AF665320

Surfaces and Interfaces of Ceramic Materials

NATO ASI Series

Advanced Science Institutes Series

A Series presenting the results of activities sponsored by the NATO Science Committee, which aims at the dissemination of advanced scientific and technological knowledge, with a view to strengthening links between scientific communities.

The Series is published by an international board of publishers in conjunction with the NATO Scientific Affairs Division

A Life Sciences
B Physics

Plenum Publishing Corporation
London and New York

C Mathematical and Physical Sciences
D Behavioural and Social Sciences
E Applied Sciences

Kluwer Academic Publishers
Dordrecht, Boston and London

F Computer and Systems Sciences
G Ecological Sciences
H Cell Biology

Springer-Verlag
Berlin, Heidelberg, New York, London, Paris and Tokyo

Series E: Applied Sciences - Vol. 173

Surfaces and Interfaces of Ceramic Materials

edited by

Louis-C. Dufour

Réactivité des Solides, UA 23 CNRS,
Université de Bourgogne, Dijon, France

Claude Monty

Physique des Matériaux, CNRS,
Laboratoires de Bellevue, Meudon, France

and

Georgette Petot-Ervas

Ingénierie des Matériaux et des Hautes Pressions,
CNRS, Université de Paris-Nord,
Villetaneuse, France

Springer-Science+Business Media, B.V.

Published in cooperation with NATO Scientific Affairs Division

Proceedings of the NATO Advanced Study Institute on
Surfaces and Interfaces of Ceramic Materials
CAES-CNRS, Ile d'Oléron, France
September 4–16, 1988

Library of Congress Cataloging in Publication Data

NATO Advanced Institute on Surfaces and Interfaces of Ceramic Materials (1988 : Ile d'Oleron, France)
Surfaces and interfaces of ceramic materials / edited by Louis-C. Dufour, Claude Monty, Georgette Petot-Ervas.
p. cm. -- (NATO ASI series. Series E, Applied sciences ; vol. 173)
"Proceedings of the NATO Advanced Institute on Surfaces and Interfaces of Ceramic Materials, CAES-CNRS, Ile d'Oléron, France, September 4-16, 1988."
"Published in cooperation with NATO Scientific Affairs Division."
ISBN 978-94-010-6957-1 ISBN 978-94-009-1035-5 (eBook)
DOI 10.1007/978-94-009-1035-5
1. Ceramic materials--Surfaces--Congresses. I. Dufour, Louis-Claude. II. Monty, Claude. III. Petot-Ervas, Georgette. IV. North Atlantic Treaty Organization. Scientific Affairs Division. V. Title. VI. Series: NATO ASI series. Series E, Applied sciences ; no. 173.
TA455.C43N37 1988
620.1'4--dc20 89-20048

Printed on acid free paper

Originally published by Kluwer Academic Publishers in 1989
Softcover reprint of the hardcover 1st edition 1989

CONTENTS

5 : INTERACTION OF CERAMICS WITH ENVIRONMENT

Tutorial Lectures

Invited Lecture

Contributed Papers

6 : PROPERTIES RELATED TO SURFACES AND INTERFACES IN SPECIFIC CERAMIC MATERIALS

Tutorial Lectures

Invited Lecture

Contributed Papers

PREFACE

This book contains the proceedings of the NATO Advanced Study Institute on Surfaces and Interfaces of Ceramic Materials, held on the Oléron island, France, in September 1988. This Institute was organized in nine months after receiving the agreement of the NATO Scientific Affairs Division. Despite this very short time, most of the lecturers contacted have accepted our invitation to prepare a specific talk.

The meeting was held at "La Vieille Perrotine" on the Oléron island. This holiday village of the French CNRS is located near the Ocean in a natural area which contributed to create a very pleasant atmosphere favourable to develop interaction between the 91 participants in this Institute.

First of all, the Institute was aimed at diffusing the foremost results on the characterization of and the role played by surfaces, grain boundaries and interfaces in preparation and overall properties of ceramic materials, mainly of oxide ceramics. Through its interdisciplinary character, the Institute was also aimed at developing interaction between scientists and engineers interested in basic and practical aspects of processing and use of ceramics. Lastly, the Institute has emphasized the fundamental importance and the key role of the external and internal surfaces in the advanced technology of these materials. 42 hours of courses were given by 23 lecturers and 13 additional hours on more limited subjects by 13 invited lecturers. 8 hours were devoted to three round table discussions on segregation, ceramic-metal bonding and sintering processes. Moreover, a poster session, preceded by a short oral introduction by the authors, enabled the students to present their own works.

Most of the lectures and papers presented during this Advanced Study Institute are gathered in this book which is divided in six sections according to the following description:

1/ STRUCTURAL AND MICROSTRUCTURAL ASPECTS OF SURFACES AND INTERFACES OF CERAMIC MATERIALS, INCLUDING ELECTRONIC AND CHEMISORPTION PROPERTIES. Although progress is still needed, many theoretical calculations are now available and reliable to predict the thermodynamical stability of both free and doped surfaces and interfaces *. Modern high resolution spectroscopy and microscopy make it possible to explore and understand the crystallographic, electronic and chemical structure of surfaces and interfaces on an atomic scale. These investigations reveal the fundamental role played by defects and additives or impurities in the properties of surfaces and interfaces of these materials. Also the theory of fractals is considered a very interesting mathematical approach to treat some specific problems, particularly wetting, elaboration by sol-gel route or sintering of ceramics **.

2/ SEGREGATION PHENOMENA AND TRANSPORT PROPERTIES. One important point emphasized here concerns the role of segregation now better analyzed and controlled in the equilibrium conditions. After thermal treatment, this phenomenon is often involved in modifying, sometimes drastically, the expected behaviour of the ceramic materials and its study therefore is to be considered a priority. Recent approaches on the role of both surface and grain boundary diffusions in the overall properties of ceramic protective layers, solid electrolytes, sensors and electrical ceramics, including new high T_c oxide superconductors, are presented.

3/ THERMODYNAMIC AND STRUCTURAL ASPECTS OF CERAMIC-METAL INTERFACES. Another question largely involved in many industrial applications concerns the ceramic-metal bonding. The way to improve the bond strength of ceramic-metal interfaces by subtle changes in chemical composition or stoichiometry at interfaces is discussed. Recent data on preparation and properties of small particles and very thin films of metals on oxides is also presented.

---

* For instance , see references in: (a) Henrich V.E., *The surfaces of metal oxides*, Rep. Prog. Physics, 48 (1985) 1481-1541; (b) Egdell R.G. and Mackrodt W.C., this book ; (c) the papers published by Stoneham A.M., Tasker P.W. and coworkers (*e.g.* , ref. in the chapter 1, *The theory of ceramic surfaces*, in 'Surface and Near-Surface Chemistry of Oxide Materials', Nowotny J. and Dufour L.C., (Eds), Elsevier, Amsterdam, 1988)

** Some references in this field may be found in: (a) Mandelbrot B., 'The Fractal Geometry of Nature', Freeman, San Francisco, 1982; (b) Cherbit G. (Ed.), 'Fractals, Dimensions Non Entières et Applications', Masson, Paris, 1987; (c) Feder J., 'Fractals', Plenum Pub. Corp., New York, 1988; (d) Botet R. and Julien R., *A theory of aggregating systems of particles: the clustering of clusters process*, Ann. Phys. Fr., 13 (1988) 153-221

4/ ROLE OF SURFACES AND INTERFACES IN ELABORATING CERAMIC MATERIALS. Current issues on processing of ceramic powders, particularly from monosized particles and on grain growth in sintered ceramics are analysed in this section.

5/ INTERACTION OF CERAMIC MATERIALS WITH ENVIRONNMENT, INCLUDING SOLID-GAS AND SOLID-LIQUID REACTIONS. Several chapters present the most recent development on the role of surfaces and interfaces in the interactions of ceramics with gas at high temperature and with liquids such as acidic solutions. The key role of the surface defects as well as the formation, in specific conditions, of thin protective amorphous films is discussed in the processes of dissolution of ceramic oxides.

6/ PROPERTIES RELATED TO SURFACES AND INTERFACES OF SPECIFIC CERAMIC MATERIALS AND COMPOSITES, INCLUDING MECHANICAL PROPERTIES. Here attention is more particularly concentrated on the relationship between microstructural properties of the interfaces and strength of composites, on the surface properties of vitreous silicates and the elaboration of ceramic membranes.

The aim of an Advanced Study Institute is to collect the knowledge gained in a scientific field and to make this information easily accessible to those interested. We trust that this book will show that the subject treated is timely and important for basic research as well as in the development of modern technology for ceramic materials. This research area is now in fast expansion and has an ever increasing impact on the way to improve ceramics and composite materials for catalysis, metallurgy, energetics, electronics, space science,.. No doubt that, in short future, this scientific branch becomes one of the most active and rapidly developing of contemporary Materials Science.

Louis-C. Dufour,
Claude Monty
Georgette Petot-Ervas

ACKNOWLEDGEMENTS

The Organizers of this NATO Advanced Study Institute gratefully acknowledge support and co-sponsorship from the following Institutions and Companies:

- NATO Scientific Affairs Division which provided the main support. Our sincere gratitude goes particularly to Dr. L.V. da Cunha and Dr. G. Sinclair for their encouragement and assistance.
- Centre National de la Recherche Scientifique (CNRS), Paris. Particular thanks are due to Prof. M. Fayard, Head of the Department of Chemistry.
- European Research Office of the US Army Research, Development and Standardization Group, London. We particularly thank Dr W.C. Simmons, Chief of the Materials Sciences Branch.
- Riber SA., Rueil-Malmaison, France

The Organizing Committee is very grateful to lecturers and participants who have contributed to the success of this meeting and cooperated in the preparation of this book

Our thanks are also addressed to:

- Prof. J-C. Colson, University of Burgundy, for his encouragement and for providing facilities
- Mr G. Lutrot and his staff, for their efficient assistance during the meeting at the CNRS Village "La Vieille Perrotine" on the Oléron island
- Drs K. Chhor and J.L. Lirhman, CNRS Villetaneuse, Dr. P. Bracconi, CNRS Dijon and Mr M. Vareille, University of Burgundy, for their kind assistance.
- the "Groupe Français de Céramique" and the French "Club Surfaces et Interfaces" which have kindly allowed us to use their address listings
- the Editors of "The Bulletin of the American Ceramic Society", "l'Actualité Chimique", "Journal de Physique", "Reactivity of Solids" and "Journal de Chimie Physique" who announced the meeting in their Journal.

The Members of the Organizing Committee:
Drs Louis-C. Dufour, Director; Alan Atkinson; Claude Monty ;
Janusz Nowotny; Georgette Petot-Ervas; Werner Weppner

PARTICIPANTS

Mr.	Jean-André	ALARY	CEGEDUR Péchiney, Centre de Recherche et de Développement, B.P. 27, 38340 VOREPPE (France)
Mr.	S.	ALPERINE	Office National d'Etudes et de Recherches Aérospatiales (ONERA), B.P. 72, 92322 CHATILLON CEDEX (France)
Dr.	Alan	ATKINSON	Materials Development Division, AERE Harwell, HARWELL, Oxfordshire OX11 ORA (UK)
Mr.	Jørgen B.	BILDE-SØRENSEN	Metallurgy Department, Risø National Laboratory, Postbox 49, DK 4000 ROSKILDE (Denmark)
Dr.	Sara A.	BILMES	Physikalisches Institut III, Universität Düsseldorf, Universitätstrasse 1, 4000 DÜSSELDORF 1 (FRG)
Dr.	Michel	BISCONDI	Ecole Nationale Supérieure des Mines, 158, Cours Fauriel, 42 023 ST ETIENNE CEDEX (France)
Prof.	Jack M.	BLAKELY	Dept of Materials Science and Engineering, Coll of Engineering, Cornell University, Bard Hall, ITHACA, NEW YORK 14853 (USA)
Dr.	Miguel A.	BLESA	Comision Nacional de Energia Atomica, Dpto "Quimica de Reactores", Avenida del Libertador, 8250, 1429 BUENOS AIRES (Argentina)
Prof.	Günter	BORCHARDT	Institut für Allgemeine Metallurgie, Technische Univ.Clausthal, Robert-Koch-Str. 42, Postfach 1253, 3392 CLAUSTHAL-ZELLERFELD (FRG)
Dr.	Gilles	BORDIER	DPC / STS, Bâtiment 125, Centre d'Etudes Nucléaires de Saclay, 91191 GIF/SUR/YVETTE CEDEX (France)
Mr.	Jeffrey H.	BOY	University of Illinois, Coll. Engng, Dept Mat.Sci. & Engng, Ceramics Division, 204 Cer.Blg, 105 South Goodwin Av., URBANA, ILLINOIS 61801 (USA)
Prof.	Richard J.	BROOK	Max-Planck Institut für Metallforschung, Pulvermetallurgisches Laboratorium, Heisenbergstrasse 5, 7000 STUTTGART 80 (FRG)
Prof.	François	BROUERS	Institut de Physique, Université de Liège, B6, Sart Tilman, B-4000 LIEGE (Belgium)
Prof.	A.J.	BURGGRAAF	Twente University of Technology, Dpt Chem. Engng /Lab Inorg. Chemistry. & Materials Science, P.O. Box 217, 7500 AE ENSCHEDE (The Netherlands)

Prof.	C. Barry	CARTER	Dept. of Materials Science and Engineering, College of Engineering, Cornell University, Bard Hall, ITHACA, NEW YORK 14853 (USA)
Mrs.	Dominique	CHATAIN	Lab. de Thermodynamique et Physico-Chimie Métallurgiques (LTPCM), Domaine Universitaire, BP 75, 38402 ST MARTIN D'HERES (France)
Dr.	David R.	CLARKE	I.B.M., Thomas J. Watson Research Center, Materials Science Department, P.O. Box 218, YORKTOWN HEIGHTS, NEW YORK 10598 (USA)
Ms.	Charlotte	CLAUSEN	Metallurgy Department, Risø National Laboratory, P.O. Box 49, DK 4000 ROSKILDE (Denmark)
Mr.	Michel	COURBIERE	42, rue Marguerite, 69100 VILLEURBANNE, (France)
Mr.	Julian	DE ANDRES	CERATEN S.A., Carpinteros 10, Pol Industrial Los Angeles, 28906 GETAFE (MADRID) (Spain)
Dr.	Duane	DIMOS	I.B.M., Thomas J. Watson Research Center, Materials Science Department, P.O. Box 218, YORKTOWN HEIGHTS, NEW YORK 10598 (USA)
Mr.	Jan	DIPHOORN	DSM Research BV, Department MT Ker, P.O. Box 18, 6160 MD GELEEN (The Netherlands)
Prof.	Arturo	DOMINGUEZ-RODRIGUEZ	Depto de Fisica de la Materia Condensata, Universidad de Sevilla, Apdo 1065, SEVILLA (Spain)
Dr.	Louis C.	DUFOUR	Réactivité des Solides, UA 23 CNRS, Université de Bourgogne, BP. 138, 21004 DIJON CEDEX (France)
Dr.	Russell G.	EGDELL	Department of Chemistry, Imperial Coll. of Science and Technology, SOUTH KENSINGTON, LONDON SW7 2AZ (UK)
Mr.	José V.	EMILIANO	R. Coronel Artur Whittacker 85413690 DESCALVADO SP (Brasil)
Mr.	Claude	ESNOUF	INSA, Bâtiment 502 / GEMPPM, 69 621 VILLEURBANNE CEDEX, (France)
Dr.	Martine	GAUTIER	Centre d'Etudes Nucléaires Saclay, DPC/ SPCN /SES, 91191 GIF-SUR-YVETTE CEDEX (France)
Mr.	Dominique	GESLIN	Centre de Recherches Péchiney, Service RD-R, B.P. 54, 13541 GARDANNE CEDEX (France)
Mr.	Jacques	GILLOT	Société de Céramiques Techniques, B.P. 1, 65 460 BAZET, (France)
Prof.	Daniele	GOZZI	Dipartimento di Chimica, Universita "La Sapienza", Piazzale Aldo Moro 5, 00 185 ROMA (Italy)

Prof.	Hans J.	GRABKE	Max-Planck Institut für Eisenforschung, Max-Planck Strasse 1, Postfach 140 260, 4000 DÜSSELDORF 1 (FRG)
Mr.	Helmut B.	GRÜBMEIER	Institut für Reaktorwerkstoffe, KFA Jülich GmbH, Postfach 1913, D-5170 JÜLICH 1 (FRG)
Mr.	Frédéric	GRUY	Ecole Nationale Supérieure des Mines, 158, cours Fauriel, 42023 SAINT-ETIENNE, (France)
Dr.	Riza	GÜRBÜZ	Metallurgical Engineering Department, Middle East Technical University, P.K. 06531, ANKARA (Turkey)
Dr.	Jean-Marc	HEINTZ	Chimie du Solide CNRS, Université de Bordeaux 1, 351, cours de la Libération, 33405 TALENCE CEDEX (France)
Prof.	Victor E.	HENRICH	Applied Physics, Yale University, Becton Center, P.O. Box 2157 Yale Station, NEW HAVEN, CONNECTICUT 06520-2157 (USA)
Mr.	Joseph	HOAGLAND	Washington State University, Chemistry Dept, PULLMAN, WASHINGTON 99164-4630 (USA)
Prof.	Jan	JANOWSKI	Director, St Staszic University of Mining and Metallurgy, ul. Szymanowskiego 1-10A, 30 047 KRAKOW (Poland)
Prof.	Stanislawa	JASIENSKA	Academy of Mining and Metallurgy, Institute of Metallurgy, al. Mickiewicza, 30, 30 059 KRAKOW (Poland)
Dr.	Halit	KALEBOZAN	Department of Metallurgical Engineering, Middle East Technical University, Inönü Bulvary, 06531 ANKARA (Turkey)
Prof.	Gretchen	KALONJI	Dpt. Mat. Science and Engineering. / Room 13 5094, Massachusetts Institute of Technology, 77, Massachusetts Avenue, CAMBRIDGE, MA 02139 (USA)
Dr.	J.T.	KLOMP	Centre for Technical Ceramics, C/O University of Technology, P.O. Box 513, 5600 MB EINDHOVEN (The Netherlands)
Prof.	Per	KOFSTAD	Department of Chemistry, University of Oslo, P.B. 1033 BLINDERN, 0315 OSLO 3 (Norway)
Mr.	Fathi	KOOLI	CNRS/ CRSOCI, 1B, rue de la Ferollerie, 45071 ORLEANS CEDEX 2, (France)
Dr.	Albert A.	KRUGER	Battelle Pacific Northwest Laboratories, P.O. Box 999, RICHLAND, WA 99352 (USA)
Dr.	Mongi	LABIDI	Faculté des Sciences de Tunis, Département de Physique, LE BELVEDERE 1001, (Tunisia)
Dr.	Sylvie	LARTIGUE	Métallurgie Structurale, Bâtiment 413, Université Paris XI, 91405 ORSAY CEDEX (France)

Dr.	Colin	LEACH	Imperial College London, Department Materials - Room B 513, LONDON SW72BP, (UK)
Mr.	Libero	LIGGIERI	Istituto. Chimica Fisica Applicata dei Materiali, C.N.R. GENOVA, Via Lungobisagno Istria 34, 16141 GENOVA (Italy)
Mr.	Pedro M.	MANTAS	Dept. Eng. Ceramica e do Vidro, Universidade de Aveiro, 3800 AVEIRO, (Portugal)
Dr.	Hj.	MATZKE	CEC - JRC, / European Inst.for TransuraniumElements, Postfach 2340, D 75OO KARLSRUHE (FRG)
Ms.	Caroline	MAUNIER	Centre d'Etudes Nucléaires Saclay, CENS /SRMP Bâtiment 20, 91191 GIF-SUR-YVETTE, (France)
Dr.	Andrea	MENNE	Max-Planck Institut Festkörperforschung, Heisenbergstrasse 1, Postfach 80 06 65, D-7000 STUTTGART 80 (FRG)
Prof.	Alain	MOCELLIN	Ecole Polytechnique Fédérale, Dépt. des Matériaux, Lab. de Céramique, 34, Chemin de Bellerive, CH-1007 LAUSANNE (Switzerland)
Mrs.	Laura	MONTANARO	Politecnico di Torino, Dipto Scienza dei Materiali, Corso Duca Degli Abruzzi 24, 10129 TORINO (Italy)
Dr.	Regina C.	MONTEIRO	Universidade. Nova de Lisboa, Fac.Ciências Tecnol., Dpto Ciências Materiais, Quinta da Torre, 2825 MONTE DE CAPARICA (Portugal)
Dr.	Claude	MONTY	CNRS, Physique des Matériaux, 1, place Aristide Briand, BELLEVUE, 92195 MEUDON CEDEX (France)
Dr.	François	MORIN	Chimie des Matériaux, Institut de Recherches d'Hydro-Québec, C.P. 1000, VARENNES, Québec, JOL 2PO (Canada)
Mrs.	Maria Luigia	MUOLO	Istituto Chim. Fisica Applicata dei Materiali, C.N.R. Lungobisagno Istria, 34, 16141 GENOVA (Italy)
Dr.	Michael G.	NICHOLAS	Metals Technology Centre, Harwell Lab., UKAEA, DIDCOT, OXON OX11 ORA (UK)
Prof.	Jean-C.	NIEPCE	THOMSON-LCC, /Centre de Saint-Apollinaire, Av.du Colonel Prat, 21850 ST APOLLINAIRE (France)
Dr.	Truls	NORBY	Department of Chemistry, University of Oslo, P.B. 1033 BLINDERN, N- 0315 OSLO 3 (Norway)
Dr.	Gérard	NOUET	LERMAT-CNRS/ISMRA, Boulevard du Maréchal Juin, 14032 CAEN CEDEX (France)
Dr.	Janusz	NOWOTNY	Max-Planck Institut Festkörperforschung, Heisenbergstrasse 1, Postfach 80 06 65, 7000 STUTTGART 80 (FRG)

Prof.	Marcel	PERDEREAU	Réactivité des Solides, UA 23 CNRS, Université de Bourgogne, BP 138, 21004 DIJON CEDEX (France)
Dr.	Georgette	PETOT-ERVAS	LIMHP/CNRS, Centre Univ.Paris-Nord, Avenue Jean-Baptiste Clément, 93430 VILLETANEUSE (France)
Dr.	Albert P.	PHILIPSE	National Ceramic Centre, NKA, Netherlands Energy Research Foundation, ECN, Postbus 1, 1755 ZG PETTEN (The Netherlands)
Dr.	Bernard	PIERAGGI	Métallurgie Physique, UA 445 CNRS, E.N.S.C. Toulouse, 118, rue de Narbonne, 31077 TOULOUSE CEDEX (France)
Prof.	P.	PIROUZ	Dept of Metallurgy and Materials Science, Case Western Reserve University, CLEVELAND, OHIO 44106 (USA)
Prof.	Pierre	QUINTARD	Lab. Spectrométrie Vibrationnelle, UA 320 CNRS/ Céramiques Nouvelles, Facultés des.Sciences, 123 Av. Albert Thomas, 87060 LIMOGES (France)
Prof.	Alexandre	REVCOLEVSCHI	Université de Paris-Sud, Chimie des Solides, Batîment 414, 91405 ORSAY CEDEX (France)
Prof.	Jean	RIEU	Ecole des Mines de Saint-Etienne, 158, Cours Fauriel, 42023 SAINT-ETIENNE CEDEX (France)
Prof.	Terry A.	RING	Ecole Polytechnique Fédérale /Lab. de Technologie des Poudres,/C/O Bâtiment de Chimie, 1015 LAUSANNE-ECUBLENS (Switzerland)
Mr.	Roberto	SANGIORGI	Istituto Chim. Fisica Applicata dei Materiali, C.N.R., Lungobisagno Istria, 34, 16141 GENOVA (Italy)
Prof.	Bernard	SAPOVAL	Physique Matière Condensée /GR 38 CNRS, Ecole Polytechnique, Route de Saclay, 91128 PALAISEAU CEDEX (France)
Dr.	Helmut	SCHUBERT	Max-Planck Institut für Metallforschung, Pulvermetallurgisches Laboratorium, Heisenbergstrasse 5, 7000 STUTTGART 80 (FRG)
Dr.	Ana Maria	SEGADAES	Departamento de Engenharia. Ceramica e do Vidro, Universidade de Aveiro, 3800 AVEIRO, (Portugal)
Dr.	Ana Maria	SENOS	Departamento de Engenharia Ceramica e do Vidro, Universidade de Aveiro, 3800 AVEIRO, (Portugal)
Mr.	Derek C.	SINCLAIR	University of Aberdeen, Department of Chemistry, Meston Walk, OLD ABERDEEN AB9 2UE (UK)
Dr.	W.	SKROTZKI	Institut Geologie und Dynamik der Lithosphäre, Universität Göttingen, Goldschmidtstrasse 3, D-3400 GÖTTINGEN (FRG)
Prof.	Walt W.	SMELTZER	Institute for Materials Science, McMaster University, 1280 Main Street West, HAMILTON, Ontario, L8S 4M1 (Canada)

Dr.	Francis	TEYSSANDIER	Institut Science et de Génie des Matériaux et Procédés, CNRS/ Université de Perpignan, Av. de Villeneuve, 66025 PERPIGNAN CEDEX (France)
Mr.	G.S.A.M.	THEUNISSEN	University of Twente, Fac. of Chemical Engineering, P.O. Box 217, 7500 AE ENSCHEDE (The Netherlands)
Dr.	Peter S.	TURNER	Division of Science and Technology, Griffith University, NATHAN, BRISBANE, Queensland 4111 (Australia)
Dr.	Colette	VIGNAUD	CNRS LP 15, Physique des Liquides et Electrochimie, 51 Bvd A. Blanqui, 75013 PARIS (France)
Dr.	Werner	WEPPNER	Max-Planck Institut Festkörperforschung, Heisenbergstrasse 1, Postfach 80 06 65, 7000 STUTTGART 80 (FRG)
Mr.	Jacques	WERCKMANN	C.N.R.S./I.P.C.M.S.,Surfaces et Interfaces, Université Louis Pasteur, 4, rue Blaise Pascal, 67000 STRASBOURG (France)
Dr.	A.R.	WEST	University of Aberdeen, Department of Chemistry, Meston Walk, OLD ABERDEEN AB9 2UE (UK)
Dr.	L.R.	WOLFF	Center for Technical Ceramics (CTK), Eindhoven University of Technology, P.O. Box 513, 5600 MB EINDHOVEN (The Netherlands)
Prof.	Eiichi	YASUDA	Tokyo Institute of Technology, Research Lab of Engineering Mat., 4259 NAGATSUTA, MIDORI, YOKOHAMA 227 (Japan)

THE MATERIALS SCIENCE OF CERAMIC INTERFACES

R.J. BROOK
Max–Planck–Institute for Metals Research
Powdermetallurgy Laboratory
Heisenbergstr. 5
D–7000 Stuttgart 80
West Germany

ABSTRACT. The significance of surfaces and interfaces for the processing and application of ceramics has long been recognised. In this review attention is given to phenomena where the existence of surfaces is responsible for important phenomena in processing (free surfaces and the sintering of powders) and properties (surface cracks and the influence on mechanical strength). Attention is also given to two subject areas where an improved understanding of interface behaviour would be beneficial. The first relates to the significance of segregation at grain boundaries and at free surfaces for the development of ceramic microstructures; the second relates to the structure and behaviour of grain boundaries in the presence of second phases, a factor important both for the processing and for the high temperature properties of ceramic materials.

1. Introduction

As preliminary to a conference where much of the discussion will involve close analysis of the chemical and physical properties of surfaces and interfaces in ceramic materials, it may be helpful to review the subject from the point of view of the ceramist, namely from the point of view of someone wishing to fabricate and use this set of materials for applications and seeking to understand surfaces and interfaces as a means of improving product performance. This approach has the potential benefit that it allows identification of the critical questions as seen from the viewpoint of someone working with the processing or use of ceramics; it should also be helpful to the wider community in revealing any prejudices or preoccupations that may have arisen among those working with the development of ceramics.

In the opening section a brief review is given of certain topics where the ceramist has long recognised that interfaces or surfaces play a critical role. These include the densification of powders by sintering and the mechanical properties of ceramics at ambient temperatures. In subsequent sections two topics are treated where the current understanding of grain boundary or surface structure is incomplete and where further progress should prove rewarding for the processing or application of these materials. The two themes are the segregation of dopants or impurities at grain boundaries in single phase ceramic materials and secondly the

L.-C. Dufour et al. (eds.), Surfaces and Interfaces of Ceramic Materials, xxiii–xxxvii.

distribution and properties of second phases in the grain boundaries of multiphase ceramics. The two themes therefore relate directly to the structure and behaviour of solid/solid grain boundaries and of solid/liquid/solid boundaries.

2. Surfaces and Interfaces and the Processing, Properties and Application of Ceramics

One approach to materials science and engineering is to recognise a causal sequence between processing, the resulting microstructure, and the properties that are then associated with this microstructure; applications must then be satisfied on the basis of the properties that have been developed. Recognition of this logical sequence allows one to argue that a given application will require a given set of material properties which must then be attained by specific processing designed to achieve a microstructure capable of yielding these properties. The significance of surfaces and interfaces in each of these stages has long been recognised by the ceramics community[1].

The processing of ceramics is most commonly based upon the heat treatment of powders. The energy present in a powder as a consequence of its high surface area provides a driving force for structural change which results in the densification of the powder and the production of a finished solid component. The central importance of this sintering process in the fabrication of ceramics has made it a much studied subject.

At the macroscopic level, a requirement for successful densification is that the surface energy in the powder be greater than the interfacial energy of the resulting polycrystalline material. This then ensures that densification, the removal of porosity and the replacement of free surfaces by grain boundaries is favourable. One explanation, e.g., for the difficulties in the sintering of SiC lies in the view[2] that this energy change may be insufficient to drive densification. At the microscopic level, it is important to recognise and control the atom movements that provide the mechanism for the microstructural change. Here attention is given to the role of curved surfaces (Fig. 1) in establishing the chemical potential gradients in the system which then drive the atom diffusion. The argument is that since the diffusion of atoms from point to point in the powder bed is likely to be the slow process in the sequence of events leading to microstructural change, all other steps in the sequence can be considered to be at virtual equilibrium. Where curved surfaces occur in the system, the ability of the surfaces to act as rapid sources or sinks for atoms ensures that they are in equilibrium with the local partial pressures in an adjacent gas phase or with the local point defect concentration in an adjacent solid phase.

The local pressures and concentrations are readily calculated[3]. The first step is to recognise the pressure difference, ΔP, that occurs across a curved surface at equilibrium. The energy balance for an infinitesimal volume change dV undergone by a spherical pore of radius r in a solid is

$$\Delta P \, dV = \gamma dA \qquad (1)$$

where γ is the specific surface energy and dA the change in pore surface area. For the spherical geometry, therefore,

$$\Delta P = \frac{2\gamma}{r} \tag{2}$$

The second step is to consider two systems, one where we have a planar interface at equilibrium between two phases, A and B (these could also be two differently oriented grains of the same phase meeting at a grain boundary), and one where we have the two phases in local equilibrium with a curved interface. Molar quantities of A and B are then transferred from the first system to the second. Since both systems are themselves at equilibrium, the energy change on transferring A must equal that on transferring B, i.e.,

$$\Omega_A \, dP_A = \Omega_B \, dP_B \tag{3}$$

where Ω is the molar volume and dP the difference in the pressures acting normal to the interfaces in the two systems.

Combining (2) and (3) and taking the example for the second system of a spherical interface concave towards phase B (Fig. 2),

$$dP_B = \frac{2\gamma}{r}\left[\frac{\Omega_A}{\Omega_A - \Omega_B}\right] \tag{4}$$

If phase A is an ideal gas and phase B a solid, then with $\Omega_A \gg \Omega_B$,

$$dP_B = \frac{2\gamma}{r} \tag{5}$$

Similarly with

$$dP_A = -\frac{2\gamma}{r}\left[\frac{\Omega_B}{\Omega_B - \Omega_A}\right] \tag{6}$$

$$dP_A = -\frac{2\gamma}{r}\frac{\Omega_B \, P_A}{RT} \tag{7}$$

From (7), local variations in surface curvature cause local variations in local gas partial pressure giving rise to diffusion and atom transport in the gas. Similarly, the differences in local normal pressures at differently curved surfaces in the solid (5), give rise to chemical potential gradients which in turn lead to atom diffusion in the solid.

Much effort has gone into the conversion of these atom fluxes into the form of the corresponding microstructural changes that are observed in the sintering of powders; notwithstanding the complexities of these changes, it must be recognised that the understanding of the process is now on a sound footing.

An important distinction that must be recognised in the resulting microstructural changes is the one between densifying and non-densifying processes. If densification is to occur then material must be removed from points between the grain centres of adjacent grains to allow shrinkage. The only reversible source for atoms that exists between the two centres is that provided by the grain boundary between the grains; diffusion processes which involve the movement of atoms from grain boundaries to adjacent pores therefore result in an approach of the grain centres and in densification (Fig. 3). Movement of atoms between other sources and sinks, say from one part of the pore surface to another part of the surface does not result in densification; the changes then produce a coarsening of the microstructure in which the surface energy is reduced by reduction of the total surface area rather than by replacement of this surface by grain boundary. A major ambition in sintering is therefore to control the processing conditions, namely, powder particle size, temperature, composition, applied pressure, and processing time, so that the densification process is favoured and the coarsening process is inhibited (Fig.4). The important covalent ceramic systems SiC and Si_3N_4 are examples where the control of this balance has been recognised as a major problem.

The role of surfaces has also been recognised as of vital importance in mechanical properties[4]. When a stress is imposed on a material, elastic strain energy is introduced as the atoms are pulled from their equilibrium positions. To a first approximation the only mechanism capable of absorbing this energy during the fracture of a ceramic is the formation of new surface. If, in the propagation of a crack, the rate of elastic strain energy release exceeds the rate of surface energy demand, then the crack will grow. That is, when

$$\frac{d}{dL}\left[\frac{\sigma^2}{2E}\right] L^3 \geq \frac{d}{dL}(2\gamma_f)\, L^2 \qquad (8)$$

Here σ is the applied stress, E the elastic modulus, γ_f the energy required to form unit area of the new fracture surfaces and L a dimension scaling with the size of the crack. The fact that the elastic strain energy scales as the volume of material containing the atoms under stress while the surface energy scales as the area of new surface means that flaws above a certain size

$$L > \frac{4E\gamma_f}{\sigma^2} \qquad (9)$$

are unstable and lead to rapid brittle failure.

The critical rôle of flaw size in determining the applied stress, σ, which can be tolerated (the strength) has placed great importance on the identification of structural flaws and on their size reduction. A variability of flaw size produces a corresponding variability in mechanical strength. In order to achieve reliable and consistent strength values, therefore, a major ambition in current processing technology is to achieve control of the flaw population in the material[5].

The importance of the γ_f term, namely the energy required to bring about extension of the fracture surface, has resulted in efforts to develop mechanisms capable of raising this energy. In addition to the purely thermodynamic surface energy, contributions can be found from local plastic deformation and from a variety of mechanisms introduced by deliberate microstructural design (microcracking[6], crack branching or deflection[7], plastic deformation in metallic inclusions[8], branching or deflection or pullout in fibre containing systems[9], additional strain energy terms as in transformation toughening[10]). For many of these mechanisms, further interfacial energy terms become significant as, for example, in the case of the interface between fibrous inclusions and a matrix where the course of the crack path through the material can be influenced by modification of the interface character[11].

A further feature is the importance of interactions between the newly formed faces of the crack and the environment; these can provide additional contributions to crack extension. The general tendency to look more closely at the detailed micromechanics of crack development places more weight on the nature and behaviour of the interfaces and surfaces involved.

The two examples which have been introduced, one from processing and one from properties, are but representative of the central rôle which surface and interface character play in ceramics. To indicate the range of problems, two other examples can be very briefly mentioned. The great refinement that has occurred in recent years in the preparation of powders (Fig. 5) and in the control of the rheology of powder suspensions in connection with ceramic forming processes[12] has placed great emphasis on the nature and properties of the powder/fluid interface and on its control with additives. A further example from the mechanical properties area would be that of high temperature creep; the existence of a viscous phase in the grain boundaries of ceramics (Fig. 6) can be a major source of weakening under load at high temperatures[13]. The nature and distribution of such second phases is therefore a major research theme.

Insofar as the above processing and properties issues relate to all ceramics, the impact of surfaces and interfaces on fabrication and applications is widespread. Specific examples can, however, be identified where surface related properties become the central aspect of the application. These would include uses which depend upon the wear resistance of the material such as cutting tools[14] or medical implant materials[15]. The role of a surface in wear resistance is self-evident but full understanding of the wear process has been difficult to achieve: in general, behaviour is extremely specific to the application concerned.

The uses of functional ceramics which find application as electrical or magnetic components also provide many examples where surfaces and interfaces play a specific role. Deliberate engineering of the interface structure is, e.g., involved in the development of boundary layer capacitors[16] and of positive temperature coefficient devices[17]. Again, the behaviour of the surface can be very specific to the system involved and generalizations are difficult.

To summarize, it is important to recognise that surfaces and interfaces have long been seen by the ceramic community as critical elements in the processing and use of these materials. There are issues which are common to all ceramics such as those discussed above related to sintering and to mechanical strength. There are, in addition, examples which specifically exploit surface phenomena. The almost universal importance of surfaces and interfaces for the use of ceramics has made the community as alert to the impact of surface phenomena for these materials[1] as for any other materials class.

3. Structure of Interfaces in Ceramics

As a basis for understanding the behaviour and properties of surfaces and grain boundaries in ceramics, attention, as in other aspects of solid state science, is first given to structure. Excellent progress has been made in the understanding of interface structure in recent years and this has been well covered in reviews[18].

The structure of surfaces has been thoroughly studied, notably with the technique of LEED, and systematic descriptions[19] of surface geometry are now available much as has long been the case for internal crystal structure. A feature of such descriptions is the careful attention given to the preparation and measurement conditions since these interact so profoundly with the detailed observations.

The structure of grain boundaries is understood at a considerably less detailed level. Proposals for specific structures[20] can be developed on the basis of molecular dynamics calculations; particularly favourable orientations can then be sought and matched with the relative frequency of their occurrence to assess the extent of agreement between experimentally determined and theoretically predicted structures.

Major advances in understanding have been achieved on the basis of the rapid developments in the instrumentation available for surface and interface studies. High resolution transmission electron microscopy and more recently scanning tunneling electron microscopy[21] allow detailed study of the local physical structures at interfaces; the growing array of surface analysis methods[22] allows exceptional depth resolution of chemical characteristics normal to an interface. The combination of methods offers remarkable opportunities for the determination of the physical and chemical character of the interfaces whose behaviour is to be controlled and utilised. As seen in the following sections, however, the very complexity of the processes wherein interfaces play a role leaves room for uncertainties in interpretation and sets major tasks for future research.

4. Segregation and Boundary Migration

The wish to control the extent of structural coarsening during the heat treatment of ceramics has placed emphasis on the methods that can be made available for controlling grain boundary mobility. Of these, the most promising exploits the segregation of dopant or impurity atoms to the grain boundaries; such segregated species then act to restrain or inhibit boundary motion (the impurity drag effect)[23]. The underlying idea is that segregation arises from the existence of low energy sites for the segregant at the boundary, either as a consequence of size differences[24] and the existence of less ordered, more accommodating sites at the boundary, or as a consequence of space charges[25] and the resulting electrostatic interactions between charged dopant/impurity species and the boundary; movement of the boundary then requires movement of the segregated atoms if they are to remain in the low energy sites. The boundary mobility is then not determined by the simple jump of atoms from one side of the boundary (the shrinking grain) to the other side (the growing grain) but by the associated long distance diffusion of the cloud of segregated atoms.

The observations of segregation are now many[26] and evidence exists both for size difference[27] and for electrostatic interaction[28] as the responsible factor. Attempts have been made at a systematic classification[29]; generally segregation is to be expected except where dopant and host atoms have both the same size and the same charge (or where a similarly sized atom of different charge is accompanied by similar sized atoms of compensating charge difference).

The basis for electrostatic segregation[25] lies in the fact that the individual formation energies for the compensating intrinsic defects in ionic systems are likely to be different, e.g., the formation energy of one vacancy type is less than that of the other in Schottky disorder in a binary ionic compound. The existence of the boundary allows this difference to be reflected in a locally enhanced concentration of the lower energy defect, the resulting space charge being offset by the compensating charge at the boundary itself. Distribution of charged dopants in the resulting electrical field in the boundary region then appears as segregation.

Studies of grain growth in CeO_2 can illustrate[30] the position. Ce^{4+} (920 pm) and Gd^{3+} (938) are of similar ion size. Segregation of Gd^{3+} is, however, to be expected on electrostatic grounds (Fig. 7). If for CeO_2, anion Frenkel disorder is taken as the intrinsic disorder type and if $E_f(V_o) < E_f(O_i)$ then a defect distribution as in (a) is expected (positive spacecharge offset by a negative surface layer rich in O^{2-}). Addition of Gd_2O_3 raises the V_o concentration (b) to compensate for Gd'_{ce} so that the vacancy concentration rises above the value which it would have were charge balance requirements relaxed (dotted line). The concentration can relax to this value at the surface establishing a positive surface layer and negative space charge within which distribution of the Gd'_{ce} takes place giving rise to segregation. Grain growth in Gd-doped CeO_2 is reduced in comparison with that in the undoped material. The effect of Nb_2O_5 addition to the doped material (Fig. 8) is to enhance grain growth at small concentrations (Fig. 7d) where the co-doping allows defect concentrations to reach the values which they would adopt in the absence of any need for charge compensation and where segregation is accordingly absent; at larger concentrations, Nb_{ce} must itself be compensated by O_i (Fig. 7c) and segregation of Nb_{ce} then occurs in the resulting space charge field. Although Nb^{5+} has a different ion size (690 pm), the evidence provided by the grain growth behaviour for the existence of charge compensation among the additives underlines the importance of electrostatic factors. Similar behaviour has been found in the more complex example of $BaTiO_3$[31].

The difficulties which currently arise when attempts are made to predict the influence of additives or of dopants on the mobility of boundaries are several. They include:

(1) Segregation can, as noted, arise from different causes. Distribution in the segregation regions can accordingly take different forms.

(2) Segregation at interfaces can be highly anisotropic with different degrees of segregation arising at crystallographically different interfaces[32].

(3) The direct link between degree of segregation and boundary kinetics[33] is difficult to make experimentally. Grain growth studies[34] indicate that the influence of segregated additives can be very considerable; direct measurements are, however, difficult and are complicated by the existence of instabilities, i.e., dependence of the mobility on the local driving force for migration.

(4) Mobility can be influenced by other factors such as second phases or attached porosity[35].

The present position is that the widespread occurrence of segregation as a phenomenon has been demonstrated as has the ability of a segregated dopant to influence boundary mobility. Systematic study of behaviour, particularly on a quantitative basis, is still very much required in order that more rational use of this method for the control of coarsening can be made. The opportunities for interesting scientific and technological contributions arising from studies of segregation and its consequences are indeed considerable.

5. Special Boundaries and Liquid Films

The influence of liquid phases at high temperature in ceramics has been of longstanding interest not only for such properties as creep[13] and (following cooling t6o ambient temperature) thermal conductivity[36] but also because of the prevalence of liquid phase sintering as a processing method (where a grain boundary liquid is used to accelerate diffusion during densification) as a processing method.

An important issue concerns the stability[37] of liquid phases in grain boundaries, i.e., the extent to which boundaries remain wetted by a liquid as the temperature is reduced from that used during processing, and the dependence of this stability on the orientation of the grains forming the boundary. Although discussions of boundary energy as a function of orientation have not taken place to the extent recognised for solid/solid boundaries, a number of indications suggest that the anisotropy can be substantial. These include:

(1) The notable structural differences occurring at solid/liquid/solid interfaces of different orientation (Fig. 6) imply energy differences for the two sides of the boundary; correspondingly, different energies are expected for boundaries where one of the grains can take up a favoured orientation (upper grain in Fig. 6).

(2) Liquid infiltrated grain boundaries are known to be favourable[38] for the occurrence of abnormal grain growth (where a few grains grow rapidly within a fine-grain matrix). As shown by computer modelling[39], the existence of special boundaries of low energy is also favourable to this development. For Al_2O_3, isotropic grain structures (Fig. 9) develop[40] either in the pure material (free from grain boundary films) or in materials in which a liquid is present but with less anisotropic interfacial energies (work on MgO-doped material or at high temperature). Otherwise, anisotropic grain structures readily develop. The indications from such work are (Fig. 10) that anisotropy of the boundary energy with grain orientation is more extreme in liquid containing systems than in those where no liquid boundary phase occurs. Where additives promote anisotropy[41], e.g. CaO, then the tendency to produce faceted grains and abnormal growth is enhanced.

The influence of temperature, of orientation and of chemistry makes the interaction between grain boundaries and liquid second phases a complex and fascinating issue. The methods for controlling second phase distribution, amount and character (additives[42], distribution of additives[43], crystallisation[44], evaporation[45], solid solution formation with the grains[46]) are numerous and the theme of grain boundary engineering has developed on the basis of the different possibilities. The widespread importance of boundary films makes this an area where intensive and systematic studies will continue to prove fruitful.

6. Concluding Remarks

These introductory notes can perhaps serve to emphasize the number of research questions in ceramics where further progress will depend upon an improved ability to understand, design, construct, and exploit surfaces and interfaces. There are many other examples as in sensors or catalysts or in wear or corrosion where this is equally true.

The special feature that unites all such themes is the opportunity that now exists for rapid advance. The combination of sophisticated instrumentation, of refined modelling, and of better characterised chemical systems provides an excellent foundation for progress in establishing a quantitative and comprehensive understanding of phenomena where surfaces and interfaces are involved. The significance of such work should not be underestimated: the search for deliberate and predictive control over the processing and properties of ceramics has been a long one; imaginative and systematic work on surfaces and interfaces is one of the most promising ways to further progress.

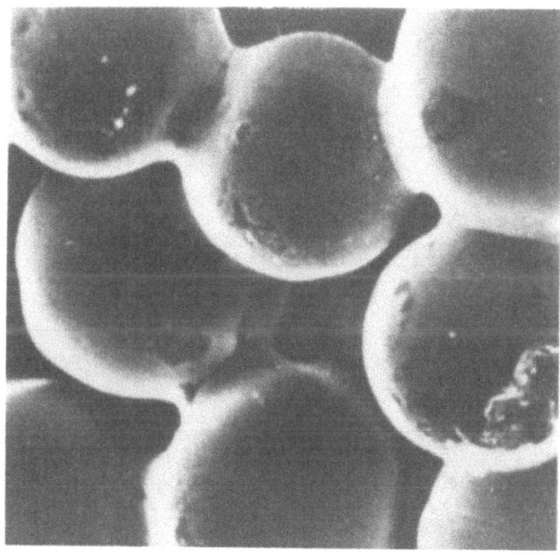

Figure 1. Differences in surface curvature drive the atomic movements which cause microstructural change during the densification of a powder. (Courtesy J.-L. Chermant).

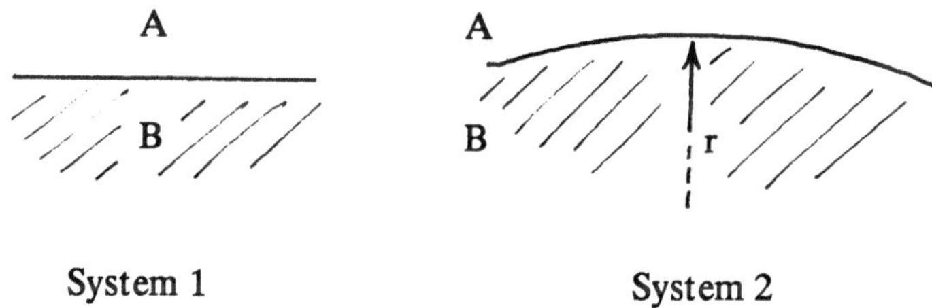

Figure 2. The gas pressure over the curved surface is raised by dP_A in system 2 relative to system 1 (Eq. 7); the normal stress in the solid is raised by dP_B for the curvature shown (Eq. 5).

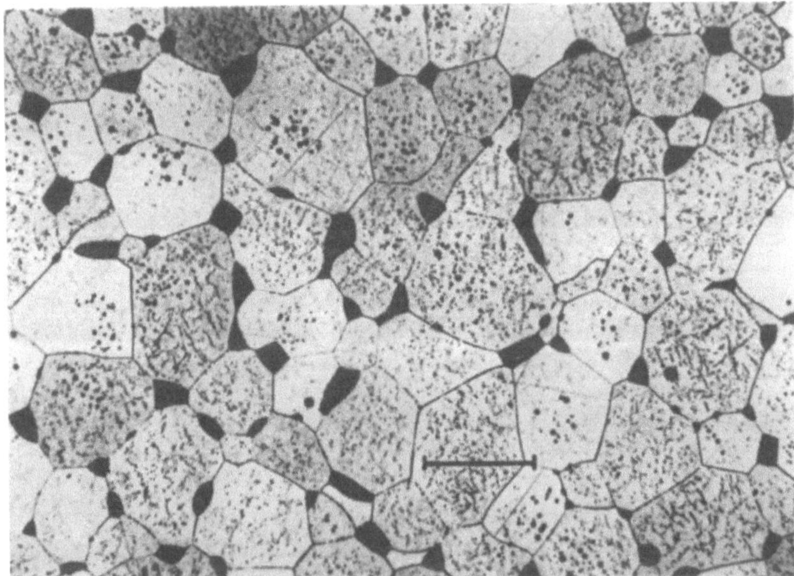

Figure 3. Atom movements from grain boundaries to pores (black) allow the shrinkage of the grain/grain inter-centre distance which represents densification. (CoO sample; 20μm marker; courtesy S.P. Howlett)

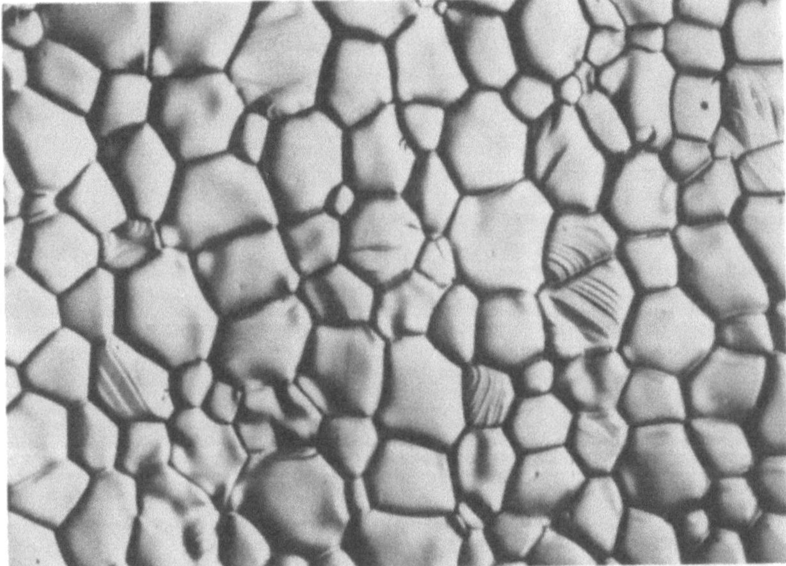

Figure 4. Completed densification in alumina (courtesy E.W. Roberts). The grain boundary curvature drives grain to grain atom movements which produce coarsening; as a result of the boundary motion, small grains disappear and the average size of the remainder increases.

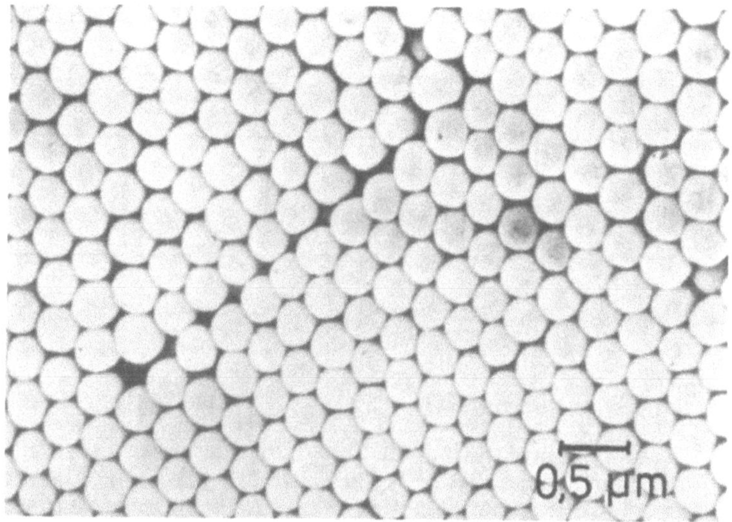

Figure 5. Chemical methods of powder production have resulted in particles specifically adapted for the sintering process. A difficulty with ordered arrays is that minor departures from size uniformity can produce large scale faulting; the existence of disordered regions between different regions is also a serious source of inhomogeneity. (SiO_2 spheres; courtesy S.J. Milne)

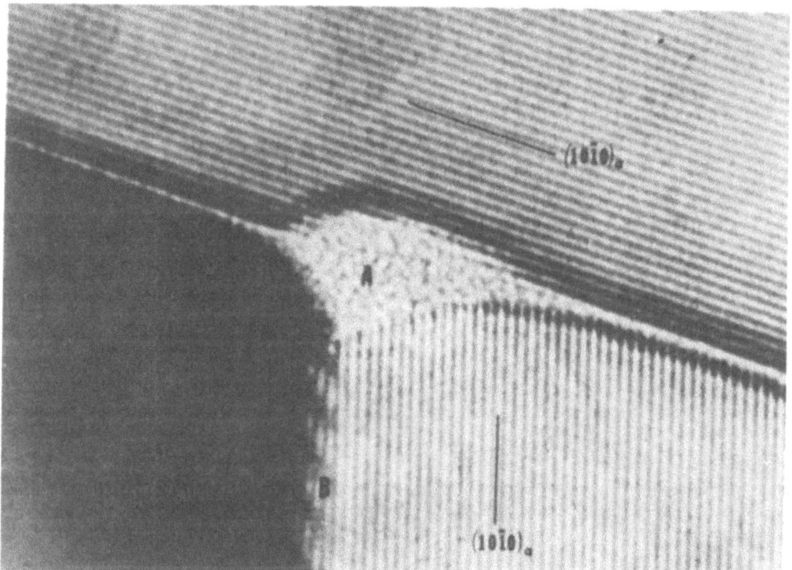

Figure 6. Interface structure between three grains of silicon nitride (courtesy D.R. Clarke). The presence of an amorphous boundary phase allows enhanced deformation under load at high temperature.

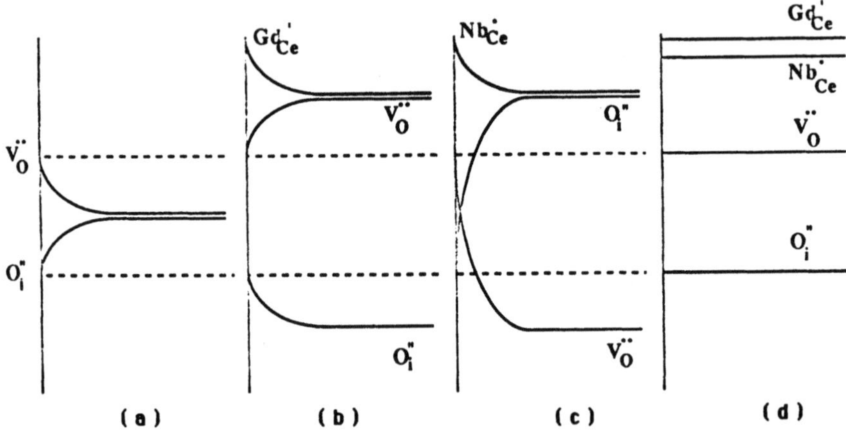

Figure 7. Defect concentrations in CeO_2 adjacent to an external surface or grain boundary (Courtesy Y.S. Zhen[30]). The schematic plots show the situation:

a) for the undoped material assuming anti-Frenkel disorder and a lower formation energy for the vacancy than for the interstitial

b) for material doped with gadolinium oxide

c) for material doped with niobium oxide

d) for co-doped material.

The dashed lines show the concentrations that would exist for the host defects on the basis of the individual formation energies, i.e., if requirements for charge balance in the bulk of the crystal could be relaxed.

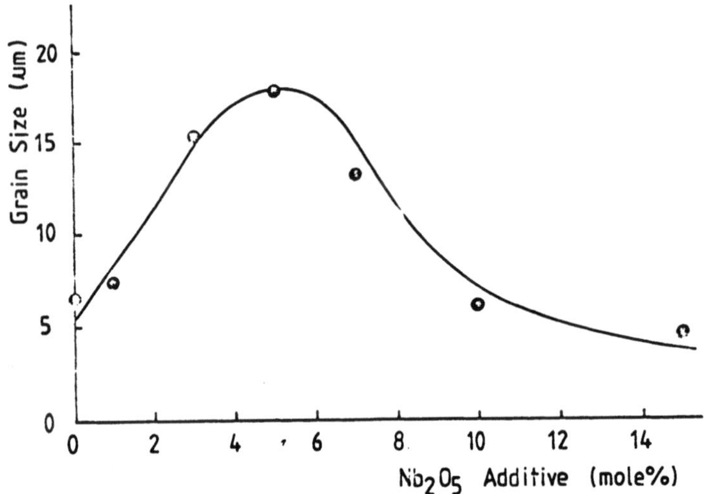

Figure 8. Grain size in cerium oxide after firing for 5 hours at 1600°C (Courtesy Y.S. Zhen[30]). The materials contain 7 mol% Gd_2O_3.

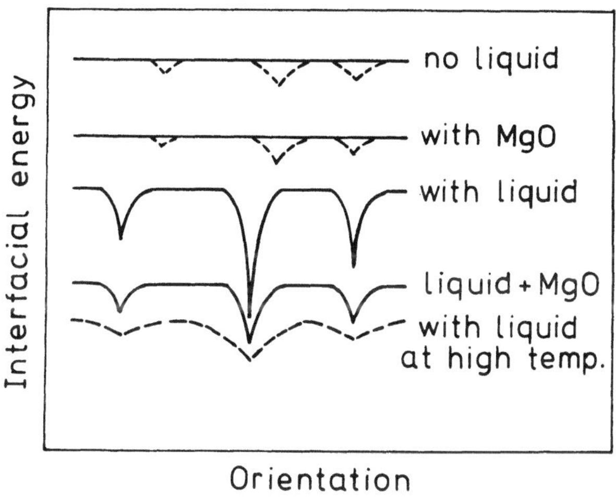

Figure 9. The dependence of grain shape on processing conditions in alumina[40].

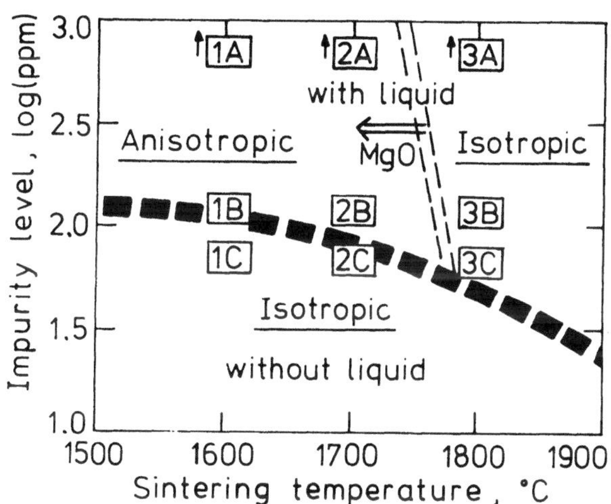

Figure 10. The pronounced shape anisotropy of grains in liquid-containing systems suggests that the orientation dependence of the solid/liquid interfacial energy is greater than that for solid/solid boundaries[40]. The degree of anisotropy is qualitatively indicated to match the experimental results shown in Fig. 9.

References

1. W.D. Kingery, J.Am.Ceram.Soc., **57** (1), 1-8 (1974).
2. S. Prochazka and R.M. Scanlan, J.Am.Ceram.Soc., **58** (1-2), 72 (1975).
3. Y.E. Geguzin, 'Physik des Sinterns', VEB Deutscher Verlag für Grundstoffindustrie, Leipzig (1973).
4. S.N. Freiman, Am.Ceram.Soc.Bull. **67** (2) 392-402 (1988).
5. F.F. Lange, J.Am.Ceram.Soc., **72** (1), 3-15 (1989).
6. N. Claussen, J. Steeb and R.F. Pabst, Am.Ceram.Soc.Bull. **56** (6), 559-562 (1977).
7. K.T. Faber and A.G. Evans, Acta Metall. **31** (4), 577-584 (1983).
8. V.D. Kostic, P.S. Nicholson and R.G. Hoagland, J.Am.Ceram.Soc., **64** (9), 499-504 (1981).
9. D.B. Marshall, B.N. Cox and A.G. Evans, Acta Metall. **33**, 2013-2021 (1985).
10. A.G. Evans and R.M. Cannon, Acta Metall. **34** (5), 761-800 (1986).
11. M.D. Thouless, O. Sbaizero, L.S. Sigl and A.G. Evans. J.Am.Ceram.Soc., **72** (4) 525-532 (1989).
12. E. Barringer, N. Jubb, B. Fegley, R.L. Pober and H.K. Bowen, pp. 315-333 in 'Ultrastructure Processing of Ceramics, Glasses, and Composites', ed. L.L. Hench and D.R. Ulrich, Wiley, New York (1984).
13. S.M. Wiederhorn, B.J. Hockey, R.F. Krause, Jr., and K. Jakus, J.Mat.Sci. **21**, 810-824 (1986).
14. J.G. Baldoni and S.T. Buljan, Am.Ceram.Soc.Bull. **67** (2) 381-387 (1988).
15. P. Vincenzini, ed., 'Ceramics in Surgery', Elsevier, Amsterdam (1983).
16. R. Mauczok and R. Wernicke, Philips Tech. Rev. **41**, 338 (1983/4).
17. H. Ihrig, Adv. in Ceram. **7**, 117 (1984).

18a. J. Nowotny and L.C. Dufour, eds., 'Surface and Near Surface Chemistry of Oxide Materials', Elsevier, Amsterdam (1988).

b. L.C. Dufour and J. Nowotny, eds., 'External and Internal Surfaces in Metal Oxides', Mat.Sci.Forum **29**, 1-303 (1988).

19. J.M. MacLaren, J.B. Pendry, P.J. Rous, D.K. Saldin, G.A. Somorjai, M.A. van Hove and D.D. Vredensky, 'Surface Crystallographic Information Service: Handbook of Surface Structures', Reidel, Dordrecht (1987).
20. D.M. Duffy and P.W. Tasker, Adv. in Ceram. **10**, 275-289 (1984).
21. W. Hirschwald, pp. 61-187 in Ref. 18a.
22. S. Hoffmann, Surface and Int. Analysis, **9**, 3-20 (1986).
23. J.W. Cahn, Acta Metall. **10**, 789 (1962).
24. D.R. McLean, 'Grain Boundaries in Metals', Oxford University Press, Oxford (1957).
25. K.L. Kliewer and J.S. Koehler, Phys.Rev., **140** (4A), 1226 (1965)
26. P. Wynblatt and R.C. McCune, pp. 247-279 in Ref. 18a.
27. R.F. Cook and A.G. Schrott, J.am.Ceram.Soc. **71** (1) 50-58 (1988).
28. W.D. Kingery, Pure Appl.Chem., **56** (12), 1703-1714 (1984).
29. A.J. Burggraaf and A.J.A. Winnubst, pp. 449-477 in Ref. 18a.
30. Y.S. Zhen, 'Oxygen Ion Conduction in Doped Rare Earch Oxides', Ph.D. Thesis, University of Leeds (1988).
31. R.J. Brook, W.H. Tuan and L.A. Xue, Ceramic Trans. **1**, 811-823 (1988).
32. S. Baik and C.L. White, J.Am.Ceram.Soc. **70** (9) 682-688 (1987).
33. A.M. Glaeser, H.K. Bowen and R.M. Cannon. J.Am.Ceram.Soc. **69** (2), 119-125 (1986). (But see in Y.M. Chiang and W.D. Kingery,

J.Am.Ceram.Soc. 72 (2) 271-277 (1989)).
34. S.J. Bennison and M.P. Harmer, J.Am.Ceram.Soc. 68, C: 22-24 (1985).
35. J. Rödel and A.M. Glaeser, J.Am.Ceram.Soc. 70 (8), C: 172 (1987).
36. Y. Kurokawa, K. Utsumi and H. Takamizawa, J.Am.Ceram.Soc. 71, 588-594 (1988).
37. D.R. Clarke, Ann.Rev.Mat.Sci. 17, 57-74 (1987).
38. F.M.A. Carpay and A.L. Stuijts, Sci.Ceram. 8, 23 (1976).
39. D.J. Srolovicz, G.S. Grest and M.P. Anderson, Acta Metall. 33, 2233-2247 (1985).
40. B. Hong, S.-J.L. Kang and R.J. Brook, for publication.
41. W.A. Kaysser, M. Sprissler, C.A. Handwerker and J.E. Blendell, J.Am.Ceram.Soc. 70 (5), 339 (1987).
42. J. White, Mat.Sci.Res. 6, 81-108 (1973).
43. R. Riedel, G. Passing and G. Petzow, Proc.2nd Int.Conf.Powd.Proc.Sci., Berchtesgaden, DKG Verlag, Köln (1989).
44. M.H. Lewis, G. Leng-Ward and S. Mason, Brit.Ceram.Proc. 39, 1-13 (1987).
45. R.W. Rice, Proc.Br.Ceram.Soc. 12, 99 (1969).
46. K.H. Jack, J.Mater.Sci. 11, 1135-1158 (1976).

ELECTRON SPECTROSCOPIC DETERMINATION OF THE ELECTRONIC, GEOMETRIC AND CHEMISORPTION PROPERTIES OF OXIDE SURFACES

VICTOR E. HENRICH
Surface Science Laboratory
Applied Physics
Yale University
New Haven, Connecticut 06520

ABSTRACT. The electronic and geometric properties of both nearly perfect and defect metal oxide surfaces have been studied by using surface sensitive electron spectroscopic techniques such as photoemission, electron energy loss spectroscopy, low energy electron diffraction, etc. and single crystal samples cleaved in ultrahigh vacuum; surface defects are created by ion or electron bombardment or thermal treatment. The use of such well characterized ceramic samples permits a determination of the physical and chemical properties of their surfaces on the atomic scale. The geometric structure of defect free oxide surfaces is nearly a termination of the bulk crystal structure, and their electronic structure is usually similar to that of the bulk. Maximal valence oxides are quite inert with respect to molecular adsorption. The interaction of sub-oxide surfaces with molecules, on the other hand, exhibits a wide range of behavior. Point defects on surfaces are the active sites for chemisorption and exhibit electronic structures very different from those of perfect surfaces.

I. Introduction

Many of the properties of ceramic materials, including grain boundary diffusion, adhesion of particles, electrical conductivity, degradation, etc., depend crucially on the properties of the surfaces of the individual crystallites. Therefore a knowledge of both the geometric and electronic properties of ceramic surfaces is important in order to understand their behavior and, ultimately, to modify it. To obtain a detailed picture of the properties of ceramic surfaces on the atomic scale from experiments on real ceramics is extremely difficult due to their complicated structure and morphology. Thus in the last few years surface scientists have begun to study the properties of single crystal surfaces of ceramic materials in order to eliminate many of the complexities present in most ceramics. In this way the fundamental electronic and geometric structure of metal oxides can be explored in a way that permits comparison with theoretical predictions.

This chapter will summarize some of the experimental and theoretical results that have been obtained on single crystal metal oxides, on the properties of defects on their surfaces, and on the manner in which various molecules interact

L.-C. Dufour et al. (eds.), Surfaces and Interfaces of Ceramic Materials, 1–28.

with both nearly perfect and defect surfaces. The emphasis will be on understanding the physics and chemistry of ceramic surfaces at the atomic level. Because of their importance in heterogeneous catalysis and the variety and complexity of the properties that result from the possibility of different cation valence states, the greatest attention has been paid to the transition metal oxides [1]. The oxides that have been best studied include TiO_2, $SrTiO_3$, $BaTiO_3$, Ti_2O_3, V_2O_3, α-Fe_2O_3, NiO, MnO, CoO, Fe_3O_4, SnO_2 and ZnO. Some work has been done on more conventional ceramics such as MgO, but not to the same level of detail as for transition metal oxides [2].

For the purpose of interpreting the surface properties of metal oxides, it is convenient to distinguish between **maximal valence oxides** such as TiO_2, $SrTiO_3$, ZnO, MgO and Al_2O_3, whose cations are in their highest valence states (i.e., fully oxidized), and **sub-oxides** such as Ti_2O_3, V_2O_3 and MnO, whose cations are in a partially reduced valence state. The distinction is particularly important with respect to chemisorption, and we will discuss the two classes of oxides separately in that section.

The methods used to determine the surface geometric and electronic properties of solids consist primarily of the very surface sensitive spectroscopies that are based on relatively low energy ($\lesssim$ 1000 eV) electrons or ions, whose inelastic scattering lengths in solids range from about 3 Å (i.e., roughly one atomic plane) to 20 - 30 Å [3]. All of the measurement techniques discussed in this chapter will be electron spectroscopies, with low energy ions used only to create point defects on surfaces in a controlled way. The most powerful technique for determining surface electronic structure in the region important for electrical conductivity, bonding and chemisorption is ultraviolet photoemission spectroscopy (UPS). In UPS, photons in the range of about 10 - 300 eV incident on a solid surface in an ultrahigh vacuum (UHV) environment excite electrons in surface orbitals or bands into normally empty states having energies above the vacuum level of the material. Some of these photoelectrons can escape from the solid into vacuum without suffering appreciable energy loss and are detected by an electron spectrometer. UPS spectra thus give, to first order, a picture of the density of occupied electron states in the surface region. When the incident photons have energies in the x-ray region ($\gtrsim$ 300 eV), electrons can be excited out of core levels of surface atoms. This is referred to as x-ray photoelectron spectroscopy (XPS) and gives information on the binding energies and valence states of the atoms. The joint density of states of electrons excited from filled orbitals to empty states within the solid can be studied by electron energy loss spectroscopy (EELS), in which the energy lost by a monoenergetic incident electron that excites an interband transition on the surface is measured after the electron has been scattered back from the surface; the process is essentially Raman scattering with electrons. While EELS spectra are not as easy to interpret as UPS or XPS spectra since the initial and final state corresponding to a particular loss energy are not unique, it is a useful method for measuring normally empty densities of states and can be used on highly insulating samples where surface charging renders the photoemission spectroscopies almost useless [4]. If the incident electron beam is highly monoenergetic and the electron spectrometer has sufficiently high resolution (FWHM $\lesssim$ 10meV = 80 cm^{-1}), then the energy lost by the incident electron in exciting surface phonons or molecular vibrations in adsorbed species can be measured; this use of EELS is referred to as high resolution electron energy loss spectroscopy (HREELS). When the atoms on a solid surface have long range order, monoenergetic electrons can be diffracted elastically from the surface and give rise to diffraction patterns on a fluorescent screen. This technique of low energy electron diffraction (LEED) is the primary method of determining the location of atoms on nearly perfect single crystal surfaces. For a thorough description of these and other surface analysis techniques, see Ref. 3.

Section II will consider the atomic geometry of oxide surfaces, with the structure of perfect surfaces as determined from LEED measurements and multiple scattering calculations discussed in Section IIA and surface defects in Section IIB. Section III discusses the electronic structure of ceramic surfaces as determined by UPS, XPS and EELS. Nearly perfect surfaces will be treated first, Section IIIA, followed by a discussion of the electronic properties of surface defects in Section IIIB. The adsorption of molecules on nearly perfect maximal valence oxide surfaces, perfect sub-oxide surfaces, and at surface defect sites will be discussed in Sections IVA, B and C, respectively.

II. Atomic Geometry of Ceramic Surfaces

Our knowledge of the position of atoms and ions on the surfaces of metal oxides comes almost exclusively from LEED [1]. (A small number of ion scattering measurements have been made; scanning tunneling microscopy has not yet been successfully applied on the atomic scale to ceramics.) LEED can only give positional information on perfect surfaces, and only a few metal oxides have been treated quantitatively by LEED. For other oxides only quantitative LEED measurements have been performed; i.e., the symmetry of the LEED pattern indicates whether there has been any reconstruction of the surface compared to a simple termination of the bulk lattice. For surfaces containing an appreciable number of defects, LEED gives no useful information and we must resort to modelling the types of defects that might be formed on a surface based upon a termination of the bulk lattice.

Since many metal oxides have a very localized electronic structure, and even conducting oxides can have some localized character to their bonds, it is useful to approach metal oxides from a localized ligand picture [5]. The O ions generally have nearly a closed shell O^{2-} configuration, even in the presence of surface defects, so the electronic structure is determined primarily by the cation electron configuration. Thus we will discuss both the geometric and electronic structure of oxides in terms of the cations and their ligands. The cations in the bulk of oxides having the rocksalt, rutile, perovskite and corundum crystal structures, which include most of the oxides that we will discuss here, all reside in sites having six O^{2-} ligands. The octahedron of O^{2-} ions is regular only in rocksalt, with the degree of distortion varying in the other structures. ZnO has the wurtzite structure, in which the Zn^{2+} ions are surrounded by a tetrahedron of four O^{2-} ions. Both the perfect surface structure and surface defect sites will be characterized in part by the reduction of the ligand coordination of the cations on the surface compared to those in the bulk.

A. Nearly perfect surfaces

The cations on the surface of a metal oxide generally have a lower coordination of O^{2-} ligands than in the bulk [1]. As an example, a model of the (100) cleavage face of the rocksalt structure is shown in Fig. 1. (This is a space filling, spherical ion model of an unrelaxed termination of the bulk lattice, with the radii of the spheres chosen to be nearly the relative ionic radii of the M^{2+} and O^{2-} ions.) Figure 1 includes a monoatomic step from one (100) terrace to another, and both cation and anion vacancies; these will be discussed in Section IIB below. On the perfect (100) surface the ligand coordination of both cations and anions is reduced to five; it is further decreased at defect sites. The five-fold coordination of the surface cations on rocksalt (100) is similar to that on the most stable surfaces of most other octahedrally coordinated ceramics; see Ref. 1 for a full discussion.

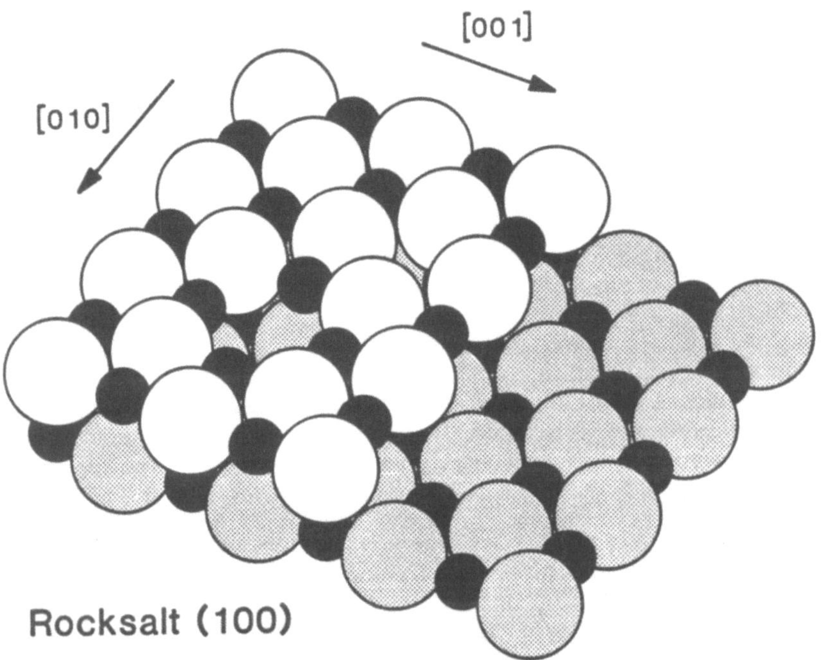

Figure 1. Model of the rocksalt (100) surface. Large circles are O anions, small circles are metal cations. A (010) step to another (100) terrace is shown, as are both missing anion and missing cation point defects.

The only metal oxides whose surfaces have been studied quantitatively by LEED in order to determine accurately the location of cations and anions on the surface are the rocksalt oxides MgO, CoO, NiO and CaO [6,7]. By far the most stable surface in the rocksalt lattice is (100), and all of those oxides cleave well along that plane (see Fig. 1.) The LEED results indicate that the (100) surface is very nearly a truncation of the bulk crystal structure, with at most a rumpling of a few percent of the interionic spacing. LEED studies of ZnO showed that the $(000\bar{1})$-O and $(11\bar{2}0)$ surfaces are very nearly a truncation of the bulk, while Zn ions on the (0001)-Zn polar face relax inward by about 0.2 Å [8,9]. The non-polar $(10\bar{1}0)$ surface exhibits significant reconstruction, with the Zn ions relaxed inward by 0.45 Å; there may also be a slight inward relaxation of the O ions.

The corundum oxides Ti_2O_3 and V_2O_3 cleave well along the $(10\bar{1}2)$ plane, which is the plane that contains the octahedral sites that are not occupied by cations in the bulk crystal structure. All cations on the $(10\bar{1}2)$ face are five-fold O coordinated. Qualitative LEED patterns indicate that the Ti_2O_3 $(10\bar{1}2)$ cleavage face has a low density of surface defects and is not reconstructed (although the actual position of the atoms is not known since quantitative LEED measurements have not been made) [10-12]. The V_2O_3 $(10\bar{1}2)$ surface contains a higher density of defects than does Ti_2O_3 [13]. Although Al_2O_3 has the same corundum structure, it does not exhibit any good cleavage planes [2].

The rutile oxides TiO_2 and SnO_2 also do not cleave, but qualitative LEED measurements show that the most stable face is the (110); it can be produced by annealing in the appropriate pressure of O_2 [14,15]. The (110) face contains equal numbers of five- and six-fold coordinated cations. Other crystal faces of the rutile lattice that have been studied are the (100), which contains all five-fold coordinated surface cations and reconstructs to give (1 x 3), (1 x 5), and (1 x 7) LEED patterns [16], and (001), which contains only four-fold coordinated cations and is very unstable, faceting into planes that have higher O coordination of the cations [17].

Two ceramics having the perovskite structure have been studied in single crystal form: $SrTiO_2$ [18] and $BaTiO_3$ [19]. The most stable surface appears to be the (100). There are two (100) planes that alternate in the perovskite structure, one has a composition of SrO (or BaO), and the other has a TiO_2 composition. Since only the Ti ions affect the electronic structure in the vicinity of the conduction and valence bands, we will consider only that surface, which contains all five-fold coordinated cations. Again no quantitative LEED measurements have been performed for perovskite oxides.

B. Surface defects

There are three general categories of defects that can (and generally are!) present on crystal surfaces [20]: point defects such as vacancies and adatoms; extended defects including steps, kinks and edges; and the intersection of dislocations and grain boundaries with surfaces. Since other chapters deal with dislocations and grain boundaries, we will not consider them here.

Little is known about the properties of steps, kinks, etc. on oxide surfaces. They are certainly present even on the best cleaved surfaces, and regular arrays of steps have been produced on some oxide surfaces [21]. The location of atoms at steps has not been determined empirically, so we must resort to idealized models such as that for rocksalt (100) in Fig. 1. A general property of atoms at steps is a reduction in their ligand coordination. For the (010) step shown on the rocksalt (100) surface, both cations and anions at step edges have only four of the six ligands that they have in the bulk; a further reduction in ligand coordination occurs at kinks and corners. Although no experimental data exist for the actual positions of atoms at steps, theoretical calculations for rocksalt monoxides indicate a significant degree of relaxation of the atoms, no doubt to compensate for the reduced ligand coordination [22-25].

The surface defects that have received the greatest amount of attention are point defects, since their electronic structure is significantly different than that of the perfect surface, and since they are generally the most active sites for chemisorption and catalysis. Two point defects are shown in Fig. 1: an O ion vacancy, or F_s center; and a cation vacancy, referred to as a V_s center. No experimental information is available on the relaxation that must certainly accompany vacancies, so they are usually considered, as in Fig. 1, in terms of the unrelaxed structure of the perfect lattice. It must be borne in mind that point defects on real surfaces are more complicated than model defects like those shown in Fig. 1.

There are two important properties of point defects on ceramics: the ligand coordination of the cations (anions) adjacent to an anion (cation) vacancy is reduced, and charge must be trapped at the vacancy site in order to maintain local charge neutrality [1]. We will return to the latter point at length in Section IIIB below. In the studies on single crystals considered in this chapter, one usually begins with cleaved or fractured surfaces, which may contain a fairly high density of extended defects such as steps but have a very low density of point defects. Point defects are then created in a controlled manner by electron or ion bombardment or by heating to a high enough temperature that a fraction of a monolayer of surface defects is in thermodynamic equilibrium and then quenching to freeze in the defect distribution. The defects that are created in this way are predominantly O ion vacancies, so we will concentrate on those. At an O vacancy on the rocksalt (100) surface (see Fig. 1), the ligand coordination of the four adjacent cations is reduced from five to four. On the corundum $(10\bar{1}2)$ surface, an O vacancy reduces the ligand coordination of two adjacent cations from five to four. The same is true of the perovskite (100) TiO_2-like plane and the rutile (100) plane. The rutile (110) surface is a bit more complicated (see Refs. 1 and 26), but again surface O vacancies reduce the coordination of adjacent cations by one. This reduced ligand coordination of surface cations at O vacancy defect sites is important in both electronic structure and in chemisorption.

III. Surface Electronic Structure

The surface electronic structure with which we are concerned in this chapter is that within a few electron volts of the Fermi level, E_F, and the vacuum level. We are restricting ourselves to that region of energy since it contains the valence band, the conduction band and low lying empty electronic levels; these are the electronic states that are responsible for chemical bonding, electrical conduction, adhesion, chemisorption and catalysis, etc. As discussed in Section I, the most useful experimental tool for determining the density of filled electronic levels is UPS, and most of what we know about surface electronic structure has been obtained by using that technique. Some information about normally empty electronic levels can be obtained by using EELS, and XPS can be useful for determining the valence state of surface ions.

A. Nearly perfect surfaces

The electronic structure of the surfaces of several metal oxides has been studied on nearly perfect surfaces prepared by cleaving single crystals in UHV [1]. One of the first oxides to be studied comprehensively, and one that has received more attention than any other ceramic, is the rutile form of TiO_2. In the bulk all Ti cations have six O^{2-} ligands and have a Ti^{4+} ($3d^0$) electronic configuration. The O 2p valence band is filled and the Ti 3d conduction band is empty, with an absolute bulk bandgap of 3.1 eV. The first theoretical calculations of the electronic structure of TiO_2 surfaces used an LCMTO slab approach and considered the effect of the reduced O^{2-} ligand coordination of the cations at the surface from six- to five-fold on the (110) surface and to four-fold on the (001) surface [27]. The (110) surface was predicted to have a slightly narrower bandgap than the bulk, but the (001) surface was calculated to have a very high density of both cation- and anion-derived states throughout the bulk bandgap region. Subsequent UPS measurements on both surfaces showed that in fact there were no intrinsic surface states in the gap for either geometry [14]. The measured surface electronic structure was strikingly similar to that of bulk TiO_2, regardless of the reduced ligand coordination of the surface cations. Thus, at least for TiO_2, the reduction of ligand coordination of cations is not in itself sufficient to alter the surface electronic structure. Subsequent theoretical calculations using a more complete scattering theoretic approach predicted that neither the (110) nor the (001) surface would exhibit any intrinsic bandgap surface states, in agreement with experiment [28-30].

Subsequent work on many oxides has shown that in general the measured density of filled electronic states on nearly perfect surfaces is indistinguishable from that of the bulk. This statement must be qualified by noting that in most cases the bulk electronic structure and density of states of metal oxides are not nearly as well understood as those for metals and semiconductors, so it is not possible to determine subtle differences between bulk and surface structure; this is an area where much work remains to be done.

The surface electronic structure of perhaps the most technologically important ceramic, Al_2O_3, is not well understood since its highly insulating nature has prevented accurate UPS measurements from being performed on single crystals [2]. Some UPS measurements have been performed on thin Al_2O_3 layers on single crystal Al [31], and XPS measurements of the valence band region on single crystal Al_2O_3 samples have been performed [32]; the only occupied electronic levels that are observed are those corresponding to the filled O 2p valence band, which also looks just like that expected for the bulk valence band in Al_2O_3 [33]. Similar XPS measurements on MgO (100) surfaces also show evidence for an O 2p valence band indistinguishable from that of the bulk [34].

The class of ceramics that is currently receiving the greatest attention from surface scientists consists of the rocksalt transition metal monoxides. These materials are interesting for reasons ranging from their basic physics to their behavior as heterogeneous catalysts and industrial ceramics. The electronic structure of the (100) cleavage faces of MnO, CoO and NiO have been measured on single crystals [35-41]; Figure 2 shows UPS spectra taken at a very surface sensitive photon energy for these three surfaces. The spectra exhibit just the features predicted for the bulk electronic structure based upon the effect of ligand field and exchange interactions on the cation d-orbitals [5,42]. All three oxides are antiferromagnetic insulators in which the majority spin 3d orbitals are all populated before any electrons occupy minority spin levels. The Mn^{2+} ion in MnO has a $3d^5$ configuration corresponding to a half filled d-band; it is thus predicted to be an insulator, as observed experimentally. In NiO, where the Ni^{2+} ion has a $3d^8$ configuration, both the majority spin band and the minority spin $t_{2g}\downarrow$ band are filled, while the minority spin $e_g\downarrow$ band, which is capable of holding the last two d-electrons, is empty. Theory thus predicts that NiO would also be an insulator, as observed. The $3d^7$ electronic configuration of the Co^{2+} ion in CoO poses a problem, however, since only two electrons occupy the $t_{2g}\downarrow$ band, which is capable of holding three electrons as in Ni^{2+}. This would normally lead to CoO being metallic, but it is in reality a fairly good insulator; theoretical work is currently underway to resolve this difference.

The electronic structure described above is clearly visible in the UPS spectra in Fig. 2. The two peaks at binding energies of about 3 and 5 eV in each spectrum correspond to emission from the O 2p π and σ orbitals, while the lower binding energy peaks arise from emission from the cation 3d orbitals; for CoO and NiO the $e_g\uparrow$ and $t_{2g}\downarrow$ orbitals have nearly the same energy and are not resolvable in the UPS spectra. So for MnO, CoO and NiO the nearly perfect surface exhibits a density of filled states that is essentially the same as that of the bulk, consistent with observations on most other metal oxides.

The electronic structure of two sub-oxides having the corundum structure, Ti_2O_3 and V_2O_3, have been rather thoroughly studied by UPS [10-13, 43-49]. Figure 3 shows the UPS spectra for UHV cleaved $(10\bar{1}2)$ surfaces of those materials. In the bulk the Ti^{3+} ions have a $3d^1$ configuration and the V^{3+} ions are $3d^2$. The two oxides are of interest since their surfaces, which are both $(10\bar{1}2)$ cleavage planes, differ only in the d-electron population on the cations. They thus constitute an excellent system in which to separate the geometric and electronic components of surface electronic structure. What is known about the electronic levels in bulk Ti_2O_3 and V_2O_3 comes primarily from optical measurements [5]. Ti_2O_3 is a semiconductor at room temperature since it has a filled a_{1g} band (which results from a pairing of nearest-neighbor cations along the c-axis of the corundum lattice) separated from the nearest empty band (e_g^{π}) by 0.1 eV. V_2O_3 is metallic at room temperature, with its two d-electrons per cation partially occupying both the a_{1g} and e_g bands. Both materials exhibit a bulk bandgap of 2 - 2.5 eV between the cation bands and the filled O 2p band.

The UPS spectra show clearly the semiconducting and metallic nature of

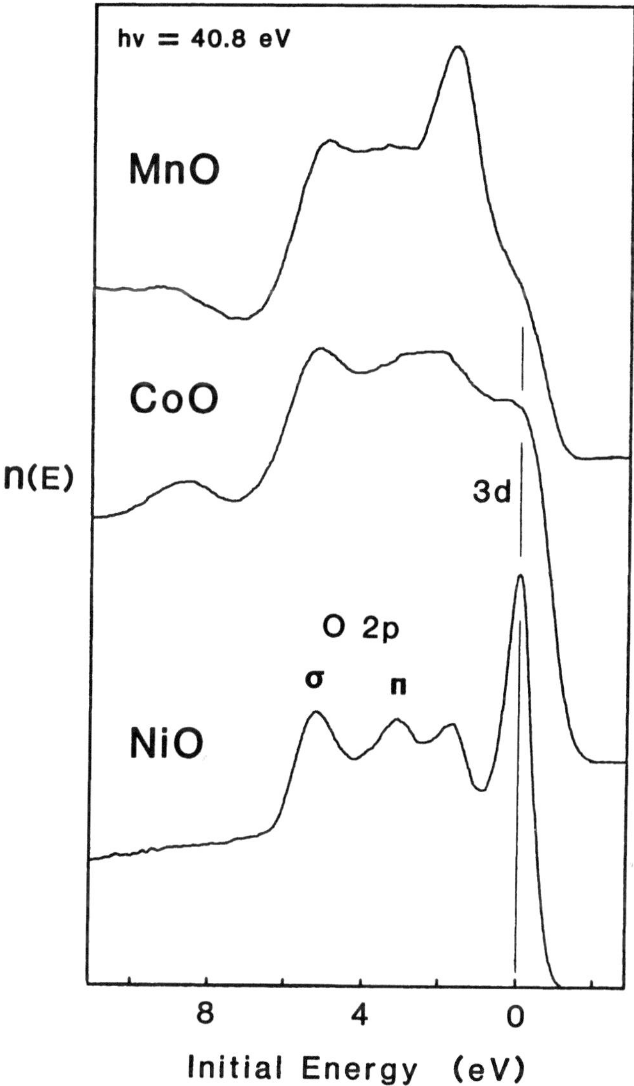

Figure 2. UPS spectra for vacuum cleaved (100) surfaces of MnO, CoO and NiO. The spectra have been aligned at the majority spin $e_g\uparrow$ band.

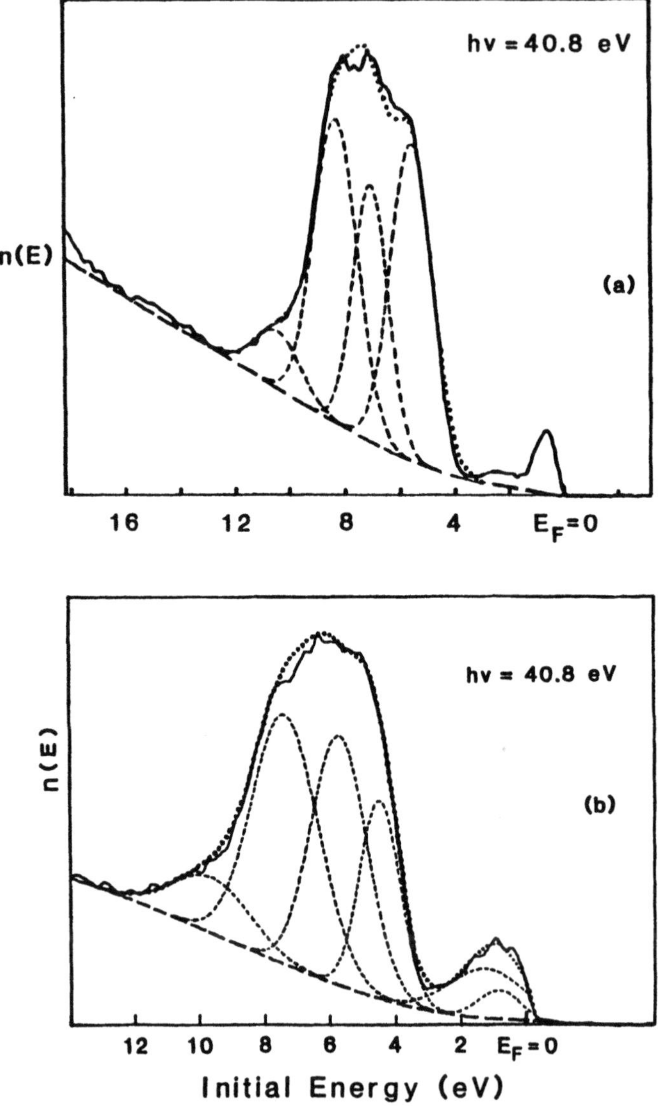

Figure 3. UPS spectra for vacuum cleaved $(10\bar{1}2)$ surfaces of (a) Ti_2O_3 and (b) V_2O_3. The dotted and dashed lines are Gaussian fits to the spectra above an extrapolated background.

Ti_2O_3 and V_2O_3, respectively, by the density of occupied electronic states at E_F; the density is very low in Ti_2O_3 since E_F lies in the narrow $a_{1g} \rightarrow e_g^{\pi}$ bandgap, and it is high in V_2O_3 since E_F is located within the bands where the density of states is large. The 2.5 eV bandgap is also apparent in Ti_2O_3, in agreement with the bulk electronic structure. For V_2O_3, however, there is no clear gap between the O 2p and V 3d bands. This is one of the few cases where the density of states of a nearly perfect oxide surface is distinguishably different from that of the bulk. Experiments involving chemisorption, the creation of defects on the surface, and angle resolved UPS measurements indicate that the density of states seen in Fig. 3 does in fact correspond to intrinsic surface states, although the separation of bulk and surface contributions to the spectra has not yet been completely sorted out [45-48].

The spectrum of empty electronic levels on the (100) surface of MgO has been studied by EELS for transitions originating on both the O 2p and Mg 2p core levels [4]. Intra-ionic excitations on the Mg^{2+} ion result in excitonic states that can be well described in terms of the electronic levels of isolated Mg^{2+} ions. By varying the incident electron energy it is possible to vary the depth into the material that is sampled; it is thus possible to separate bulk and surface contributions to the spectra. The bulk losses agree well with the known excited state spectra of the free Mg^{2+} ion, while the surface loss features are shifted to lower energy than those of the bulk. The surface spectra can be explained by a Stark shift of the free ion levels due to the strong, localized electric fields at the surface of an ionic material that arise from the gradient of the Madelung potential that exists when the bulk lattice is terminated.

Inter-ionic transitions between the O 2p levels and empty Mg orbitals give a loss peak at about 6 eV, which is below the bulk bandgap of 7.78 eV in MgO [4]. The 6 eV peak is an intrinsic surface loss corresponding to a lowering of the energy of surface Mg^{2+} levels due to the change in the Madelung potential at the surface mentioned above. Theoretical cluster calculations are in agreement with the observation of this intrinsic surface state [50,51].

B. Surface Defects

Whenever defects are present on ceramic surfaces, one must consider possible changes in the electronic structure of the ions at or near the defect. As discussed in Section IIB, the ions at step edge sites have a lower ligand coordination than those on the terraces of the perfect surface. Since it is difficult to produce regular step arrays on most oxide surfaces, little is known about the details of the electronic structure at steps. But UPS, EELS, etc. measurements on UHV fractured TiO_2 and $SrTiO_3$ surfaces, where the LEED patterns indicate that as many as 10% of the surface cations occupy step sites, indicate that there are no major differences between the electronic structure of steps and terraces [14,18]. This is in qualitative agreement with the results described in Section IIIA for the similarities in the electronic structure of TiO_2 surfaces having different cation ligand coordinations.

Striking changes in electronic structure occur when point defects, such as the predominant O vacancies, are present on a surface. The properties of point defects have been explored most thoroughly on transition metal oxides since the ability of the cations to assume different valence states can result in large changes in electronic structure [1,26]. As mentioned in Section IIB above, the electronic structure of point defects is determined primarily by the necessity of maintaining local charge neutrality in the vicinity of the defect. For O-vacancy defects such as that in Fig. 1, which we shall primarily consider here since they are the predominant type observed on ionic oxides, removal of an O ion having a valence close to -2 requires that roughly two electrons be trapped at each vacancy. A few theoretical cluster calculations have been performed for O-vacancy defects [1,51], and the trapped charge is found to be localized in the vacancy but to partially overlap the adjacent cations which, as discussed in Section IIB, have fewer O^{2-} ligands than do cations on perfect terraces. For simplicity it is customary to describe the electronic structure of a surface defect in terms of an increased population of the lowest lying normally empty orbitals on adjacent cations. For example, O-vacancy defects on TiO_2 are said to create Ti^{3+} ($3d^1$) ions adjacent to it, although the trapped charge is in reality only partially occupying those normally empty cation levels. The O^{2-} ions near an O-vacancy defect do not play a major role in the charge distribution since they already have filled 2p shells and cannot accept any additional charge.

The effects of O-vacancy point defects on the electronic structure as measured by UPS are most striking in maximal valence transition metal oxides such as TiO_2 and $SrTiO_3$. When O-vacancy point defects are created on those surfaces, a new band of surface states appears in the bulk bandgap just below the bottom of the bulk conduction band [52]; for large defect densities that band may overlap the bulk conduction band (see the clean surface spectrum for Ar^+-ion bombarded $SrTiO_3$ in Fig. 10 below). When the surface defect density is large enough (roughly one-half monolayer) the surface of the ceramic becomes electrically conducting due to the ability of charge to hop from one defect state to another; this is visible in the UPS spectrum as a finite density of states at E_F.

For oxides that already have partially filled d-bands, the increased electronic charge associated with an O-vacancy defect may either appear in UPS spectra either as an increase in the amplitude of the d-band emission or, if the sub-bands are filled in the stoichiometric oxide (e.g., NiO), as a shoulder at lower binding energy as electrons begin to occupy the next lowest lying unoccupied d-orbitals. This can be seen for CoO in Fig. 4, where the three curves correspond to the cleaved surface, that surface after bombardment to steady state by 500 eV Ar^+ ions, and the (100) surface after bombardment by 5,000 eV Ar^+ ions. The 500 eV bombarded surface is missing about 4% of the O^{2-} ions in its top layer, and the electrons at the defect sites give rise to a broadening of the low energy edge of the valence band (compare with Fig. 2). After 5,000 eV ion bombardment, about 16% of the surface O^{2-} ions are missing, resulting in the appearance of a shoulder on the upper edge of the valence band. CoO is unique in that its surface becomes <u>more insulating</u> (as determined by the charging observed in UPS spectra) when defects are created on the (100) surface; all other oxides studied become more conducting. This increase in resistivity as electrons are trapped at defect sites probably arises from the fact that the $3d^8$ electronic configuration of a Co^+ ion is more stable (i.e., the $t_{2g}\downarrow$ band is now filled) than the $3d^7$

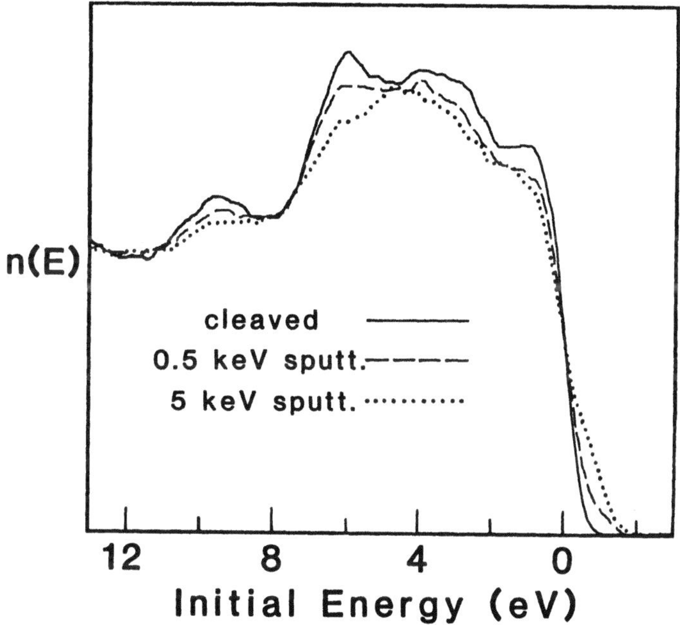

Figure 4. UPS spectra of vacuum cleaved CoO (100) (solid curve), after bombardment with 500 eV Ar^+ ions (dashed curve), and after subsequent bombadment with 5 keV Ar^+ ions (dotted curve). $h\nu = 40.8$ eV.

configuration of the Co^{2+} ion [40].

Extrinsic surface defect states have been observed in the joint density of states as measured by EELS in MgO [4,53]. When the nearly perfect MgO (100) surface is bombarded by ions or electrons, a new loss feature appears at about 2 eV that can be eliminated by either O_2 exposure or annealing. The surface defect responsible for the 2 eV loss peak has been interpreted as both an O^{2-} vacancy [4] and a Mg^{2+} vacancy [53]; the energy of the loss peak compared to calculations of surface defect transitions suggests that the latter interpretation is more likely correct.

The changes in cation electronic structure that accompany surface point defect creation can also be observed by using XPS to measure core level binding energies. Figure 5 shows the Fe $2p_{3/2}$ region of the core level XPS spectrum for an α-Fe_2O_3 (0001) surface that had been Ar^+-ion bombarded at room temperature (which reduces the surface and creates both Fe^{2+} and F^0 ions and atoms) and that surface after annealing in UHV at two different temperatures [54,55]. Annealing restores most of the missing O^{2-} ions to the surface via diffusion from the bulk, eventually eliminating all of the zero-valent Fe and greatly reducing the number of Fe^{2+} ions on the surface. The Fe^{2+} ions could never

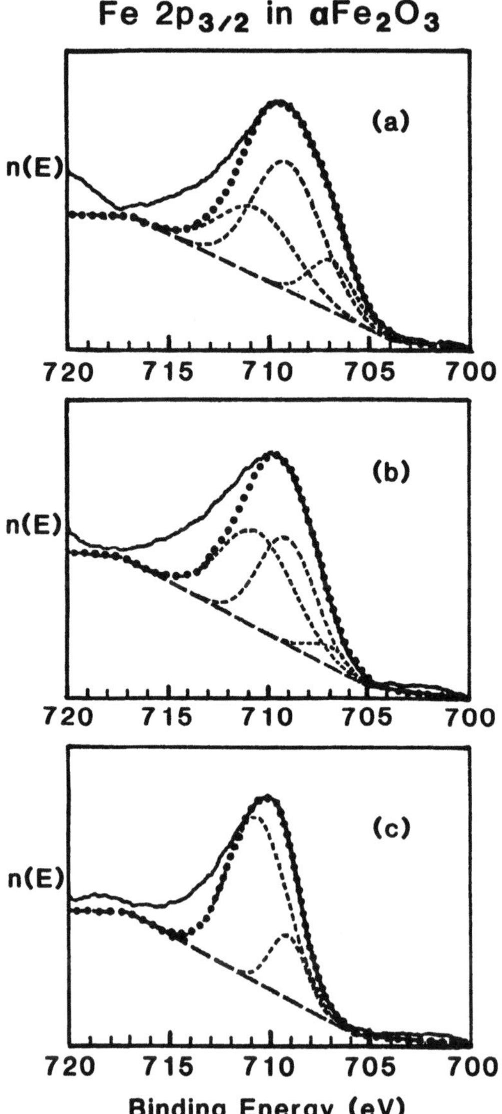

Figure 5. Gaussian fits to the Fe $2p_{3/2}$ core level in α-Fe_2O_3 for surfaces that have been (a) Ar^+ ion bombarded, (b) annealed to 973 K, and (c) annealed to 1093 K. The three Gaussians correspond, from higher to lower binding energy, to Fe^{3+}, Fe^{2+} and Fe^0 contributions to the spectra. $h\nu$ = 1487 eV.

be completely eliminated from the α-Fe_2O_3 (0001) surface by annealing, however, and LEED patterns give a clue as to what is occurring on the surface [54,56]. Figure 6 shows LEED patterns taken for three different annealing conditions. Figures 6(a) and (b) show reconstructions of the (0001) surface compared to the bulk termination of α-Fe_2O_3, with the complex pattern in Fig. 6(b) indicating the presence of an incommensurate overlayer (i.e., a monolayer having a different lattice constant) on top of the bulk oxide. Comparisons of LEED patterns with the known crystal structures of Fe_xO, α-Fe_2O_3 and Fe_3O_4 suggest that for certain conditions of temperature and O_2 partial pressure a monolayer of Fe_xO (111) on α-Fe_2O_3 (0001) occurs, while for different conditions an Fe_3O_4 (111) monolayer is stable on the α-Fe_2O_3 substrate [56]. The surfaces of iron oxides can thus be extremely complicated and depend strongly on processing conditions. The surface phases that are observed to be stable on α-Fe_2O_3 are in general different than those predicted by the bulk Fe-O phase diagram.

IV. Molecular adsorption on ceramic surfaces

The way in which molecules interact with ceramic surfaces depends upon whether or not the ceramic is a maximal valency oxide and whether or not surface defects are present; we will consider the two classes of oxide surfaces, and surface defects, separately. The most useful tool for studying chemisorption has been UPS, since the UPS spectra of a surface are modified upon adsorption by the addition of emission from the molecular orbitals of the adsorbed species. A technique for determining the nature of adsorbed species, referred to as "photoemission difference spectroscopy," consists of subtracting the clean surface UPS spectrum from the spectrum for the same surface after molecular adsorption [1]. The resulting difference spectrum contains the molecular orbital emission from the adsorbate and any changes caused in substrate electronic structure by interaction with the adsorbate. It is often possible to separate the molecular orbital emission from substrate changes, thus permitting the identification of the adsorbed species. Other electron spectroscopic techniques are useful for furnishing complementary information; only XPS and HREELS will be discussed here.

A. Perfect maximal valence oxide surfaces

Nearly perfect surfaces of maximal valence oxides, whose surface cations have no electrons to donate to adsorbed molecules, are generally found to be inert to most molecules at room temperature [1], with no observed changes in any spectra for exposures of 10^6 L or more (L = Langmuir = 10^{-6} Torr-sec). Since a 1 L exposure would result in roughly one adsorbed monolayer if every molecule or atom that struck the surface were to stick (i.e., if the sticking coefficient were 1), sticking coefficients on nearly perfect maximal valence oxide surfaces are small indeed. The number of molecules and cleaved single crystal oxide surfaces studied to date is not large, but enough systems have been examined in order for trends to become apparent.

One interesting exception to this inertness is for CO exposure of NiO, which is inert to other simple molecules studied [37]. No interaction is observed until

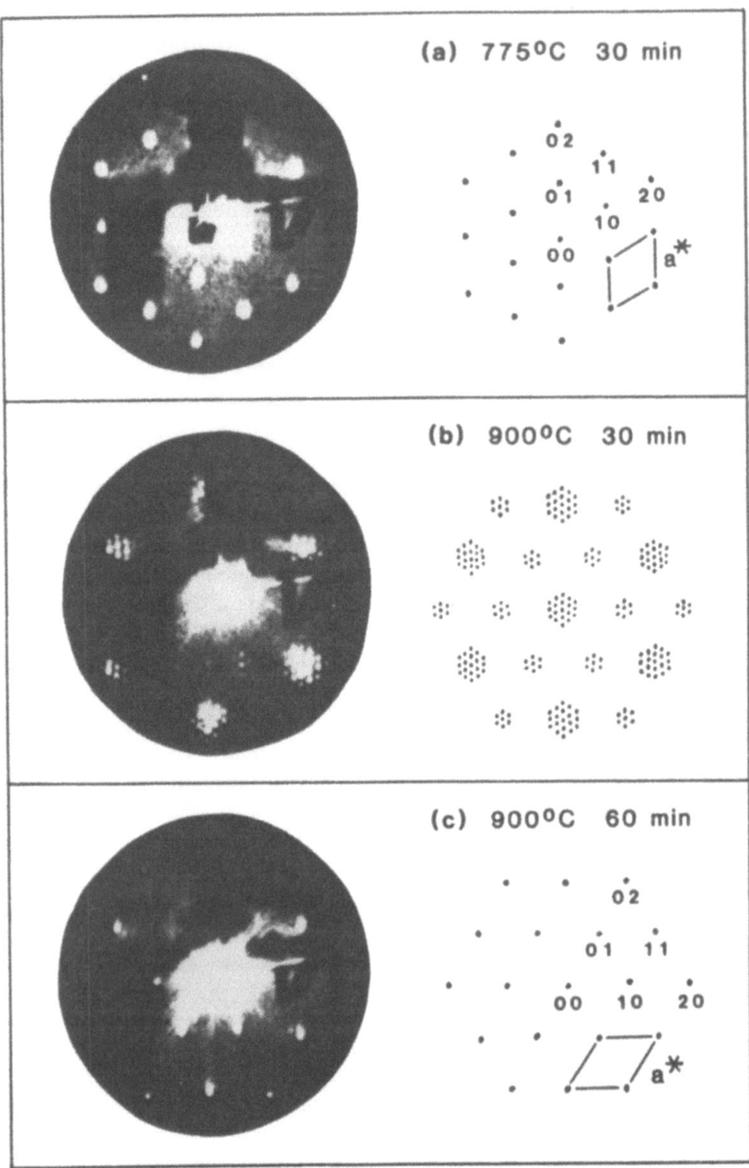

Figure 6. LEED patterns for 70 eV incident electrons from the α-Fe_2O_3 (0001) surface after annealing at different conditions in 10^{-6} Torr O_2. The indices in the schematic LEED patterns on the right correspond to the diffracted beams generated by (a) Fe_3O_4 (111), (b) an incommensurate overlayer, and (c) α-Fe_2O_3 (0001) surfaces.

about 10^9 L CO exposure, at which point the CO appears to reduce the NiO surface by abstracting O^{2-} ions to form CO_2 and leaving an O-vacancy defect behind.

One transition metal oxide that does not obey the pattern at all is MnO. The cleaved MnO (100) surface interacts strongly with O_2, H_2O and CO [38]. H_2O is adsorbed as OH^- radicals at low exposures, while CO is adsorbed molecularly. O_2 interacts more strongly with MnO (100) than either H_2O or CO and will even displace adsorbed OH^- or CO. This behavior is demonstrated in Fig. 7, which shows the MnO (100) + O_2 spectrum (bottom trace) compared to MnO (100) + H_2O and MnO (100) + CO both before and after subsequent exposure to O_2. The O_2 appears to virtually eliminate all features in the spectra due to OH^- or CO, replacing them with the MnO (100) + O_2 spectrum.

HREELS spectroscopy has been used to study the interaction of H_2O with annealed $SrTiO_3$ (100) surfaces [57,58]. In HREELS the vibrational frequencies of adsorbed species are determined by measuring the energy lost by an incident electron as it excites one quantum of vibrational excitation in the adsorbate. Figure 8 shows HREELS spectra of $SrTiO_3$ (100) before and after room temperature exposure to H_2O. The clean surface spectrum contains peaks due to the excitation of Fuchs-Kliewer surface phonon modes, but the overtones in the spectrum, which occur in the energy range of the molecular vibrations and complicate their interpretation, have been removed by a deconvolution procedure [58]. The two loss peaks that appear after H_2O exposure correspond to the symmetric O-H stretching mode and the H-O-H scissors bending mode. H_2O thus adsorbs molecularly on this surface.

B. Adsorption on perfect sub-oxide surfaces

While maximal valence oxide surfaces react weakly if at all with most molecules at room temperature, a much wider range of molecule-surface interactions is exhibited by sub-oxides of the transition metals. This is presumably due in part to the availability of electrons in weakly bound cation orbitals that are available to interact with adsorbed species. For example, when nearly perfect Ti_2O_3 $(10\bar{1}2)$ surfaces are exposed to O_2, the electrons are removed from the Ti 3d a_{1g} orbital (see Fig. 3) of the first monolayer and a half of Ti^{3+} ions, oxidizing them to Ti^{4+} $3d^0$ [10]. The electrons are transferred to the O_2, creating a negative adsorbed oxygen species. The reaction stops after a few L exposure, with no subsequent interaction of the surface with O_2. A similar interaction occurs for exposure to H_2O, although the interaction with the Ti 3d electrons is much weaker and only a slight depopulation of the a_{1g} band is observed [11]. UPS difference spectra show that the perfect Ti_2O_3 surface adsorbs H_2O molecularly, with the interaction ceasing after about a monolayer coverage.

A very different behavior is observed when Ti_2O_3 is exposed to SO_2 [43]. Figure 9 shows a series of UPS spectra for a nearly perfect Ti_2O_3 $(10\bar{1}2)$ surface clean (bottom curve) and for SO_2 exposures up to 10^4 L. The SO_2 - Ti_2O_3 interaction is extremely strong, and large changes are apparent in the UPS spectra after only 0.1 L exposure (not shown). The reaction does not saturate, with the UPS spectra continuously changing for exposures as large as 10^4 L (the top curve in Fig. 9). XPS measurements of the S 2p binding energy show that the SO_2 completely dissociates upon adsorption, with no S-O bonding remaining. The Ti 3d a_{1g} band is completely depopulated as the SO_2 oxidizes the surface to TiO_2 and TiS_2. The reaction continues as long as there are Ti^{3+} ions with which the SO_2 can interact, and the absence of saturation suggests that the microscopic TiO_2 - TiS_2 scale formed does not adhere well to the Ti_2O_3 substrate. SO_2 is thus a much stronger oxidizing agent for Ti_2O_3 than is O_2, with S catalyzing the reaction.

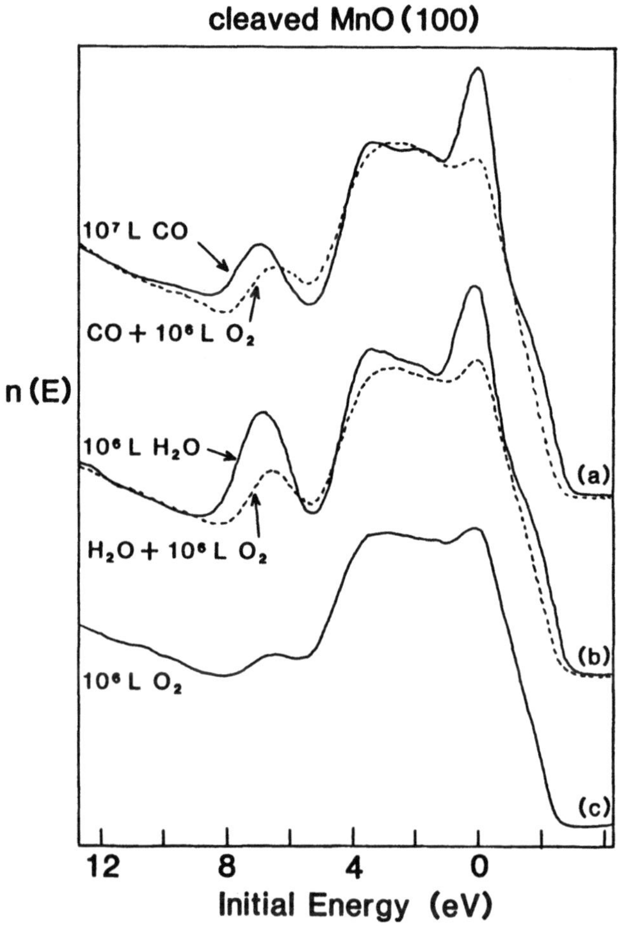

Figure 7. UPS spectra for vacuum cleaved MnO (100) surface (a) exposed first to 10^7 L CO (solid curve) and then to 10^6 L O_2, (b) exposed first to 10^6 L H_2O and then to 10^6 L O_2, and (c) exposed only to 10^6 L O_2. $h\nu = 40.8$ eV.

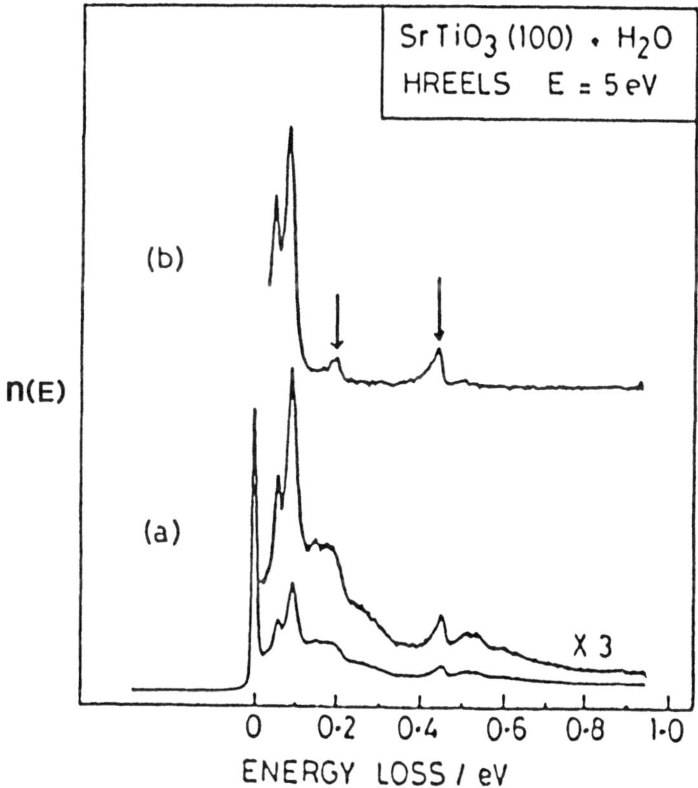

Figure 8. (a) HREELS spectrum of $SrTiO_3$ (100) dosed with H_2O. (b) Deconvoluted and smoothed loss function with surface phonon overtones removed; adsorbate induced losses are indicated by arrows. (From Ref. 58.)

No real trends have been found in the way in which molecules interact with sub-oxide surfaces. For example, the V_2O_3 (10$\bar{1}$2) surface behaves very much like Ti_2O_3 (10$\bar{1}$2) when exposed to O_2 or H_2O [13]. But V_2O_3 interacts only weakly with SO_2, with the reaction saturating by 10 L exposure [59]. The S 2p core level spectra show that both S-O and S- cation bonding occur. Thus each molecule - surface system must be examined separately.

C. Chemisorption at defect sites on oxides

The strongest and most important interactions between molecules and oxide surfaces occur at point defect sites. Electron spectroscopic studies of point defects on single crystals are usually performed by creating the defects on a cleaved or nearly perfect surface by electron or ion bombardment or by heating and quenching (see Section IIB). The first two methods have been used much more frequently than the last, although bombardment no doubt produces many types of defects, not just simple O vacancies. So it has not been possible to get as detailed a microscopic picture of the interaction of molecules with point defects as with perfect surfaces; one is really seeing effects that average over a variety of defects.

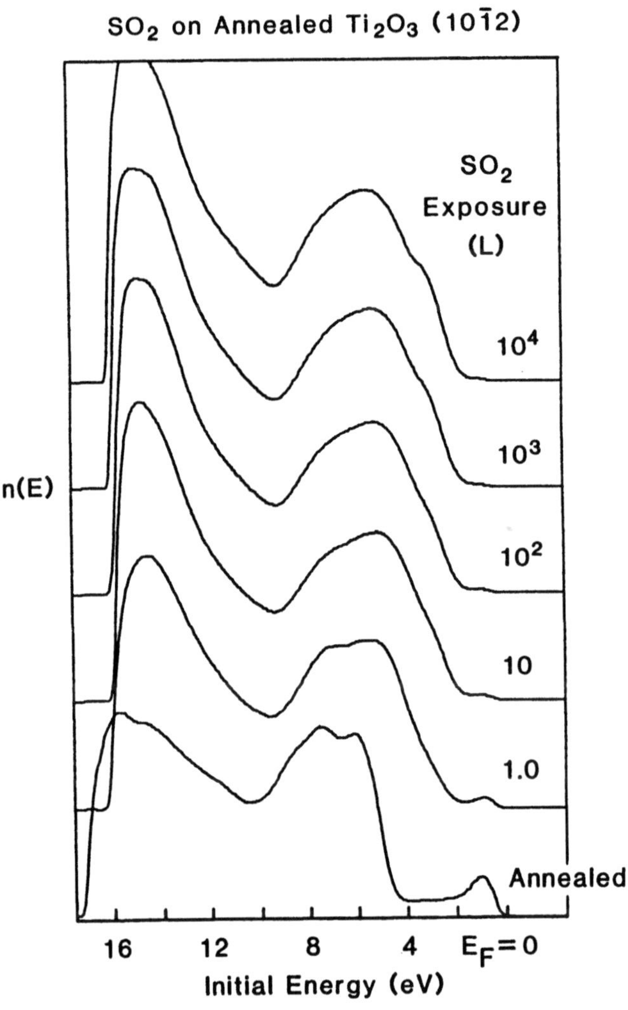

Figure 9. Series of UPS spectra for Ti_2O_3 (10$\bar{1}$2) exposed to SO_2 up to 10^4 L. $h\nu$ = 21.2 eV.

Since defective oxide surfaces are almost always O deficient, it is not surprising that O_2 interacts strongly with defects. O_2 generally dissociates for low exposures, removing electrons from cation orbitals to create a negative adsorbed O species, presumably nearly O^{2-}. An example of this is shown in Fig. 10, which shows a series of UPS spectra for an ion-bombarded $SrTiO_3$ (100) surface exposed to O_2 [60]. The strong interaction with the Ti^{3+} defect states is apparent, with the states essentially depopulated (i.e., all of the reduced Ti cations at defect sites re-oxidized) by about 10 L exposure. An increase in emission intensity in the O 2p band can be seen, along with other peaks appearing in the spectra for large exposures. Analysis of the difference spectra for these data indicates that the

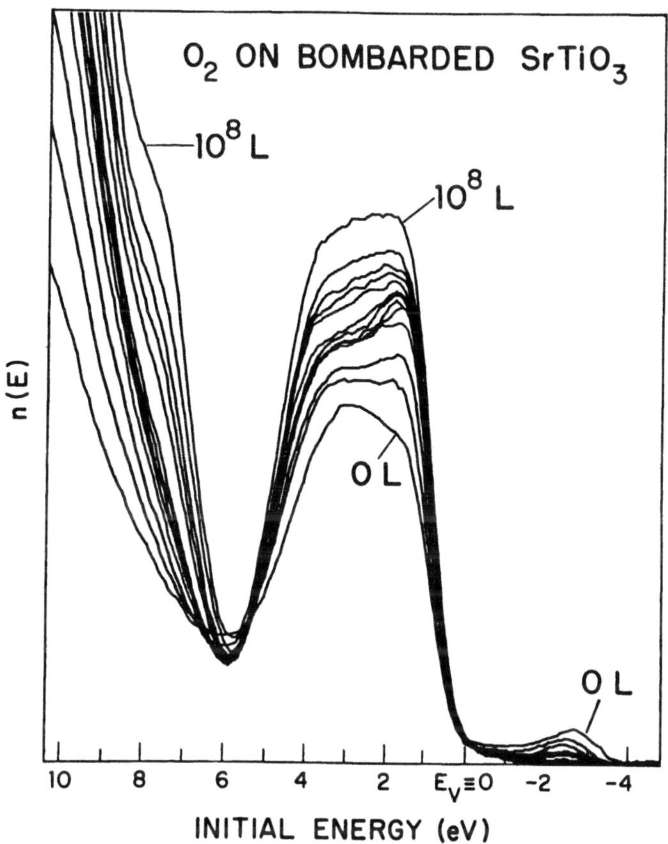

Figure 10. UPS spectra for Ar^+ ion bombarded $SrTiO_3$ (100) after successive exposures to O_2. $h\nu$ = 21.2 eV.

initial stage of adsorption consists of dissociation of the O_2 molecules to form O^{2-} ions. After this reaction has saturated the defect sites, a second species, which has been identified as O^{2-}, adsorbs much more weakly on the surface. This interaction with O_2 is characteristic of defects on most oxide surfaces, whether maximal valence or sub-oxides.

With the exception of NiO, defects on oxide surfaces interact strongly with H_2O, dissociating it and causing hydroxylation of the surface. Emission from the 3σ and 1π molecular orbitals of OH^- gives a two-peaked structure in UPS difference spectra that is easy to distinguish from the three-peaked structure of the b_2, a_1 and b_1 orbitals of H_2O [1]. After the defect surface is fully hydroxylated, some molecular H_2O adsorption is often observed for large exposures. An interesting aspect of the dissociative adsorption of H_2O is that there is only a weak interaction of the adsorbed species with the cation orbitals at the defect site; it is similar to the molecular adsorption of H_2O on nearly perfect Ti_2O_3 $(10\bar{1}2)$ mentioned above. Thus the electrons trapped at point defects do not play the central role in H_2O adsorption that they do in O_2 adsorption, even though both molecules give rise to negatively charged adsorbed species. The origin of this different behavior is not yet understood.

A very interesting exception to the above rule is NiO [36]. The perfect NiO (100) surface is inert to both H_2O and O_2. The surface containing O-vacancy defects interacts strongly with O_2 in the manner of most other oxides, but it does not interact at all with H_2O at room temperature. However, if the defect NiO surface is first exposed to about 10 L of O_2, so that negatively charged atomic O species are present at or near defect sites, H_2O does adsorb fairly strongly, dissociating to form OH^- radicals. Thus it appears that an activated O species (i.e., not lattice O^{2-}) is necessary in order to rupture the O-H bond in H_2O. Similar behavior has been seen on Ni metal and on GaAs, but it has not been observed on any other oxide surfaces.

In Section IVB we discussed the strong interaction between SO_2 and the Ti^{3+} ions on Ti_2O_3 $(10\bar{1}2)$. Nearly perfect TiO_2 does not interact at all with SO_2 since there are no Ti^{3+} ions either in the bulk or on the surface. However when a TiO_2 surface containing O-vacancy defects is exposed to SO_2, a strong dissociative adsorption similar to that on Ti_2O_3 occurs, but only until all of the surface Ti^{3+} ions have been re-oxidized to Ti^{4+}; the reaction then ceases, regardless of the amount of SO_2 exposure [44,61]. As for Ti_2O_3, SO_2 is a stronger oxidizing agent for defect TiO_2 surfaces than is O_2, and the reaction results in complete dissociation of the S-O bond and the formation of only S - cation bonds. Of course, defect Ti_2O_3 surfaces, which contain some cations that are more reduced than Ti^{3+} (i.e., Ti^{2+} and Ti^0), also react very strongly with SO_2, with no saturation observed [42,44].

The interaction of SO_2 with defects on V_2O_3 surfaces is very different than that for Ti_2O_3 and TiO_2 [59], as was the case for SO_2 interacting with the perfect surfaces. Figure 11 shows UPS spectra for Ar^+-ion bombarded V_2O_3 $(10\bar{1}2)$ exposed to up to 5 L of SO_2. The largest changes in the spectra occur during the first 1 L of exposure, and after 5 L there are no further indications of SO_2 adsorption. This is perhaps more surprising than the behavior of the stoichiometric V_2O_3 surface, since the complexity of V-O chemistry suggests that many different types of surface defect sites should be present. The reason that SO_2 interacts so weakly with any of them is not yet understood.

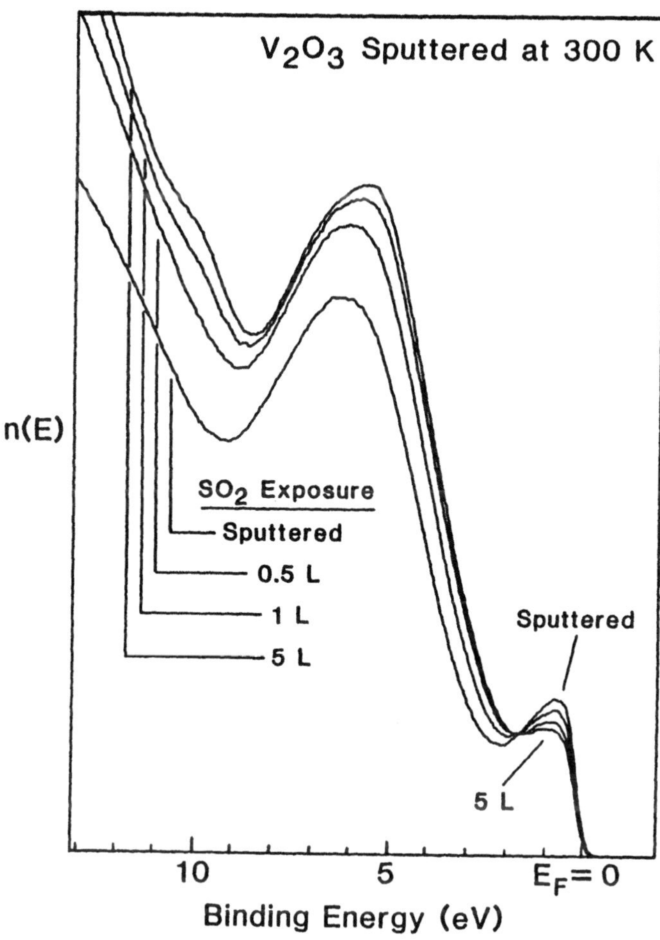

Figure 11. UPS spectra for V_2O_3 $(10\bar{1}2)$ Ar^+ ion bombarded at 300 K and that surface exposed to SO_2 up to 5 L. $h\nu$ = 21.2 eV.

V. Conclusion

The use surface sensitive electron spectroscopic techniques to study the properties of ceramic surfaces has been extended in recent years to nearly perfect surfaces produced by cleaving in UHV, to point defects on those surfaces produced in controlled ways, and to the interaction of both perfect and defect surfaces with simple molecules. Comparison of quantitative LEED measurements with multiple scattering calculations has made it possible to identify the positions of atoms in the surface plane and to determine the differences between the actual surface structure and a simple termination of the bulk. The electronic structure of ceramic surfaces in the energy range near the Fermi level and the vacuum level has been explored by using UPS, and most stoichiometric, nearly perfect oxide surfaces exhibit filled electronic levels very much like (often indistinguishable from) that of the bulk. In many cases, however, the details of the bulk electronic structure are not well enough known to permit the separation of bulk and surface structure. Some clearly identifiable surface transitions have been observed on MgO (100) which indicate the existence of empty electronic states in the bulk bandgap. Point defects on oxides have very different electronic structure than the bulk, due primarily to the presence of charge that must be trapped at defect sites in order to maintain local charge neutrality. Such point defects are the active sites for most chemisorption on metal oxides, with perfect surfaces being inert (particularly for maximal valence oxides) or relatively weakly interacting compared to defect surfaces. There are exceptions to these trends, however; for example, the interaction between the perfect Ti_2O_3 $(10\bar{1}2)$ surface and SO_2 is a stronger one than most molecular interactions with surface defects. The experimental results that are obtained on well characterized single crystal surfaces can be compared in a detailed way with theoretical calculations of surface structure. This combination of theory and experiment is yielding a great deal of insight into the microscopic processes that affect the physics, chemistry and materials science of ceramic materials.

Acknowledgements

The author is indebted to S.-P. Jeng, R.L. Kurtz, R.J. Lad, J.L. Mackay, J.M. McKay, M.H. Mohamed, H.R. Sadeghi, K.E. Smith and Z.M. Zhang for their imaginative, tireless and profound contributions to the understanding of oxide surfaces and for making the field an enjoyable one in which to work. Grants from the Solid State Chemistry Division of the National Science Foundation and from the Chemical Processes and Techniques Branch, Office of Basic Energy Sciences, Department of Energy, made the work at Yale possible.

References

1. 'The Surfaces of Metal Oxides,' V.E. Henrich, Rep. on Prog. in Physics **48**, 1481 (1985).

2. 'The Surface Electronic Structure of MgO and Al_2O_3 Ceramics,' V.E. Henrich, Adv. in Ceramics **10**, 205 (1984).

3. Modern Techniques of Surface Science, D.P. Woodruff and T.A. Delchar (Cambridge University Press, 1986).

4. 'Energy-Dependent Electron-Energy Loss Spectroscopy: Application to the Surface and Bulk Electronic Structure ofr MgO,' V.E. Henrich, G. Dresselhaus, and H.J. Zeiger, Phys. Rev. B **22**, 4764 (1980).

5. 'Metallic Oxides,' J.B. Goodenough, in Progress in Solid State Chemistry, Vol. 5, ed. H. Reiss (Pergamon, Oxford, 1972), p. 145.

6. 'LEED Studies of the Structures of the (100) Surface of Divalent Metal Oxides,' M. Prutton, J.A. Walker, M.R. Welton-Cook and R.C. Felton, Surf. Sci. **89**, 95 (1979).

7. 'Surface Structure of MgO (001) Surface Studied by LEED,' T. Urano, T. Kanaji and M. Kaburagi, Surf. Sci. **134**, 109 (1983).

8. 'Atomic Geometry of the Low-Index Surfaces of ZnO: LEED Analysis,' A.R. Lubinsky, C.B. Duke, S.C. Chang, B.W. Lee and P. Mark, J. Vac. Sci. Technol. **13**, 189 (1976).

9. 'Calculation of Low Energy Electron Diffraction Intensities from ZnO $(10\bar{1}0)$: II. Influence of Calculational Procedure, Model Potential, and Second Layer Structural Distortions,' C.B. Duke, R.J. Meyer, A. Paton and P. Mark, Phys. Rev. B **18**, 4225 (1978).

10. 'The Surface Electronic Structure of Corundum Transition-Metal Oxides: Ti_2O_3,' R.L. Kurtz and V.E. Henrich, Phys. Rev. B. **25**, 3563 (1982).

11. 'Chemisorption of H_2O on the Surface of Ti_2O_3: Role of d-Electrons and Ligand Geometry,' R.L. Kurtz and V.E. Henrich, Phys. Rev. B **26**, 6682 (1982).

12. 'Preparation and Properties of the Ti_2O_3 (047) Surface,' J.M. McKay and V.E. Henrich, Surf. Sci. **137**, 463 (1984).

13. 'Surface Electronic Structure and Chemisorption on Corundum Transition-Metal Oxides: V_2O_3,' R.L. Kurtz and V.E. Henrich, Phys. Rev. B **28**, 6699 (1983).

14. 'The Surface Electronic Structure of TiO_2: Atomic Geometry, Ligand Coordination, and the Effect of Adsorbed Hydrogen,' V.E. Henrich and R.L. Kurtz, Phys. Rev. B **23**, 6280 (1981).

15. 'Rh on TiO_2: Model Catalyst Studies of the Strong Metal-Support Interaction,' H.R. Sadeghi and V.E. Henrich, Appl. Surf. Sci. **19**, 330 (1984).

16. 'Low Energy Electron Diffraction and Electron Spectroscopy Studies of the Clean (110) and (100) Titanium Dioxide (Rutile) Crystal Surfaces,' Y.W. Chung, W.J. Lo and G.A. Somorjai, Surf. Sci. **64**, 588 (1977).

17. 'Thermal Faceting of the Rutile TiO_2 (001) Surface,' L.E. Firment, Surf. Sci. **116**, 205 (1982).

18. 'Surface Defects and the Electronic Structure of $SrTiO_3$ Surfaces,' V.E. Henrich, G. Dresselhaus and H.J. Zeiger, Phys. Rev. B **17**, 4908 (1978).

19. 'Photoemission Study of $BaTiO_3$ (100) Surfaces,' B. Cord and R. Courths, Surf. Sci. **152/153**, 1141 (1985).

20. 'The Nature of Defects on Solid Surfaces as Studied by Electron Spectroscopy,' V.E. Henrich, in Defects in Solids - Modern Techniques,' ed. A.V. Chadwick and M. Terenzi (Plenum, New York, 1986), p. 311.

21. 'Interaction of CO, CO_2 and O_2 with Nonpolar, Stepped and Polar Zn Surfaces of ZnO,' W.H. Cheng and H.H. Kung, Surf. Sci. **122**, 21 (1982).

22. 'Defective Non-Planar Surfaces of MgO,' E.A. Colbourn, J. Kendrick and W.C. Mackrodt, Surf. Sci. **126**, 550 (1983).

23. 'Irregularities at the (001) Surface of MgO: Topography and Other Aspects,' E.A. Colbourn and W.C. Mackrodt, Solid State Ionics **8**, 221 (1983).

24. 'Surfaces of Magnesia and Alumina,' P.W. Tasker, Adv. Ceramics **10**, 176 (1984).

25. 'The Structure and Properties of the Stepped Surfaces of MgO and NiO,' P.W. Tasker and D.M. Duffy, Surf. Sci. **137**, 91 (1984).

26. 'The Nature of Transition-Metal-Oxide Surfaces,' V.E. Henrich, Prog. Surf. Sci. **14**, 175 (1983).

27. 'Theoretical Electronic Properties of TiO_2 (Rutile) (001) and (110) Surfaces,' R.V. Kasowski and R.H. Tait, Phys. Rev. B **20**, 5168 (1979).

28. 'Surface Electronic Structure of Rutile-Type Semiconductors: SnO_2 (110) and TiO_2 (110),' S. Munnix and M. Schmeits, Surf. Sci. **126**, 20 (1983).

29. 'Electronic Densities of States of Defect-Free TiO_2 (110) and TiO_2 (001) Surfaces,' S. Munnix and M. Schmeits, Phys. Rev. B **28**, 7342 (1983).

30. 'Electronic Structure of Ideal TiO_2 (110), TiO_2 (001) and TiO_2 (100) Surfaces,' S. Munnix and M. Schmeits, Phys. Rev. B **30**, 2202 (1984).

31. 'Al-Al_2O_3 Interface Study Using Surface Soft-X-Ray Absorption and Photoemission Spectroscopy,' A. Bianconi, R.Z. Bachrach, S.B.M. Hagstrom and S.A. Flodstrom, Phys. Rev. B **19**, 2837 (1979).

32. 'Valence and Conduction Band Structure of the Sapphire ($1\bar{1}02$) Surface,' W.J. Gignac, R.S. Williams and S.P. Kowalczyk, Phys. Rev. B **32**, 1237 (1985).

33. 'Electronic Structure of α-Al_2O_3', I.P. Batra, J. Phys. C **15**, 5399 (1982).

34. 'The Electronic Structure of $SrTiO_3$ and Some Simple Related Oxides (MgO, Al_2O_3, SrO, TiO_2),' S.P. Kowalczyk, F.R. McFeely, L. Ley, V.T. Gritsyna and D.S. Shirley, Solid State Commun. **23**, 161 (1977).

35. 'Structure of Valence and Conduction Levels in NiO,' J.M. McKay and V.E. Henrich, Phys. Rev. Lett. **53**, 2343 (1984).

36. 'Surface Electronic Structure of NiO: Defect States, O_2 and H_2O Interactions,' J.M. McKay and V.E. Henrich, Phys. Rev. B **32**, 6764 (1985).

37. 'Adsorption and Interaction of O_2, H_2O and CO on the NiO (100) Surface,' J.M. McKay and V.E. Henrich, J. Vac. Sci. Technol. A **5**, 722 (1987).

38. 'Surface Electronic Structure and Chemisorption Properties of MnO (100),' R.J. Lad and V.E. Henrich, J. Vac. Sci. Technol. A **6**, 781 (1988).

39. 'Electronic Structure of MnO Studied by Angle-Resolved and Resonant Photoemission,' R.J. Lad and V.E. Henrich, Phys. Rev. B (in press).

40. 'Electronic Properties of CoO (100) Surfaces: Defects and Chemisorption,' J.L. Mackay and V.E. Henrich, Phys. Rev. B (submitted).

41. 'Magnitude and Origin of the Bandgap in NiO,' G.A. Sawatzky and J.W. Allen, Phys. Rev. Lett **53**, 2339 (1984).

42. 'Transition-Metal Monoxides: Band or Mott Insulators?,' K. Terakura, A.R. Williams, T. Oguchi and J. Kubler, Phys. Rev. Lett. **52**, 1830 (1984).

43. 'Interaction of SO_2 and CO with the Ti_2O_3 $(10\bar{1}2)$ Surface," K.E. Smith and V.E. Henrich, Phys. Rev. B **32**, 5384 (1985).

44. 'Adsorption and Reaction of SO_2 on Titanium Oxide Surfaces,' K.E. Smith, J.L. Mackay and V.E. Henrich, J. Vac. Sci. Technol. A **5**, 689 (1987).

45. 'Valence Band Structure of V_2O_3 $(10\bar{1}2)$,' K.E. Smith and V.E. Henrich, J. Vac. Sci. Technol. A **6**, 831 (1988).

46. 'Molecular Orbital Effects in Resonant Photoemission from Ti_2O_3 and V_2O_3,' K.E. Smith and V.E. Henrich, Solid State Commun. **68**, 29 (1988).

47. 'Bulk Band Dispersion in Ti_2O_3 and V_2O_3,' K.E. Smith and V.E. Henrich, Phys. Rev. B **38**, 5965 (1988).

48. 'Resonant Photoemission in Ti_2O_3 and V_2O_3: Hybridization and Localization of Cation 3d Orbitals,' K.E. Smith and V.E. Henrich, Phys. Rev. B (in press).

49. 'Photoemission Study of the Interaction of SO_2 and H_2S with Titanium and Vanadium Oxides,' K.E. Smith and V.E. Henrich, J. Vac. Sci. Technol. (in press).

50. 'Discrete Variational Xα Cluster Calculations. II. Application to the Surface Electronic Structure of MgO,' C. Satoko, M. Tsukada and H. Adachi, J. Phys. Soc. Jpn. **45**, 1333 (1978).

51. 'Theory of Electronic Structure of Oxide Surfaces,' M. Tsukada, H. Adachi and C. Satoko, Prog. Surf. Sci. **14**, 113 (1983).

52. 'Observation of Two-Dimensional Phases Associated with Defect States on the Surface of TiO_2,' V.E. Henrich, G. Dresselhaus and H.J. Zeiger, Phys. Rev. Lett. **36**, 1335 (1976).

53. 'The Surface Defect Peak in the Electron Energy Loss Spectrum of MgO (100),' P.R. Underhill and T.E. Gallon, Solid State Commun. **43**, 9 (1982).

54. 'Geometric Structure of α-Fe_2O_3 (001) Surface: A LEED and XPS Study,' R.L. Kurtz and V.E. Henrich, Surf. Sci. **129**, 345 (1983).

55. 'Surface Electronic Structure and Chemisorption on Corundum Transition Metal Oxides: α-Fe_2O_3,' R.L. Kurtz and V.E. Henrich, Phys. Rev. B **36**, 3413 (1987).

56. 'Structure of α-Fe_2O_3 Single Crystal Surfaces Following Ar^+ Ion Bombardment and Annealing in O_2,' R.J. Lad and V.E. Henrich, Surf. Sci. **193**, 81 (1988).

57. 'HREELS Studies of Adsorbates on Polar Solids: Water on $SrTiO_3$ (100),' P.A. Cox, R.G. Egdell and P.D. Naylor, J. Electron Spectros. **29**, 247 (1983).

58. 'Application of Fourier Transform Techniques to Deconvolution of HREEL Spectra,' P.A. Cox, W.R. Flavell, A.A. Williams and R.G. Egdell, Surf. Sci. **152/153**, 784 (1985).

59. 'Surface and Bulk Electronic Structure and Chemisorption Properties of Titanium and Vanadium Oxides,' K.E. Smith, Ph.D. Thesis, Yale University (unpublished).

60. 'Chemisorbed Phases of O_2 on TiO_2 and $SrTiO_3$,' V.E. Henrich, G. Dresselhaus and H.J. Zeiger, J. Vac. Sci. Technol. **15**, 534 (1978).

61. 'Interaction of SO_2 with Nearly Perfect and Defect TiO_2 (110) Surfaces,' K.E. Smith, J.L. Mackay and V.E. Henrich, Phys. Rev. B **35**, 5822 (1987).

STRUCTURE & MICROSTRUCTURE OF INTERFACES IN CERAMIC MATERIALS

C. BARRY CARTER
Department of Materials Science & Engineering,
Bard Hall,
Cornell University,
Ithaca,
NY 14853, USA

ABSTRACT. The present understanding of internal interfaces in ceramic materials is reviewed by considering several recent studies. The emphasis of the review is experimental with illustrations, obtained using transmission electron microscopy, of interfaces in alumina, spinel and MgO/NiO. The structure of twin boundaries are discussed in some detail emphasizing the different types of interface which are commonly grouped together under this topic. Some of these interfaces do not involve any major distortion of the anion sublattice while others are more closely related to twin boundaries in metals or semiconductors. In the first group, the position of the interface is determined solely by the distribution of the cations. This feature makes these interfaces similar, in important respects, to certain heterophase boundaries, which are therefore also discussed.

1. Introduction

The purpose of this tutorial paper is to review some of the fundamental concepts of internal interfaces in ceramic materials by considering recent experimental observations. Although most of the analysis will concentrate on homophase boundaries (i.e., interfaces between grains of the same structure and composition - grain boundaries), examples of heterophase boundaries will also be discussed (i.e., where the structure and/or chemistry is different for the adjoining grains). The primary experimental tool used for the analysis of the structure of these interfaces is the transmission electron microscope (the abbreviation, TEM, is used to denote both the technique and the machine), both in the conventional diffraction contrast mode and at high (<2Å point-to-point) resolution. In practice, a wide range of TEM imaging techniques are employed and these will be discussed briefly. The importance of selected-area diffraction for determining local misorientations will also be emphasized.

The materials discussed will include those with the structures of, or related to, alumina, magnesia and spinel. Details of the perfect crystal structures are given in textbooks (e.g., 1,2) and articles concerned with the specific materials (e.g., 3-5). From a research point of view, this choice of materials allows the effects of changing crystal structures and chemistry, both separately and together, to be examined and also covers a wide range of technologically important ceramics; it also allows direct comparison with the much greater and more detailed body of knowledge concerning grain boundaries in metallic systems.

The particular grain boundaries included in the discussion are i) low-angle grain boundaries, where the structure of the interface can be related to the allowed lattice dislocations with both perfect and partial Burgers vectors, ii) twin boundaries, where either

L.-C. Dufour et al. (eds.), Surfaces and Interfaces of Ceramic Materials, 29–55.

the whole of the crystal is affected by the presence of the interface or where, to a first approximation, only one sublattice is affected, and iii) high-angle grain boundaries. In the latter case, the importance of the grain boundary plane will be emphasized. This factor can be related to both the structure of surfaces in these materials and the structure of other grain boundaries which may be wet by an amorphous phase; these two topics will be dealt with more extensively elsewhere in these proceedings. The heterophase boundaries discussed include those occurring between pairs of these same materials (alumina/spinel and spinel/ "magnesia"). Finally the relationship between the faceting of interfaces parallel to certain favored planes and the structures of steps on these interfaces will be considered. In particular, factors affecting the height of such steps will be discussed.

It should be stressed that many questions relating to the structure of internal interfaces in ceramic materials cannot be answered satisfactorily at this time. In fact, in some respects, as new results are obtained, the answers to some questions appear to recede further. The object of the talk on which this paper is based was "to present an introduction and summary of the present status of understanding." The talk therefore posed questions to those researchers who are more familiar with interface studies in metals. The fundamental question is, of course, "do interfaces in ceramic materials have any properties which are fundamentally different from those found in metals?" If the answer is yes, as the evidence increasingly suggests, then a second question arises: "how can technology take advantage of these differences?"

Throughout this discussion, the motivation for this study should be remembered, *viz.*, in most applications of ceramic materials, interfaces play an important, often controlling, role, and it is therefore necessary that our understanding of these planar defects be improved. For the future, such studies will lead to improved interface engineering - an old subject in ceramics. A knowledge of the geometry and composition of interfaces in ionic and covalent materials is important for a better understanding of the processing and properties of many modern ceramic materials [6]. The types of interface commonly found in ceramic materials range from situations where the distance between the grains is $\geq 0.1\mu m$ wide and is filled with a second phase (e.g., 7), to the basal twin boundary in alumina which is atomically abrupt and almost certainly very clean [8]. It is also now clear that the types of grain boundary present in the sintered (or hot-pressed) compact may be significantly influenced by the surface characteristics of individual particles before and during sintering [9]. The behavior of the surface at high temperatures under conditions appropriate to sintering is also relevant to such studies [10,11]. The chemistry of the grain boundary region is clearly very important but will not be discussed in detail here.

The discussion of many of the illustrations as been intentionally limited. Reference should be made to the original articles for details. The inclusion of so many illustrations is a direct consequence of the nature of the paper; they are intended to promote thought on the complexities of interfaces in ceramic materials.

2. Grain Boundaries: Notation, Concepts and Theories

The definition of a grain boundary is simply that it is the interface between two grains. This definition is becoming a little blurred for ceramic materials where second phases are present and may have a significant thickness, in which case, two interfaces will actually be present. In the present discussion, therefore, the term "grain boundary" will be taken to denote an internal interface between two crystalline grains which are either in intimate contact or are separated by a second phase which is so thin that its properties are significantly influenced by the properties of the interface (e.g., it has a new imposed periodicity - see e.g., ref. 12). Grain boundaries in ceramic materials can thus be divided into three groups:

i) Interfaces where the two grains are in intimate contact over their entire area; such interfaces correspond most closely to the grain boundaries which have been extensively characterized in metals and semiconductors.
ii) Interfaces where the two grains are separated by a layer of a second phase which may be crystalline or amorphous.
iii) Interfaces which are a combination of the first two kinds.

2.1 FORMATION

Grain boundaries and phase boundaries are formed in most ceramic processing procedures. It is likely that the process used to form the ceramic will not only influence the type and structure of the grain boundary but also its chemistry. Situations in which grain boundaries or phase boundaries will form are:

A. The Sintering Process. Grain boundaries in ceramic materials are formed during the sintering (or hot-pressing, etc.) process. Thus the surface which is stable at the high temperatures necessary for sintering is one important factor in deciding which grain boundaries will form.

B. Phase transformations and reactions invlolving a ceramic precursor (e.g., alumina from boehmite). In this case the process is more complex, since heterophase boundaries are then involved. These processes may be summarized by the following:

- gel -> crystalline, glass -> crystalline,
- 2-phase -> 1 phase, n-phase -> (n-r) phase!

C. Thin-layer ceramics, epi-growth. Deposition of a ceramic film on a substrate is growing in importance with the development of thin-film ceramic materials (e.g., the high-Tc superconductor).

2.2 GRAIN BOUNDARIES -- MORE DETAILED CONCEPTS

Grain boundaries are commonly divided into two groups:

1. Interfaces between two grains of the same materials -- a homophase boundary.
2. Interfaces between two grains of different materials -- a heterophase boundary.

The first of these interface types is then commonly referred to as a grain boundary, while the second is known as a phase boundary. However, in ceramic materials these distinctions are becoming less well defined. There are also important variations within this categorization:

i) Both "grains" may be crystalline.
ii) Both "grains" may be glass.
iii) 1 grain may glass while the other is crystalline

Then there are special combinations of such boundaries. Perhaps the best "known" of these is one which may be referred to as a simple sandwich, where there is a glassy film present at the "grain boundary." The term "grain boundary" is used here to emphasize that these interfaces are not well understood (or known) at present, and that the interface may not actually be a single interface if the "film" thickness is greater than a certain value. This interface variation is often included as type 3 in the above classification. The problem involving the limiting case (see the lecture by David Clarke), is that the film may actually take on characteristics of the grains, e.g., a periodicity. Or if it is thicker, there will actually be two interfaces present. As an additional complication, it is possible for the "intergranular film" to be crystalline.

Continuing with the concept of increasing complexity of interfaces, it is also possible for a material to contain periodically repeating grain boundaries. The result will then be a new structure. This is related to the phenomenon of chemical twinning. Illustrations of such structures would include $YBa_2Cu_3O_{6+x}$ and β'''-alumina. In the latter, for example, the crystal structure can be described as blocks of spinel separated by defective twin planes.

2.3 GRAIN BOUNDARY THEORIES

A problem with certain theories of grain boundaries is that they tend to gives methods for cataloguing, not understanding, the structure of these interfaces.

The geometrical description is necessary and quite straightforward. Define 2 infinite lattices -- then introduce a plane and remove all the point corresponding to one lattice from one side and all the points corresponding to the other lattice from the other side. Then let the two crystals rigidly move with respect to one another, if necessary, and finally allow local relaxations. Clearly, for each misorientation of two infinite crystals, there is an infinite number of possible grain boundary structures. (Reminder: these concepts are not limited to homophase boundaries.)

The basic concepts which have been developed are listed below:

a) Misorientation	Rotation about a common axis
b) CS	The lattice of coincident individual-lattice sites
c) Σ	The inverse of the fraction of individual-lattice sites which are common to the two adjoining grains
d) O-lattice	The lattice of "origins"
e) DSCL	The displacement-shift-complete lattice
f) Translation	The rigid-body translation
g) Translations	Local atomic relaxations

Having taken the above notation into account, there are then a number of special grain boundary "types" which are encountered.

A) Twist B) Tilt C) Mixed

Although the intended meaning of these terms is clear, their has been much confusion over the importance of this type of categorization as noted, particularly clearly in the recent work of Wolf [13]. The terminology arises from the historical development of grain boundary analysis. Low-angle grain boundaries are composed of arrays of dislocations. When the rotation axis is normal to the grain boundary plane, the grain boundary is called a twist boundary and is composed of at least two arrays of screw dislocations. When the rotation axis is contained within the grain boundary plane, the boundary is called a tilt boundary, and contains at least one array of edge dislocations. When neither of these special situations holds, the boundary is referred to as a mixed boundary. The question is: Is this terminology equally relevant to high-angle grain boundaries?

The difficulty illustrated by the faceting twin boundary (found in cubic metals, semiconductors and insulators) shown in Figure 1. Is this boundary a twist or tilt boundary? The diagram shows that both descriptions are correct. More complex twin boundaries also occur in nature. Such interfaces can be easily generated by simply mirroring the structure across a plane other than the $\{111\}/\{11\bar{2}\}$ plane (or the $(000\bar{1})$ plane in hcp materials). Examples include the $\{11\bar{2}3\}$ and $\{1\bar{1}01\}$ twin boundaries in alumina and will be discussed below.

2.4 METHODS FOR STUDYING GRAIN BOUNDARIES

The methods available for studying grain boundaries are too numerous and diverse to be covered here, and will therefore only be noted selectively. The approaches can be separated into two groups: the direct and indirect methods of observation.

2.4.1 Direct methods:

i) Visible-Light. Extremely valuable in studying interfaces in ceramics. Many ceramics are transparent, measurements are often made on grain boundary grooves formed by thermal or chemical etching (e.g., in order to estimate grain boundary energies). In comparison with the techniques that follow, visible-light microscopy has the important advantage that the sample does not need to be coated to increase electrical conductivity. The technique is also used during specimen preparation for TEM.

ii) Scanning electron microscopy (SEM) and transmission electron microscopy (TEM - also EDS, EELS, STEM etc.). SEM is a surface sensitive technique which is particularly valuable for scanning large areas at reasonably high magnification and giving a much greater depth resolution than is possible using visible-light. TEM is, of course, the essential tool for any study of interface structure. The resolution can be as high as 1.7Å for structural detail and ~10Å for chemical detail. The major disadvantage of this technique is that it is almost always destructive.

2.4.2 Indirect Approaches

i) Fracture is used to study segregation, etc. The sample is broken along grain boundaries, and the morphology of the fracture surface examined. If the process is carried out in a UHV system, the chemistry of the fracture surface can also be examined to give information on grain boundary segregation. One drawback with this technique (although a strong point from another perspective) is that it selects the weakest interfaces. The other major difficulty is that no information can subsequently be obtained on the crystallographic parameters for the interfaces involved.

ii) Surface grooving studies. The approach of producing thermal grooving(see above) in order to estimate the energy of interfaces is a well established method for grain boundary studies. The objections to the method are equally well known. No information is

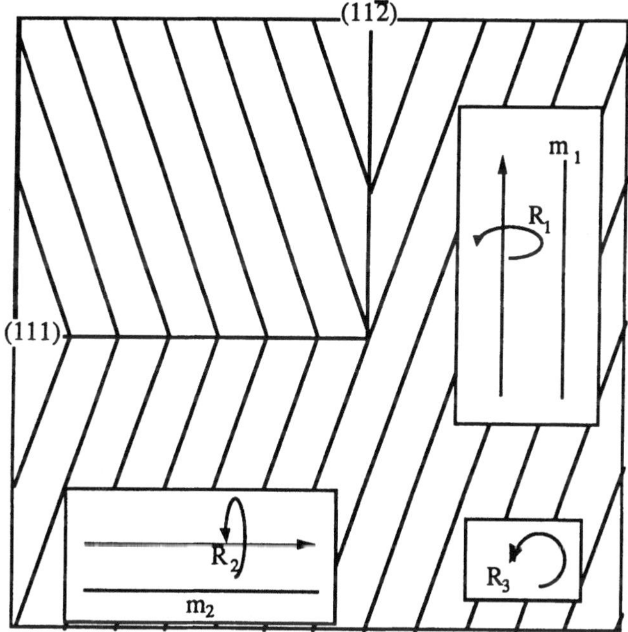

Fig. 1 The rotation axes (R_1, R_2 and R_3) and mirror planes (m_1, m_2) which can be used to produce the "coherent" {111} and "incoherent" (or lateral) {112} twin boundaries in fcc materials.

usually available regarding the plane of the grain boundary, the anisotropic nature of the etching process has too often been neglected. Surface structures, and therefore energies, may depend on the temperature and the atmosphere used in the experiment. However, in spite of these drawbacks, it is still one of the few techniques available for estimating energies of internal interfaces.

iii) New techniques. Examples of the new, recently developed techniques which are now being applied to the study interfaces include surface force measurements and scanning tunneling microscopy. At this time, the impact of these new techniques has not been widely felt but should again give important information on the region where the grain boundaries intersect the surface.

3. Interfaces in ceramic materials - Special Considerations

The structures of defects in ceramic materials can be more complex than those of similar defects in the more thoroughly studied pure metals (e.g., Al [14]) because of the presence of two, or more, ionic species on two sub-lattices. For example, one of the most striking structural phenomena which results from the presence of two types of ions, is that of dissociation by climb of both lattice and grain boundary dislocations. Such a dissociation has been recorded for many materials including spinel, alumina and garnet (for reviews see refs 15-17). Actually, this phenomenon can also occur in the more complex metallic, ordered alloys [see e.g. ref. 18,19], for precisely the same reason. In the ceramic materials, the partial dislocations created by the dissociation process are "perfect" dislocations of the oxygen sub-lattice; the stacking fault thus formed is therefore only a fault on the cation sublattice.

So ... What is different about grain boundaries in ceramics? A guide to the answere(s) to this question can be found by considering the materials themselves.

1) Ceramic materials always have atoms present which are not on lattice sites. This property is not unique to ceramic materials, c.f., the hcp metals, Si, etc. However, most ceramics have $\geq$ 2 atomic species present. Si and Au have the same fcc crystal structure, but Si has atoms which do not lie on the lattice sites; Si can be regarded as a ceramic material. Another example would be MgO, which also has the fcc crystal structure.

2) Ceramic materials necessarily involve ionic and covalent bonding (also others, e.g., van der Waals), and this property does have identifiable, though still not well understood, consequences. In particular the covalent component introduces directional bonds, while the ionic character allows local density variations which are potentially large.

3) The importance of impurity effects in ceramic materials cannot be understated but the details are certainly not fully understood. Actually, very little is known at present about clean boundaries - ceramic materials are almost invariably "dirty" even when they are "clean." The solubility of these impurities in the lattice is often low (lower at lower temperatures - higher at the higher temperatures used for processing these materials. The classic example is the intentional addition of $\geq$ 200ppm MgO in Al_2O_3. It is still not clear what the role of the MgO is in aiding densification but it is presumably a grain boundary effect.

4. The Structure of Grain Boundaries in Ceramic Materials

The approach followed in the research described here is to choose groups of materials which have particular features in common while differing in other respects. The materials chosen include oxides, carbides, nitrides, etc., but reference is also made to metallic materials. Examples of materials used for such a comparison include the following:

-- compare	NiAl, CuAu, Ni_3Al etc.	complex ordered alloys
-- compare	β-SiC or GaAs with Si & Ge	Sphalerite v. diamond-cubic
-- compare	α-SiC, AlN, ZnO, BeO	wurtzite structure
-- compare	ZnO and Al_2O_3	"hcp" oxygen sublattices
-- compare	NiO and Al_2O_3	fcc v. "hcp"
-- compare	Ni0 or Mg0 with $NiFe_2O_4$ or $MgAl_2O_4$	fcc oxygen sub-lattice

The experimental approach used then is to choose special boundaries to examine particular features. Ideally the program will include computer modelling and the experimental study of these chosen interfaces.

When discussing the structure of grain boundaries in metallic systems and in such model ceramic materials as MgO [20], NiO [21-23] and alumina [24-26], the O-lattice, the coincident-site lattice (CS) and DSC lattice have been extensively used. The background to these concepts has been extensively discussed by Bollmann and others [27,28-29]. Since the DSC lattice (with respect to which the Burgers vectors of allowed secondary, i.e., grain boundary, dislocations are defined for a given grain boundary) is a geometric construction, it can be used to derive the allowed Burgers vectors of grain boundary dislocations in ceramic materials [30,31]. However, as has often been noted [32,33], this construction does not, in general, predict the actual Burgers vectors.

It has also recently been reported [13,34] that the plane adopted by the interface is a very important factor in any discussion of grain boundaries. It has often been observed that grain boundaries in ceramic materials show a particularly strong tendency to facet. The facet plane is almost invariably parallel to a low-index plane in one or both grains. Reasons for this phenomenon have been discussed by Wolf [13]; the basic argument is that low-index planes are the most widely separated planes and thus involve less energy when a "stacking error" is present in the crystal. A particularly striking illustration is provided by the Σ=99 and Σ=41 boundaries in spinel where, by chance, the interface can facet parallel to several pairs of (different) low-index planes in both grains simultaneously. A similar situation exists for phase boundaries.

It is well known that liquid phases are important in polycrystalline materials (e.g., 35,36). The evidence that most interfaces in some materials (e.g. Si_3N_4) usually contain a very thin (~10-50Å) amorphous film is strong [37,38], while in other materials it is the subject of controversy [39,40]. It is sometimes assumed that all grain boundaries (other than special twin boundaries) in alumina contain such a thin amorphous film. Whether or not this is the case may have a strong effect on the usefulness, or use, of the material. In some materials the same interface may or may not contain an amorphous film depending on how the material was processed.

The following question then arises: Is this film a necessity or can the same interface (i.e., one with the same relative grain misorientation and the same grain boundary plane) exist in either a structured state or with an amorphous film present, depending on such factors as the cleanliness of the sample? If this is so, then can the interface change from one configuration to another by varying the processing conditions, or the post- or pre-processing treatment?

An additional complication is that none of the techniques presently employed for the identification of these films (the Fresnel fringe technique, the diffuse scattering technique, EDS analysis and lattice-fringe imaging) can, at present, unambiguously identify very thin grain-boundary films without the availability of other information on the material. The method proposed to overcome this difficulty in this program is to fabricate special samples where the presence of the glassy film will be pre-determined. In particular, interfaces will be prepared which contain both structured regions and amorphous films in the same sample. The technique is well illustrated in Fig. 2 which shows a grain boundary in a hot-pressed Si bicrystal; dislocations can easily be seen in the structured region.

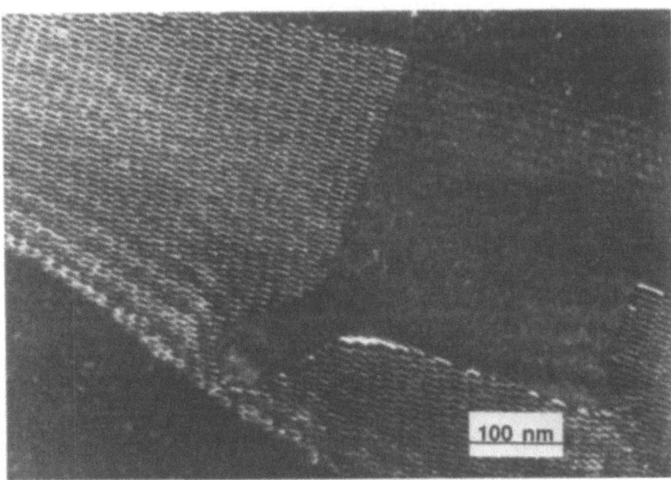

Fig. 2 An interface in a Si bicrystal prepared by hot-pressing, where both an amorphous film and a structured region are present (e.g., Si (41,42)). A "weak-beam" image.

The theoretical aspects of thin films at grain boundaries have recently been extensively reviewed by Clarke [43]. Marion et al. [44] have proposed that the glassy film may actually be itself structured at the amorphous/crystalline interface. A related proposal has been made for the Si and Ge crystalline/amorphous interface [45].

5. Grain boundary chemistry

The structure of grain boundaries in ceramic materials should not be considered in isolation from their chemistry [46], even in so-called pure materials. It has also been suggested that the wetting of a surface may be altered by the doping of this layer [47]. It is therefore also important to identify and characterize the segregation of impurities and additives to grain boundaries in these materials. The problem here is two-fold: the nature of the grain boundary must be identified (the misorientation, the plane of the interface, etc.) and the distribution of "foreign" elements must then be accurately measured. It has recently been demonstrated that the last part of this process is actually even more difficult than has been generally appreciated (see §4). The added complexity arises from the methods which are, at present, routinely used to prepare samples of ceramic materials for examination in the TEM. Whenever it is desired to relate the chemical profile of an interface to its structure, and thence to the overall structure of the ceramic composite, the specimen is usually thinned by ion-milling (crushing destroys all such crystallographic information). This thinning process has been shown to result in cross-contamination of the specimen and in the formation of a groove at the intersection of the interface with the surface. The degree of contamination depends on a large number of factors including the type of oil used in the diffusion pump of the ion-miller (many ion-millers use a silicone-based diffusion pump oil since it is less expensive and more resistant to cracking than are Si-free oils), and may even be influenced by the cleanliness of the system used to carbon-coat the thinned TEM specimen [48]. It has also been conclusively shown that the Ar-ion thinning process preferentially removes the lighter atoms and implants Ar into loose, or open, structures [49].

The methods used in the detection of very thin, amorphous films can be ambiguous if no other independent information is available. Furthermore, it has been proposed that a continuous, thin, amorphous film at an interface may be thermodynamically unstable although, since little is known about the amorphous/crystalline interface, it is difficult to take this factor into account. It is experimentally extremely difficult to determine unambiguously by TEM whether a grain boundary in Al_2O_3 does or does not contain a glassy phase when the width of this phase (if present) is less than ~30 Å.

The reasons for these conclusions have been detailed by Simpson et. al [50]. The primary problem is directly attributable to the result of ion-thinning techniques which are currently used to prepare samples for examination in the TEM. The older method of crushing the bulk sample is not appropriate for this type of study particularly when the relation of the grain boundary to the microstructure as a whole may be important. It has been shown that the surface of NiO can be chemically reduced to such an extent by ion-thinning that the NiO may subsequently spontaneously reduce to Ni metal [51]! Similar problems have recently been demonstrated for semiconductors by Chew and Cullis [52].

One of the many experiments used to examine the implications of this preferential thinning and contamination is illustrated in fig. 3 where a grain boundary has been examined using the "diffuse-scattering "technique in both the edge-on and tilted configurations. This type of experiment has clearly demonstrated that the amorphous material may be associated with the surface rather than with the grain boundary as a whole. Often only one bright line can be seen when the sample is tilted; this observation can be explained by the asymmetry of the grooving at the two surfaces. The result of this grooving can be observed directly by high-resolution microscopy [48]. In the study of the grooving of grain boundaries in Ge, the specimen was prepared from a Czochralski-grown bicrystal; plan-view imaging showed that all the interface was structured.

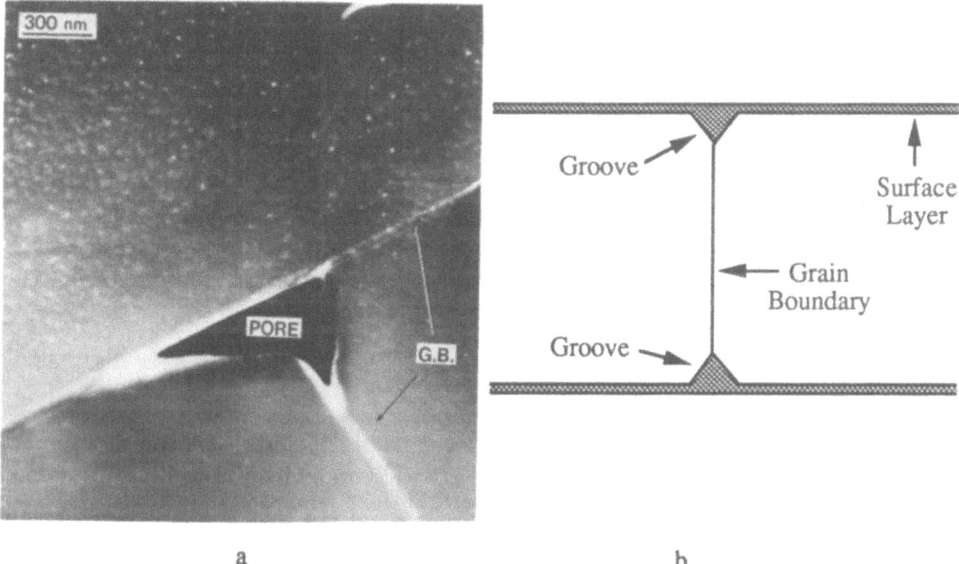

Fig.3a A grain boundary imaged using the diffuse-scattering technique with the interface inclined to the electron beam.

Fig.3b The sample geometry which can be produced by ion-thinning.

The schematic diagram shown in fig. 3b illustrates a possible configuration for an ion-thinned sample. The amorphous and/or heavily damaged material coats the two surfaces and fills the groove. If this structure is tilted over, the grooves on the surface will give rise to two white lines, as can be seen in "a."

6. Relationship between surfaces, grain boundaries, and sintering

The relationship between these different topics can be appreciated from the following gedanken experiment: consider a powder composed of a large number of identical small spheres of a crystalline ceramic. When this powder is heated to the sintering temperature, the particles will facet [49,53]. In alumina, it has been shown that the basal plane is a strongly preferred plane [54,55], although the $\{11\bar{2}0\}$ and $\{\bar{1}102\}$ planes are also favored. When two planes join with their basal planes parallel, either a single crystal or twinned bicrystal may form (Σ=7, 13, 21, and 39 will also be favored [56,57]). The twin interface is a low-energy grain boundary and the bicrystal "grain" will effectively have twice the volume of the initial particles; if the grains are slightly rotated away from the exact twin misorientation, secondary dislocations will form in the resulting interface (similarly, dislocation arrays in the form of low-angle grain boundaries may form when the Σ=1 interface is produced). It has been suggested that, following standard sintering models, such an anomalously large particle may initiate exaggerated grain growth and that this phenomenon may explain the occurrence of twin boundaries in many anomalously large grains in sintered materials.

A new approach for studying the geometry of annealed surfaces and the movement of surface steps in ceramic materials has recently been developed [58]. The TEM sample is annealed after it has been thinned to perforation so that no further ion-thinning is necessary. The technique has been extended to give a direct method of studying the growth of amorphous particles in a crystalline matrix.

7. Low-Angle Grain Boundaries

Low-angle grain boundaries, twin interfaces and special, high-angle grain boundaries are usually interfaces of the first kind although not necessarily so. There have been many studies of low-angle grain boundaries in materials such as alumina [59], NiO and spinel [60-64]; such studies not only provide new understanding of grain boundaries themselves but can also give new insight into the structure and behavior of dislocations in these materials [65]. Low-angle grain boundaries are rather neglected in studies of ceramic materials because most such studies are aimed at characterizing sintered materials. Do they exist ? -Yes, but often not as nice planar interfaces.

In this area one of the more extensively studied materials is NiO, much of which was initiated by the program of Dr. Norman Peterson in growing bicrystals of NiO; illustrations of such interfaces in the work of Merkle, Rühle, Sass, Eastman, Vaudin, et al., (20-22) show that faceting, due to elastic interactions, is ubiquitous. If the dislocations forming the grain boundary are in equilibrium they must be regularly spaced. However, the Burgers vectors of the dislocations present in the low-angle grain boundary, and the spacing of these dislocations, will be different for the different planes. This effect is shown in fig. 4.

To appreciate the complexities of grain boundaries in ceramic materials, consider the same $(1\bar{1}0)$ tilt boundary in spinel and NiO (ignoring any difference in lattice parameter and choose the same misorientation, θ). The magnitude of the Burgers vector of the dislocations in spinel is twice that of the dislocation in NiO and hence, according to the formula in fig. 4, the dislocations must be twice as far apart in spinel as they are in the NiO. However, the dislocation in spinel can dissociate into two partial dislocations with

Burgers vector, a/4[110] which are separated by a stacking fault on the (110) plane. The resulting configuration is that shown in fig. 5. The dislocations in the spinel are then said to be climb dissociated (since their Burgers vector does not lie in the plane of the stacking

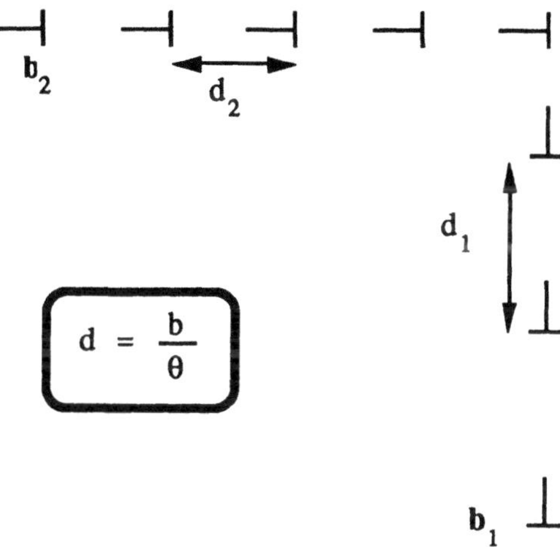

Fig. 4 The change in dislocation spacing with a change in boundary plane for a given pure-tilt boundary. The equation relating d, θ and b is the same for both facet planes.

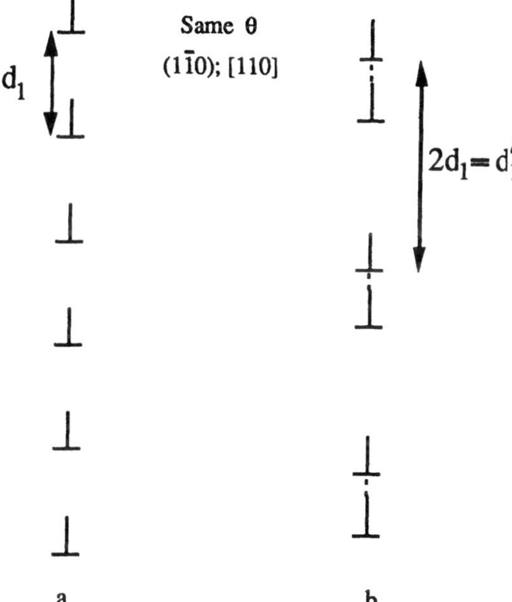

Fig. 5 The spacing of dislocations in NiO (a) and spinel (b) for pure-tilt boundaries with the same misorientation.

fault). Other clear illustrations are the dissociation of an a<1$\bar{1}$00> dislocation in Al_2O_3 into 3 a/3<1$\bar{1}$00> partial dislocations and the dissociation of the dislocations in the low-angle twist boundary in spinel (see fig. 6). In this example, the situation actually even more complex since there are two different planes for the stacking fault, and these two planar defects have a different energy per unit area. (The stacking fault is, of course, another "special" interface which involves only a translation.)

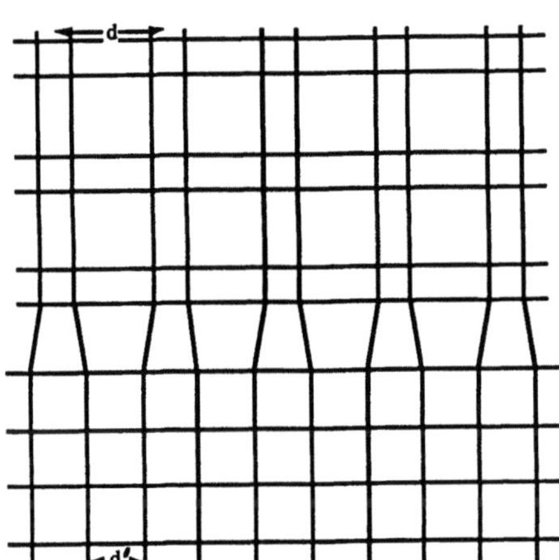

Fig. 6 A network of orthogonal screw dislocations producing a small-angle twist boundary in spinel. d is the spacing of the perfect, lattice dislocations; d' is the periodicity of the screw partial dislocations.

8. Twin Boundaries

Twin boundaries are found in many ceramic materials including MgO [66,67], spinel [68-71], Al_2O_3 [72,73], Fe_2O_3 [74], and the rare earth oxides [75]. Indeed, it seems likely that such interfaces will occur in all ceramic materials. The observation of so many twin boundaries is interesting because their presence in a crystal can introduce new polyhedral sites which may be occupied by cations which could not fit so well into the perfect crystal lattice. This ability of such interfaces to accommodate impurity ions has been developed [76] into the model of chemical twinning, wherein apparently different crystal structures can be related to one another by the periodic repetition of a pair of twin interfaces. It was subsequently [43] demonstrated that this concept could then be used to understand the actual mechanism of a phase transformation in a ceramic system. This latter study also emphasized the close relation between grain boundaries and phase boundaries.

The first-order twin boundary in spinel has been particularly interesting since, crystallographically, the misorientation of the adjoining grain is then the same as occurs for twin boundaries in fcc metals and Si. This grain boundary has thus been studied in greater detail than any other. (The work of Clark, Pond, Smith and Vitek is particularly relevant here.) In spinel the structure has been found to be very different from that found in metals and Si/Ge, as is illustrated in Fig. 7. The interface structure consists of an array of so-called "white-spot defects" which actually correspond to triangular pyramids of less dense material running in <110> directions. The size of the pyramids is such as to translate the interface normal to its original plane by three {111} spinel planes. This defect is essentially dislocation free and causes no long-range shearing of the sample. In fact as shown by our work, the "coherent twin interface" in this material is also interesting since its structure is not unique, i.e., unlike the coherent twin interface in Al, Si, etc., it can exist with several different translation states.

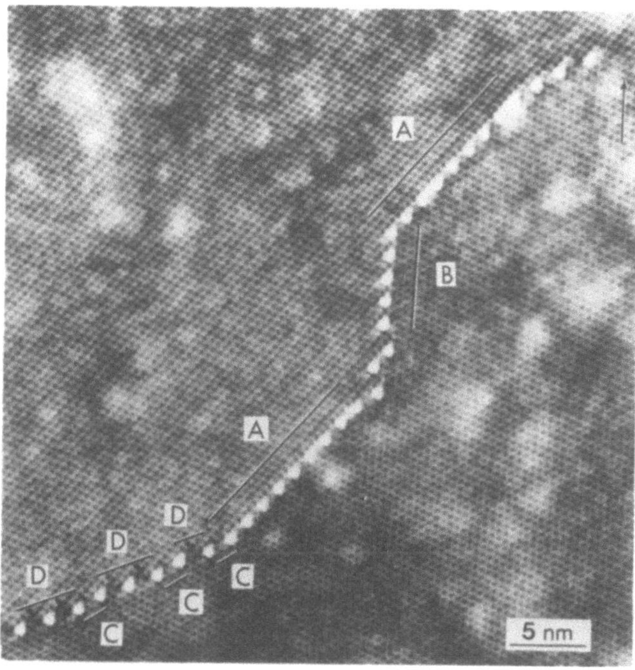

Fig. 7 A segment of a lateral twin boundary in spinel showing that the interface structure is actually a series of "white-spot" defects which can be arranged in different arrays to give the impression of micro-faceting.

This interface in Ge, together with the corresponding interface in Si, is possibly the interface which has been most thoroughly studied by TEM. This attention has, of course, been primarily due to its common occurrence in lower grade Si (e.g. the polycrystalline solar-cell material), and its effect on the electrical performance of those materials. In a recent study by Bourret's group using high-resolution TEM for two different zone axes, both being parallel to the plane of the interface, it was shown that the structure was that of a translation (as opposed to mirror) twin boundary and that bond reconstruction had occurred at the interface.

The image shown in fig. 8 shows a small region of a lateral twin boundary. The Ge was grown as a bicrystal using the Czochralski method. Although dislocations, which accommodate a small deviation from exact coincidence, are present in the boundary, they do not obscure the important result which is that two crystallographically distinct configurations are present: one with, and the other without, a rigid-body translation parallel to the interface.

8.1 The basal Twin Boundary in Al_2O_3

The basal twin boundary in Al_2O_3 bares an interesting resemblance to the stacking fault in that material: there is no detectable translation and the oxygen sublattice is not (to a very good approximation) disturbed. However, there are 3 configurations for the interface when it lies parallel to the common (0001) plane: a mirror configuration and two glide configurations. A schematic diagram the perfect Al_2O_3 structure and of the basal twin boundary lying parallel to the (0001) and ($1\bar{1}00$) plane are shown in fig. 9. Figure 10 then shows this same interface faceting from the (0001) plane to the {$11\bar{2}0$} plane; for simplicity, only one pair of oxygen planes parallel to {$1\bar{1}00$} is shown. It is found

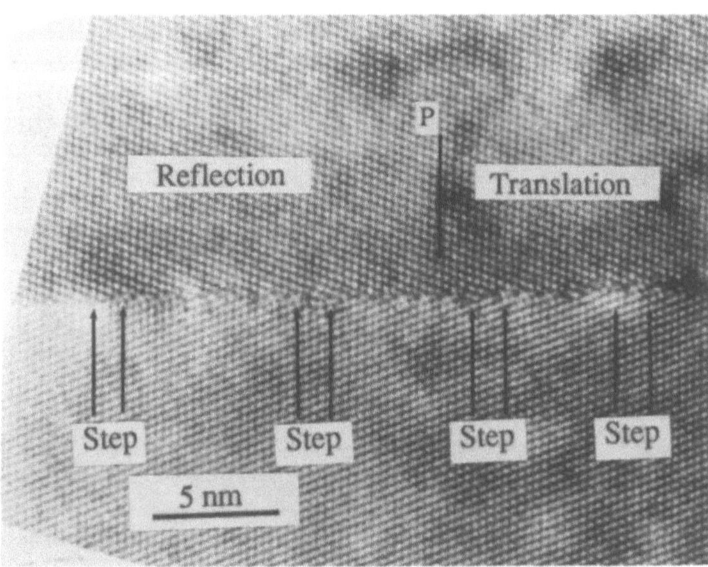

Fig. 8 A segment of a lateral twin boundary in Ge showing the existence of two different translation states on the same interface.

invariably to be faceted even though, to a first approximation, only the cation sublattice is affected by its presence. The faceting of the grain boundary can therefore be attributed directly to the influence of the cations.

In most earlier discussions of this interface, attention was focussed on the (0001) facet, since it was this facet which appeared to be most relevant to deformation twinning. However, facets on the {$1\bar{1}00$}, {$11\bar{2}0$} and {$\bar{1}012$} planes have also been observed experimentally. Questions which have not yet been answered include: what is the favored configuration for the ($1\bar{1}00$) boundary plane? What is the preferred configuration for steps on either boundary plane? It was suggested earlier in our study, that by analogy with the structure observed for the first-order twins in spinel, the basal twin would step over six planes when stepping from one (0001) facet to another, parallel facet, as illustrated in fig. 10. Since such a step is one unit cell high, and thus dislocation-free, it need not cause the oxygen sublattice to distort or to shear and only requires local rearrangement of the cations. Steps on such interfaces may thus show many interesting features. For example, i) the distributions of cations at the first and second bends in a unit-cell-high step are different, ii) If the step is not a unit-cell-high step, the relative orientation of the "matrix" (lower) and

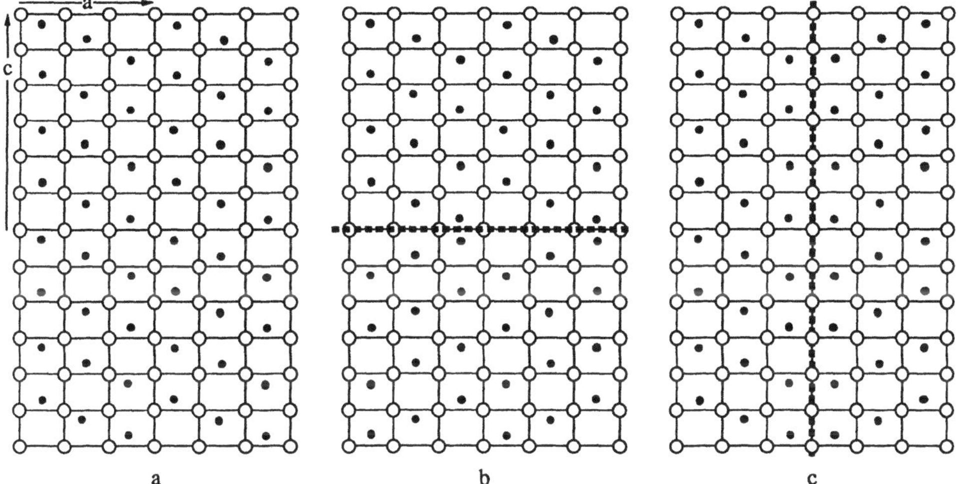

Fig. 9 Schematic diagrams of the basal twin boundary in alumina: a) the perfect material; b) parallel to the (0001) plane; c) parallel to the {1$\bar{1}$00} plane.

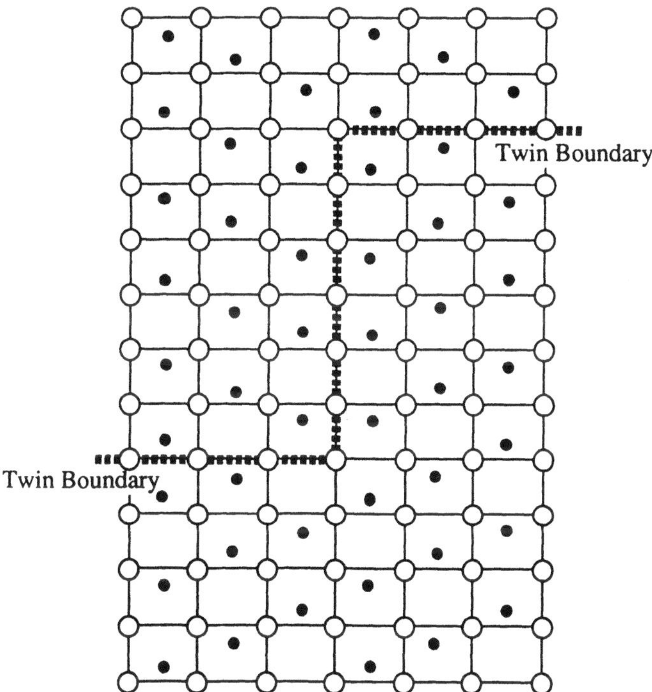

Fig. 10 A schematic of a basal twin boundary faceting from the (0001) plane onto the (1$\bar{1}$00) plane and back to the (0001) plane. The step is one unit cell high.

twin grains may be the same on one segment, but translated on the other.

Although not discussed in any detail here, an example of the value in comparing related interfaces in similar materials can be seen by comparing the above structures with those of twin boundaries in ZnO or BeO.

There may be a long-range coulombic interaction between the facets. This suggestion may also be used to explain the absence of small facets on long (0001) facets. The similarities between these situations and the interactions between steps on the external surface, or between jogs on a dislocation line, is striking. Actually the faceting of surfaces of ceramic materials is even more closely related to the subject of grain boundaries since surface-faceting can directly influence which grain boundaries form during sintering.

The observations made on the basal twin interface in Al_2O_3 are of course special in the sense that these interfaces had sufficient time to reach an equilibrium, or low-energy, configuration. We have monitored the migration of a basal twin boundary in Al_2O_3 by repeatedly annealing a polycrystalline TEM sample at 1200°C. The importance of this result is not only that it shows that grain boundary migration in an oxide can be monitored at relatively low temperatures but also that the technique may be applicable to other low energy interfaces in this material.

8.2 Twin Boundaries in MgO

MgO serves as a model cubic material and will provide an informative comparison with the related interface in spinel. Consider firstly the boundary lying parallel to the common (111) plane - the distortion of the lattice is not too serious (other than the mirroring). The structure of this interface is shown schematically in fig.11. The magnesium ions all lie in octahedral sites; only one stable configuration would be expected for this interface. Actually this interface has rarely been observed, and then only when water vapor has been present during processing [77].

If instead, the lateral twin boundary lying parallel to the {112} plane is considered, the distortion of the oxygen sublattice becomes much more significant, as is shown in fig. 12. The question to be answered is clearly: what happens in the triangular pyramid? One obvious conclusion is that such an interface will be an excellent sink for point defects and impurities. The figure is not intended to represent the actual configuration, since there must be a distribution of cations along the interface plane to restore charge balance, and some of the anions are clearly closer than would be possible in reality.

8.3 Twin Boundaries in Spinel

Now, to make the situation a little more complex, although staying with a relatively simple oxide, the second cation can be introduced. As shown in fig. 13, when the grain boundary is parallel to the common (111) plane, the oxygen sublattice is not distorted too much. However, unlike the corresponding interface in MgO, there are now 12 different structures possible for the (111) twin boundary in spinel, only one example is shown in figure 13 [78]. In some of these situations, there would appear to be a charge imbalance at the interface which must be compensated. Possibly impurities will segregate preferentially to this interface - an unexpected result for the "coherent" twin boundary.

Again, the situation is different for the {112} lateral twin boundary as shown in fig.14. The oxygen sublattice is now distorted. Examples of tetrahedra and octahedra are denoted by shading. The dark-line triangles denote the regions of distortion as seen in the experimental image. Multiple structures, some of which produce non-stoichiometric interfaces, are also possible in this situation.

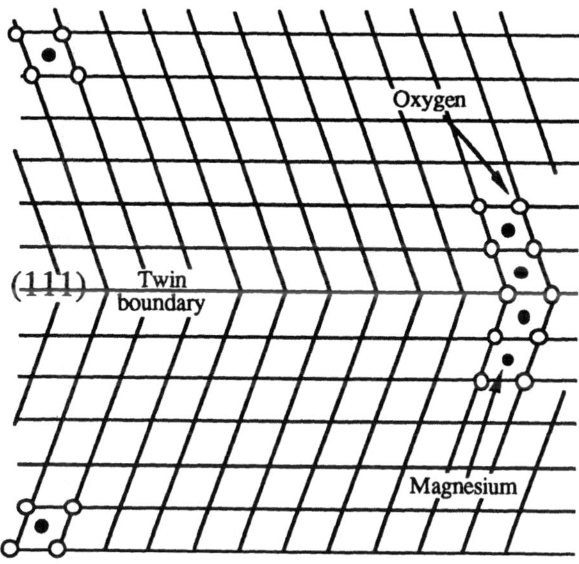

Fig. 11 Schematic of a twin boundary in MgO showing the lattice of oxygen ions with the magnesium ions in the octahedra. Most of the cations have been omitted to emphasize the similarity to the corresponding interface in fcc metals.

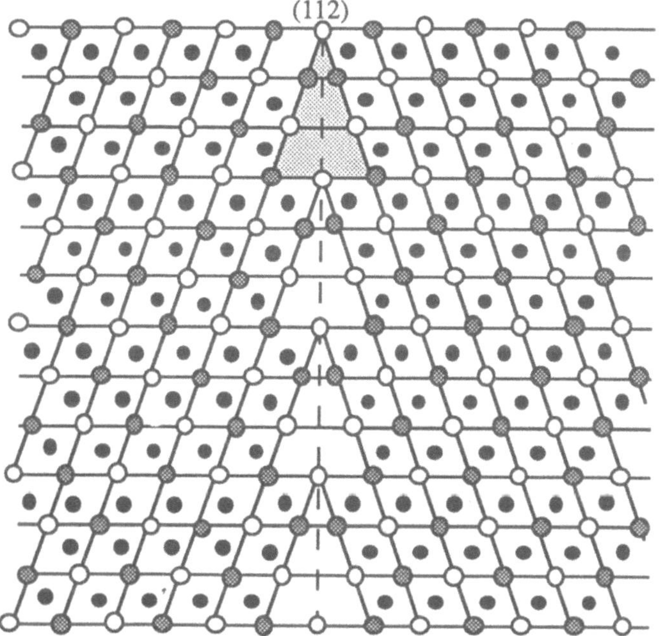

Fig. 12 The {112} lateral twin boundary in MgO. Notice the triangular pyramids lying along the boundary plane (one is shaded).

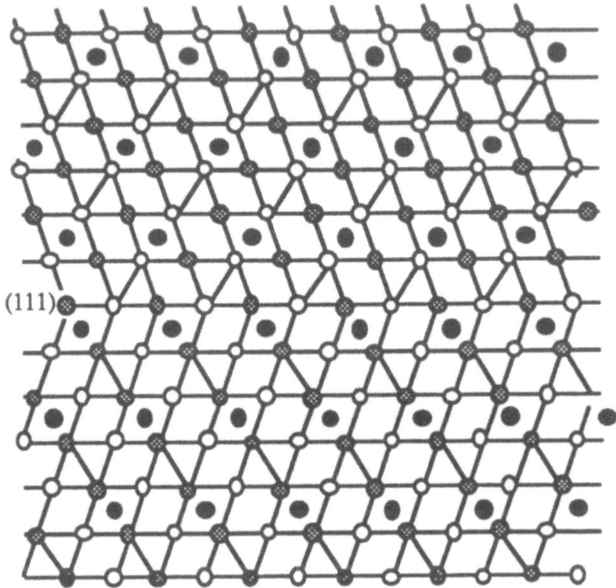

Fig. 13 The twin boundary in spinel lying parallel to the common {111} plane.

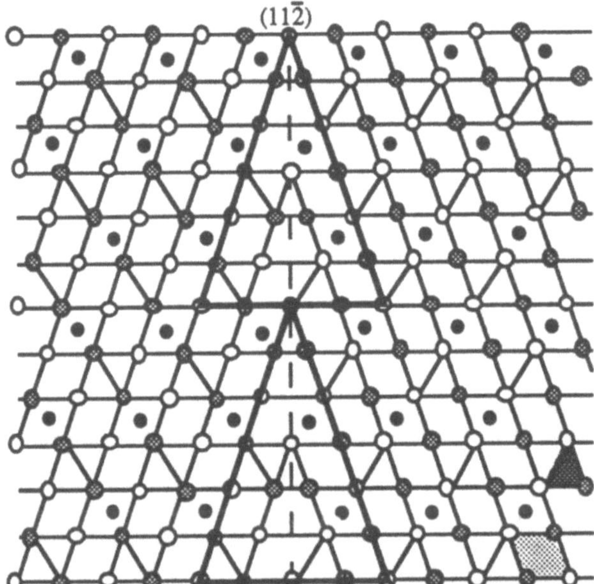

Fig. 14 A schematic representation of the {112} lateral twin boundary in spinel. Note that the configuration shown is for the interface which does not have a mirror plane along the boundary. The triangle and rhombus in the lower right corner identify tetrahedra and octahedra, respectively.

8.4 OTHER TWIN BOUNDARIES IN Al_2O_3

A second special interface which had not previously been identified is the $\{11\bar{2}3\}$ "twin boundary" in Al_2O_3. This interface has been chosen to illustrate how the size of available polyhedra in a particular grain boundary may vary depending on the size of the dopant or impurity atoms which must be accommodated at the grain boundary (i.e., the structure of the grain boundary may itself depend on the chemistry as well as the reverse). Fig. 15a shows an experimental image of such an interface which was found in a sintered polycrystalline alumina sample. The interface has been referred to as a $\{11\bar{2}3\}$ twin interface because it can be created by mirroring the alumina structure across the $\{11\bar{2}3\}$ plane, as can be directly appreciated from the schematic representation shown in fig. 15b. A rigid-body translation could bring the alternate set of ions on the interface into exact coincidence at the grain boundary and thus decrease the size of the available polyhedra. Other special boundaries in Al_2O_3 identified in this study include the $\{10\bar{1}2\}$ and $\{10\bar{1}1\}$ twin boundaries.

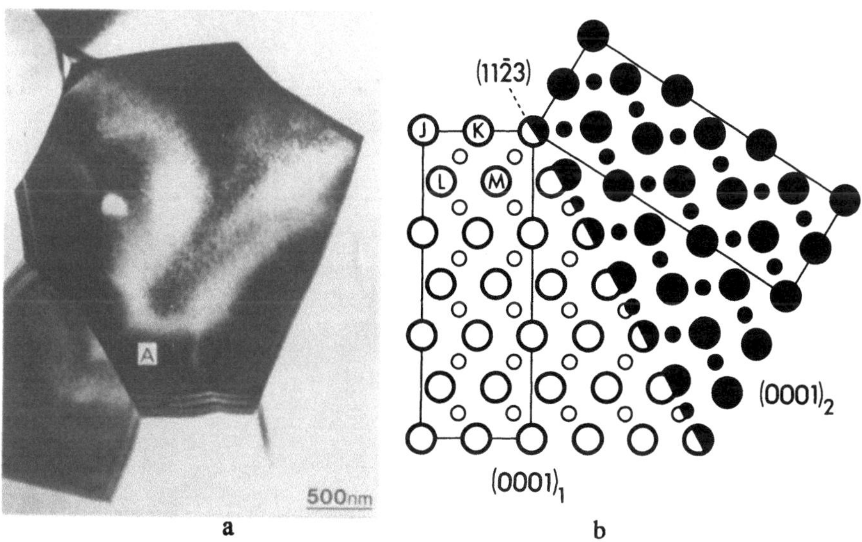

Fig.15 The $\{11\bar{2}3\}$ twin boundary in Al_2O_3. a) Bright-field image; b) schematic of the structure.

9. Special Heterophase Boundaries

Before continuing with grain boundaries, it is instructive to consider the closely related phase boundaries, $NiO/NiFe_2O_4$, i.e., the spinel/wustite interface [79]. If the lattice misfit is neglected, then, to the usual good approximation, the oxygen sublattice is undisturbed, but the chemistry changes at the interface. The situation is illustrated in fig. 16 by a schematic and in fig. 17 by an experimental image. In the schematic, the NiO can be recognized by pairs of numbers within an ellipse, denoting octahedral sites; the spinel contains both octahedra and tetrahedra (c.f., fig. 14). The experimental image was produced by imaging in a spinel reflection which was not common to the NiO. Actually, the location of the individual ionic species is not known from the high-resolution image, only the distribution of occupied polyhedra within the structure are identified.

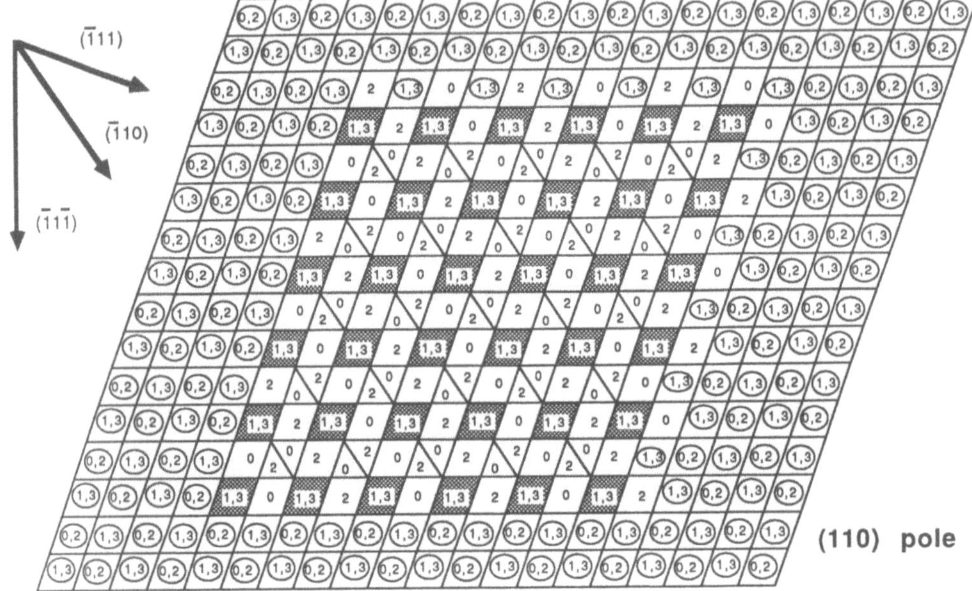

Fig. 16 A schematic showing a spinel particle embedded within a NiO matrix. In this (110) projection, two sets of {$\bar{1}$11} planes are seen edge-on [80].

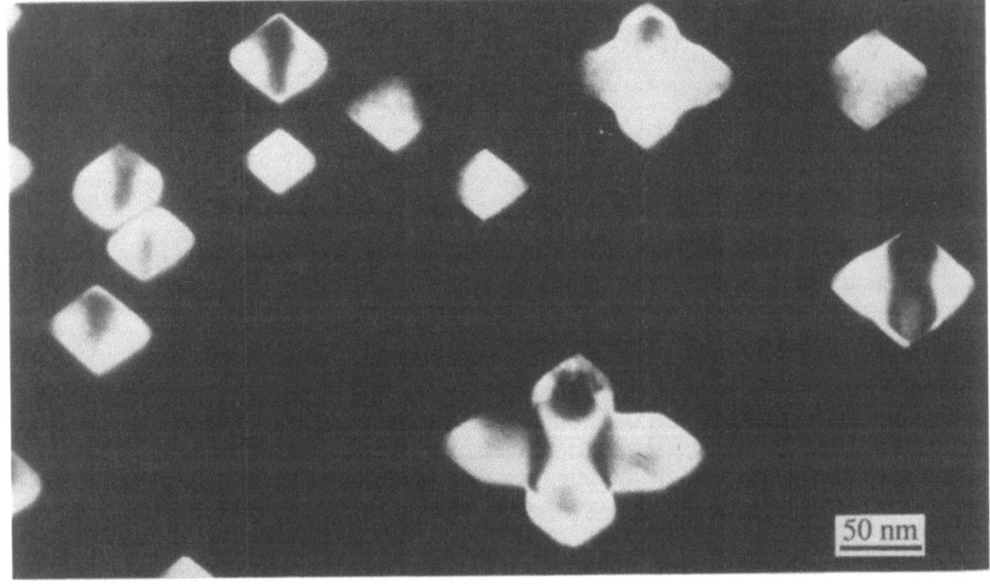

Fig. 17 A dark-field image of spinel particles embedded within a NiO matrix. The same (110) view as shown in fig. 16.

10. High-Angle Grain Boundaries

Before discussing high-angle grain boundaries in detail, the important point regarding the nomenclature should again be noted. The terms twist and tilt were introduced to explain low-angle grain boundaries, not high-angle ones. These terms are used extensively because they are so "ingrained" in this subject. The main point of difference for the present discussion is that a "tilt" boundary is one where a low-index direction is common to the two grains, and the interface is therefore suitable for high-resolution (lattice) microscopy.

High-angle grain boundaries, other than the twin boundaries, have been the subject of far fewer studies, in part because of the experimental difficulties in characterizing such interfaces. This lack of information is unfortunate since these interfaces are ubiquitous in sintered materials. High-angle grain boundaries formed by hot-pressing together two single crystals of MgO appeared to behave as would have been anticipated from studies on the corresponding interfaces in fcc metals and semiconductors (c.f. refs 81 and 82), and supported the earlier interpretation of experiments on MgO "smoke" [83]. However such interfaces have more recently been the subject of extensive computer-modelling studies which tend to give a contradictory view [84,85].

One reason which has been suggested as a possible cause for this disagreement between experiment and modelling is that it is believed to be difficult to prepare ideal clean interfaces in ceramic materials. However, recent work has shown that, for example, the $\Sigma=5$ grain boundary in NiO can be stabilized by adding Schottky defects to the interface, i.e. by decreasing the density at the grain boundary while keeping it "pure." Experimental observations do show that the density at interfaces can indeed be lower than that of the bulk material [86-89]. From the experimental point of view, it has recently been shown that for grain boundaries in alumina and, by implication, other oxides, the techniques used to obtain chemical information at high-resolution from actual grain boundaries may themselves give misleading results.

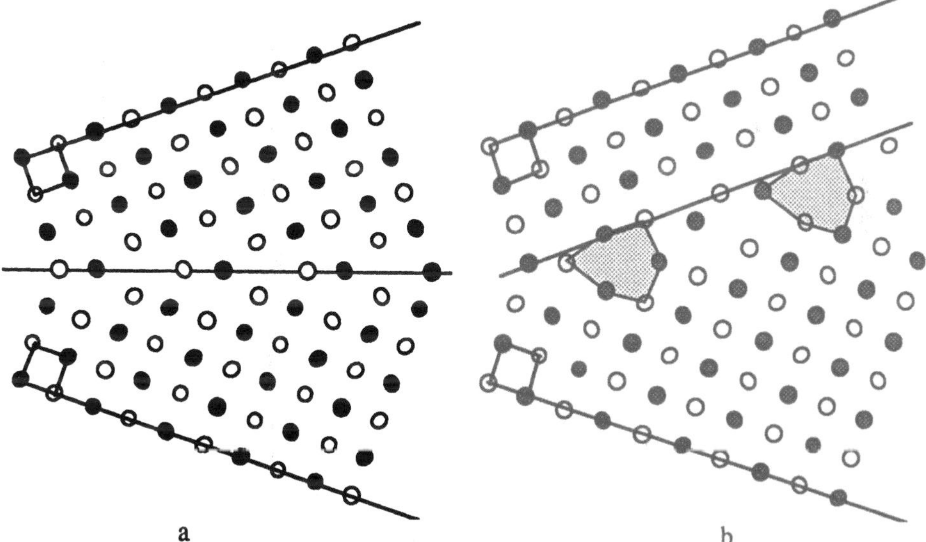

Fig.18 Different possible configurations of the $\Sigma=5$ tilt boundary in NiO: b) A symmetric configuration; b) an asymmetric configuration with the interface faceting parallel to the {001} plane in one grain.

Structured high-angle grain boundaries can certainly exist in ceramic materials such as MgO, Al_2O_3 and spinel. The study of grain boundaries in MgO by Sun and Balluffi, was made on specially prepared bicrystals and revealed periodic arrays of secondary dislocations. The high-angle grain boundaries in Al_2O_3 were found in polycrystalline alumina samples and the presence of secondary dislocations was a rare occurrence although line defects were observed on the $\{11\bar{2}3\}$ and $\{\bar{1}102\}$ twin interfaces. It was noted, however, that a characteristic feature of such interfaces was that they tended to facet parallel to the basal plane, or other densely packed planes, in one grain or the other. It has been proposed by other researchers that such faceting only occurs because a thin, glassy phase is present along the interface. The results of our program show that this statement is not correct. Theoretical arguments given by Wolf also show that faceting will be expected for these high-angle grain boundaries. An illustration of the type of faceting which has now been experimentally observed [90]. The Σ=5 boundary in NiO can facet parallel to {120} or {130} planes producing a symmetric configuration (fig. 18a) or it may facet parallel to one of the low index planes in one grain or the other. In the example shown in fig.18b, the interface is faceting parallel to an {001} plane in the upper grain. Similarities (in terms of special polyhedra) can be recognized, but differences are also apparent.

11. Other Asymmetric Interfaces

A special grain boundary is shown in fig. 19. This boundary is the Σ=99 (!) boundary in spinel. The interface is faceted both along the edge-on segments and, apparently, on the inclined segments. There was very little amorphous material present at the triple junctions in this material as was expected in view of the preparation technique used; also the interface does display periodic contrast features. It is therefore unlikely that an amorphous film was present along the interface causing the two grains to align so closely to the Σ=99 orientation. It is believed that this interface is truly a structured, high-angle, grain boundary. This particular grain is important because it directly supports the proposal (also made by Wolf) that the grain boundary plane may be more important than previously realized and that asymmetric boundary planes may be particularly low energy

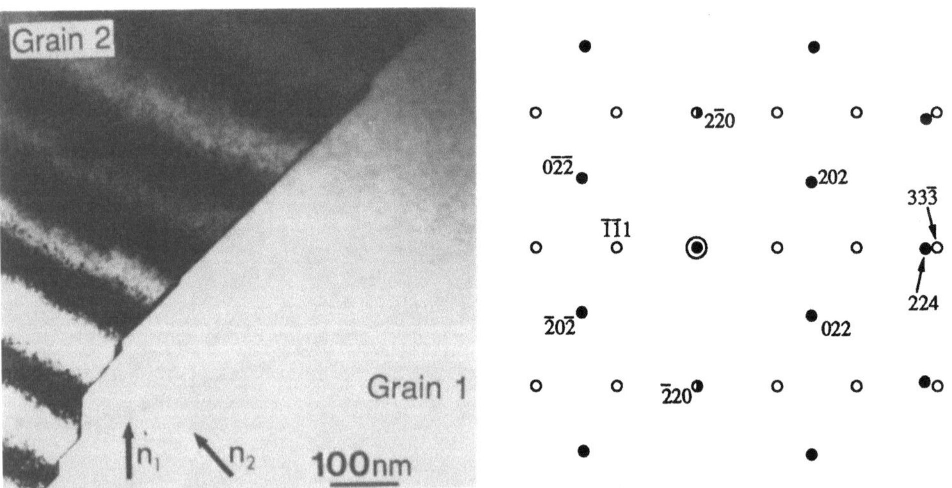

Fig. 19 Strong-beam, bright-field image and corresponding diffraction pattern of a Σ=99 boundary in Mg-Al spinel.

configurations. This interface has recently been observed in aluminum, where it was reported to be able to form in the symmetric orientation but to prefer the asymmetric orientation.

The importance of the grain boundary plane has also been recognized from observations on the $\Sigma=13$ boundary in Al_2O_3, and the $\Sigma=41$ in spinel. (The $\Sigma=41$ boundary has recently been manufactured intentionally, in gold bicrystals, and shown to facet preferentially in the asymmetric orientation.). Similar conclusions have been deduced from observations of special phase boundaries in the spinel/Al_2O_3 system. A particularly striking example is shown in fig. 20, where the phase boundary can be seen to lie precisely along a $\{111\}$ plane in the spinel and a $\{11\bar{2}3\}$ plane in the alumina [91].

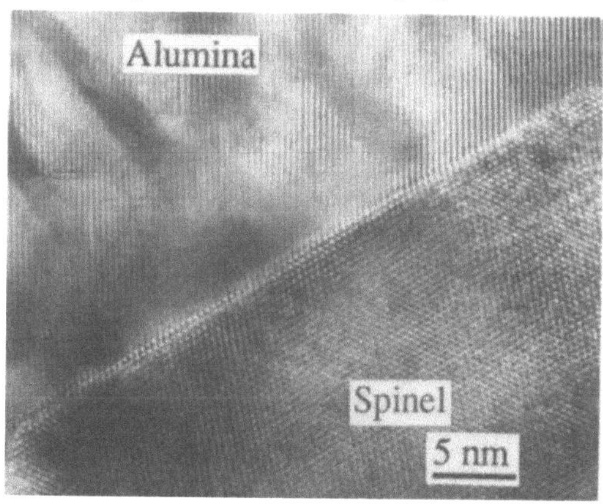

Fig. 20 A high-resolution image of the $\{111\}/\{11\bar{2}3\}$ spinel/alumina phase boundary.

12. Steps in Grain Boundaries

When discussing grain boundaries in most materials and ceramics in particular, the importance of steps and their relationship to grain boundary faceting cannot be overstated. Examples of steps in twin boundaries in Al_2O_3 and Spinel have been noted in previous sections. It is particularly instructive to examine figs 7 and 10.

A final illustration is given in fig. 21, to show that even the discussion of steps may become complex when the detailed structure is considered. The figure shows the interface

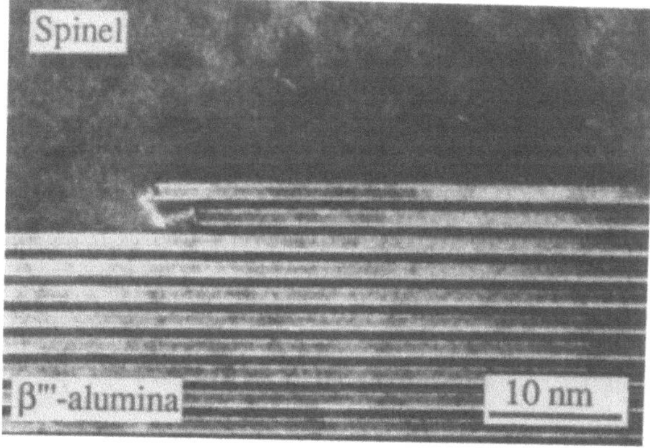

Fig. 21 A step propagating along the interface between spinel and a particle of β'''-Al_2O_3.

between a grain of spinel and a growing grain of β'''-Al_2O_3. The interesting question is then: is the transformation taking place by the movement of a step along the surface of the β'''-Al_2O_3, or is a microtwin propagating in the spinel close to the interface between the spinel and the β'''-Al_2O_3?

13. Summary

The present situation in our understanding of grain boundaries in ceramic materials can be summarized briefly:

A. What are the important parameters ?
- a) Misorientation
- b) Grain boundary plane
- c) Chemistry at the interface
- d) Energy
- e) Impurity effects
- f) Others !

B. How well are these parameters understood?
Not well enough but progress is being made.

C. What are the problems ?
1) Theory ... interface energies are very elusive both experimentally and from a theoretical viewpoint.
2) Experiment ... the purity of ceramic materials is always questioned. The techniques used for determining the possible presence of a second phase at an interface characterizing it, must be improved, if possible.

D. What facts are actually known about grain boundaries in ceramic oxides? At this time, the following statements can be made:
i) Structured grain boundaries do exist in ceramic materials. This conclusion applies to both high- and low-angle grain boundaries.
ii) Interfaces in ceramic materials do exist which are not "structured" but are, instead, wet over their entire area by a thick amorphous phase [92]. (c.f., §6.1, example D.)
iii) The structure of grain boundaries in ceramic materials tends to be relatively more open (less dense) than is found for interfaces in metals which are otherwise crystallographically similar.
iv) The effects of several ions being present in the unit cell can be recognized. They can produce new interfaces where only one sublattice is affected or can cause a modification of the interface structure itself.
v) The structure of an interface may not be unique because of the presence of two ions or because two or more structures are possible which do not have significantly different energies. It is also likely that the presence of impurities at the interface will modify the structure by their favoring different polyhedral sites [93,94].

As was noted above, the extensive analysis of the structure of grain boundaries in ceramic oxides using computer modelling does not appear yet to explain the experimental observations. One limitation of this approach is that the grain boundary plane is, de facto, fixed. Furthermore, computer modelling of asymmetric interfaces has not yet been possible; constructing the unit cell for asymmetric grain boundaries is difficult. Most of this computer modelling has been directed at understanding the structure of grain boundaries in materials with the rock-salt structure. In these materials (primarily NiO and MgO), the oxygen-ion sublattice is fcc while the cations are located at the octahedral sites. The situation is more complex for most other oxides when different interstices are occupied; for example in spinel two-thirds of the cations occupy octahedral interstices while one- third occupy tetrahedral interstices. In a sintered material, it cannot be assumed that all the ions have found their "ideal" sites (e.g. some Al^{3+} ions in $MgAl_2O_4$ may be on tetrahedral sites). The structure predicted by computer modelling for the {112} lateral twin interface in NiO [95] contains a rigid-body translation. Such a translation is not observed experimentally for the same type of interface in spinel. It may be that the reason for this difference is that the translation-free configuration is that which is present on a migrating grain boundary and not that which is predicted on the basis of minimum energy calculations. So the history of the interface is also an important factor.

Acknowledgments

I would like to acknowledge the contributions to this work form my research students and postdoctoral associates, past and present, namely Dr. Z. Elgat, Dr. K. J. Morrissey, Dr. K. M. Ostyn, Dr. Y. K. Simpson, Mr S. R. Summerfelt, Dr. D. W. Susnitzky and Ms L.A. Tietz. I am particularly grateful to Ms Tietz for carefully criticizing this manuscript. Mr Ray Coles has consistently maintained the microscopes; these are part of a Facility supported, in part, by NSF through the Materials Science Center at Cornell. Ms M. Fabrizio has carried out our photographic processing. This research is presently supported by the US National Science Foundation (DMR-8521834; Dr. J. Hurt) and the US Department of Energy (DE-FGO2-84ER45092; Dr. J. Routbort and Dr. R. J. Gottschall).

References

1. Megaw, H. D., 1973, "Crystal Structures: A Working Approach" (Saunders, Philadelphia).
2. Wells, A. F., 1984, "Structural Inorganic Chemistry" (Oxford U. Press, Oxford).
3. Kronberg, M., 1957, Acta Met. **5**, 507.
4. Hornstra, J., 1960, J. Phys. Chem. Solids **15**, 311.
5. Kingery, W. D., Ed., 1984, Adv. Ceram **10**. (Interfaces in Alumina and MgO.)
6. See Advances in Ceramics vols **5** and **6** (1983).
7. Bongers, P. F. and Franken, P.E.C., 1980, Advances in Ceramics, **1**, 38.
8. Morrissey, K. J. and Carter, C. B., 1984, J. Am. Ceram. Soc. **67**(4), 292.
9. Morrissey, K. J. and Carter, C. B., 1984, Mat. Res. Soc., Symp. Proc. **24**, 121.
10. Susnitzky, D. W., Kouh, Y. M., De Cooman, B.C. and Carter, C. B., 1985, Mat. Res. Soc. Symp. Proc. **60**, 219.
11. Susnitzky, D. W. and Carter, C. B., 1986, J. Am. Ceram. Soc. **69** (9), C217.
12. Marion, J. E., Hseuh, C. H. And Evans, A. G., 1987, J. Am. Ceram. Soc. **70**, 708.
13. D. Wolf, 1985, J. de Phys. **46**, C4, 197.
14. Smith, D.A., Vitek, V. and Pond, R. C., 1978, Acta Metal. **25**, 475.
15. Mitchell, T. E., Donlon, W. T., Lagerlof, K.P.D. and Heuer, A. H., 1984, Mat. Sci. Res. **18**, 125.
16. Heuer, A. H. and Castaing, 1984, J., Adv. Ceram. **10**, 238.
17. Carter, C. B., 1984, Proc. Int. Colloq. on Dislocations, Aussois, France (Eds P. Veyssière, L.P. Kubin and J. Castaing; CNRS Editions), 227.
18. Veyssière, P., 1984, Phil. Mag. A**50**(2), 189.
19. Veyssière, P., 1988, Revue Phys. Appl. **23**, 431.
20. Sun, C. P. and Balluffi, R. W., 1982, Phil. Mag. A**46**, 49, 63.
21. Liou, K. Y. and Peterson, N. L., 1981, Mater. Sci. Res. **14**, 189.
22. Schmidt, H., Rühle, M. and Peterson, N. L., 1981, Mater. Sci. Res. **14**, 177.
23. Eastman, J., Schmückle, F., Vaudin, M.D. and Sass, S. L., 1984, Adv. Ceram. **10**, 324.

24. Morrissey, K. J. and Carter, C. B., 1984, J. Am. Ceram. Soc. **67**(4), 292.
25. Morrissey, K. J. and Carter, C. B., 1983, Adv. Ceram. **6**, 85.
26. Carter, C. B. and Morrissey, K. J., 1984, Adv. Ceram. **10**, 303.
27. Bollmann, W, 1970, "Crystal Defects and Crystalline Interfaces" (Springer, NY)
28. Smith, D.A. and Pond, R. C., 1976, Int. Met. Rev. No. 205.
29. Warrington, D., 1979, in "Grain Boundary Structure and Kinetics" (ASM, Metals Park, Ohio), p1.
30. Sun, C. P. and Balluffi, R. W., 1982, Phil. Mag. A**46**, 49, 63.
31. Carter, C. B. and Morrissey, K. J., 1984, Adv. Ceram. **10**, 303.
32. Clark, W.A.T. and Smith, D.A., 1978, Phil. Mag. A**38**, 367.
33. Skrotzki, W., Wendt, H., Carter, C. B. and Kohlstedt, D. L., 1984, Mat. Res. Soc. Symp. Proc. **25**, 299.
34. C. B. Carter, 1988, Acta Met. **36**(10), 2753.
35. Flaitz, P. F., 1981, Ph. D. Thesis, UC Berkeley.
36. Moya, J. S., Kriven, W. M. and Pask, J. A., 1981, Mater. Sci. Res. **14**, 317.
37. Clarke, D. R., 1979, Ultramicroscopy **4**, 33.
38. Clarke, D. R., 1987, J. Am. Ceram. Soc.**70**(1), 15.
39. Kouh, Y. M., Carter, C. B., Sklad, P. and Bentley, J., 1986, Mat. Res. Soc. Symp. Proc. **62**, 427.
40. Simpson, Y. K., 1989, Ph.D. Thesis, Cornell University, USA.
41. Carter, C. B., Rose, J. and Ast, D. G., 1981, Proc. **39**th Annual Meeting EMSA, Atlanta, GA, 160.
42. Carter, C. B. and Elgat, Z., 1983, Adv. Ceram. **6**, 192.
43. Clarke, D. R., 1979, Ultramicroscopy **4**, 33.
44. Marion, J. E., Hseuh, C. H. And Evans, A. G., 1987, J. Am. Ceram. Soc. **70**, 708.
45. Spaepen, F., 1978, Acta Met. **26**, 1167.
46. Kingery, W. D., 1984, Pure and Appl. Chem. **56**(12), 1703.
47. Flaitz, P. F., 1981, Ph. D. Thesis, UC Berkeley.
48. Kouh, Y. M., Carter, C. B., Morrissey, K. J., Angelini, P., and Bentley, J. 1986, J. Mat Sci. **21**, 2689.
49. Morrissey, K. J. and Carter, C. B., 1983, Mat. Sci. Res. **15**, 297.
50. Simpson, Y. M. K., Carter, C. B., Sklad, P., and Bentley, J. 1986, Mat. Res. Soc., Symp. Proc. **62**, 427.
51. Ostyn, K. M. and Carter, C. B., 1982, Surf. Sci. **121**, 360.
52. Chew, N. and Cullis, A. G., 1988, Ultramicroscopy.
53. Segall, R. L., Smart, R.St.C. and Turner, P. S., 1978, J. Chem. Soc. (Lond) Farad. Trans. I, **74**(12), 2907.
54. Susnitzky, D. W. and Carter, C. B., 1987, Mat. Res. Soc. Symp. Proc., **82**, 463.
55. Morrissey, K. J., Czanderna, K. K., Carter, C. B. and Merrill, R. P., 1984, J. Am. Ceram. Soc. 67 (5), C88.
56. Morrissey, K. J. and Carter, C. B., 1984, J. Am. Ceram. Soc. **67**(4), 292.
57. Morrissey, K. J. and Carter, C. B., 1984, Mat. Res. Soc., Symp. Proc. **24**, 121.
58. Simpson, Y. K. and Carter, C. B., 1988, Proc. **46**th Annual Meeting EMSA P570.
59. Barber and Tighe, N, 1966, Phil. Mag. **14**, 531.
60. Lewis, M. H., 1966, Phil. Mag. **14**, 1003.
61. Veyssière, P., Rabier, J., Garèm, H. and Grihlé, J., 1976, Phil. Mag. **33**, 143.
62. Mitchell, T. E., Hwang, L. and Heuer, A. H., 1976, J. Mat. Sci. **11**, 264.
63. Ostyn, K. M. and Carter, C. B., 1983, Adv. Ceram. **6**, 44.
64. Wendt, H., Ostyn, K. M. and Carter, C. B., 1984, Mat. Res. Soc. Symp. Proc. **24**, 87.

65. Carter, C. B., 1984, Mat. Res. Soc. Symp. Proc. **24**, 267.
66. Cowley, J. M., Segall, R. L., Smart, R. St. C. and Turner, P. S., 1979, Phil. Mag. A**39**(2), 163.
67. Newcomb, S. B., Smith, D. J. and Stobbs, W. M., 1983, J. Microsc. **130**, 137.
68. Shaw, T. M. and Carter, C. B., 1982, Scripta Met. **16**(12), 1431.
69. Carter, C. B., Elgat, Z. and Shaw, T. M., 1986a, Phil. Mag. A**55**(1), 1.
70. Carter, C. B., Elgat, Z. and Shaw, T. M., 1986b, Phil. Mag. A**55**(1), 21.
71. Elgat, Z., 1985, Ph. D. thesis, Cornell University.
72. Morrissey, K. J. and Carter, C. B., 1984, J. Am. Ceram. Soc. **67**(4), 292.
73. Hansen, S.C. and Phillips, D. S., 1983, Phil. Mag. A**47**(2), 209.
74. Bursill, L. A. and Withers, R. L., 1979, Phil. Mag. A**40**, 213.
75. Yangui, B., Boulesteix, C., Bourret, A., Nihoul, G. and Schiffmacher, G., 1982, Phil. Mag. A**45**, 443.
76. Hyde, B.G., Andersson, S, Bakker, M, Plug, C. M. and O'Keeffe, M., 1974, Prog. Solid St. Chem. **12**, 273.
77. Turner, P. S., priv. communication (1988).
78. Carter, C. B., Elgat, Z. and Shaw, T. M., 1986, Phil. Mag. A**55** (1), 1.
79. Summerfelt, S. R. and Carter, C. B., 1988, Proc. **46**th Annual Meeting EMSA P766.
80. Summerfelt, S. R. and Carter, C. B., 1989, to be published.
81. Schober, T. and Balluffi, R. W., 1970, Phil. Mag. **21**, 109.
82. Carter, C. B., Föll, H, Ast, D. G. and Sass, S. L., 1981, Phil. Mag. A**43**, 441.
83. Chaudhari, P. and Matthews, J. W., 1971, J. Appl. Phys. **42**, 3063.
84. Wolf, D., Mater. Sci. Res. **14**, 13 (1981).
85. Duffy, D. M. and Tasker, P. W., 1986, Phil. Mag. A**53**(1), 113.
86. Merkle, K. L., Reddy, J. F. and Wiley, C. L., 1985, Ultramicroscopy **18**, 281.
87. Merkle, K. L., Reddy, J. F., Wiley, C. L., Smith, D. J. and Wood, G. J., 1987, Mat. Res. Soc. Symp. Proc. **82**, 383.
88. Shaw, T. M. and Carter, C. B., 1982, Scripta Met. **16**(12), 1431.
89. Carter, C. B., Elgat, Z. and Shaw, T. M., 1986b, Phil. Mag. A**55**(1), 21.
90. Merkle, K., priv. communication (1988).
91. Simpson, Y. Kouh and Carter, C. B., 1987, Proc. **45th** Ann. Meeting EMSA P294.
92. Stenton, N. and Harmer, M.P., 1983, Adv. Ceram. **7**, 156.
93. Hyde, B.G., Andersson, S, Bakker, M, Plug, C. M. and O'Keeffe, M., 1974, Prog. Solid St. Chem. **12**, 273.
94. Shackelford, J. F., 1983, Adv. Ceram. **6**, 96.
95. Duffy, D. M. and Tasker, P. W., 1983, Phil. Mag. A**48**(1), 155

INTERGRANULAR PHASES IN POLYCRYSTALLINE CERAMICS

D. R. Clarke
IBM Research Division
Thomas J. Watson Research Center
Yorktown Heights, NY 10598

ABSTRACT. The majority of polycrystalline ceramics contain phases, often noncrystalline, along their grain boundaries. These intergranular phases affect the overall electrical and high temperature mechanical properties of the ceramics. This tutorial describes the microstructure, the conditions for wetting, and dewetting, of the boundaries by the phase and the factors affecting the equilibrium thickness of the phase. Finally, how the connectivity of the intergranular phase may be exploited in post-fabrication processing is described.

I. Introduction

One of the major new insights into the behavior of polycrystalline ceramics has come from the examination of their microstructures using the techniques of high resolution transmission electron microscopy. The microscopy reveals that many different ceramics contain a thin glass phase along their grain boundaries both at the three grain junctions and, significantly, at many of the two grain junctions. On the basis of such microscopy evidence the prevailing picture of the microstructure of polycrystalline ceramics has changed from consisting of simply an assemblage of grains joined by crystalline grain boundaries to one consisting of crystalline grains with a continuous glass phase threading between them. Such a two phase microstructure model provides the basis for interpretating many of the mechanical and electrical properties of ceramics in terms of the properties of the both the crystalline phases and the intergranular phases. (In many respects the microstructures in many ceramics may thus be likened to the liquid layer grain boundary model for metals [1,2] introduced by Quiacke and current during the early decades of this century).

Intergranular phases have been observed in almost all structural ceramics including the silicon nitride based materials [3-7], the majority of transformation toughened zirconia alloys [8,9], nuclear waste ceramics [10-12], the multicomponent electronic substrate ceramics [13,14], and in many dielectric, magnetic and electrical ceramics [15-20]. Furthermore, the thickness of the intergranular glass phase can be quite small, being 0.8 nm to 2 nm, in some silicon nitride and zirconia ceramics. Electron microscopy also reveals that, in general, the intergranular phase is siliceous and contains impurities that cannot be taken into solid solution within the crystalline grains [21]. Common impurities include Al, Ca, Na, impurities known to be glass modifiers and to be particularly effective in decreasing the viscosity of glasses at elevated temperatures and thus promoting high temperature degradation.

L.-C. Dufour et al. (eds.), Surfaces and Interfaces of Ceramic Materials, 57–79.

In all cases the intergranular phases are believed to be the remnants of a liquid phase present during the liquid phase densification of the ceramics. As presently understood, the liquid phase can be formed for a number of reasons. In many ceramics, the phase results from the liquid phase sintering process [22,23] used to densify them. Examples of these include the silicon nitride alloys, the zinc oxide varistor materials, and alumina substrate materials. (In these materials, particularly the covalent ceramics, additives are deliberately added to the powders before sintering so as to promote or facilitate densification during firing). In others, such intergranular films are present because the materials are prepared by the controlled but incomplete crystallization of a glass (glass-ceramics). A third, but practically important category, is that in which the phase forms from the impurities present in the starting materials. For instance, a number of the ceramics developed for nuclear waste encapsulation and certain of the polycrystalline, tetragonal zirconia ceramics contain such intergranular films. In all these instances, the existence of the intergranular phase is the result of the inability of the crystalline phases to accommodate into solid solution the constituents of the intergranular phase and the inability of the constituents to crystallize out as distinct phases. The former appears to be particularly marked, again, in the case of the strongly covalent materials.

The presence of such a glass phase can have a profound effect on the properties of such polyphase ceramics. For instance, the unique electrical properties of a number of electrical ceramics, such as the barrier layer capacitors, results from the presence of an intergranular phase and processes have been developed to deliberately form the phase. It is also well established now that many of the high temperature properties (creep deformation, creep fracture, oxidation and corrosion resistance) of the silicon nitride based ceramics are determined by the properties of the intergranular glass phase at these temperatures [24-30]. This growing recognition of the roles of the intergranular phase has spurred interest in how the wetting behavior of such films can be altered and, perhaps, controlled. In the liquid phase sintering process it is envisaged that the crystalline grains precipitate and grow from out of the liquid phase. As they grow their surfaces approach one another with a consequent decrease in thickness of the intergranular phase until an equilibrium value is attained [31]. In discussing the intergranular phase it has generally been assumed that it behaves as a liquid at high temperatures, and that the pertinent microstructural features of the phase are frozen in when the materials are cooled for subsequent examination at room temperature by, for instance, transmission electron microscopy.

This tutorial reviews the current concepts concerning the stability of intergranular phases in polycrystalline ceramics, and in particular addresses the questions as to why grain boundaries are wet, what determines the observed thicknesses, the structure of the glass phase, and how a boundary might dewet. In addition, the role of the intergranular phase in providing a short circuit diffusion path through the microstructure and its consequences for the macroscopic properties of ceramics will be described. Finally, a number of ways in which the presence of an intergranular phase might be exploited in altering the properties of a polycrystalline ceramic -- post-fabrication processing -- will be outlined.

II. Observation of Intergranular Phases

It has long been argued, on the basis of optical and scanning electron microscopy observations of apparent zero contact angles measured in polished sections, that second phases in polycrystalline ceramics exist as thin intergranular films at grain boundaries. However, it was not until the advent of high resolution electron microscopy techniques that it was unequivocally demonstrated that they existed along grain boundaries or their thicknesses measured directly [5-7]. Observations reveal that the major fraction of intergranular phase is located at the intersections of three grains (triple junctions) and at four grain junctions. Higher resolution techniques [7] further indicate that the majority of grain boundaries (two-grain junctions) contain a thin film of intergranular material. (Figure 1).

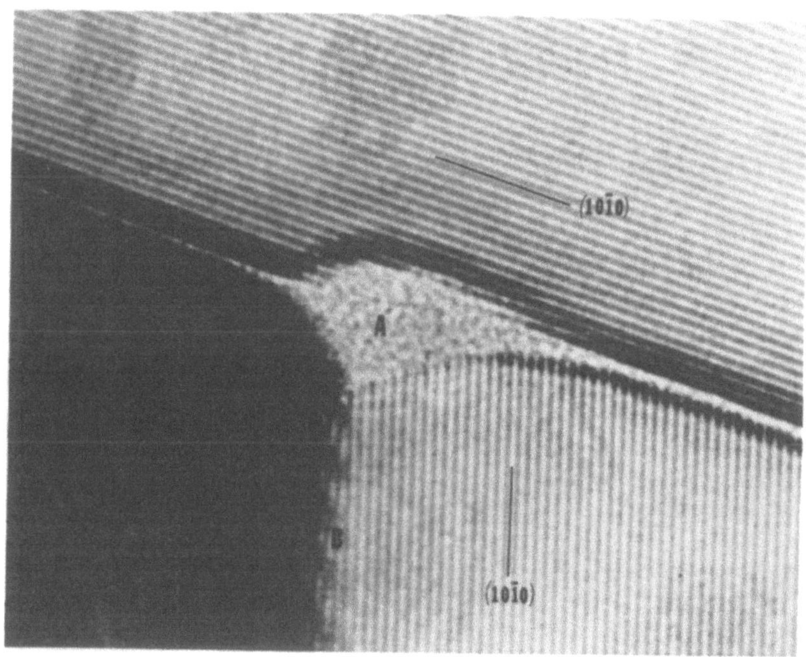

Figure 1. Lattice fringe image of a grain boundary region showing the presence of an intergranular phase at the triple grain junction and between the adjacent crystalline grains. The lattice fringe spacing corresponds to 6.6 Å. The thickness of the intergranular phase is typical of the thicknesses observed in many polycrystalline ceramics.

Selected area electron diffraction techniques have confirmed that the intergranular phase material is also non-crystalline (figure 2).

Figure 2. Selected area electron diffraction pattern recorded from the intergranular phase at a three-grain junction in a silicon nitride ceramic. The circular halo pattern is indicative of the material being non-crystalline.

As the boundaries in more and more ceramics are scrutinized by transmission electron microscopy techniques a greater number of materials are found to contain an intergranular phase. Such thin intergranular phases have been reported not only in the silicon nitride alloys in which their presence was first confirmed but also in a number of zinc oxide varistor ceramics, certain silicon carbides, capacitor dielectrics, (single phase tetragonal and partially stabilized) zirconia ceramics, glass-ceramics, nuclear waste ceramics and aluminas, including many substrate grade aluminas. Intergranular films have also been detected in particularly high purity materials, such as in Lucalox-types of aluminum oxides, where they were thought not to be present.

Perhaps the most surprising observation is that the thickness of the intergranular material ($\simeq$1nm) is found to be relatively constant from one boundary to another in any given sample of material. In other words, within the accuracy of the observations, the thickness does not appear to be dependent on the angle of misorientation across the boundary. The exception being that in a number of boundaries, which are widely believed to be low-angle boundaries and boundaries of "special" orientations such as a coherent twin boundary, no intergranular phase is detectable. In polyphase ceramics, such as in many silicon nitride and Sialon materials, the thickness of the intergranular phase appears to have one value at boundaries between like phases and a different value at boundaries between dis-similar phases. An example of this is shown in figure 3, a micrograph of a yttria-alumina fluxed silicon nitride [32].

Analysis of the elemental constituents of the intergranular phases in silicon nitride based materials using electron microscope techniques (X-ray microanalysis and electron energy loss spectroscopy) suggest that they are highly siliceous [21]. Such an interpretation is consistent with the observations made by Auger electron spectroscopy of both ion-beam sputtered fracture surfaces [33,34] and, more recently, of high temperature in-situ fracture surfaces [35]. Only one polyphase ceramic, a developmental nuclear waste ceramic, has been reported to contain an intergranular phase that is not siliceous and in that case it was found to be phosphate based [12]. In all materials examined to date the intergranular phase is also found to contain the impurity elements and, in those cases where a sintering aid is used, the elements from the aid. In those few cases where a complete analysis has been obtained, the composition is close to that of silica. It is important to note that because of the extremely fine scale of the intergranular phase the microscope based analyses have been restricted to three and four grain regions rather than at the grain boundaries themselves.

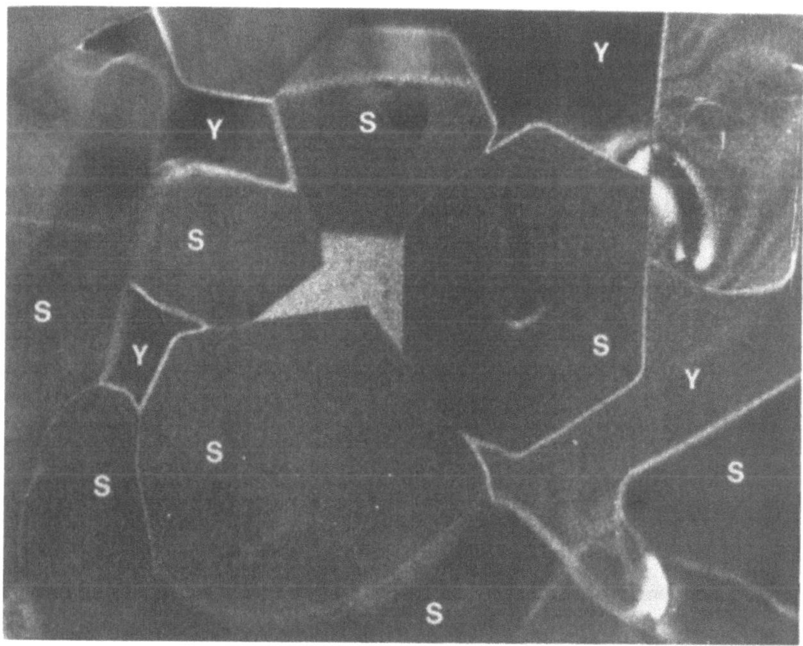

Figure 3. The distribution of the intergranular glass phase in a yttria-alumina fluxed silicon nitride ceramic as revealed by diffuse dark field imaging in the transmission electron microscope. The micrograph, in which the intergranular phase appears bright, illustrates a characteristic difference in thickness of the intergranular phase between silicon nitride (S)-silicon nitride (S) grains and that between silicon nitride (S)-yttrium silicon oxy-nitride (Y) grains. (Micrograph courtesy of T.M. Shaw, and originally published in Journal Of The American Ceramic Society 62(11-12), 585-590, 1979 as figure 4d).

III. Microstructural Stability of Intergranular Glass Phases

CONDITIONS FOR WETTING

Why should a grain boundary be coated with an intergranular film ? The simplest statement that can be made concerns the energies involved: a grain boundary will be wet by a liquid intergranular film if its interfacial energy is thereby reduced. This leads to a necessary condition for wetting that can be stated as:

$$2\gamma_{SL} < \gamma_{SS} \qquad [1]$$

where γ_{SL} and γ_{SS} are the energies of a crystal-liquid interface and a crystal-crystal interface respectively. This condition can be established by comparing two end-point states, shown schematically in figure 4, one in which the boundaries A,B,C are all wet, and the other where they have crystal-crystal contact with all the liquid localized at the three grain junction. Such a comparison of states has been used by Raj [36] and more recently by the author [31,37] in discussing the stability of intergranular phases.

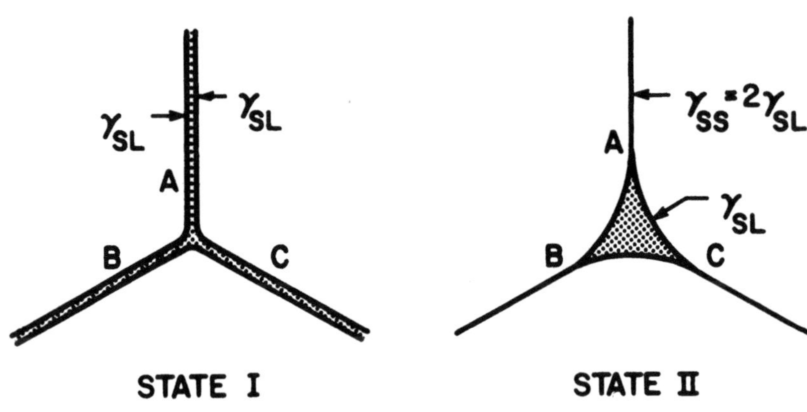

Figure 4. End point configurations for comparing the energies of boundaries wet (state I) and not wet (state II) by an intergranular phase. Redrawn after Raj.

ORIENTATIONAL DEPENDENCE

A number of observations, for instance those of debased aluminas [13,14], hot-pressed silicon nitrides [5-7,21], zinc oxide varistors [17-20], and recently in a Sialon material by Schmid and Ruhle [38], all indicate that whilst the majority of grain boundaries in these materials contain an intergranular phase some do not. The suggestion made by these authors has invariably been that the boundaries are crystallographically "special", although the detailed crystallography of the boundaries has not always been presented. This raises the question as to why all boundaries are not wet ? One way of rationalizing the observations lies in comparing the orientational dependence of the energy of a boundary.

In making the end-point comparison of figure 4, it is implicitly assumed that the interfacial energies are isotropic such that the relative orientation of the boundaries A, B and C need not be specified. This is the simplest possible situation and thus is useful in gaining some insight into the problem as presented in the previous section. However, the interfacial energies are generally dependent on the orientation of the surface in contact with a liquid (the basis of the Wulff plot), and this is expected to be especially marked for the majority of ceramics of interest since most are crystallographically anisotropic. In determining whether a given grain boundary is wet or not the parameter of interest is not the orientation of the surfaces forming the boundary per-se but rather the <u>relative</u> misorientation of the grains forming the grain boundary. Thus the starting point for any discussion is a boundary having a particular misorientation. Then, by comparing the interfacial energy of that grain boundary with the same boundary but in a wetted state for a variety of relative misorientations one can establish an orientational dependence. This rationalization is shown schematically in figure 5.

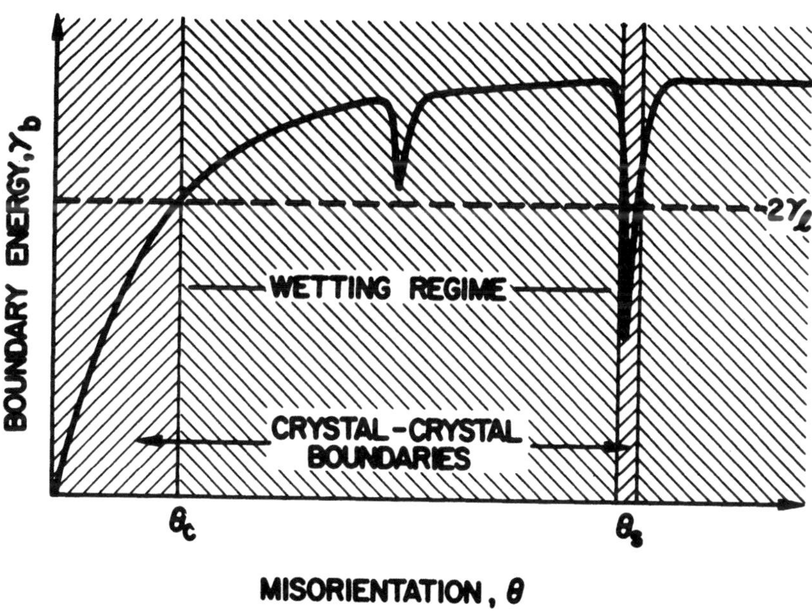

Figure 5. Grain boundaries will only be free of an intergranular film if their energy is lower than that of a wetted boundary, ie for misorientations for which the crystal-crystal boundary energy (solid line) is lower than that of the wetted boundary (dashed line). Schematic.

On a diagram of boundary energy versus misorientation, are graphed both the form of the energy curve for a crystal-crystal grain boundary (the solid line) and that for a wetted boundary (dashed horizontal). From this construction and the condition established in the previous section (equation 1), follows whether a boundary will be wet or not; for all those boundary misorientations for which the solid curve has the lowest energy the boundary will not be wet, and vice versa. The energy curve sketched for the crystal-crystal boundary has the characteristic features one expects of such a boundary and which have recently been predicted by detailed atomistic calculations: there is the region of small misorientation in which the energy can be described in terms of the elastic energy associated with an increasing density of dislocations (the Read-Shockley model, [39]); the monotonically increasing energy as a function of further misorientation; and the energy cusps associated with "special" boundaries of one sort or another, for instance, coherent twin boundaries. On the basis of such a description low angle grain boundaries (up to a critical value Θ_c) will be free of glass whereas all high angle boundaries, with the exception of deep "cusp" orientations, will be wet.

EFFECT OF TEMPERATURE

The situation described so far pertains to that at any arbitrary temperature, for instance the liquid phase sintering temperature. At such a high temperature most boundaries are expected to be coated with a liquid film, and indeed according to the above only those boundaries that do not satisfy the inequality of equation 1 will be free of such a film. As the material is cooled down to room temperature after sintering the wetting of the boundaries will be altered provided that there is a different temperature dependence of the interfacial energies of the wetted and unwetted boundaries and the surfaces of the adjoining grains remain unaltered. Since, in general, the entropic contribution to the free energy of a liquid is expected to be larger than that for a crystalline interface the overall energy of a wetted grain boundary will have a greater temperature dependence than that of its crystalline counterpart. Then, as the temperature is reduced from, say, the sintering temperature the curve for the wetted boundary on figure 5 will rise relative to that of the purely crystalline boundary. The result will be that over a range of orientations it will no longer be energetically favorable for the boundary to be coated with a glass film and it will proceed to de-wet. (There will however be, as will be seen later, a nucleation barrier to be overcome). Observation of such a dewetting of grain boundaries has been reported in zinc oxide varistor materials [19,20] and in a LiF liquid phase sintered spinel [40].

IV. The Equilibrium Thickness

Observations such as those of figures 1 and 3 pose the intriguing question of what dictates the equilibrium thickness of an intergranular glass film and what the detailed atomic arrangement of the interfacial region is.

CONTINUUM APPROACHES

Two approaches, one based on a conventional, but modified, diffuse interface theory and the other on the force balance normal to a flat grain boundary, have recently been presented [37,31] in a first attempt to establish the reason for the observed thickness. Both approaches indicate that a stable thickness of an intergranular liquid phase will result as a consequence of the interplay of an attractive van der Waals- dispersion interaction and a repulsive long range structural interaction. In this respect there is a striking but predictable similarity with the findings of the DLVO theory (41) of colloid stability where a balance is sought between attractive van der Waals forces and repulsive electrostatic forces.

The necessary condition for the mechanical equilibrium of an intergranular liquid film is that the net force acting normal to the film is zero:

$$P_c + \Pi + P_A = 0 \qquad [2]$$

Here the combined effects of the capillarity force (due to curvature at the three and four grain junctions) P_c , and any externally applied stress, P_A , which would tend to draw adjacent grains together, are balanced by an equal force which, following Derjaguin (42), may be termed the disjoining pressure, Π. This is, in reality, a combination of forces given by:

$$\Pi = \Pi_{vdw} + \Pi_{st} + \Pi_{ad} + \Pi_{edl} \qquad [3]$$

where the individual terms here are the contributions from the London-van der Waals interaction energy between the grains, the structural interactions of the molecules in the liquid (a steric effect), the preferential solute adsorption on the grain surfaces, and finally that due to any electrical double layer interactions. Little is known from experiment about whether the latter two are of any significance for the type of system under consideration, so the analysis performed to date has been limited to considering the contributions of the van der Waals and structural disjoining terms alone.

The van der Waals contribution is the standard force resulting from the dispersion interaction between two media having differing dielectric permittivity, $\varepsilon(i\zeta)$ as a function of frequency ζ. The structural disjoining force is less well known, but can be visualized through the following thought experiment. Consider two solids having flat, parallel surfaces rotated with respect to one another and well separated by a liquid having a random structure. The first layer of liquid molecules is assumed to have a preferred orientation on the solid surfaces. To accommodate this alignment the liquid molecules in the gap arrange themselves to match the orientation of the monolayers on either solid. When the size of the gap is large this is a minor concern, the bulk of the liquid remains random, and the distortion adds little to the overall energy of the system. The solid surfaces are now brought closer together by allowing free flow of the liquid out from the gap. As the gap gets smaller and smaller the configurational matching becomes progressively more severe, and the orientation of the liquid molecules less random. Correspondingly, the energy of the system rises. This is manifest as a force acting on the grains to resist their being brought together, a force that steadily increases in magnitude as the distortion increases as the grains are brought closer together. Clearly, when the mismatch has to be accommodated by just a molecular units, the distortion is particularly extreme, and the repulsive force correspondingly large.

Analysis of the two contributions leads to an expression for the net force (per unit area) acting on the thin liquid film of thickness, h, being:

$$P_A + P_c + \frac{\hbar}{8\pi^2 h^3}\int_0^\infty \left[\frac{\varepsilon_\alpha(i\zeta) - \varepsilon_\beta(i\zeta)}{\varepsilon_\alpha(i\zeta) + \varepsilon_\beta(i\zeta)}\right]^2 d\zeta - 4a\eta_o^{\,2}\exp\left(-h/\xi\right) = 0 \qquad [4]$$

In the colloid science literature, the integral expressing the van der Waals term is written in the form of a Hamaker constant, H, as $H/6\pi h^3$. The other terms in the equation, ξ and $a\eta_o^2$, refer to an orientational correlation length and to the free energy difference between an ordered and random liquid structure.

An interesting feature of this equation is that the van der Waals attractive term depends only on the dielectric constant of the crystalline grain (normalized to that of the intergranular phase) whereas the repulsive, structural term depends on the conformational distortion of the molecules in the intergranular phase and not on the nature of the adjoining grains. It is thus to be expected that the equilibrium thickness will vary from one ceramic material to another even when they contain the same intergranular liquid phase. This is seen for the simplest case, in which both the applied stress and the capillarity stresses are zero, in the calculations listed in Table I. The stable thicknesses all lie in approximately the same regime, with the exception of silicon carbide system where the van der Waals attraction is larger at all distances than the structural disjoining term. This is a consequence of the exceptionally high refractive index of silicon carbide. It is also interesting to note that to date no silicon carbide ceramic has been observed to contain a siliceous intergranular phase along the two grain boundaries.

Table 1

Estimated Equilibrium Thicknesses

Ceramic System	Hamaker Constant $10^{-21}\,J$	Equilibrium Thickness ξ
$Al_2O_3 - SiO_2 - Al_2O_3$	22	8.0
$BeO - SiO_2 - BeO$	17	8.3
$ZrO_2(c) - SiO_2 - ZrO_2(c)$	103	5.0
$ZrO_2(t) - SiO_2 - ZrO_2(t)$	93	5.2
$AlN - SiO_2 - AlN$	85	5.5
$Si_3N_4 - SiO_2 - Si_3N_4$	76	5.6
$SiC - SiO_2 - SiC$	233	0

The thicknesses are expressed in terms of the orientational correlation length between molecules, approx. $0.3nm$

The analysis also provides a ready explanation as to why the thickness of the intergranular phase is found to be of the order of 1nm rather than tens or hundreds of nanometers. That is, the equilibrium thickness is a natural consequence of the fact that both the structural disjoining pressure and the van der Waals- dispersion force have only appreciable magnitude over distances of a few nanometers. Furthermore, the analysis provides some insight as to why the intergranular thickness has a different value at boundaries between like phases and a different value at boundaries formed between dissimilar phases, as illustrated for instance in figure 3. In such asymmetric cases, there is still an attractive van der Waals interaction but since it is weaker the net equilibrium separation will be wider. (Since the dielectric constant of silica is generally lower than that of other ceramic phases, the dispersion interaction is likely to be always attractive).

A further prediction of the continuum models is that, contrary to expectation for liquids, grain boundaries containing a liquid film of equilibrium thickness will be able to support a normal stress. Such a situation has not hitherto been contemplated but is expected to be of importance to high temperature deformation, for instance under creep conditions. Since the equilibrium thickness corresponds to the thickness at which the net force acting normal to the film is zero (equation 2), the application of an external force has the effect of altering the equilibrium thickness, with a compressive force acting to further thin the film and a tensile force acting to thicken it. However, because of the strong thickness dependence of both the van der Waals and the structural terms in equation 4 the application of a compressive stress has little effect on the equilibrium thickness

The analysis outlined in the preceding paragraphs has been formulated in terms of a continuum description of the the forces acting across a liquid intergranular film. The magnitude of the calculated equilibrium thickness however lies near the lower limit at which such continuum models might be expected to apply. At these distances an additional force, a solvation force, which oscillates with distance between being attractive and repulsive, and is due to an osmotic pressure term related to the size of the molecules making up the liquid may well begin to be significant (43).

STRUCTURAL CONSIDERATIONS

A molecular unit model embodying many of the features of the disjoining term has been advanced (26) for the specific case of a silica intergranular phase in silicon nitride ceramics, based on the similarity of the structural units of silica and silicon nitride and the nature of their chemical bonding. Silicon nitride is a covalently bonded material that can be considered to consist of SiN_4 , tetrahedra joined at their corners, with a Si-N bond length of approximately 1.74 A. Silica glass is also a covalently bonded material and can be thought of as a network of SiO_4 tetrahedra with an average approximate Si-O bond length of 1.62A. On the basis of the similarity of these two materials it has been proposed that the SiO_4 tetrahedra adjacent to the silicon nitride grains bond to the SiN_4 tetrahedra forming an epitaxial-like partially ordered monolayer of silica on the surface of each grain of silicon nitride. This interfacial region can alternatively be described as corresponding to a layer of the (known) compound Si_2N_2O . The recent high resolution transmission electron microscopy observations by Thorel etal (44) provide support for such a structural interpretation.

Figure 6. Structure image of the $\beta -$ silicon nitride- glass interface revealing the presence of a thin, crystalline transition phase between the silicon nitride structure and the featureless glass phase. Transmission electron micrograph recorded at a defocus value of 800 angstroms. (Micrograph reproduced courtesy of Drs. Thorel and Laval).

Figure 6, taken from their work, shows a change in atomic positions at the grain boundary indicating the existence of a crystalline transitional zone. By comparing their images with computer simulated ones they conclude that the material in the zone is isostructural with the silicate mineral known as Dalyite. One structural feature of crystal structure of this mineral is that the bonding arrangement is similar to that found in silicon oxynitride. On the basis of this observation they suggest that the intergranular phase itself can be considered to consist of repeating units of the Dalyite phase. This implies that the thickness of the intergranular phase should correspond to an integral number of unit cells of dalyite. There is thus an analogy with the molecular models of liquids trapped between solid walls that exhibit the oscillatory solvation force mentioned above. Unfortunately, there exists insufficient experimental data at the time of writing to test such a hypothesis.

V. Dewetting of Grain Boundaries

The simple picture of figure 5 indicates that grain boundaries wet by a liquid film at high temperatures may be de-wet at lower temperatures provided the free energy of the material can be lowered. Such a situation is consistent with what is intuitively expected in ceramics prepared by liquid phase sintering where the boundaries are wet during sintering but have non-wetted boundaries when examined at room temperature. Two processes are believed to be involved, one the homogeneous thinning of the intergranular phase as the grains grow together by the precipitation from the liquid phase as the temperature is lowered and the other a heterogeneous process leading to the actual de-wetting.

NUCLEATION BARRIER TO DEWETTING

As described above there will be boundaries, wet at high temperatures, that can lower their energy by the retraction of the liquid phase when the temperature is reduced below a critical value. Contrary perhaps to expectation, the dewetting is not a spontaneous process and requires that a nucleation barrier be overcome. In consequence, the material has to be "supercooled" for the liquid to retract. As will be seen below the nucleation barrier, in the absence of elastic stresses, although small in absolute value is nevertheless finite and significant.

The existence of an energy barrier to the nucleation of a region of crystal-crystal grain boundary can be understood by comparing the energy of a completely wetted boundary (figure 7a) to that of the same boundary containing (for simplicity of expression) a circular, dewetted region (the nucleus) of radius, r, (figure 7b).

As in the standard analysis of the nucleation and growth of a coherent second phase precipitate in a matrix, nucleation of the dewetted region involves two contributions that scale differently with the size of the nucleus. Formation of the nucleus in the case of interest here involves the replacement of two crystal-liquid surfaces of area πr^2 with one crystal-crystal grain boundary having the same area, but a different interfacial energy (of γ_{SS} rather than γ_{SL}). At the same time, formation of the nucleus increases the area of the solid-liquid interface by an amount proportional to the thickness of the liquid phase and corresponding to the circumferential area of the nucleated island. The excess energy of the nucleated island can thus be written as:

$$\Delta E = -(\gamma_{SS} - 2\gamma_{SL}).\pi r^2 + \gamma_{SL}.2\pi.r.h \qquad [5]$$

At equilibrium, the excess energy associated with the formation of the nucleated island is a stationary value corresponding to:

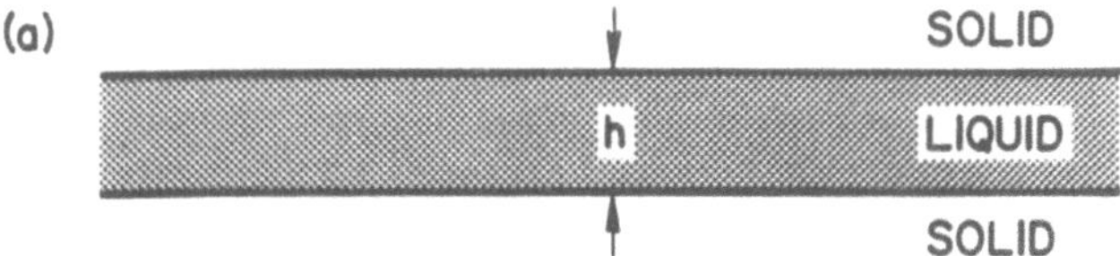

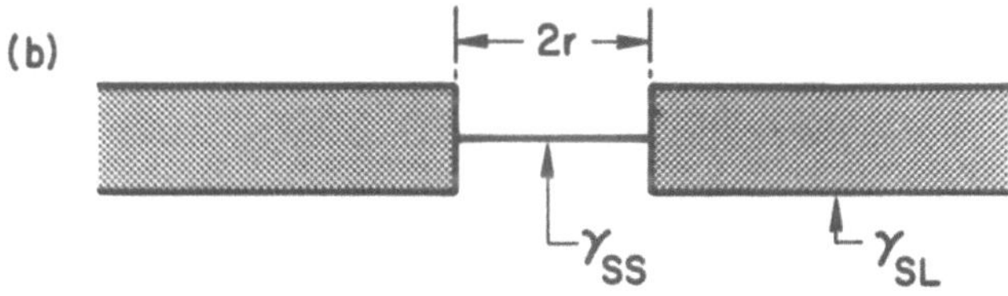

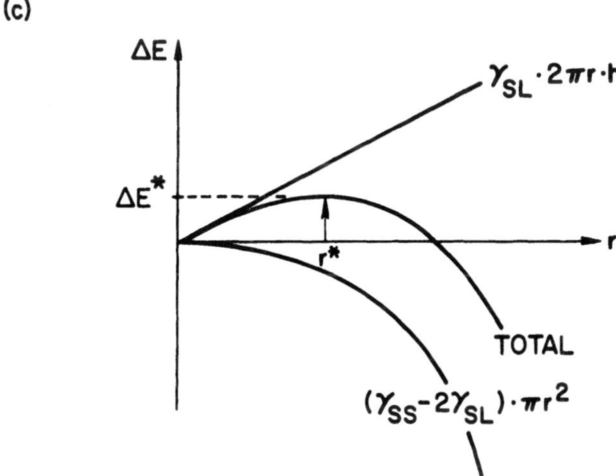

Figure 7. Nucleation of a dewetted, "pill-shaped", region of radius, r, requires the replacement of two solid-liquid interfaces of area $\pi . r^2$ by one solid-solid interface of the same area. However, this is at the expense of the creation of a solid-liquid interface having a circumferential area equal to that of the nucleus. The contribution of the interfacial energies associated with these two terms as a function of the radius is shown schematically in (c).

$$\frac{d\Delta E}{dr} = -(\gamma_{SS} - 2\gamma_{SL}) \,.\, 2\pi r + \gamma_{SL} \,.\, 2\,\pi \,.\, h = 0 \qquad [6]$$

Evaluation of this condition leads to a value for the critical nucleus size, r^*, given by:

$$r = r^* = \frac{\gamma_{SL}}{\gamma_{SS} - 2\gamma_{SL}} \,.\, h \qquad [7]$$

Substitution of this value into equation 5 provides a value for the energy barrier to the nucleation of an island of dewetted boundary:

$$\Delta E^* = \frac{\pi . \gamma_{SL}^2}{2\gamma_{SL} - \gamma_{SS}} \,.\, h^2 \qquad [8]$$

dewetting has not been established. An alternative mechanism by which an intergranular phase might be removed, and one that does not require heterogeneous nucleation in the grain boundary, is the nucleation of growth ledges at the three-grain junctions and subsequent propagation along the grain surfaces in the grain boundary displacing liquid into the triple junction region.

The removal of liquid phase from the three grain channels represents a more formidable problem in that a larger volume of material is involved and because it requires mass transport over considerably larger distances commensurate with the size of the sample as distinct to the grain size.

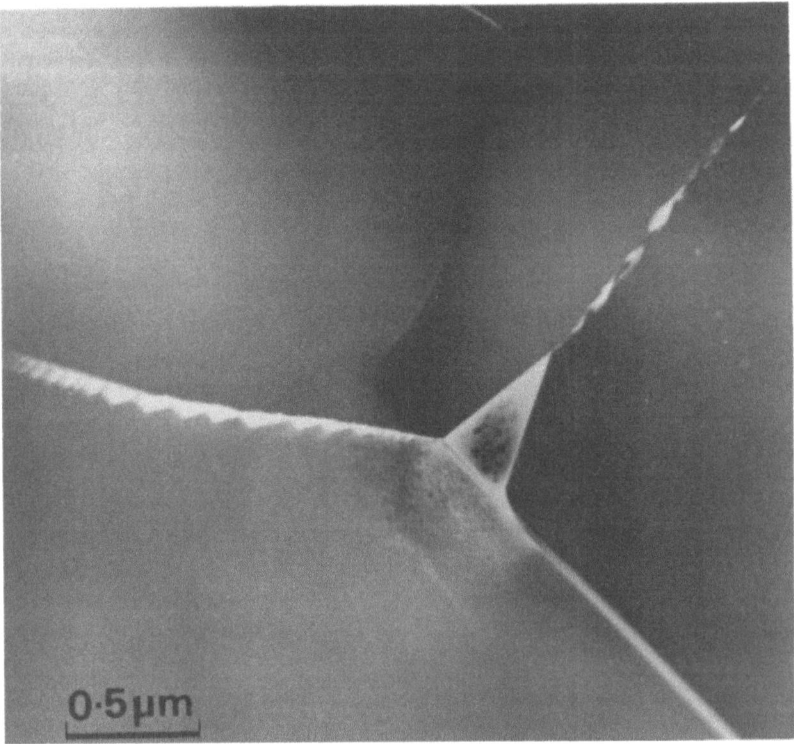

Figure 8. Partially retracted intergranular phase in a debased alumina material. The intergranular phase, appearing bright in this micrograph, is seen to completely wet two of the boundaries but to have partially dewet at the third.

VI. Post-Fabrication Processing

In view of the often detrimental role the intergranular phase plays in determining high temperature mechanical (and electrical) properties, a rational materials design strategy would be to avoid the formation of an intergranular phase, or at least, to make materials having a

Examination of this simple equation provides a number of interesting insights. The driving force for dewetting is represented by the denominator, indicating that the larger this term the lower is the nucleation barrier. Hence, the necessity for the material to be "supercooled" to overcome the nucleation barrier. In the limit at which the driving force goes to zero, i.e. at that particular temperature at which a wetted and dewetted boundary has the same energy, the nucleation barrier becomes infinitely large. The second salient feature of the equation is the quadratic dependence of the nucleation barrier on the thickness of the intergranular liquid film. For the thicknesses appropriate to those observed in many ceramics, the quadratic dependence on the thickness serves to make the magnitude of the energy barrier rather small, but it can nevertheless be appreciably larger than any thermal fluctuation. As in the way of an example consider the dewetting of a 2 nm thick intergranular liquid phase at a temperature of $1200°C$. If the driving force, the difference between the surface energies of the grains and the wet boundary, has risen to a value of $0.5\,J\,m^{-2}$ dewetting requires that an energy barrier of $300kT$ be overcome. The nucleation barrier only falls to the value of a thermal fluctuation (kT) when the driving force exceeds $154\,J\,m^{-2}$! (In these order of magnitude estimates the quite reasonable value for the solid--liquid interfacial energy of $0.5\,J\,m^{-2}$ has been assumed).

This estimate of the size of the nucleation barrier assumes a particularly simple geometry for the nucleus (equivalent to a contact angle of ninety degrees between the solid and the liquid). However, when a nucleus shape is assumed that incorporates a wetting angle of less than ninety degrees an even larger nucleation barrier is calculated than that above.

The conclusion that a wetted boundary will not spontaneously dewet because there is a finite nucleation barrier to be overcome is consistent with the results of an alternative, continuum, analysis presented some years ago by Mullins (45). He showed that a thin, planar liquid film, of infinite lateral extent, is stable to all (small scale) fluctuations in thickness.

NUCLEATION MECHANISM

The withdrawal of the glass phase from a two-grain boundary as its energy is lowered requires either that the phase uniformly thins until it is of a monolayer thickness (and is no longer a distinguishable phase) or that a process occurs in which the adjacent grains contact at one point and the liquid then retracts (figure 8).

The numerical example below serves to illustrate that the energy barrier to dewetting is considerably larger than any thermal activation for any realistic values of a dewetting force. The implication then is that homogeneous nucleation of dewetting is unlikely, and that some form of heterogeneous nucleation is required. Such heterogeneous nucleation also appears to be most likely on the basis of the relatively few observations made of partially dewetted films. However, the nucleation mechanism itself has not been identified. One possible mechanism by which a dewetted region may nucleate is by faceting of one of the two adjacent grains with the facet growing across the intergranular phase until it makes contact with the other grain surface. This requires nucleation of a preferred facet and its subsequent growth by surface diffusion accommodated by volume diffusion in the glass phase. For it to occur on cooling also requires that the faceting take place not at the sintering temperature but rather be most probable at intermediate temperatures. Facetting across an intergranular phase has been observed in magnesium-Sialon materials but whether this is responsible for

reduced volume fraction of glass of the highest possible viscosity. However, in doing so one is faced with a dilemma -- what might be termed the sintering dilemma --the very formation of the ceramics is promoted by the presence of a liquid phase during sintering. Moreover, the factors that enhance liquid phase sintering, namely the volume fraction of the liquid phase, its viscosity and solubility, are the very ones that exaggerate the degradation in high temperature mechanical properties.

The resolution of the sintering dilemma requires that the liquid phase be present during densification but then disappear after fabrication. One method traditionally used in the ceramic community is to incorporate an additive that forms a transient liquid phase during sintering, much as LiF is a transient sintering aid in the densification of $MgAl_2O_4$. Unfortunately few transient liquid phases have been identified for the sintering of ceramics, especially for ceramics of technological interest, such as the silicon nitride based ceramics and the zirconia ceramics.

In the absence of an identifiable transient liquid phase, an alternative is to adopt a post-fabrication processing step; a process, after fabrication and shaping, specifically chosen to alter the characteristics of the intergranular phase without affecting the properties of the crystalline phases in the material. Such an approach has two objectives. One is to draw glass (as liquid at high temperatures) out of the material, thereby decreasing the volume fraction of the intergranular material remaining in the ceramic. The other is to alter the composition of the intergranular glass phase, thereby altering the viscosity of the remnant glass phase, and in some cases, its connectivity and surface energy. In the following, one class of post-fabrication processing -- processing under an imposed chemical potential gradient -- is described.

PROCESSING USING CHEMICAL POTENTIAL GRADIENTS

The essential idea of chemical potential gradient processing is to establish a chemical potential difference across the sample at high temperatures so as to drive impurities out of the sample. This represents an example of the general diffusion problem: species migrate under the influence of a negative chemical potential gradient, with the diffusive flux being related to the potential gradient by an effective diffusion constant. (Fick's first law). In order that this processing approach can be exploited in practice to "purify" the intergranular phase, the intergranular phase must indeed be continuous through the material so that it can act as a rapid transport path and the temperature must be sufficiently high that the mobility of the diffusing species is large enough for an appreciable number of impurities to migrate within the time scale of the processing. This is particularly important since, in the majority of ceramics the cross-sectional area represented by the intergranular phase is quite small, since it typically is only 1-10 nm in thickness at the grain boundaries. In addition, the temperature cannot be too high otherwise bulk diffusion through the crystalline phase may occur.

As in other diffusion processes, different driving forces can be used to cause the diffusing species to migrate. In the sections below examples of the concepts of processing using three distinct driving forces are mentioned, compositional gradients, an electric field gradient and a surface tension gradient. Each is described, together with an example of their application to modifying the dielectric loss tangent -- a sensitive measurement [46] -- of a hot-pressed silicon nitride ceramic. The heat treatments described below may simply be characterized as "Impurity Pump" methods in that impurities are drawn or pumped out of the material.

The use of other potential energy gradients, such as a temperature gradient (leading to the Soret effect), or a stress gradient may also be applicable but have not, as far as the author is aware, been explored.

COMPOSITIONAL GRADIENT PROCESSING

Since the composition of the intergranular glass phase in silicon nitride ceramics is largely silica based, heating any of the materials in intimate contact with pure silicon dioxide will establish a compositional gradient for the impurities between the intergranular phase and the surrounding silica.[1] This can be done either by surrounding the sample with packed silica glass, or by heating the silicon nitride in air so as to produce, by oxidation of the silicon nitride itself, a surface layer of silica. This latter case is shown schematically in figure 9.

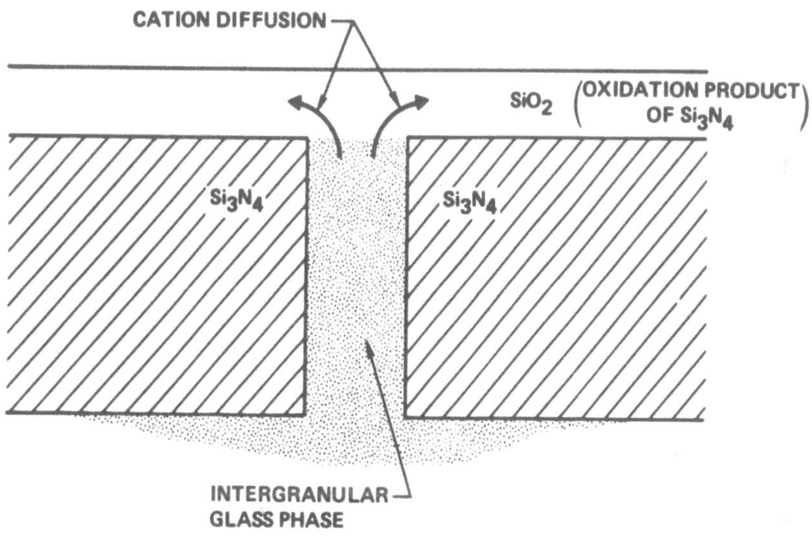

Figure 9. Schematic diagram illustrating the diffusion of cations to the surface under the influence of a compositional gradient.

The impurities drawn to the surface collect in the silica layer which can then be removed. The impurity diffusion is the basis of the property improvements reported for silicon nitride based ceramics [48,49] and for post-fabrication processing of yttria densified silicon nitrides

[1] It is equally possible to use a chemical potential gradient to drive impurities into a material, and thereby degrade its high temperature properties [47]. Such migration of impurities into the intergranular phases, and the consequent degradation of the properties of the phase, is an important mechanism in the environmental attack of many structural ceramics.

[50]. Another example of effectiveness of this approach is shown in figure 10, in which the loss tangent at 35GHz of a MgO hot-pressed silicon nitride is compared with that of the same material but after heating in contact with fused silica glass for 200 hrs at 1400°C.

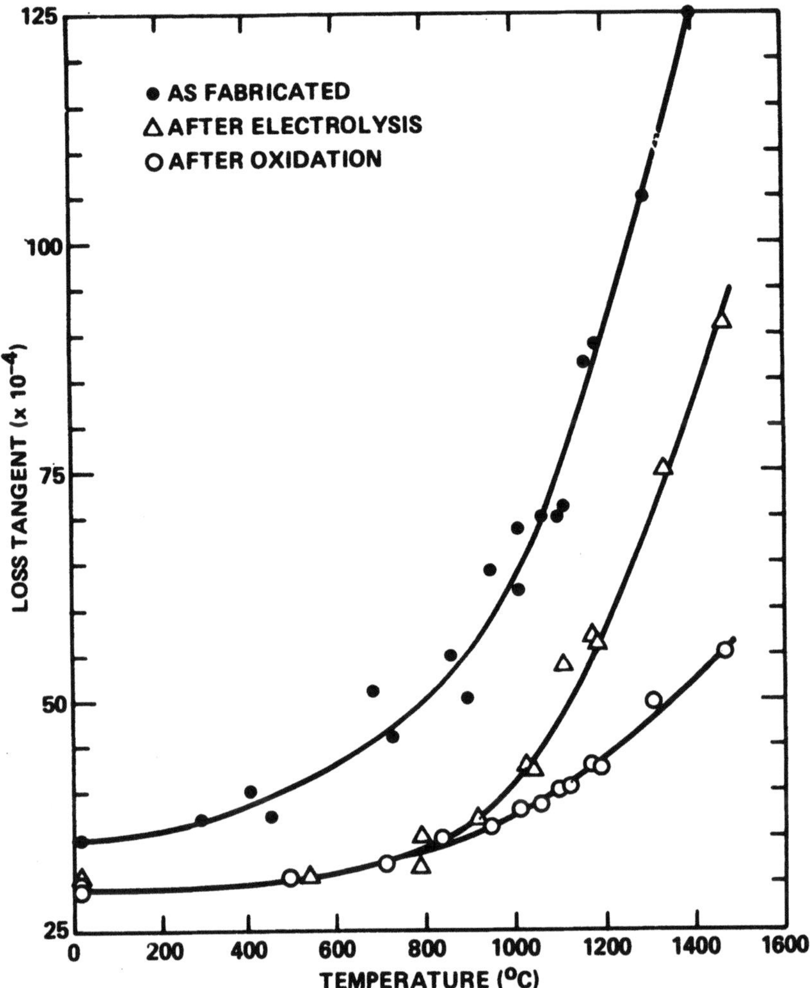

Figure 10. Dielectric loss tangent measured at 35 GHz as a function of temperature for as-fabricated silicon nitride and after an electrolysis treatment and after an oxidation treatment.

An applied electric field may also be used to drive impurities from the intergranular phase if they have a net charge. The model [51] proposed for utilizing an applied electric field is represented schematically in figure 11.

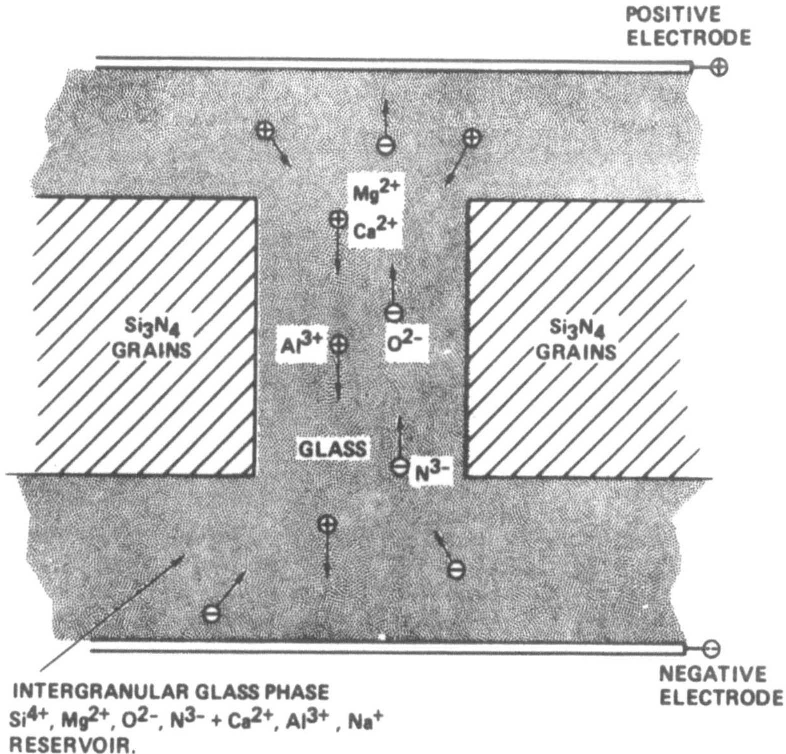

Figure 11. Schematic representation of model for migration of impurity and additive ions through a continuous intergranular phase under the influence of an externally applied electric field.

A d.c. electric field applied across the sample causes cations in the intergranular phase to migrate through the phase to the negative electrode and anions to the positive electrode (where in principle, at least, they may be collected), leaving the intergranular material depleted of the mobile ions. For appreciable mass transport to occur, the driving force on an ion must exceed the resistance due to viscous forces and any electro- viscous retardations.

The magnitude of the mass transport attainable will also be limited by the internal electric field developed by polarization phenomena. An example of the use of an applied electric field to drive impurities and additives out of a hot-pressed silicon nitride ceramic is shown in figure 10. In this example, the material was subject to an electric potential gradients of the order of 10^4V/m, for 94hrs at a temperatures of 1100 °C. The principal impurities drawn to the surface were Ca and Mg.

SURFACE TENSION GRADIENT PROCESSING

Gradients in surface tension are known to be capable of driving fluid flow in liquids, and such flows are collectively referred to as the Marangoni effects [52] following their identification by J.J. Thomson and Marangoni in the nineteenth century. (Perhaps the best known example of the Marangoni effect is the formation of "tears" of wine, the phenomena in which wine and brandy in a glass is seen to creep up the walls of the glass. The surface tension of wine, a solution of alcohols in water, increases with decreasing concentration of alcohol. Thus, as alcohols evaporate from the meniscus the surface tension increases drawing the wine up the walls of the glass). The same surface tension driven flow of impurities out of a ceramic containing an intergranular glass phase has yet to be unequivocally demonstrated. However, it is suggested that the work of Drennan and Hannink [53], in which additions of SrO to certain partially stabilized zirconias led to silica glass being extruded to the surface, may well be an example of such surface tension driven flow of impurities.

CONCLUDING REMARKS

In the decade since the existence of thin glass phases at the grain boundaries in ceramics was first established there has been a growing realization that their presence can have a controlling influence on the properties of technical ceramics, particularly their electrical and high temperature properties. Considerable insight into the microstructural stability of the intergranular phase has been gained from simple wetting arguments, but more detailed analysis is required before one can predict the conditions under which such phases are stable, and those under which they can be encouraged to retract. In many respects this is a relatively neglected area of condensed matter physics since most "wetting" phenomena investigated have been between fluids, or of liquids on a free surface rather than fluids confined between solid interfaces. There is also a lack of experimental data in the area of characterization and in studies of the penetration and retraction of liquids along grain boundaries. The advent of higher resolution electron microscopes (transmission and tunneling) promises to provide much of the necessary structural and chemical information. Despite these limitations considerable progress has been made. For instance, we now have an appreciation of the conditions for wetting of the grain boundaries, the existence of a nucleation barrier to retraction of a liquid film from a grain boundary, the forces controlling the equilibrium thickness of a liquid film, and finally the fact that very thin liquid films can support a normal stress. Furthermore, based on the concepts outlined here a number of ideas have been explored for new ceramic processing approaches.

REFERENCES

1. C. Quincke, 'Berliner Monats', **139**, (1869), Proc. Roy. Soc. A, **76**, 431 (1905).
2. McLean, D., *'Grain Boundaries In Metals'* Oxford, Oxford University Press (1957).
3. Lange, F.F., 'Silicon Nitride Polyphase Systems: Fabrication Microstructure and Properties', Int'l. Metals. Revs. **247**, 1-20 (1982)
4. *Progress In Nitrogen Ceramics*, edited F.L. Riley, Martinus Nijhoff Publishers, Boston (1983).
5. Clarke, D.R. and Thomas, G., 'Grain-Boundary Phases in Hot-Pressed MgO Fluxed Silicon Nitride', J. Am. Ceram. Soc. **60**, 491-495 (1977).
6. Lou, L.K.V., Mitchell, T.E. and Heuer, A.H., 'Impurity Phases in Hot-Pressed Si_3N_4', J. Am. Ceram. Soc. **61**, 392-396 (1978).
7. Clarke, D.R., 'On the Detection of Thin Intergranular Films by Electron Microscopy', Ultramicroscopy **4**, 33-44 (1979).
8. *Science and Technology Of Zirconia*, Heuer, A. and Hobbs, L. eds., Am. Ceram. Soc., Columbus (1981).
9. *Advances In Ceramics*, Ruhle, M., Claussen, N. and Heuer, A. eds., Amer. Ceram. Soc., Columbus (1984).
10. Clarke, D.R., 'Ceramic Materials for the Immobilization of Nuclear Waste', Ann. Rev. Mat. Sci. **13**, 191-218 (1983).
11. Morgan, P.E.D., Clarke, D.R., Jantzen, C.M. and Harker, A.B., 'High Alumina Tailored Nuclear Waste', J. Am. Ceram. Soc. **64**, 249-258 (1981).
12. McCarthy, G.J., Pepin, G.J., Pfoertsch, D.E. and Clarke, D.R., in *Ceramics In Nuclear Waste Management*, DOE (1979).
13. Hansen, S.C. and Phillips, D.S., 'Grain-Boundary Microstructure in a Liquid-Phase-Sintered Alumina (γ - Al_2O_3)', Phil. Mag. A**47**, 209-234 (1983).
14. Clarke, D.R., 'High Temperature Deformation of a Polycrystalline Alumina Containing an Intergranular Glass Phase', J. Mater. Sci. **20**, 1321-1332 (1978).
15. Yan, M. and Heuer, A., *Additives and Interfaces In Electronic Ceramics*, Am. Cer. Soc., Columbus (1983).
16. *Grain Boundary Phenomena In Electronic Ceramics*, Levinson, L.M., Hill, D.C. eds., Am. Ceram. Soc., Columbus (1981).
17. Clarke, D.R., 'The Microstructural Location of the Intergranular Metal Oxide Phase in a Zinc Oxide Varistor', J. Appl. Phys. **49**, 2407-2411 (1978).
18. Kingery, W.D., Vander Sande, J.B. and Mitamura, T., 'A Scanning Transmission Electron Microscopy Investigation of Grain-Boundary Segregation in a $ZnO - Bi_2O_3$ Varistor', J. Am. Ceram. Soc. **62**, 221-222 (1979).
19. Gambino, J., Phd Thesis, MIT (1984).

20. Olsson, E., 'Interfacial Microstructure in ZnO Varistor Materials', Phd Thesis, Chalmers University (1988).

21. Clarke, D.R., Zaluzec, N.J. and Carpenter, R.W., 'The Intergranular Phase in Hot-Pressed Silicon Nitride Alloys. I. Elemental Composition', J. Am. Ceram. Soc. **64**, 601-607 (1981).

22. Kingery, W.D., 'Densification During Sintering in the Presence of a Liquid. I. Theory.', J. Appl. Phys. **30**, 301-310 (1959).

23. Hampshire, S. and Jack, K.H., in *Progress In Nitrogen Ceramics*, Boston, Martinus Nijhoff (1983).

24. Lange, F.F., Clarke, D.R. and Davis, B.I., 'Compressive Creep of Si_3N_4/MgO Alloys: Effect of Composition', J. Mater. Sci. **15**, 601-610 (1980).

25. Lange, F.F., 'Evidence for Cavitation Crack Growth in Si_3N_4', J. Am. Ceram. Soc. **62**, 222-223 (1979).

26. Clarke, D.R. and Lange, F.F., 'Oxidation of Silicon Nitride Alloys: Relationship to Phase Equilibria. I $Si_3N_4 - SiO_2$—MgO Systems', J. Am. Ceram. Soc. **63**, 586-593 (1980).

27. Karunaratne, B.S.B. and Lewis, M.H., 'High Temperature Fracture and Diffusional Deformation Mechanisms in Si-Al-O-N Ceramics', J. Mater. Sci. **15**, 449 (1980).

28. Clarke, D.R., 'High Temperature Environmental Strength Degradation of Hot-Pressed Silicon Nitride: An Experimental Test.', J. Am. Ceram. Soc. **66**, 156-158 (1983).

29. Clarke, D.R., 'Compositional Modification of Hot-Pressed Silicon Nitride by High Temperature Electrolysis', J. Am. Ceram. Soc. **67**, 455-459 (1984).

30. M. Billy and J. G. Desmaison, 'High Temperature Oxidation of Silicon-Based Structural Ceramics', *High Temperature Technology*, **4**(3) 131-139 (1986).

31. Clarke, D.R., 'On the Equilibrium Thickness of Intergranular Glass Phases in Ceramic Materials', J. Am. Ceram. Soc. **70**, 15-22 (1987).

32. Krivanek, O.L., Shaw, T.M. and Thomas, G. 'The Microstructure and Distribution of Impurities in Hot-Pressed and Sintered Silicon Nitrides', J. Am. Ceram. Soc. **62**, 585-590 (1979).

33. Powell, P.D. and Drew, P., 'The Identification of a Grain-Boundary Phase in Hot-Pressed Silicon Nitride by Augers Electron Spectroscopy', J. Mater. Sci. **9**, 1867-1870 (1974).

34. Hofmann, S. and Gauckler, L.J., 'A Study of Fracture Surfaces of Hot-Pressed Silicon Nitride by Auger Electron Spectroscopy', Powder. Metall. Int. **6**(2), 90-92 (1974).

35. Clarke, D.R., 'The High Temperature Microstructure of a Hot-Pressed Silicon Nitride', J. Am. Ceram. Soc., Submitted (1988).

36. Raj, R., 'Morphology and Stability of the Glass Phase in Glass-Ceramic Systems', J. Am. Ceram. Soc. **64**, 245-248 (1981).

37. Clarke, D.R., 'Grain Boundaries in Polyphase Ceramics', J. de Physique. **C4**, 51-59 (1985).

38. Schmid, H. and Ruhle, M., 'Structure of Special Grain Boundaries in SiAlON Ceramics', J. Mater. Sci. **19**, 615-628 (1984).

39. Read, W.T. and Shockley, W., 'Dislocation Models of Crystal Grain Boundaries', Phys. Rev. **78**, 275-289 (1950).

40. Clarke, D.R. and Lange, F.F., 'Morphological Changes of an Intergranular Thin Film in a Polycrystalline Spinel', J. Am. Ceram. Soc. **65**, 502-506 (1982).

41. Israelachvili, J.N., *Intermolecular and Surface Forces*, Orlando, Academic Press.

42. Derjaguin, B.V., Rabinovich, Y.I. and Churnaev, N.Y., 'Direct Measurement of Molecular Forces', Nature **272**, 313-318 (1978).

43. Ninham, B.W., Chemica Scripta. **25**, 3-6 (1985).

44. Thorel, A., Laval, J.Y., Broussaud, D., deSagey G. and Schiffmacher, G., Proc. of JIMIS-4, Tokyo (1986).

45. Mullins, W.W., 'Flattening of a Nearly Plane Solid Surface Due to Capillarity', J. Appl. Phys. **30**, 77-83 (1959).

46. Clarke, D.R. and Ho, W.W., 'Effect of Intergranular Phases on Dielectric Losses of Silicon Nitride Ceramics', Adv. Ceram. **7**, 246 (1983).

47. Clarke, D.R., 'High Temperature Environmental Strength Degradation of Hot-Pressed Silicon Nitride: An Experimental Test', J. Am. Ceram. Soc., **66**, 156 (1983).

48. Clarke, D.R. and Lange, F.F., 'Oxidation of Silicon Nitride Alloys: Relationship to Phase Equilibria. I. $Si_3N_4 - SiO_2$—MgO Systems', J. Am. Ceram. Soc. **63**, 586 (1980).

49. Karunaratne, B.S.B. and Lewis, M.H., 'High Temperature Fracture and Diffusional Deformation Mechanisms In Si-Al-O-N Ceramics', J. Mater. Sci. **15**, 449 (1980).

50. Clarke, D.R., Lange, F.F. and Schnittgrund, G.D., 'Strengthening of Sintered Silicon Nitride by a Post-Fabrication Heat Treatment', J. Am. Ceram. Soc. **65**, C51 (1982).

51. Clarke, D.R., 'Compositional Modification of Hot-Pressed Silicon Nitride by High Temperature Electrolysis', J. Am. Ceram. Soc. **67**, 455 (1984)

52. Scriven, L.E. and Sternling, C.V., 'The Marangoni Effects', Nature **187**, 186 (1960).

53. Drennan, J. and Hannink, R.H.J., 'Effect of SrO Additions on the Grain-Boundary Microstructure and Mechanical Properties of Magnesia-Partially-Stabilized Zirconia', J. Am. Ceram. Soc. **69**, 541 (1986).

INVESTIGATION OF THE FRACTAL STRUCTURE OF THE PORE-GRAIN INTERFACE IN ALUMINA CERAMICS

F.BROUERS+ and A.RAMSAMUGH*

\+ Institute of Physics, Liege University, Sart-Tilman, Liege 4000 Belgium.

* Department of Physics, University of the West Indies, Mona, Jamaica.

ABSTRACT. At low frequency, the conductivity of fluid saturated porous media varies with porosity as ϕ^m where m is the Archie's exponent. At higher frequencies, the real part of the conductivity exhibits a dispersive (non Gaussian) behaviour characterized by a power law $\sigma'(\omega) \propto \omega^x$. The real part of the dielectric constant which can attain very large value at low frequency varies in the dispersive region as $\varepsilon'(\omega) \propto \omega^{-y}$ with $x + y = 1$. We obtained the values of these exponents from data in brine saturated porous alumina ceramics presented in previous papers.

The conclusion of our analysis is that the low frequency conductivity of porous alumina ceramics saturated with saline water yields an Archie's exponent compatible with the theory of percolation. The conductivity frequency exponents in the dispersive region are consistent with the model of anomalous conduction on fractals and the high value of the real part of the dielectric constant can be accounted for if the microgeometry of the grain-pore interface and multipole effects are incorporated into the Maxwell-Garnett theory of composite media.

1. INTRODUCTION

In this lecture, we discuss how simple electrical and impedance measurements can yield information on the microstructure of porous ceramics in particular on the nature of the pore-grain interface.

L.-C. Dufour et al. (eds.), Surfaces and Interfaces of Ceramic Materials, 81–92.

For a long time this problem has been addressed to by geophysicists. One very important application of the theory of the dielectric response of mixtures is in oil exploration. From a measurement of the complex dielectric function, dielectric permittivity and conductivity of a rock where pore space is saturated with brine and/or oil, and from the knowledge of porosity one would like to know the relative quantities of oil and brine.

We are convinced that, although the formation mechanisms of ceramics and sedimentary rocks are quite different, some of the concepts introduced in the physics of rocks may be useful to describe aspects of the physics of ceramics.

The features common to the dielectric functions of many sedimentary rocks systems are (Keynon, 1984; Stroud et al., 1986;Thompson et al., 1987)

1) the ratio of the dc conductivity of the mixture to that of the conducting components (brine) has a power-law behaviour with respect to the brine filling fraction ϕ : $\sigma \sim \sigma_w \phi^m$ where $m \approx 2.$ and σ_w is the conductivity of the brine (Archie (1942)'s law);

2) the real part of the dielectric function of the composite increases dramatically with decreasing frequency, surpassing the corresponding values of both components and showing a crossover to saturation at both low and high frequency ends. The low frequency saturation corresponds to gigantic values of the order 10^6 or 10^7 ;

3) the ac conductivity of the composite increases with frequency with a crossover to saturation at both low-frequency and high frequency ends.

Such a behaviour has been predicted by studying random resistors networks in the framework of the percolation theory. Heuristic arguments based on the theory of phase transition as well as simulation Monte-Carlo calculations predict a scaling both with respect to the filling fraction, static scaling, and with respect to frequency, dynamic scaling. Static scaling has indeed been observed below percolation in systems of metal films (Abeles et al., 1975, Deutscher et al., 1978) and above percolation in a system of small Ag particles randomly dispersed throughout a KCl host (Granan et al., 1981). Dynamic scaling has been reported above percolation for a system of gas evaporated gold black

(Niklasson and Granqvist, 1986) and both below and above percolation in thin Au films. (Laibowitz and Gefen, 1984).

It was tempting to see if porous ceramic would exhibit such static and dynamic scalings. We made a study of the frequency variation of the dielectric permitivity and conductivity of brine saturated porous alumina ceramics. From such measurements previously reported (Ramsamugh and Brouers, 1987), an empirical scaling law for the real part of the frequency-dependent electrical conductivity $\sigma'(\omega)$ to its dc limit was identified.

This scaling law can be written as

$$\sigma'(\omega)/\sigma(0) = f\,(A\epsilon_0\omega/\sigma(0)) \qquad (1)$$

where $f(x) \approx x^{0.75}$ over the main part of the dispersive region. A is a

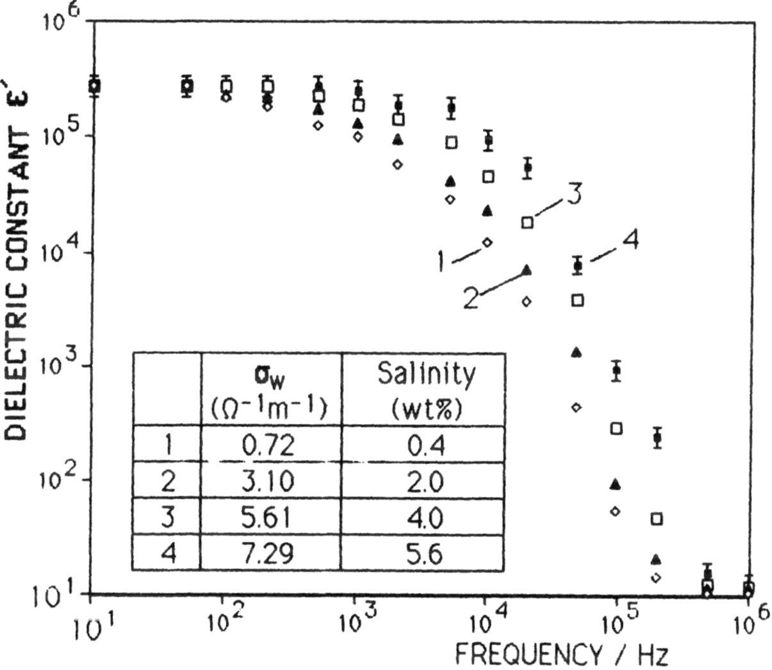

Figure 1. The frequency dependence of the real part of the dielectric constant of brine saturated alumina ceramic (From Brouers and Ramsamugh, 1988).

constant which depends on salinity, porosity and possibly on temperature. It was found that this result is very similar to what is found in hopping semiconducting systems. This scaling law fits the data

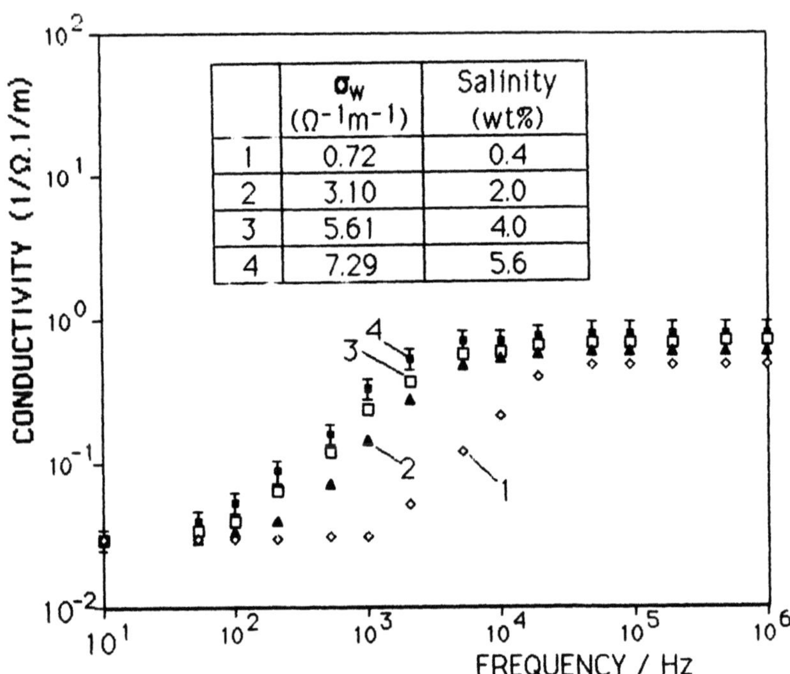

Figure 2 Frequency dependence of the elec-trical conductivity of brine saturated alumi-na ceramics (From Brouers and Ramsamugh, 1988).

obtained for a number of samples with different porosity and various brine conductivity. At the same time (Brouers and Ramsamugh, 1986), it was found that the same samples where obeying a static scaling law similar to the Archie's law for the rocks. The value of the dynamical scaling law exponent is one of the closest to what the metal-insulator percolation theory predicts (Efros and Shklovskii, 1976). Some of the results are shown in Fig 1 and 2. The reader is invited to consult F.Brouers et al.(1987) and Brouers and Ramsamugh (1988) for a detailed presentation of these data. In the next section we summarize the arguments which lead to the derivation of the scaling laws in systems presenting an insulator-metal transition.

2. STATIC AND DYNAMIC SCALING LAWS

Anomalous behaviour of the ac. conductivity in mixture of insulators and conductors is a subject of recent interest. In contrast, the power law behaviour of the dc conductivity as the concentration of conductor approaches a critical value p_c, the percolation threshold, has long been recognized. In a macroscopic composite system made of two constituents characterized by the dc conductivities $\sigma_A = \sigma$ and $\sigma_B = 0$, the medium conductivity in the vicinity of the percolation threshold scales as:

$$\sigma^* = \sigma(p - p_c)^t \tag{2}$$

while the static dielectric constant scales as:

$$\epsilon^* \sim (p - p_c)^{-s} \tag{3}$$

Although p_c depends on the type of disorder and the type of lattice or of continuous model, the <u>exponents</u> s and t are supposed to be <u>universal</u> i.e. to depend only on the space dimension. Percolation is a second order phase transition and analogies with the thermodynamic phase transition formalism where $(T - T_c)/T_c$ is replaced by $(p - p_c)/p_c$ have proved to be useful and powerful.

In particular if σ_B is not zero but much smaller than σ_A ($\sigma_B/\sigma_A \ll 1$) the quantity $h = \sigma_B/\sigma_A$ plays the role of an external magnetic field h and as in the para-ferromagnetic phase transition the static scaling law for the sharp transition is replaced by a smooth transition which can be described by the use of a scaling function. In that case the medium conductivity can be written as (Straley, 1976; Efros and Shlovskii, 1976)

$$\sigma^* = \sigma_A |p - p_c|^t F_{\pm} [h |p - p_c|^{-(t+s)}] \tag{4}$$

The function $F_{\pm}$ is a scaling function which has different asymptotic values above and below p_c for different extremes of its argument. In the limit where $h \to 0$, one recovers the previous results. Straley's relation provides predictions concerning the system's response to an ac source if one explicitly substitutes $\sigma_A \rightarrow \sigma$ and $\sigma_B \rightarrow i\omega\epsilon\epsilon_0$ and one writes the complex conductivity σ^* as:

$$\sigma^* = \sigma \mid p - p_c \mid^t F_{\pm} \left[(i\omega/\omega_0) \mid p - p_c \mid^{-(t+s)} \right] \qquad (5)$$

where $\omega_0 \equiv \sigma/\epsilon\epsilon_0$. For frequencies much less than the crossover frequency $\omega^* \equiv \omega_0 \mid p - p_c \mid^{(t+s)}$, Equation (5) will produce the conventional limiting forms discussed above : Equations (2) and (3). For $\omega^* \ll \omega \ll \omega_0$ one has an anomalous scaling of the conductivity:

$$\sigma^* \sim \sigma \left(\omega/\omega_0 \right)^u \qquad \text{where } u = t/(s+t) \qquad (6)$$

A Kramers-Kronig analysis gives for the dielectric permittivity:

$$\epsilon^* \sim \epsilon \left(\omega/\omega_0 \right)^{-(1-u)} \qquad (7)$$

The theoretical results we have just reported strictly refer only to ramdom systems. Such systems always exhibit a non zero percolation threshold p_c below which the dc conductivity vanishes. In rocks and presumably in ceramics the pore space remains connected even in samples with very low porosity. Presumably the analog for porous media of the limit $p \rightarrow p_c$ is the low porosity limit $p \rightarrow 0$ and the Archie's law is the analog of Eq. (2), the static scaling law, if $p = \phi$ and $p_c = 0$.

The values of the Archie's, dielectric constant and conductivity exponents m, s/(s+t) and t/(t+s) measured on a series of alumina samples with porosities varying from 0.179 to 0.433 and for various salinities (see Figures 1 and 2) are respectively 1.94, 0.284 and 0.746. The value of m is very close to the percolation theory values of Balberg et al. (1983), Balberg and Bozowski (1982), Derrida et al. (1983). If the Archies's exponent m is identified with the percolation theory conductivity exponent t, our data yield a value of $s = 0.73 \pm 0.06$ in agreement with theoretical evaluations ($s_3 = 0.75 \pm 0.04$) of Herrmann et al. (1984) and the experimental measurement ($s_3 = 0.73 \pm 0.07$) of Bowman and Stroud (1984).

These results raise the question of whether or not Archie's law can be explained by ordinary percolation in the continuum (i.e., is the theory consistent with the very small critical porosity ϕ_c and with the observed values of m?)

Although a corresponding universality would explain the wide applicability of Archie's law with m ~ 2, it remains to be proved that such universality exists with respect to variations such as grain shape and packing geometry. If percolation theory could be applied to interpret Archie's law this would have the advantage of being based on universal behaviour and of not requiring *a priori* assumption regarding the microstructure of the pores in the materials. This would be extremely useful for geophysical and materials science studies if there exist universal laws governing dielectric and conductivity properties of fluid saturated porous media.This seems to be the case for the ceramic samples used for our study. They all yield a value of m close to 1.95. This is not the case for data related to rocks where (Sen et al., 1981) values close to 1.5 and 4 have been reported. However as we have argued (Brouers, 1988), the measurements are generally done for non zero frequencies and this could explain the dispersion of the exponent values. The response to the question of universality must wait until more dc data are available.

3. FRACTAL DIMENSION

A <u>fractal dimension</u> d_f can be determined using theoretical expressions derived for percolating clusters. Near percolation threshold p_c, the random character of the mixture (alumina and filled pores) leads to <u>self-similarity</u> over a significant range of length scales, as predicted by the scaling theory of percolation. A series of micrograph of the porous ceramics taken at varying magnifications show qualitatively similar pictures, i.e. the observer is unable to tell in what order of magnifications they were taken. However, below a certain magnification gross features become apparent and the order of magnification is easily established. According to our current understanding of percolation (Stauffer, 1985) this occurs when the frame size is of the order of a characteristic length, the "<u>correlation length</u>" ξ_p, which diverges at p_c as:

$$\xi_p = a \mid p - p_c \mid^{-\nu} \tag{8}$$

where a is a microscopic length and ν is is a critical exponent $\nu_3 = 0.85$ in three dimensions. Near p_c, ξ_p can become quite large. At p_c, it is strictly infinite for an infinite system. The property of self-similarity is then observable over a sizeable range of scales L, i.e. $a < L < \xi_p$.

Above threshold, the study of the infinite conducting cluster shows that it has an anomalous scale dependent density. If one measures the density ρ as a function of the length scale L, one obtains a power law density decrease $\rho(L) \propto L^{-(d-d_f)}$. In other words, the mass is not proportional to the volume L_d, where d is the usual dimensionality., but rather varies as L^{d_f} where d_f is an anomalous, or fractal, dimensionality. This result is the logical consequence of the assumption made in the percolation theory that there is only one relevant length scale ξ_p in the problem, i.e. on scales smaller than ξ_p there is no observable characteristic length scale and the object must therefore be self-similar and fractal. This self similarity property can be visually identified by the fact that the infinite cluster has in it holes (here the insulating rock or ceramic) of all sizes up to ξ_p . At $L > \xi_p$, which appears as the characteristic length scale of the cluster, the density is size independent and changes in magnification can be detected. In three dimensions, the fractal dimension for percolating clusters ($L > \xi_p$) is $d_f = 2.5$.

4. ANOMALOUS DIFFUSION ON FRACTALS

As first shown by Gefen et al. (1983), diffusion on a fractal is such that

$$\langle r^2(t) \rangle \propto t^{d_{rw}} \tag{9}$$

where d_{rw} is the randomwalk critical exponent. In a normal diffusion process in a euclidean 3-dimension space, d_{rw} is equal to 1. Above percolation which is the the physical situation we are interested in the porous ceramic, this anomalous diffusion gives rise to a frequency dependent conductivity and dielectric constant in a frequency region delimited by ω_ξ and ω_a ,the frequencies corresponding to the correlation length ξ and the microscopic length limit a. The knowledge of the brine diffusion constant and of the values of ω_ξ allows us to determine an estimation of the correlation length i.e. ξ = 5–12 μm.

Although models of percolating clusters allow us to gain some insight into the problem of electrical conductivity or transport processes in inhomogeneous materials, we do not know today of any exact relation between the exponent of the conductivity and the other exponents of the percolation theory, except in dimensions higher than 6. This failure is the consequence of the fact that, in the transport on a percolation network, both the diffusion process of a particle and the percolating network itself are random.

If we use the relations between the exponents t, s, ν, d, d_f and the random walk dimension d_{rw} established by Alexander and Orbach (1982) and Coniglio and Stanley (1984) (see also Niklasson and Granqvist 1986) i.e.:

$$t/\nu = 1 + d_{rw} - d_f \tag{10a}$$

and

$$s/\nu = (d_f - 1)/2 \tag{10b}$$

we can calculate d_f and d_{rw} noting however than (10a) is exact and (10b) still a conjecture. The exponent ν can be first determined using the "hyperscaling law" proposed by Straley (1980) :

$$t + s = d\nu \tag{11}$$

The average value ν = 0.89 ± 0.02 is in agreement with the value of 0.88 ± 0.02 given by Gaunt and Sykes (1983) and the value of 0.87 ± 0.02 given by Hermann et al. (1984).

The value of d_f and d_{rw} are given in Table 1. In the last column we give the values of the spectral dimension

$$d_s = 2d_f/d_{rw} \tag{12}$$

which are slightly higher than the Alexander and Orbach (1982) conjecture (4/3).

Porosity ϕ	t	s	ν	d_f	d_{rw}	d_s
0.179	1.94	0.72	0.89	2.64	3.79	1.39
0.193	1.94	0.72	0.89	2.64	3.79	1.39
0.253	1.94	0.79	0.91	2.74	3.87	1.41
0.275	1.94	0.67	0.87	2.54	3.77	1.35
0.433	1.94	0.76	0.90	2.69	3.84	1.42
Average value	1.94	0.73	0.89	2.65	3.81	1.39

TABLE 1. Conductivity t and superconductivity s exponents. The correlation length exponent ν, the fractal, random walk and spectral dimensions d_f, d_{rw} and d_s are obtained from Equations (10-12).

5. DISCUSSION

We have found that the critical exponents s, t, ν and the relations between the dimensions d_f, d_{rw} and d_s of brine saturated porous alumina ceramics are consistent with percolation theory. This may strengthen the recent proposal by Balberg (1986) that "ordinary percolation provides a realistic model for the pore space " .

However it must be kept in mind that the porosities (up to 40%) are far from the percolation threshold (ϕ_c = 0). The fact that ϕ_c = 0 presumably means that the pore network geometry is dominated by the formation mechanism and is not at all random. It is now known that fractal growth models can explain the fractal nature of ceramics and that diffusion-limited aggregation and other related models may have a fractal dimension and other exponents rather close to those of percolation clusters.

On the other hand, if one follows the controversial suggestion (Katz and Thompson, 1985) that the pore volume behaves as a fractal with the fractal dimension of the pore-grain interface, the value obtained d_f = 2.65 ± 0.10 is very close to the value of the alumina surface fractal dimension, d_f= 2.79±0.03, obtained by Avnir et al (1984) and Pfeifer (1986) from an analysis of molecular adsorption on porous alumina ceramics. It also agrees with the fractal dimension of a rough surface determined by Monte Carlo simulations, d_f = 2.8 ± 0.03, (Mon, 1986) for a solid-on-solid model above the roughening transition.

In our case the value of the correlation length is practically equal to the grain size as measured by electron microscopy. This result supports the Katz and Thompson hypothesis. One can therefore imagine that ,as the porosity decreases, a cross-over from a fractal structure dominated by the pore-grain interface to a percolation cluster fractal structure is taking place. This might not be easy to detect since the fractal dimension of a 3-dimension percolation cluster (< ξ_p) is equal to 2.5 (Stauffer 1985).

In this lecture we have shown how low frequency conductivity and dielectric measurements can become a potentially rich method to provide useful information on the the fractal structure of porous media. However to make definite statements on the validity of the various competing models (Wong 1987) more data and a deeper understanding of their relation to the morphology of the materals are necessary.

ACKNOWLEDGEMENT

We would like to thank Dr. Duvigneaud of the Laboratoire de Chimie Industrielle, Université Libre de Bruxelles, who has prepared the samples. This work was supported by a grant of the Third World Academy of Science (Trieste).

REFERENCES

Abeles B, Pinsh H.L. and Gittelman , 1975 , Phys.Rev.35, 247
Alexander S. and Orbach R.J., 1982 , Phys.Lett. 43, L625
Archie G.E. , 1942, Trans.AIME (Am.Inst.Min.Metall.Pet.Eng.) 46, 54
Avnir D.,Farin D. and Pfeifer P., 1984 , Nature 308, 261
Balberg I. , 1986 , Phys Rev B 33, 3618
Balberg I., Bienenbaum N. and Anderson C.H., 1983, Phys.Rev.Lett. 51, 1605
Balberg I.and Bozowski S., 1982, Solid State Commun.44, 551
Bowman D.R. and Stroud D., 1984, Phys.Rev.Lett. 52, 299
Brouers F., 1988, in " Universalities in Condensed Matter", Ed. R.Jullien Springer, Berlin.
Brouers F. and Ramsamugh A., 1988, J.Phys.C 21, 1839
Brouers F., Ramsamugh A. and Dixit V.V., 1987, J.Mat.Sc.22, 2759

Coniglio A. and Stanley H.E., 1984, Phys.Rev.Lett.52 , 1068

Derrida B.,Stauffer D., Herrmann H.J. and J.Vannimenus, 1983, J.de Phys.Lett. 44, L701

Deutscher G., Rappaport M.L. and Ovadyahu Z., 1978, Solid State Comm. 28, 593

Efros A.L. and Shklovskii B.I., 1976, Phys.Status Solidi (b) 76, 475

Gaunt D.S. and Sykes M.F., 1983, J.Phys.A 16, 783

Gefen Y, Aharony A. and Alexander S., 1983, Phys. Rev. Lett. 50, 77

Granan D.M., Garland J.C. and Tanner D.B., 1981, Phys. Rev. Lett. 46, 375

Herrmann H.J., Derrida B. and Vannimenus J., 1984, Phys.Rev. B 30, 4080

Katz A.J. and Thompson,1985, Phys.Rev.Lett. 54, 1325

Kenyon W.E., 1984, J.Appl.Phys.55, 3153

Laibowitz R.B. and Gefen Y., 1984, Phys.Rev.Lett.53, 380

Mon K.K., 1986, Phys.Rev.Lett.57, 866

Niklasson G.A. and Granqvist C.G.,1986, Phys.Rev.Lett.56, 256

Pfeifer P., 1986, in "Fractal in Physics" Ed. L. Pietronero and E. Tosatti Elsevier Science Pub., p.47.

Ramsamugh A. and Brouers F., 1987, Phil. Mag. Lett. 55, 301

Sen P.N. , Scala C. and Cohen M.H., 1981, Geophysics 46, 781

Stauffer D.,1985, Introduction to Percolation Theory , Taylor and Francis, London.

Straley J.P.,1980, J.Phys.C.13, 819

Stroud D. and Bergman D.J., 1982, Phys.Rev. B25, 5145

Thompson A.H., Katz A.J. and Krohn C.E., 1987, Adv.Phys. 36, 625

STRUCTURE AND PROPERTIES OF GRAIN BOUNDARY IN MgO BICRYSTALS

Eiichi YASUDA & Shiushichi KIMURA*

Res. Lab. of Eng. Mater., Tokyo Institute of Technology,
4259, Nagatsuta, Midoriku, Yokohama, 227, Japan.
*:Faculty of Engineering, Tokyo Institute of Technology
2-12-1 Oookayama, Meguroku, Tokyo, 152 Japan.

Abstract

Bicrystals with (100) and (110) symmetrical tilt boundaries and with (100) twist boundaries were obtained by joining two single crystals at about 2480°C without pressure. Boundary energies were measured by the thermal grooving technique at 1200 to 1600°C. It was clear that small angle boundaries lower than 10° are composed of an array of dislocations, and it was also confirmed from observation by transmission electron microscopy (TEM) in (100) twist and (110) tilt boundaries. The critical angle estimated from diffusional properties in the (100) tilt boundary was also 10°. On the other hand, the critical angle of the (100) tilt boundary following the Read-Shockley model was measured to be 20°. Up to 10° the (100) tilt boundary is composed of a straight dislocation arrangement. Above 10° boundary is composed of a crisscrossed dislocation arrangement in a kink and jog fashion. For high angle boundaries, all of the low energy boundaries observed in the (100) twist boundary and (110) tilt boundary correspond to the high density coincident site lattice (CSL) boundary. In the (100) tilt boundary, no low energy boundary was observed at higher angles because of the higher atomic density on the (100) plane than on the (110) and the introduction of dislocations with different slip systems.

1. Introduction

Research on grain boundary structure in ceramics is more complicated than with metals because of the difficulty of obtaining pure materials without secondary phases at the grain boundary. Many experimental and theoretical efforts have been devoted to studying the structure and properties of grain boundaries in ceramic materials; however, their characterizations have not been clarified sufficiently.

Grain boundaries are an interface between two crystals of different orientations. Boundaries are classified into two groups from the viewpoint of (1) symmetry and (2) orientation of the two crystals and

L.-C. Dufour et al. (eds.), Surfaces and Interfaces of Ceramic Materials, 93–108.

boundary planes. In the latter group there are tilt boundaries and twist boundaries.

In studying the structure and properties of boundaries, the use of bicrystals becomes necessary. It is very difficult to synthesize bicrystals because of the high melting temperature or high vapor pressure of ceramics. However, several fabrication methods using single crystals have been developed in some metallic oxides.

In this paper the following are discussed: (1) the fabrication technique of MgO bicrystals, (2) the structure of grain boundaries and the chemical properties of MgO grain boundaries, and (3) the discrepancy between the maximum angle estimated from the Read-Shockley relationship and the angle measured from diffusional behavior.

2. Fabrication Technique of Bicrystals

Magnesium oxide, which has a highly symmetrical crystal structure, is used to study various properties of grain boundaries. However, few studies have been performed on making MgO bicrystals because the melting point and vapor pressure are higher than for other materials. Wuensch and Vasilos[1] and Roshko and Kingery[2] reported on methods of making bicrystals. According to Wuensch and Vasilos, the (001) plane is hot pressed for 2 hours at 1800°C under a pressure of 2500psi (17.2MPa). Roshko and Kingery's method involves hot-pressing at 31MPa in Ar at 1500°C for 2 to 6 hours.

Kimura et al.[3] fabricated bicrystals by joining two single crystals at 2480°C for 10 minutes using a graphite resistance furnace without applying pressure, as shown in Figure 1. The assembly was heated slowly (about 6°C/min) up to 1500°C to sinter the powder uniformly without cracks. It was heated quickly from 1500 to 2480°C and was held for 10 minutes at 2480°C. After this, it was cooled slowly (9°C/min) to 2250°C and then the heating power supply was shut off. The bicrystals were separated from the

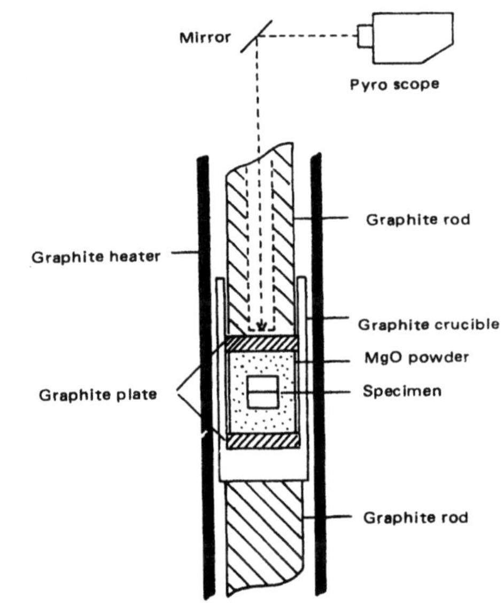

Figure 1. Schematic illustration of the setting assembly of joining for MgO bicrystals.

surrounding polycrystalline MgO and the boundaries were observed through the single crystal layer with optical microscopy.

After being heated according to the schedule described above, the interface could not be recognized with the naked eye because the greater part of the single crystal surface was well joined. However, many rectangular particles were observed on the outer part of the joined boundary with optical microscopy. These rectangular particles indicate that the boundary formation of MgO is mainly controlled by vapor deposition. The preferred growth plane and direction should be (001) and <001> respectively. However, diffusion should also take part in the final stage to remove the pores formed by contact of rectangular particles at the interface.

3. Microstructure of the Grain Boundary

Bicrystals were cut perpendicular to the grain boundary, polished to a thickness less than 200 μm, chemically thinned with hot orthophosphoric acid, and finally ion thinned. The thin sections were observed in a TEM at 100 and 1000 kV.

Observation of the microstructure by electron microscopy revealed many 0.8x3μm ellipsoidal pores as shown in Figure 2(a). When examined in detail, there were flat planes (facets) on the surface of the pores. Kinks could also be seen at the grain boundary, which indicated that there were regions of symmetric and asymmetric boundary. The same symmetric and asymmetric boundaries have also been reported for Si[4]. These structures result from the boundary formation process of vapor deposition. Figure 2(b) shows a (020) dark-field image of the 6.8°tilt boundary. We can recognize many criss-crossed edge dislocations at the grain boundary and parallel to the <100> direction, as shown in Figures 2(c) and (d). These results suggest that two different edge dislocations are present. This will be discussed in the following section.

4. Grain Boundary Energy

Atoms at the grain boundaries are displaced from the positions they would occupy in the perfect crystal lattice. Hence, their free energy may be higher than that of the atoms on an undistorted crystal lattice. Theoretical interpretation of the thermal groove developed on the grain boundaries was performed by Mullins[5]. His analytical method has been widely used to investigate the energy of the grain boundaries relative to the surface energy. According to the theory, the driving force of material transportation from the grain boundaries to the surface is a reduction of the chemical potential near the grain boundaries, which develops the thermal groove. There are some important conditions to be satisfied for thermal grooving experiments before we can apply Mullin's equation for analyzing boundaries:

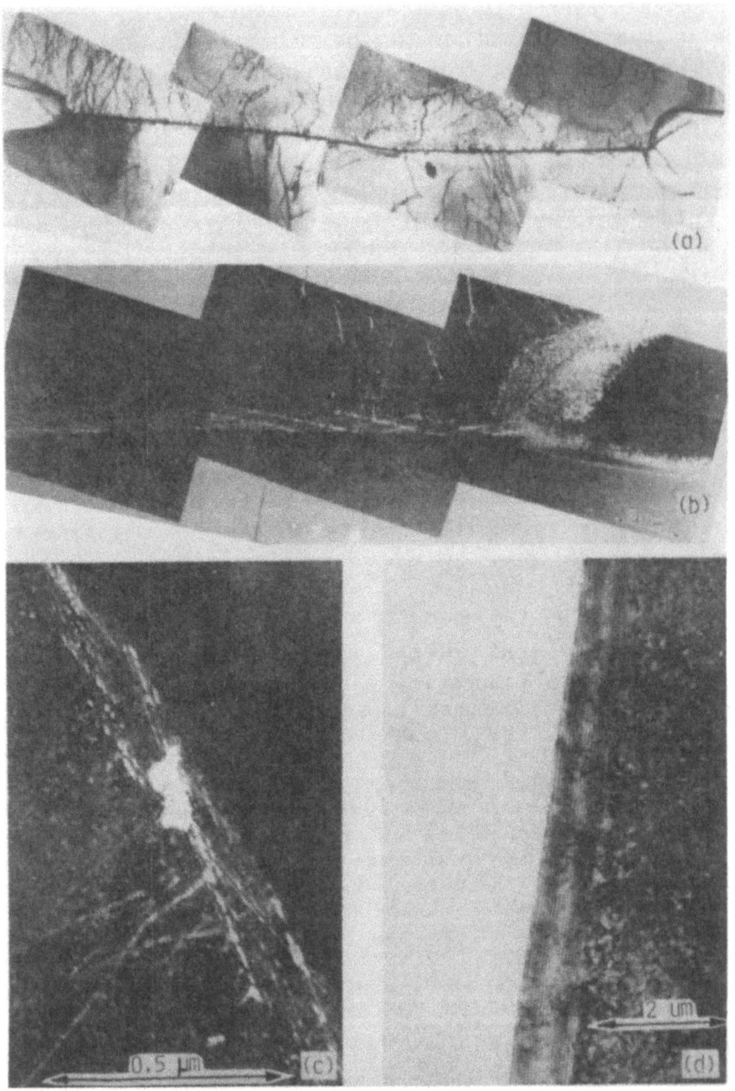

Figure 2. TEM observation of the grain boundary in 6.8° tilt (001) symmetric bicrystals.
(a):bright-field image, (b):(002) dark-field image.
(c):crossed edge dislocations on grain boundary in b.
(d):parallel edge dislocations on grain boundary.

(1) The specimen should be in a closed system, which means mass transpotation must not occur out of the system and the vapor pressure of the material is in equilibrium in the system.
(2) Development of the grooves is governed by a simple mechanism. When development of boundary grooves is dominated by surface diffusion, lattice diffusion, or evaporation/condensation, the time exponent m of width and depth of the groove is 1/4, 1/3, or 1/2, respectively.
(3) The inclination of the groove (dihedral angle) should be large enough.

Thermal grooving of MgO bicrystals was carried out at 1290 to 1600°C in a MgO crucible to satisfy condition (1). MgO bicrystals were cut perpendicular to the grain boundary, and the cut planes were (001) for a (001) tilt boundary, (110) for a (110) tilt boundary, and (f(0),1,0) for a (001) twist boundary. The dihedral angle was measured by an interference microscope. Condition (2) was satisfied by plotting the logarithms of width or depth of the groove against logarithms of the time as shown in Figure 3.

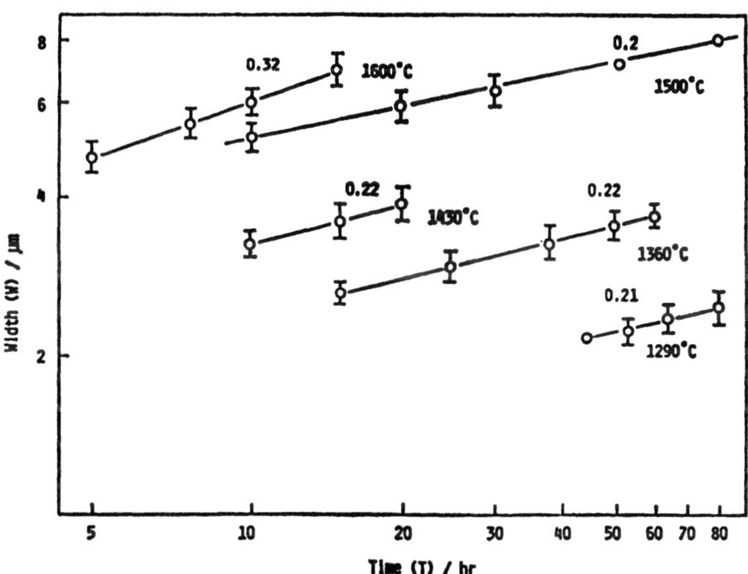

Figure 3. Groove width vs. time for (100) tilt bicrystals of MgO (Θ=5.4°) at various temperatures.
The numbers attached above each line indicate the slope.

Two different types of curves are obtained between relative grain boundary energy (σ_b/σ_s) and misorientation angle as shown in Figures 4, 5, and 6. The curves in Figure 4, which are obtained from the (001) tilt boundary of MgO, show that relative boundary energy (σ_b/σ_s) increases in accordance with an increase in tilt angle and then it becomes constant. Usually this type of curve is observed in (100) tilt boundaries in NiO[6], (110) tilt boundaries in Al_2O_3 , and (100) tilt boundaries in NaCl[8].

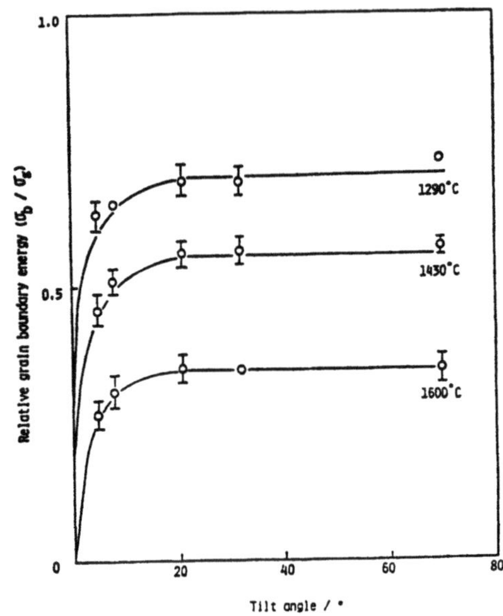

Figure 4. Relative grain boundary energy (σ_b/σ_s) of (100) tilt boundary of MgO as a function of tilt angle.

The curves in Figures 5 and 6, which are obtained from the (110) tilt boundary and (001) twist boundary of MgO, respectively, show the existence of cusps with low energy. Cusps usually appear in the (110) tilt and twist boundaries of ceramic materials such as NiO[9]. An increase in relative boundary energy at low angle has been explained by the dislocation model proposed by Read and Shockley[10], which will be discussed in the following section. In the case of high angle tilt boundaries, it is difficult to apply the dislocation model because the spacing of dislocations is so small that each dislocation is arra nged independently in a wide area[11]. When this coincides with atomic arrangement at the boundaries, cusps in the energy curve as shown in Figures 5 and 6 can be explained as a CSL boundary. The structure of the low angle boundary and also the reason for the flatness in the boundary energy at higher tilt angles in the (001) tilt boundary will be discussed in the following section.

5. Temperature Dependence of Grain Boundary Energy

The groove widths of the (100) tilt boundary were plotted at various temperatures as shown in Figure 3. As can be seen, the time exponent m was about 0.22 below 1500°C and 0.32 above 1600°C. This indicates that the development of the groove was controlled by surface diffusion below 1500°C and by volume diffusion above 1600°C. The shape of the relationship between the relative grain boundary energy and tilt angle of the (001) tilt boundary did not change and no cusps were observed

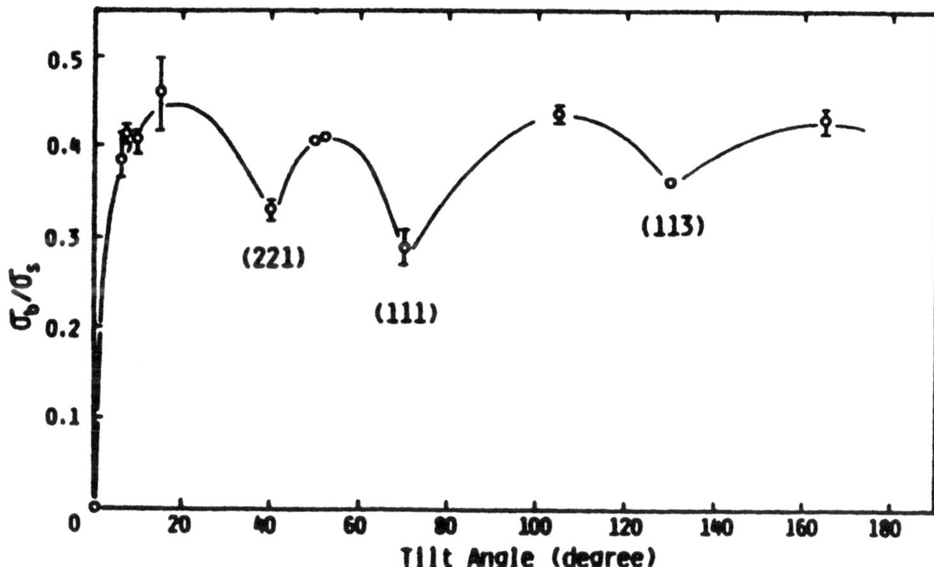

Figure 5. Relative boundary energy (σ_b/σ_s) of (110) tilt boundary of MgO as a function of tilt angle.

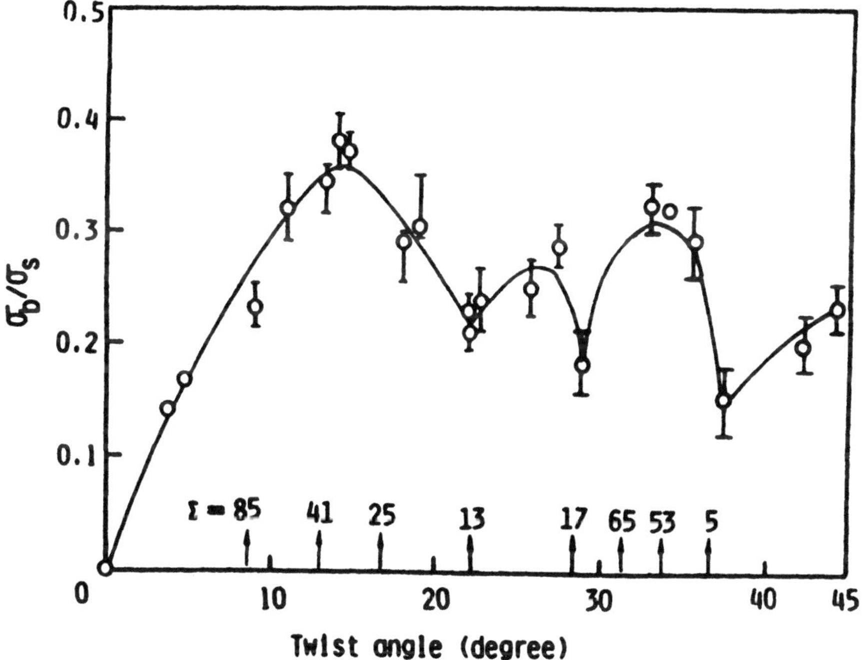

Figure 6. Relative boundary energy (σ_b/σ_s) of (100) twist boundary of MgO as a function of twist angle.

during the experiment, as shown in Figure 4.

The relative boundary energy decreased with an increase in temperature. The temperature dependency of the relative grain boundary energies is shown in Figure 7 for a 5.4° and 69° inclined (100) tilt boundary. Although the vertical axis is not pure boundary energy but relative boundary energy, which is affected by surface energy, it showed a linear relationship. Both the surface energy and boundary energy can be expressed as U-TS; however, the relative boundary energy decreased with an increase in temperature. This result suggests that the entropy of the boundary S_b will be larger than that of the surface S_s. The same results are reported in NiO (100) and (110) tilt boundaries by Ready & Jech[12] and Dhalenne et al.[13]

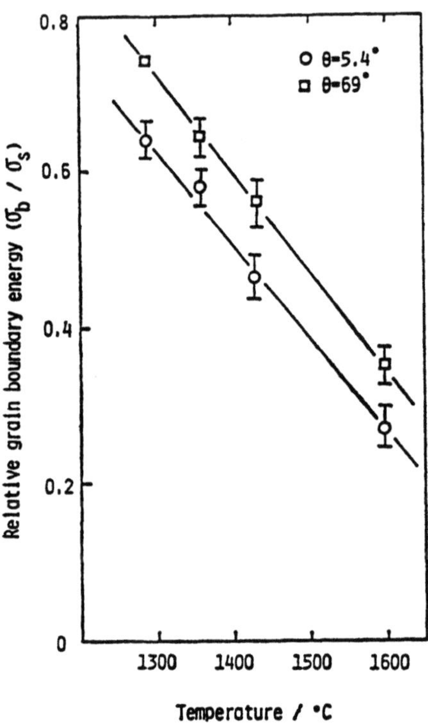

Figure 7. Relative boundary energy (σ_b/σ_s) of (100) tilt boundary as a function of temperature.

When the 5.4° line in Figure 7 is extrapolated to higher temperatures, the relative boundary energy will cross zero at about 1850°C. The 69° line will cross zero at a higher temperature. The meaning of this is not clear, but it can be speculated that the misorientation of the atoms at the grain boundary is similar to the movement by thermal vibration of the atoms near the boundary.

6. Grain Boundary Diffusion

The grain boundary diffusion of Mn in MgO bicrystals with a (100) tilt boundary was measured at 1200 to 1500°C[14]. As shown in Figure 8(a), grain boundary diffusion perpendicular to the rotation axis of the tilt boundary increased about 10 times from 5° to 10° and became constant. On the other hand, the diffusion parallel to the axis increased only 40% from 5° to 10° and also became constant as shown in Figure 8(b).

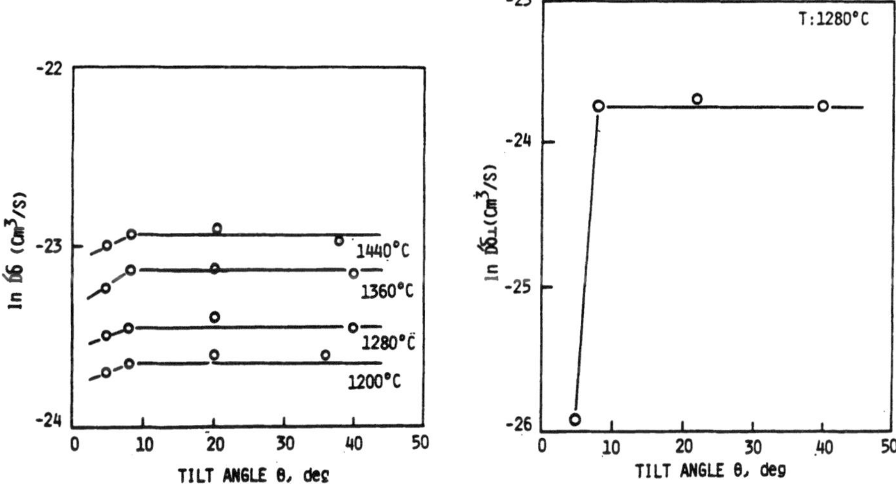

Figure 8. Grain boundary diffusion products Do of Mn in (100) tilt bounary of MgO as a function of tilt angle.
(a): along rotation axis, (b):perpendicular to it.

A large degree of anisotropy was found. The ratio of the grain boundary diffusion parallel and perpendicular to the rotation axis was 10 at about 5°, and above 10° it decreased to about unity. This strong anisotropy of diffusion at low tilt angle supports the pipe diffusion and can be explained as a dislocation arrangement. Same results were reported by Osenbach and Stubic an[15] for the diffusion of Cr in MgO bicrystals with a (100) tilt boundary.

In the case of metals, anisotropy was also observed, but it decreased continuously to 45°, and no critical angle for diffusion was observed in the (100) tilt boundary of Ag[16].

7. Structure of Grain Boundaries

In the early days, the structure of grain boundaries was considered an amorphous film analogous to a liquid. If so, all boundaries should have isotropic properties. However, there is now much experimental evidence that the properties of boundaries show anisotropy varying with misorientation. This theory has been revised to include a transition of lattice layer. Many models have appeared since these theories. The lattice dislocation model for the low angle boundary was originally suggested by Bragg[17].

Van der Merwe[18] also modified the related equation between misorientation and boundary energy, considering the atomic bonding near the boundaries. It was reported by Gjostein and Rhines[19] that this equation was applicable for tilt and twist boundaries in copper. Read and Shockley[10] developed it and deduced related equations between misorientation and boundary energy as follows:

$$E = E_o.\theta\,(A-\ln\theta) \quad ......(1)$$
$$E_o = G\cdot a/4\pi(1-\mu) \quad ......(2)$$
$$A = 1+\ln(a/2\pi r_o) \quad ......(3)$$

where
G = rigidity modulus
a = lattice parameter
μ = poisson's ratio
r_o= radius of dislocation core

Boundary energies versus tilt angles of (100) and (110) low angle tilt boundaries are plotted in Figure 9, which indicates that the low angle tilt boundaries are composed of dislocations. This was also confirmed by TEM observation. A sphere model of a low angle (100) tilt boundary is shown in Figure 10. One dislocation is inserted in every eight planes which forms about an 83°tilt boundary. The critical angles following the Read-Shockley relationship are 22° and 10°for the (100) tilt and (110) tilt boundaries, respectively. The difference in the critical angle will be discussed later.

Many models have been proposed extending the dislocation model for high angle tilt boundaries. Brandon et al.[20] succeeded in explaning

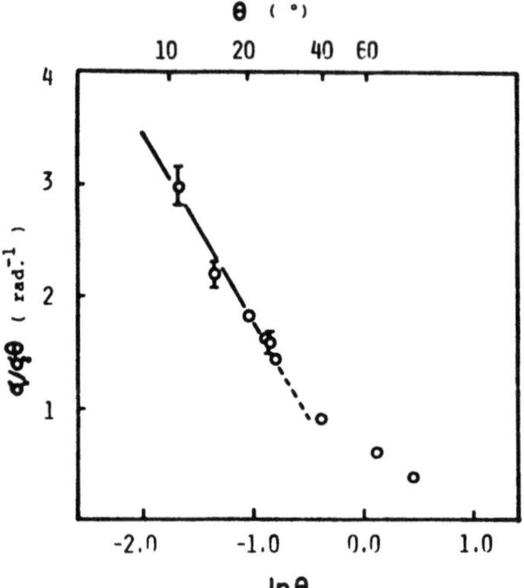

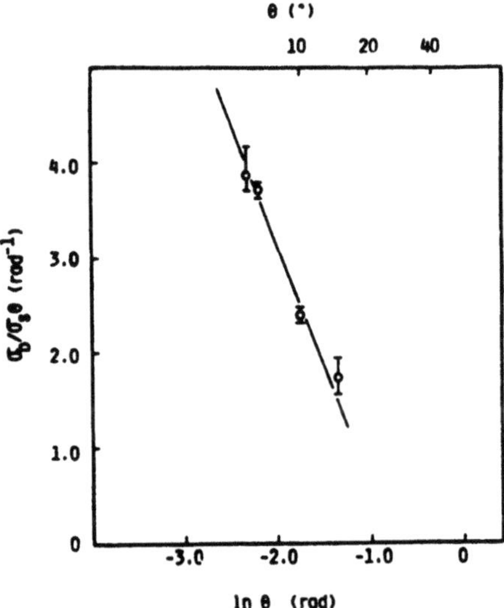

Figure 9. Read-Shockley plots of tilt boundaries of MgO.
(a):(100) tilt boundary,
(b):(110) tilt boundary

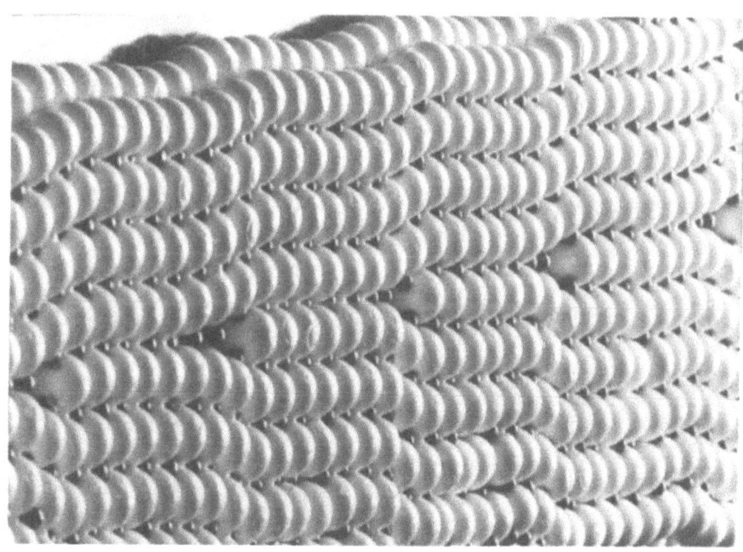

Figure 10. Sphere model of low angle (110) tilt boundary.

high angle boundaries using the concept of coincident site lattice (CSL) theory combined with the dislocation model. Sun and Balluffi[21] examined it by using (100) twist boundaries fabricated by welding MgO under high temperature and high pressure. Similar results were observed that slightly deviated from Σ =5(36.9°) and Σ =13(22.6°) in (100) twist boundaries in NiO[22]. Other TEM experiments[23,24] have also shown that boundaries deviating slightly from high density CSL misorientation often consist of a grain boundary dislocation network.

Duffy and Tasker[25] have also performed computer simulation in (001) twist boundaries in NiO and have found the existence of cusps at Σ =5(53.1°). Experimental evidence for the existence of cusps in the relationship between boundary energy and crystal misorientation with high density CSL has been obtained in (100) twist boundaries[26,27]. All of the low energy angles in the (100) twist boundary agreed with the angles obtained from experiments using small MgO crystals, which were deposited on the (100) plane of a MgO single crystal[27].

Fig. 11 gives a schematic illustration of (110) tilt boundaries with high density CSL. As can be seen, the density of the atoms in the (110) plane is not too high, and there are many spaces to relax the position of the atoms. In addition, the arrangement of the atoms at the grain boundary in the (110) tilt boundary with Σ =3 and Σ =9 is not too distorted and the coordination number in the plane of the opposite ion at the boundary is the same as that of the bulk. The arrangement

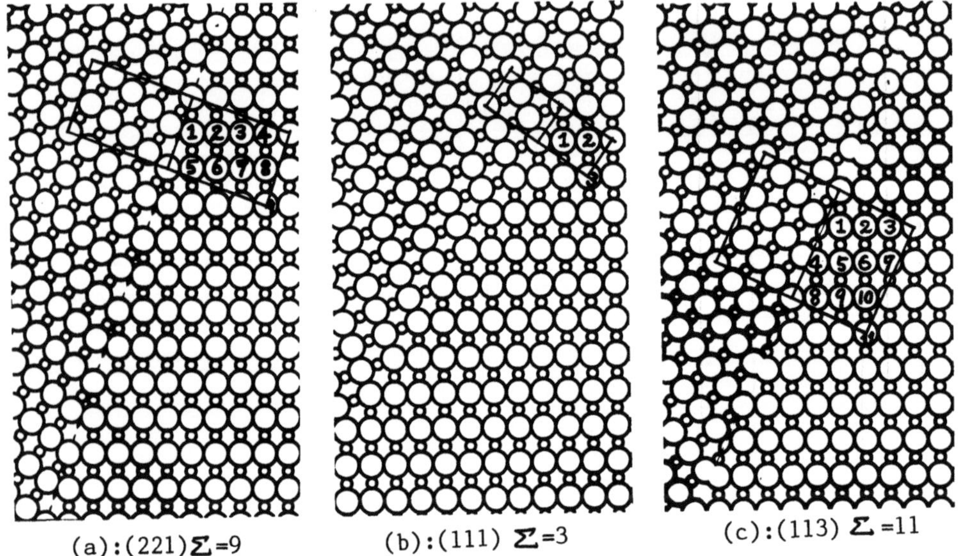

Figure 11. Sphere model of (110) symmetrical tilt bounary.

of the atoms in the (110) tilt boundary with Σ =11 is distorted but, when one of the overlapped atoms and one of the opposite site ions (one pair) are taken out in every cycle, the coordination numbers of the large atoms are easily maintained at the boundary.

For tilt boundaries, the existence of cusps in (100) boundaries is definitively inconclusive[7,9]. Hasson & Goux[28] reported that in metals of FCC structure such as Al, (100) tilt boundaries showed many cusps in the energy curve of the (100) tilt boundary. The cusps appeared at (017), (013), (012), and (023), which corresponded well to the high density CSL boundary[29]. The arrangement of spheres in the (100) tilt boundary is different from the (110) tilt boundary. The coordination number of the opposite ions is 4 in the (100) plane. As can be seen in Figure 12, the density of the atoms in this plane is very high.

The computational models for the boundary structure that allow three-dimensional relaxation are either molecular dynamic or static. Molecular dynamic calculations involving large numbers of atoms are complex; therefore, most of the calculations have been made using static procedures[30]. Duffy and Tasker[25] have recently carried out molecular static calculations of the structures and energies of NiO (001) symmetric tilt boundaries. The boundaries are stable and possess much lower energies than the energies of the two free surfaces. The difference in the boundary energy in (100) tilt boundaries between the measured and calculated values or Al metal will mainly depend on the

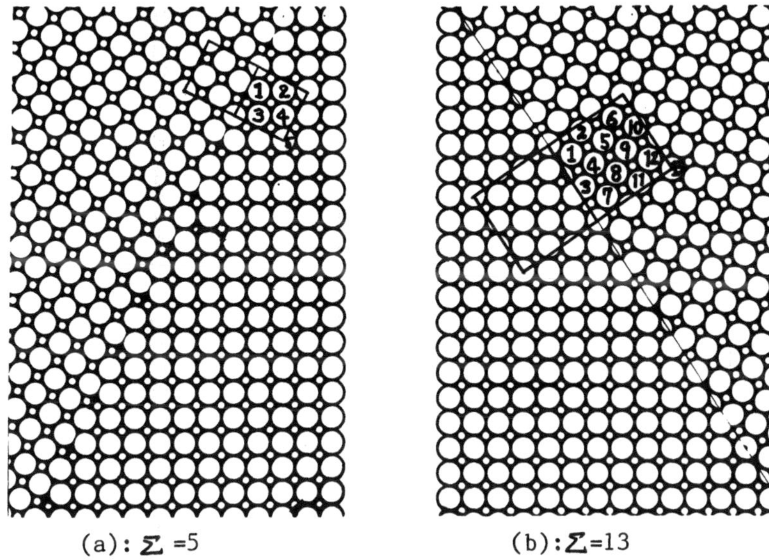

(a): Σ =5 (b): Σ=13

Figure 12. Sphere model of (100) symmetrical boundary.

large effect of Coulomb force and the low capability of adjustment of the atoms.

8. Discrepancy Between Grain Boundary Energy and Diffusion

As mentioned above, strong anisotropy in diffusion was observed up to 10° in (100) tilt boundaries and was the result of pipe diffusion in the dislocation arrangement; above 10° the anisotropy disappeared and became nearly isotropic. On the other hand, the Read-Shockley plot suggests that the (100) tilt boundary is constructed by dislocation arrays up to 20°. The discrepancy in the critical tilt angle in the (100) boundary suggests the differences in dislocation arrangement between diffusional properties and boundary energy.

When the (100) small angle tilt boundary is composed of (100) dislocation arrangements, all of the Burgers vectors are parallel to each other. Then all of the dislocations are arranged at the same distance and never cross each other[31] because of the repulsion force of the dislocations. In metals, the minimum distance for dislocation arrangements with the same Burgers vector is thought to be about five lattices. This distance corresponds to 11.3°. The critical tilt angle following the Read-Shockley relationship in (110) tilt boundaries is about 10°, and the angle corresponds well to the critical angle of the closest dislocations with a parallel arrangement.

As shown in Figure 13, the dislocation is generally inserted into the

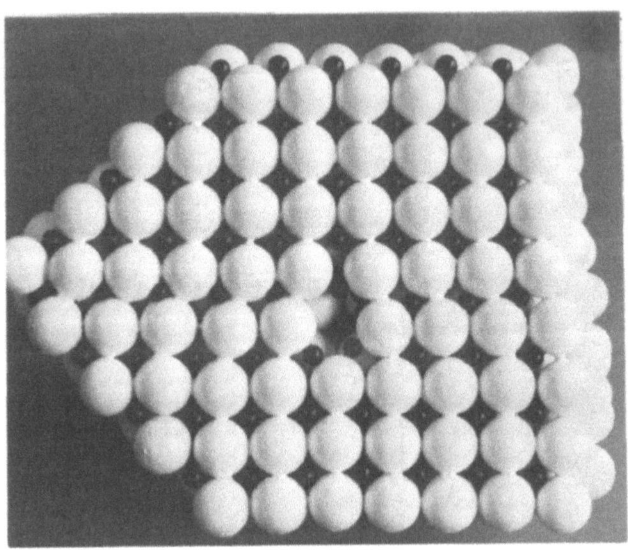

Figure 13. Sphere model of dislocation where dislocation core passes through <001> direction; jogging to <110> is also possible.

(110) plane and the dislocation core passes into the <001> direction. Of course the dislocation can be also introduced into the (100), but it is not an active slip system below 1700°C[32]. It is easy to introduce dislocations into the (110) first, then the dislocations in the (100) can be substituted for (110) dislocations. As can be seen in Figure 2(b), which is a (020) dark-field image of the low angle tilt boundary, there are many crisscrossed dislocations of a (110) <110> system. When the (100) tilt boundary is composed of (110) dislocations, the Burgers vectors of dislocation in the (100) tilt boundary are normal to each other and the dislocations attract each other and can crisscross. In this case the critical angle for the dislocation arrangement will not exist. It might be the reason why no cusps exist in the (100) tilt boundary.

Looking at the sphere model (Figure 13) closely, the dislocation core not only passes through in the <100> direction but also in the <110> direction. This arrangement is one type of a kink. The fact that anisotropic diffusion cannot be observed above 10° in the (100) tilt boundary will depend on the crisscrossed edge dislocations with the kinked arrangement.

9. Summary

(1) Bicrystals of MgO were fabricated by joining two single crystals at 2480°C. The boundary was formed by evaporation/condensation and diffusion.

(2) Boundary energy was measured by the thermal grooving technique at 1200 to 1600°C. From the Read-Shockley plot it was suggested that the (100) and (110) low angle tilt boundaries up to 20° were constructed by dislocation arrangements.

(3) The (110) high angle tilt boundary followed the CSL theory, and three low energy boundaries were observed at (221), (111) and (113). A low energy boundary was not observed in the (100) tilt boundary. There are two possibel reasons for these differences: (1) the high density of the atoms on the (100), which causes a large distortion of the atoms; or (2) substituting (100) dislocations into the (110), <110> system, which enables crisscrossing the dislocations.

(4) Grain boundary diffusion at low angles showed large anisotropy, which confirmed the dislocation model. The difference is the critical angle between the estimate from the Read-Shockley relationship and diffusional properties will be derived from the crisscrossed arrangement (kink and jog type) of the dislocation core.

Acknowledgment

The authors express their gratitude to Dr.Y.Moriyoshi, Dr.Y.Tanabe, and Dr.H.Kim for useful discussions.

References

1) B.J. Wuensch and T. Vasilos, J. Am. Ceram. Soc., **47** [2] 63 (1964)
2) A. Roshko and W.D. Kingery, J. Am. Ceram. Soc., **68** [12] C331 (1985) Am. Ceram. Soc., (1985), pp.347
3) S.Kimura, et al., Advances in Ceramics Vol.10, Edt.by W.D.Kingery, Am.Ceram.Soc., (1985), pp.347
4) B.Cunningham & D.Ast, "Grain boundaries in semiconductors" Edt. by Pike et al., Elsevier, Amsterdam pp.21-26 (1982).
5) W.W. Mullins, J. Appl. Phys., **28** [3] 333 (1957)
6) D.W.Readey and R.E. Jech, J. Am. Ceram. Soc., **51** [4] 201 (1968) 201
7) G. Dhalenne et al., Phys. Stat. Sol., **(a)** **56** [1] 267 (1979)
8) R.L. Moment and R.B. Gordon, J. Am. Ceram. Soc., **47** [11] 570 (1964)
9) G.Dhalenne et al., J. Am. Ceram. Soc., **65** [1] C11 (1982)
10) W.T. Read and W. Shockley, Phys. Rev.,**78** [3] 275 (1950)
11) S.Kimura et al., Yogyo Kyokai Shi, **94** [8] 795 (1986)
12) D.W.Ready & R.E.Jech, J.Am.Ceram.Soc., **51** 201-8 (1968)

13) G.Dhalene, M.Dechamps & A.Revcolevschi, "Advances in Ceramics. vol.6" edt.by W.Kingery, Am.Ceram.Soc.,(1983) pp139-149
14) H.Kim, Doctor thesis (1987) Tokyo Institute of Technology.
15) J.W.Osenbach & V.S.Stubican, J.Am.Ceram.Soc., **66** 191-5 (1983)
16) R.Hoffman, Acta Met., **4** 98 (1956).
17) W.L. Bragg, Proc. Phys. Soc., A52, 54 (1940)
18) J.H. Van der Merwe, ibid, **A63**, 616 (1950)
19) N.A. Gjostein and F.N. Rhines, Acta Met., **7** [5] 319 (1959)
20) D.G. Brandon et al., Acta Met., **12** [7] 813 (1964)
21) C.P. Sun and R.W. Balluffi, Phil. Mag., **A46** [1] 49 (1982)
22) K.Y.Liou and N.L.Peterson, Surface and Interface in Ceramic and Ceramic/Metal Systems, Ed. by J.Pask and A.Evans, Plenum Press, New York, (1981), pp.189
23) R.W. Balluffi et al., Surf. Sci., **31** [1] 68 (1972)
24) D.J. Dingley and R.C.Pond, Acta Met., **27** [4] 667 (1979)
25) D.M. Duffy and P.W. Tasker, Phil. Mag., **A47** [6] 817 (1983)
26) C.P. Sun and R.W. Balluffi, Phil. Mag., **A46** [1] 63 (1982)
27) H.Mykura et al., Phil. Mag., **A42** [2] 225 (1980)
28) G. Hasson and C. Goux, Scripta Met., **5** [10] 889 (1971)
29) G.Hasson, M.Biscondi, P.Lagarde,J.Levy & C.Goux, "The nature and Behavior of Grain Boundaries" edt.by H.Hu, Plenum N.Y.pp.3-40 (1972).
30) R.W. Balluffi et al., J. Am. Ceram. Soc., **64** [1] 23 (1981)
31) "Theory of Dislocations" edt.by Jap.Soc.Met., Maruzen Tokyo (1971) pp.59
32) A.G.Evans & T.G.Langdon, Proc.Mater.Sci., **21** 171 (1976)

INTERFACES IN DIRECTIONALLY SOLIDIFIED OXIDE-OXIDE EUTECTICS

G. DHALENNE and A. REVCOLEVSCHI
Laboratoire de Chimie des Solides, UA CNRS 446,
Université Paris-Sud, Bât. 414,
91405 Orsay Cédex, France.

1. INTRODUCTION

During the last three decades, an extensive research and development effort in the aera of composite materials has taken place. Among the various materials which have been studied, oxide - oxide aligned eutectics were considered because of their anisotropic properties, high strength- to - weight ratios at high temperatures, high melting points, and also resistance to oxidation [1-2].

Interfaces obviously are largely responsible for the behaviour of these materials, and quite a large number of studies have been devoted to a better understanding of the structure and properties of such interphase boundaries.

Among these aligned materials, we shall particularly point out those recently grown in several NiO or cubic ZrO_2 or MgO - based binary eutectic systems with the object of yielding interfaces associating a cubic structure (NiO, ZrO_2 or MgO) with other phases of cubic or lower symmetry [2,3]. The subject has been recently reviewed [3] and we shall briefly mention here some of the main data.

2. METHODS OF GROWING OXIDE-OXIDE EUTECTICS AND RESULTING MICROSTRUCTURES

According to Mollard and Flemings, well - aligned regular eutectic structures are produced when solidification proceeds under plane-front conditions, that is, when the solid-liquid interface is planar, with the two phases growing at the same rate [4] . Such conditions are met for high G/R values (G is the temperature gradient at the solid - liquid interface and R is the solidification rate). It was also shown that when the G/R ratio is sufficiently high, non - dendritic composite structures can be grown at off-eutectic compositions.

L.-C. Dufour et al. (eds.), Surfaces and Interfaces of Ceramic Materials, 109–114.

Several methods can be used to grow ceramic eutectics by directional solidification from the melt :

- *Floating zone technique*

A molten zone produced by local heating (thermal imaging, R. F, electron beam) is established between two cylindrical vertical rods and moved along one of them [5-7].

- *Czochralski technique*

The technique consists of pulling samples of eutectic composition from a charge melted in a crucible, by a conventional heating method or by the "Skull Melting" technique [8-9].

- *Bridgman technique*

A crucible, made of refractory metal and containing a molten charge is lowered through a RF-heated graphite susceptor [2,6,10-11].

The microstructures of the oxide - oxide samples unidirectionally solidified by these techniques,generally consist of rods of one of the phases distributed uniformly in a matrix of the second, or of alternating plates of the two eutectic phases in a lamellar structure.

The occurrence of any of these two morphologies is generally in good agreement with the calculation relative to lamellar - to - rod transition in eutectic structures, which establishes that when the energy of the surface separating the two phases is isotropic, a minimum surface energy criteron favors rods, when the volume fraction of the minor phase is less than 30% ; for higher volume fractions, a lamellar morphology is expected [12] . Stubican and Bradt have reviewed, for a large number of non-metallic eutectics, the correlation existing between microstructure and volume fraction [2].

It was also established, that the interphase spacing λ should decrease with increasing growth rate R so that the product $\lambda^2 R$ remains constant [13]. This relationship was verified in many systems [2].

A list of the various oxide -oxide systems which have been studied to date will be found in reference 3. Figure 1 shows the two types of aligned structures which can be obtained.

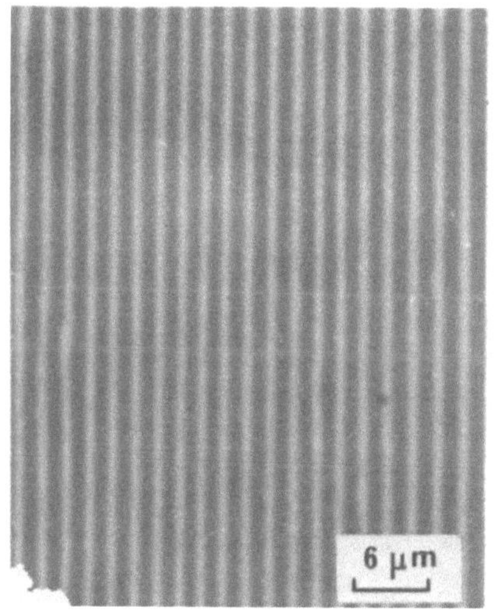

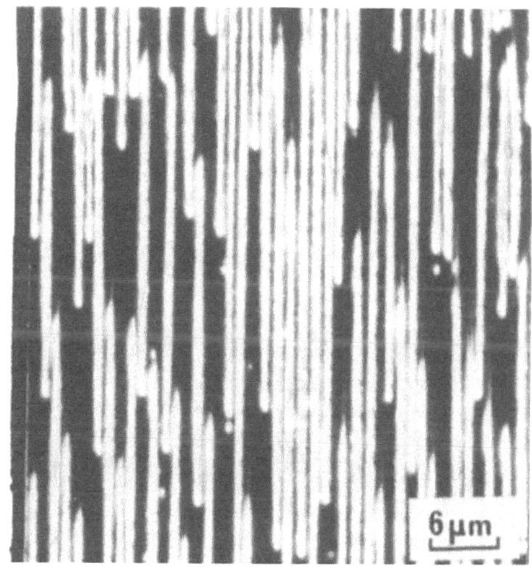

Figure 1.

Longitudinal section of lamellar $NiO-ZrO_2(CaO)$ eutectic

Longitudinal section of fibrous $NiO-La_2NiO_4$ eutectic

3. CRYSTALLOGRAPHY OF UNIDIRECTIONALLY SOLIDIFIED OXIDE - OXIDE EUTECTICS

X - ray and electron diffraction techniques can be used to determine the crystallographic features of the interfaces.

X-ray diffractometry experiments carried out on transverse sections of ingots exhibiting aligned structures, allow the determination of the preferred growth directions of each phase.

X-ray techniques (Weissenberg,De Jong and Buerger methods) applied to microfragments, provide the accurate growth directions of each phase and the orientation relationships which exist between them [7,14-16]. Optical or scanning electron microscope observations are used in addition in order to specify the interface plane in lamellar eutectics [7,14].

Similar information can be obtained by transmission electron microscopy (TEM) of lamellar or fibrous type specimen thinned by argon ion - milling [17]. TEM observations have revealed planar interfaces even in fibrous composites for which curved parts of the interfaces can be resolved at higher magnification into facets corresponding to well defined crystallographic planes with low Miller indices [18-19].

The main crystallographic data from these studies can be summarized as follows :

- In lamellar oxide - oxide eutectics, the apparent growth direction can vary over a small range, but is contained in the interface plane which is unique,
- Interfaces are low Miller indices. Generally, these planes correspond to dense arrangements of atoms,
- A unique crystallographic orientation relation generally exists between the two phases and is due, as we shall see in next paragraph, to a good degree of atomic matching at the interface planes.

Table I summarizes the main crystallographic features relative to lamellar or fibrous oxide-oxide eutectics.

a)

Coupled structures	Phases	Growth direction	Interface plane
Rocksalt-Rocksalt	$NiO-CaO$	$[1\bar{1}0]NiO//[1\bar{1}0]CaO$	$(111)NiO//(111)CaO$
Rocksalt-Ln_2O_3 (C)	$NiO-Y_2O_3$	$[1\bar{1}0]NiO//[001]Y_2O_3$	$(111)NiO//(100)Y_2O_3$
Rocksalt-Ln_2O_3 (B)	$NiO-Gd_2O_3$	$[1\bar{1}0]NiO//[010]Gd_2O_3$	$(111)NiO//(20\bar{1})Gd_2O_3$
Fluorite-Rocksalt	ZrO_2-MO (M=Ni,Mg)	$[001]ZrO_2//[1\bar{1}0]MO$	$(100)ZrO_2//(111)MO$
Fluorite-Ln_2O_3 (A)	$ZrO_2-Nd_2O_3$	$[1\bar{1}0]ZrO_2//[2\bar{1}\bar{1}0]Nd_2O_3$	$(111)ZrO_2//(0001)Nd_2O_3$
Fluorite-Ln_2O_3 (B)	$ZrO_2-Sm_2O_3$	$[1\bar{1}0]ZrO_2//[010]Sm_2O_3$	$(111)ZrO_2//(20\bar{1})Sm_2O_3$
Fluorite-Ln_2O_3 (C)	$ZrO_2-Dy_2O_3$	$[110]ZrO_2//[110]Dy_2O_3$	$(hkl)ZrO_2//(hkl)Dy_2O_3$

b)

Coupled structures	Phases	Growth direction	Orientation relation
Rocksalt-Fluorite	$MgO-ZrO_2$	$[1\bar{1}0]MgO//[1\bar{1}0]ZrO_2$	$(111)MgO//(111)ZrO_2$
Rocksalt-K_2NiF_4	$NiO-La_2NiO_4$	$[5\bar{5}4]NiO//[210]La_2NiO_4$	$(110)NiO//(100)La_2NiO_4$
Rocksalt-Spinel	$MO-MAl_2O_4$ (M=Ni,Mn)	$[111]MO//[111]MAl_2O_4$	$(hkl)MO//(hkl)MAl_2O_4$
Fluorite-Corundum	$ZrO_2-Al_2O_3$	$[010]ZrO_2//[0001]Al_2O_3$	$(100)ZrO_2//(2\bar{1}\bar{1}0)Al_2O_3$

Table I

Crystallographic relations at the lamellar (a) or fibrous (b) oxide-oxide eutectics

4. STRUCTURE OF EUTECTIC INTERFACES

Interface models have been suggested for several lamellar eutectics on the basis of crystallographic data and structure of the two eutectic phases.

Oxide - oxide interfaces can be described by a coincidence - or pseudo coincidence- lattice with a more or less complex interface structure (misfit dislocations, facets) which accomodates the residual lattice mismatch. Despite the quite different structures and symmetry of the coupled phases, the orientation relations give rise to a low lattice mismatch which amounts to a few per cent only. In particular in the interface plane, good atomic matching along the growth direction of each phase should be found.

Another important feature observed in this description is a similar atomic stacking on both sides of the interface. The sequence of planes perpendicular to the boundary is the same in each phase and generally corresponds to an alternate stacking of cationic and anionic planes with comparable atom density. The distances between two consecutive metal-oxygen layers along the direction perpendicular to the interface are not very different and this situation satisfies fairly well electroneutrality and bond length requirements at the interface [7,16-17,20-22].

Current developments in high resolution electron microscopy (HREM) make possible the observation of interfaces between oxides at the atomic scale. This technique has been recently used to examine the structure of eutectic interfaces between zirconia and several refractory oxides (Al_2O_3, MgO) [11,17-19,23]. These observations indicate that transition from one phase to the other takes place indeed over one atomic distance.

REFERENCES

1) R. L. Ashbrook, J. Am. Ceram. Soc. **60** (1977) 428.

2) V. S. Stubican, R. C. Bradt, Ann. Rev. Mater. Sci. **11**, 267. Ed. R. A. Huggins, R. H. Bube and D. A. Vermileya (Ann. Rev. Inc., Palo Alto, 1981).

3) A. Revcolevschi, G. Dhalenne and D. Michel, Mat. Sci. Forum **29**, 273. Ed. L. C. Dufour and J. Nowotny (Trans Tech Pub., 1988).

4) F. R. Mollard and M. C. Flemings, Trans. AIME **239** (1967) 1526.

5) D. Rowcliffe, W. Warren, W. Elliot and W. Rothwell, J. Mat. Sci. **4** (1969) 902.

6) C. O. Hulse and J. A. Batt, Final Report UARL N910803-10 NITIS AD-781995 (Springfield, USA, 1974)
7) A. Revcolevschi, G. Dhalenne and F. d'Yvoire, J. Phys. Colloq. **C4-46** (1985) 441.
8) D. Michel, Y. Rouaux and M. Perez Y Jorba, J. Mat. Sci. **15** (1980) 61.
9) D. Michel, L. Mazerolles and M. Perez Y Jorba, Studies in Inorganic Chemistry **3**, 841. Ed. R. Metselaar, H.J. M. Heijligers and J. Schoonman (Elsevier Sciences, Amsterdam, 1983).
10) F. Schmid and D. Viechnicki, J. Mater. Sci. **5** (1970) 470.
11) J. Echigoya, Y. Takabayashi, K. Sasaki, S. Hayashi and H. Suto, Trans. Jpn Inst. Metals **27** (1986) 102.
12) D. J. Cooksey, D. Munson, M. P. Wilkinson and A. Hellawell, Phil. Mag. **10** (1964) 745.
13) J. D. Hunt and K. A. Jackson, Trans AIME **236** (1966) 1129.
14) A. Revcolevschi, Materials Science Research **20**, 115. Ed. R. E. Tressler, G. L. Messing, C. G. Pantano and R. E. Newnham (Plenum Press, New York, 1986).
15) M. Fragneau and A. Revcolevschi, J. Am. Ceram. Soc. **66** (1983) 162.
16) M. Fragneau, A. Revcolevschi and D. Michel, Advances in Ceramics **6** , 110. Ed. M. F. Yan and H. A. Heuer (The American Ceramic Society, Columbus, 1983).
17) D. Michel, L. Mazerolles and R. Portier, Zirconia Ceramics **6**, 45. Ed. S. Somiya (Uchida Rokakuho, Tokyo, 1986).
18) L. Mazerolles, D. Michel and R. Portier, J. Am. Ceram. Soc. **69** (1986) 252.
19) L. Mazerolles, Thesis Paris 1988.
20) M. Fragneau, D. Michel and A. Revcolevschi, Studies in Inorganic Chemistry **3**, 413. Ed. R. Metselaar, H. J. M. Heijligers and L. Schoonman (elsevier sciences, Amsterdam, 1983).
21) B. Dubois, G. Dhalenne, F. d'Yvoire and A. Revcolevschi, J. Am. Ceram. Soc. **69** (1986) C6.
22) A. Revcolevschi, B. Dubois, G. Dhalenne and F. d'Yvoire, Grain Boundary Structure and Related Phenomena, proceedings of JIMIS **4**. Supplement to Trans. Japan Inst. Met., (1986) 1045.
23) J. Echigoya and H. Suto, Grain Boundary Structure and Related Phenomena, Proccedings of JIMIS **4**. Supplement to Trans. Jpn Inst. Met., (1986) 213.

GRAIN BOUNDARIES IN HEXAGONAL CARBIDE CERAMICS

J. VICENS, S. LAY, E. LAURENT-PINSON, G. NOUET
LABORATOIRE D'ETUDES ET DE RECHERCHES SUR LES MATERIAUX, URA CNRS N°1317, INSTITUT DES SCIENCES DE LA MATIERE ET DU RAYONNEMENT, BOULEVARD DU MARECHAL JUIN - 14032 CAEN Cedex, FRANCE.

ABSTRACT. Fundamental definitions of the geometrical description of grain boundaries are firstly recalled. Then the particularities of the non-cubic systems are emphasized and the results obtained in a hexagonal material, tungsten carbide-cobalt composite are presented. Different techniques such as TEM, HREM, EDX were used to characterize the grain boundaries according to the orientation relationship, the nature of the grain boundary planes and the chemical composition of the grain boundaries.

1. INTRODUCTION

Carbide ceramics are materials of high interest regarding their high thermomechanical properties. Among the different materials which can be classified as ceramics, cermets constitute a special group since they are not made only of light elements. Moreover, their properties are due to their composition, i.e. a mixture of brittle (hard) and ductile (soft) materials, each of which is used because of its specific properties : hardness to resist wear, softness to increase toughness. The most typical cermet is the composite material : tungsten carbide-cobalt (WC-Co).

As for any polycrystalline materials, grain boundaries interact with dislocations during the deformation process either as sinks or sources,and play an important role. For instance, creep tests can involve dislocation motion and diffusion in grain boundaries. Such a good knowledge of the properties of a material requires a precise description of these crystallographic and chemical defects. In the case of grain boundaries, the model describing low-angle grain boundaries in terms of individual dislocations is no longer valid for high angle grain boundaries which are the main topic of the present paper. Thus we recall the fundamental definitions of different geometrical elements (Coincidence

L.-C. Dufour et al. (eds.), Surfaces and Interfaces of Ceramic Materials, 115–140.

Site Lattice, O-Lattice, primary and secondary dislocations, coincidence index Σ) which are necessary to describe some special grain boundaries in terms of dislocations or periodic patterns. Since tungsten carbide and several other ceramics are hexagonal, the particularities related to the value of the parametric ratio c/a are also outlined. After a short introduction to the concept of coincidence in hexagonal systems, we deal with tungsten carbide : elaboration condition, crystallographic data, frequency of coincidence grain boundaries, second phase at the grain boundaries, investigation of the grain boundary structures with transmission electron microscopy.

2. COINCIDENCE ORIENTATION: APPLICATION TO HEXAGONAL SYSTEM

2.1. Coincidence site lattice and O-lattice

The concept of coincidence site lattice (CSL) was mathematically developed by Bollmann with the O-lattice theory (1, 2). This approach was aimed to calculating the crystallographic structure of interfaces between two adjacent grains and to deducing the nature of the crystallographic defects associated with these grain boundaries.

The interest of the O-lattice theory is to generalize the concept of coincidence site lattice to equivalent points with the same internal coordinates. Let us consider two interpenetrating lattices related point-to-point by the transformation A such as :

$$x^2 = A\ x^1 \qquad [1]$$

where x^2 and x^1 are any vectors in the coordinate systems associated with the crystal lattices 1 and 2, respectively. A is a linear homogeneous transformation such as a rotation, a shear, an expansion ...

If the difference vector $x^2 - x^1$ is a lattice vector b of crystal 1 then x^2 is defined as a O-point x^o :

$$b = x^2 - x^1 \qquad [2]$$

$$x^o = A\ x^1 \qquad [3]$$

$$b = (I - A^{-1})\ x^o \qquad [4]$$

I : identity matrix.

According to the rank of $(I - A^{-1})$, the set of O-vectors may be imaged by a point (rank 3), a line (rank 2) or a plane lattice (rank 1). If the determinant of $(I-A^{-1}) \neq 0$, the O-lattice vectors are given by :

$$x^o = (I - A^{-1})^{-1} b \qquad [5]$$

This O-lattice is a set of invariant elements for which the fit between the two lattices is maximum. The Wigner-Seitz cells associated with these O-elements are areas where the misfit between these two lattices is concentrated.

When the grain boundary plane intercepts this O-lattice and the walls of the associated Wigner-Seitz cells, this intersection defines dislocations which are primary dislocations since they correspond to a discontinuity between two crystal lattices which is equal to Burgers vectors b.

2.2. Displacement Shift Complete Lattice

For two interpenetrating lattices the O-points are the set of the origin of the rotations leading to the coincidence site lattice. So, if the origin of the rotation is shifted from one O-point to another, the coincidence orientation is not modified and the interpenetrating pattern is conserved. In a similar way, other displacement vectors can recreate the same pattern provided that they belong to the Displacement Shift Complete Lattice (DSC lattice). By definition, this DSC lattice is the common lattice containing the two crystal lattices 1 and 2 as superlattices. If one crystal is translated by such a vector, the interpenetrating pattern is not modified but the origin of the rotation is shifted and all the O-elements are also shifted. The interest of the DSC lattice is obvious since it enables the CSL to be retained. Examples of coincidence orientations are given in figure 1 with the corresponding DSC lattices.

If an experimental rotation deviates from a coincidence orientation the supplementary rotation can be compensated by a new dislocation network or secondary dislocation network for which the Burgers vectors will be vectors of the DSC lattice. These Burgers vectors are no longer vectors of the crystal. It is worth noticing that if the O-lattice is a function of the transformation A, the CSL and DSC lattices are independant of this choice.

As for the determination of the primary dislocation network, a new O-lattice or O2-lattice can be calculated to determine the secondary dislocation network. The O2-lattice is the O-lattice associated with the transformations A and A'. A' is the transformation corresponding to the orientation relationship including the slight deviation from the exact coincidence. So, the same calculation procedure can be used and the vectors of the DSC lattice are given by the following equation :

$$b^{(DSC)} = (I - B^{-1})^{-1}\ x^{02} \qquad [6]$$

with $\quad B = A'.A^{-1} \quad \text{and} \quad (I - B^{-1}) \neq 0 \qquad [7]$

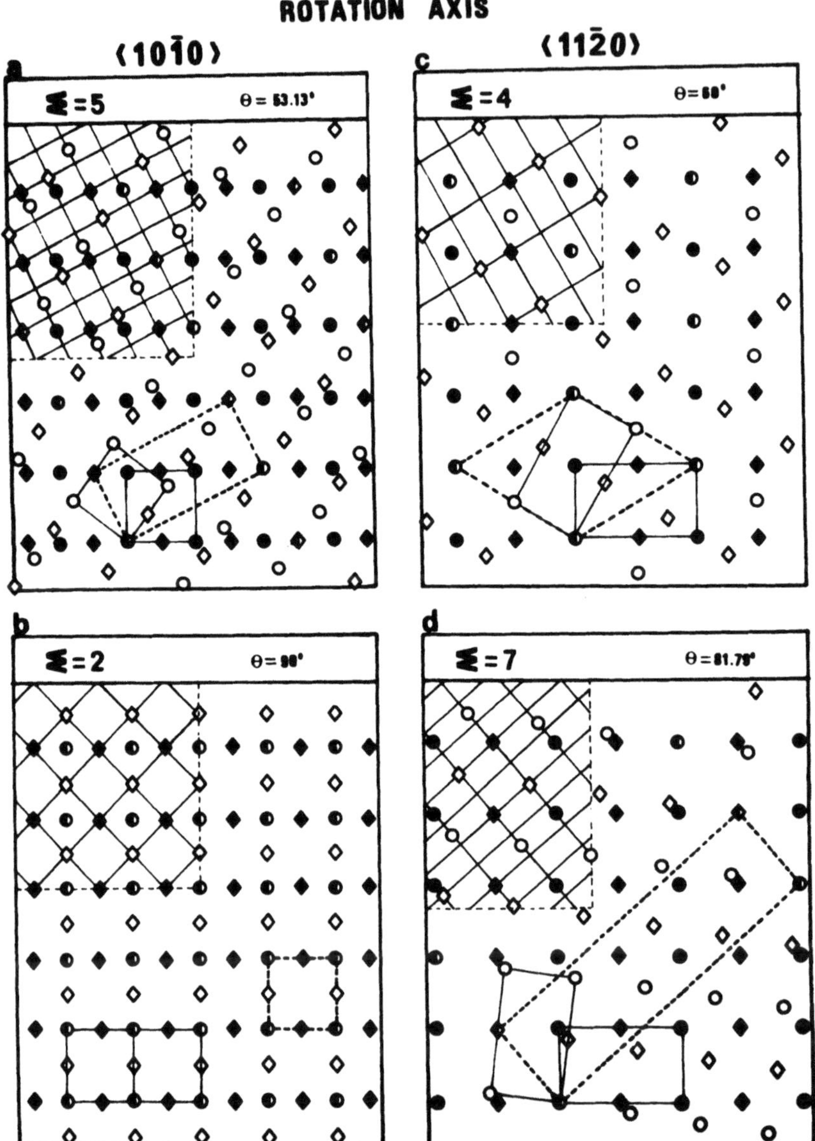

Fig. 1 : Rotations around <10$\bar{1}$0> θ=53.13°(Σ5), θ=90° (Σ2) and <11$\bar{2}$0> θ=60°(Σ4), θ=81.79°(Σ7) axes for c/a=1

○ ● : 0

◇ ◆ : 1/2

crystal 2 1

dotted lines : CSL insert : DSC lattice

Similarly the misfit areas or secondary dislocations are the cell walls around the O2-elements and the directions of the dislocation lines are the intersections of the boundary plane with these cell walls.

Until now, no assumption has been made regarding the nature of the crystallographic system to define the CSL and DSC lattices. Except for the cubic system for which exact coincidences can be systematically obtained, the parametric ratio must be taken into account and the notion of near-coincidence or constrained coincidence must be introduced (3-6). In the following, attention will be focused on the hexagonal system.

2.3. Rotation matrix in the hexagonal coordinate system

In an orthonormal coordinate system the matrix describing any rotation is given by equation 8 where p_1, p_2 and p_3 are the director cosines of the rotation axis and θ is the value of the rotation angle :

$$R_0 = \begin{bmatrix} p_1^2(1-\cos\theta)+\cos\theta & p_1p_2(1-\cos\theta)-p_3\sin\theta & p_1p_3(1-\cos\theta)+p_2\sin\theta \\ p_1p_2(1-\cos\theta)+p_3\sin\theta & p_2^2(1-\cos\theta)+\cos\theta & p_2p_3(1-\cos\theta)-p_1\sin\theta \\ p_1p_3(1-\cos\theta)-p_2\sin\theta & p_2p_3(1-\cos\theta)+p_1\sin\theta & p_3^2(1-\cos\theta)+\cos\theta \end{bmatrix} \quad [8]$$

with $p_1^2 + p_2^2 + p_3^2 = 1$

The most general rotation matrix R_0 can also be written as follows [eq. 9] with $t = \tan(\theta/2)$ and $\theta \neq 180°$:

$$R_0 = (1+t^2)^{-1} \times \begin{bmatrix} 2t^2p_1^2+1-t^2 & 2t^2p_1p_2-2tp_3 & 2t^2p_1p_3+2tp_2 \\ 2t^2p_1p_2+2tp_3 & 2t^2p_2^2+1-t^2 & 2t^2p_2p_3-2tp_1 \\ 2t^2p_1p_3-2tp_2 & 2t^2p_2p_3+2tp_1 & 2t^2p_3^2+1-t^2 \end{bmatrix} \quad [9]$$

The hexagonal structure is determined by the three unit vectors a_1, a_2 and c, the lengths of which are :

$$|\vec{a}_1| = |\vec{a}_2| = a \quad ; \quad |\vec{c}| = c$$

and the three angles :

$$(\vec{a}_1, \vec{c}) = (\vec{a}_2, \vec{c}) = 90° \quad ; \quad (\vec{a}_1, \vec{a}_2) = 120°$$

The coordinates of these three unit vectors are given in the orthogonal coordinate system by the columns of the structure matrix S :

$$S = \begin{pmatrix} 1 & -1/2 & 0 \\ 0 & \sqrt{3}/2 & 0 \\ 0 & 0 & c/a \end{pmatrix} \qquad [10]$$

The rotation matrix may be expressed in the hexagonal coordinate system (a_1, a_2, c) by the relation :

$$R_H = S^{-1} R_0 S \qquad [11]$$

The matrix elements of R_H are then given in equation 12.

$$R_H = (1+t^2)^{-1} x \qquad [12]$$

$$\begin{bmatrix} 2t^2p_1^2+2t^2p_1p_2/\sqrt{3} +2tp_3/\sqrt{3}+1-t^2 & t^2p_2^2-t^2p_1^2+ +2t^2p_1p_2/\sqrt{3}-4tp_3/\sqrt{3} & (c/a)(2t^2p_1p_3 +2t^2p_2p_3/\sqrt{3}+2tp_2 -2tp_1/\sqrt{3}) \\ 4t^2p_1p_2/\sqrt{3} +4tp_3/\sqrt{3} & 2t^2p_2^2-2t^2p_1p_2/\sqrt{3} -2tp_3/\sqrt{3}+1-t^2 & (c/a)(4t^2p_2p_3/\sqrt{3} -4tp_1/\sqrt{3}) \\ (a/c)(2t^2p_1p_3 -2tp_2) & (a/c)(-t^2p_1p_3 +\sqrt{3}t^2p_2p_3+tp_2+\sqrt{3}tp_1) & 2t^2p_3^2+1-t^2 \end{bmatrix}$$

By use of the hexagonal coordinates of the vector parallel to the rotation axis, U, V and W, and a new parameter M such as $M^2 = [(a/c)^2 \ (U^2+V^2-UV)+W^2]/\ 3t^2$, the rotation matrix R_H can be written in the hexagonal coordinate system [eq. 13] :

$$R_H = (U^2 + V^2 - UV + (c/a)^2 \ (W^2 + 3M^2))^{-1} x \qquad [13]$$

$$\begin{bmatrix} U^2-V^2 +(c/a)^2(3M^2+2MW-W^2) & (2V-U)U -4(c/a)^2MW & 2(c/a)^2UW +2(c/a)^2M(2V-U) \\ (2U-V)V +4(c/a)^2MW & -U^2+V^2 +(c/a)^2(3M^2-2MW-W^2) & 2(c/a)^2VW +2(c/a)^2M(2U-V) \\ (2U-V)W -3(a/c)^2MV & (2V-U)W +3(a/c)^2MU & -U^2-V^2+UV +(c/a)^2(3M^2+W^2) \end{bmatrix}$$

For special values of the rotation angle and of the rotation axis a three dimensional coincidence site lattice can be obtained (7-12).

2.4. Coincidence orientation in the hexagonal symmetry

A rotation matrix leads to a lattice of common nodes or coincidence site lattice (CSL) for special conditions : the nine elements of the rotation matrix must be integers with the following property gcd $(R_{ij}) = 1$.

In a hexagonal coordinate system this condition is only fulfilled for any value of the ratio $(c/a)^2$ if the rotation axes are :

i) <001> with the rotation angle θ given by $\cotan^2(\theta/2) = 3M^2$ (M : rational number)

ii) <UV0>, $\theta = 180°$

On the other hand, the rotation can lead to a CSL only when the ratio $(c/a)^2$ has a rational value. By writing that $(c/a)^2 = \mu/\nu$ and by use of the parameters m and n, such as $t = (n/m)(d/3\mu)^{1/2}$ with $d = \nu(u^2+v^2-uv) + \mu w^2$, the rotation matrix is given by the equation [14].

$$R_H = (3\mu m^2 + dn^2)^{-1} x \tag{14}$$

$$\begin{bmatrix} \nu(u^2-v^2)n^2 + \mu(3m^2-w^2n^2)+2\mu mnw & \nu(2v-u)un^2 - 4\mu mnw & 2\mu uwn^2 - 2\mu mn(2v-u) \\ \nu(2u-v)vn^2 + 4\mu mnw & \nu(-u^2+v^2)n^2 + \mu(3m^2-w^2n^2)-2\mu mnw & 2\mu vwn^2 - 2\mu mn(2u-v) \\ \nu(2u-v)wn^2 - 3\nu mnv & \nu(2v-u)wn^2 + 3\nu mnu & \nu(-u^2-v^2+uv)n^2 + \mu(3m^2+w^2n^2) \end{bmatrix}$$

The main differences between the matrices (13) and (14) are :

- $(c/a)^2 = \mu/\nu$ is rational where gcd $(\mu/\nu) = 1$

- u, v, w are integers, gcd (u, v, w) = 1

- m, n are integers, gcd (m, n) = 1

Consequently, the CSL orientation can be determined by using the two parameters μ and ν related to the crystallographic structure and the five parameters m, n, u, v, w associated with the orientation relationship (13).

Coincidence index Σ defined as the ratio of the CSL primitive cell volume to the crystal primitive cell volume is, actually, the least positive integer such that ΣR_H and ΣR_H^{-1} are integral matrices (14) :

$$\Sigma = [\mu \ (3m^2 + w^2 n^2) + \nu (u^2 - uv + v^2) \ n^2] \ / \ \alpha$$

2.5. Equivalent rotations

Applying of one of the twelve symmetry elements to one grain leaves unchanged the orientation relationship, only the rotation axis-rotation angle will be modified. Moreover, the choice of the reference crystal which remains unchanged with the rotation can be interchanged, therefore 2.12^2 hexagonally equivalent rotations can be obtained (15). Table 1 gives the twelve equivalent rotations in function of the parameters μ, ν , u, v, w , m, n .

1	un	vn	wn	m
2	2vn	-2(u-v)n	wn-3m	wn+m
3	-2(u-v)n	-2un	-wn-3m	wn-m
4	-(u-2v)n	-(2u-v)n	-3m	wn
5	2(u+v)n	-2(u-2v)n	3(wn-m)	wn+3m
6	-2(2u-v)n	-2(u+v)n	-3(wn+m)	wn-3m
7	-2μ(wn+m)	-4μm	ν(2u-v)n	νvn
8	-4μm	2μ(wn-m)	ν(u-2v)n	νun
9	2μ(wn-m)	2μ(wn+m)	$-\nu$(u+v)n	ν(u-v)n
10	4μwn	2μ(wn+3m)	-3νun	ν(2u-v)n
11	2μ(wn-3m)	4μwn	-3νvn	ν(2u-v)n
12	-2μ(wn+3m)	2μ(wn-3m)	3ν(u-v)n	ν(u+v)n

Table 1- The twelve hexagonally equivalent descriptions for the same coincidence orientation.

Now it is important to point out that the twelve mathematical descriptions of the same orientation relationship are not of the same nature. The descriptions 1 to 6 are independent of the crystallographic parameters μ and ν whereas the six other descriptions include these parameters. Thus, the crystallographic indices of the six last rotation axes are parallel to irrational directions of the hexagonal crystal characterized by the parameters a and c. In fact, in the numerical determination of coincidence orientations, the ratio $(c/a)^2$ is replaced by the rational number μ/ν. In these conditions the indices of these six last rotation axes are depending on the choice of the μ and ν values.

Owing to the symmetry of the hexagonal system the twelve rotation axes lie on two meridians of the stereographic projection : one contains all the low indice rotation axes (rational indices) and the second the high indice rotation axes (irrational indices). So, strictly speaking, the twelve rotation axes lie on two different planes, P and P', perpendicular, in the direct space. Consequently, for a description with the ratio c/a, only the P plane contains coincidence sites along six rotation axes and this plane defines a two dimensional coincidence site lattice.

In the P' plane, the rotation axes are close to rational directions but not exactly parallel to them. But since experimental results show that three dimensional coincidence lattices exist in real materials a local deformation of the adjacent grains near the grain boundary must be taken into account. This deformation can be relaxed by dislocation networks in the grain boundary plane. Thus two types of deviation from exact coincidence must be considered :

- parametric deviation introduced by the difference between the ratios μ/ν and $(c/a)^2$
- angular deviation due to the superposition of a low angle rotation on the exact coincidence relationship.

In fact, these two deviations are accommodated by the same secondary dislocation network as shown by Bollmann (2).

In the second part, examples of studies of grain boundaries in a hexagonal material will be given.

3. GRAIN-BOUNDARIES STUDY IN TUNGSTEN CARBIDE-COBALT COMPOSITES

3.1. Specimen preparation

The tungsten carbide-cobalt materials are prepared by a liquid sintering process. They belong to the cemented carbide alloys where cobalt is mainly used as the binder phase. In the case of tungsten carbide and cobalt alloys the very low interfacial energy between the two phases result in a nearly perfect wetting and consequently in a very good adhesion in the solid state (16,17).

A tungsten carbide and cobalt powder mixture (with cobalt volume ratios of 5 up to 37 %) compacted under high pressure is heated at about 1400°C under vacuum. At this temperature the liquid cobalt phase can dissolve some amount of tungsten carbide crystals. The solubility of

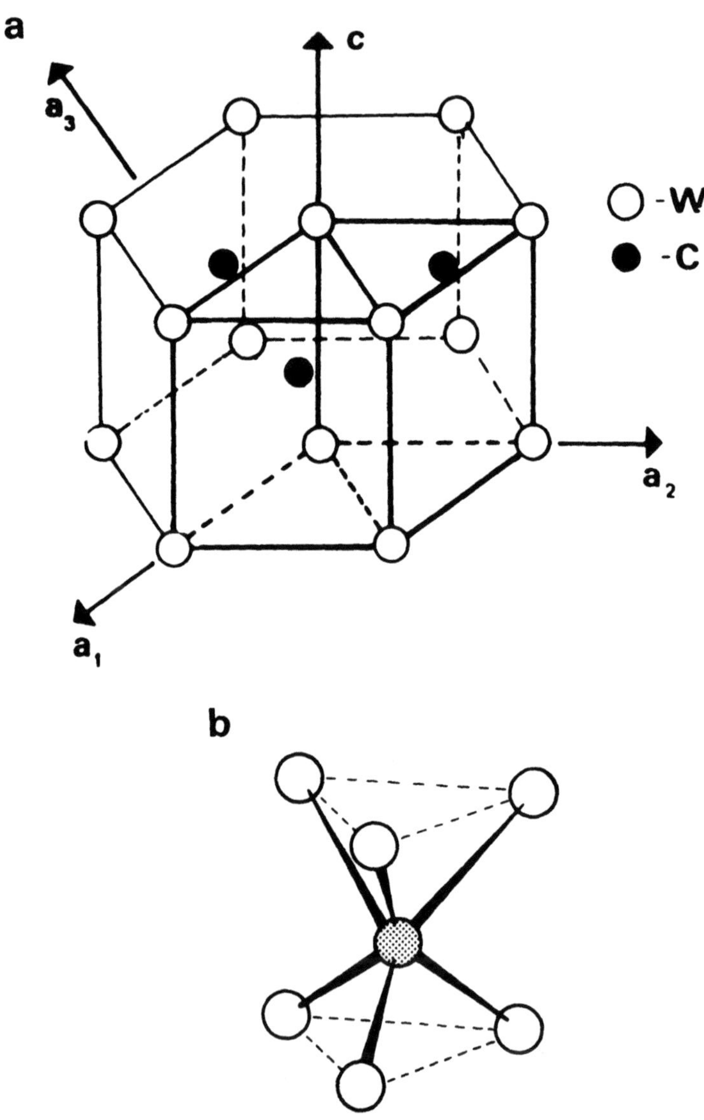

Fig. 2a : Tungsten carbide crystal structure. The lattice parameters are a = 0.2906 nm and c = 0.2837 nm with a c/a ratio equal to 0.976

b : Trigonal prismatic coordination of carbon atoms in WC

tungsten carbide in the liquid phase is important and reaches about 30 wt % near 1400°C (18). After saturation of the liquid cobalt phase and during the cooling process, precipitation, redeposition and growing of tungsten carbide crystals on preexisting carbide crystals occur giving rise to the well characterized microstructure of such specimens.

3.2. Structural considerations

Tungsten carbide has a simple hexagonal structure with two atoms per unit cell (Fig. 2a). Its space group is $P\bar{6}m2$ with tungsten and carbon atoms in the following positions 1(a) 000 and 1(f) 2/3 1/3 1/2 respectively. The characteristic trigonal prismatic coordination of carbon atoms in WC is illustrated in Fig. 2.b. The lattice parameters are a=0.2906 nm and c=0.2837 nm with a c/a ratio close to 1 (0.976). The tungsten-carbon distance is d_{WC} = 0.2197 nm, tungsten and carbon radii are r_W=0.1368nm and r_C=0.0829 nm. Consequently there are no contacts between tungsten atoms but only contacts between tungsten and carbon atoms. Due to the lattice parameters of WC, tungsten atomic rows (parallel to the electron beam) can easily be imaged by HREM (Jeol 200 CX, point resolution 0.22 nm) for two crystal orientations <0001> and <11$\bar{2}$0>.

The equilibrium shape of tungsten carbide crystals can be deduced from some crystallographic considerations. In that structure there are two sets of three equivalent {10$\bar{1}$0} planes which induce a typical, flat prismatic with a triangular base crystal shape, with facets parallel to {10$\bar{1}$0} planes limited in the perpendicular direction by the basal plane (19).

3.3. Microstructure

The optical micrograph of Fig. 3a depicts the typical WC-Co microstructure which appears simply as a two-phase material : the cobalt phase (white) and the tungsten carbide phase (grey). The sample has a cobalt volume ratio equal to 22% and a mean WC grain diameter D_{WC}=2.2μm. The grain boundaries of the cobalt phase have not been revealed on this micrograph. Other experiments have shown that they have a mean diameter larger than 20 μm (20). During the sintering process the tungsten carbide crystals develop well defined facets and tend to adopt the equilibrium shape. The optical micrographs show typical grain shapes with rectangular or triangular faces suggesting {10$\bar{1}$0} or {0001} crystallographic orientations. This fact also appears on the transmission electron micrograph (Fig.3b) which shows three carbide grains with planar grain boundaries. It also suggests an important microstructural aspect of such alloys which could accordingly be described by means of a tungsten carbide

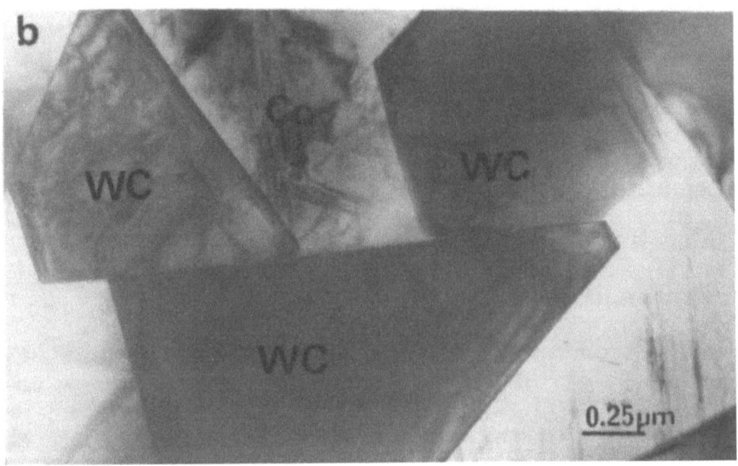

Fig. 3a : Optical micrograph of a WC-Co composite with a cobalt volume ratio equal to 22%

b : Transmission electron micrograph of the same material

skeleton in a soft cobalt matrix. Defects also appear inside the two phases. They are partly due to the sintering process that induces stresses during the preparation which are accommodated by plastic deformation. The corresponding dislocation mechanisms have been studied in details for the carbide phase (21, 22). The cubic cobalt phase contains a high density of stacking faults which can result from a local transformation into the hexagonal cobalt structure. The hexagonal cobalt structure is the equilibrium phase at room temperature. In sintered alloys containing WC the cubic cobalt phase can be stabilized at low temperature (below 417°C) by mechanical constrains or, more likely, by dissolved tungsten and carbon atoms (16).

Different microstructural models have been proposed to explain the mechanical properties of such materials.One of them tends to describe the microstructure not only by a carbide skeleton but also by carbide crystal aggregates (23). In all of these models, grain boundaries between the carbide crystals play an important role and many studies have been focused on the description of their crystallographic and chemical nature by the use of several techniques.

3.4. Chemical nature of grain boundaries in WC-Co composites

During the sintering process of the tungsten carbide-cobalt alloys, the Co-W-C solid solution may be squeezed out of the intergranular region or retained as a continuous film between the carbide grains. TEM observations of many grain boundaries have shown that both cases can be found. An effective dewetting of the cobalt phase at tungsten carbide grain boundaries have often been observed (Fig. 3b). On the contrary, well defined intergranular films of cobalt with a thickness of 1 nm (Fig. 4a) and 10 nm (Fig. 4b) are also observed between carbide grains. The presence of a small amount of cobalt has been deduced by STEM - EDX analysis and Fig. 5 corresponds to X-Ray spectra taken inside a WC grain (Fig. 5a) and at a grain boundary (Fig. 5b). The two spectra reveal the presence of cobalt. A noticeable cobalt content has been found at that grain boundary (13 at % Co) as compared to the quantity found inside the grain where the solubility of cobalt can be estimated to be 3 at % Co (24, 25).

Many authors have already discussed the different possible configurations of tungsten carbide grain boundaries with a continuous cobalt film or not, or with a small amount of segregated cobalt phase. Several physical analysis methods have been put to work and the most recent investigations have been made by using HREM, STEM-EDX and field ion microscopy.Grain boundary studies by lattice fringe imaging (26) and HREM (27) confirm the contiguity of

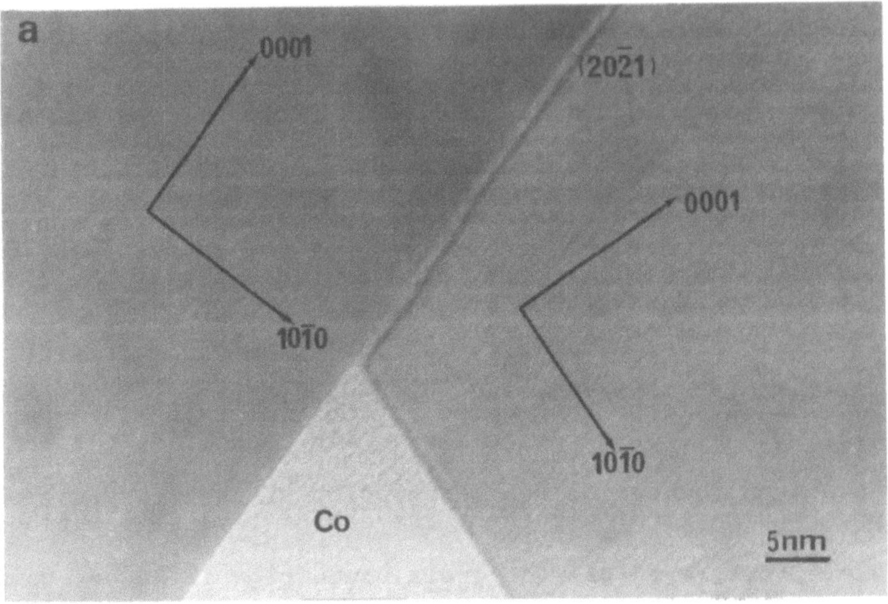

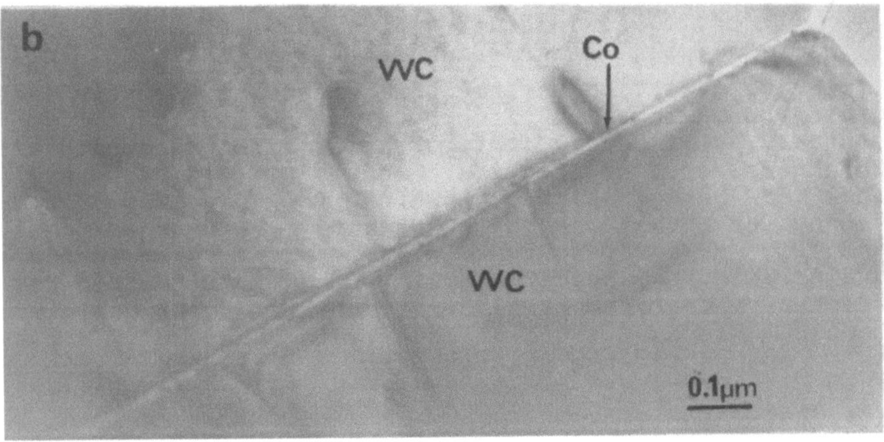

Fig. 4 : Well defined film of cobalt at grain boundaries with a thickness of 1nm (a) and 10 nm (b)

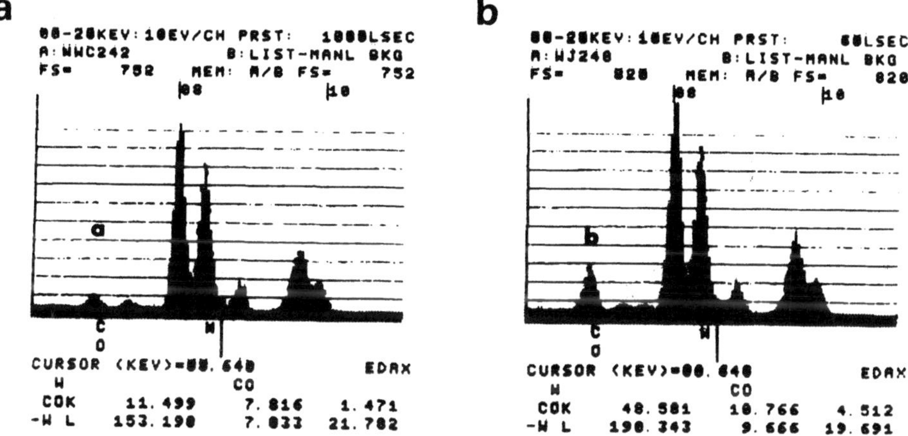

Fig. 5 : X-Ray spectra from a) WC grain, b) general grain boundary in an as-sintered sample.

some grain boundaries which do not contain a cobalt phase. By STEM-EDX analysis (24, 25, 28), the cobalt phase was found to segregate at grain boundaries and field ion microscopy (29) confirmed that cobalt is present as a segregate with half a monolayer coverage in analyzed grain boundaries.

In such a material the relationship between the orientations of the carbide grains and the cobalt content at these interfaces is of interest. So, cobalt segregation profiles have been obtained by STEM-EDX analysis performed on tungsten carbide grain boundaries previously orientated by TEM. The data confirm that many near - coincidence grain boundaries showed no cobalt segregation and results have emphasized two points playing a role on the cobalt segregation : the crystallographic nature of the grain boundary plane and the presence of defects (dislocations) at the interface (24, 25). More detailed results will be published elsewhere.

3.5. Coincidence orientations

The geometric descriptions of grain boundaries based on the principle of coincidence orientations have been applied to tungsten carbide with the approximation $c/a = 1$. The list of the calculated coincidence orientations up to $\Sigma=100$ has been obtained. Experimental orientations of carbide grains with respect to each other have shown that many grain boundaries are very close to low index coincidence orientations (Σ = 2, 4, 5, 7a, 7b, 10 ...). Some low angle grain boundaries between carbide grains (Σ = 1) can also be found in this material. The grain boundary orientations can be classified in three main families. The first family $\Sigma = 13a$ corresponds to a rotation around the [0001] axis and therefore is independent of the c/a ratio. In the second family (Σ=2, 5, 10...) the grain boundary orientations can be described as rotations around the $\langle 10\bar{1}0\rangle$ axis perpendicular to the usual $\{10\bar{1}0\}$ boundary plane. The third family (Σ = 4, 7b ...) corresponds to rotation around the $\langle 11\bar{2}0\rangle$ axis.

Fig.6 illustrates four diffraction patterns corresponding to coincidence or near coincidence orientations : $\Sigma = 2$ with an exact orientation (a) and with a deviation angle of 9° around the $[0001]_1$, or $[11\bar{2}0]_2$ axis (b), a $\Sigma = 4$ orientation (c) and a $\Sigma = 13a$ orientation (d). It is an example of the three main grain boundary families previously defined.

3.6. Nature of the grain boundary plane

Two crystallographic planes play an important role in the formation of tungsten carbide interfaces, the $\{10\bar{1}0\}$ prismatic plane and the $\{0001\}$ basal plane. As previously shown these two planes are the two main facets of WC crystals.

Most grain boundaries between two crystals rotated around the $\langle 10\bar{1}0\rangle$ axis are parallel to a prismatic plane. So they can be classified as twist grain boundaries. Due to the crystal structure of WC and to the shape of the crystal, the prismatic plane is highly promoted as a boundary plane during the sintering process. These facts can also explain the very frequent observation of $\Sigma = 2$ or near $\Sigma = 2$ orientations in this material.

All the grain boundaries described by a $\langle 11\bar{2}0\rangle$ rotation axis can be described as asymmetric tilt grain boundaries. All of them actually have a grain boundary plane parallel to the basal plane of one crystal. Consequently, the indices of the grain boundary plane in the other crystal have high crystallographic indices. This asymmetric configuration is very commonly observed in many systems (30).

Some grain boundaries described by a $\langle 0001\rangle$ rotation axis can be defined as twist boundaries.

Fig. 7a and 7b illustrate examples of the main families of grain boundaries observed in WC. Fig. 7a presents twist grain boundaries (Σ = 2 orientations) where the grain noted 2 is in a $\Sigma = 2$ orientation with respect to the crystals noted 1 and 3. The two interfaces are both parallel to $\{10\bar{1}0\}$ planes, the crystal 2 having its c axis parallel to the electron beam in the orientation shown. Fig.7b is a Σ=4 orientation with an asymmetric tilt grain boundary plane.

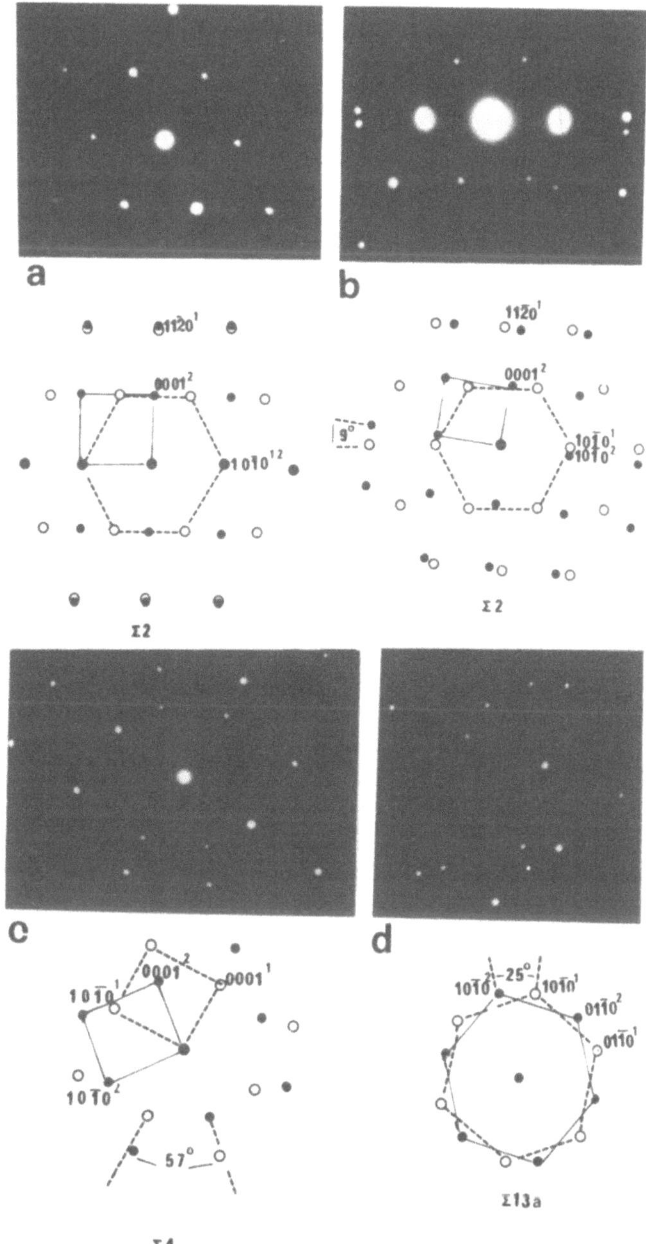

Fig. 6 : Four examples of diffraction patterns corresponding to coincidence or near coincidence orientations a) Σ2 b) Σ2 with a 9° deviation angle around the $[0001]_1$ or $[11\bar{2}0]_2$ directions c) Σ4 d) Σ13a

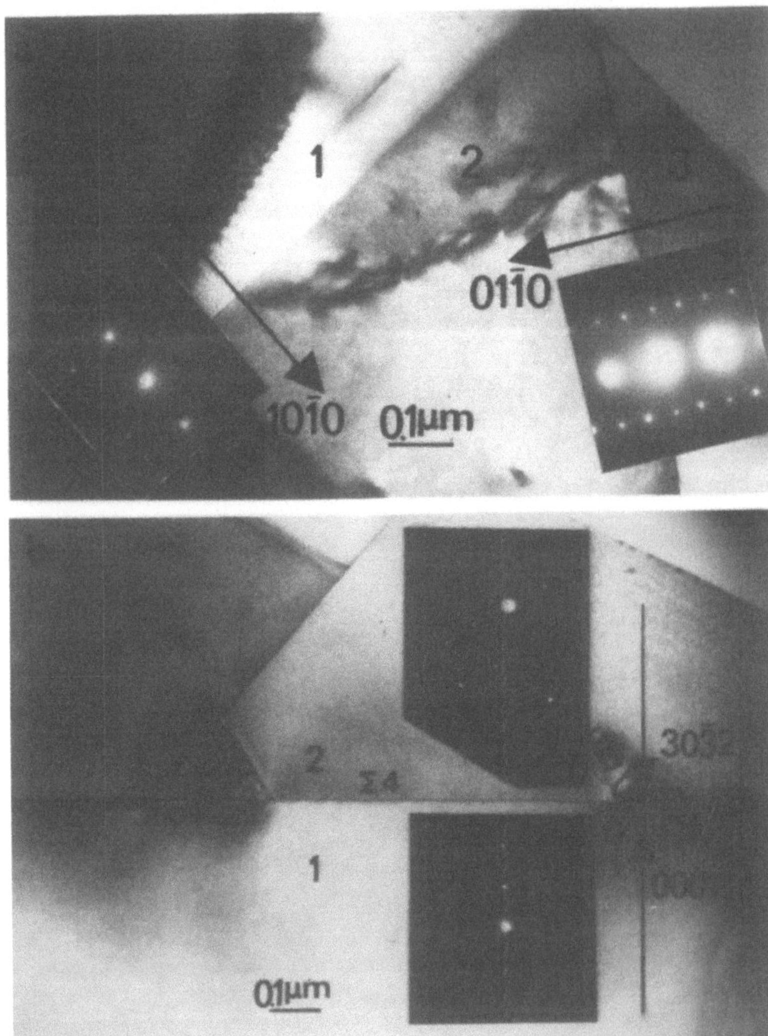

Fig. 7a : Two examples of $\Sigma 2$ orientations ; the crystal 2 has its c axis parallel to the electron beam and has been found in a $\Sigma 2$ orientation with respect to crystals 1 and 3; both interfaces are parallel to $\{10\bar{1}0\}$ planes

b : $\Sigma 4$ orientation with an asymmetric tilt grain boundary plane $(0001)_1$ and $(30\bar{3}2)_2$

3.7. HREM images of grain boundaries

HREM images of interfaces have been obtained in the WC-Co specimens. They give precise information about the defect structure of the grain boundaries and interfaces previously discussed.

Fig. 8 is a HREM micrograph between a tungsten carbide crystal in a $[11\bar{2}0]$ orientation and the cobalt phase illustrating the microfaceting of the tungsten carbide crystal. As expected, microfacets can grow along the $\{10\bar{1}0\}$ and $\{0001\}$ densest planes of the WC structure. On this micrograph a lattice image in the cobalt phase has been obtained but no clear relationship between the two phases has been found.

Many grain boundaries belonging to the main families found in this system have been imaged in HREM. We have reported here two examples of grain boundaries the first one corresponding to a $\langle 10\bar{1}0\rangle$ rotation ($\Sigma = 2$, Fig. 9), the second ($\Sigma = 37$) being described by a $\langle 11\bar{2}0\rangle$ rotation (Fig. 13).

Fig. 9 is a section of an exact $\Sigma = 2$ grain boundary and show the defect structure at the interface. That particular orientation can have been resolved in the two crystals because the grain boundary plane is exactly parallel to the electron beam. One crystal is in a $\langle 0001\rangle$ orientation and the other one in a $\langle 11\bar{2}0\rangle$ orientation. The two drawings in Fig. 10a and b show two crystals in a $\Sigma = 2$ orientation for a=c. In the first drawing, the two crystals are represented by two hexagonal cells, and the second drawing shows the atomic projection (W and C) in a plane perpendicular to the grain boundary. The DSC-vectors associated with a $\Sigma 2$ orientation have been determined in order to be correlated to the experimental observations. For a 90° rotation around $[10\bar{1}0]$ axis the unit cells of the CSL, DSC and O-lattices are given in the coordinate system of crystal 1 :

CSL :	$X_1 = \frac{1}{3}[\bar{1}2\bar{1}0]$	$X_2 = \frac{1}{3}[\bar{1}2\bar{1}3]$	$X_3 = [10\bar{1}0]$
O-lattice :	$X_1^0 = \frac{1}{6}[\bar{1}2\bar{1}3]$	$X_2^0 = \frac{1}{6}[1\bar{2}13]$	$X_3^0 = \frac{1}{2}[10\bar{1}0]$
DSC-lattice :	$d_1 = \frac{1}{6}[\bar{1}2\bar{1}3]$	$d_2 = \frac{1}{6}[1\bar{2}1\bar{3}]$	$d_3 = \frac{1}{2}[10\bar{1}1]$

The CSL unit cell is drawn on crystal 1 in figure 11. An other representation is given in figure 12 with the boundary plane parallel to the foil plane. In a real crystal there is a mismatch between the two parameters a and c :

$$41\,(a/2) = 21c$$

consequently a good matching of the two $\{11\bar{2}0\}_1$ and $\{0001\}_2$

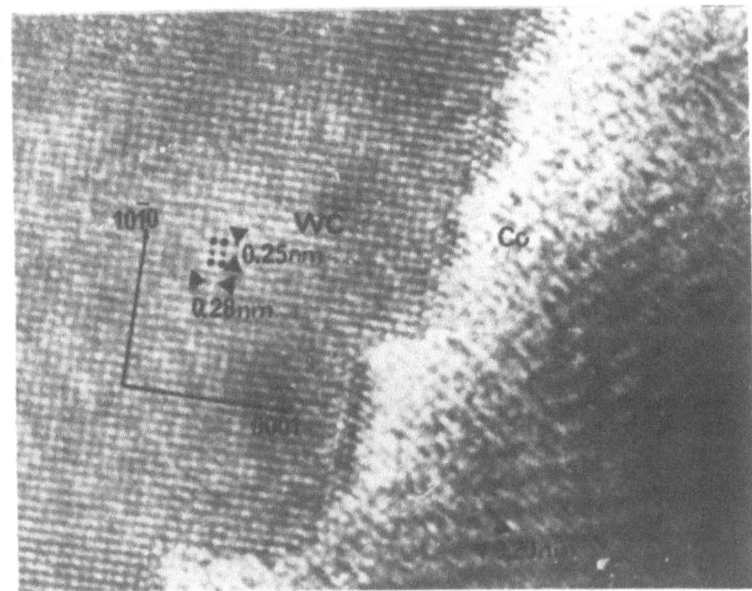

Fig. 8 : Microfaceting of the tungsten carbide-cobalt interface.

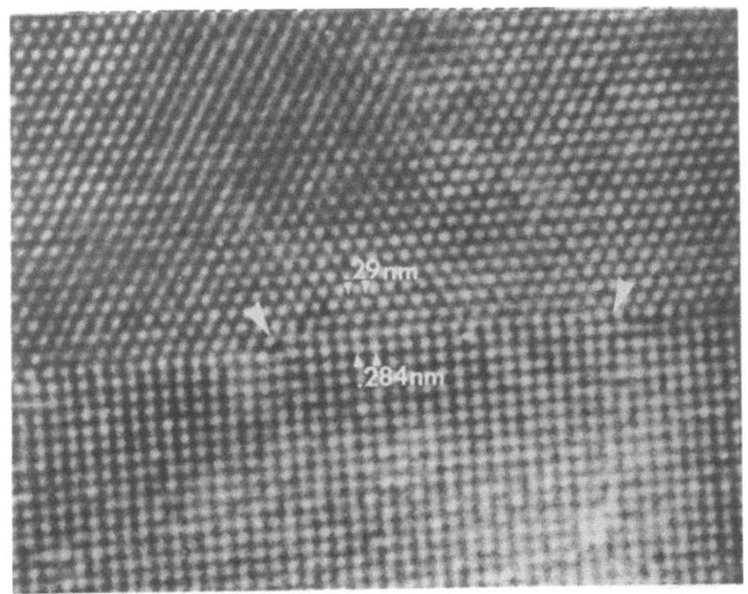

Fig. 9 : HREM image of a Σ2 grain boundary. Regular steps are imaged in the grain boundary plane.

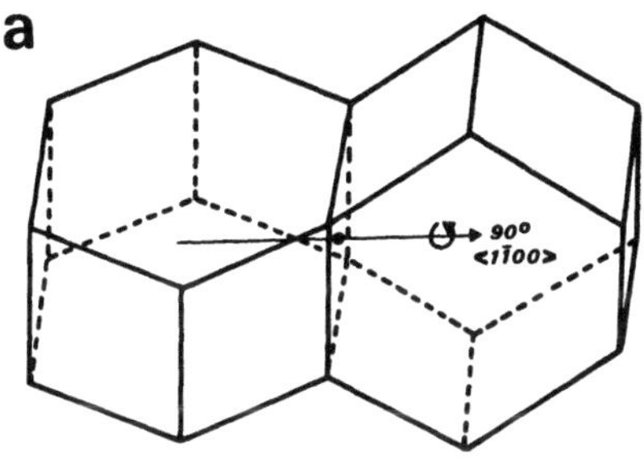

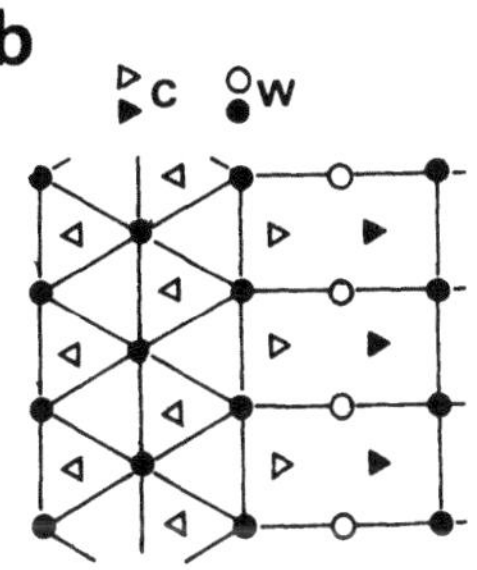

Fig. 10a : a) two crystals in a Σ2 orientation. The two crystals are schematically represented by two hexagonal cells.
b) projection of both crystals in a Σ2 orientation in a plane perpendicular to the grain boundary plane. Atomic positions of W and C are indicated on the drawing

planes is obtained when this condition is fulfilled. Moreover a good matching of the atomic sites in both crystals is obtained when the accommodation has been realized in the two perpendicular directions $[0001]_2$ or $[11\bar{2}0]_1$ and $[0001]_1$ or $[11\bar{2}0]_2$. Two regions of good matching are separed by an atomic step. Therefore the grain boundary can be described as a double stepped interface. The HREM image (Fig. 9) is a projected view of this interface. Regular steps (arrowed on the micrograph of Fig. 9) associated with vectors of the DCS-lattice previously determined are imaged with a separation equal to 5.5 nm, in good agreement with the calculated value 5.95 nm.

An asymmetric tilt grain boundary has been imaged and is showed in Fig. 13. In this case (Σ = 37) the grain boundary adopts two different orientations parallel to the two densest planes $(10\bar{1}0)_1$ and $(0001)_2$ respectively. This fact leads to a microfaceted interface which could not be detected at lower magnification.

CONCLUSION

The concept of coincidence site lattice (CSL) and the O-lattice theory previously developed for the cubic symmetry have been applied to the hexagonal symmetry. It has been shown that a lattice of common nodes or CSL between two lattices is obtained if the nine elements of the rotation matrix are integers. In a hexagonal coordinate system, where the two parameters are a and c, the rotation can generate a CSL only if the $(c/a)^2$ ratio has a rational value

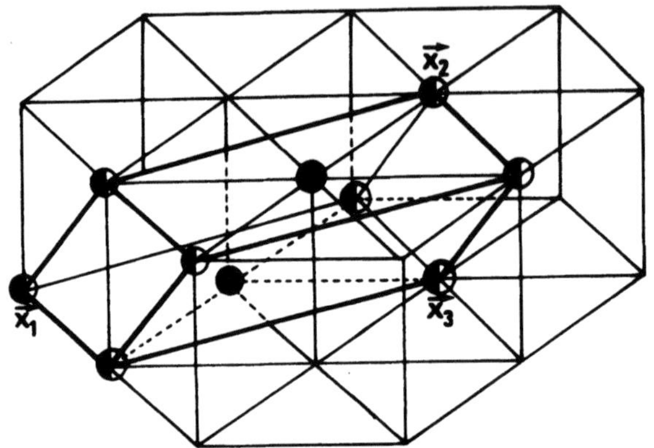

Fig. 11 : CSL unit cell of a Σ2 orientation

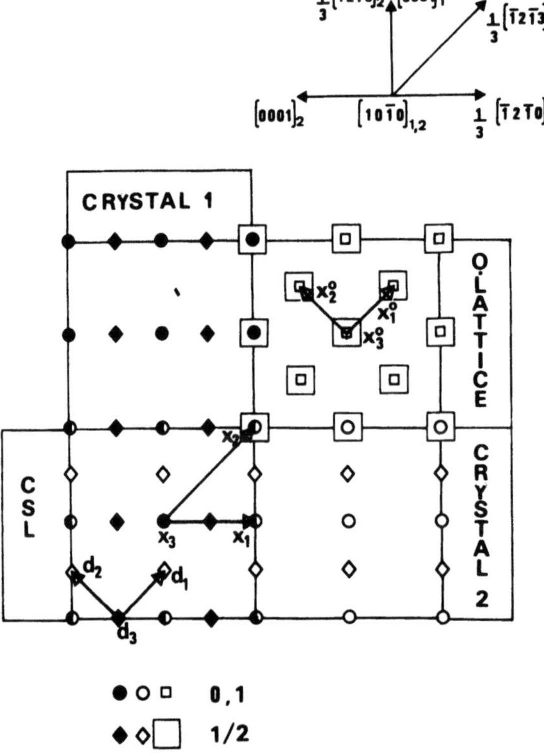

Fig. 12 : Crystals 1 and 2 projected on the common $(10\bar{1}0)$ plane with their corresponding CSL (x_1, x_2, x_3) DSC lattice (d_1, d_2, d_3) and O.lattice (x_1, x_2, x_3)

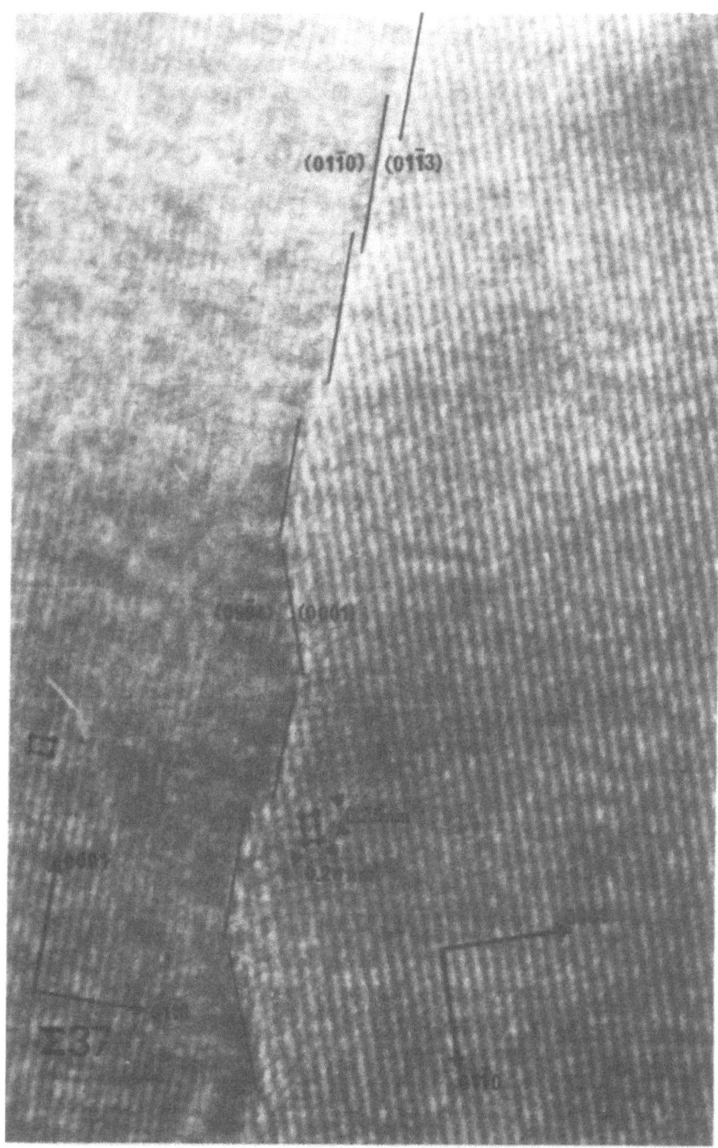

Fig. 13 : HREM image of an asymmetric tilt grain boundary where the grain boundary plane adopts two different orientations parallel to the two densest planes of WC respectively and a list of coincidence orientations can be written down for any $(c/a)^2 = \mu/\nu$ rational values. Experimental results have shown that a three dimensional coincidence lattice exists in a real material. Due to the real value of the ratio c/a, the two crystals have to be accommodated at the grain boundary even if they are in an exact coincidence orientation. For a near coincidence orientation, the relaxation at the grain boundary has to compensate the parametric deviation resulting from the difference between the ratios μ/ν and $(c/a)^2$, and an angular deviation due to the superposition of a low angle rotation on the exact coincidence relationship. As shown these two deviations are accommodated by the same secondary dislocation network.

A material of interest with respect to this theoritical approach is the tungsten carbide-cobalt composite in which a large number of near coincidence orientations have been found between the carbide crystals. This is partly due to the particular sintering process of this cermet. As illustrated in this work, the saturation of the liquid cobalt by the carbide phase at the sintering temperature gives rise to the precipitation and to the growth of the carbide crystals during the cooling of the specimen. This process leads to the formation of a high number of planar interfaces parallel to the densest planes and many of them have minimized their grain boundary energy by approaching a coincidence orientation. Very acute grain boundary description could be obtained in this material by HREM. Two cases have been shown concerning the $\Sigma = 2$ orientation and the $\Sigma = 37$ asymmetric tilt grain boundary. The knowledge of the exact grain boundary structure is of interest to explain, for instance, segregation phenomena or plastic accommodation in grain boundary. Other experiments are now in progress in tungsten carbide-cobalt alloys in these two fields to correlate segregation or dislocation interactions with well defined grain boundaries between carbide crystals.

REFERENCES

1 - BOLLMANN W. Crystal Defects and Crystalline Interfaces (1970) Springer - Verlag, Berlin

2 - BOLLMANN W. Crystal Lattices Interfaces, Matrices (1982) Bollmann, Genève

3 - BONNET R. Acta Cryst. (1980) **A36**, 116

4 - DELAVIGNETTE P. J. Microsc. Spectrosc. Electron. (1982) **8**, 111

5 - EROCHINE S., NOUET G.
Scripta. Metall. (1983) **17**, 1069

6 - DONI E.G., FANIDES Ch., BLERIS G.L.
Cryst. Res. Technol. (1986) **21**, 1469

7 - WARRINGTON D.H.
J. Phys. (Paris) (1975) **36**, **C4**, 87

8 - FORTES M.A., SMITH D.A.
Scripta Metall. (1976) **10**, 575

9 - BONNET R., COUSINEAU E., WARRINGTON D.H.
Acta Cryst. (1981) **A37**, 184

10 - BLERIS G.L., NOUET G., HAGEGE S., DELAVIGNETTE P.
Acta Cryst. (1982). **A38**, 550

11 - GRIMMER H., WARRINGTON D.H.
Acta Cryst. (1987) **A43**, 232

12 - HAGEGE S., NOUET G.
Acta Cryst. (1989) **A45**, 217

13 - HAGEGE S., NOUET G.
Script. Metall. (1985) **19**, 11

14 - GRIMMER H.
Acta Cryst. (1976) **A32**, 783

15 - GRIMMER H.
Acta Cryst. (1980) **A36**, 382

16 - EXNER H.E.
International Metals Reviews (1979), **4**, 149

17 - EXNER H.E. and FISCHMEISTER H.
Z. Metalkd (1966) **57**, 187

18 - SCHWARTZKOPF P. and KIEFFER R.
Cemented carbides, Mac Millan Company, New York (1960)

19 - FRENCH D.N. and THOMAS D.A.
Anisotropy in single crystal refractory compounds Eds. (VALLDIEK F.V. and MERSOL J.), (1968), **1**, 55, New York, Plenum Press.

20 - HAGEGE S.
These Caen - France (1979)

21 - HAGEGE S., VICENS J., NOUET G., DELAVIGNETTE P.
Phys. Stat. Sol. (a) (1980), **61** 675

22 - LAY S., DELAVIGNETTE P., VICENS J.
Phys. Stat. Sol. (a) (1985), **90**, 53

23 - LAY S., VICENS J., OSTERSTOCK F.
J. Mat. Sci. (1987), **22**, 1310

24 - LAVAL J.Y., VICENS J., NOUET G.
J. Phys., (1984), **45**, **C2**, 695

25 - VICENS J., NOUET G., DUBON A., LAVAL J.Y.
11th International Congress on X-Ray optics and microanalysis ICXOM (1986), 356, LONDON, CANADA

26 - JAYARAN V., SINCLAIR R.
J. Am. Ceram. Soc., (1983) **66**, C137

27 - VICENS J., LAURENT-PINSON E., CHERMANT J.L., NOUET G.
Interface Science and Engineering 87 LAKE PLACID N.Y. USA, J. of Phys. (1988), **49**, **C5**, 271

28 - SHARMA N.K., WARD I.D., FRASER H.L., WILLIAMS W.S.
J. Am. Ceram. Soc. (1980), **63**, 194

29 - HENJERED A., HELLSING M., ANDREN H.O., NORDEN H.
Mat. Sc. and Tech. (1986), **2**, 847

30 - LAURENT-PINSON E., NOUET G., VICENS J.,
Inst. Phys. Conf. Sci. N°93, (1988), **2**, 543 EUREM 88, ENGLAND

INTERFACES BETWEEN PIGEONITE, CLINOAMPHIBOLE AND AUGITE

W. SKROTZKI and W.F. MÜLLER*
Institut für Geologie und Dynamik der Lithosphäre, Universität Göttingen, Goldschmidtstr. 3, D–3400 Göttingen, FRG
**Institut für Mineralogie, Technische Hochschule Darmstadt, Schnittspahnstr. 9, D–6100 Darmstadt, FRG*

ABSTRACT. Pigeonite and clinoamphibole exsolution lamellae in augite have been studied by conventional and high resolution transmission electron microscopy. The pigeonite lamellae lie close to the (100) and (001) planes of augite with a deviation of about 15°. With increasing width of the lamellae the interphase boundary becomes semicoherent. The nucleation of pigeonite occurs preferentially at dislocations. Clinoamphibole lamellae lie parallel to the (010) plane of augite. Its formation is via stacking faults with a 1/2 [101] displacement vector. The stacking faults are bounded by partial dislocations with Burgers vectors of the same type. The minimum width of the lamellae is 9Å, that is one double chain. Broadening of the lamellae is by the motion of partial dislocations (steps) along the interface. With increasing thickness, the clinoamphibole lamellae become semicoherent. Nucleation of clinoamphibole also occurs at augite-orthopyroxene interfaces. The deviation of the pigeonite exsolution lamellae from (100) and (001) can be explained in terms of the differences in lattice parameters of the matrix and the precipitate phase. The reason for the formation of amphibole lamellae along (010) planes in augite is given by the dislocation aided exsolution mechanism.

1. Introduction

A detailed study of the orientation and microstructure of exsolution lamellae in silicate minerals is important in understanding the nature of nucleation and growth processes in crystalline materials. In the past, much work has been done on pyroxenes showing that the phase distributions that have been revealed by X-ray diffraction and transmission electron microscopy (TEM) are the most diverse of any mineral group (for a review see Buseck et al, 1980). Several augites and pigeonites from lunar and terrestrial basalts were found to have microstructures consistent with exsolution by spinodal decomposition whereas those from plutonic environments show microstructures that are consistent with nucleation and growth. In general, it is considered that within clinopyroxenes, where structural differences are slight, the operation of these mechanisms changes with decreasing cooling rate. It is also found that the nature of the phase boundaries depends on the respective structures of the parent and daughter phases and the conditions of formation of the pyroxene crystals. This correlation has been used to unravel the cooling history of pyroxene containing rocks (Robinson et al., 1977; Rietmeijer and Champness, 1982). So far only limited laboratory

L.-C. Dufour et al. (eds.), Surfaces and Interfaces of Ceramic Materials, 141–153.

work has been done to explore the exsolution phenomena in pyroxenes (Ried, 1982). Therefore, our present knowledge of subsolidus phase relations in pyroxenes is mainly based on investigations of natural materials, the thermal history of which is the subject of much speculation.

The present work gives a detailed characterization of different types of interphase boundaries between pigeonite, clinoamphibole and augite. The precipitation of pigeonite and clinoamphibole in augite has taken place under slow cooling metamorphic conditions.

2. Materials and Methods

The material investigated is augite from a pyroxenite dike of the Balmuccia massif (NW-Italy). Suitably oriented augite grains were selected from petrographic thin sections after characterization by optical microscopy. Electron transparent foils were prepared by ion beam thinning. To avoid charging effects by the electron beam the foils had to be coated with a thin carbon film. The specimens were examined using both conventional bright field (BF) and dark field (DF) TEM and high resolution transmission electron microscopy (HRTEM). A Siemens Elmiskop 101, a Philips EM 420 ST and a Philips CM 12 operating at an acceleration voltage of 100 and 120 kV, respectively, were used for the investigations. The chemical composition of the different phases, which were obtained by energy dispersive X-ray microanalysis, is $Wo_{46}En_{38}Fs_{16}$ for the augite and $Wo_2En_{56}Fs_{42}$ for the pigeonite (Wo = Wollastonite $CaSiO_3$, En = Enstatite $MgSiO_3$, Fs = Ferrosilite $FeSiO_3$). The scatter of each component is about ±2 mole percent. Both phases also contain some Al and Ti with a lower concentration in the pigeonite. The elements found in the clinoamphibole suggest this mineral phase to be hornblende, the exact structural formula has not been determined.

3. Results

3.1. PIGEONITE-AUGITE INTERFACE

Precipitation of pigeonite in augite has been generally observed as lamellae close to (100) and (001), termed "100" and "001" lamellae, respectively. The small precipitates in the material investigated are mainly of the "100" type having a lens shape with their longest axis inclined to (100) by about 15° (Fig. 1a). Often these lamellae change their morphology towards (001) as is shown in Fig. 1b. Less frequently "001" lamellae are also observed. These lamellae are inclined to (100) by an angle of about 60° (Fig. 2) and change their morphology towards (100) with the same angle as the "100" lamellae described above. Morphology changes of "001" lamellae have been also reported by Kitamura et al. (1981), Rietmeijer and Champness (1982) and Müller and Müller (1988) for different materials which experienced different cooling histories.

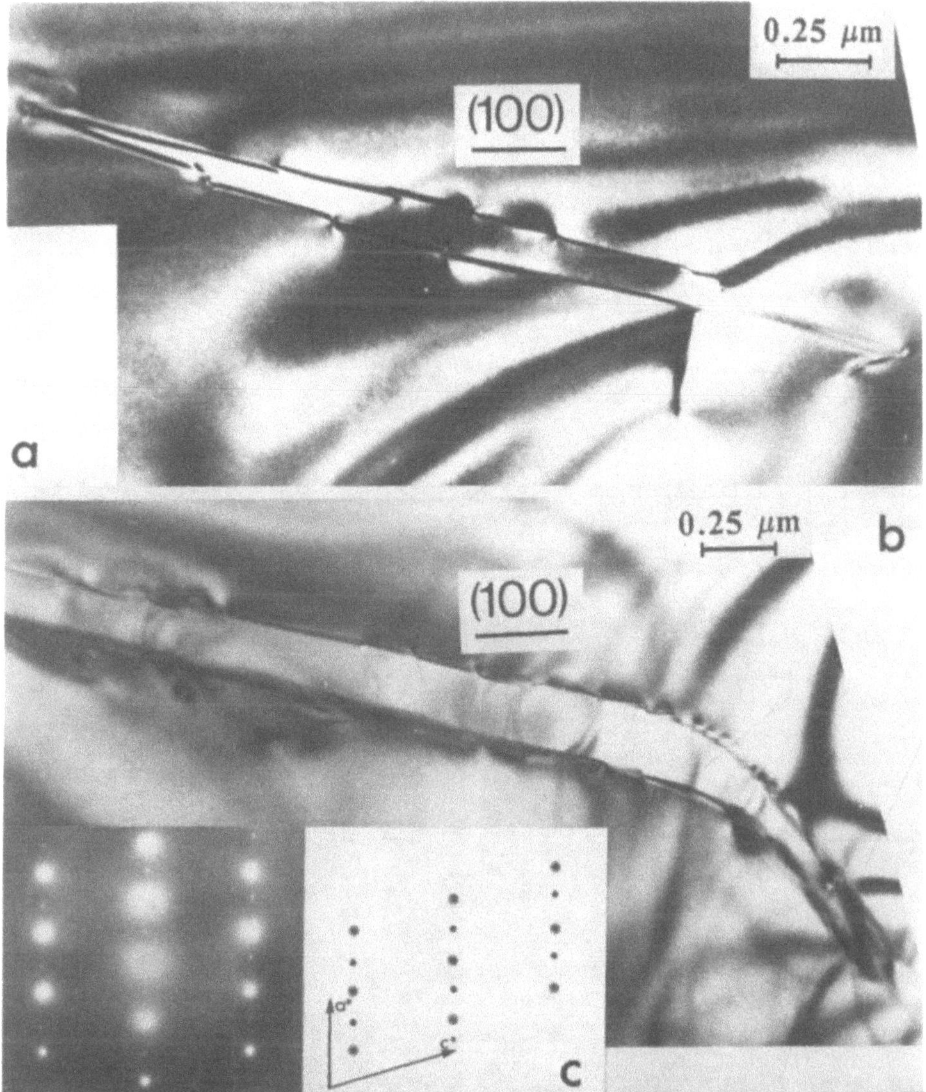

Figure 1. "100" pigeonite lamellae in augite shown in the [010] direction. Lamella (b) changes its morphology towards (001). BF images taken with g = 002 reflection. Insert (c) is the diffraction pattern of both phases with the strong spots belonging to the augite.

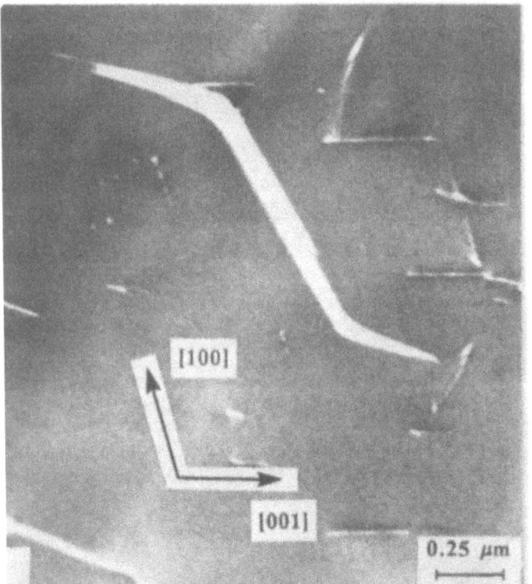

Figure 2.
"001" pigeonite lamella changing its morphology towards (100). DF image taken with $g = 002$ reflection.

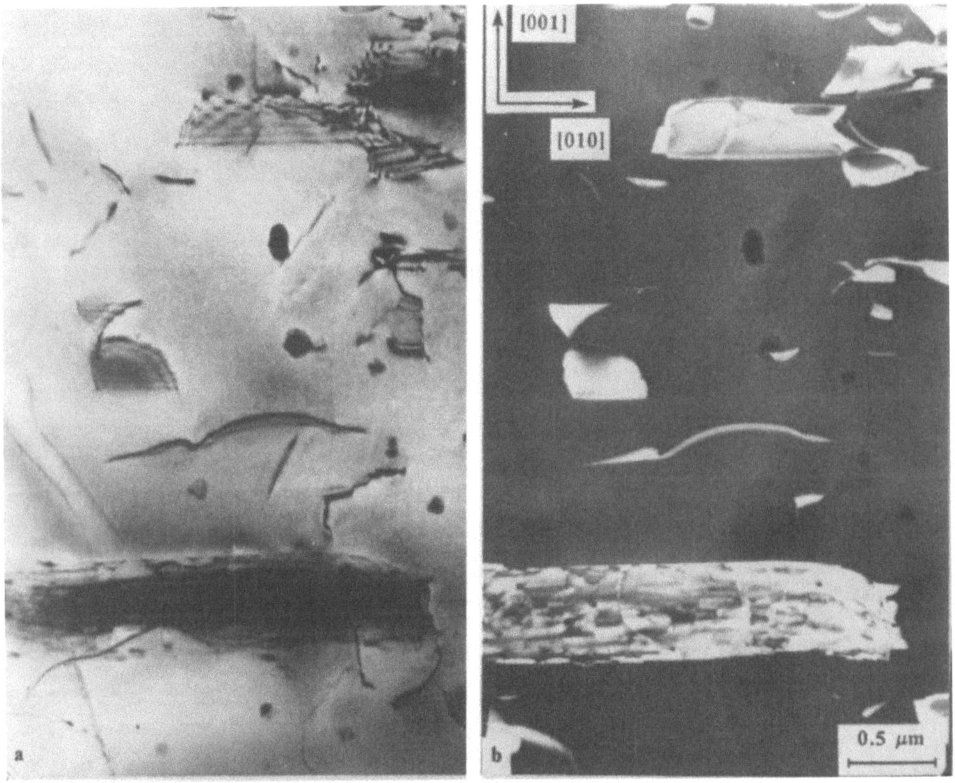

Figure 3. Flat-on view on pigeonite lamellae. The nucleation is preferrentially at dislocations. BF (a) and DF (b) images taken with $g = 002$ reflection.

The lamellae have ledges in their long interfaces (Figs. 1 and 2). The motion of these ledges may be an operating growth mechanism normal to this interface. The ledges are generally connected with dislocations, which are imaged in Fig. 4b. In the flat-on view of Fig. 3, it is clearly seen that the nucleation of the precipitates is preferentially at dislocations. HRTEM reveals that the interface follows an nonrational plane and does not show any micro-facetting along low index planes (Fig. 5). It is also found that there is a small relative lattice rotation of about 0.5° of pigeonite with respect to augite.

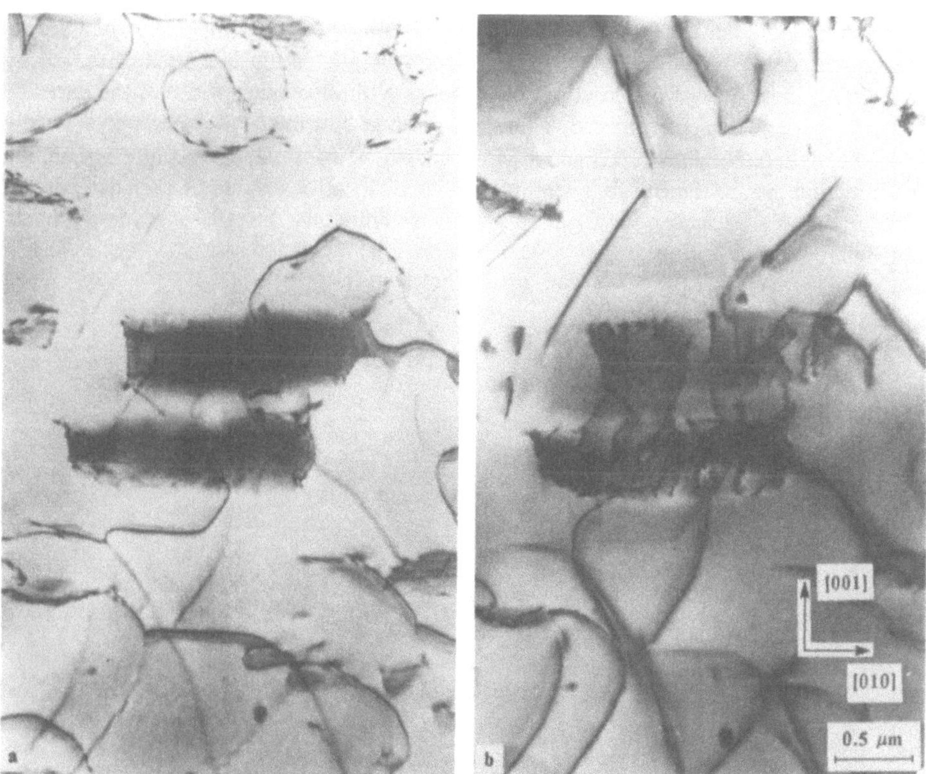

Figure 4. Dislocation microstructure of the augite matrix and the pigeonite-augite interphase boundary. BF images taken with $\boldsymbol{g} = 002$ (a) and $\boldsymbol{g} = 060$ (b) reflection.

3.2. CLINOAMPHIBOLE-AUGITE INTERFACE

Thin coherent hornblende lamellae lie parallel to the (010) plane of augite (Fig. 6). The formation of these lamellae, which exhibit stacking fault-like contrast, is via partial dislocations (Fig. 7a). The Burgers vector of the partial dislocations has been analyzed to be 1/2 [101]. The partial dislocations are most probably the result of a dissociation of perfect

[001] dislocations according to the reaction [001] → 1/2 [101] + 1/2 [$\bar{1}$01], where [001] is the main slip vector in pyroxenes (Nicolas and Poirier, 1976). The displacement vector of the stacking fault is also 1/2 [101] or integer multiples of it. The minimum width of the lamellae is 9Å characterized by one dark line in Fig. 6 corresponding to one double chain in a single chain pyroxene matrix. Broadening of the lamellae occurs by the motion of partial dislocations or pairs of it along the interface (point A in Fig. 6b, Fig. 7b). Similar results on the nature of thin clinoamphibole lamellae have been reported by Yamaguchi et al. (1978) and Crawford et al. (1983). In areas with orthopyroxene lamellae, the nucleation of the clinoamphibole lamellae has been found at the semicoherent orthopyroxene-augite interphase boundary (Fig. 8). The same heterogeneous nucleation has been observed by Smith (1977) in augites of different plutonic rocks. With increasing width of the lamellae the interface becomes semicoherent. There exist two sets of misfit dislocations, a narrow spaced set aligned along the [001] and a wider spaced set approximately aligned along the [100] direction. So far the Burgers vectors of these dislocations have not been determined unambiguously. An edge-on view on the interphase boundary reveals ledges in the interface (Fig. 10).

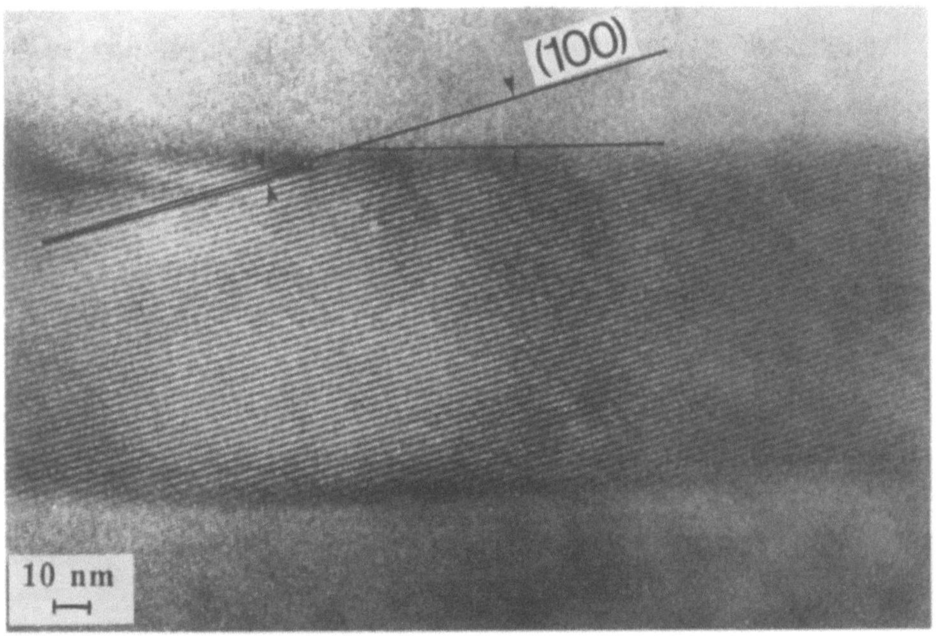

Figure 5. HRTEM image of an edge-on pigeonite lamella. The deviation of the coherent interface from (100) and the small relative lattice rotation are marked.

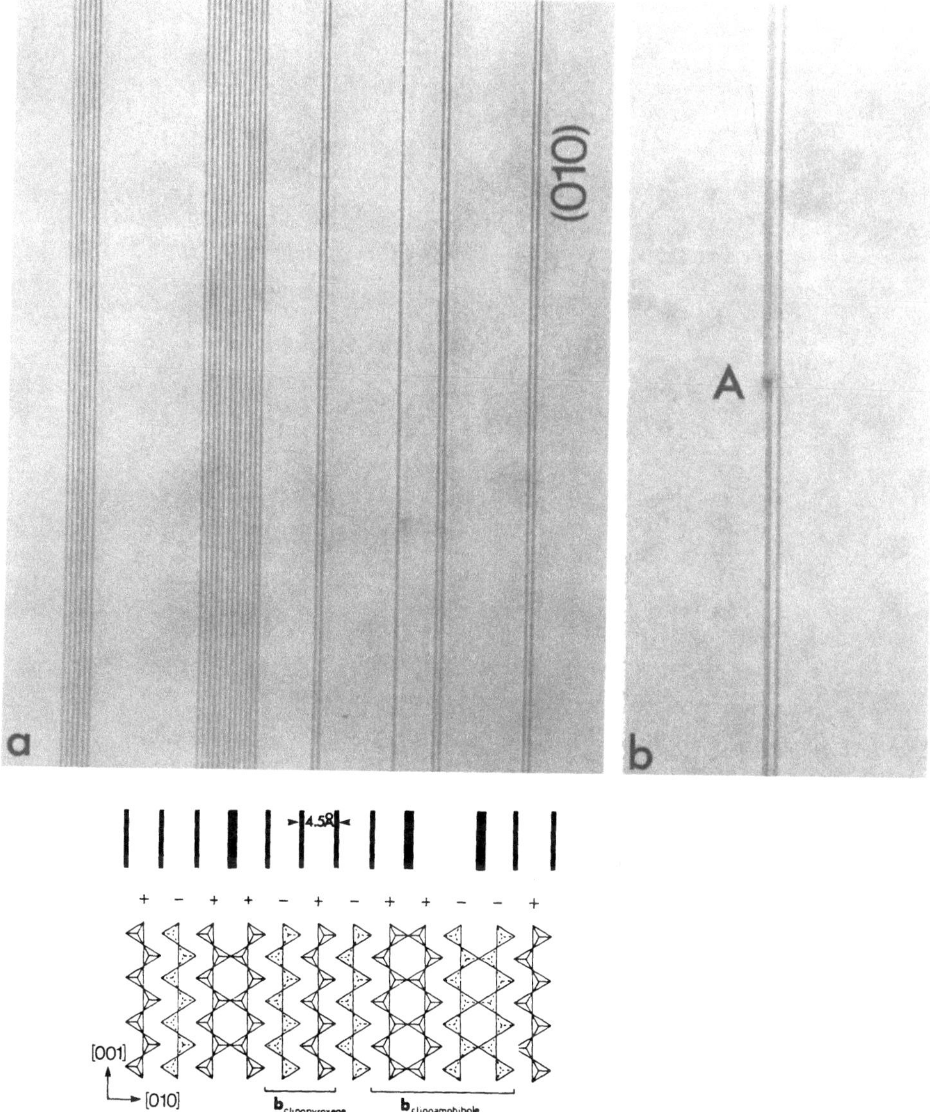

Figure 6. Thin coherent clinoamphibole lamellae viewed edge-on. Broadening of the lamellae (b) is by propagation of partial dislocations along the interface. Schematic explanation of (010) faults in the clinopyroxene structure projected on the (100) plane (c). Tops of SiO_4-tetrahedra pointing along + and - **a**-direction are indicated by full and dashed lines, respectively.

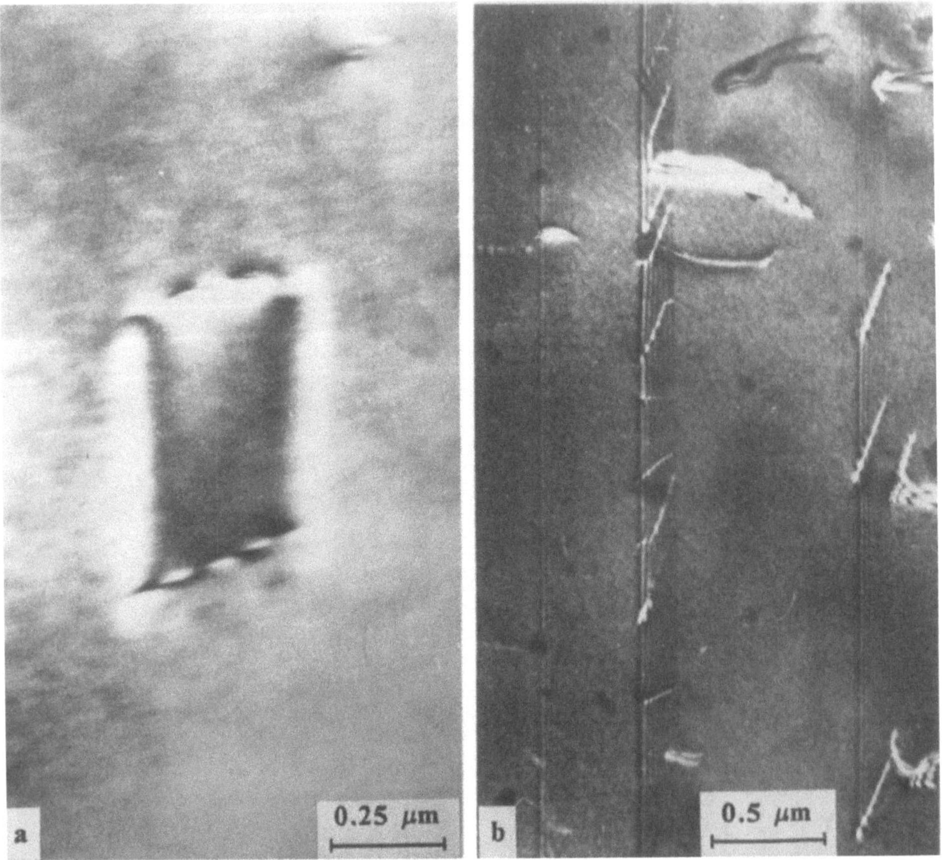

Figure 7. (010) faults viewed flat-on. The partial dislocations bounding and broadening the faults are imaged in (a) and (b), respectively. (a) BF g = 400, (b) DF g = 002.

4. Discussion

4.1. EXSOLUTION MECHANISMS

During slow cooling from a single-phase region heterogeneous nucleation of precipitates at interfaces and dislocations is preferred over homogeneous nucleation because of the reduction of strain energy which can be released at these sites. The nucleation of the pigeonite investigated occurs preferentially at dislocations. No precipitation has been found at the orthopyroxene-augite interfaces.

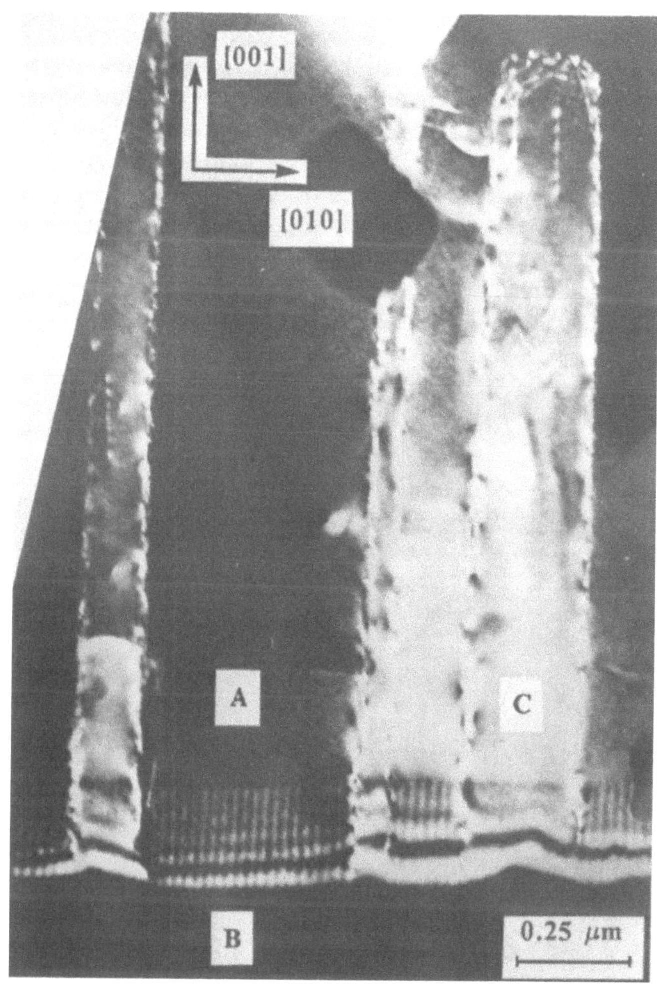

Figure 8. Edge-on clinoamphibole lamellae (C) nucleating at a semicoherent (100) augite-orthopyroxene (A-B) interphase boundary. The clinoamphibole-augite interface is semicoherent. DF image taken with $\boldsymbol{g}$ = 002 reflection.

The nucleation of clinoamphibole is quite different. In this case, unit dislocations that probably have a [001] Burgers vector dissociate into 1/2 [101] partials; the motion of these dislocations produces a shear transformation (Chisholm, 1973). Simultaneous or succeeding cation and anion exchange with the stacking fault completes the phase transformation. This process, which may be aided by deformation, circumvents the classical nucleation barrier and permits exsolution at lower solute supersaturations. A similar mechanism also leads to exsolution of clino- from orthopyroxene (Champness and Lorimer, 1973; Kirby and Etheridge, 1981). It is assumed that the clinoamphibole lamellae nucleating at the orthopyroxene-augite interfaces have been produced by the same process with the partial dislocations being generated at the semicoherent interface. In this case, diffusion along the interphase boundary is faster than in the bulk yielding thicker lamellae. Exsolution of hornblende from augite implies the existence of finite solid solution between pyroxenes and clinoamphiboles. However, from the present observations it cannot be excluded that the latter type of lamella may be also due to alteration, that is, due to the action of a hydrothermal solution diffusing preferentially along interfaces. Both possibilities are discussed in more detail by Smith (1977) and Yamaguchi et al. (1978).

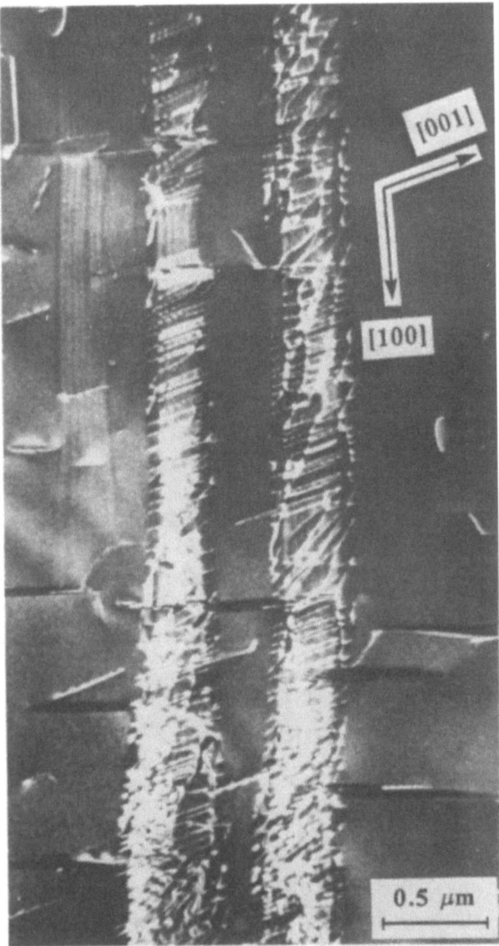

Figure 9.
Flat-on view on a coherent (left) and two semicoherent amphibole lamellae. DF image taken with $\mathbf{g}$ = 002 reflection.

4.2. ORIENTATION AND MICROSTRUCTURE OF THE INTERFACE

It is generally found that the pigeonite lamellae in augite initially form with coherent boundaries, the orientation of which is correlated with the mismatch of the lattice parameters (Jaffe et al., 1975; Robinson et al., 1977). With decreasing Δa and Δc and increasing $\Delta\beta$, $\Delta b \approx 0$ for all augite compositons, the inclination of the lamellae with the [100] and [001] axis goes to zero. To calculate the orientation of the interphase boundary, the above-mentioned authors developed a purely geometrical model which basically is identical to the O-lattice theory of Bollmann (1970). An application of the O-lattice theory to exsolved alkali feldspars has been given by Bollmann and Nissen (1967). According to this theory the optimal boundary plane is chosen in such a way that it passes through as many O-points as possible. To also provide structural continuity across the interface the boundary plane has to be symmetrical with respect to both lattices yielding an additional small relative lattice rotation.

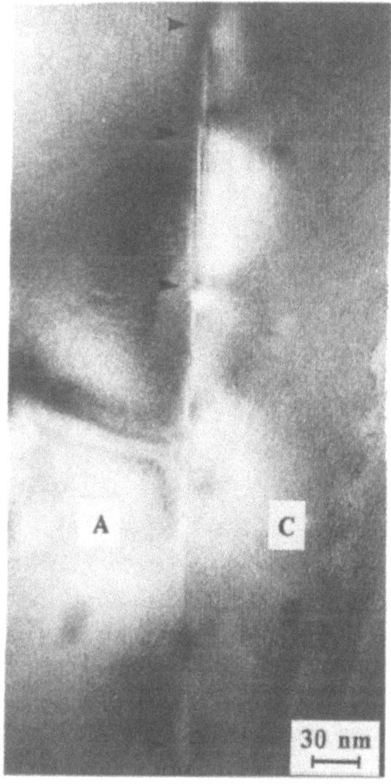

Figure 10.
HRTEM lattice fringe image of an edge-on semicoherent (010) augite-clinoamphibole (A-C) interphase boundary. Ledges in the interface are marked by arrows.

Taking the lattice parameters of Turnock et al. (1973) measured for synthetic clinopyroxenes (augite: a = 9.77Å, b = 8.95Å, c = 5.25Å, β = 105.9°; pigeonite: a = 9.65Å, b = 8.91Å, c = 5.20Å, β = 108.6°) a calculation of the deviation of the interphase boundary of "100" and "001" lamellae from [100] according to Robinson et al. (1977)

gives 14° and 57°, respectively. The relative lattice rotation is 0.4°. There is good agreement with the observed values which are about 15°, 60° and 0.5°, respectively. However, this agreement should not be overstressed for several reasons. (i) The calculations are very sensitive to the accuracy of the lattice parameters. To check the model applied the accuracy has to be better than 0.1%. The lattice parameters used have been measured with this accuracy, but not for the same material. As has been emphasized by Turnock et al. (1973) the data of synthetic Ca-Mg-Fe clinopyroxenes are not directly applicable to natural pyroxenes, since other elements in the latter may affect the unit cell parameters. However, preliminary X-ray powder diffraction measurements of the lattice parameters in the augite matrix show good agreement between the synthetic and natural material. Because of the small volume fraction of the pigeonite this could not be checked for the latter phase. (ii) The precipitates do not have ideal lamellar shape especially in the early stage of exsolution. As a result there is a larger spread of boundary orientations. (iii) Exsolution of pigeonite probably has taken place at elevated temperatures; therefore the lattice parameters at the exsolution temperature have to be used provided the initial orientation of the interface is frozen-in. It should be noted, that a good agreement between measured and calculated pigeonite interface orientations in metamorphic augites have also been reported by Jaffe et al. (1975) using room temperature lattice parameters. However, for faster cooled materials this does not hold any longer (Robinson et al., 1977).

The precipitation of clinoamphibole in clinopyroxene is via stacking faults formed by the motion of partial dislocations with [001] line direction and 1/2 [101] Burgers vector. These parameters determine the (010) slip plane and hence the orientation of the clinoamphibole lamellae.

ACKNOWLEDGEMENTS. Thanks are due to the Deutsche Forschungsgemeinschaft for support.

References

Bollmann. W.,1970. *Crystal defects and crystalline interfaces.* Springer.

Bollmann. W. and Nissen, H.-U., 1967. 'A study of optimal phase boundaries: The case of exsolved alkali feldspars'. *Acta Cryst.* **A 24**: 546-557.

Buseck, P.R., Nord, Jr., G.L. and Veblen, D.V., 1980. 'Subsolidus phenomena in pyroxenes'. In: Prewitt, C.T. (ed.), *Pyroxenes, Reviews in Mineralogy*, **17**, Mineralogical Society of America, 117 - 211

Champness, P.E. and Lorimer, G.W., 1973. 'Precipitation (exsolution) in an orthopyroxene'. *J. Mater. Sci.* **8**: 467 - 474.

Chisholm, J.E., 1973. 'Planar defects in fibrous amphiboles'. *J. Mater. Sci.* **8**: 475 - 483.

Crawford, E.S., Folkes, J.A., Williams, J.O., Barnicoat, A.C. and O'Hara, M.J., 1983. 'Electron microscope studies of minerals: phase boundaries in an extremely slowly cooled clinopyroxene (augite)'. *Proc. R. Soc. Lond.* **A 387**: 21 - 30.

Jaffe, H.W., Robinson, P., Tracy, R.J. and Ross, M., 1975. 'Orientation of pigeonite exsolution lamellae in metamorphic augite: Correlation with composition and calculated optimal phase boundaries'. *Am. Mineral.* **60**: 9 - 28.

Kirby, S.H. and Etheridge, M.A., 1981. 'Exsolution of Ca-clinopyroxene from orthopyroxene aided by deformation'. *Phys. Chem. Minerals* **7**: 105 - 109.

Kitamura, M., Yasuda, M. and Morimoto, N., 1981. 'Morphology change of exsolution lamellae of pigeonite in Bushveld augite: An electron microscopic observation'. *Proc. Japan Acad.* **57**, Ser. B: 183 - 187.

Müller, W.F. and Müller, G., 1988. 'Widths of coherent exsolution lamellae as a function of unit cell parameters'. Submitted to *Phys. Chem. Minerals.*

Nakajima, Y. and Hafner, S.S., 1980. 'Exsolution in augite from the Skaergaard intrusion'. *Contrib. Mineral. Petrol.* **72**: 101 - 110.

Nicolas, A. and Poirier, J.P., 1976. *Crystalline plasticity and solid state flow in metamorphic rocks.* John Wiley & Sons.

Rietmeijer, F.J.M. and Champness, P.E., 1982. 'Exsolution structures in calcic pyroxenes from Bjerkreim-Sokndal lopolith, SW Norway'. *Min. Mag.* **45**: 11 - 24.

Ried, H., 1982. 'Über den Realbau von Pyroxenen und Pyroxenoiden: Transmissionselektronische Untersuchungen insbesondere von Kettenperiodizitätsfehlern und Entmischungen'. Ph. D. Thesis, Univ. Frankfurt am Main.

Robinson, P., Ross, M., Nord, Jr., G.L., Smyth, R.J. and Jaffe, H.W., 1977. 'Exsolution lamellae in augite and pigeonite: fossil indicators of lattice parameters at high temperature and pressure'. *Am. Mineral.* **62**: 857 - 873.

Smith, P.P.K., 1977. 'An electron microscopic study of amphibole lamellae in augite'. *Contrib. Mineral. Petrol.* **59**: 317 - 322.

Turnock, A.C., Lindsley, D.H. and Grover, J.E., 1973. 'Synthesis and unit cell parameters of Ca-Mg-Fe pyroxenes' *Am. Mineral.* **58**: 50 - 59.

Yamaguchi, Y., Akai, J. and Tomita, K., 1978. 'Clinoamphibole lamellae in diopside of garnet lherzolite from Alpe Arami, Bellizona, Switzerland'. *Contrib. Mineral. Petrol.* **66**: 263 - 270.

A VIBRATIONAL STUDY OF TETRACYANOETHYLENE ADSORBED ON MAGNESIA.

J.J. HOAGLAND and K.W. HIPPS
Department of Chemistry
Washington State University
Pullman, WA 99164-4630
USA

ABSTRACT. The saturation adsorption of tetracyanoethylene (TCNE) from the vapor phase onto magnesia activated at 200 and 400 C is studied. Fourier transform infrared and electron spin resonance spectroscopy are used to characterize the chemisorbed material. FT-IR spectra of authentic samples of TCNE, $TCNE^{-1}$, $TCNE^{-2}$, and the tricyanovinylalcoholate anion (TVA) are also reported for comparison purposes. We find that the uptake of TCNE as a function of activation temperature is most closely related to the surface hydroxyl coverage. It does not correlate well with the BET surface area. We also find that the primary chemisorption product is a TVA like species (perhaps polymeric). Radical signals are detected, presumably from $TCNE^{-1}$, but correspond to less than 5% of the chemisorbed TCNE.

1. Introduction

There is a continuing interest in the redox chemistry of TCNE with various surfaces[1-25]. The early work was directed toward the detection and identification of radicals formed by chemisorption on oxides[1-12]. Only recently has surface vibrational spectroscopy been applied extensively to this area of research[13-25]. In the case of TCNE adsorption on alumina, it was shown that the radical anion of TCNE, $TCNE^{-1}$, was not the major chemisorption product[13-17]. The non-radical dianion, $TCNE^{-2}$, and the tricyanovinylalcoholate anion, TVA^{-1}, are also formed in significant quantities.

FT-IR, Raman, and EELS studies of TCNE adsorbed on single crystal metals and on electrode surfaces have yielded results that are surface dependent[18-25]. On carbon electrodes both the dianion and TVA^{-1} are observed[20]. In contrast, the radical anion is the principal species reported on platinum electrodes[21-23]. Both the dianion and the radical anion are reported to be formed on Cu(111)[18,19], but only the radical is reported on Ni[25].

The surface dependence of the mechanism and extent of chemisorption of TCNE on various surfaces is interesting but not well understood. The present paper is the first of a series in which we probe the redox adsorption of TCNE.

L.-C. Dufour et al. (eds.), Surfaces and Interfaces of Ceramic Materials, 155–163.

Here we will focus on the adsorption of TCNE from the vapor on magnesia activated at 200 and 400 °C. Below ca. 300 °C the sample is formally magnesium hydroxide. Above ca. 360 °C it is properly thought of as MgO with bound hydroxyl groups. During the decomposition of $Mg(OH)_2$ the surface area of the sample rises markedly. Thus, adsorption at these two temperatures should provide insight into the role of hydroxylation and surface area on the chemisorption of TCNE.

2. Experimental

2.1 MATERIALS: TCNE was purchased from Aldrich Chemical Company and purified by sublimation through Alpha Products 28 mesh activated charcoal at 100 °C. The $Mg(OH)_2$ was used as purchased from Alpha Products. Authentic samples of KTCNE, CsTVA, and Na_2TCNE were prepared by previously described methods[13-16].

2.2 SAMPLE PREPARATION AND HANDLING: Oxide activation and TCNE chemisorption were carried out under vacuum on a standard grease-free double (UHP N_2 / LN_2-trapped vacuum) manifold in a demountable 2-chamber cell. Each chamber of the glass cell was isolated by a greaseless Teflon valve. First the TCNE (ca. 0.lg) was loaded and pump-purged and then sealed under vacuum. The amount of TCNE always exceeded the maximum possible uptake by the magnesia sample. Then, the $Mg(OH)_2$ (ca. 1.0g) was loaded in the second chamber and the adsorption cell was again attached to the manifold. Typically, 4 activations were carried out simultaneously by heating 4 samples to the desired temperatures for >12 hrs under vacuum ($<10^{-5}$ torr). A TCNE free activated sample (blank or reference magnesia) was always prepared with the TCNE adsorbed sample. Following activation, the vacuum valve was closed and the oxide was cooled to 100 °C and the TCNE chamber was heated to the same temperature. The valve separating the chambers was then opened. Both chambers and valves were maintained at 100 °C for 4 hours. At the end of this period, the TCNE sidearm was cooled to LN_2 to remove the residual TCNE vapor from the chamber containing magnesia. After condensation of the TCNE the TCNE chamber was closed and the entire cell was brought to room temperature. At this point 1 atmosphere of nitrogen was introduced into both chambers. Each chamber was closed and the cell was transported to a Vacuum Atmospheres inert atmosphere glove-box. Post adsorption handling occurred within the glove-box.

The amount of residual TCNE was collected and weighed to determine (by difference) the total TCNE chemisorption. Because condensation of TCNE occurred over most of the surface of the chamber and on the valve stem, the possible error in this number was large, of the order of 50%.

The magnesia was removed from the second chamber; then, BET and spectroscopic samples were prepared. ESR samples were sealed in 4 mm tubes and used without further handling. In the case of IR spectral samples, both diffuse reflectance and transmission samples were made. Diffuse reflectance IR spectra of TCNE/magnesia samples and of the magnesia references were run without dilution in a lmm micro-cup. Transmission IR spectra were obtained by dilution of sample and reference each in a fixed amount of KBr and forming a pressed pellet.

2.3 TGA AND SURFACE AREA: Thermal gravimetric analysis was performed in flowing nitrogen on a 1.060 mg sample of $Mg(OH)_2$. A commercial Perkin-Elmer Delta series TGA7 was used. Scan rates of 1 and 5 °C/min gave similar

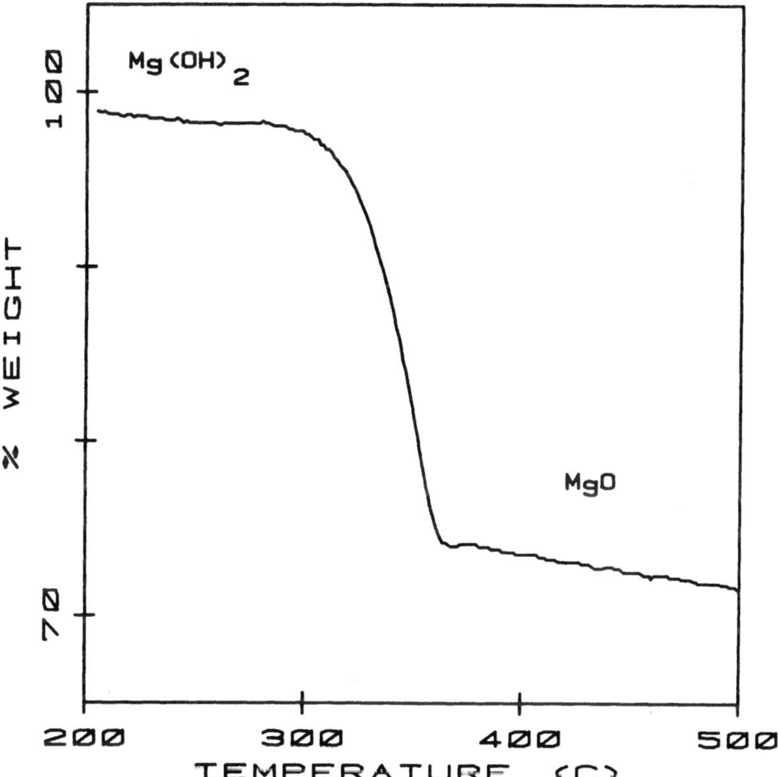

Figure 1. Percentage weight loss with heating of an $Mg(OH)_2$ sample heated at 5 °C/minute.

results. The BET (N_2) surface areas of the 200 and 400 °C activated samples were measured on a Micromeritics Flowsorb 2300 surface area analyzer.

2.4 ESR AND FT-IR SPECTRA: ESR spectra were obtained at 9.5 GHz on a Varian E-3 spectrometer. All spectra were measured at room temperature (283 ± 3 K) and intensities were determined by comparison to DPPH reference samples. FT-IR spectra were measured on an IBM instruments IR-98 spectrometer. This is a vacuum bench spectrometer. A liquid nitrogen cooled MCT detector was used and all data were collected at 4cm^{-1} resolution. Typically, 512 scans were added to yield a single spectrum. Authentic samples of TCNE, KTCNE, Na_2TCNE and CsTVA were studied by transmission and diffuse reflection in the IR region. All samples were prepared in the glove box under inert conditions.

3. Results and Discussion

Figure 1 shows the thermal gravimetric analysis (TGA) data obtained from our $Mg(OH)_2$ sample. Our observed transition temperature is higher than that reported by Che et.al.[7], of the order or less than the (broad) decomposition temperature reported by El-Salaam[26], and by Anderson *et. al.*[27]. Chen and Fong[28] have reported differential scanning calorimetry results which place the transition temperature near 400 °C. Apparently the physical and chemical history of the sample greatly effects the observed

TABLE 1 Magnesia and adsorption parameters as a function of activation temperature. [a]

Activation Temperature °C	Surface Area m^2/g	TCNE [b] molec/Å^2	mole % TCNE-	H_2O [c] molec/Å^2	Color
200±4	20±2	0.18±0.07	1.7±0.7	0.11	light brown
400±5	294±20	0.01±0.005	3.0±1.0	0.05	dark brown

(a) Adsorption of TCNE from the vapor phase at 100 °C on $Mg(OH)_2$ and MgO.
(b) Amount of TCNE taken up from the vapor phase. It is not this composition on the surface
(c) Water released from surface by activation. Actually present as hydroxyl

transition temperature. In the case of our sample, 200 °C corresponds to $Mg(OH)_2$ while 400 °C activation results in MgO with some retained hydroxyls. The number of water molecules per Angström^2 (molec. /Å^2) of $Mg(OH)_2$ at 200 °C can be estimated from the lattice spacing in magnesium hydroxide. Using O.21nm as the Mg-O spacing[27] a value of 0.11 water molecules/Å^2 of surface is obtained. At 400 °C, the number of water molecules per Å^2 of MgO can be estimated from Figure 1. The values obtained in this manner are in good agreement with those of Anderson[27], and they are presented in Table I. Also shown in Table I are the BET surface areas. The dramatic increase in surface area with activation observed here is similar to that reported in the decomposition of magnesium carbonate[29].

Figure 2 contrasts the diffuse reflectance IR spectra obtained from TCNE, KTCNE, Na_2TCNE, and CsTVA plotted to correct relative absorbance. While

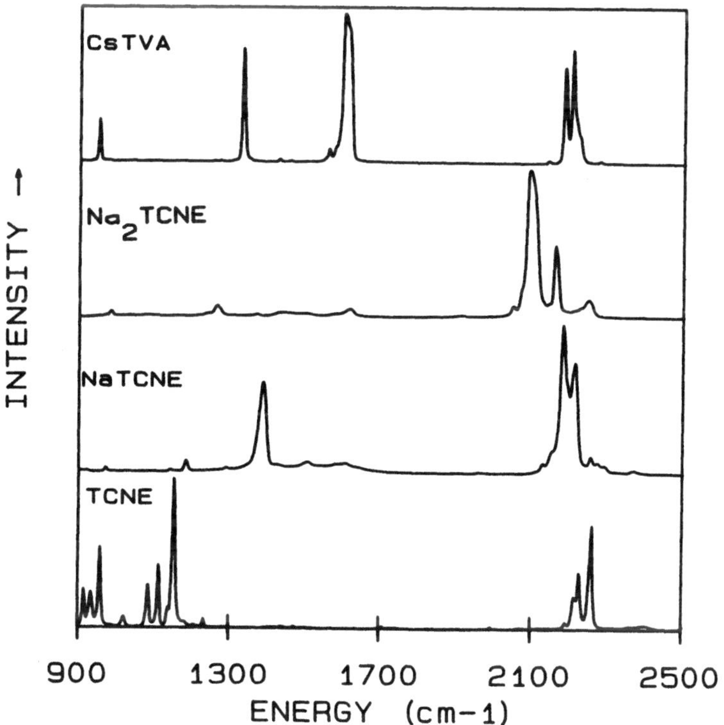

Figure 2. Diffuse reflectance spectra of the indicated compounds

much of this data is available piece-meall[3-16,30-36], the spectra have never been collected in this manner. Figure 2 is important because the IR spectrum of each ion provides a "fingerprint" for identifying it in the presence of the others. Furthermore, these easily distinguished patterns fall within the transmission window of most oxides and the high reflectivity region of most metals. IR spectroscopy, therefore, is a nearly ideal method for identifying the TCNE/surface chemisorption products. TCNE itself also has a characteristic frequency distribution in this spectral region, but its relatively weak CN band would be difficult to observe if TCNE were a minor component in the product distribution. It is also noted that depending on the surface cation, it might be difficult to identify small amounts of TCNE- in the presence of TVA- because of the cation induced variation in the bands near 1350 cm^{-1}. EPR spectroscopy is the technique of choice for identifying the radical anion.

The double integrals of the sample EPR spectra with the EPR spectra of DPPH standards indicates that about 1.7 and 3.1 mole % of the radical is formed on TCNE adsorption on 200 and 400 °C activated magnesia,

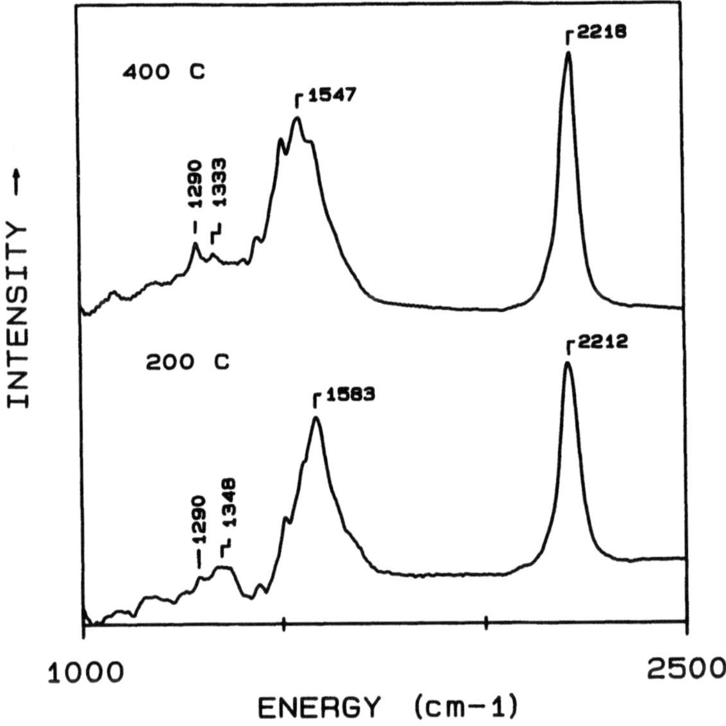

Figure 3. Difference FT-IR spectra (TCNE chemisorbed on magnesia minus magnesia) for 200 and 400 °C activated oxides. Adsorption occurred from the vapor at 100 °C.

respectively. The area of oxide increases by a factor of about 15 with 400 °C activation. Thus, the number of radicals per unit surface area is significantly less for the 400 °C activated magnesia than for the 200 °C activated material. Surface spin densities are reported in Table I.

The amount of TCNE uptake (total weight of TCNE removed from the vapor phase per $Å^2$ of oxide surface) by the activated magnesia is also reported in Table I. The TCNE uptake for MgO is considerably smaller than for $Mg(OH)_2$. This seems rather curious unless one assumes that the TCNE uptake is related to the available hydroxyl. The surface hydroxyl coverage drops by more than 2 between 200 and 400 °C. Further, the 200 °C activated material is essentially $Mg(OH)_2$; thus, hydroxyl groups from the near surface region might also play a role in the chemisorption process. Overall, if a TCNE/OH^- reaction/interaction dominates the chemisorption for 200 and 400 °C oxides, we would expect the 200 °C material to be much more effective in acquiring TCNE.

The absorbance spectra obtained from TCNE adsorbed on activated magnesia were compared to the spectra acquired from the corresponding reference magnesia. The significant points were : 1) A significant loss of intensity near 1650 and 3700 cm^{-1} on adsorption indicating a strong TCNE/OH^- interaction, 2) New structure near 2200 and 1500 cm^{-1} in the sample containing TCNE. The TCNE induced structure can be better observed by taking the difference of the sample and reference spectra, as shown in Figure 3. Comparing the spectra in Figure 3 with those in Figure 2, we see that all of the features in the chemisorption spectra can be accounted for by broadening the features of the TVA- spectrum. Thus, the dominant chemisorption product observed under these adsorption conditions is an alcoholate like species.

By "alcoholate-like" we leave open the possibility of the presence of dimers, trimers, and polymers of TCNE. Storbeck and Starke[37] synthesized very long chain cyanovinylalcohol (TVAH) polymers haveing much broader IR bands placed in positions similar to those of TVA-. Unfortunately, they identified this poly-cyanovinylalcohol as a "TCNE polymer". Several years ago, Flockhart[9] observed the IR spectrum of material washed from TCNE treated alumina. He identified this material as a "TCNE polymer" because of similarities between his spectrum and that of Storbeck's polymer. Undoubtedly, Flockart was observing a cyanovinylalcoholate species. Almost certainly it was not polymerized to the extent of Storbeck's polymer since Storbeck's is essentially insoluble and Flockhart was able to wash product from the alumina surface. Our own spectral studies of TCNE adsorption on alumina also indicate that a TVA- "like" species is always a major product on alumina activated below 500 °C[13-15].

4. Conclusions

Under the conditions of vapor phase adsorption at 100 °C, TCNE adsorbes on $Mg(OH)_2$ or highly hydroxylated MgO to form a (possibly polymeric) cyanovinylalcoholate species as the principal chemisorption product. A radical anion is also formed, but (under these conditions) it never amounts to more than about 5% of the adsorbed TCNE. Physisorbed TCNE is not observed.

We find that the amount of TCNE chemisorbed on $Mg(OH)_2$ or highly hydroxylated MgO depends on both the surface area and the available hydroxyl groups. The TCNE molecule, laying flat would form a monolayer at a surface density of 0.03 molecule/$Å^2$. The observed TCNE uptake on 400 °C activated MgO 0.3 monolayer, a reasonable value. On the other hand, the computed uptake on magnesium hydroxide corresponds to 6 monolayers! Since no TCNE is observed on $Mg(OH)_2$, we conclude that both surface and sub-surface hydroxyl are combining to react with TCNE.

Acknowledgement

We gratefully acknowledge the National Science Foundation and the Division of Materials Research for their support in the form of grant DMR-8414566.

References

1) Che, M.; Dyrek, K.; Louis, C. *J. Phys. Chem.* (1985) **89** 4531.
2) Flockhart, B.D.; Liew, K.Y.; Pink, R.C. *J.C.S. Faraday I* (1980) **76** 2026.
3) Marczewki, M.; Malinowski, S. *Bull. Acad. Polonaise Sci.* (1976) **XXV** 187.
4) Davidson, R.S.; Slater, R. *J. Chem. Soc. Faraday 1*(1976) **72** 2416.
5) Kijenski, J.; Malinowski, S.; Zielinski, B. *React. Kinet. Catal. Lett.* (1976) **4** 251.
6) Fiedorow, R.; Wieckowski, A. *React. Kinet. Catal. Lett.* (1975) **2** 355.
7) Che, M.; Naccache, C.; Imelik, B. *J. Catal.* (1972) **24** 328.
8) Wong, P.K.; Allen, A.O. *J. Phys. Chem.* (1970) **74** 774.
9) Flockhart, B.D.; Leith, I.R.; Pink, R.C. *Trans. Faraday Soc.* (1968) **65** 542.
10) Terlecki, A.; Nicolau, C.; Cristea, A.; Angelescu, E.; Nicolescu, I. *Rev. Roum.Chim.* (1968)**13** 997.
11) Naccache, C.; Kodratoff, Y.; Pink, R.; Imelik, B. *J. Chim. Phys.* (1966) **63** 341.
12) Flockhart, B.D.; Naccache, C.; Scott, J.A.; Pink, R.C. *Chem. Comm.* (1965) 238.
13) Hipps, K.W.; Mazur, U. *J. Phys. Chem.* (1982) **86** 5105.
14) Hipps, K.W. *J. Elect. Spect. Rel. Phenom.* (1983) **30** 175.
15) Mazur, U.; Hipps, K.W. *J. Phys. Chem.* (1984) **88** (1984).
16) Hipps, K.W.; Mazur, U. *Rev. Sci. Instrum.* (1984) **55** 1120.
17) Mazur, U.; Hoagland, J.J.; Hipps, K.W. *work in progress.*
18) Erley, W. *J. Phys. Chem.* (1987) **91** 6092.
19) Erley, W. Ibach, H. *J. Phys. Chem.* (1987) **91** 2947.
20) Datta, M.; Jansson, R.E.; Freeman, *J.J. Spectrosc. Lett.* (1986) **19** 129.
21) Daschbach, J.; Heisler, D.; Pons, S. *Appl. Spect.* (1986) **40** 489.
22) Pons, S.; Khoo, S.; Bewick, A.; Datta, M.; Smith, J.; Hinman, A.; Zachmann, G. *J. Phys. Chem.* (1984) **88** 3575.
23) Pons, S. *J. Electroanal. Chem.* (1983) **150** 495.

24) Chou, Y.C.; Liang, N.T. *Chem. Phys. Lett.* (1984) **106** 472.
25) Pan, F.; Hemminger, J.; Ushioda, S. *J. Phys. Chem.* (1985) **89** 862.
26) El-Salaam, K.M.; Babr, R.M.; Said, A.A. *Surface Technology* (1979) **9** 427.
27) Anderson, P.J.; Horlock, R.F.; Oliver, J.F. *Trans. Faraday Soc.* (1965) **61** 2754.
28) Chen, D.T.; Fong, P.H. *J. Thermal Analysis* (1977) **12** 5.
29) "Physical and Chemical Aspects of Adsorption and Catalysis", Ed. Fortuin,J.M.; Okkerse, C.; Steggerda, J.J. (1970) Academic Press. London
30) Hipps, K.W.; Poshusta, R.D. *J. Phys. Chem.* (1982) **86** 4112.
31) Jeanmaire, D.L.; Suchanski, M.R.; VanDuyne, R.P. *J. Amer. Chem. Soc.* (1975) **97** 1699
32) Michaelian, K.H.; Rieckhoff, K.E.; Voigt, E.M. *J. Molec. Spec.* (1982) **95** 1.
33) Chaplot, S.I.; Mierzejewski, A.; Pawley G.S. *Molec. Phys.*(1985) **56** 115.
34) Hinkel, J.J.; Devlin, J.P. *J. Chem. Phys.* (1973) **58** 4750.
35) Moore, J.C.; Smith, D.; Youhne, Y.; Devlin, P.J. *J. Phys. Chem.* (1971) **75** 325.
36) Khatkale, M.S.; Devlin, J.P. *J. Phys. Chem.* (1979) **83** 1636.
37) Storbeck, I; Stake, M. *Ber. Bunsen. Physik Chem.* (1965) **69** 343.

INVESTIGATION OF SURFACE HYDROXYLS ON CORDIERITE AEROGEL BY FT-IR SPECTROSCOPY

M.I. BARATON, T. MERLE-MEJEAN, P. QUINTARD*
Faculté des Sciences, Laboratoire de Spectrométrie Vibrationnelle
(U.A. 320 du CNRS), F-87060 LIMOGES (France)
V. LORENZELLI•
• Istituto di Chimica, Facolta di Ingegneria
Ple Kennedy, I-16129 GENOVA (Italy)

ABSTRACT - *A high surface area cordierite aerogel was studied by Fourier Transform Infrared Spectroscopy. After activation , treatments with methanol or isotopic exchange with deuterium have evidenced that the silanol groups are all located on the surface. Methanol reaction with hydroxyl, gives typical methoxyl groups covalently grafted on the surface and very heat resistant.*

1. INTRODUCTION

Cordierite ceramics could replace alumina as an integrate circuit substrate, because of its low dielectric constant and its low thermal expansion coefficient (1, 2). In addition, its high refractoriness (3) allows to use it practically as heat exchanger for gas turbine engines, industrial furnaces and also as a support of catalysts, e.g. for automobile exhaust gas control (4), in relation with its resistance to thermal shock. Regarding the catalytic use, a computerized literature survey over the last fifteen years pointed out that 93 % of the analyzed documents were patents. This clearly indicates the promising industrial interest of this material but also the lack of fundamental studies in this field, whereas a thorough surface study, on the first stage of monolithic cordierite gel, may provide information on the main surface groups which control catalytic properties and sintering ability. We managed an investigation by Fourier-Transform Infrared (FT-IR) spectroscopy, to characterize the surface of a sol-gel prepared cordierite.

2. EXPERIMENTAL

Several authors (1,5,6) have reported unconventional routes to synthesize powders in the cordierite system : SiO2-Al2O3-MgO, through a sol-gel process. Our sol-gel processing method (7) starts with aluminium sec-butoxide and silicon tetra-ethoxide. Hydrolysis of the precursors gives a reaction with silanol groups and polycondensation takes place. Magnesium nitrate is added

* *to whom correspondence should be addressed*

L.-C. Dufour et al. (eds.), Surfaces and Interfaces of Ceramic Materials, 165–171.

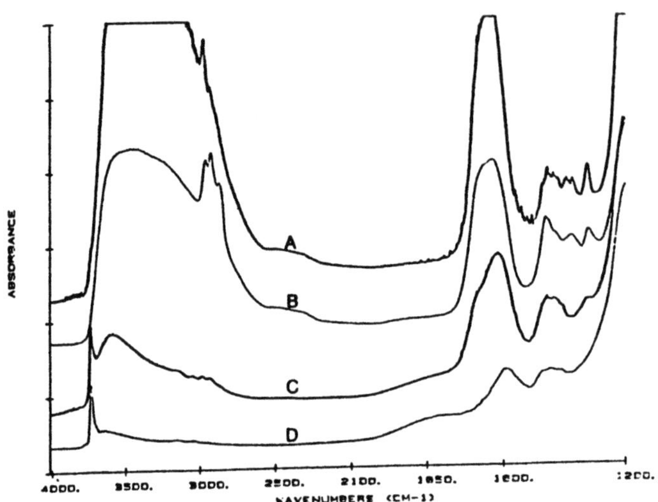

Fig. 1 - Absorption spectra, in the surface-sensitive zone of a grid-supported pellet of cordierite aerogel after in-situ treatments under dynamic vacuum **(A)** *as received;* **(B)** *470K, 30 mn ;* **(C)** *570K, 30 mn;* **(D)** *670K, 30 mn. Complete activation needs 670K, 2 hours (see Fig. 3A)*

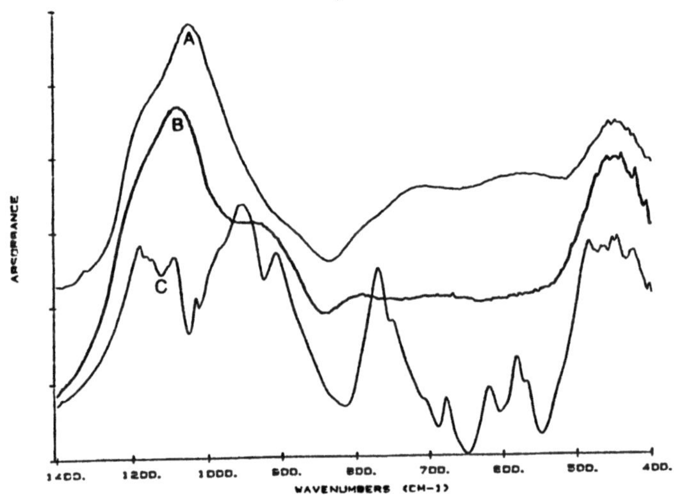

Fig. 2 - Absorption spectra in the bulk absorption zone, of cordierite aerogel **(A)** *as received, and after treatments at* **(B)** *1230 K, 1 hour and* **(C)** *1520K, 10 hours.*

to the solution and then, this homogeneous solution is hypercritically dried. The monolithic aerogel has a high microporosity, an apparent density of 340 kg/m^3 and so, a high specific surface area (400 m^2/g).

For bulk characterization the sample is pressed with KBr and for surface studies as a fine grid supported pellet without any KBr. The material is then investigated by FT-IR transmission/absorption spectroscopy (4600-400 cm^{-1}) with a Nicolet 5DX. It is put inside a greaseless cell where it undergoes thermal treatments under vacuum (1.10^{-6} mbar) or specific atmospheres. First of all the surface is activated (cleared out of physisorbed species) by heating up to 720K under vacuum, and then cooled down to room temperature.

3. RESULTS AND DISCUSSION

3.1. Evidencing the hydroxyls

The infrared spectra of " as received" cordierite recorded under vacuum (Fig. 1A and 2A) and after different treatments (Fig. 1B, C, D, and Fig. 2B, C) exhibit two main zones of interest. The first one below 1400 cm^{-1} is mainly due to absorptions of the bulk vibrations with two broad bands near 1050 and 450 cm^{-1} respectively associated with stretching and bending modes of SiO bonds in crystalline or amorphous silicates, while AlO and MgO are observed at 950 cm^{-1} and in the range 550-700 cm^{-1} respectively. Their behaviour when heating up to 1520K (Fig. 2B, 2C) is related to the transformation from amorphous form of cordierite to a highly ordered structure in the orthorombic form (Beta cordierite) through the mu and alpha forms; this has been discussed elsewhere (8,9) in connexion with X-Rays, ATD and ATG results.

When pure cordierite powder is pressed into grid supported pellet, the region where the bulk absorptions fall becomes obscure, because more amount of material is needed. On the other hand, the range above 1400 cm^{-1} is still accessible and very sensitive to surface treatments (Fig 2). We can observe a broad band centered at 3420 cm^{-1} due to stretching modes of hydrogen bonded OHs. When heating at higher temperatures this absorption is progressively replaced by a sharp band at 3745 cm^{-1} , close to the value observed in silica (10,11,12,13,14). So it is assigned to the stretching vibration of free surface silanols. It can be noticed that the frequency is almost the same as those we measured on oxidized silicon nitride and carbide (15,16).

However, cordierite is a "mixture" of Si, Al, Mg cations from which we can expect other OH group stretching frequencies e.g. at 3785, 3760, 3745 and 3700 cm^{-1} for Al-OH on different sites of gamma alumina or 3752 and 3610 cm^{-1} for Mg-OH on magnesia (11,12,13). This lack of the expected bands has already been noticed with other oxides, for instance with silica-alumina (12) and $MgAl_2O_4$ (17). Could it be related to differences in chemical surface properties of a given oxide which are substantially changed by the presence of added cations ? For example catalytic properties of alumina are modified by Mg^{2+} (18), and the presence of inorganic salts on the surface influences the sintering (19).

3.2. Deuteration

When a few mbar dose is added at 670K to the activated sample two new absorption bands appear concomitantly with a decrease in intensity of the

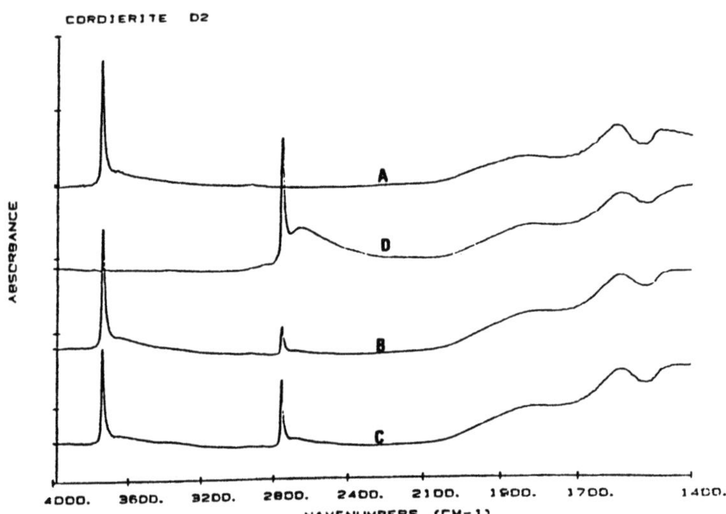

Fig. 3 - Evolution of the cordierite spectrum after activation **(A)** *and after addition at 670K during 10 mn of 4 mbar D2* **(B)**; *6 mbar D2, 10 mn* **(C)**; *350 mbar D2, 10 mn* **(D)**

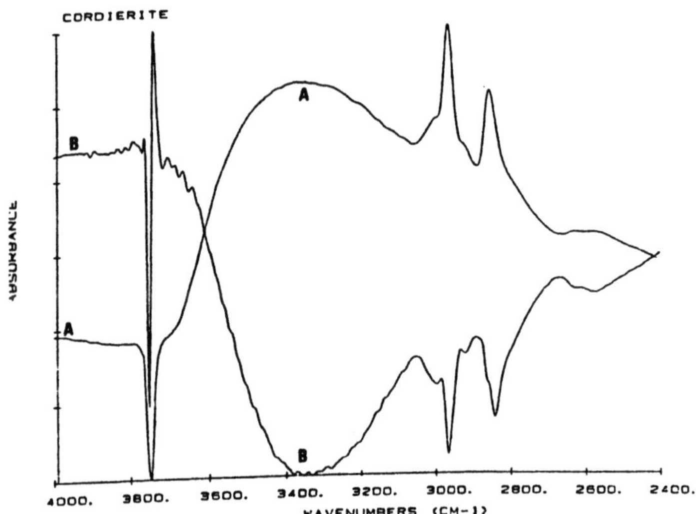

Fig. 4 - Difference spectra between activated cordierite, cooled down to r.t. and **(A)** *addition of a few mbar methanol at r.t.;* **(B)** *addition of methanol at r.t. followed by heat treatment at 370K. The shift of the 3745 cm^{-1} band is partly due to temperature effect (14). Bands downwards are due to species disappearing.*

3745 cm^{-1} (vOH) : an intense and sharp one at 2760 cm^{-1} and a broad and weak one at 2660 cm^{-1} (Fig 3). The first new absorption can be assigned to surface vOD stretching mode because the rate of the isotopic exchange is incompatible with bulky hydroxyls and the lowering of the OH frequency by a factor 1.35 which is close to the square root of 2 as expected for a simple harmonic oscillator. The 2660 cm^{-1} band arises from deuteration of remaining H bonded silanols.

Successive additions of deuterium doses (Fig. 3B, C, D) magnify the vOD with a simultaneous decrease of vOH. This process is completely reversible by adding hydrogen. The integrated intensities (Surface of the bands) of some steps of deuteration are given in table I.

Dose/mbar	4	6	10	18	27	101	350	6*	350*
S(OH)	15,2	13,7	7,2	7,1	7,0	5,0	1,28	8,6	0
S(OD)	0,37	6,1	20,8	23,2	23,1	26	40	15,4	49
% OH	99	90	50	47	46	33	8	57	0
% OD	1	12	43	47	47	53	82	31	100

Table I. Integrated intensities of the stretching vibration of OH (S(OH)) and OD (S(OD)), measured after 10 mn addition of successive doses of D2. Excess deuterium was evacuated before introducing a new dose. Values with the (*) mark are measured after 1.5 h contact of the cordierite pellet with D2.

There is a one-to-one correspondance between the bands in the two regions considering the difference in molar absorption coefficients of OH and OD which are nearly in the ratio 3.2.

So, all silanols are located on the surface of the cordierite aerogel and easily accessible to deuteration; a complete exchange can be obtained with a pressure superior to 200 mbar leading to complete vanishing of vOH.

3.3. Reaction with methanol

When methanol is added (Fig. 4A) on the cordierite surface, absorptions become visible in the range of stretching (vCH : 3700- 2800 cm^{-1}) and bending (δCH : 1500-1300 cm^{-1}) vibrations. At room temperature we observed sharp bands at 2966 and 2853 cm^{-1} with shoulders at 2997 and 2921 cm^{-1}, while δCH are at 1465 and 1446 cm^{-1} . The latter are superimposed to some bands of the bulk, which change in intensity and position with temperature (8). They will not be discussed here.

Adsorption of methanol also leads to a decrease in intensity of the absorption band of surface hydroxyl groups at 3745 cm^{-1} while a broad band nearly centered at 3375 cm^{-1} grows. This value is close to that of the liquid methanol (3350 cm^{-1}) and corresponds to molecules in a capillary-like condensed phase (20).

The sharp νCH stretching bands, which are slightly perturbed compared with the free molecule are connected with the formation of hydrogen bonds between surface silanols and the methanol OH group (20) as proposed for the same phenomenon in reference 16.

The remaining free OH (3745 cm^{-1}) could be due to the large size of the methyl groups which provides a high coverage of the surface and so, a steric hindrance to penetration of more methanol.

Under evacuation at 370K (Fig. 4B), most of the free hydroxyl groups are restored. Sharp CH bands and shoulders decrease in intensity, so does the 3375 cm^{-1} band which shifts to 3350cm^{-1} .

After treatment of cordierite at 670K, under vacuum, for one hour the spectrum still shows (Fig. 5) two weak bands at 2964 and 2856 cm^{-1} and possibly a slight shoulder near 3000 cm^{-1} visible on ratioed spectra with high methanol doses. At this temperature the H bond no longer exists. By comparison with silica and silica-like surfaces (11,16,20), we assigned these very heat resistant species to methoxyl groups linked to silicon according to the substitution reaction

$$Si\text{-}O\text{-}H + CH3OH \longrightarrow Si\text{-}O\text{-}CH3 + H2O$$

These bands which are not removed from the surface under high temperature treatment, are respectively assigned to CH stretching vibrations, symmetric (2964) and assymetric (2856).

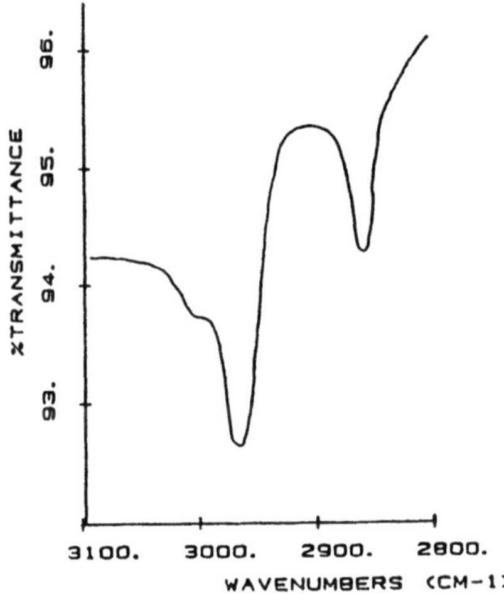

Fig. 5 - The two bands characteristic of heat resistant methoxy groups. This methoxylation is obtained at r.t., after activation. Cordierite is then heated at 670K under dynamic vacuum.

4. CONCLUSION

We evidenced by FT-IR spectroscopy, the presence of OH surface groups on Si sites in cordierite aerogel: i.e. silanols. Such hydroxyls are all located on the surface and all available to isotopic exhange with deuterium. They are able to undergo condensation reactions with methanol giving typical methoxy species. No band characteristic of Al-OH or Mg-OH groups has been observed yet; further experiments are required and call for new probe molecules.

The authors are indebted to Dr H. VESTEGHEM (ENSCI, Limoges, France) for providing them with cordierite aerogel .

References

1-G.GENSSE and U.CHOWDRY - Proceed. Symp. on Electronic Packaging Materials Science (April 1986), M.R.S. pub. 1986, K.A.Jakson ed.

2-Y.HIROSE, H.DOI and O.KAMIGAITO - J. Mater. Sci. Lett., 3, p 95-96 and 3, p 153-155 (1984).

3-F.Y.KHARITONOV and D.A.DAHIR - Study of the thermal conductivity of heat-resistant ceramics - Vopa. Povysh. Nadez. Optim. 81, p138-143(1982).

4-I.M.LACHMAN and R.N.Mc NALLY - High temperature monolithic supports for automobile exhaust - Ceram. Eng. Sci. Proc., 81, 2, p 337-351 (1982).

5-B.ZELINSKI, B.FABES and D.R.UHLMANN - Crystallization behavior of sol-gel derived glasses - J. Non-Crystal. Solids 82, p 307-313 (1986).

6-H.SUZUKI, K.OTA and H.SAITO - Preparation of cordierite ceramics from metal alkoxides - Yogyo-Kyokai-Shi, 95, 2, p 163-169 and 170-175 (1987).

7-H.VESTEGHEM, A.R.DI GIAMPOLO and A.DAUGER - Low temperature cordierite glass from autoclave-prepared gel - J. Mater. Sci.Lett., 6, p 1187-1189 (1987).

8-T.MERLE-MEJEAN, H.VESTEGHEM, P.QUINTARD and V.LORENZELLI -FT-IR investigation of cordierite aerogel-J.Mol.Struct. 175,p401-406 (1988).

9-K.LANGER and W.SCHREYER - Infrared and Powder X-Ray diffraction studies on the polymorphism of cordierite-Amer.Mineral.,54,p1442-1459(1969).

10-M.I.BARATON, T.MERLE-MEJEAN, P.QUINTARD and V.LORENZELLI - FT-IR study of the OH surface groups on cordierite aerogel - J. de Phys.,Colloque (in print).

11-L.H.LITTLE - Infrared spectra of adsorbed species - Academic Press, London (1966).

12-M.L.HAIR - Infrared spectroscopy in surface chemistry, M. Dekker pub., New York (1967).

13-H.P.BOEHM and H.KNOZINGER - Nature and estimation of functional groups on solid surfaces. In CATALYSIS, Science and technology Springer Verlag, Berlin (1983); Chap. II p 39-20

14-B.A.MORROW - Infrared studies of reactions on oxide surfaces II Methanol on silica- J. Chem. Soc. Faraday Trans., 70, p 1527-1545 (1974).

15-G.RAMIS, G.BUSCA, V.LORENZELLI, M.I.BARATON, T.MERLE-MEJEAN and P.QUINTARD - FT-IR spectroscopy of high surface area silicon nitride and carbide : in this NATO-ASI proceeding.

16-G.BUSCA, V.LORENZELLI, G.PORCILE, M.I.BARATON, and P.QUINTARD - FT-IR study of the surface properties of silicon nitride - Mat. Chem. Phys.,14, p 123-140 (1986).

17-C.MORTERA, G.GHIOTTI, F.BOCCUZZI and S.COLUCCIA -An infrared spectroscopic investigation of the surface properties of magnesium alumi nate spinel - J. catalysis 51, p 299-313 (1978).

18-K.JIRATOVA and L.BERANEK - Properties of modified aluminas Appl. Catalysis, 2, p 125-138 (1982).

19-R.S.McDONALD - Surface functionality of amorphous silica - J. Phys. Chem., 62, p 1168 (1958).

20-A.V.KISELEV and V.I.LYGIN - Infrared spectra of surface compounds- John Wiley , New York (1975).

FT-IR CHARACTERIZATION OF HIGH SURFACE AREA SILICON NITRIDE AND CARBIDE

Gianguido RAMIS, Guido BUSCA and **Vincenzo LORENZELLI**
Istituto di Chimica, Facolta di Ingegneria Ple Kennedy,
I-16129 GENOVA (Italy)
Marie Isabelle BARATON, Thérèse MERLE-MEJEAN and **Pierre QUINTARD***
Laboratoire de Spectrométrie Vibrationnelle (U.A. 320 du CNRS)
Faculté des Sciences, 123 Ave. A. Thomas, F-87060 LIMOGES (France)

ABSTRACT - *The surface chemistry of SiC and Si3N4 prepared by laser-driven gas-phase reactions has been characterized by transmission/absorption FT-IR spectroscopy. Different surface sites, such as surface silanols, silyl amino- and imido- groups, alkyl groups, silane bonds, surface carbonyls and siloxanes bridges have been evidenced. Their amounts depend on surface treatments : evacuation, oxidation, surface nitridation .*

1. INTRODUCTION

The synthesis of ceramic powders by laser-driven gas-phase reactions is a promising procedure to obtain ultrafine and ultrapure monodispersed powders, whose chemical and morphological properties are expected to be ideal for further sintering processes (1-3). In addition, these powders are of potential interest in the field of heterogeneous catalysis, as mechanically strong supports and perhaps also for their own catalytic properties (4).

The knowledge and the control of the state of the surface of these materials are therefore highly requested . Infrared (IR) spectroscopy, whose performances have been emphasized in this decade by the commercialization and diffusion of Fourier transform (F.T.) spectrometers, has proved very useful in the field of heterogeneous catalysts (5,6) and may be also used for characterizing the advanced engineering ceramics.

In this paper we report a characterization of the surface properties of silicon nitride and carbide powders prepared by laser-driven reactions, using FT-IR spectroscopy. The properties of these materials are compared with those of silica which was the object of many investigations in the past, and represents an useful reference material.

2. EXPERIMENTAL

The preparation method of the nitride and carbide samples has been reported in references 2 and 3. The surface areas are 80 and 65 m^2/g for the carbide and the nitride, respectively. XRD analysis shows the crystallization

* *to whom correspondence should be addressed.*

L.-C. Dufour et al. (eds.), Surfaces and Interfaces of Ceramic Materials, 173–184.

of SiC in the beta form and the presence of Si (small amounts), while Si3N4 appears to be substantially amorphous. Aerosil silica (from Degussa, Hanau, West Germany) has been used for comparison.

The IR spectra were recorded using a Nicolet MX-1 Fourier transform spectrometer (1 cm^{-1} resolution, 96 scans, no smoothing) equipped with greaseless gas-manipulation/evacuation line and infrared cells. For the surface properties experiments, the samples were pressed into self-supported disks (1 cm diameter; 20-30 mg weight; 10^8 Pa pressure) and evacuated at different temperatures before treatments in controlled atmospheres.

3. RESULTS AND DISCUSSION

The IR spectra of SiC and Si3N4 powders pressed in KBr disks (Fig. 1, a and b) display very strong and complex absorptions due to the skeletal vibrations. These absorptions are at 820 cm^{-1} with a shoulder at 900 cm^{-1} in the case of SiC and in the 1100-850 cm^{-1} range for the nitride. Figure 1,c shows, for comparison, the spectrum of amorphous silica. There is a broad band at 1110 cm^{-1} with a pronounced shoulder at 1220 cm^{-1} , as well as other weaker bands near 810 and 470 cm^{-1} . The spectrum of the nitride exactly corresponds neither to that of the alpha form nor to that of the beta form (7), but looks similar to that of amorphous Si3N4 (8). Moreover characteristic absorptions show that silica is present in this preparation. The IR spectrum points out that this high-surface-area sample is more oxidized and even less crystalline than the high-surface-area sample prepared by nitridation of Aerosil silica, investigated previously (10, 13). The IR spectrum of the SiC sample instead does not indicate a significant amount of silica. It corresponds to the spectrum of beta-SiC (9), but silicon is not detected (sharp band at 610 cm^{-1}). In all three cases no other strong absorption is observed at higher frequencies using these KBr pressed disks.

If the pure powders are pressed, the regions where the skeletal absorptions fall are obscured, due to the larger amount of material . However, if the particle size of the powders is sufficiently small to limit light scattering (as in the present cases), the samples transmit the IR radiation out of these cut-off regions of the skeletal absorptions. So, the spectral ranges 4800-1300 cm^{-1} and 4800-1000 cm^{-1} are available for transmission/absorption studies of silicon nitride and silicon carbide respectively. Several absorption bands are observed. Some of them are modified by surface treatments and consequently ascribed to surface vibrational modes. Other bands are not sensitive to surface phenomena, so, they can be assigned to overtones of the skeletal absorptions.

The IR spectra in the 4000-2800 cm^{-1} region of a pressed disk of pure Si3N4, recorded after evacuation at different temperatures are given in figure 2. The general picture is very similar to that of another silicon nitride sample prepared by a classical chemical reaction, invetigated previously (10), although some features appear more clearly resolved here.

For comparison, the spectrum of silica in similar conditions is shown in figure 3,e : it only exhibits a very sharp band (11) at 3745 cm^{-1} , wich is due to the OH stretching of surface silanol groups free from hydrogen bonding, together with a broad shoulder in the lower frequency side, due to H-bonded mainly internal (12) OH groups. The same sharp band is also observed in the silicon nitride spectra, although it clearly presents a second component at

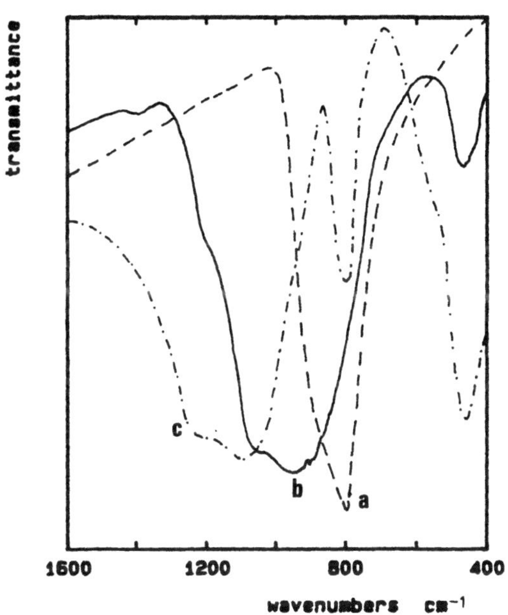

Fig. 1 - FT-IR spectra of KBr pressed disks of SiC (**a**), Si_3N_4 (**b**) *and amorphous silica* (**c**)

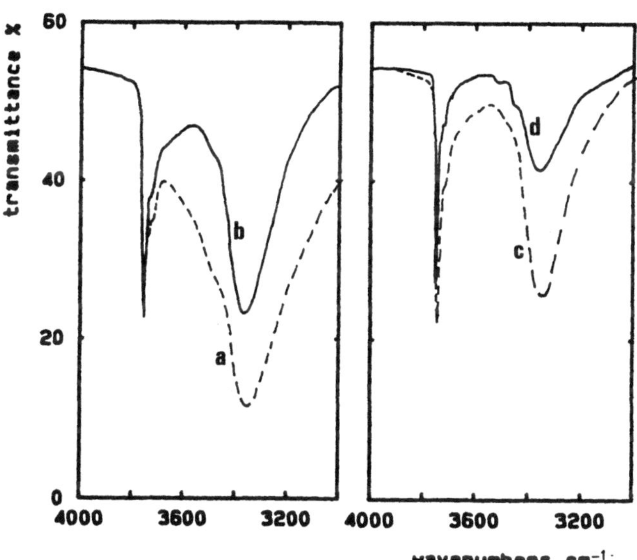

Fig. 2 - FT-IR spectra of self supported pellets of Si_3N_4 *(νOH and νNH regions) recorded after evacuation at 300K* (**a**), *550K* (**b**), *630K* (**c**) *and 1070K* (**d**)

3717 cm^{-1}. Moreover, the main maximum, assigned to O3SiOH groups on silica rafts, significantly shifts towards higher frequencies when the evacuation temperature is increased, i.e. 3740 cm^{-1} at r.t., 3743 cm^{-1} at 773 K and 3746 cm^{-1} at 1073 K. The component at 3717 cm^{-1} necessarily arises from a second type of surface silanol groups (vOH), possibly N3SiOH.

Another band, very intense and broad, also occurs in the silicon nitride spectra near 3350 cm^{-1}; this band, observed on silicon oxinitride samples too (13), is assigned to the NH stretching of surface imido-groups Si-NH-Si. Correspondingly, a sharp band at 1550 cm^{-1} was attributed in a previous work to the Si-N-H bending of the same groups (10). However, this assignment seems unsatisfactory. In fact, this frequency coincides with that of the scissoring mode of amido-groups Si-NH2 formed on silica by reaction with ammonia, whose N-H stretchings are observed at 3530 and 3450 cm^{-1} (asymmetric and symmetric modes, respectively (11)). Nevertheless, in the spectra reported in figure 2, some more features are clearly observed. In particular, a shoulder on the high-frequency side of the vNH band, near 3470 cm^{-1}, clearly splits into two bands, at 3510 and 3450 cm^{-1} after evacuation at 773-1073K (Fig. 2,d). In the same conditions, a weak band is also observed at 1470 cm^{-1} (see Fig. 4,d). It seems then more satisfactory to assign the stronger deformation band at 1550 cm^{-1} to Si-NH2 (scissoring mode) connected to the two weak stretching bands at 3510 and 3450 cm^{-1}. The weak deformation mode at 1470 cm^{-1} can instead be assigned to the deformation mode of the Si-NH-Si groups, whose stretching is very intense at 3350 cm^{-1} .

The spectra of SiC in the same spectral region after different evacuation treatments are reported in figure 3. The band of surface silanol groups is also observed in these spectra, but it is much less intense. The spectra 3,c and 3,d relate to the same powder after treatments with oxygen and ammonia respectively. It is evident that these treatments modify the surface, with the consequent appearance of silanol groups and silyl-amine or imine groups, respectively, whose bands are almost at the same position as discussed above for the nitride surface.

In figure 4 the evolution of the spectrum of silicon nitride in the region 1800-1300 cm^{-1} upon evacuation treatments is shown in detail. The sample exposed to air shows a rather strong band near 1630 cm^{-1}, due to adsorbed molecular water (scissoring mode). To this species is also due a broad and strong OH stretching band centered near 3300 cm^{-1}. The sample evacuated at r.t. (Fig. 4,b) still shows bands at 1665, 1625, 1590 and 1455 cm^{-1} . They probably arise from adsorbed water (1625 cm^{-1}), amonia (1590 cm^{-1}), deformation mode,and ammonium ion (1665 and 1455 cm^{-1}), symmetric and asymmetric deformation modes respectively. These bands completely disappear after evacuation at 550K (Fig. 4,c). By evacuation at higher temperatures, the band at 1550 cm^{-1} decreases progressively, in parallel to the vNH stretching bands, while the weak band at 1470 cm^{-1} appears more clearly. These bands have been previously assigned to the deformation modes of amine and imine groups respectively. It is worth noticing that these absorptions are superimposed on a broad absorption in the region 2000-1400 cm^{-1} having some resolved complexity near 1800 cm^{-1}, and which is due to overtones of the skeletal bands discussed above. These overtone absorptions are very frequently observed in the spectra of covalent compounds such as silica (11), silicon oxinitride (13) and the carbide as well.

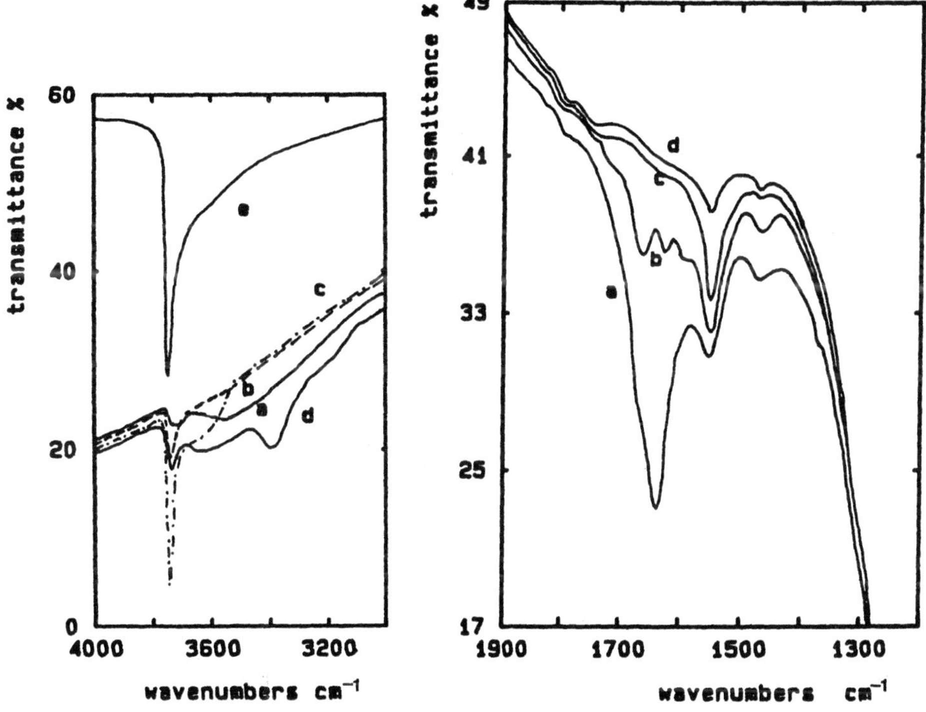

Fig. 3 - FT-IR spectra of SiC (νOH and νNH regions) after evacuation at 300K (**a**) and 770K (**b**) and after contact with oxygen (**c**) and ammonia (**d**) gases : 200 Torr at 770K and evacuation at 300K. Spectrum of amorphous silica (**e**) evacuated at 870K.

Fig. 4 - FT-IR spectra of Si_3N_4 (1900-1200 cm^{-1}) in contact with air (**a**) and evacuated at 300K (**b**), 550K (**c**) and 1073K (**d**).

The spectra of SiC in fact show a strong couple of bands at 1620 and 1540 cm^{-1} (Fig. 5) that, together with weaker absorptions at 1400 and 1310 cm^{-1} appear to be not sensitive to surface phenomena. They correspond to overtones and combinations of the fundamental optical and acoustical modes of beta-SiC (14). At higher frequencies a band near 1730 cm^{-1} is observed on the fresh sample (Fig. 5,a) but disappears after evacuation at 770K (Fig. 5,b) and reappears more intense by oxidation (Fig. 5,c). It is assigned to vC=O of surface carbonyl groups.

Figure 6 shows the evolution of the spectrum of silicon nitride in the vSiH region. The spectrum of the sample evacuated at 550K (Fig. 6,b) points out a sharp band at 2285 cm^{-1} , with a weaker and broader band poorly resolved near 2180 cm^{-1} . When adsorbed molecular water is present (Fig. 6,a), the main band is almost absent, indicating that the sites responsible for this absorption are perturbed or no longer exist. Further evacuation at 630K increases again the intensity of both absorptions, while successive heating in vacuum at temperatures between 600 and 1070K causes the band at 2285 cm^{-1} (that shifts to 2290 cm^{-1}) to decrease and the broader one at lower frequencies, to grow and split into two components at near 2200 and 2170 cm^{-1} (Fig. 6, c-f). Successive contacts with water vapour do not regenerate the band near 2290 cm^{-1}.

The spectrum of silicon carbide in the same region and after similar treatments shows only a very weak band near 2280 cm^{-1} and a sharp and intense band near 2120 cm^{-1} (Fig. 7, a and b). Treatment with ammonia produces a new couple of bands at 2295 and 2220 cm^{-1} (Fig. 7,c).

All bands in this region fall in a range typical of Si-H stretchings. The band near 2290 cm^{-1} is similar to that which occurs on silica after reaction with hydrogen at high temperature (15,16). This frequency is slightly higher than that of silane groups bonded to oxygen (near 2200 cm^{-1} for trialkoxy-silanes, above 2250 cm^{-1} for carboxysilanes (17,18)). We assign this band to H-SiOxN3-x species. Silane groups bonded to alkyl or aryl groups are characterized by a lower Si-H stretching frequency (near 2120 cm^{-1} (17,18)) close to the frequency observed on SiC that we have already assigned to H-SiC3 surface group. The band near 2220 cm^{-1} appearing on silicon carbide after ammonia treatment is assigned to H-SiNxC3-x species (x = 1 or 2).

The bands near 2200 and 2160 cm^{-1} which arise when heating silicon nitride in vacuum, are assigned to "bulk" silane bonds, produced by migration of hydrogen into the bulk. In fact, bands at the same position are observed in the bulk of this compound (8,19).

As we have already mentioned, if the silicon carbide sample is heated in oxygen, bands appear or grow near 3745 cm^{-1} (vOH of surface silanol groups) and 1730 cm^{-1}(vCO of carbonyls). Correspondingly, bands at 1090 and 1190 cm^{-1} also grow very intense (Fig. 8), in connection with the Vas Si-O-Si modes of silica. Instead, the same treatment on the nitride only causes the appearance, of a band near 1260 cm^{-1} very weak and detected only in the ratioed spectra. This band may be associated to surface Si-O-Si stretchings or to a deformation mode of H-bonded Si-OH groups.

After isotopic exchange with D2, the shift of the bands related with the stretching of bonds involving H atoms, indicate that they belong to species located on the surface. Our results agree with our theoretical expectations. (10, 21).

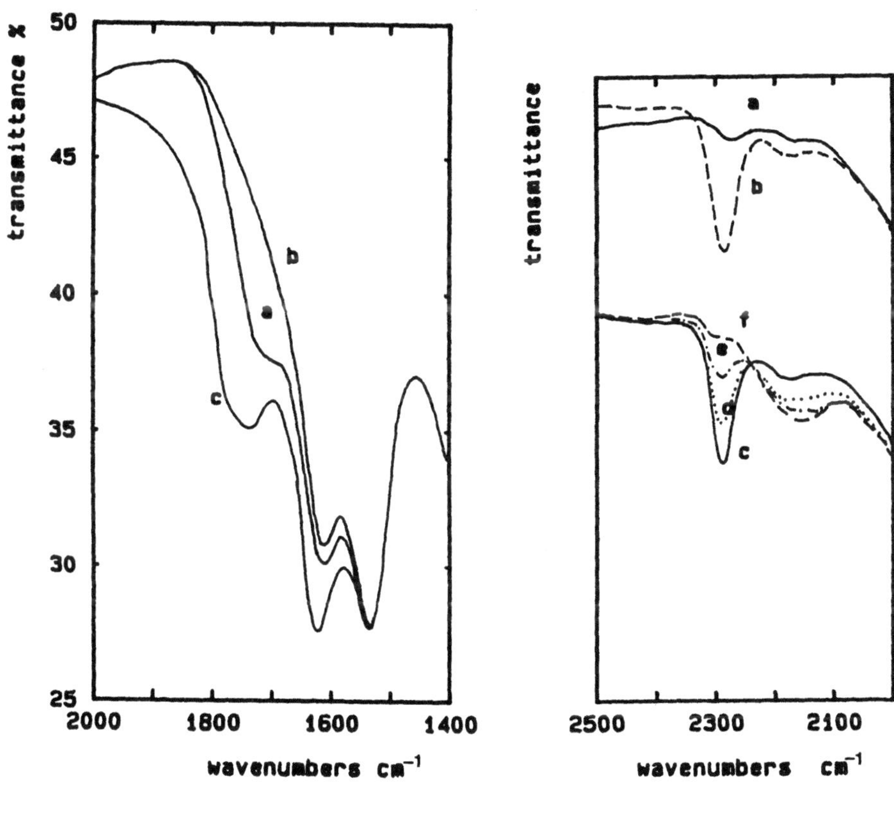

Fig. 5

Fig. 6

Fig. 5 - FT-IR spectra of SiC (2000-1400 cm^{-1}) after evacuation at 300K (**a**) *and 770K* (**b**), *and after contact with oxygen* (**c**) *(200 Torr) at 670K followed by evacuation at 300K.*

Fig. 6 - FT-IR spectra of Si3N4 (νSiH region) after evacuation at 300 K(a), 550K (**b**), *630K* (**c**), *740K* (**d**), *900K* (**e**) *and 1070K* (**f**)

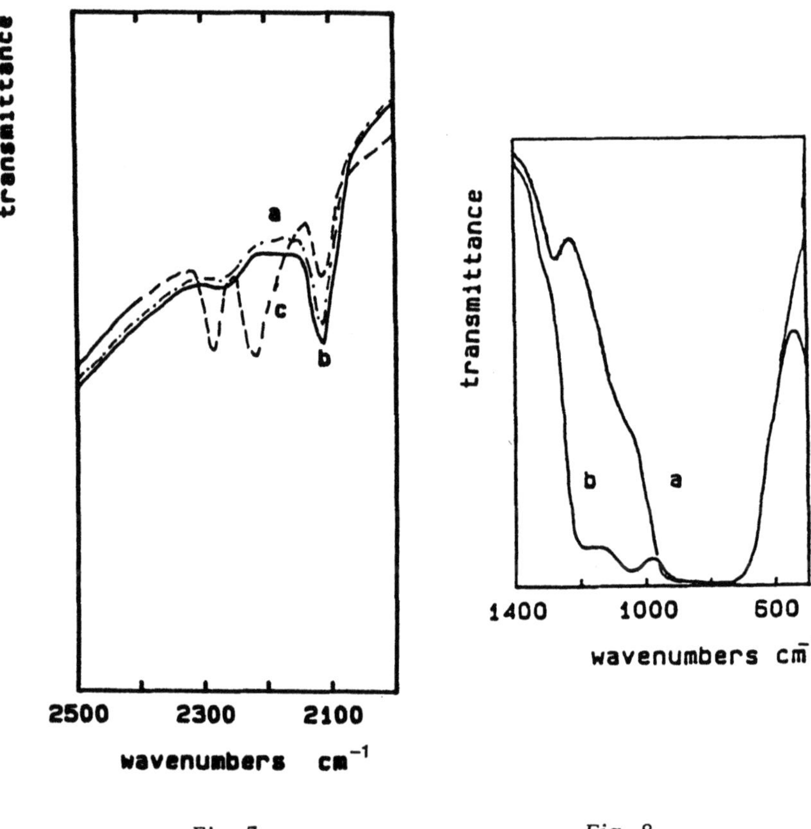

Fig. 7

Fig. 8

*Fig. 7 - FT-IR spectra of SiC (νSiH region) after evacuation at 300K (**a**) and 770K (**b**), and after contact with ammonia gas (**c**) (200 Torr) at 770K followed by evacuation at 300K.*

*Fig. 8 - FT-IR spectra of SiC (νSiO and νSiC regions) evacuated at 300K (**a**), and after contact with oxygen (**b**) 200 Torr at 770K.*

CONCLUSIONS

The present study shows that the FT-IR spectroscopy technique in the transmission/absorption mode, applied to submicrometric powders, allows unique inspection of the surface of ceramic materials. Such surfaces, whose systematic investigation has been undertaken only very recently (20,21), appear to be very complex, regarding the presence and the activity of a number of different surface sites.

However, the surface structure looks similar to what we may foresee considering the surface as obtained by cutting a crystal along one or more faces and by saturating the residual valences with the atoms from water (adsorbed from the environment) or of the reactants used to prepare the materials (silane, ethylene, ammonia). Figure 9 shows for example a representation of the (0001) face of beta silicon nitride (22) that may be taken as a model for the surface of this compound. Closed symbols refer to coordinatively saturated atoms near the surface (circles Si, rhombi N type 1, triangles N type 2), while the corresponding open symbols refer to the same atoms when they are exposed on the surface. Open circles represent Si atoms bonded to three nitrogen atoms that may produce silanol or Si-NH2 groups by reacting with either water or ammonia, respectively. Open rhombi are type 1N atoms that, when exposed, remain bonded to two silicon atoms only, so producing silyl-imido Si-NH-Si groups by taking a proton. Finally, open triangles are type 2N atoms, already bonded to three Si atoms but exposing the lone pair towards the surface, so acting potentially as Lewis basic sites.

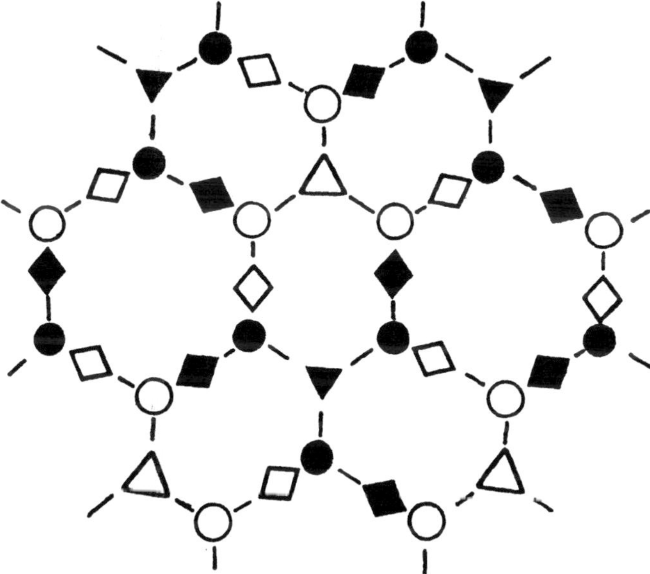

Fig.9 - A model for the (001) face of beta -Si3N4 (see text)

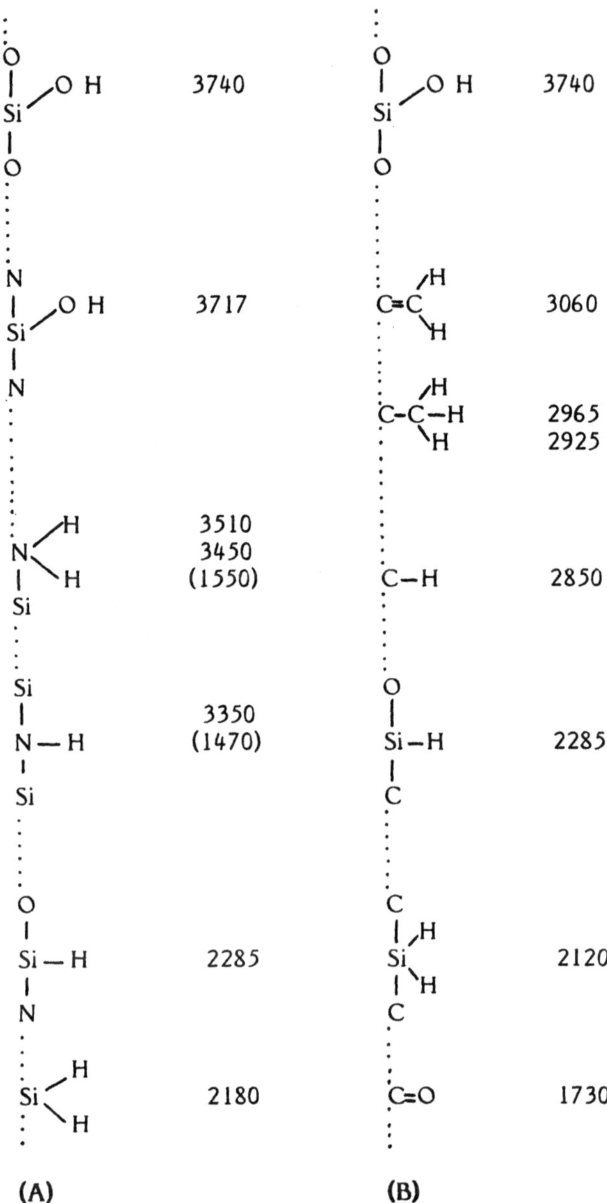

Fig. 10 - Main surface groups on silicon nitride (A) and silicon carbide (B) after high temperature treatment under evacuation (frequencies in cm^{-1})

Almost all that sites we have directly detected are then foreseen by this model, considering also that part of the surface is possibly modified by the presence of silica.

The presence of reactive groups such as, for example, silanols and silyl-imido-groups also suggests the possibility to modify selectively these surfaces by anchoring organic chains or organo-metallic complexes, in order to prepare surface-active materials for different purposes.

The main surface groups on silicon nitride and carbide, as they can be obtained after high temperature treatment and under evacuation, in order to eliminate physisorbed species, are given on figure 10.

The authors are indebted to Dr M. CAUCHETIER (C.E.N. - Saclay - France) for the preparation of the samples.

REFERENCES

1) W. R. Cannon, S. C. Danforth, J. H. Flint, J. S. Haggerty and R. A. Marra, "Sinterable Ceramic Powders from Laser-Driven Reactions: I, Process Description and Modeling", J. Am. Ceram. Soc., 65 (7) 324-330 (1982)
2) M. Cauchetier, O. Croix, M. Luce, M. Michon, J. Paris and S. Tistchenko, "Laser Synthesis of Ultrafine Powders", Ceram. Intern., 13 13-17 (1987)
3) M. Cauchetier, O. Croix and M. Luce, "Laser Synthesis of Silicon Carbide Powders from Silane-Hydrocarbon Mixtures", J. Amer Ceram. Soc. in press
4) R. B. Levy, "Properties of Carbides, Nitrides and Borides: Implications for Catalysis" in Advanced Materials in Catalysis (J.J. Burton and R.L. Garten ed.), Academic Press, New York, 1977, p. 101.
5) A. Zecchina, E. Garrone and E. Guglielminotti, "Structural Characterization of Surface Species and Surface Sites by Conventional Optical Spectroscopies" in Catalysis, Royal Society of Chemistry, London, Vol. 6, 1983, p. 90.
6) J. B. Peri, "Infrared Spectroscopy in Catalytic Research", Catalysis Science and Technology, Vol. 5, 171-220, J. R. Anderson and M. Boudart eds., Springer-Verlag (1984)
7) J.P. Luongo, "Infrared Characterization of α- and β-Crystalline Silicon Nitride", J. Electrochem. Soc. 130, 1560 (1983)
8) J. P. Luongo, "IR study of Amorphous Silicon Nitride", Appl. Spectrosc. 38, 195 (1984)
9) W. W. Pultz and W. Hertl, "Perturbations in the Infrared Spectra of Submicroscopic Size, Single Crystal, Fibrous SiC", Spectrochim. Acta 22, 573 (1966)
10) G. Busca, V. Lorenzelli, G. Porcile, M. I. Baraton, P. Quintard and R. Marchand, "FT-IR Study of the Surface Properties of Silicon Nitride", Mater. Chem. Phys., 14 123-140 (1986)
11) J. B. Peri, "Infrared Study of OH and NH_2 Groups on the Surface of Dry Silica Aerogel", J. Phys. Chem. 70, 2937 (1966)
12) S. Kondo, H. Yamagouchi, Y. Kajiyama and T. Ishikawa, J. Chem. Soc. Faraday Trans. I, 80, 2033 (1984

13) M. Sorlino, G. Busca, V. Lorenzelli, R. Marchand, M.I. Baraton and P. Quintard, "Propriétés de la Surface de l'Oxynitrure de Silicium Si_2ON_2: Etude Preliminaire par Spectrométrie infrarouge par Transformée de Fourier", Ann. Chim. Sci. Mater. 10, 105 (1985)
14) L. Patrick and W. J. Choyke, "Lattice Absorption Bands in SiC", Phys. Rev., 123 813-815 (1961)
15) C. Morterra and M. J. D. Low, "Chemisorption of Molecular Hydrogen on Highly Dehydrated Silica", Chem. Comm. 1491 (1968)
16) M.J.D. Low, "Reactive Silica. XVII: The nature of Reaction Center", J. Catal. 103, 496 (1987)
17) L. Smith, "Infrared Group Frequencies for Structure Determination in Organosilicon Compounds", in Chemical, Biological and Industrial Applications of Infrared Spectroscopy (J.R. Durig ed.), Wiley, New York, 1985, p. 73.
18) P. Quintard, G. Ramis, M. Cauchetier, G, Busca and V. Lorenzelli, "FT-IR Characterization of Si-H Bands on the Surface of Silicon Carbide", J. Mol. Struct., 174, 369 (1988)
19) H.J. Stein and H.A.R. Wegener, "Chemically Bound Hydrogen in CVD Si_3N_4: Dependence on NH_3/SiH_4 Ratio and on Annealing", J. Electrochem. Soc. 124, 908 (1977)
20) M. H. Rahaman, Y. Boiteux and L. C. De Jonghe, "Surface Characterization of Silicon Nitride and Silicon Carbide Powders", Am. Ceram. Soc. Bull., 65 1171-1176 (1986)
21) V. Lorenzelli, G. Ramis, G. Busca and P. Quintard, "Surface Structure and Stability of Advanced Engineering Ceramics: Fourier Transform Infrared Study of High-surface Silicon Nitride and Carbide Powders in Different Atmospheres", Proc. 31st International Congress of Pure and Applied Chemistry, Invited Lectures Section 5, 40-52, Sofia, Bulgaria (1987)
22) H. M. Jennings, J.O. Edwards and M.H. Richman, "Molecular Structure, Microstructure, Macrostracture and Properties of Silicon Nitride", Inorg. Chim. Acta 20, 167 (1976)

THEORY OF DOPANT SEGREGATION IN CERAMIC OXIDES

R. G. EGDELL
Department of Chemistry
Imperial College,
London
U.K.

W. C. MACKRODT
I.C.I. Polymers and Chemicals Group
P.O. Box 8
The Heath
Runcorn
Cheshire
U.K.

ABSTRACT. The basic approaches to the computer simulation of ionic solids are briefly reviewed, leading to a strategy for the calculation of the heat of segregation of cation dopants in metal oxides. It is shown that equilibrium surface coverages of dopant cations can be derived from simulation studies if due account is taken of the variation of segregation energy with surface coverage. Isovalent dopants in MgO are considered in detail and comparison is made with experimental work on Ca/MgO(001) and Ba/MgO(001). The behaviour of isovalent dopants in Al_2O_3 is more complex owing to major surface relaxation effects in the pure material and the existence of several different low index surfaces of comparable energy. However comparison with experimental work on Y/Al_2O_3 is encouraging. The behaviour of aleovalent dopants is illustrated with respect to Ti in MgO and Mg in Al_2O_3. In both cases the simulation studies treat a neutral dopant-vacancy complex.

1. Introduction

It is now widely recognised that the mechanical and electrical properties of ceramic materials are governed to a large extent by the presence of impurities and by the structure and stability of surface and grain boundary interfaces. Surprisingly then there has been little systematic experimental work directed toward the study of dopant segregation in metal oxides under well defined conditions. On the other hand, theoretical simulation procedures based on the ionic model [1] have been used extensively in recent years to explore the surface and defect structure of metal oxides. It is the purpose of this article to review theoretical work on dopants in MgO and Al_2O_3.Theoretical modelling has now reached the level of maturity where it is possible to make direct contact with experimental

L.-C. Dufour et al. (eds.), Surfaces and Interfaces of Ceramic Materials, 185–203.

work and it is hoped that the approaches developed to deal with dopants in magnesia and alumina will be extended to a wider range of oxide systems. It is beyond the scope of this short paper to deal with the technical aspects of simulation procedures and the reader is referred to [2] for further details.

2. An outline of basic strategy in simulation studies of dopant segregation.

The simulation studies to be described here involve calculation of the energy U_L of a static lattice using simple two body potentials incorporating a long range Coulomb term and a short range component $\Theta_{ij}(r_{ij})$:

$$U_L = \sum_{i>j} q_i q_j / r_{ij} + \sum_{i>j} \Theta_{ij}(r_{ij}) \tag{1}$$

Here i and j index ions with charge q_i and q_j and the r_{ij} are interionic separations. The bulk substitution energy $\Delta U_{sub}(\mathrm{bulk})$ corresponding to the process:

$$MO + D(\infty) \longrightarrow DO + M(\infty) \tag{2}$$

(where M is the host cation and D is the dopant cation) then appears as the energy difference between the fully relaxed bulk substituted lattice and the perfect host lattice. The corresponding surface substitution energy $\Delta U_{sub}(\mathrm{surf})$ is defined by the energy difference between the fully relaxed surface substituted lattice and the relaxed surface lattice. The segregation energy ΔU_{seg} is then given by :

$$\Delta U_{seg} = \Delta U_{sub}(\mathrm{surf}) - \Delta U_{sub}(\mathrm{bulk}) \tag{3}$$

Note that in general $\Delta U_{sub}(\mathrm{surf})$ may vary with surface coverage of the dopant. In the work to be described here it is generally assumed that the bulk dopant concentration is so small that interactions between dopant ions can be ignored.

2.1. INTERIONIC POTENTIALS

The long range Coulomb potentials in (1) use fully ionic charges for the Mg^{2+}, Al^{3+} and O^{2-} host ions and for the dopants. Despite contrary indications from some experimenal techniques (notably helium diffraction which [3] probes the corrugation in potential at the surface), recent theoretical work on MgO clusters, slabs and infinite arrays supports the idea that full ionicity is retained at the surface with ionic charges in excess of 1.9e emerging from Mulliken population analyses [4,5].

Treatment of the short range potentials is more problematic. Two fundamentally different approaches have been applied in simulation studies. The first uses a simple functional form for the short range potential with one or more empirical parameters to be fitted by reference to experimental data such as the lattice energy, elastic constants and dielectric constant. One popular form:

$$\Theta(r) = A \exp(-r/\sigma) - C/r^6 \tag{4}$$

involves three parameters A,C and σ for each class of ion pair [6]. Empirical parameterisation effectively samples interionic potentials close to the equilibrium interionic separations of the perfect crystal, but in low symmetry structures this may involve a range of different interionic distances for each pair type.

Alternatives rely on a theoretical description of short range potentials. One method [6] that has been particularly popular in segregation studies uses a density functional approach in which the energy of closed shell ions is expressed in a form depending only on the electron density $\rho(r)$. The electron densities associated with two ions, say A and B, are assumed to be simply superimposable (the so-called rigid electron gas approximation) and the energy of the two interacting ions then depends on the superimposed electron density:

$$\rho_{AB}(r_{AB}) = \rho_A + \rho_B \tag{5}$$

$$E[\rho_{AB}(r_{AB})] = E[\rho_A + \rho_B] \tag{6}$$

The interaction potential Θ_{AB} is given by the difference:

$$\Theta_{AB}(r_{AB}) = E[\rho_{AB}(r_{AB})] - E[\rho_A] - E[\rho_B] \tag{7}$$

In general the electron densities may be obtained from ab-initio Hartree Fock calculations on the appropriate ions. A particular problem in applying this approach to metal oxides is that free O^{2-} ion is unstable and the O^{2-} must therefore be surrounded by an appropriate spherically symmetrical stabilising positive potential. This can conveniently be taken as the Madelung potential at the oxide site in the crystal of interest. Because the site potential differs from one oxide to another, the oxide ion wavefunctions and hence the interaction potentials are different for different oxides. Because they are non-empirical, electron gas potentials do not necessarily reproduce the observed lattice parameter particularly well. Moreover there are some interesting qualitative differences between the non-Coulombic components of electron gas and empirical potentials for the O^{2-}-O^{2-} interaction. Thus whereas the electron gas model predicts the expected short range repulsive interaction, empirical potentials for both MgO and Al_2O_3 are attractive [6].

The model developed so far takes no account of the electronic polarisation responsible for the high frequency dielectric behaviour of ionic crystals. In a crude way the high frequency dielectric constant can be reproduced by introducing the idea of point polarisable ions. However this approach fails to take account of the interplay between the polarisation of the ion charge clouds and the short range repulsive interaction. The so called *shell model* [6] illustrated schematically in figure 1 provides a simple mechanical model that both reproduces high frequency dielectric behaviour and treats in a realistic way the interaction between short range repulsion and polarisation. Even within the electron gas model, shell model parameters are usually derived empirically by reference to low and high frequency dielectric constants. The cations Mg^{2+} and Al^{3+} are presumed to be non-polarisable, but it is not satisfactory to ignore the polarisability of large dopant cations such as Ba^{2+}. In general terms the shell model represents a great improvement over the point polarisable ion model in dealing with relaxation of ions around impurities and defects and at surfaces.

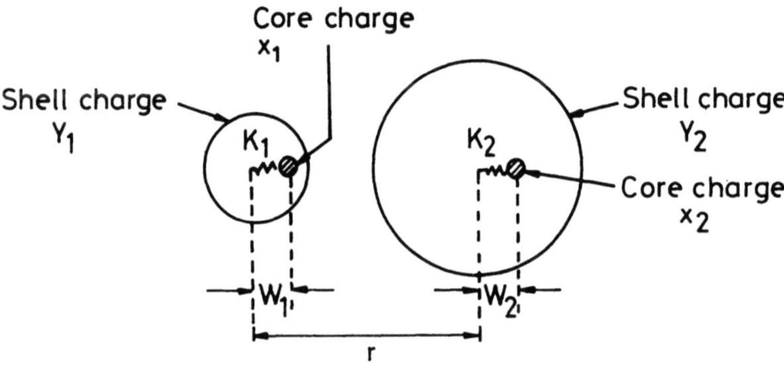

Figure 1 . A schematic representation of the shell model.

2.2. A THERMODYNAMIC MODEL FOR DOPANT SEGREGATION

We now turn to the problem of relating the equilibrium surface coverage of a dopant cation to the heat of segregation derived from simulation studies. A fuller account of the model developed here is given in reference [7]. Let us suppose cations of type 1 are doped into a lattice where the dominant cations are of type 2. Let $g_s(i)$ and $g_b(i)$ be the free energy of ions of type i in surface and bulk sites respectively. The total free energy of the system is G where:

$$G = n_b(1)g_b(1) + n_b(2)g_b(2) + n_s(1)g_s(1) + n_s(2)g_s(2) - k_BT\log_e\Omega \qquad (8)$$

Here $k_B\log_e\Omega$ is the configurational entropy. If we assume that the host and dopant cations compete for the same sites both in the bulk and at the surface and that there is a random occupation of the available sites we have:

$$\Omega = [n_b(1) + n_b(2)]!\,[n_s(1) + n_s(2)]! \,/\, [n_b(1)!\, n_b(2)!\, n_s(1)! n_s(2)!] \qquad (9)$$

It is now convenient to define a free energy of segregation Δg

$$\Delta g = g_s(1) - g_s(2) - g_b(1) + g_b(2) \qquad (10)$$

Differentiating (8) with respect to $n_s(1)$ subject to conservation relationships on the number of bulk and of surface sites and on the total number of host and dopant cations we arrive after a little manipulation at the relationship:

$$dG/dn_s(1) = \Delta g + n_s(1) d\Delta g/dn_s(1) + k_B T \log_e (x_s/x_b) \qquad (11)$$

Here x_s is the surface cation ratio $n_s(1)/n_s(2)$ and x_b is the bulk cation ratio $n_b(1)/n_b(2)$. Since we have that :

$$n_s(1) d\Delta g/dn_s(1) = x_s(1+x_s)\, d\Delta g/dx_s \qquad (12)$$

It follows that the free energy is minimised when the surface atom ratio is related to the bulk atom ratio by:

$$x_s = x_b \exp [-\{ \Delta g + x_s(1+x_s)\, d\Delta g/dx_s \} / k_B T] \qquad (13)$$

Given values for the free energy of segregation as a function of coverage, equation (13) must be solved numerically to derive the equilibrium surface coverage as a function of temperature and bulk doping level.

A major problem in relating results from simulation studies to experimental surface coverages is that the former yield internal energies rather than free energies and thus entropy terms are ignored when comparing with experimental data. There have been few studies of vibrational entropies of segregation, although Masri *et al* [8] considered the particular case of Ca segregated MgO at half coverage of the (001) surface. They calculated phonon dispersions in a 15 layer MgO slab both with and without the Ca surface dopant and then evaluated vibrational partition functions and entropies of substitution in both the surface and in the bulk. The surface entropy of substitution was only $0.06k_B$ in the temperature range 500K to 1000K as compared with a bulk entropy of substitution of $0.30k_B$. Thus the vibrational entropy of segregation $-0.24k_B$ arises mainly from a loss of bulk entropy. At 1000K the entropic contribution to the free energy of segregation was only 0.03eV as compared with an enthalpy contribution -0.48eV so that the neglect of the vibrational entropic term is not too serious in this particular case. Masri *et al* also considered the problem of using computed internal energies in place of the enthalpy term left in (13) after neglect of vibrational entropy, and concluded that this should usually be a reasonable approximation [8].

In experimental work it is usual to attempt to measure enthalpies for a reaction by making a plot of $\log_e$ (equilibrium constant) against 1/T. In the present context it will be seen from equation (13) that the slope of $\log_e (x_s/x_b)$ against k_B/T is given not by Δg but by $\Delta g + x_s(1 + x_s)\, d\Delta g/dx_s$. Since the energy of segregation often varies strongly with coverage, the second term can not in general be ignored. Linear behaviour in experimental work will be found either when the heat of segregation is independent of coverage or when the variation with coverage is of the form in (14) over a range of coverages:

$$\Delta g(x_s) = g_c - \Lambda\,(1 + 1/x_s) \qquad (14)$$

where g_c and Λ are constants independent of coverage.

3. Isovalent dopants in alkaline earth oxides.

The surface energetics of the alkaline earth oxides contrasts with that of alumina (see below) in that the surface energies for (001) surfaces are very much lower than for other low index surfaces such as (110) and (111), which has infinite surface energy in the simplest model. Thus the (100) surface energies for MgO, CaO and SrO are calculated to be 1.07, 0.80 and 0.65J/m^2 respectively whereas the corresponding energies for the (110) surfaces are 2.28, 1.90 and 1.49J/m^2. Thus most experimental work on single crystal surfaces has focussed on low energy (001) cleavage planes. One of the interesting contributions of simulation studies has been to make comparison with more open (110) surfaces. The results of simulation studies [9,10] on dopants in MgO are summarised in table I. Throughout the calculations use electron gas potentials appropriate to MgO , together with shell model parameters for O^{2-}. Polarisation of dopant cations have however been ignored. In all cases ions in the surface plane are allowed to relax along the direction of the surface normal.

Several qualitative features of table I deserve comment. Firstly we see a strong correlation between the excess size of the dopant cation and the segregation energy. Thus the segregation energy for Ni , which has almost exactly the same ionic radius as Mg, is negligible. In the series Ca,Sr,Ba we see increasingly negative heat of segregation (at zero surface coverage of the dopant) with increasing ionic size. This reflects the fact that the surface provides the possibility for relief of elastic strain associated with bulk substitution by relaxation of dopant cations away from the surface. However, as the coverage of large cations at the surface increases, the heat of segregation becomes increasingly less favourable owing to steric overcrowding. In the particular case of Ba at the (001) surface of MgO, the energy of segregation at full coverage is positive if the surface Ba and O ions are individually constrained to the same plane so that the (1x1) surface structure is preserved. Steric overcrowding is relieved to some extent if the surface ions are allowed to split into two layers, producing a (√2x√2) surface reconstruction. Details of the relaxed reconstructed surface structure are shown in figure 2 with distances appearing in this figure defined in table II. The reconstructed surface structures are seen to involve a strongly rumpled surface monolayer. This contrasts with the intrinsic (001) surface which is an almost perfect termination of the bulk structure with little relaxtaion of ions away from their ideal positions

TABLE I

Calculated energies of segregation of isovalent impurities at (001) and (110) surfaces of MgO
All values in kJ/mole

Coverage	Be	Ni	Ca	Sr	Ba
		(001) surfaces			
Isolated impurity	+20.26	+0.97	-104.19	-215.14	-376.25
1/4 coverage	-	+1.93	-81.04	-162.08	-288.46
1/2 coverage		+2.89	-46.31	-62.71	-107.09
full coverage (unreconstructed)	+38.59	+2.89	-38.59	-25.08	+28.94
full coverage (reconstructed)			-46.31	-108.85	-136.03
		(110) surfaces			
Isolated impurity	-290.39	-0.97	-137.96	-307.75	-541.22
full coverage (unreconstructed)	-36.66		-69.46	-134.10	-209.35
full coverage (reconstructed)				-208.39	-389.76

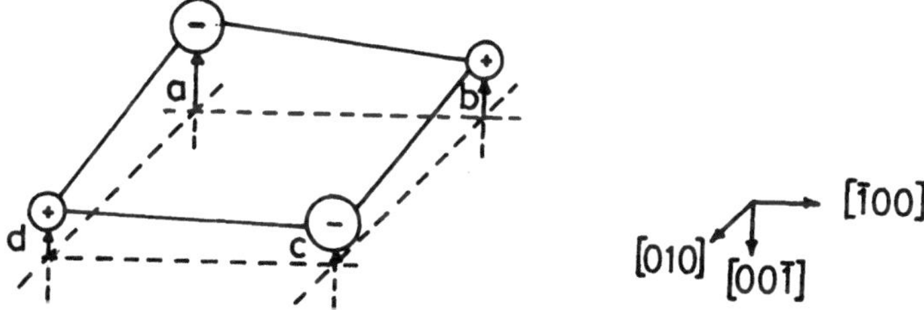

Figure 2. Reconstruction of MgO(001) with monolayer substitutional incorporation of large dopant cations. Values of the displacements a-d are given in table II. Dashed line shows the unrelaxed surface plane.

TABLE II
Perpendicular displacements of ions in reconstructed segregated surface of MgO

displacement	Ca	Sr	Ba
a	0.39	0.75	0.79
b	0.21	0.68	1.21
c	0.05	0.14	0.79
d	0.21	0.24	0.38

a and c are anion displacements; b and d are cation displacements out of the surface. All values given in units of a_0, the host cation-anion separation.

Returning to table I we see that large dopant cations are more easily accomodated at the open (110) surface of MgO than at the (001) surface, particularly at high surface coverage. In the case of monolayer Ba coverage the difference in segregation energy between the two surfaces is sufficiently large that the (110) surface energy is lowered below that for the (001) surface. Thus if we could experimentally achieve full monolayer coverage this would be expected to lead to facetting of the (001) surface. Moreover the large heat of segregation actually leads to a negative surface energy for both MgO(001) and MgO(110), which implies a thermodynamic barrier to sintering [10].

We turn now to consider more detailed comparison between simulation studies and experiments on Ca/MgO and Ba/MgO.

3.1 Ca on MgO(001)

The large heat of segregation for Ca in MgO means that even at low bulk doping levels there may be an accumulation of the dopant at free surfaces sufficient that it is possible to probe surface Ca concentrations with spectroscopic techniques such as Auger electron spectroscopy (AES), ion scattering spectroscopy (ISS) or X-ray photoelectron spectroscopy (XPS). The technique of ISS is usually presumed to sample only the topmost ionic layer, whereas both AES and XPS probe further into the solid over a length range determined by the electron pathlength. In a series of elegant experiements, McCune and Wynblatt equilibrated a 200ppm Ca-doped MgO(001) crystal at elevated temperatures over the range 900°C to 1450°C in a UHV system and measured Ca/Mg intensity ratios in ISS and AES [11]. At the lower end of the temperature range studies it was estimated that hold times of the order of 2,000 days would be necessary to equilibrate the surface Ca with that in the bulk. In fact maximum hold times of 9 days were used. In both AES and ISS the Ca concentration at the surface increased with decreasing temperature, as expected from an equilibrium thermodynamic model. One problem with both surface analytical techniques was that they appeared to remove the surface layer being analysed, so that the Ca peak intensity decreased with time of the spectroscopic measurement.

The results of ISS experiments are summarised in figure 3, where the log of the peak area ratio is plotted against 1/T. For comparison surface coverages derived from simulation studies [7,10] are also shown. At high temperatures (low surface coverage) both theory and experiment find linear behaviour that can be interpreted in terms of a heat of segregation of order 45kJ/mole. With increasing surface coverage, however, nonlinear behaviour sets into the computed curve due to the steric overcrowding encapsulated in the second term of equation (13). Non linear behaviour is also found experimentally, although this was originally interpreted in terms of a failure to reach the full surface equilibrium coverage of Ca [11]. The only disappointment of figure 3 is that the calculated absolute surface coverage is somewhat higher than found experimentally. This may be due to the surface erosion during the ISS measurements alluded to above.

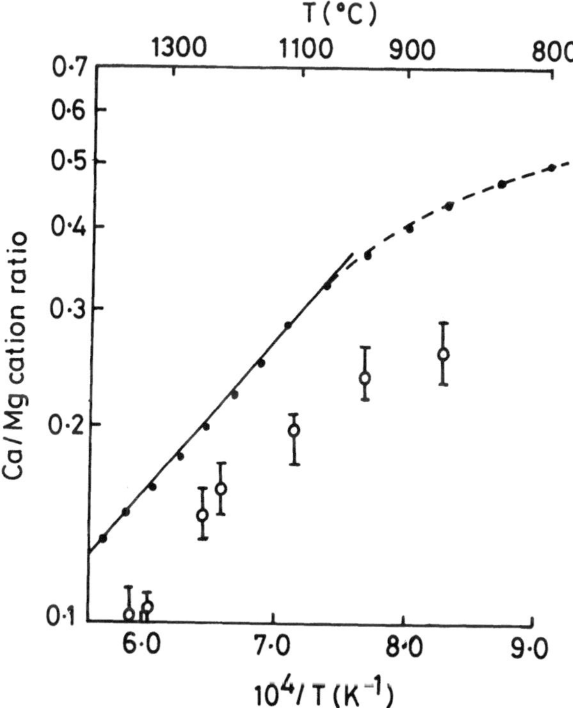

Figure 3. Surface cation ratio Ca/Mg for 200ppm Ca-doped MgO(001) as a function of inverse temperature. Full points result from simulation studies [7]. Open circles are from experiments of McCune and Wynblatt [11].

The excellent agreement between theory and experiment for Ca doped MgO is encouraging and the work of McCune and Wynblatt stands as a benchmark for further experimental efforts in this area. However the need to hold samples at elevated temperatures for prolonged periods of time in expensive UHV systems has probably discouraged widespread adoption of their methodology.

3.2 Ba on MgO

Cotter *et al* [12,13] have studied the behaviour of Ba at the surface of MgO using both polycrystalline and single crystal material. The principal surface analytical technique was XPS. The Ba was introduced either as a bulk dopant [12], or by deposition of Ba onto the MgO(001) surface from a Ba getter source [12,13].

In the first series of experiments, doped ceramic pellets were equilibrated outside the UHV system at 1360°C for 48 hours and were then rapidly quenched before transfer into the XPS system [12] The Ba:$3d_{5/2}$ to Mg:1s intensity ratio was then measured as a function of bulk doping level, with results summarised in figure 4. The remarkable feature of these experiments is that the Ba/Mg intensity shows very little change with changing bulk doping level over a range of three orders of magnitude.

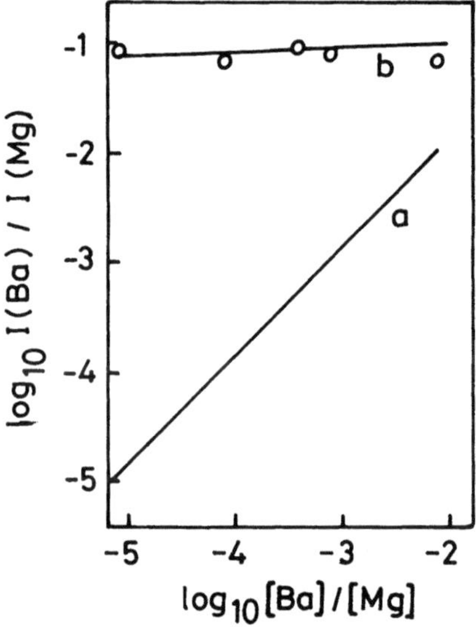

Figure 4. XPS intensity ratio between Ba:$3d_{5/2}$ and Mg:1s peaks in Ba-doped MgO ceramics equilibrated at 1360°C. Circles are experimental points. Line a is calculated assuming zero heat of segregation whilst line b is derived from heat of segregation calculated assuming 'intermediate potentials' (see text)

In these experiments care was taken to establish that the Ba ions reside only in the topmost ionic layer. Additionally an internal calibration to the experiments was provided by careful measurements of peak intensities in MgO and BaO. The surface coverage of Ba was calculated using the procedures outlined in section 2 and the Ba/Mg XPS intensity ratio was then derived from a model taking proper account of electron pathlengths for the various core levels involved. This approach leads to calculated peak intensity ratios in excellent absolute agreement with observed values. There are however a number of technical aspects of the calculations that need to be considered.

The first problem concerns potentials. When dealing with bulk substitutional replacement of Mg by Ba it is reasonable to use interionic potentials appropriate to the MgO host lattice. These potentials will also be adequate to deal with surface substitution of Ba at low surface coverages. However when we have complete coverage by a BaO monolayer it is more appropriate to use BaO potentials. In the intermediate coverage regime, we must envisage a changeover between the two limiting cases. In deriving the theoretical curve of figure 4 a linear variation of the heat of segregation with coverage has been assumed which incorporates a switchover of this sort [12]. The second major problem concerns surface structure. The idealised structure of figure 2 is based on crude potentials that do not allow for Ba polarisation and treat interactions involving the oxide ion with MgO potentials. If cation polarisation is first introduced, a more complex surface reconstruction emerges with four distinct surface cation sites in the surface BaO monolayer [14]. However with full BaO potentials, no stable reconstruction is found at full monolayer coverage if the surface ions are constrained to on-top sites. This suggested that more radical surface reconstruction should be considered. One possibility emerges from the following considerations. The lattice parameter a(BaO)=5.52Å is much greater than a(MgO)=4.21Å, but there is an accidental near match with $\sqrt{2}$a(MgO)=5.95Å This suggests that half a monolayer of BaO can be accomodated not by substitutional replacement but by formation of an 'on-top' layer. The on-top monolayer can sit either on the substrate cation sublattice or the anion sublattice or above hollow sites. Despite the fact that they involve near neighbour repulsive interactions between ions of the same charge, the former structures are found to be favourable, with an energy of 0.73eV per Ba ion lower than for bulk substitutional replacement.The energy of segregation for surface substitutional replacement is only -0.42eV, so that the commensurate overlayer structure is favoured. A final possibility is that BaO sits as an overlayer on the substrate at its own natural lattice spacing. Unfortunately simulation studies cannot deal easily with this non-commensurate structure as it would involve use of a prohibitively large surface unit cell. However one can calculate the energy per Ba ion in a free BaO layer, and this turns out to be 1.67eV lower than for bulk substitution. The energy must be further lowered by interaction with the surface, but probably not by much. Thus simulation studies suggest that as Ba builds up on the MgO surface, the lowest energy structure involves BaO at its own lattice spacing. These considerations were all incorporated in setting up figure 4. Direct study of the structure of BaO on MgO confirms these ideas. Both in Ba segregated MgO(001) at half monolayer coverage (here defined in terms of substitutional replacement) or for BaO layers deposited onto MgO(001) it is found that the spot positions in low energy electron diffraction (LEED) correspond to the lattice parameter of bulk BaO.

The temperature variation of Ba coverage on MgO has not been studied experimentally, although simulation studies have been performed: some results are shown in figure 5.

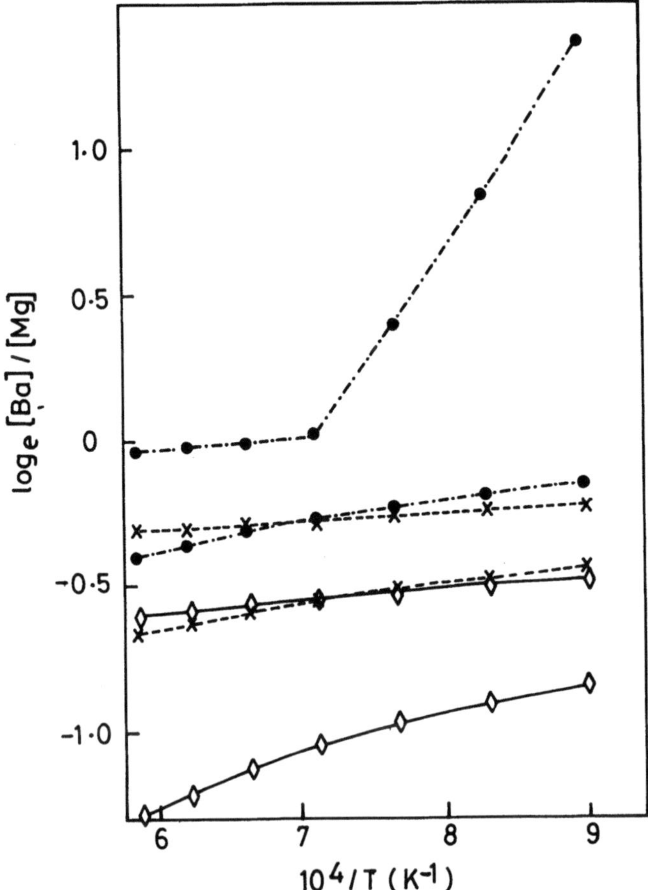

Figure 5. Calculated surface cation ratio Ba/Mg for Ba-doped MgO as a function of inverse temperature. Solid lines and diamonds: BaO potentials.Crosses and dashed lines; intermediate potentials. Dash-dot lines and circles; MgO potentials. In each case the lower of the curves corresponds to 7ppm bulk doping and the upper curve to 7000ppm doping.

The temperature variation of surface coverage is remarkably muted in these calculations owing to competition between a large negative heat of segregation term and a positive term arising from variation of the heat of segregation with coverage. The latter is accentuated by the switchover in potentials so that the intermediate potentials case has the least pronounced temperature variation of coverage. Interpreted in terms of simple Langmiur behaviour, the data for 7ppm doping has a slope corresponding to a heat of segregation of only 0.06eV. This is of course very much lower than the atomistic heat of segregation.

4. Isovalent dopants in α-Al_2O_3.

As mentioned previously, Al_2O_3 differs from MgO in that there are several different surfaces of comparable energy. Moreover the surface energies are strongly influenced by major surface relaxtions as shown in table III. Relaxation is particularly important for the basal plane where 59% inward relaxation of the outermost cation plane leads to almost $4J/m^2$ stabilisation

TABLE III
Calculated surface energies of low index surfaces of α-Al_2O_3
All values given in J/m^2

Surface	Unrelaxed energy	Relaxed energy
(0001)	5.95	2.03
$(10\bar{1}0)$	6.46	2.23
$(10\bar{1}2)$	3.63	2.29
$(11\bar{2}0)$	4.37	2.50
$(10\bar{1}1)$	5.58	2.52

Segregation behaviour at the basal surface of alumina differs from that for MgO(001). The ions Fe^{3+}, Y^{3+} and La^{3+} are all larger than Al^{3+}, but surprisingly for Fe^{3+} and Y^{3+} the heat of segregation to the basal surface is positive (figure 6a) and only the very large dopant La^{3+} is expected to accumulate at this surface This is because the large inward relaxation of the outer Al plane leaves little room to accomodate large ions. By contrast, at the $(11\bar{2}0)$ surface there is more room for large dopant cations and large negative heats of segregation are found for all three trivalent dopants (figure 6b). The only experimental work on alumina which may be directly compared with the simulation studies relates to Y^{3+} in alumina [18]. It has been confirmed that there is no increased concentration of the dopant at the basal surface but surface enrichment is found in polycrytalline material. The observed variation of Y/Al Auger intensity ratio with temperature corresponds closely to that expected for the $(11\bar{2}0)$ surface, but the absolute value is lower than expected for the single crystal surface. This must be largely due to the fact that surfaces other than $(11\bar{2}0)$ are present in the polycrystalline material.

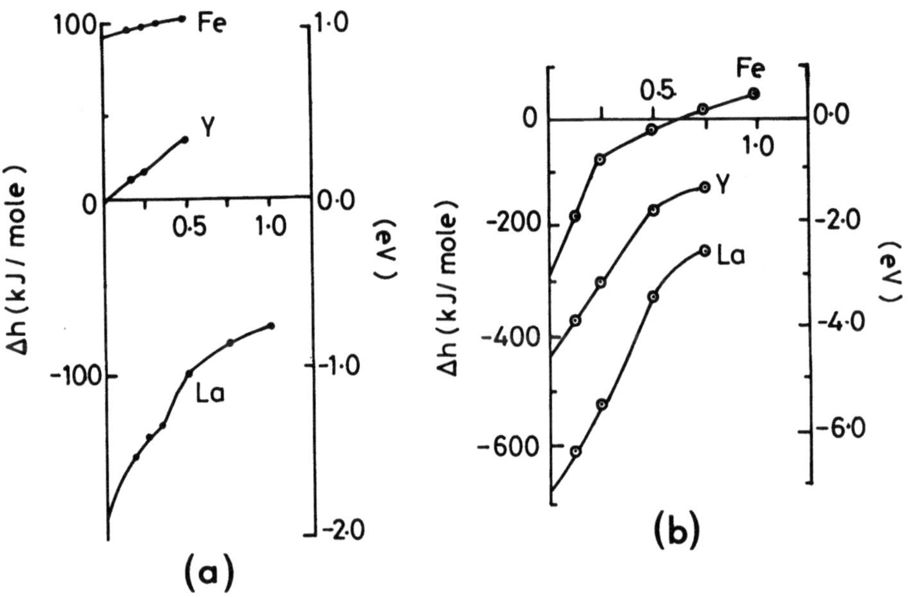

Figure 6. Heats of segregation of Fe,Y and La as a function of surface coverage to (a) the relaxed (0001) surface of α-Al_2O_3 (b) the relaxed $(11\bar{2}0)$ surface.

5. Aleovalent dopants.

5.1 Ti in MgO

In general aleovalent dopants are more difficult to study than isovalent dopants because space charge effects may influence segregation behaviour. The strategy that has been used so far in simulation work is to treat segregation of a suitable neutral donor-vacancy complex. In the case of Ti doped MgO, the appropriate complex is a Ti^{4+}-Mg^{2+}(vacancy) pair [19]. In the bulk of MgO, the association energy between dopant and vacancy is

1.5eV, so that there is justification for this procedure. The bound neutral vacancy has a positive heat of segregation to the topmost ionic plane at zero coverage, although there is a negative enthalpy of the order of 0.5eV for the next ionic layer in from the surface. Both in the bulk planes and at the two surface planes the enthalpy becomes increasingly negative with increasing coverage (figure 7), so that there is a thermodynamic driving force for aggregation of the defect impurity pairs. This may be understood in terms of incipient formation of a second phase, presumably $MgTiO_3$.

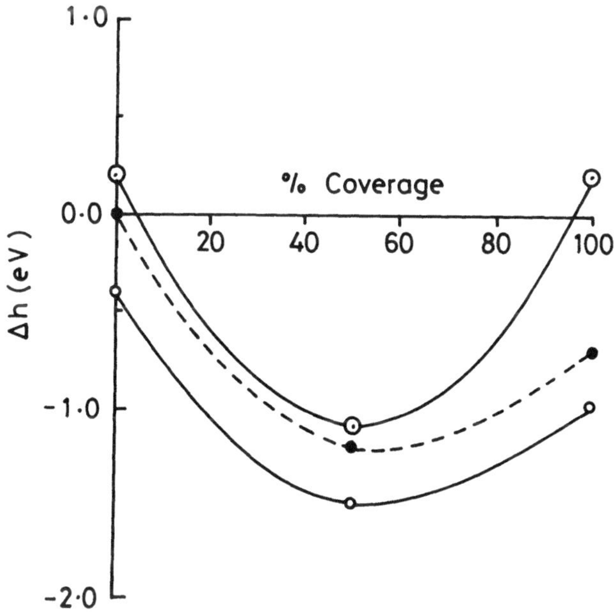

Figure 7. The heat of segregation of Ti to MgO(001) as a function of coverage. Circles with dots correspond to Ti-Mg(vacancy) complex in surface ionic layer; open circles correspond to impurity-vacancy complex in the next ionic layer down. The dashed line describes bulk aggregation behaviour in a (001) plane.

5.2 Mg on Al_2O_3(0001)

There has been a longstanding interest in magnesium doped alumina because of the practical importance of alumina ceramics.The difference in charge between Mg^{2+} and Al^{3+} can be accomodated by formation a neutral Mg^{2+}-O^{2-}(vacancy)-Mg^{2+} complex.as shown schematically in figure 8. In the bulk there is an association energy of 1.64eV, so that again the vacancy is strongly bound to the impurity. At the surface, one must consider alternative impurity-vacancy complexes, such as those of figure 8. Of these the third has the lowest energy at zero coverage. As with the case of Ti-doped Al_2O_3, the segregation energy passes through a minimum at a coverage apparently corresponding to a surface

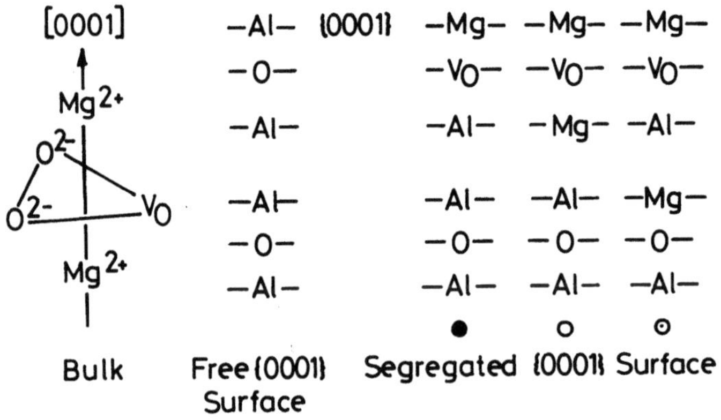

Figure 8. Structure of the Mg(dopant)-O(vacancy)-Mg(dopant) complex suggested by simulation studies together with different possible surface complexes.

phase of stoichiometry $MgAl_2O_4$.

Experimentally, Mg-doped alumina has proved more difficult to study than the other systems discussed so far because Mg evaporates from the alumina surface at the temperatures required to achieve sufficiently fast bulk diffusion to equilibrate the surface with the bulk. However in some experiments on a quasi-closed system in which two alumina crystals were held in contact inside a UHV system during equilibration at high temperature, Blakely and co-workers found the expected variation of surface coverage with equilibration temperature [20]. Their experimental Auger intensity ratios are shown in figure 9, along with computed surface coverages. The two sets of data have been arbitarily scaled to coincide in an absolute sense. It will be seen that there is good agreement over the temperature range of the experiments, although the interesting regime where a distinct surface phase appears in the calculations was not accessible experimentally owing to the low diffusion rate at low temperatures.

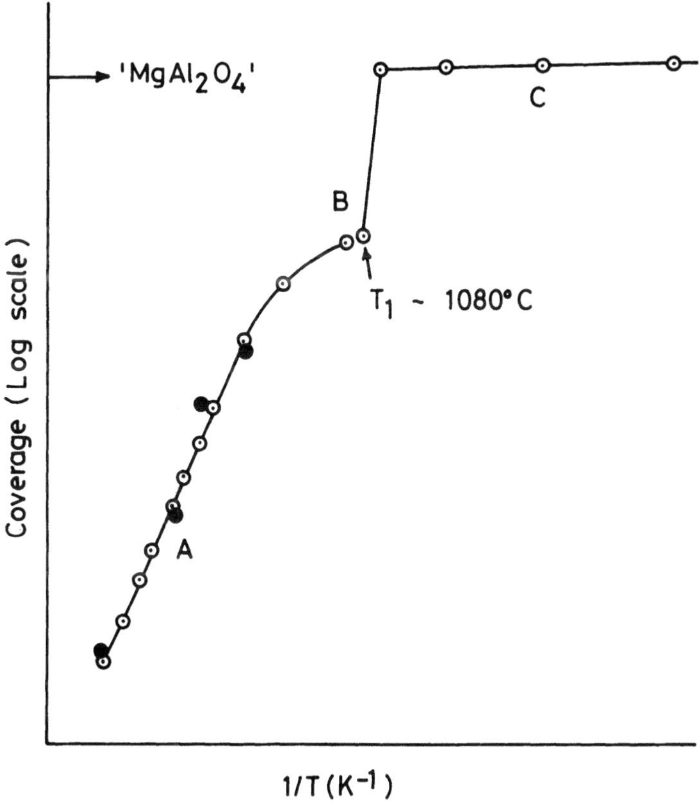

Figure 9. Surface cation ratio Mg/Al at the (0001) surface of 40ppm doped α-Al_2O_3 as a function of temperature. The solid line is from simulation studies. Experimental data points from AES have been scaled in an absolute sense to lie close to the theoretical curve so that visual comparison of the slopes may be made.

6. Some concluding remarks.

Simulation procedures have now explored a wide range of problems in relation to the behaviour of dopants at surfaces of magnesia and alumina. The scope of experimental work is more restricted, although in a few cases there is now a pleasing tie up between theory and experiment. This should give some confidence in the application of simulation procedures to study situations that are more difficult to approach experimentally e.g. grain boundary segregation. The theoretical work on surface coverages and on the structure of segregated surfaces should serve as a stimulus to well-designed experiments in this area.It is a particular problem that it is not always possible to locate true global energy minima in simulation studies and guidance from experimental data will often prove to be invaluable here.

7.Acknowledgements

We are deeply indebted to Dr. Alan Atkinson, who delivered the text of this lecture at the NATO ASI in September 1988.

REFERENCES

1. C.R.A. Catlow, 'Compututational techniques and simulation of crystal structure' in *Solid State Chemistry, Techniques*, eds. A.K. Cheetham and P.Day, Clarendon Press, Oxford, 1987, p.231.
2. *Computer Simulation of Solids*, eds. C.R.A. Catlow and W.C. Mackrodt, Springer Verlag, Berlin, 1982.
3. K.H. Rieder, 'Helium diffraction from MgO(001)', *Surf. Sci.*, **197** (1982) 57
4. M. Causa, R. Dovesi, C. Pisani and C. Roetti, 'Ab-initio Hartree Fock studies of MgO(001), Surf. Sci., **175** (1986) 551
5. E.A. Colbourn and W.C. Mackrodt, 'A theoretical study of CO chemisorption at (001) surfaces of non-defective doped MgO', *Surf. Sci.*, **143** (1984) 391
6. C.R.A. Catlow, M. Dixon and W.C. Mackrodt, 'Interionic potentials in ionic solids', ref. 2, p.130.
7. W.C. Mackrodt and P.W. Tasker, 'Segregation isotherms at the surfaces of oxides' ,*J.Amer. Ceram. Soc.*, in press.
8. P. Masri, P.W. Tasker, J.P. Hoare and J.H. Harding, 'Entropy of segregation of isovalent impurity cations at the surface of an ionic crystal: MgO(001)Ca^{2+}', *Surf. Sci.*, **173** (1986) 439.
9. E.A. Colbourn, W.C. Mackrodt and P.W. Tasker, 'The segregation of calcium ions at the surface of magnesium oxide: theory and calculation', *J. Mater. Sci.*, **18** (1983) 1917.
10. P.W. Tasker, E.A. Colbourn and W.C. Mackrodt, 'Segregation of isovalent dopant cations at the surfaces of MgO and CaO', *J. Amer. Ceram. Soc.*, **68** (1985) 74.
11. R.C. McCune and P.W. Wynblatt, 'Calcium segregation to a MgO(001) surface', *J. Amer. Ceram. Soc.*, **66** (1983) 111.
12. M. Cotter, S. Campbell, L.L. Cao, R.G. Egdell and W.C. Mackrodt, 'Surface segregation of Ba in MgO', *Surf. Sci.*, in press.
13. M. Cotter, S. Campbell, R.G. Egdell and W.C. Mackrodt, 'Growth of ordered BaO overlayers on MgO(001)', *Surf. Sci.*, **197** (1988) 208.
14. E.A. Colbourn, W.C. Mackrodt and P.W. Tasker, 'The structure of Ba segregated MgO(001)', *Physica B*, **131** (1985) 41.
15. W.C. Mackrodt and P.W. Tasker, 'Calculated impurity segregation at the surfaces of α-Al_2O_3', *Mat. Res. Soc. Symp. Proc.*, **60** (1986) 291.

16.W.C. Mackrodt, 'The calculated equilibrium segregation of Fe^{3+}, Y^{3+} and La^{3+} at low index surfaces of α-Al_2O_3', *Advances in Ceramics*, **23** (1987) 293.

17.W.C. Mackrodt, 'Atomistic lattice simulations of the structure, energetics and impurity segregation behaviour of ceramic boundaries', in 'Ceramic microstructures '86: role of interfaces' eds. J.A. Pask and A.G. Evans, *Materials Science Research*, **21** (1987) 271.

18.R.C. McCune, W.T. Donlan and R.C. Ku, 'Yttrium segregation and YAG precipitation at surfaces of Y-doped alpha alumina', *J. Amer. Ceram. Soc.*, **69** (1986) C196.

19.W.C. Mackrodt, P.W. Tasker and E.A. Colbourn, 'The surface segregation of titania in MgO', *Surf. Sci.*, **152** (1985) 940.

20.S. Baik, D.E. Fowler, J.M. Blakely and R. Raj, 'Segregation of Mg to the (0001) surface of doped sapphire', *J. Amer. Ceram. Soc.*, **68** (1985) 281.

SURFACE AND GRAIN BOUNDARY SEGREGATION IN METAL OXIDES

J. Nowotny
Max-Planck-Institut für Festkörperforschung
7000 Stuttgart 80, F.R.G.

ABSTRACT. Experimental approaches in studies of surface and grain boundary segregation in ionic solids are considered. The experimental material on both intrinsic and extrinsic segregation is discussed from the viewpoint of predominant driving forces of segregation. The direct effect of segregation involves an increased solubility limit within the interface region and resulting formation of bidimensional structures of extraordinary properties such as transport, magnetic and optical properties. The effect of segregation on properties of materials is considered in terms of phase diagrams of fine powder ceramics and sintering mechanism of ceramics. Also the catalytical aspect of the problem is briefly discussed.

Experimental data on surface and grain boundary segregation are reviewed for several oxide materials such as ZrO_2, Al_2O_3, MgO, SnO_2, NiO, CoO and their solid solutions. Possible effect of interfaces on critical parameters of high T_c oxide superconductors is also considered. Most urgent research pathways in the surface and grain boundary chemistry are discussed.

1. INTRODUCTION

It is well known that chemical composition of solid interfaces such as surfaces and grain boundaries differs from that of the bulk as a result of segregation. An awareness is also growing that the interface composition of ceramic materials has a substantial effect on their reactivity and many properties of practical importance.

The present knowledge on segregation in solids is practically limited to metals. It has been recently realized, however, that our understanding of the phenomenon of segregation for non-stoichiometric compounds is far behind application requirements. Accordingly, an interest has been recently generated in segregation studies of ionic solids.

The purpose of the present paper is to describe several aspects of surface and grain boundary segregation and its effect on processing and properties of oxide materials.

L.-C. Dufour et al. (eds.), Surfaces and Interfaces of Ceramic Materials, 205–239.

Experimental approaches to the study of segregation in metal oxides will be briefly reviewed and the extent of segregation of both intrinsic and extrinsic defects will be considered for several oxide materials such as NiO, CoO, Al_2O_3, SnO_2, MgO, ZrO_2 and their solid solutions. Moreover, the available experimental data on the effect of interfaces and the interface segregation on properties of oxide materials, including high T_c oxide superconductors, will be considered.

2. DEFINITIONS OF TERMS

Considerations in the present work are limited to the equilibrium segregation. The conceptual approach to describing this type of segregation involves essentially the same terms and relationships which have been derived for adsorption of gases. The main difference in both these phenomena involves the temperature at which the equilibrium state may be reached. Thus, in contrast to adsorption the equilibrium segregation, which is usually rate controlled by the lattice diffusion through the solid phase, may be established at elevated temperatures at which lattice diffusion becomes relatively fast. Accordingly, experimental studies relevant to the segregation equilibrium state should be performed in equilibrium conditions. Studies of solid interfaces in these conditions (high temperature and controlled gas atmosphere) are the subject of severe difficulties.

Most of the data in the literature on segregation are reported for quenched materials. Since the surface composition of metallic solids may be relatively well quenched the picture of segregation for metals and alloys is relatively well known. On the other hand the effect of segregation for ionic solids is much more awkward for experimental studies. Quenching of these materials may lead to as many changes (in both composition and structure) as many quenching procedures are applicable. Therefore, the main effort in segregation studies of ionic solids should be directed to 'in situ' investigations. Unfortunately, most of the available modern surface sensitivite techniques, which have been used in segregation studies, operate at high vacuum and also at room temperatures.

The experimental material in this work is essentially limited to metal oxides. Their properties depend on the type and the extent of non-stoichiometry. Fig. 1 illustrates possible types of the deviation from stoichiometric composition and resulting predominant defects as well as semiconducting properties.

3. THEORETICAL APPROACHES

Extensive description of the theory of segregation in solids is given by Blakely [1] and Edgell [2] (present proceedings) as well as by Hondros and Seah [3]. Therefore, theoretical considerations in the present paper are limited to a short discussion of fundamental relationships which have been used in considering segregation in solids.

1. p-TYPE SEMICONDUCTORS

DEFICIT OF METAL - $M_{1-y}O$ - ACCEPTORS (V_M)
NiO, CoO, MnO, FeO
Cu_2O, Fe_3O_4, Cr_2O_3

EXCESS OF OXYGEN - MO_{1+y} - ACCEPTORS (O_i)

2. n-TYPE SEMICONDUCTORS

EXCESS OF METAL - $M_{1+y}O$ - DONORS (M_i)
ZnO, Fe_3O_4

DEFICIT OF OXYGEN - MO_{1-y} - DONORS (V_O)
ZrO_2, UO_2 , TiO_2

3. EXTRINSIC SEMICONDUCTOR

Al_2O_{3-y}, $Mg_{1-y}O$

Figure 1. Deviation from stoichiometry and related semiconducting properties of transition metal oxides

The most general approach to describe both adsorption and segregation (adsorption from the bulk phase) is based on the Gibbs approach which gives the relationship between the surface enrichment Γ, surface tension Υ and the bulk composition X^b. For the i-th component we have:

$$\Gamma_i = \frac{-1}{RT}\frac{\partial \Upsilon}{\partial \ln X_i^b} \tag{1}$$

where Γ is the quantity expressed in moles of solute per unit surface. Thus the enrichment factor of the i-th element can be expressed as:

$$f = \frac{X^s}{X^b} = \left(\frac{1}{RT\Gamma_i^o}\frac{d\Upsilon}{dX^b}\right)_{T,\ X_j,\ X_i\ =\ const} \tag{2}$$

where Γ^o corresponds to the monolayer coverage, X_j is the composition of j-th element (j = i) and X^s denotes the surface composition. One should emphasize that f can be considered as a characteristic quantity of materials only for the identical impurity spectrum.

From the Langmuir adsorption isotherm we receive the following relation between the bulk and the surface composition (X^b and X^s, respectively):

$$\frac{X^s}{(1-X^s)} = \frac{X^b}{(1-X^b)}\exp\left(-\frac{\Delta G_{seg}}{RT}\right) \tag{3}$$

where ΔG_{seg} is the free energy of segregation. The regular solution approximation leads to a similar relationship which for a binary system assumes the form:

$$\frac{X_2^s}{X_1^s} = \frac{X_2^b}{X_1^b} \exp\left(-\frac{\Delta H_{seg}}{RT}\right) \quad (4)$$

where ΔH_{seg} is the enthalpy of segregation. According to Wynblatt and McCune [4, 5] ΔH_{seg} comprises three principal components resulting from a surface energy contribution (ΔH_γ), a binary solution interaction (ΔH_w) and an elastic solute strain energy (ΔH_e). In the case of segregation of aliovalent ions, an additional electrostatic component (ΔH_{el}) may result from the space charge in the boundary layer. Hence:

$$\Delta H_{seg} = \Delta H_\gamma + \Delta H_w + \Delta H_e + \Delta H_{el} \quad (5)$$

Both elastic strain energy and electrostatic components are usually predominant in ionic solids.

4. EXPERIMENTAL APPROACHES

Evaluation of segregation in solids is usually based on a comparison between the surface and bulk chemical composition data resulting either from a direct chemical analysis or the determination of parameters sensitive to chemical composition.

The most common approach in determination the surface composition and segregation-induced concentration profiles within the near-surface or the grain boundary region is based on depth profiling. Ion sputtering techniques combined with surface analysis using such methods as Secondary Ion Mass Spectrometry (SIMS), Auger Electron Spectrometry (AES), X-Ray Photoelectron Spectroscopy (XPS), Analytical Electron Microscopy (AEM) and Low Energy Ion Scattering (LEIS) provide information on changes in concentration of elements vs. distance from the surface. These methods are characterized by different depth resolutions varying between 0.5 and 10 nm. Elaboration of the depth profile intensity data in terms of concentrations requires a calibration by using appropriate standards. Another difficulty in studies of insulating surfaces such as Al_2O_3 or ZrO_2 results from a charging effect during the ion bombardment. Recent reports indicate a possibility to minimize this effect by the use of neutralized ions for sputtering [6, 7].

The sputtering technique leads to a destruction of the surface structure. Also the sputtering may be selective to certain lattice ions thus leading to changes in composition of the analyzed spot.

Another, non-destructive approach for the depth resolved analysis consists in electron spectrometry in which different electron kinetic energies are used. For the kinetic energy between 5 to 10^3 eV the mean free path (penetration or escape depth) may vary between 0.5 and 10 nm [8]. Also varying the incident angle one may change the depth resolution [8].

In contrast to several approaches in studies of surface properties depth profiling allows to perform a direct chemical analysis. The main disadvantage of the depth profile analysis is its limitation to study quenched specimens.

In these conditions only quenched solids may be studied. Thus, obtained experimental data may not be relevant to the state of the surface in segregation equilibrium. Accordingly, the depth profile analysis data should always be analyzed with a great caution regarding possible effect of quenching.

Another experimental approach in segregation studies involves the measurements of electrical parameters such as electrical conductivity (EC), Seebeck effect (SE), thermoelectric power of solid state galvanic cells (TEP) and work function (WF). Some of these methods (EC, SE and TEP) are essentially sensitive to the bulk while WF is entirely sensitive to the outer surface layer. Also the impedance spectroscopy has found a wide application in studies of grain boundaries. The main advantage of electrical methods involves their high sensitivity to structure and composition. Moreover, these methods can operate at elevated temperatures and under controlled gas atmospheres. However, these methods provide only indirect information about chemical composition and structure and require a verification with another methods.

Different approaches to evaluate segregation and its effect on structure and composition involve measurements of diffusion data independently for the bulk and the surface layer. It has been observed that segregation and resulting concentration profiles have a substantial effect on the transport of charged defects across the nearsurface layer [9, 10]. Recently proposed models of transport processes within the near-surface region enable to evaluate segregation from the re-equilibration kinetic data, if the bulk diffusion is known [9]. Several other experimental approaches to evaluate segregation are based on the oxygen pressure dependence of a composition related parameters and studies of materials of different grain size. A description of of these approaches is given in ref. [11, 12].

5. INTRINSIC SEGREGATION

Studies of several oxide materials have indicated that external surfaces and grain boundaries are enriched in intrinsic lattice defects such as vacancies and interstitials of host lattice elements. There are several reasons for the enrichment which are caused by the tendency of the surface to decrease its energy:

1) diffusion of defects from the bulk as a result of a segregation driving force (see § 3),
2) formation of defects which are not stable in the bulk phase,
3) formation of low energy bidimensional interface structures also termed as overlayer structures or 'surface phases'.

It has been observed that the outer crystal layer of metal deficient oxides ($M_{1-y}O$) are enriched in cation vacancies [12].

The reported enrichment coefficient for NiO, for example, may vary between three for the grain boundary [13] and five for the outer surface layer [14].

It has also been reported that the surface of such oxides as NiO and CoO may be enriched with oxygen vacancies [15] or cation interstitials [16] which were not observed in the bulk. This kind of segregation results from specific energetical conditions of the surface (the excess of surface energy) giving rise to different free energy of defects formation ΔG_f . The concentration of defects is the following function of both p_{O_2} and T:

$$[D] = p_{O_2}^{1/n} \exp\left(-\frac{\Delta G_f}{RT}\right) \tag{6}$$

where [D] corresponds to the deviation from stoichiometry and n is the parameter sensitive to the defect structure. ΔG_f is composed of both enthaply and entropy terms:

$$\Delta G_f = \Delta H_f - T(\Delta S_c + \Delta S_v) \tag{7}$$

where ΔH_f is the enthalpy of formation and ΔS_c and ΔS_v denote the configuration and vibration entropy terms, respectively. All these quantities should be considered as a function of the distance from the surface x. Accordingly, in considering the deviation from the stoichiometric composition we have to realize that this parameter and resulting defect concentrations within the interface layer are a function of x.

So far, the function $\Delta G_f(x)$ is not known even for simple oxide systems.

6. EXTRINSIC SEGREGATION

In contrast to intrinsic segregation there are many experimental approaches to the study of segregation of extrinsic lattice defects. Their concentration resolution varies between about 1% for SIMS and 5% for such methods as XPS or AES. The term 'extrinsic defects' involves solutes introduced into the lattice both intentionally (dopant) and unintentionally (impurities). Unfortunately, most of available data on this type of segregation are not well defined concerning the segregation conditions and thus cannot be considered as material data.

First signals on the extrinsic segregation in metal oxides resulted from studies of oxide catalysts involving the effect of doping on their catalytic properties. It has been observed that parameters of catalytic reactions in most cases do not exhibit a correlation with bulk properties of catalysts. The studies of Bielanski et al. [15, 17, 18] have shown that the activation energy of CO oxidation, performed over Li-doped NiO as a catalyst, change non-monotonously with composition (Fig. 2).

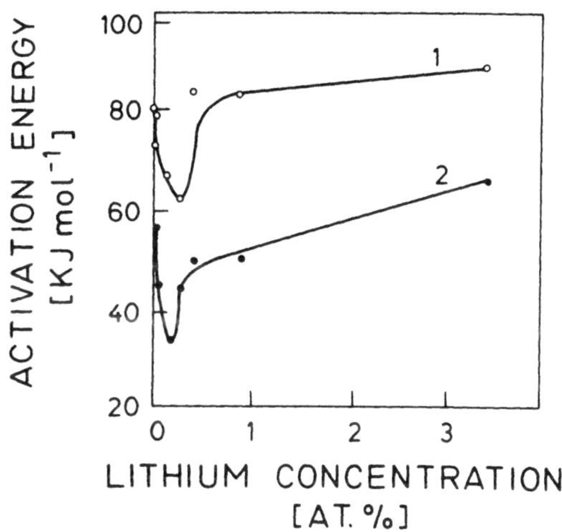

Figure 2. Activation energy of CO oxidation on Li-doped NiO catalyst as a function of composition. 1 - stoichiometric CO - O_2 mixture, 2 - excess of oxygen [15].

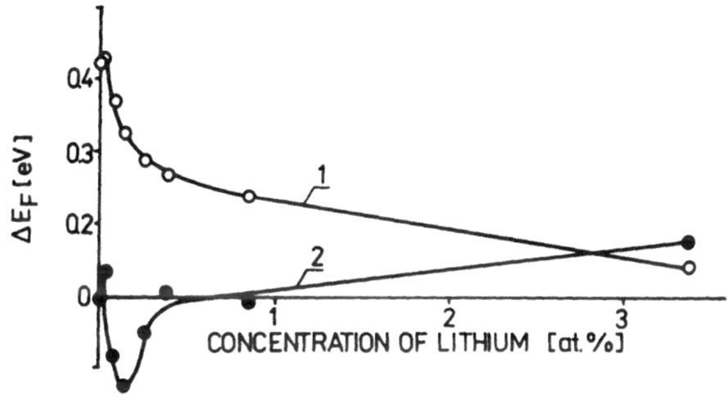

Figure 3. Changes in E_F for Li-doped NiO in the bulk (1) and at the surface (2) [15]

It has been observed that several bulk parameters such as the lattice constant [15], magnetic susceptibility [18] and bulk Fermi energy (curve 1 in Fig. 3) [15] do not exhibit a correlation with catalytical properties. On the other hand, there is an excellent agreement between catalytical parameters and work function changes (curve 2 in Fig. 3). The effect has been explained in terms of different mechanisms of lithium incorporation into the crystalline bulk (Eq. 8) and the surface layer of NiO (Eq. 9):

$$Li_2O + 1/2\ O_2 \rightleftarrows 2Li'_{Ni} + 2h^{\bullet} + 2O_O^{x} \qquad (8)$$

$$Li_2O \rightleftarrows 2Li_i^{\bullet} + V''_{Ni} + O_O^{x} \qquad (9)$$

Extensive studies were also performed for Cr-doped NiO [9, 19, 20]. It has been observed that Cr incorporates into the bulk of NiO according to the following substitutial mechanism:

$$Cr_2O_3 \rightleftarrows 2Cr^{\bullet}_{Ni} + V''_{Ni} + 3O_O^{x} \qquad (10)$$

Thus generated donor centres ($Cr^{\bullet}_{Ni}$) are responsible for the rise of the bulk Fermi energy level (E_F) what has been observed experimentally by measurements of the Seebeck effect (curve 1 in Fig. 4). Convers ely the surface electronic effect recorded by work function measurements does not confirm the mechanism (10) for the surface layer. It has been shown that the outer surface layer of $NiO-Cr_2O_3$ solid solutions is enriched in Cr and assumes a spinel-type structure [20]. SIMS studies have confirmed Cr segregation (Fig. 5) [8].

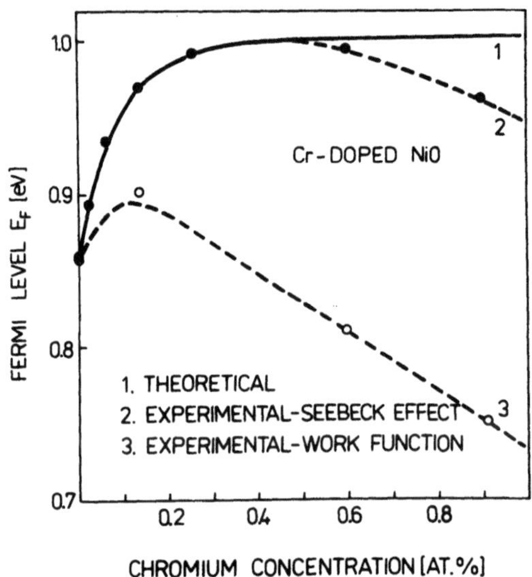

Figure 4. Changes in E_F for Cr-doped NiO. 1 - as predicted by the mass action law, 2 - in the bulk, 3 - at the surface [12]

Based on the above considerations one can propose the segregation model for undoped NiO as well as Li- and Cr-doped NiO. The outer surface layer of NiO enriched with cation vacancies assumes a negative surface charge which is compensated by the positive space charge formed by electron holes (Fig. 6). In the case of Li-doped NiO the negative surface charge of cation vacancies in compensated by the space charge composed of both Li interstitials and electron holes (Fig. 7).

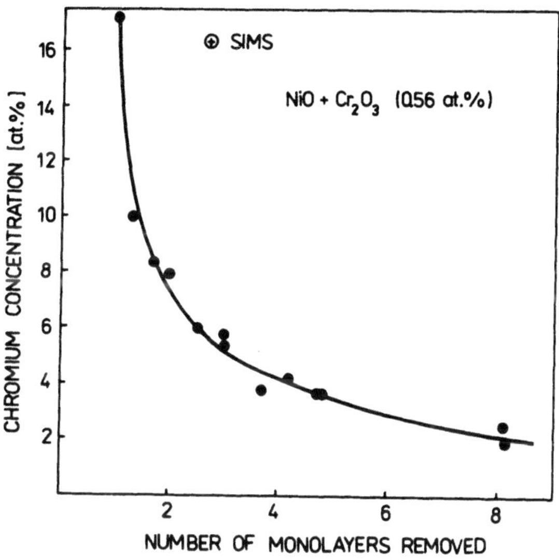

Figure 5. The concentration profile of Cr in Cr-doped NiO [8]

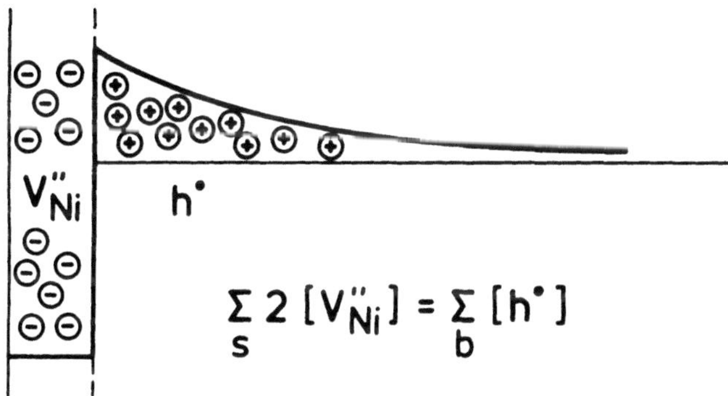

Figure 6. The surface model of undoped NiO

V_{Ni}''

$Li^{\bullet}, h^{\bullet}$

$$\sum_s 2[V_{Ni}''] = \sum_b [Li_i^{\bullet}] + [h^{\bullet}]$$

Figure 7. The surface model of Li-doped NiO

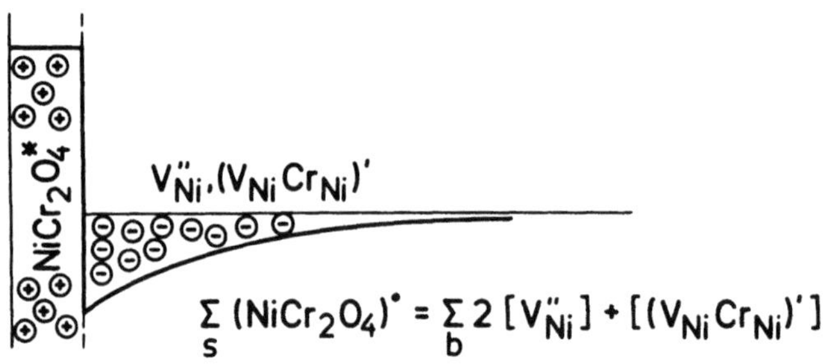

Figure 8. The surface model of Cr-doped NiO

Reduction of the strain energy may be considered as a predominant segregation driving force for Li segregation in the form of $Li_i^{\bullet}$ centres. Small amounts of Cr result in a drastical change in the NiO segregation model involving the formation of a bidimensional core layer of the spinel-like structure, which is positively charged, and the negative space charge layer formed of both cation vacancies and complexes $(V_{Ni}Cr_{Ni})'$ (Fig. 8). Comparison of ionic radii of Cr^{+3} and Ni^{+2} (0.072 and 0.069 nm, respectively) indicates that the elastic strain energy term cannot be responsible for segregation of Cr in NiO, if the same structure in the bulk and at the surface is taken into account. In this case, however, the outer surface layer is the subject of structural transitions leading to the formation of the spinel-type ordering.

Apparently, the tendency in the formation of this structure at the surface is responsible for Cr segregation.

Segregation of extrinsic defects has been the subject of many reports mainly for such refractory oxides as Al_2O_3 and MgO. Intensive segregation studies in alumina have been generated since Coble has reported that small additions of MgO into Al_2O_3 have a substantial effect on the mechanism of sintering which results in a high density alumina [21, 22]. The beneficial role of MgO was ascribed to the inhibition of a discontinuous grain growth and the enhancement of sintering. The studies of Jorgensen and Westbrook [23] have shown that the addition of MgO causes an increase in the grain boundary microhardness (Fig. 9).

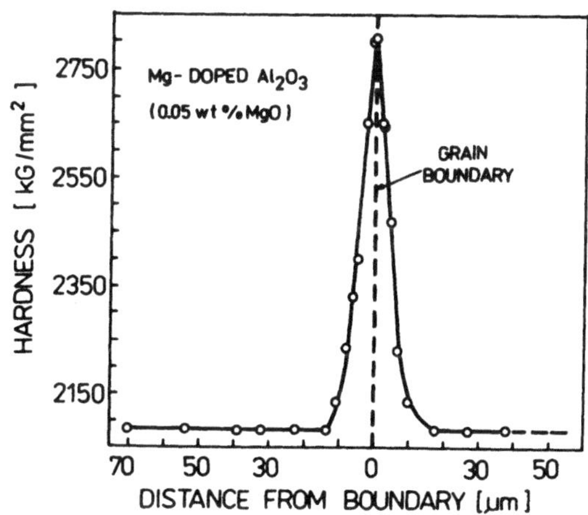

Figure 9. Hardness profile across a grain boundary in Mg-doped Al_2O_3 [23]

This effect has been ascribed to Mg segregation. Later surface analyses performed by AES seemed to indicate that segregation in Mg-doped Al_2O_3 is limited to Ca which is present in alumina at the level of several ppm while Mg segregation was not observed [24-26]. Fig. 10 illustrates segregation depth profiles of Ca in undoped (but not pure) alumina. The first evidence of Mg segregation was reported by Taylor et al. [27, 28]. They have shown that Mg segregates though the enrichment coefficient for Ca is substantially higher. These results seemed to be very consistent with values of ionic radii of the host and segregating cations (0.05, 0.065 and 0.099 nm for Al^{+3}, Mg^{+2} and Ca^{+2}, respectively). One may thus expect that the decrease in the elastic strain energy term is responsible for segregation. Recent works of Baik, Blakely et al. [29-32] have shown that Mg segregates as effectively as Ca, however, in contrast to Mg, Ca exhibits a substantial segregation anisotropy in alumina.

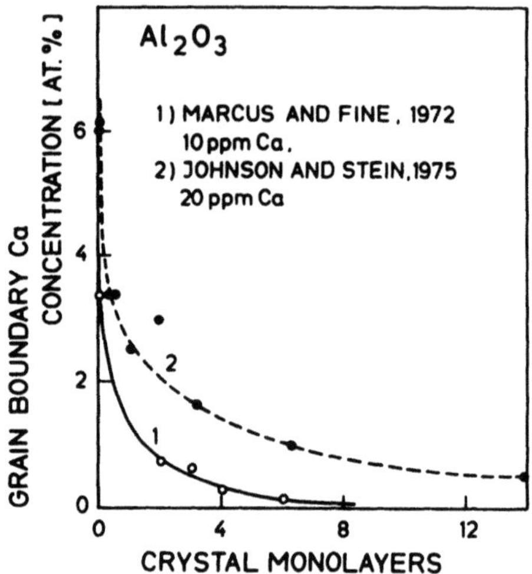

Figure 10. Grain boundary concentration profiles of Ca in alumina after Marcus et al. [24] and Johnson et al. [25].

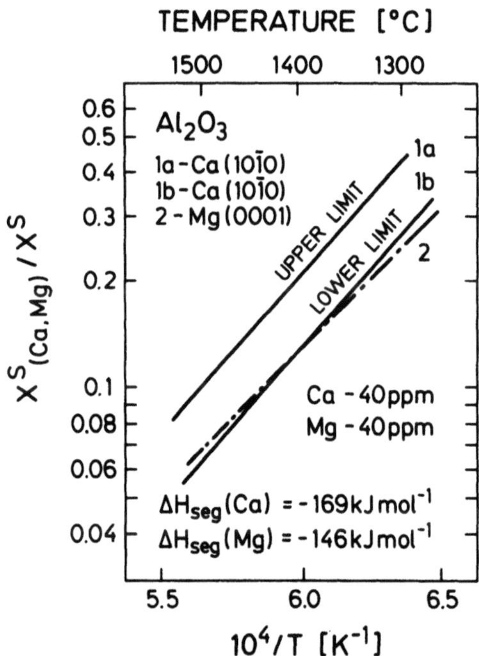

Figure 11. Grain-boundary concentration profiles of Ca and Mg in Al_2O_3 [32]

The enrichment coefficient of Ca is 10^3 and 50 for $(10\bar{1}0)$ and (0001) planes, respectively [31] (Fig. 11). Mg segregates to both planes with the enrichment about 10^3, leading to the formation of an overlayer bidimensional structure [29, 31]. It has been also shown that the presence of Mg is limited to grain boundaries while MgO evaporates from external surfaces [29, 30]. The evaporation increases with the increase of oxygen pressure. This explains the absence of Mg at free surfaces as it has been reported earlier [24-26]. Finally, it has been shown that at the $(10\bar{1}0)$ plane Mg can succesfully compete with Ca for surface sites [32]. A similar competitive effect has been observed for Ca and Y (Fig. 12).

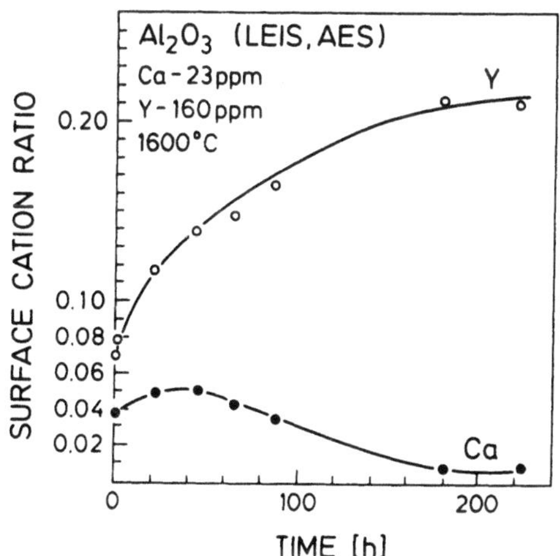

Figure 12. Changes in Ca/Y ratio at the surface of Y-doped alumina during the thermal treatment [26]

Pae k et al. [33] report that the advanced densification of alumina is favoured by the increase in oxygen partial pressure. This is in agreement with the observed effect of Mg evaporation with p_{O_2} [28, 29].

It has also been reported that the addition of Y results in a similar way on sintering of Al_2O_3 as it is the case for MgO [26] while Ca causes a negative effect.

Li and Kingery [34] have studied segregation of various elements in polycrystalline alumina. The enrichment of the elements measured at different temperatures were analyzed from the viewpoint of the misfit strain energy as a predominant driving force. The considerations are based on an assumption that the measured enrichment values are characteristic for the elements in the material.

However, as was indicated above (Eq. 2) the enrichment may be considered as a characteristic parameter for a given element, if several parameters having an effect on the enrichment, such as concentration of all lattice components and temperature, are fixed. Accordingly, the reported linear dependence between the enrichment and the strain energy parameter [34] seems to be accidental.

According to Baik [30] the outer core layer for Mg-doped Al_2O_3 is enriched by negatively charged Mg ions and the positive space charge layer (Fig. 13).

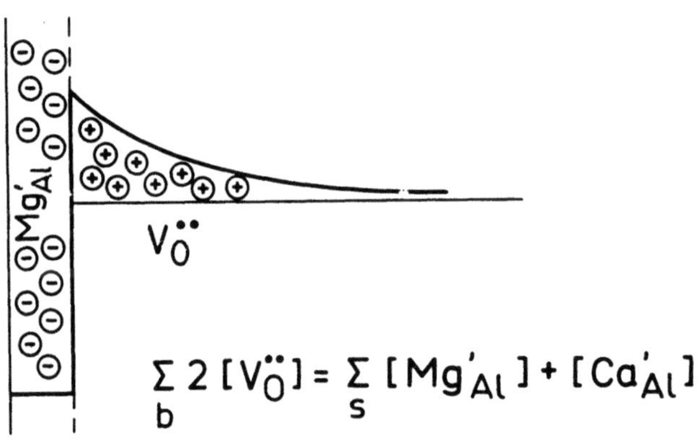

Figure 13. The surface model of Mg-doped alumina

Another material which has been the subject of segregation studies is MgO. Extensive investigations of Ca segregation in MgO have been reported by McCune and Wynblatt [4, 5]. The results of both AES and LEIS spectroscopy suggest that segregation is limited to the outermost layer. Thus, the ΔH_{seg} determined for (100) surface is about 50 kJ mol^{-1} [35]. Turner et al. [35] have shown that Ca segregation has an effect on the initial dissolution rate of MgO. Kingery et al. [36, 37] have reported that the grain boundary structure does not have an effect on Fe segregation. No explanation of this effect was given. Black and Kingery [38] have also reported the effect of p_{O_2} on both Cr and Fe segregation in MgO (see § 7).

An important ceramic material which should at least be mentioned here is zirconia. Axelsson et al. [39] have shown that segregation of Si in undoped ZrO_2 results in the formation of the $ZrSiO_4$ overlayer. Burggraaf et al. [40] have demonstrated that yttria segregates to the surface of yttria stabilized zirconia with the enrichment coefficient of about 1.5. The segregation of yttria is blocked by Bi which exhibits much higher enrichment (f=10).

It has been shown recently that the real surface model of yttria stabilized zirconia is composed of an overlayer, essentially enriched in Ca, and the sublayer enriched with yttria. In this case the surface enrichment in Ca, which is present in the bulk at the level of 10 ppm, is about 10^4 [41].

7. EFFECT OF COMPOSITION

The surface layer formed as a result of segregation remains in contact with both the bulk phase and and the gas phase. Therefore, the segregation equilibrium should be considered as an equilibrium between the gas phase and the surface layer on one side and the surface layer and the bulk phase on the other side.

In the case of oxide crystals the most important parameter of the gas phase is oxygen activity. Since oxygen is a constituent of the lattice its activity in the gas phase has a direct impact on the deviation from stoichiometric composition and thus resulting concentration of defects in both cation and anion sublattices. One should expect that the extent of non-stoichiometry and resulting defect structure has a substantial effect on segregation although, so far, little is known on the dependence between concentration of intrinsic defects and the extent of segregation of extrinsic defects. The observed effect of p_{O_2} on the near-surface segregation profile has been considered either in terms of different rate of evaporation of segregating elements [12, 38] or in terms of different driving forces as a result of changing their valency with p_{O_2} [38]. In the first case the segregation equilibrium cannot be reached as it has been described for Mg segregation to open surfaces in vacuum [28, 29]. In the second case the effect consists in changing the valency of segregating ions as it has been demonstrated for Fe-doped MgO [38]. At low p_{O_2} Fe^{2+} ions are isovalent and do not segregate in the surface space charge. The increase in p_{O_2} leads to the formation of Fe^{3+} ions which segregate effectively to the MgO surface (Fig. 14).

The effect of composition may be considered in terms of lateral interactions between segregated species. Very dilute systems may be considered within an ideal approximation. However, as concentration increases then lateral interactions may lead to substantial deviations from ideality. This results in strong changes in ΔH_{seg} as it has been observed for the Sb_2O_5-SnO_2 system (Fig. 15) [42]). For Cr-doped NiO ΔH_{seg} remains relatively constant [43]. Changes in lateral interactions may also lead to major changes in the structure leading to the formation of bidimensional overlayers [12, 16, 29, 31, 39, 44].

The observed tendency of a constant surface coverage results in a strong effect of bulk composition on the enrichment factor. As is seen from Fig. 16, the enrichment factor for Cr-doped CoO varies between 15 and 100 at high and low Cr concentration, respectively.

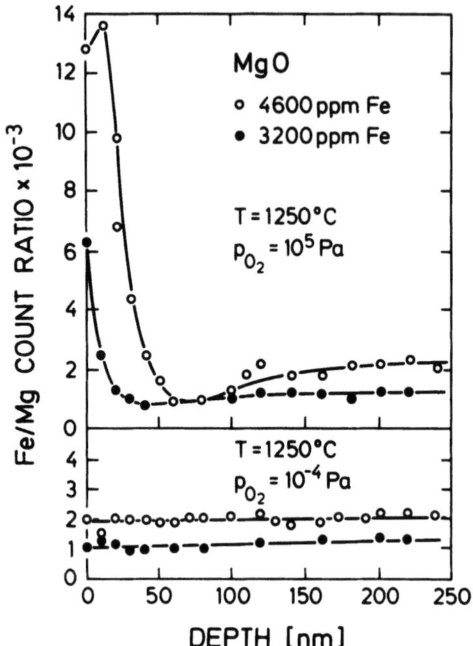

Figure 14. Effect of p_{O_2} on Fe segregation in MgO [37]

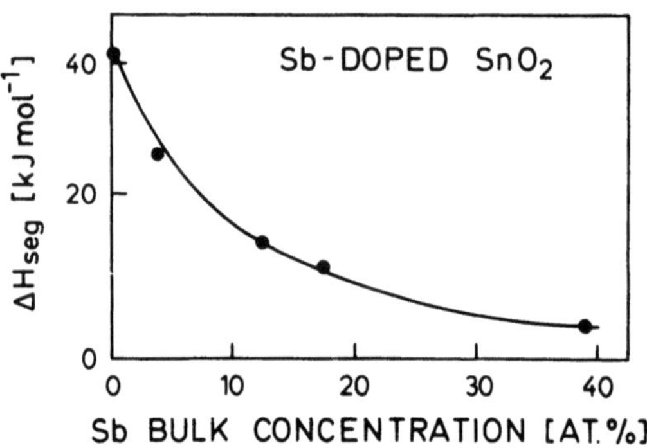

Figure 15. ΔH_{seg} for Sb-doped SnO_2 as a function of composition [42].

For very dilute solutions and high values of segregation driving forces the enrichment coefficient may reach 10^4 - 10^5 as has been reported for Ca segregation in yttria stabilized zirconia [41] and alumina [25, 29-32].

8. SURFACE VS. BULK SOLUBLITY LIMIT

The enrichment at the surface and within the near-surface region produced as a result of segregation may be considered in terms of a local solubility limit which is a function of the distance from the surface or a grain boundary (x). Accordingly, the relation between the bulk and the surface concentration ratios of the i-th element (solute) to the concentration of the host element (solvent) in a binary metal oxide may be written in the form:

$$\frac{c_i^b}{c_h^b} < \frac{c_i^s}{c_h^s} \tag{11}$$

where h refers to the host element. The upperscripts b and s correspond to the bulk and the surface, respectively, The ratio c_i^s/c_h^s strongly increases as x decreases and assumes maximum at the outer layer (x=0).

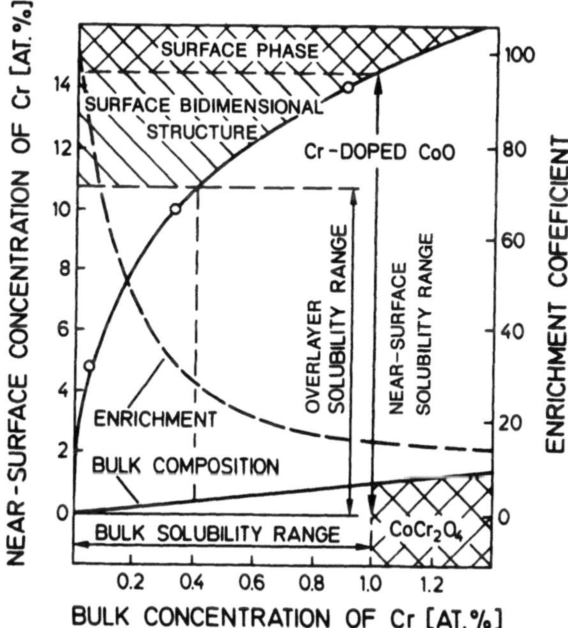

Figure 16. Near-surface vs. bulk composition of Cr-doped CoO [12]

Accordingly:

$$c_i^o = \lim_{x \to 0} c_i(x) \tag{12}$$

where $c_i(x)$ is the concentration function within the interface region. For extrinsic defects $c_i(x)$ can be determined by depth profiling.

The solubility limit is usually considered as an overall, maximum solubility in a crystal. This value can be termed as the 'bulk solubility'. Within the near-surface layer the solubility is usually enhanced as a result of segregation. The solubility limit in the regions close to the surface can be termed as the 'near-surface solubility'. This quantity is a function of x. Moreover, there has been an accumulation of experimental data indicating that overlayer structures may be formed within the bulk solubility limit [12, 16, 29, 31, 39, 44]. Thus, formed bidimensional structures cannot be termed as phases from the thermodynamics viewpoint as the system in these conditions is still bivariant. The highest concentration of the solute corresponding to the formation of the overlayer structures will be termed as the 'overlayer solubility'. These three types of solubility can be illustrated for the Cr_2O_3-CoO solid solution. The bulk solubility limit for Cr in CoO at 1273 K is about 1 at%. [44]. The corresponding near-surface solubility is over 14 at% Cr and is thus over one order of magnitude higher than in the bulk. As is seen from Fig. 16 the overlayer saturation for Cr- doped CoO is over 10 at% and corresponds to 0.4 at% Cr in the bulk [44]. Accordingly, the overlayer saturation is higher from bulk solubility by the factor of 25. It has been observed that the formation of bidimensional surface layers results in dramatic changes in both lattice transport mechanism [44] and the enthalpy of segregation [43]. This effect is of importance in the interpretation of phase diagrams for very fine grain materials. One should therefore expect a large dependence on the parameters corresponding to the phase transition and the grain size, if the grain size is lower than a certain critical value.

9. NEAR-SURFACE AND BULK EQUILIBRATION KINETICS

Re-equilibration in the interface region can be considered in terms of two quasi-equilibria: (1) between the gas phase and the interface region, and (2) between the interface region and the bulk phase.
1) When oxygen partial pressure is suddenly changed over an initially equilibrated oxide crystal then a new oxygen activity is imposed at the surface which corresponds to a change in oxide non-stoichiometry. The re-equilibration of the oxide crystal thus involves the propagation of the new level of non-stoichiometry y_2 from the surface into the bulk. For a metal deficient oxide we may, therefore, write:

$$2\,\frac{1-y_2}{\Delta y}\,M_{1-y_1}O + O_2 \rightleftarrows 2\,\frac{1-y_1}{\Delta y}\,M_{1-y_2}O \tag{13}$$

where subscripts 1 and 2 correspond to two p_{O_2} values.

where:

$$y_2 = y_1 + \Delta y \tag{14}$$

The re-equilibration process is rate controlled by the diffusive transport of defects from the surface into the bulk or vice versa. Accordingly, the re-equilibration kinetics has been frequently used in determination of the chemical diffusion coefficient [9, 45]. In the case of $M_{1-y}O$ type oxides, the re-equilibration involves the transport of cation vacancies which are the predominant lattice defects. Since the near-surface crystal layer differs from the bulk in many aspects such as lattice parameter [11], lattice ordering [44] and composition, the re-equilibration process should be considered independently for this layer and the bulk. Therefore, assuming that at elevated temperatures cation vacancies are doubly ionized, the interaction involves the formation of the vacancies at the surface and their transport into the bulk:

$$1/2\ O_2 + MO \rightleftarrows (V_M'')_s + O_O^x + 2h^{\bullet} \tag{15}$$

$$(V_M'')_s \rightleftarrows (V_M'')_b \tag{16}$$

where subscripts s and b correspond to the near-surface (or a grain boundary) region and bulk, respectively.
2) The re-equilibration process should be considered involving all kinds of bulk defects such as impurities and dopants which segregate to the interface region during the thermal treatment. Considering only one type of i-th solute the equilibrium between the bulk and the near-surface region may be expressed as:

$$c_i^b \rightleftarrows c_i^s \tag{17}$$

Segregation leads to a shift of Eq. (17) into the right. Even assuming that the bulk concentration of the i-th element is very low, its changes in the bulk as a result of segregation can be neglected, if the surface to volume ratio is low. On the other hand changes of surface concentration may be significant and cannot be neglected. Accordingly, the considerations on the re-equilibration process in the bulk can be limited to intrinsic defects formed according to Eq. (15), while for the near-surface layer the picture is more complex. The near-surface re-equilibration also involves all kinds of solutes which segregate from the bulk to the surface. If the lattice transport kinetics of solutes is very rapid then both near-surface and bulk equilibration processes are rate controlled by diffusion of cation vacancies. Conversely, when the transport of solutes is low then the re-equilibration process of the near-surface layer is determined by the kinetics of segregation.

It has been a general assumption that at high temperatures the surface process is always rapid in comparison with bulk transport processes. However, it has been recently demonstrated that the rate of the near-surface re-equilibration may be low and rate controlling. This has been observed for undoped NiO [46] and yttria-doped zirconia [41]. In both cases the bulk diffusion of intrinsic defects is relatively rapid while the equilibration of the near-surface layer is rate controlled by Ca segregation which is present in the bulk as an impurity at the level of 10 ppm [41, 46]. The segregation of Ca corresponds to the drift of the reciprocal of the oxygen pressure exponent (Eq. 6) for the surface of undoped (but not pure) NiO while in the bulk this parameter remains constant [46]. Similar effect has been observed for yttria stabilized zirconia [41].

From the above considerations it results that the mechanism and the kinetics of the re-equilibration for the near-surface layer on one side and the crystalline bulk on the other side can be entirely different and should be considered independently.

10. THICKNESS OF THE BOUNDARY LAYER

Segregation leads to the formation of the surface charge which is compensated by the space charge in the boundary layer. According to Frenkel [47] the surface charge in Schottky-type defected crystals is formed as a result of different formation energy of defect pairs. This results in a tendency to form one kind of defects at the surface in an excess concentration.

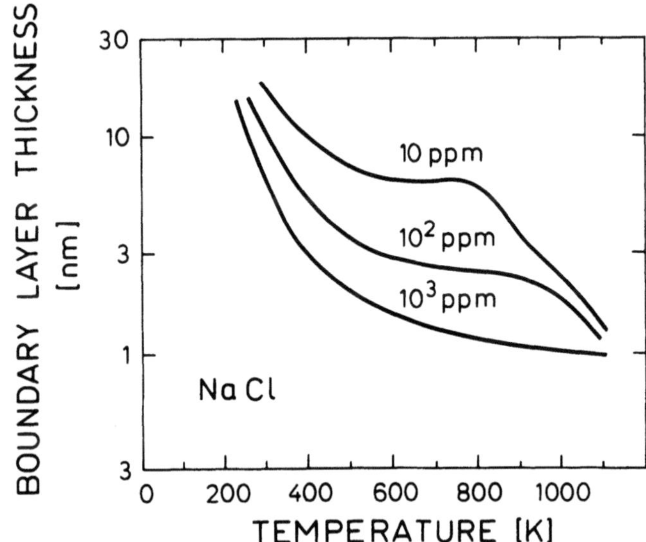

Figure 17. The boundary layer thickness for NaCl calculated by Kliewer and Koehler [48]

Kliewer and Koehler [48] have developed the Frenkel model for NaCl- type crystals. They have determined the depth of the space charge layer at elevated temperatures at different concentration of divalent impurities. As can be seen from Fig. 17 the thickness of the boundary layer according to Kliewer and Koehler [48] varies between about 0.1 nm at 1000 K and 10 nm or more around room temperature.

The model of Kliewer and Koehler has been derived making certain assumptions. The model is limited to Schottky-type defects in NaCl-type stoichiometric crystals. Therefore, the model is not applicable to non-stoichiometric compounds. As it has been widely demonstrated the extent of the segregation gradient for oxide ceramic materials may be entirely different from that predicted by the model of Kliewer and Koehler.

Assuming that the surface charge is limited to the outer layer the depth of the space charge layer is characterized by the Debye length:

$$L = \left(\frac{\varepsilon\varepsilon_o kT}{8\pi q^2 c}\right)^{1/2} \tag{18}$$

where ε and ε_o denote the dielectric permittivity of the sample and of vacuum, respectively, and c is the concentration of charge carriers. Thus calculated values of the Debye length for undoped and Cr-doped NiO are shown in Fig. 18. It can be seen that the values differ strongly from those in the model of Kliewer and Koehler [48].

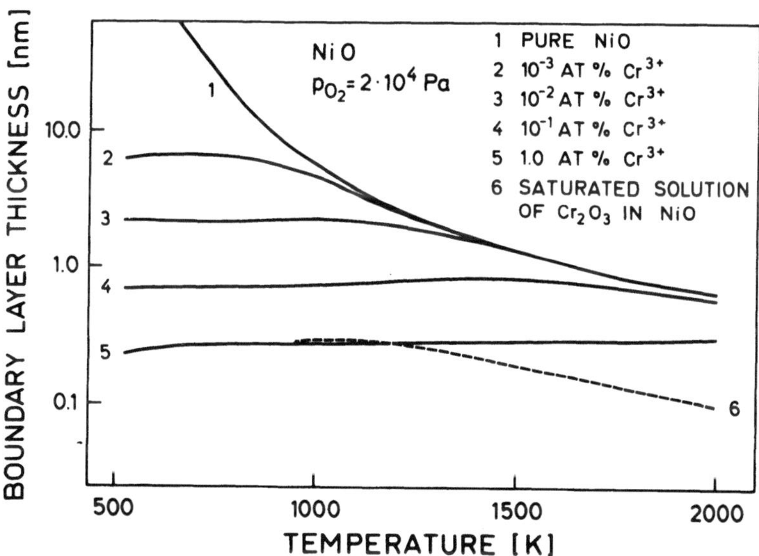

Figure 18. The Debye length for Cr-doped NiO

Table I shows the boundary layer thickness for several oxide materials determined experimentally by Komatsu et al. [49].

TABLE I
Surface layer thickness for binary oxides after Ikuma and Komatsu [49]

Oxide	Temperature [^oC]	Thickness [nm]	Method
CoO	253 - 723	4.0 - 60	kinetic
Cu_2O	280 - 380	7.5 - 92	kinetic
TiO_2(rutile)	883 - 965	4.5 - 20	kinetic
α-Al_2O_3	900 - 1000	3.0 - 13	kinetic
NiO	300 - 600	0.3 - 2.8	isotope ex.
Y_2O_3	350 - 450	2.8 - 4.6	isotope ex.
MgO	600 - 900	0.3 - 0.5	isotope ex.

11. INTERFACE TRANSPORT KINETICS

As can be seen from the schematic illustration in Fig. 19 the surface transport kinetics (D_s) is much faster than the bulk lattice diffusion (D_l). The grain boundary diffusion (D_{gb}) is usually faster than D_l. Generally, the activation energy of these three types of diffusion increases in the following order:

$$E_l > E_{gb} > E_s \qquad (19)$$

Accordingly, at lower temperatures, both surface and grain boundary transport mechanisms predominate, while at high temperatures the lattice transport prevails.

Both surface and grain boundary diffusion have been generally considered in terms of the transport along interfaces. In contrast to this type of transport also the transport across the interface region should be distinguished. This transport has been usually assumed as very rapid with comparison to the lattice diffusion. Recently, it has been shown, however, that this assumption may not be valid and that the overall heterogeneous kinetics may be rate controlled by the transport kinetics across the interface region [9].

11.1. Grain Boundary Diffusion

The transport along grain boundaries or within the grain boundary region, in the direction parallel to the interface, is usually termed as the grain boundary diffusion.

The grain boundary diffusion is very sensitive to both composition and structure as it has been observed for the bulk.

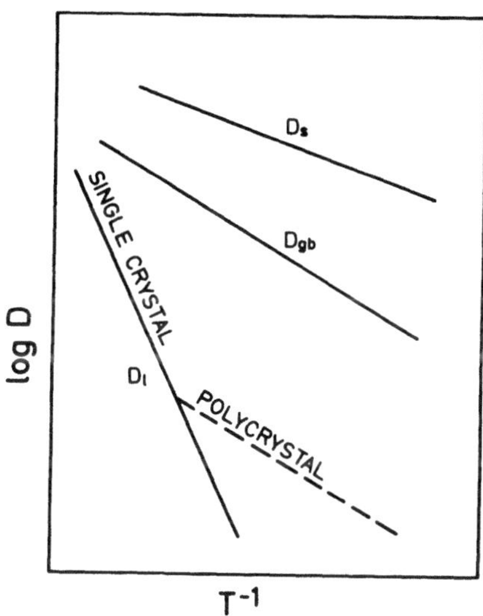

Figure 19. Illustration of the temperature dependence of different types of diffusion; surface diffusion (D_s), grain boundary diffusion (D_{gb}), lattice diffusion (D_l)

Unfortunately, both composition and structure for the interface region are poorly defined and, therefore, the grain boundary diffusion data available in the literature should be interpreted with great caution concerning their physical meaning. Recent studies of Barbier and Dechamps [50-52] have shown that segregation of impurities to grain boundaries has a substantial effect on the measured diffusion coefficient which should be interpreted along with the picture of segregation. Generally, the grain boundary segregation of defects results in an enhanced grain boundary transport. However, for pure specimens, when segregation is low, the grain boundary transport kinetics may be comparable with the lattice diffusion as it has been observed for NiO [50, 51].

It seems that the lack of characterization of grain boundaries concerning their structure and composition is responsible for an apparent disagreement between the reported experimental data.

Extensive description of the grain boundary transport kinetics is given by Atkinson [53].

11.2. Transport Across Interfaces

In contrast to the grain boundary diffusion considered above, the transport of defects across interfaces may be rate controlling the entire heterogeneous kinetics even at high temperatures [9]. Despite the importance of this transport mechanism little is known about its influence.

Segregation of solutes causes the formation of concentration gradients across the interface region and accompanying potential drop. Thus resulting electric fields may either retard or enhance the transport of charged defects (Fig. 20).

EFFECT OF ELECTRIC FIELD ON DIFFUSION

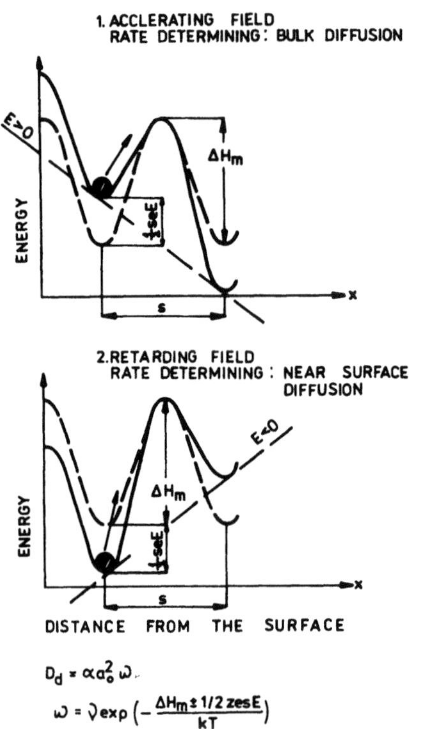

Figure 20. Illustration of the effect of an electric field on diffusion

For an enhanced potential barrier, normal to the interface, the overall equilibration kinetics is rate controlled by lattice diffusion and thus resulting kinetic data are suitable for determination of bulk-related chemical diffusion. Conversely, when the near-surface potential barrier retards the transport of defects then the overall heterogeneous kinetics is rate controlled by the near-surface diffusive resistance. The retarding effect depends essentially on temperature and the valency of diffusing species [9].

The effect of segregation on the transport kinetics across interfaces has been considered for the gas-solid heterogeneous system [9]. Solution of Fick's second law together with the Poisson's equation enables calculation the effect of the near-surface electric field on the transport kinetics across the boundary layer. Fig. 21 and 22 illustrate the effect of the segregation induced potential barrier on the apparent value of the chemical diffusion coefficient for undoped NiO [9]. As can be seen the barrier may produce a substantial retarding effect which results in changes of $\tilde{D}$ by several orders of magnitude. The effect becomes negligible above 1600 K. As a matter of fact, most of the available diffusion data in the literature correspond to lower values of T. These data require a verification from the viewpoint of a possible contribution of the segregation-induced diffusive resistance. Determination of this contribution is necessary for the correct understanding of the physical meaning of chemical diffusion data.

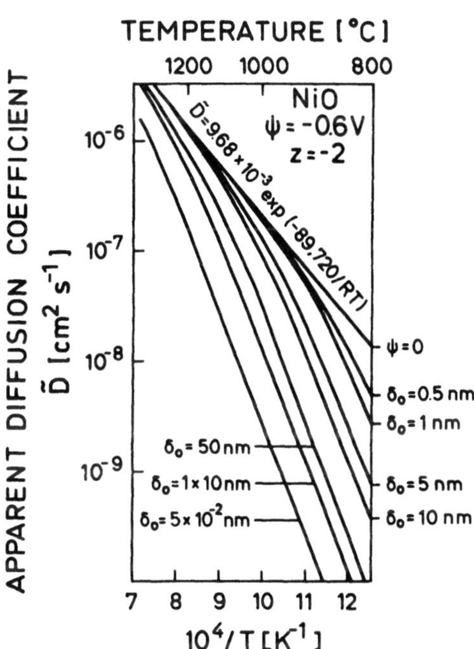

Figure 21. Effect of the surface potential barrier on the apparent value of chemical diffusion coefficient of NiO [9]

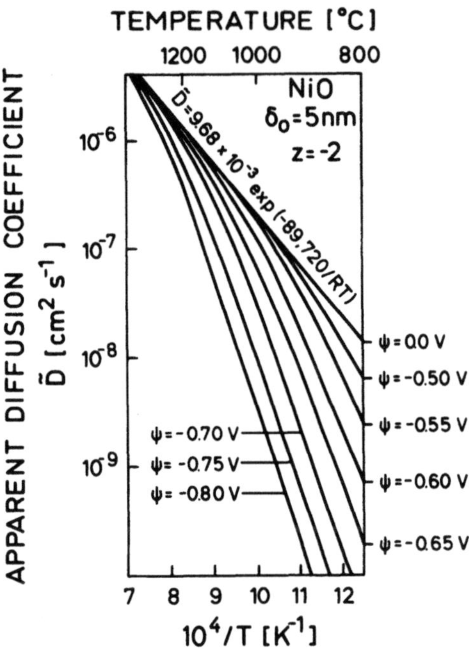

Figure 22. Effect of the extent of the potential barrier on the apparent value of chemical diffusion coefficient of NiO [9]

12. HIGH T_c OXIDE SUPERCONDUCTORS

So far, little is known on segregation in high T_c oxide superconductors. The available data on surface and grain boundary properties of these materials are frequently controversial and do not allow to derive a consistent picture of the effect of segregation on properties of these important oxide materials. There are many reports, however, which indicate that superconductivity is not homogeneous in ceramic materials [54-59].

It has been already reported by Seyama [60] that cations may segregate in perovskite-type oxides. The segregation depends on the redox pre-treatment. Wen et al. [61] have shown that the outer surface layer of $YBa_2Cu_3O_x$ exhibits a reconstruction and is depleted in yttria. Ginley et al. [54] have even claimed that the superconductivity in the Y-Ba ceramic cuprates is limited to thin shells while the bulk of grains remains metallic. Although it has been later documented that the superconductivity is also a bulk property there are many signals of evidence that grain boundaries and other dislocations play an important role. Cai et al. [59] have demonstrated that the critical temperature at grain boundaries of Y-Ba-Cu oxides may reach 167 K while the T_c for the crystalline bulk remains at the level of about 95 K.

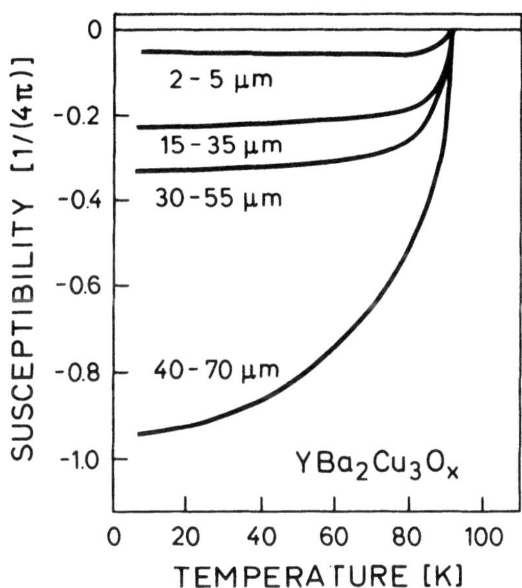

Figure 23. Effect of grain size on magnetic susceptibility of $YBa_2Cu_3O_x$ [63]

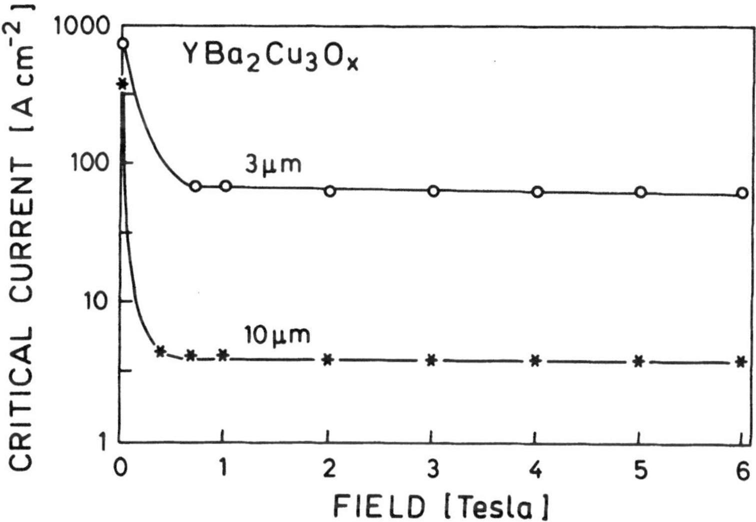

Figure 24. Effect of grain size on the critical current of $YBa_2Cu_3O_x$ [64]

Recent work reported by Dimos et al. [62] has shown a strong effect of the crystal misorientation on the critical current density. It has also been shown that the grain boundary density has a strong effect on the magnetic susceptibility (Fig. 23) and the critical current (Fig. 24) of Ba-Y-Cu oxides [63, 64]. Recently Duan et al. [65] have documented that the magnetic susceptiblity (below T_c) substantially decreases after the outer crystal layer is removed (Fig. 25).

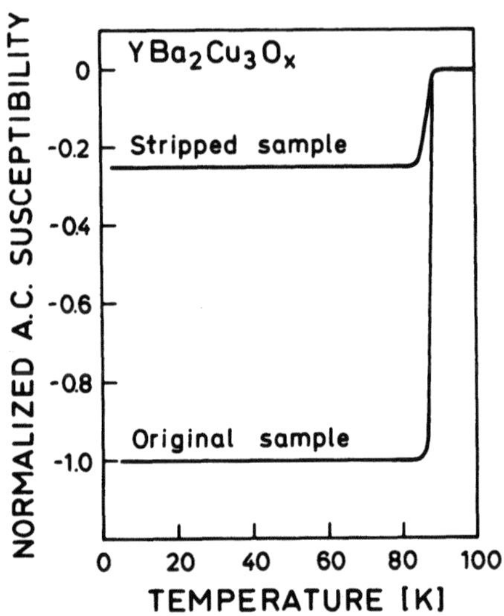

Figure 25. Magnetic susceptibility of both entire grain and its bulk (after removal of the outer layer) [65]

The main difficulty in studying the effect of interfaces on superconductivity arises due to its poor reproducibility. This seems to result from the variety of impurity spectra of studied materials. As has been shown above, segregation may lead to a high concentration of impurities at interfaces even for materials of a high purity, if the segregation driving force is high enough. Therefore, in the discussion of the effect of interfaces on superconductivity the segregation of both intrinsic and extrinsic defects should be taken into account. However, so far little is known about segregation in Ba cuprates. The reported experimental material is limited to the evaluation of segregation of Y [61] and Cu [64, 66].

One should expect that conditions of preparations such as gas phase composition, temperature of annealing and quenching procedures have an effect on the picture of segregation involving not only host ions such as Y, Ba, Cu and oxygen but also impurities.

Even if the redox treatment is easily reproduced it is difficult to control the impurity level. Thus resulting concentration gradients may lead to changes in several properties within the interface region as it is illustrated in Fig. 26.

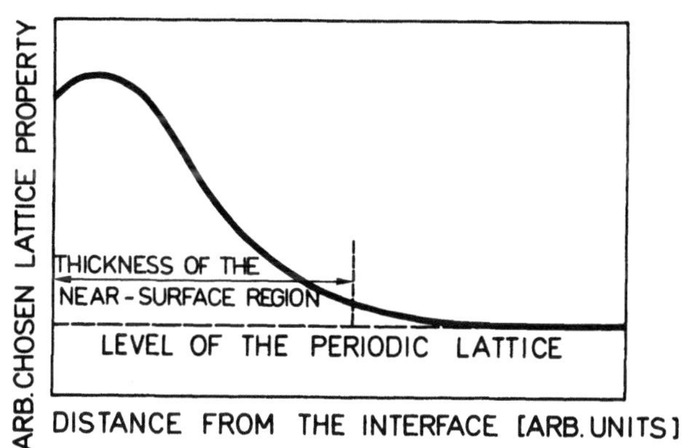

Figure 26. Changes of an arbitrary chosen lattice property within the interface region as a function of the distance from the interface

Both T_c and other parameters of superconductors may change accordingly [67]. One should, therefore, expect that besides the geometric factor (misorientation) the chemical concentration gradient may be responsible for extraordinary properties of materials within the interface region including their superconductivity.

13. URGENT RESEARCH ORIENTATIONS

From the experimental material presented in this work, it can be seen that the phenomenon of segregation has a strong impact on properties of materials. Segregation leads to changes in chemical composition of the interface region. The solubility of both intrinsic and extrinsic lattice defects within this region may be essentially different from that in the bulk. It has also been shown that solutes have a tendency to achieve a constant surface concentration, independent of bulk composition. Consequently, the enrichment coefficient of defects increases with the decrease of their bulk concentration. Within the solubility limit ionic crystals exhibit a tendency to form overlayer structures which can only be stable in specific conditions of the interface region.

So far, the effect of segregation on the interface defect chemistry is not known even for simple oxide materials. Accordingly, it is urgently required that systematic studies on the effect of segregation on the defect structure of the interface region be undertaken.

As segregation is a thermodynamic phenomenon, the relevant composition data for single and multicomponent segregation should be studied 'in situ' i.e. in the conditions of the thermodynamic equilibrium. As long as the results for quenched speciments are considered we have to realize that the effect of quenching on the interface defect structure can be significant. Therefore, new experimental approaches to study surface properties at elevated temperatures and under controlled gas atmospheres are required.

Conceptual approaches in evaluation of interface properties may be based on the defect structure approach similar to that applicable in the bulk defect chemistry. A general description of this approach should involve specific conditions of the interface region and, consequently, electroneutrality conditions based on particular solutions of the Poisson's equation. In general, the parameters for the interface region should be considered as a function of the distance from the interface. Thus, resulting 3 D - type dependences for Cr-doped CoO, as an example, are illustrated in Fig. 27 and 28.

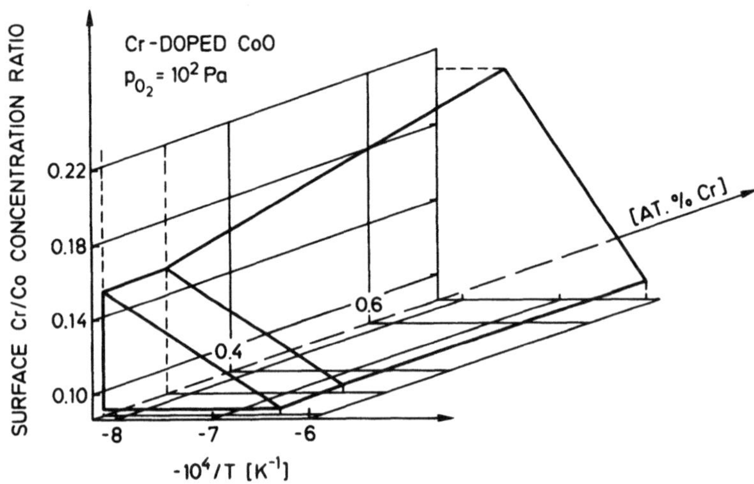

Figure 27. 3-D illustration of the effect of T on surface vs. bulk composition of Cr-doped CoO [68]

Further accumulation of 'in situ' experimental data on segregation are needed to verify the available theories of segregation and to adopt them for ionic solids. Also further theoretical studies oriented on non- stoichiometric compounds would be advisable. Unfortunately, the present material in this matter is practically limited to quasi-perfect and stoichiometric crystals such as MgO.

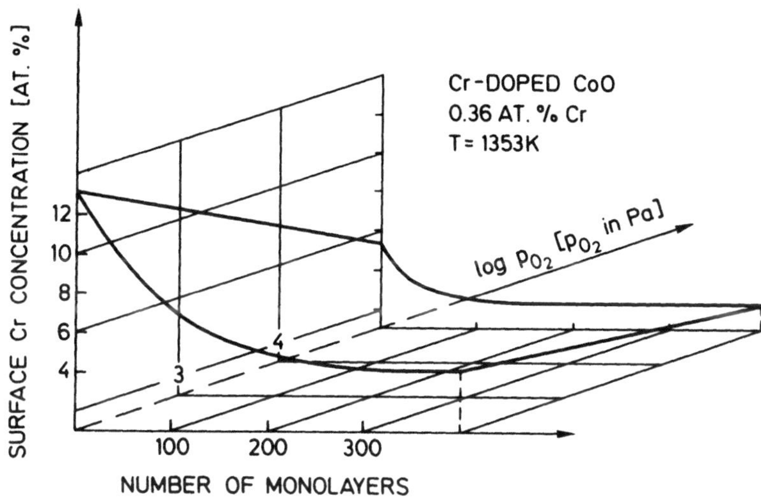

Figure 28. 3-D illustration of the effect of p_{O_2} on surface vs. bulk composition of Cr-doped CoO [68]

14. CONCLUSIONS

It has been shown that the excess of surface (interface) energy leads to several complications at and near the interface region of non-stoichiometric compounds. The complications involve mainly:

1. Segregation of intrinsic lattice defects and resulting formation of a gradient of non-stoichiometry within the interface region.
2. Segregation of extrinsic defects and resulting formation of concentration gradients of defects within the interface region.
3. Formation of defects which can only be stable in specific conditions of the interface region.
4. When the interface concentration surpasses a certain critical value then lattice reordering occurs leading to the formation of interface superstructures of extraordinary properties such as enhanced electronic and ionic transport and catalytical properties.
5. Segregation may have a substantial effect on the rate of heterogeneous processes even at high temperatures. The effect is caused by the interface diffusive resistance produced by the formation of electric fields within the interface region. The effect has a severe impact on the interpretation of the diffusion data determined by the re-equilibration kinetics.

REFERENCES

1. J.M. Blakely and S.M. Mukhopadhyay, 'Segregation at Ceramic Surfaces and Effect on Mass Transport', this book
2. R.G. Edgell and W.C. Mackrodt, 'The Theory of Dopant and Impurity Segregation in Ceramic Oxides, this book
3. E.D. Hondros and M.P. Seah, 'Segregation to Interfaces', Intern. Metals Rev., 222 [12], 262-301 (1977)
4. R.C. McCune and P. Wynblatt, 'Calcium Segregation to a Magnesium Oxide (100) Surface', J.Am.Cer.Soc., 66 [2] 111-117 (1983)
5. P. Wynblatt and R.C. McCune, 'Surface Segregation in Metal Oxides' in: 'Surface and Near-Surface Chemistry of Oxide Materials', J. Nowotny and L.C. Dufour, Eds., Elsevier, Amsterdam, 1988, p. 247-279
6. G. Borchardt, 'Use of SIMS to Study Surface Segregation in Insulating Materials', this book
7. G. Borchardt, 'SIMS analysis of Poorly Conducting Surfaces', Solid State Ionics, in print
8. W. Hirschwald, 'Selected Experimental Methods in the Characterization of Oxide Surfaces', in: Surface and Near-Surface Chemistry of Oxide Materials', Elsevier, Amsterdam, 1988, p. 61-- 187
9. Z. Adamczyk and J. Nowotny, 'Effect of Segregation on Near-Surface and Bulk Transport Phenomena in Ionic Crystals', J.Phys.Chem. Solids 47 [1] 11-24 (1986)
10. Z. Adamczyk and J. Nowotny, 'Effect of Surface Charge on the Defect Transport Kinetics Through Interfaces of Ionic Solids, Reactivity of Solids 4 139-150 (1987)
11. J. Nowotny, 'Surface Segregation of Defects in Oxide Ceramic Materials', Solid State Ionics, in print
12. J. Nowotny, 'Certain Aspects of Segregation in Oxide Materials', Mater.Sci.Forum 29, 99-126, 1988
13. A. Atkinson and R.I. Taylor, 'The Diffusion of ^{63}Ni Along Grain Boundaries in Nickel Oxide', Phil.Mag., 43, [4], 979-998, 1981
14. J. Nowotny, M. Sloma and W. Weppner, 'Work Function in Studies of the Defect Structure of Near-Surface Layers of Oxide Ceramics', Adv.Ceram., 23 159-173 (1987)
15. A. Bielanski and J. Deren, 'Relations Between Electronic and Catalytic Properties of Semiconducting Oxide Catalysts', in: 'Electronic Phenomena in Chemisorption and Catalysis on Semiconductors', K. Hauffe and F.F. Volkhenstein, Eds., W. de Gruyter, Berlin, 1969, p.149-165
16. J. Nowotny, M. Sloma and W. Weppner, 'Defect Structure of CoO in the Vicinity of CoO/Co_3O_4 Phase Boundary', in: 'Non-Stoichiometric Oxides', J. Nowotny and W. Weppner, Eds., Kluwer Acad., Amsterdam, in print
17. A. Bielanski, K. Dyrek and Z. Kluz, 'Effect of Doping of Nickel Oxide Catalysts with Altervalent Metal Additives. I. Composition and Defect Concentration in Li-Doped NiO', Bull.Acad.Polon.Sci., Ser.Sci.Chim., 12 [9] 657-661 (1964)
18. A. Bielanski, K. Dyrek and Z. Kluz, 'Effect of Doping of Nickel Oxide Catalysts with Altervalent Metall Additives'. II. Magnetic Properties of Li and Fe-Doped NiO', Bull.Acad.Polon.Sci., Ser-Sci. Chim., 13 [4] 285-290 (1965)

19. J. Deren, B. Russer, J. Nowotny, G. Rog and J. Sloczynski, 'Relation Between Physicochemical, Electronic and Catalytic Properties of NiO Doped with Chromium', J.Catal., 34 124-131 (1974)
20. J. Nowotny and J.B. Wagner, Jr., 'Influence of the Surface on the Equilibration Kinetics of Non-Stoichiometric Oxides', Oxid.Metals 15 [1/2] 169-198 (1981)
21. R.L. Coble, 'Sintering of Crystalline Solids. I. Intermediate and Final State Diffusion Modes', J.Appl.Phys., 32 [5] 787-792 (1961)
22. R.L. Coble, 'Sintering of Crystalline Solids. II. Experimental Test of Diffusion Models in Powder Compacts', J.Appl.Phys., 32 [5] 793-799 (1961)
23. P.J. Jorgensen and J.H. Westbrook, 'Role of Solute Segregation at Grain Boundaries During Final-Stage Sintering of Alumina', J.AmCera.Soc., 47 [7] 332-338 (1964)
24. H.L. Marcus and M.F. Fine, 'Grain Boundary Segregation in MgO-doped Al_2O_3', J.Am.Ceram.Soc., 55 [11] 568-570 (1972)
25. W.C. Johnson and D.F. Stein, 'Additive and Impurity Distribution at Grain Boundaries in Sintered Alumina', J.Am.Ceram.Soc., 58 [11-12] 485-488 (1975)
26. P. Nanni, C.T.H. Stoddart and E.D. Hondros, 'Grain Boundary Segregation and Sintering of Alumina', Mater.Chem., 1 297-320 (1976)
27. R.I. Taylor, J.P. Coad and R.J. Brook, 'Grain Boundary Segregation in Al_2O_3', J.Am.Ceram.Soc., 57 [12] 539-540 (1974)
28. R.I. Taylor, J.P. Coad and A.E. Hughes, 'Grain Boundary Segregation in Mg-Doped Al_2O_3', J.Am.Ceram.Soc., 59 [7-8] 374-376 (1976)
29. S. Baik, D.E. Fowler, J.M. Blakely and R. Raj, 'Segregation of Mg to the (0001) Surface of Doped Shapphire', J.Am.Ceram.Soc., 68 [5] 281-286 (1985)
30. S. Baik, 'Segregation of Mg to the (0001) Surface of Single-Crystal Alumina': Quantification of AES Results', J.Am.Ceram.Soc., 69 [5] C-101 - C-103 (1986)
31. S. Baik and C.L. White, 'Anisotropic Calcium Segregation to the Surface of Al_2O_3', J.Am.Ceram.Soc., 70 [9] 682-688 (1987)
32. S.M. Mukhopadhyay, A.P. Jardine, J.M. Blakely and S. Baik, 'Segregation of Magnesium and Calcium to the (1010) Prismatic Surface of Magnesium-Implanted Sapphire', J.Am.Ceram.Soc., 71 [5] 358-362 (1988)
33. Y.-K. Pae k, K.-Y. Eun and S.-J.L. Kang, 'Effect of Sintering Atmosphere on Densification of MgO-Doped Al_2O_3', J.Am.Ceram.Soc., 71 [8] C-380 - C-382 (1988)
34. Chien-We Li and W.D. Kingery, 'Solute Segregation at Grain Boundaries in Polycrystalline Al_2O_3', Adv.Cer., 10 368-378 (1984)
35. R.L. Segall, R.St.C. Smart and P.S. Turner, 'Oxide Surfaces in Solution', ref. 5, p.527-576
36. T. Mitamura, E.L. Hall, W.D. Kingery and J.B. Vander Sande, 'Grain Boundary Segregation of Iron in Plycrystalline Magnesium Oxide Observed by STEM", Ceramurgia Intern., 5 [4] 131-136 (1979)

37. A. Roshko and W.D. Kingery, "Segregation at Special Boundaries in MgO", J.Am.Ceram.Soc., 68 [12] C-331 - C-333 (1985)
38. J.R.H. Black and W.D. Kingery, "Segregation of Aliovalent Solutes Adjacent Surfaces in MgO", J.Am.Ceram.Soc., 62 [3-4] 176-178 (1979)
39. K.-O. Axelsson, K.-E. Keck and B. Kasemo, "Surface Composition of ZrO_2 in H_2O, H_2 and Atomic Hydrogen, Investigated by AES and EELS", Appl.Surf.Sci., 25, 217-230 (1986)
40. A.J. Winnubst, P.J. Kroot and A.J. Burggraaf, AES/STEM Grain Boundary Analysis of Stabilized Zirconia Ceramics, J.Phys.Chem.Solids 44, 955-960 (1983)
41. J. Nowotny, M. Sloma and W. Weppner, "Surface Relaxation of Y_2O_3-Stabilized Zirconia", Solid State Ionics, 28 (1988), in print
42. Y.M. Cross and D.R. Pyke, "An X-Ray Photoelectron Spectroscopy Study of the Surface Composition of Tin and Antimony Mixed Metal Oxide Catalysts", J. Catal., 58 61-67 (1979)
43. W. Hirschwald, I. Sikora, F. Stolze and J. Oblakowski, "Effect of Temperature on Chromium Segregation to CoO (100) and NiO (100) Surfaces", Surf.Interf.Anal., in print
44. J. Nowotny, I. Sikora and J.B. Wagner, Jr., "Segregation and Near-Surface Diffusion for Undoped and Cr-Doped CoO", J.Am.Ceram.Soc., 65 [4] 192-196 (1982)
45. P.E. Childs, L.W. Laub and J.B. Wagner, Jr., "Chemical Diffusion in Non-Stoichiometric Compounds", Proc.Brit.Ceram.Soc., 19, 29-53 (1971)
46. J. Nowotny, M. Sloma and W. Weppner, "Equilibration Processes at NiO Surface", Surf.Interf.Anal., 12, 269-272 (1988)
47. J. Frenkel, "Kinetic Theory of Liquids", Oxford Univ.Press, New York, 1946, p.36
48. K.L. Kliewer and J.S. Koehler, "Space Charge in Ionic Crystals". I. General Approach with Application to NaCl", Phys.Rev., 140, [15], A 1226-A 1240 (1965)
49. Y. Ikuma and W. Komatsu, "Oxygen Surface Diffusion and Surface Layer Thickness in Oxides Measured by ^{18}O-Exchange Reaction", Ikutoku Kogyo Daigaku Kenkyu Hokoku, B-8, 187-200 (1984)
50. F. Barbier, PhD thesis, "Studies on Certain Properties of Grain Boundaries of NiO: Diffusion, Energy and Structure of Dislocations", Université de Paris-Sud, Orsay, 1986
51. F. Barbier, C. Monty and M. Dechamps, "On the Grain Boundary Diffusion of Co in NiO Bicrystals", Phil.Mag.A, 58, in print
52. M. Dechamps and F. Barbier, "Interface Transport in Monoxides", in ref. 16
53. A. Atkinson and C. Monty, "Grain Boundary Diffusion in Ceramics", in "Surfaces and Interfaces of Ceramics", this book
54. D.S. Ginley, E.L. Venturini, J.F. Kwak, R.J. Banghman, M. Morosin, and J.E. Schirber, "Semiconducting Shells in Ceramic $YBa_2Cu_3O_7$", Phys.Rev.B, 36 [1] 829-832 (1984)

55. D.A. Esparza, C.A. Dovidio, J. Guimpel, E. Osquiguil and F. de la Cruz, 'The Granular Nature of Bulk Superconductivity at 40 K in $La_{1.8}Sr_{0.2}CuO_4$', Solid State Comm., 63 [2] 137-140 (1987)
56. D.C. Larbaleister, M. Daeumling, X. Cai, J. Seuntjens, J. McKinnel, D. Hampshire, P. Lee, C. Meingast, T. Willis, H. Muller, R.D. Ray, R.G. Dillenburg, E.F. Hellstrom and R. Joynt, 'Experiments Concerning the Connective Nature of Superconductivity in $YBa_2Cu_3O_7$', J.Appl.Phys., 62 [8] 3309-3313 (1987)
57. P.M. Grant, S.S.P. Parkin, V.Y. Lee, E.M. Engler, M.L. Ramirez, J.E. Vazguez, G. Lim and R.O. Jackowitz, 'Evidence of Superconductivity in La_2CuO_4', Phys.Rev.Lett., 58 [23] 2482-2485 (1987)
58. J.R. Cooper, L.W. Zhou, B. Dunn, C.T. Chu, B. Alavi and G. Grüner, 'Traces of Superconductivity in Sintered La_2CuO_{4-y}', Solid State Comm., 64 [2] 253-255 (1987)
59. X. Cai, R. Joynt and D.C. Larbalestier, 'Experimental Evidence for Granular Superconductivity in Y-Ba-Cu-O at 160 K', Phys.Rev.Lett., 58[26] 2798-2801 (1987)
60. T. Seyama, 'Surface Reactivity of Oxide Materials in Oxidation-Reduction Environment', in ref. 5, p. 189 - 218
61. S. Wen, J. Feng, X. Song, L. Lu, Chenhong Li, Chengen Li and D. Li 'Grain Boundaries and Lattice Defects in $YBa_2Cu_3O_{7-x}$ Ceramics', Mater.Lett., 5[11-12] 425-428 (1987)
62. D. Dimos, P. Chaudhari, J. Mannhardt and F.K. Le Goues, 'Orientation Dependence of Grain Boundary Critical Currents in $YBaCu_3O_{-x}$ Bicrystals', Phys.Rev.Lett., 61[2] 219-222 (1988)
63. Duan Hongmin, Lu Li and Zhang Dianlin, 'The Role of Surface Shell of the $YBa_2Cu_3O_x$ Single Crystals in High T_c Superconductivity', Chinese Phys.Lett., 5[6] 253-256 (1988)
64. Y.M. Chiang, J.A.S. Ikeda and A. Roshko, 'Grain Boundary Segregation and Critical Current Density in $YBa_2Cu_3O_{7-x}$ Superconductors', Adv.Ceram., to be published
65. Duan Hongmin, Lu Li, Zhang Dian Lin, '$YBa_2Cu_3O_7$: A Bulk Superconductor?', Solid State Comm., 67[8] 809 - 813 (1988)
66. M. Sarikaya and B.L. Thiel, 'Identification of Intergranular Cu_2O in Polycrystalline $YBa_2Cu_3O_{7-x}$ Superconductors', J.Am.Ceram.Soc., 71[6] C305-C309 (1988)
67. J. Nowotny, M. Rekas, D.D. Sarma and W. Weppner, 'High T_c Oxide Superconductors. Possible Effect of Interfaces', in ref. 5, p. 669-699
68. I. Sikora, F. Stolze and W. Hirschwald, 'Segregation of Chromium in CoO-Cr_2O_3 Solid Solutions and $CoCr_2O_4$ Spinel Phase Studied by SIMS abd ESCA', Surf.Interf.Anal., 10, 424-429 (1987)

SURFACE DIFFUSION AND SURFACE ENERGIES OF CERAMICS with application to the behavior of volatile fission products in ceramic nuclear fuels.

Hj. Matzke
Commission of the European Communities, Joint Research Centre
Karlsruhe Establishment, European Institute for Transuranium Elements
Postfach 2340, D-7500 Karlsruhe, Federal Republic of Germany

ABSTRACT

The available data on surface diffusion and matter transport driven by surface energy in ceramics are reviewed. The competing effects of matter transport by evaporation-condensation and by volume diffusion are treated. The materials considered are mainly oxides (MgO, NiO, Al_2O_3, Fe_3O_4, UO_2 etc.), but other ceramics (UC, UN) are also dealt with. The knowledge on surface energies of such ceramics is reviewed as well. A number of different techniques has been developed to measure both properties, and the effects of deviation from stoichiometry on surface diffusion or surface energy have been measured for $UO_{2\pm x}$ and $Fe_{3-x}O_4$. The data are applied to discuss the technologically important aspect of the formation and the behavior of gas-filled bubbles in UO_2 nuclear fuel.

1. INTRODUCTION

Surface diffusion and matter transport driven by surface tension or surface free energy of solids are phenomena which are well known to material researchers. For example, the change in surface energy provides the driving force for material transport during sintering which, in some stages, can be rate-controlled by surface diffusion. The formation of equilibrium surface

L.-C. Dufour et al. (eds.), Surfaces and Interfaces of Ceramic Materials, 241–272.

states at high temperatures can be influenced by surface diffusion, though evaporation-condensation processes and volume diffusion in surface-near layers are always competing and often dominating mechanisms. Also, surface diffusion of matrix atoms can provide bulk mobilities of gas-filled bubbles. Finally, most materials researchers will recall the visual impression of actual jumps of atoms by surface diffusion from field emission microscopy on metal tips.

Whereas diffusion processes in the *bulk* of crystals have been measured and studied extensively for the past half century, the diffusion of adsorbed atoms on surfaces, and *surface self-diffusion* have been investigated quantitatively only since the sixties. This development parallels that of surface science as a whole: surfaces of oxides had long been considered to be just the external boundaries of solids without intrinsic character of their own. Recently, a large number (of about one hundred) modern surface-sensitive techniques became available, and understanding of surface diffusion has made significant progress. Unfortunately, despite of the importance of surface properties of ceramics, most applications were oriented towards *metal surfaces*. A recent Nato Advanced Study Institute on "Surface Mobilities on Solid Materials, Fundamental Concepts and Applications" (1), for instance, treats metals in detail without mentioning ceramics. Knowledge on surface self-diffusion of metals was, in fact, advanced sufficiently as early as in 1972 to motivate the edition of a monograph on this topic (2). Fortunately more recently, significant progress on *surface properties* of ionic solids, compounds and ceramics has been achieved as is evidenced by a book (3) on "External and Internal Surfaces in Metal Oxides" edited by two of the organizors of the present NATO ASI. This book includes some data on surface diffusion on NiO and Al_2O_3.

The oldest of the modern methods for observing surface diffusion are field electron emission and field ion emission microscopy which achieves resolution at the atomic level. The field emission microscope, invented by E.W. Müller in 1937, yielded the first direct observation (4) of single atom diffusion at surfaces and is most illustrative to demonstrate surface adatom- and surface self-diffusion down to the level of individual jumps of single atoms. Such data and observations are only available for metal tips and not for ceramics. For metals and metallic supports, much additional evidence has been obtained on such phenomena as formation of equilibrium shapes of e.g. tips, on surface mobilities of adsorbed atoms in chemical reactions and catalysis, and on crystal growth.

In addition to the direct microscopic observation of atom motion on surfaces, there are basically two more sources for obtaining information on surface self-diffusion of solids: semi-macroscopic investigation of mass transport of tracer atoms or of matrix atoms in transport processes related to surface diffusion, and theoretical treatments of surface potentials and surface mobility. The present article deals with the available knowledge for ceramics (mainly oxides); this knowledge is nearly exclusively due to the second of the above sources of information, i.e. tracer studies and mass-transport related to surface diffusion.

The second topic treated here is surface energy of ceramics, γ_S. Values of γ_S can a priori also be deduced from experiments similar to those for surface diffusion. Grooving and surface relaxation measurements yield usually the product $\gamma_S D_s$. Other techniques, such as wetting or the multi-phase equilibration technique yield independent values of γ_s, and indentation methods (Hertz or Vickers) yield supporting results on (fracture) surface energies. Calculational methods are also available and were applied to some ceramics. Again, as for D_s, available data on γ_s are critically discussed.

The available knowledge is applied to a description of the operational performance of ceramic nuclear fuels, and in particular to the behavior of gaseous fission products in these fuels. Mobility of rare-gas bubbles rate-controlled by surface diffusion is indicated for specific conditions in all types of ceramic nuclear fuels, oxides, carbides and nitrides. Phenomena interfering with the resultant fast mobility of gas bubbles exist during most irradiation and operation conditions and are discussed in detail as well.

2. SURFACE DIFFUSION OF CERAMICS

In this section, experimental methods to measure surface diffusion coefficients, D_S, or surface diffusivity products, $D_S\delta$ applicable and applied to ceramics are described (δ is the thickness of the high diffusivity surface layer; for impurity diffusion, $\alpha\delta$ has to be used with α=segregation factor). Then, a brief summary of basic aspects of surface diffusion and of results on metals is given in an attempt to provide a basis for understanding fundamental mechanisms of surface diffusion. Subsequently, available data for ceramics are reviewed.

2.1 Experimental Techniques to Measure Matter Transport by Surface Diffusion.

Two different types of techniques exist to measure D_S or the surface diffusivity product, $(\alpha)D_S\delta$: use of radiotracers as distinguishable atoms and mass-transfer observing the migration of indistinguishable matrix atoms.

Tracer techniques: Whereas the application of tracers to study *volume* diffusion has largely advanced this area of research, tracer techniques are not as straightforward for surface diffusion, the main reason being loss of tracer atoms into the crystal lattice due to volume diffusion and spreading of tracer atoms on the surface by evaporation-condensation. Three principle geometrical configurations are used to apply the tracer (see Fig. 1): i) the edge source from which surface diffusion takes place onto another surface via the edge; ii) the half-plane source, covering half of a solid assumed to be semi-infinite, from which diffusion takes place onto the other part; and iii) the point source, with a needle containing the radio-active tracer being placed onto the solid. Mathematical solutions to the relevant diffusion equation exist for all three configurations of tracer sources, based on solutions of grain boundary diffusion where simultaneous fast diffusion along the grain boundary and slow volume diffusion away from the interface occur as well (e.g. 5-8). The exact solutions provided by Whipple (7) and Suzuoka (8) are difficult to apply to experimental data. LeClaire (9) has shown that for most experimental conditions, Whipple's infinite-source and Suzuoka's finite-source solution can be expressed in a more easily applicable way. A factor of $^1/_2$ is added to allow for diffusion on one surface only. As an example, the solution used by Stubican et al. (10) for evaluating data of Cr-surface diffusion on MgO, Al_2O_3 and $MgAl_2O_4$ with the edge source tracer geometry, is given

$$D_s\delta = \frac{1}{2}\left(\delta \ln\bar{c}/\delta y^{6/5}\right)^{-5/3}(4D_v/t)^{1/2}\ 0.78^{\ 5/3} \qquad (1)$$

where D_s is the desired surface diffusion coefficient, δ the thickness of the high diffusivity layer, $\bar{c}$ the average specific activity in a section of thickness Δy at distance y from the original surface, D_v the lattice or volume diffusion coefficient and t the annealing time. (Note that δ should be replaced by $\alpha\delta$ for impurity diffusion where α is a solute segregation factor). Besides the application of the edge source tracer method for different oxides (10, 11), the point source method has been used to study surface diffusion on a ceramic as well, i.e. surface self-diffusion of U and impurity surface diffusion of Pu on UO_2 (12, 13). More frequently, however, use was made of different mass-transfer techniques not involving tracers, partly because of the relative simplicity of these methods, and partly because suitable radioactive tracers do not exist for important ceramics, as is the case for Mg, Si and Al.

Mass-transfer techniques: these techniques observe the migration of indistinguishable matrix atoms, and are based on investigating changes in the geometry of surfaces caused by the net flow of material along crystal planes. This flow is the consequence of the attempt of the crystal to minimize its surface free energy, and *may* be rate-controlled by surface diffusion.

a) Sintering and neck-growth: The formation of a neck between well-defined bodies (e.g. sphere-to-sphere, sphere-to-plane, wire-to-wire, wire-to-plane) during sintering follows a relation of the type

$$x^n/d^m = Ct \qquad (2)$$

where x=neck diameter, d=size of sphere or wire, t=sintering time, C=emphirical constant. Possible mechanisms competing with surface diffusion are viscous or plastic flow (for which n=2, m=1), evaporation-condensation (n=3, m=1), and volume diffusion (n=5, m=2). If surface diffusion is rate-controlling: n=7, m=3 (14). The method was frequently applied to metals. A difficulty often encountered are changes in surface orientation in the vicinity of the neck. Shrinkage during neck growth was also measured in oxides, e.g. in CeO_2 where, however, kinetics were found indicating that Ce-grain boundary diffusion rather than surface diffusion was rate-controlling for the experimental conditions used (T-range 1000-1100 °C) (15).

b) Grain boundary grooving: At the intersection of a grain boundary and a free surface, a groove develops upon annealing because the system can locally lower its total free surface energy by forming a dihedral angle (2β) thus equilibrating grain boundary and surface energy. Evaporation-condensation, volume diffusion, surface diffusion, and viscous flow are competing mechanisms for groove formation. A complete equation including all possible diffusion paths for grain boundary grooving has been derived (16). The same problems exist for the methods of

c) single scratch decay and d) relaxation of periodic profiles. These are the two most common profiles applied to solids. They are schematically shown in Table 1 which also shows the development of width w and depth d with annealing time (17, 18). For more details, the reader is referred to the book of ref. (2) or the review by Bonzel (19) on surface diffusion in metals in the proceedings of a previous Nato ASI (1). Single scratches are e.g. made with a diamond indentor, and periodic profiles are usually made by applying a photoresist on the specimen, exposing through a line mask of the desired grating, processing the photoresist and subsequently etching grooves into the uncovered surface region. When the residual photoresist is removed, the specimen is ready for annealing. The development and the relaxation of sinusoidal profiles can be followed stepwise with interference microscopy or continuously by monitoring the diffraction pattern of a laser beam.

The general equation to describe the time-dependence of the amplitude a(t) of a sinusoidal profile under the above four competing transport processes (assumed to occur independently) is

$$a(t) = a(o)\exp(-Kt) = a(o)\exp[-(F\omega + A'\omega^3 + C\omega^3 + B\omega^4)t] \quad (3)$$

(a factor sin ωx has to be added if a(x,t) is wanted). The constants F, A', C and B result from the four mechanisms and are defined as follows

viscous flow	$F = \gamma_s/2v$	γ_s = (isotropic) surface energy, v = viscosity coefficient
evaporation-condensation	$A' = pD_g\gamma_s\Omega^2/kt$	p = equilibrium vapor density over a plane surface, D_g = diffusion coefficients of the vapor phase (molecules) in the experimental atmosphere (e.g. inert gas)
volume diffusion	$C = D_v\gamma_s\Omega/kT$	

Table 1: Mass transfer methods to measure surface diffusion

starting condition	schematic profile	growth parameters for rate-control by	
		surface diffusion	volume diffusion
grain-boundary grooving		$w = 4.6\,(Bt)^{1/4}$ $d = 0.97\,m\,(Bt)^{1/4}$	$w = 5.0\,(Ct)^{1/3}$ $d = 1.01\,m\,(Ct)^{1/3}$
single surface scratch		$w = 6.9\,(Bt')^{1/4}$ $d = 0.049\,M_2\,(Bt')^{-3/4}$ $M_2 = 0.063\,w^3d$	$w = 6.22\,(Ct')^{1/3}$ $d = 0.053\,M_2\,(Ct')^{-1}$ $M_2 = 0.078\,w^3d$
periodic profile		$\ln(a_o/a) =$ $(2\pi/\lambda)^4\,Bt$	$\ln(a_o/a) =$ $(2\pi/\lambda)^3\,Ct$

Here, w = width and d = depth of surface feature; λ = wave length and a = amplitude of sinusoidal profile (or of multiple surface scratches). The mass transfer constants are

$$B = D_s \cdot N_s \cdot \gamma_s \cdot \Omega^2/kT$$
$$C = D_v \cdot \gamma_s \cdot \Omega/kT$$

where N_s = surface atom density, γ_s = surface free energy, Ω = atomic volume, and k = Boltzman constant, $m = \cos\beta = \gamma_b/2\gamma_s$ with β = dihedral angle at the base of the groove, γ_b = grain boundary energy, and t' = variable allowing for development of the initial profile.

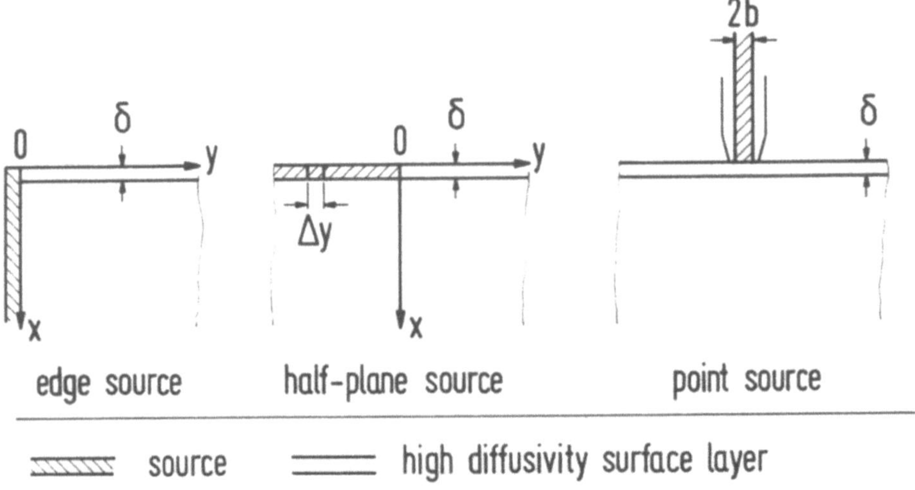

Fig. 1: Tracer methods to measure surface diffusion

<u>surface diffusion</u> $B = D_S \gamma_S \Omega^2 N_S / kT$ (see also Table 1)

Experimentally, the predominant surface relaxation mechanism can be identified from the wavelength dependence ($\lambda = 2\pi/\omega$) of the decay constant K. If a log-log plot of K vs. λ yields a slope of -4, surface diffusion is rate-controlling for relaxation; if the slope is -3, diffusion in the lattice or in the gas-phase are rate-determining; and a slope in between is obtained when the relaxation occurs by a combination of the various diffusion processes.

All three mass-transfer methods b) to d), i.e. grain boundary grooving, single scratch decay and relaxation of sinusoidal profiles have been applied to ceramics, e.g. to UO_2 (20-22).

2.2 Basic Aspects of Surface Diffusion

Surface diffusion coefficients, D_S, may be described by an Arrhenius-type relation

$$D_S = D_{S,o} \exp(-\Delta H_S / kT) \tag{4}$$

just as is conventionally done with volume and grain boundary diffusion. In contrast to volume diffusion, however, surface diffusion is not necessarily just a process involving one mechanism, say adatom migration on an otherwise flat surface. It rather is usually a very complex process involving several transport mechanisms. This is best illustrated by regarding the terrace-ledge-kink (TLK) model of a surface, developed independently by Kossel and by Stranski at around 1930 (see Fig. 2). There are different possible surface sites of an atom:

- in the terrace plane, where also surface vacancies (terrace vacancies) can exist
- on the terrace plane as adatom
- in the ledge or on the ledge, as ledge-adatom. The ledge can also have vacancies.

Among these sites, the kink positions (half-crystal positions) are important. They are the equilibrium sites during evaporation-condensation. The terraces are atomically flat if they correspond to low index planes. High index planes tend to roughen, exhibiting a structure of monatomic ledges and kinks.

The consequences for surface diffusion are that a number of types of atomic jumps with different jump lengths and different formation and migration energies exist. Some authors call the diffusion of single atoms on otherwise defect-free flat terraces *intrinsic surface self-diffusion*. The ΔH_S of equ. (4) is then only a migration term for the adatom, $\Delta H_{m,a}$. In contrast, whenever larger diffusion distances are involved, some authors speak of mass transfer surface diffusion, involving additionally trapping and emission of adatoms on and from kink sites. In this case, ΔH_S includes also a formation term, $\Delta H_{f,a}$, somewhat in analogy to bulk diffusion.

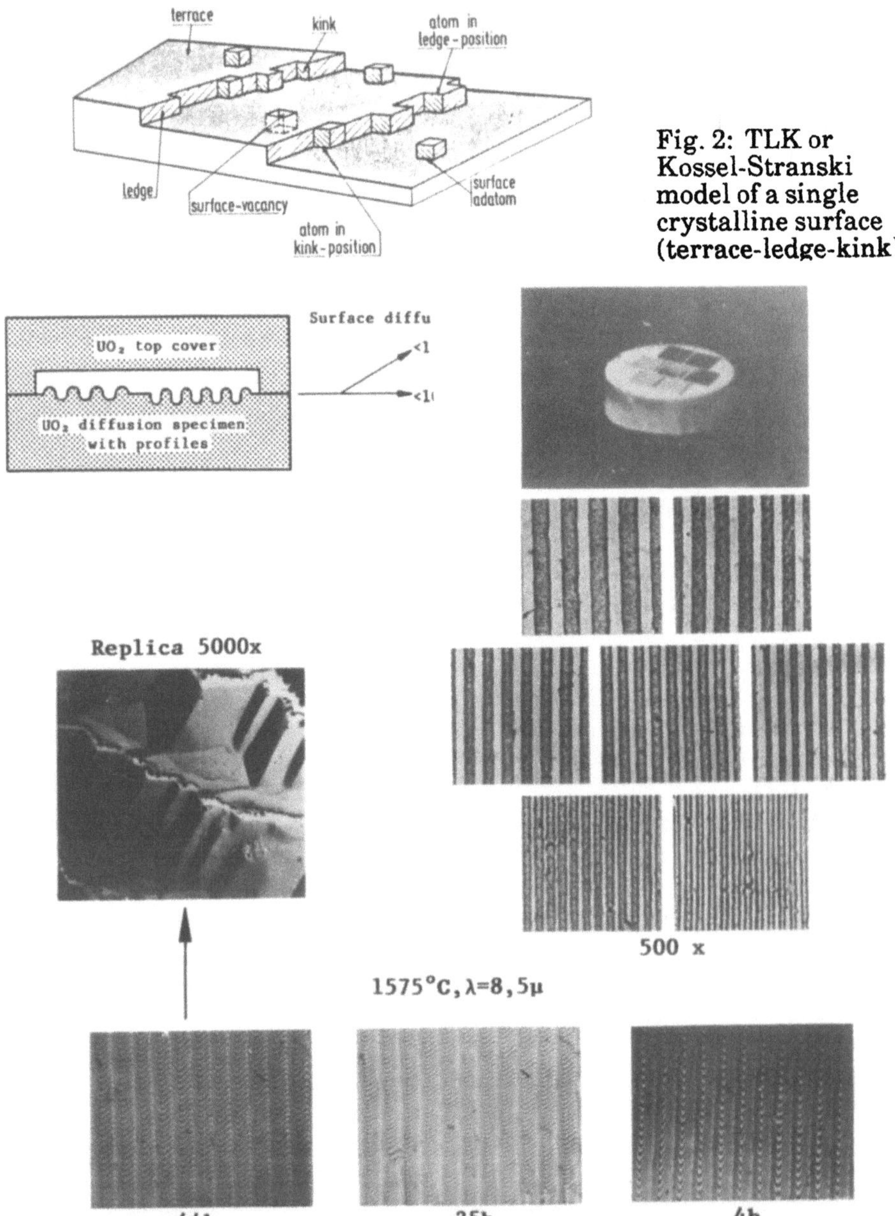

Fig. 2: TLK or Kossel-Stranski model of a single crystalline surface (terrace-ledge-kink)

Fig. 3: Attempt to measure surface diffusion on UO_2 single crystals: Profiles with 7 different wave lengths are produced in a way that crystal surface, direction of the profiles and surface of the UO_2 top cover were all oriented along low index planes or orientations of the same UO_2 crystal. Interference microscopy of one profile and a surface replica are also shown (annealing temperature 1575 °C).

There are at least five more complicating difficulties: i) The two-dimensional potential of the surface which determines migration energies for intrinsic surface diffusion is a function of surface orientation: surface diffusion is anisotropic, both orientationally and directionally. ii) Surface diffusion depends on surface cleanliness: adsorbed impurities can strongly affect D_S; experimental evidence shows that they may either suppress or enhance it. iii) The phenomenological equations for tracer diffusion are (and have to be) derived by taking into account simultaneous surface and volume diffusion. They yield $D_s\delta$, where δ is the effective thickness of the surface layer in which surface diffusion takes place (see above). For metals, and based on our current knowledge of the atomistics of their surfaces, δ should be one atom layer, i.e. of the order of 2×10^{-8} cm. For ceramics, however, larger values for δ increasing with temperature have been postulated (though the present author questions these results, see below) (23). iv) For mass transport experiments (Table 1), the product of $D_S\gamma_S$ is obtained. The surface energy is also anisotropic and may display cusps or singularities at some planes. This leads to facetting of profiles making an evaluation impossible (see Fig. 3). v) For diatomic ceramics or ceramics consisting of 3 or more elements, mass transport experiments to measure surface diffusion leave the moving species undefined. Since for true mass transport, *all* components have to be transported, otherwise fast surface diffusion of one component may become suppressed. Also, in contrast to metals, ceramics can often not be annealed in UHV as one can do with metals, because of their incongruent evaporation. In particular for non-stoichiometric oxides, annealing in CO/CO_2 or H_2O/H_2 atmospheres is needed to maintain their O/M-ratio. Therefore, surface cleanliness is more difficult to achieve.

2.3 Some Results for Surface Diffusion on Metals.

Before discussing results for ceramics, it is instructive to mention a few typical results for metals for which much more data exist, and for some of which field ion emission microscopy, FIM, data exist for random walk measurements of individual adatoms at low temperatures. The corresponding (intrinsic) $D_{S,a}$-values, if extrapolated into the temperature range of mass transfer measurements (yielding D_S), are much larger than D_S, typically $D_{S,a}/D_S > 10^3$ (* see next page). $D_{S,a}$ was found to show pronounced directional anisotropy, e.g. on (110) surfaces of fcc metals. For instance, for (110) Pt, $\Delta H_s = 1.70$ and 3.16 eV for <110> and <001> directions, respectively, with D_S for <110> being higher by a factor of $\sim 10^2$ at 1200 K (24). D_S can also be tremendously influenced by pretreatment: hydrogen etched (110) Ni-surfaces show an increase of about 10^7 in $D_{S,a}$ at around 100 K compared with thermally etched surfaces.

The mass transfer surface self-diffusion data for fcc and bcc metals fall into two relatively narrow scatter bands, yielding roughly $D_S \sim 10^{-7}$ to 10^{-8} cm^2/s

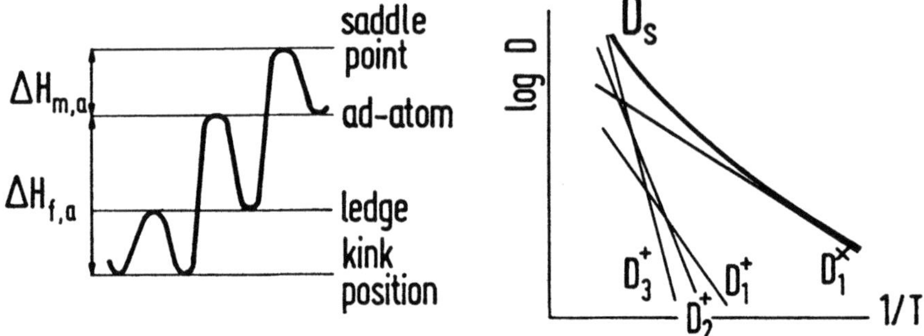

Fig. 4: Schematic presentation of potentials for surface diffusion of ad-atoms (or of evaporation-condensation) and schematic Arrhenius diagram for composite diffusion due to localized ad-atoms and (excited) ad-atoms, dimers and trimers in the non-localized state. (The latter can be stabilized by additional degrees of freedom, e.g. of rotation.)

* The intrinsic surface diffusion coefficient of adatoms, $D_{S,a}$, and the mass transfer coefficient, D_S, are related via

$$D_S = (n/N_S)D_{S,a}$$

with N_S = number of adsites/cm^2, n = number of adatoms/cm^2
Since $n/N_S = \exp(-\Delta H_f/RT)$, D_S has an additional formation term in ΔH, in addition to the migration term of $D_{S,a}$, i.e. to $\Delta H_{m,a}$.

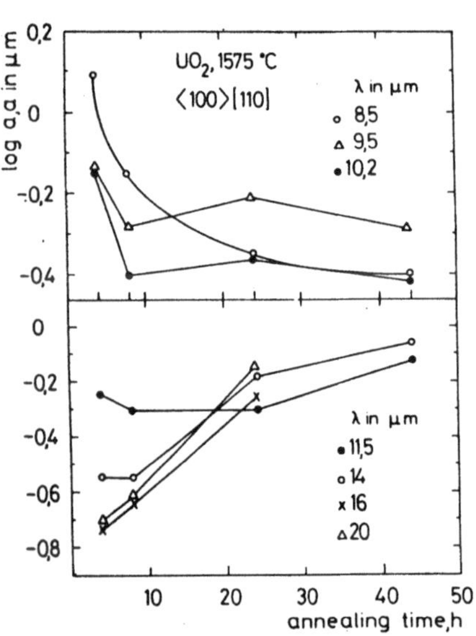

Fig. 5: Development of the amplitude of the profiles shown in Fig. 3 with time for annealing at 1575° C (22)

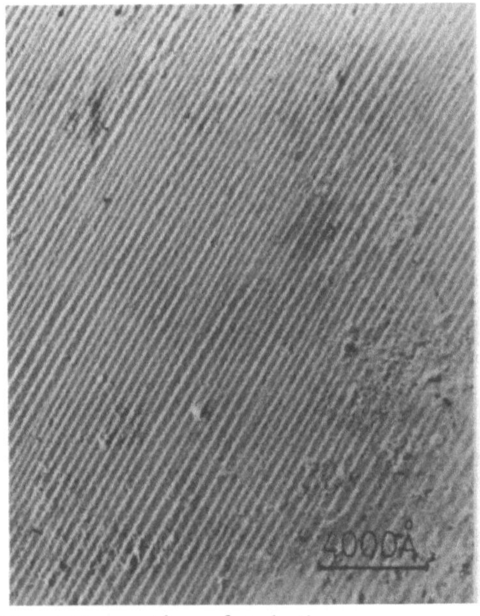

Fig. 6: Formation of striations on an originally flat surface of UO_2: surface replica following annealing at 1950° C in H_2 for 30 min.

for fcc metals (Cu, Ni, Au, Pt) and $D_s \sim 10^{-8}$ to 10^{-9} cm^{2}s^{-1} for bcc metals (Mo, Nb, W, Re) at 0.5 T_m, and (extrapolated) D_s-values at the melting point T_m of 10^{-4} to 10^{-5} cm^2 s^{-1}. However, an unexpected increase in D_s is observed at high temperatures ($T/T_m \geq 0.7$) leading to actual D_s-values of about 10^{-3} cm^{2}s^{-1} at T_m. This is about 100 times larger than D-values in a liquid. Bonzel (25) explained these high D_s-values with a non-localized surface diffusion process in which adatoms diffuse by a 2-dimensional gas-like flight in addition to the normal hopping mechanism (localized diffusion), somewhat resembling the older "rolling-stone" model. These "excited adatoms" have a life-time long enough to yield a mean-square displacement which is very large compared to the low temperature jump length. Also pairs, trimers etc. of non-localized adatoms are suggested to exist, stabilized by additional degrees of freedom. This model also explains the high pre-exponential factors D_0 which often are higher by 10^4 to 10^5 as compared with the Zener and Wert predictions and the experimental results for lattice diffusion (see Fig. 4).

Pronounced effects of impurities were frequently observed in the presence of metallic vapors: D_s can be significantly increased and can become as high as 0.1 to 1 cm^{2}s^{-1} (example Cu or Au surface self-diffusion in the presence of Pb-vapor which is thought to create a two-dimensional liquid at the surface (26) or to cause a modified diffusion process involving metal-impurity complexes (27)). Decreased D_s-values have also been found, e.g. by adsorbed carbon on Cu. Another example will be given below for gas bubbles in nuclear fuels (see section 4). To indicate the progress achieved in surface physics of metals, the recent use of tunneling microscopy to observe surface diffusion at the atomic level should also be mentioned (28).

2.4 Results on Surface Diffusion on Ceramics.

a) Difficulties, high diffusivity surface layer. As mentioned before, much less data exist for ceramics than for metals, and the experimental difficulties are much more severe. Some of these difficulties are mentioned at the end of section 2.2, and they are indicated in Figs. 3 and 5. Due to vapor pressures of ceramics (or of components of ceramics) generally being significantly higher than those of metals (at a given fraction of the melting point), mass transport by evaporation-condensation frequently interferes with surface diffusion. Fig. 3 shows that though crystal surface, direction of the profiles and surface of the top cover were all orientated along low index planes and directions of the same UO_2 single crystal, the profiles developed a crystallographic substructure rather than being sinusoidally flattened (22). We have shown above that theory claims that the seven profiles on UO_2 should decay with time yielding a straight line in a plot of log amplitude, a, versus time, t. Fig. 5 shows that small profiles (wave-length 8.5 to 10.2 µm) did shrink, but they neither showed a sinusoidal shape nor the expected time dependence of decay. The bigger profiles were growing, in contrast to theory. The anisotropy of surface energy, γ, and simultaneous occurrence of mass transport by surface diffusion and by evaporation-condensation transfer even originally flat surfaces into a different high temperature state (Fig. 6). This well known phenomenon affects volume diffusion measurements but should not be mistaken to be a proof for the existence of a thick high diffusivity layer. As shown in Fig. 7, the tracer of an originally thin (U-223)O_2 layer on a UO_2 crystal for volume diffusion measurements becomes embedded to a depth of the order of 0.07 to 0.15 μm by this fast initial surface relaxation. This process involves both annealing of polishing damage and formation of the high temperature equilibrium shape, e.g. in form of striations. Once this surface state has been reached, no further tracer mobility can be seen (Fig. 7).

Recently, Ikuma, Komatsu et al. (e.g. 23) have postulated similarly thick high diffusivity layers (thickness δ) to exist on oxides and to affect surface diffusion. δ was reported to increase with temperature, and to reach values of up to 50 nm (CoO, 640°C). The δ-values were calculated from changes in shape of weight gain curves during anneals of the oxide in 0-18 atmosphere or from the positive intercept of reaction curves during the formation of ternary oxides. The powders used were assumed to be spherical with a concentric core in which diffusion is governed by D_V and an outer shell of thickness δ where D_S is rate-determining. The implications of the large postulated δ-values are significant: Many published values of surface diffusion are deduced from experimentally determined products of $D_S\delta$, which are then converted to D_S-values by assuming δ to be one atomic layer. If δ would be much larger and, in addition, temperature dependent, the reported D_S-values would be much too high and the ΔH-values would need correction as well.

The present author feels that the evidence for such high δ-values is not convincing. The model of spherical solids with concentric shell and core seems to be an oversimplification. Powders usually do not consist of spheres and there is no proof for the existence of the postulation that the faster reacting fraction is really a thin layer covering all of the substance. Furthermore, and more importantly, there is no physical model that could explain a fast mobility far below the surface. Certainly, all suggested and observed mechanisms for surface diffusion cannot be operative at such depths. Also, there is ample recent theoretical evidence for ceramics that bulk properties are dominant some 5 atom layers below the surface. As a matter of fact, due to surface relaxation, polarization of oxygen and the related *decrease* in lattice parameter, vacancy formation energies are *higher* in the second and third atom layer than in the bulk (29).

A rather convincing, easy to measure proof of bulk properties being dominant just a few atom layers away from the surface consists in introducing (insoluble) rare gas ions such as Kr or Xe into different depth near to the

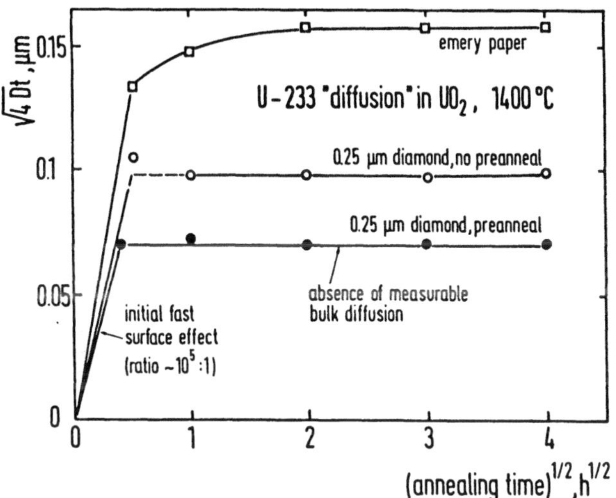

Fig. 7:
Evolution of penetration of U-233 tracer into UO_2 as function of time in a plot of $(4\ D_v t)^{1/2}$ vs. $t^{1/2}$. Three different surface treatments are used before applying the tracer layer and annealing at 1400° C. Only surface effects are operative, no penetration due to volume diffusion is seen.

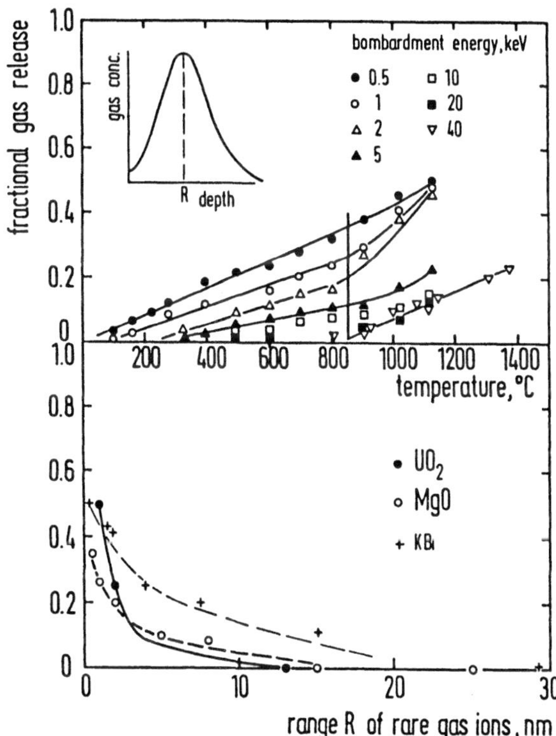

Fig. 8:
Release of heavy rare gases (Kr, Xe) from ceramics (MgO, UO_2, but also KBr) following ion implantation. The upper part shows release at low temperatures from MgO, the lower part gives the fraction of gas released at low temperatures vs. the range R (see inset in upper part) of the ions (30-33).

surface by using ion implantation with different energies. Fig. 8 shows that surface-near heavy rare gas ions can indeed be released by a fast mobility process at low temperatures. The examples are MgO, UO_2 (and also KBr) (30-33). However, even at a range of the gas of 2-3 atom layers, only a fraction of the gas is involved in this fast mobility. The data show that gas located deeper than about 5 atom layers behaves as being located in the bulk. Note that this is true though implantation occurs at moderate accelerator vacua (say 10^{-4} mbar) where 1-2 atom layers of water and other gases are still adsorbed, and release is measured in laboratory air (MgO, KBr) or in $H_2(UO_2)$, hence again under conditions not guaranteeing good surface cleanliness.

In summary, the concept of a "high diffusivity layer" many atomic distances thick can safely be rejected for well prepared high-density sintered or single-crystalline ceramics. Enough is known about crystal surfaces including surfaces of ceramics to state that the bulk crystal structure persists to the surface, with only slight relaxation of the surface planes of atoms. It seems also safe to reject any correction of published D_s-values of ceramics because of high δ-values, both because of experimental, physical and theoretical reasons. The general picture of relative diffusion rates as schematically shown in Fig. 9 has thus not to be changed. As an example, recent results for the system Co/NiO yield $\Delta H_s = 139$ kJ/mol, $\Delta H_{gb} = 180$ kJ/mol, $\Delta H_v = 227$ kJ/mol and, for 800°C, $D_s \sim 10^2 D_{gb} \sim 10^7 D_v$ (42).

b) Uranium oxide, UO_2. UO_2 is a fluorite-type oxide with a very large range of nonstoichiometry at elevated temperatures (e.g. from O/U = 1.65 to O/U = 2.25 at ~2500° C (34)). Very many diffusion measurements have been performed with UO_2. Bulk diffusion of O and U depend strongly on O/U-ratio (34, 35). As mentioned above, all three mass-transfer methods to measure D_s-values, i.e. grain boundary grooving, single scratch decay and relaxation of sinusoidal profiles (e.g. 20-22) were applied to UO_2, in addition to tracer techniques with U and Pu-tracers (12, 13, 36, 37). The data existing in 1969 were summarized in the review on surface diffusion of oxides by Robertson (38), those published until 1975 in the review on lattice defects and irradiation damage in ThO_2, UO_2 and $(U,Pu)O_2$ by the present author (39). Some of the experimental difficulties encountered were already discussed above and are illustrated in Figs. 3 and 5.

The main *experimental difficulties* with measuring D_s in UO_2 are its high vapor pressure and the anisotropy of its surface energy (see Section 3). Depending on experimental details and atmosphere (pressure) used, evaporation-condensation frequently interferes with surface diffusion or dominates surface mass transport. For the mass-transfer techniques, surface diffusion dominates usually up to about 1400 °C, evaporation-condensation above about 1700 °C, and both mechanisms are important between 1400 and

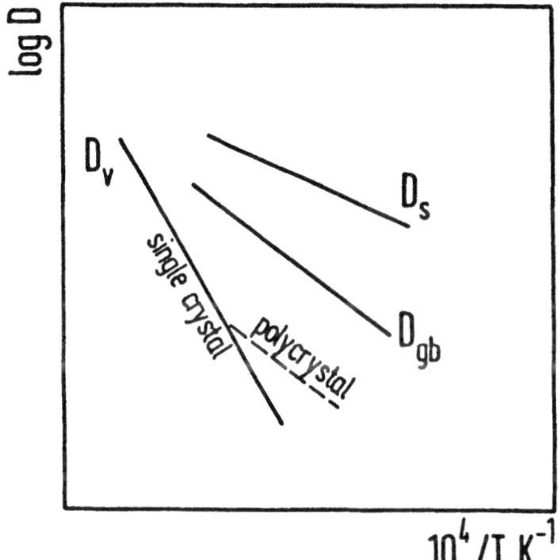

Fig. 9: Schematic Arrhenius diagram showing approximate relative rates of surface, grain boundary and volume diffusion in ceramics

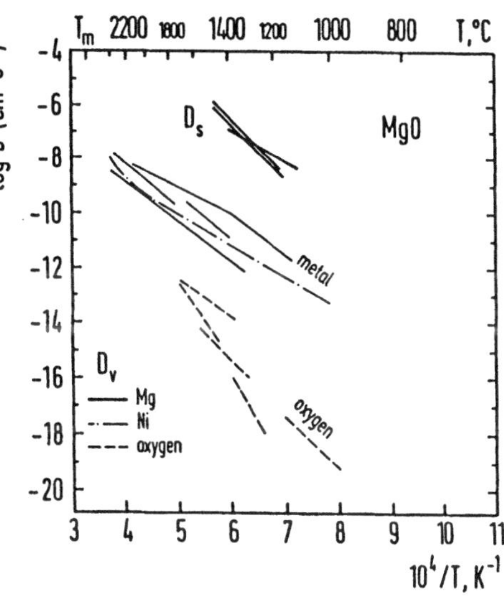

Fig. 11: Arrhenius diagram for diffusion processes in MgO (10,11,38,41-44)

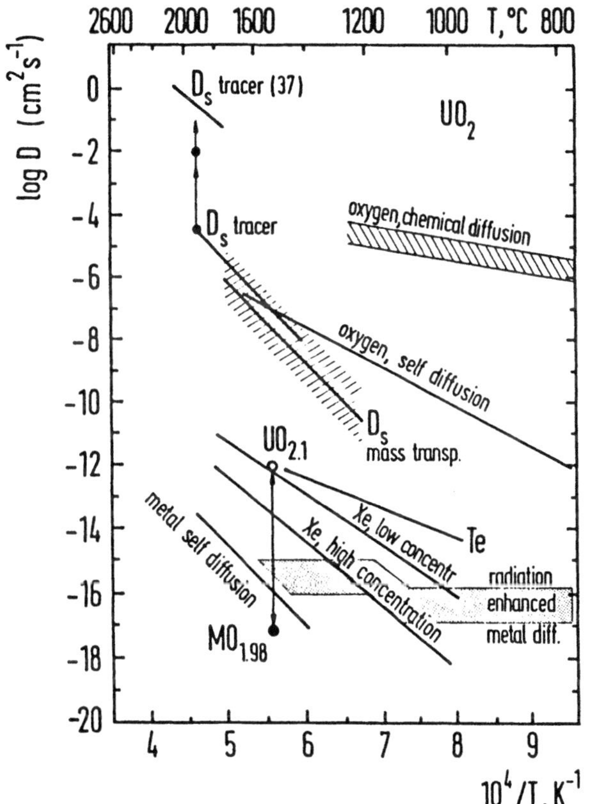

Fig. 10: Arrhenius diagram for diffusion processes in UO_2 (see text and ref. (35) for explanation).

1700 °C. The results of 9 different investigations (for refs. see (39)) are contained in Fig. 10 in the band labelled D_s, mass transport. The line within this band is described (39) by

$$D_s = 5 \times 10^5 \exp(-450.000/RT)\ cm^2s^{-1}$$

Since ΔH_s with 450 kJ/mol (or 4.7 eV/atom) is smaller than the heat of vaporization (5.4eV/atom), the change in mechanism with increasing temperature becomes understandable. (The term "atom" used here indicates that U-atoms are transported , but does not imply that U-atoms move by themselves. Rather, mobility of UO_2- or UO_3-molecules is included. This remark holds also for the other materials described below).

Fig. 10 shows, as expected, that metal self-diffusion in the volume is much slower ($\Delta H_v \sim$ 540 kJ/mol) than D_s determined from mass transport methods. Also, important fission products diffuse in the bulk much more slowly than U on the surface. This is indicated for Xe at low concentration (diffusion of single gas atoms) and at high concentration (Xe-atoms trapped in defects), as well as for Te. Oxygen, on the other hand, as fast species in the bulk of UO_2, diffuses faster than U-mass transport at the surface, in particular whenever a chemical gradient exists (see band labelled oxygen, chemical diffusion in Fig. 10, describing diffusion coefficients effective following a change in the oxygen potential, $\Delta G(O_2)$, of the annealing atmosphere, hence during oxidation or reduction within the $UO_{2\pm x}$ phase field). Radiation and fission during operation of UO_2 fuel in a reactor increase the diffusion rates of U very significantly (see band labelled radiation enhanced metal diffusion). Attempts to measure possible radiation enhancement of D_s were performed in the laboratory of the author but did not yield unique results.

The arrow at 1500 °C in the lower part of Fig. 10 shows that metal diffusion (U in UO_2, Pu in MO_2, (for $(U,Pu)O_2$ as fast breeder reactor fuel)) varies significantly with $\Delta G(O_2)$ or O/M-ratio. The difference between D_v for U diffusion in $UO_{2.1}$ and that of D_v for Pu diffusion in $(U, Pu)O_{1.98}$ is rather dramatic, i.e. more than 4 orders of magnitude (39). A dependence of D_s on $\Delta G(O_2)$ or O/U-ratio is likely as well but has not been measured yet.

The main *mathematical (theoretical) difficulties* with measuring D_s in UO_2 were found with tracer experiments. The first experimental data of Marlowe and Kaznoff (12) obtained by measuring the spread of U-235 tracer from a $(U\text{-}235)O_2$ needle pressed on a UO_2 specimen coincide in their as-reported form with the band for mass transfer D_s in Fig. 10, as shown by their reported D_s-value at 1915 °C (see full dot labelled D_s, tracer in Fig. 10). The raw data on spreading of U-235 around the tracer-needle need, of course, corrections for evaporation-condensation as positive contribution, and volume diffusion as negative contribution. It is apparent that the spreading tracer layer is similar to the initial condition of a volume diffusion experiment. Similar to the better known case of grain boundary diffusion, there is a continuous loss of tracer from the surface into the interior since bulk or volume diffusion acts as sink for surface atoms. Such a correction is not trivial and necessitates the existence of reliable values of D_V. The first correction of the original data by Robertson (38) caused the reported D_s values to increase by a large factor of 300, and a very recent re-analysis by Olander (36) yielded a further increase in the corrected D_S, the correction factor now being 10^4. Though Olander used too high D_V-values, the final corrected D_S-value will be very high appoaching the order of a D in gases. Recent sophisticated experiments by Zhou and Olander (37) yield also very high D_S-values (see line D_S, tracer in Fig. 10) with a pre-exponential factor $D_0 \sim 5 \times 10^6$ cm^2/s and a ΔH_S of 300$\pm$ 60 kJ/mol. In the above nomenclature, this would be a $D_{S,a}$-value in accord with the non-localized surface diffusion model of Bonzel described above.

The implication is that large mean free paths are involved for the diffusing species, which is likely be an activated UO_2 molecule with unhindered rotational freedom. This is consistent with the knowledge on UO_2 evaporation. Though UO_2, UO_3, UO, U and O_2 are all present in the gas phase, UO_2 is the dominant species in slightly substoichiometric UO_2 as it would be expected to exist in high temperature experiments. However, the UO_2 molecule in the gas phase is significantly different from the UO_2 in the bulk. U-O distances in the bulk are 0.237 nm and bonding is largely ionic, whereas the gaseous UO_2 molecule is covalent with U-O distances of 0.179 nm (40). It is thus not surprising, that a gas-like surface-UO_2 molecule has rotational degrees of freedom and is not readily trapped in molecular surface vacancies. (Similar arguments hold for stoichiometric UO_2 above which the dominant gas phase is UO_3.)

The ratio of $D_{s,a}/D_s$ for UO_2 is in the range known for metals (see Section 2.3), i.e. $\sim 10^3$ to 10^4, thus giving further confidence to the suggested mechanism. Despite the many experiments performed with UO_2 so far, however, a careful study with controlled oxygen potential is still needed to measure the dependence of $D_{S,a}$ and D_S on $\Delta G(O_2)$ or O/U-ratio.

c) Magnesium and nickel oxide MgO and NiO. These two oxides were also investigated extensively in their diffusion properties. In these NaCl-type monoxides and in contrast to the fluorite-type dioxides, oxygen is the slow species. Lack of space does not allow to give all literature references for the Arrhenius diagrams shown in Figs. 11 and 12. The reader is referred to a previous review (41) for refs. on volume diffusion; the ones for surface self-diffusion are refs. (10, 11, 38, 42 - 44) for MgO and refs. (11, 16, 45) for NiO. Dhalenne et al. (16) have provided a very extensive study and a complete mathematical analysis for all possible diffusion paths for grain boundary grooving in NiO, and they conclude that all three processes of volume and surface diffusion as well as evaporation-condensation contribute to surface mass transport, and that volume and surface diffusion are controlled by Ni-ions, the transport of oxygen occurring through the gas phase. Their results are included in Fig. 11 labelled D_S (at 1400-1520°C). At 1520°C, the ratio of D_S/D_V for Ni on NiO is ~7000.

Tracer data on surface diffusion exist also for MgO and NiO due to recent work of Stubican et al. (10, 11, 42, 44). These data are for impurities such as Co and Cr, and they confirm the general picture very convincingly (see line for Co surface diffusion on NiO in Fig. 12.). These data have already been mentioned above as a typical example to explain the schematic presentation in Fig. 9.

The tracer data on MgO show knees in the Arrhenius diagrams at about 1100 °C, possibly indicating a transition to delocalized transport at higher temperatures. Comparison of surface diffusion of divalent Co and trivalent Cr on MgO shows similarities to bulk diffusion: Co diffuses faster than Cr in both cases, probably due to a larger impurity-vacancy bonding energy for Cr.

d) Non-cubic oxides: Al_2O_3,Fe_3O_4. Very many mass transport experiments have been performed with Al_2O_3. Since no suitable radioactive tracers exist for either Al or O, these experiments are the only source of information on surface self-diffusion on Al_2O_3, though tracer data exist for impurity diffusion (Cr) on Al_2O_3. The data for self-diffusion (46, 47) and typical data for surface diffusion (48-50) are shown in Fig. 13. Grain boundary grooving of pure and impure Al_2O_3 (type Morganite or Lucalox), crack healing, scratch smoothing, initial stage sintering etc. have all been investigated in about a dozen studies. A representative equation (49) is

$$D_S = 4 \times 10^5 \exp(-452.000/RT)\ cm^2s^{-1}$$

which is obtained from 17 data points for boundaries between pure single crystals from four investigations (see thick line labelled D_S in Fig. 13). Impure (e.g. 99.7% pure sinters) Al_2O_3 give a ΔH about 30% lower and D_S-values about a factor of 10 higher.

Recent data (42) on surface diffusion of Co on magnetite $Fe_{3-x}O_4$ under different oxygen potentials indicated a minimum of D_S (at $p(O_2) = 10^{-14}$ bar at 750°C). Similar minima occur for volume diffusion of Fe and of impurities as first shown by Dieckmann and Schmalzried (e.g. 51). This indicates that the mechanism of surface diffusion is similar to that of volume diffusion. It should also be mentioned that the recent studies on oxides have yielded rather high ΔH values for surface diffusion.

e) Carbides and nitrides, UC and UN. A number of mass transport studies have been performed on UC (grain boundary grooving, relaxation of sinusoidal profiles (see e.g. (52-54), and also monograph on "advanced nuclear fuels, hence carbides and nitrides of U and Pu" (55))). Fig. 14 shows that for suitable experimental conditions, surface diffusion can be rate-controlling. Otherwise, fast carbon diffusion through the bulk and U transport through the vapor phase can control surface mass transport. Since no tracer data exist for surface diffusion on either UC or UN, identification of the rate-controlling species and of the rate-controlling mechanism is not trivial. Fig. 15 shows that volume diffusion of carbon in UC is very fast, and the dominant species in gas

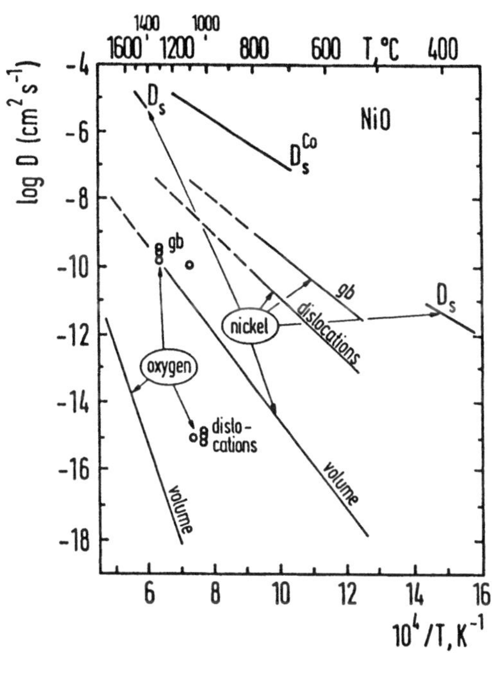

Fig. 12 (left):
Arrhenius diagram for diffusion processes in NiO (11, 16, 41, 45)

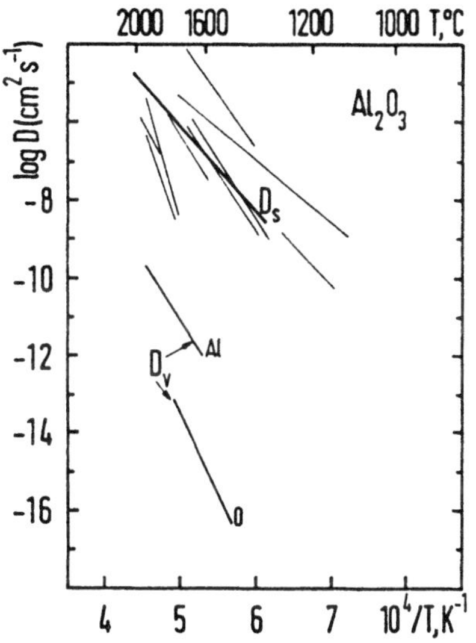

Fig. 13 (below right):
Arrhenius diagram for diffusion processes in Al_2O_3 (38, 46-50)

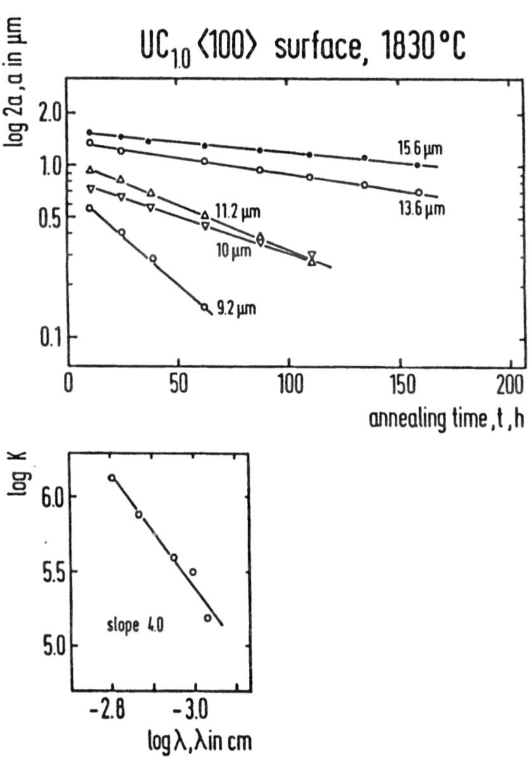

Fig. 14 (below left):
Relaxation of surface profiles such as those shown in Fig. 2 for UO_2. The data are for UC and they follow the kinetics of surface diffusion (see equ. (3))(52).

phase is known to be U (55). Data on volume diffusion and grain boundary diffusion of U are also included (e.g. 55-58). Both C and U volume diffusion depend on C/U-ratio, as indicated by the vertical arrows in Fig. 15. D_S may depend on C/U-ratio as well, but again no experimental results exist to answer this question.

The available results on surface matter transport on UC may be interpreted (55) as showing a change-over in rate-controlling mechanism at about 1800°C. The most reliable data (53) are included in Fig. 15 and are described by

$$D_S = 5{,}6 \times 10^{-4} \exp(-130.000/RT) cm^2 s^{-1}$$

Both D_0 and ΔH are surprisingly low. This may be due to a change in C/U-ratio at different temperatures. More data are definitely needed.

An interesting alternative of measuring D_S is observing the mobility of rare gas-filled bubbles in a temperature gradient (see Section 4). Such experiments were performed with both UC and UN (see Fig. 16)(e.g. 59). A more detailed discussion is given in Section 4. It suffices here to note that the corresponding D_S-values for UN are shown in Fig. 15, upper part, and that the indicated ΔH-values are again surprisingly low (~175 kJ/mol for UN).

3. SURFACE ENERGY OF CERAMICS

As mentioned in the introduction, surface energies of ceramics, γ_S, are of importance in surface diffusion processes. Matter transport measurements usually yield the product $D_S \gamma_S$ (see e.g. equ. (3) and Table 1), and anisotropy of γ_S is the reason for the surface striations shown in Figs. 3 and 6. In the following, experimental techniques to determine γ_S are briefly mentioned and experimental as well as calculational results are discussed for UO_2, a ceramic for which many results exist.

3.1 Experimental Techniques to Measure Surface Energies

Basically, two different approaches have been used to determine γ_S in ceramics experimentally, so-called "thermodynamic methods" and "fracture methods". The first ones determine the "thermodynamic surface energy", i.e. the γ_S corresponding to *relaxed* high temperature surfaces which, however, are often contaminated with adsorbed impurities or impurities which segregated at the surface. Segregation of impurities can easily change transport kinetics and surface energies significantly (60). The latter techniques have the advantage of creating new fresh surfaces without any segregation, dirt, adsorption etc. but their drawback is the partly empirical nature of the relations used to deduce γ_S, and the fact that *unrelaxed* fracture surfaces are formed. Experimental difficulties exist also, e.g. with crack branching.

a) Wetting experiments and the multi-phase equilibrium technique.

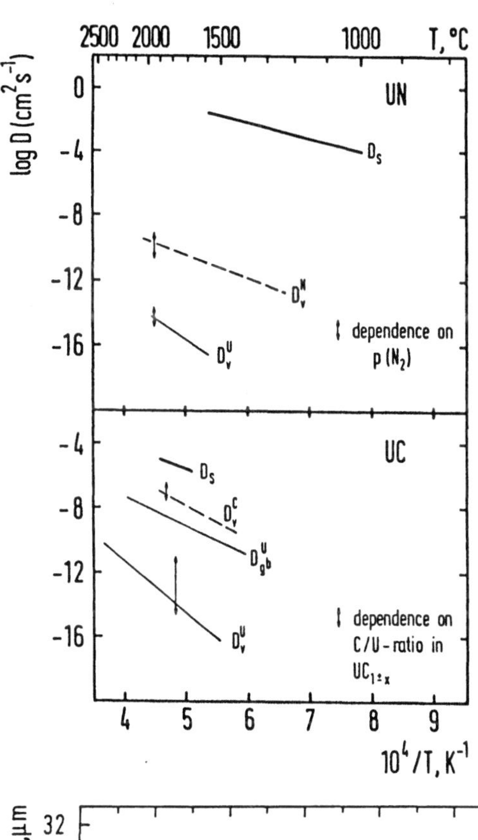

Fig. 15:
Arrhenius diagrams for diffusion processes in UC and UN. Vertical arrows at 1950°C for UN and at 1850 or 1800°C for UC show that volume diffusion of both metal and non-metal atoms depend on C/U or N/U-ratio (55-58). Similar experiments for D_s are still missing.

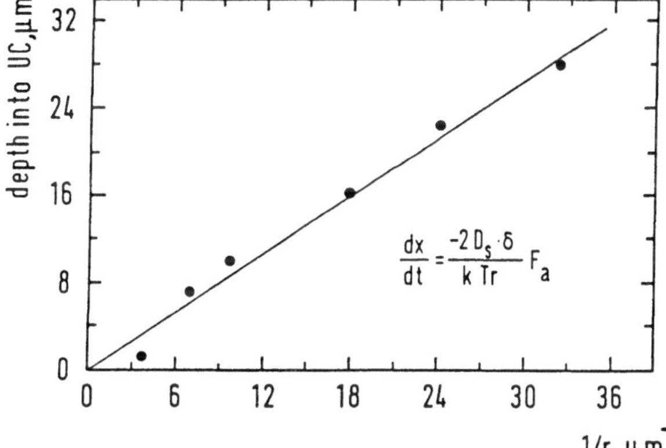

Fig. 16:
Diffusion of He-bubbles in UC (in a temperature gradient (59)).
The distance migrated is proportional to 1/r, (r = bubble radius), as expected for surface diffusion as rate-controlling process. The equation relates the bubble velocity dx/dt with D_s and the driving force F_a, i.e. the temperature gradient.

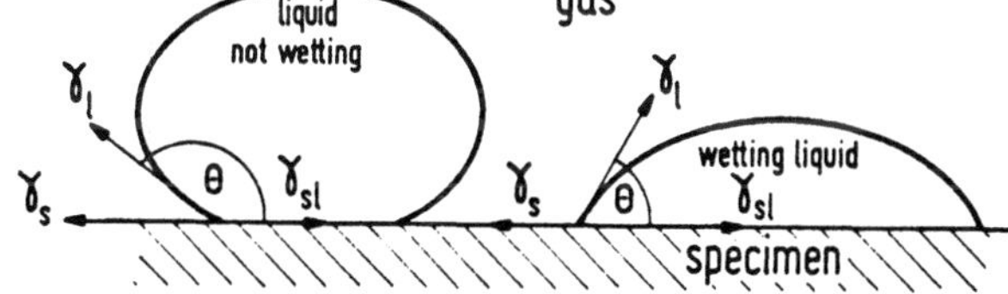

Fig. 17:
Schematic illustration of a wetting experiment with a liquid metal drop on a ceramic surface.

The wetting of a ceramic by liquid metals can be used to get a first indication for the value of γ_s. The degree of wetting is measured by the contact angle Θ which, as shown in Fig. 17 is the angle through the liquid medium which the tangent to the contour of the liquid surface makes with the solid surface at the point of contact with the solid. A contact angle $<90°$ gives wetting. The spreading of the liquid over the solid surface is determined by the condition that the free energy of the system as a whole should decrease. The balance is given by the relation

$$\gamma_s = \gamma_{sl} + \gamma_l \cos\Theta \tag{5}$$

where γ_l is the surface energy of the liquid and γ_{sl} the interfacial energy between solid and liquid (see e.g. ref. (61)).

The *multi-phase equilibration technique* yields, in principle, very accurate results of γ_s. It was applied to UO_2 by 3 groups (62-64). It consists of a series of 4 experiments (at constant, high temperature)

- measuring grain boundary groove angles ψ following annealing in inert atmosphere
- measure similarly ψ^* in the vapor of a metal
- measure the dihedral angle ϕ at a grain boundary under the liquid metal
- measure the contact angle Θ (as in Fig. 17) of a metal drop (in metal vapor).

The surface energy of the solid γ_s (also often called γ_{sv}) is then given by

$$\gamma_{sv} = \gamma_{lv} [\cos\Theta \cos 0.5\,\phi \cos 0.5\,\psi^* / \cos 0.5\,\psi\,(\cos 0.5\,\phi - \cos 0.5\,\psi^*)] \tag{6}$$

The difficulty in exact determination of γ_s(or γ_{sv}) consists in the fact that the denominator will be small (errors will be large) for $\phi \approx \psi^*$. An error of one degree can be significant, a large set of values (preferably up to 100) are desirable, given the fact that the grain boundaries are not necessarily perpendicular to the investigated surface.

b) Fracture and indentation techniques.

Such techniques determine γ_s from crack nucleation, formation or growth. Indentation techniques have been extensively applied to UO_2 and are increasingly used to measure fracture toughness, K_{Ic}, and (fracture) surface energy of nonmetals.

The Hertzian technique (see Figs. 18, 19) uses spherical indentors to produce ring cracks which extend conically into depth. It is a physically based method not necessitating any empirical relations, at least in principle. In practice, evaluation of γ is based on the knowledge of a number of quantities in addition to that of the critical load P_c for crack formation and the radius, r, of the ring crack formed, i.e. the two quantities measured directly in the experiment, since

$$\gamma = f[P_c, r, v, E, \beta, a/c''(\phi'')^{-2}] \tag{7}$$

where v = Poisson ratio, E = elasticity or Young's modulus, β = a constant and $a/c''(\phi'')^{-2}$ a crack extension function for which values and graphical presentations are available (e.g. 65, 66).

Vickers indentation of brittle materials such as ceramics leads to cracks emerging from the corners of the edges of the diamond indentation provided

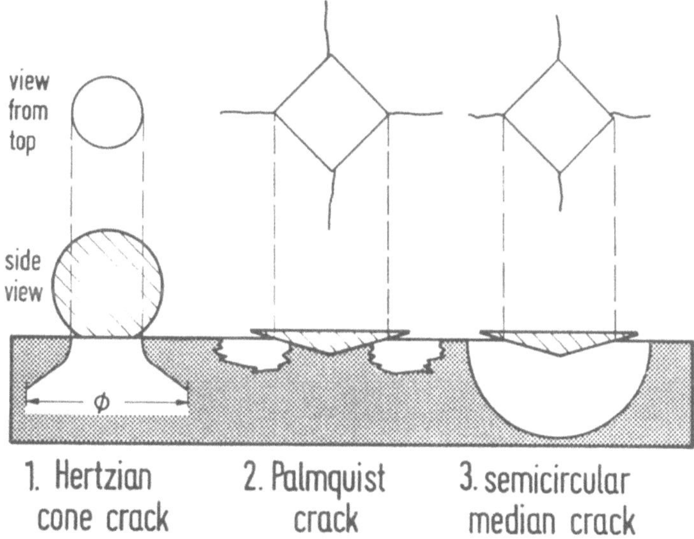

Fig. 18: Three types of indentation cracking, viewed from top or sectioned vertically in a median plane.

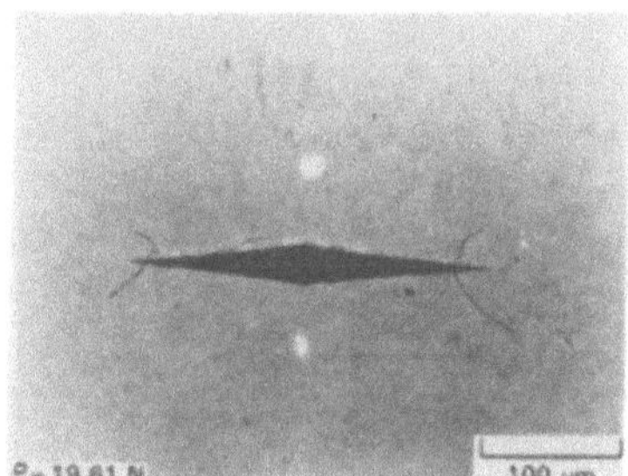

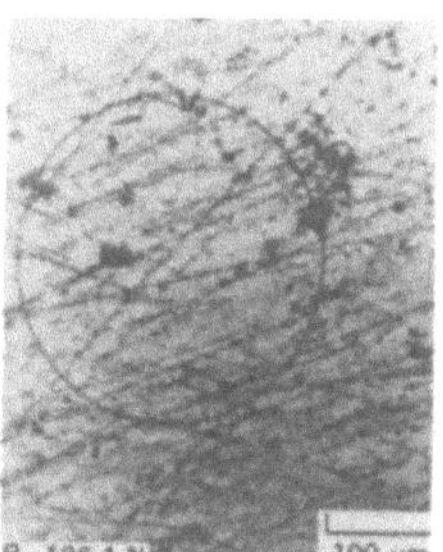

Fig. 19: Fotographs of a Vickers indentation (left) and a Knoop indentation (middle) in glass and of a Hertzian ring crack in ThO_2.

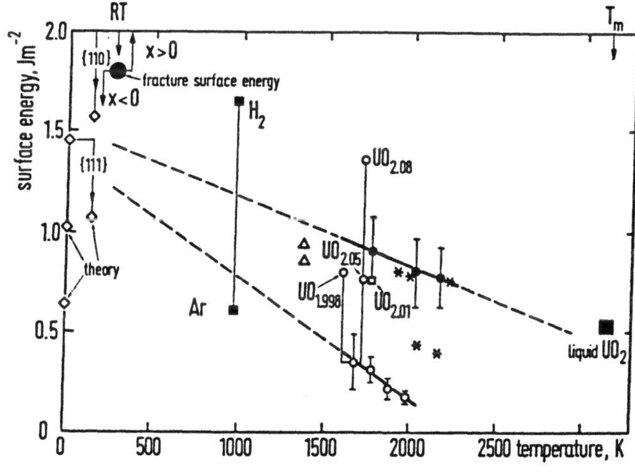

Fig. 20: Results on surface energy of UO_2 (indentation techniques, multi-phase equilibration and calculations) (see (65, 67) for detailed references).

the load is high enough. These cracks extend into depth and can be of the Palmqvist or the half-penny type (median cracks) as shown in Fig. 18. (In practice, often more complex crack systems are formed including lateral cracks underneath the surface, evidenced by the interference pattern in the left photo of Fig. 19). An empirical relation based on a number of determinations of fracture properties of different materials and with different techniques is used to determine K_{Ic} and γ via a relation of the type.

$$\gamma = [(1-v^2)/2E]K_{Ic}^2 \text{ with } K_{Ic} = \text{const } H\sqrt{a}\,(E/H)^{0.4}(c/a)^{-3/2} \qquad (8)$$

where the constant is an empirical number (=0.057) whereas the other quantities are measured in the test (H = hardness, a = half diagonal of Vickers indentation, c = length of radial crack) or can be measured separately (E-modulus).

The drawback of fracture techniques mentioned above, i.e. the fact that the energy of *unrelaxed* surfaces is measured, is not necessarily an argument against those techniques which are simple, fast and necessitate small specimens only. Measurements in the laboratories of the author [at increased temperatures (67) or of (Hertzian) crack extension (in contrast to crack nucleation) (68)] have yielded results in agreement with those expected for relaxed surfaces, i.e. the surfaces encountered in practical work at elevated temperatures. Even at room temperature and with measurements at the surface only, relative trends in γ (e.g. with deviations from stoichiometry in a given substance, etc.) can easily be obtained.

3.2 Results on Surface Energies for $UO_{2\pm x}$

Not very many data on γ_s exist for the materials treated here, and for other ceramics. Most results have been published for UO_2 for which also critical reviews (e.g. 69) and theoretical values (70) exist. These results will therefore be summarized here as being typical for ceramics though on UO_2 values for some other ceramics have also been published, e.g. for Al_2O_3, ThO_2, UC and UN.

Fig. 20 shows published results as function of temperature. Fracture and indentation techniques in polycrystals yield average values of unrelaxed surfaces: $\gamma_s = 1.8$ J/m^2 for UO_2 at room temperature (see big black circle in Fig. 20), with a decrease of γ_s in UO_{2-x} and an increase in UO_{2+x}, as indicated by arrows at the black circle. This γ_s represents the energy to create a unit area of new surface. Since every system tends to reduce its total free energy to a minimum, the fresh surface tends to reduce its energy, e.g. by re-adjusting the location of the surface atoms, or by their polarization. Such relaxed surfaces will, for instance, have the large ions (oxygen in oxides) in the surface. A study of Hertzian indentation on ThO_2 using preheated specimens (71) showed that the room temperature γ_s-value decreased by $\sim$ 35% if the ThO_2 is preheated to between 200 and 300°C, i.e. the temperature range in which oxygen becomes mobile. A similar effect in UO_2 would decrease γ_s for UO_2 from 1.8 to about 1.2 J/m^2.

Calculations by Tasker (70) showed a similar decrease (from 1.48 to 1.07 J/m^2 for {111} surfaces) between an unrelaxed and a relaxed UO_2 surface. On relaxation, the first oxygen plane of the {111} surface is predicted to move substantially (~ 14pm). The {111} plane is the energetically favored plane in UO_2; it is also the natural cleavage plane of UO_2. In agreement with earlier experimental work (72) on γ-plots for UO_2 (i.e. plots of γ as function of direction) showing a pronounced anisotropy of γ, higher surface energies are calculated for other planes. For instance, $\gamma = 1.54$ J/m^2 is calculated for a relaxed {110} plane. A random value will thus be in between the extreme values. The strong anisotropy in γ_s for UO_2 explains also the difficulties with surface diffusion measurements using mass transfer techniques and the formation of striations on UO_2 surfaces during annealing (see Section 2.4 and Figs. 3, 5 and 6).

Fig. 20 shows a number of results obtained at higher temperatures, indicates the scatter obtained and the effect of O/M-ratio (or annealing atmosphere: Ar, H_2).Recommendable data are indicated by the upper dashed line connecting the low temperature values with the results for liquid UO_2. Those latter results are based on 4 independent studies (for refs. to all data see (65, 69)).

For the purpose of better understanding surface diffusion on ceramics, the above discussion on anisotropy in γ_s, scatter in the data and dependence of γ_s on deviations from stoichiometry serves to show the difficulty in deducing meaningful D_s-values from measurements yielding $D_s\gamma_s$, as for the mass transfer methods in table I.

4. APPLICATION TO THE BEHAVIOR OF VOLATILE FISSION PRODUCTS IN CERAMIC NUCLEAR FUELS

In the above description of surface diffusion and surface energy, ceramics of interest in nuclear energy were frequently referred to. UO_2 is the most common fuel of today's power reactors, and the so-called advanced fuels, carbides and nitrides, are candidates for advanced breeder reactors or space reactors (e.g. 66). In nuclear fission, large amounts of gaseous or volatile elements are formed as fission products. For instance, for 100 fissions of Pu-239, about 25 rare gas atoms (2 Kr and 23 Xe atoms) and 22 other volatiles (1 Rb, 2 I and 19 Cs atoms) are formed. Though Rb and Cs can occupy substitutional sites in UO_2 (73), the solubility of Kr and Xe is very low. Therefore, the rare gases precipitate into gas bubbles even at low temperature whenever the concentration is very high, or, at lower concentrations, whenever the temperature is high. In equilibrium, the pressure in the bubble is compensated by the surface energy of the matrix, i.e. $p = 2\gamma/r$. Typical

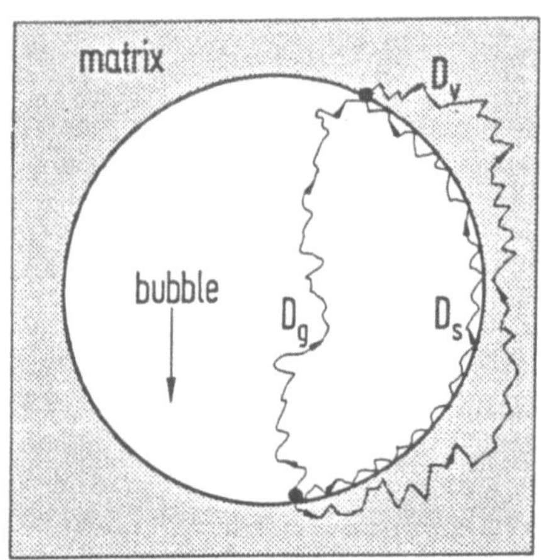

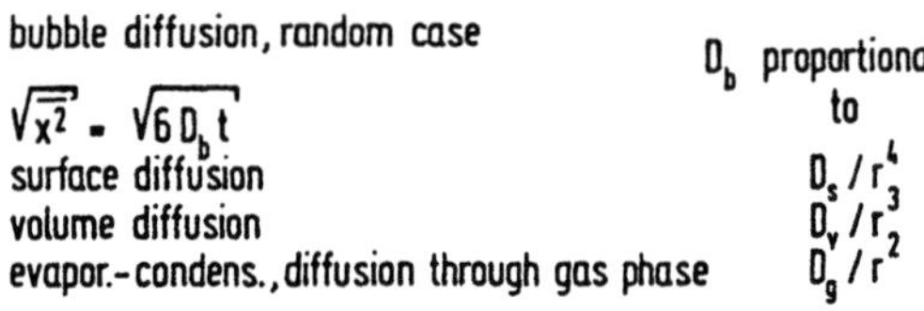

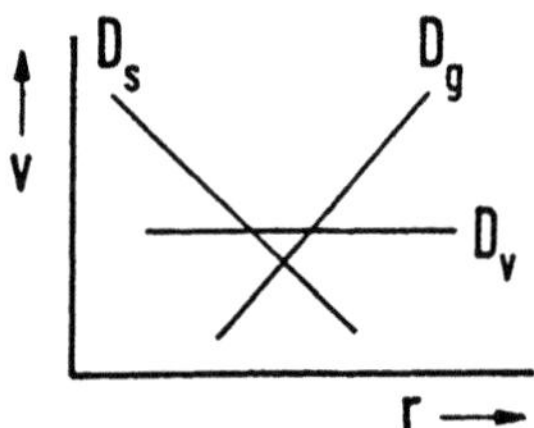

Fig. 21: The three mechanisms of inert gas bubble diffusion via surface diffusion, volume diffusion or evaporation-condensation of matrix atoms.

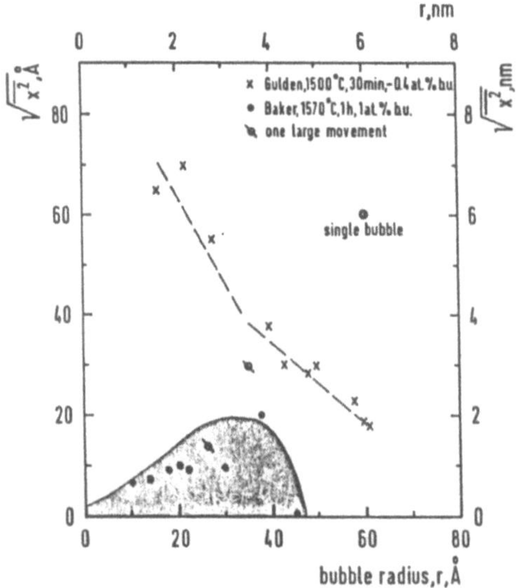

Fig. 22: Measured migration distances versus bubble radius for fission gas bubbles in oxide fuels (76-78)

Fig. 23: D_b-values measured and predicted for UO_2, as a function of the bubble radius for T = 1843 K (77-83)

bubble radii, r, are in the range of 1 to 20 nm. With r = 10 nm and at 1500 K, a bubble in UO_2 contains about 30000 rare gas atoms. At high temperatures, bubbles can move by random Brownian motion or directed in the temperature gradient existing in the fuel. Different dependences of the bubble diffusion coefficient, D_b, on bubble radius are predicted, as shown in Fig. 21. The figure shows also the three basic mechanisms of motion. These are the same as those competing for surface mass transfer, i.e. surface diffusion, volume diffusion and evaporation-condensation (with diffusion through the gas phase within the bubble, coefficient D_g) of matrix atoms. The centre of gravity of the bubble shifts (hence the bubble moves) if matrix atoms move from one side of the bubble to the other.

Obviously, bubble formation and mobility are very important phenomena since they determine the volume increases, i.e. the swelling of the ceramics. If bubbles grow the pressure decreases ($p = 2\gamma/r$) and swelling increases; if they diffuse they may reach grain boundaries or surfaces and be vented by releasing their gas. Fig. 21 shows in its right part that very large bubbles, or rather pores, will move fast by evaporation-condensation, a well known phenomenon for restructuring of UO_2 in a temperature gradient. Small bubbles should also move fast by a surface diffusion process. We see here an immediate application of results for D_s and γ_s to an important technical question.

Small rare gas bubbles have indeed been found to move by a fast surface diffusion mechanism. Examples are He-bubbles in UC and UN (see Fig. 16, section 2, and refs. (59, 66)) or in Al (74). However, this fast mechanism is very frequently *not* operative. What are the reasons?

Even for such simple systems as He-bubbles in Au or Cu, the bubble velocities were observed to be only 10^{-5} to 10^{-2} of those predicted for a surface diffusion mechanism (75). We have here a case which is illustrative for surface diffusion. Basically, the case of bubble mobility should be ideal: we deal with a new surface created by precipitation of rare gas, hence also with a very pure atmosphere. Yet, surface diffusion has slowed down. For He-filled bubbles in Au or Cu, this could be attributed to facetting of the bubble surfaces (75). Nucleation of steps on these facets determines atomic transport on the surface and decreases the effective D_s. Anisotropy in γ_s is thus affecting and decreasing bubble mobility.

Another mechanism decreasing bubble mobilities is their interaction and pinning at dislocation lines or networks. In the case of fission gas bubbles, additional mechanisms slowing down bubbles are: i) presence of other volatiles (Cs, I, Te, etc.) in the bubble decreasing D_s, ii) interaction of the bubble with precipitates of solid fission products, e.g. the often observed metallic precipitates of the (insoluble) metallic fission products Pd, Ru, Rh,Tc and Mo, iii) blocking of adatom jumps in small gas bubbles because of the high pressure of rare gas within them.

Fig. 22 shows measured migration distances of very small fission gas bubbles in irradiated UO_2 (76 - 78). Some data indicate an increase in distance with decreasing r, others do not. Apparently, bubble mobility is slow and irregular (at $T \leq 1800$ °C). Individual bubbles may move a certain distance but a general trend is not followed. Fig. 23 finally shows the same fact in another presentation (79): D_b-values are plotted as function of r. The theoretical predictions for bubble diffusivities by surface diffusion (with an uncertainty band corresponding to an error in T of ± 150 K) and for evaporation-condensation are shown as lines, experimental results are labelled with numbers (1 to 5 corresponding to refs. (77, 78, 80 - 83)). Obviously, near to all observed bubbles have effective D_b-values very significantly below those calculated for surface diffusion, in particular if they are of the size most frequently found in irradiated fuel, i.e. $r \leq 50$ nm. Trails in transmission electron micrographs of transient tested fuel have recently been seen (80) which, if representing the migration distance of bubbles, may be compatible with a fast surface diffusion mechanism (see square labelled (1) in Fig. 23). However, this is only an indication and needs verification.

Most bubbles obviously are not moving by the fast surface diffusion mechanism. In addition to facetting or contamination of bubble surfaces and blocking of adatom mobility, interaction with dislocations and precipitates have been mentioned above as processes slowing down bubble mobilities and/or changing the rate-controlling mechanism for their mobility. Calculations on binding forces for those two possibilities have already been made in the sixties (84, 85). Those calculations show that for temperature gradients and temperatures normally found in UO_2 nuclear fuel (e.g. $dT/dx \approx 1000$ °C/cm, $T \leq 1700$ °C), bubbles with $r \leq 100$ nm tend to remain pinned at dislocations or precipitates.

A large amount of electron microscope work has recently been performed on reactor irradiated UO_2, including transient tested fuel, and on UO_2 containing rare-gases and volatile fission products (I, Rb, Cs, Te) introduced by controlled ion implantation (e.g. 86 - 89). These studies yielded clear evidence for bubble facetting and bubble pinning and interaction with dislocation lines and precipitates, as shown in Figs. 24 a -c.

5. SUMMARY AND CONCLUSIONS

The present manuscript describes surface diffusion and surface diffusion controlled phenomena in ceramics (MgO, NiO, Al_2O_3, Fe_3O_4, UO_2, UC, UN, etc.). Brief reviews on surface diffusion on metals and on surface energies in ceramics (example UO_2) are also given. The available knowledge is applied to the technologically important problem of the mobility of rare gas-filled bubbles.

For surface diffusion on ceramics, additional difficulties exist that are not encountered to the same extent on metals. Those include the difficulty of determining which of the elements of the compound are moving and rate-controlling, strong anisotropy in surface energies, high vapor pressures favoring evaporation-condensation to become rate-controlling for mass transport at ceramic surfaces and often existing non-stoichiometric compositions. D_S may depend on deviation from stoichiometry and the fact of a finite width of a single phase field itself often prevents UHV techniques to be used; consequently, ceramic surfaces are generally less clean than metal surfaces in diffusion experiments.

Nevertheless, a reasonably clear picture for the existence of both adatom (or "ad-molecule") mobility (non-localized diffusion) and mass transport surface diffusion (localized diffusion) has evolved. The recently postulated existence of a thick and temperature-dependent high diffusivity layer on oxides is rejected as being not compatible with theoretical and experimental evidence of modern surface physics.

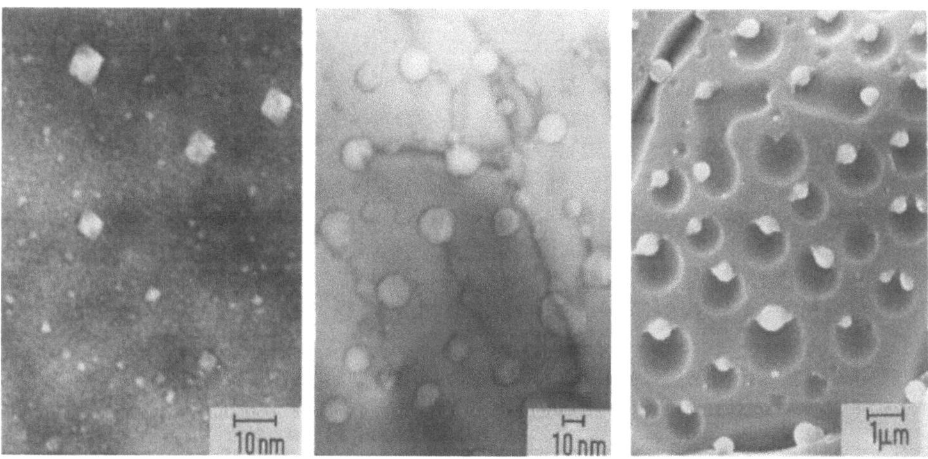

Fig. 24: Electron microscope examination of UO_2. a) Facetted fission gas bubbles in reactor irradiated and annealed UO_2, b) rare gas bubbles at dislocations in ion-implanted and annealed UO_2, c) interaction of metal precipitates and fission gas bubbles in reactor-irradiated UO_2.

References

1) *Surface Mobilities on Solid Materials, Fundamental Concepts and Applications,* Ed. Vu Thien Binh, Nato ASI Series B: Physics, Vol. **86**, Plenum Publ. Corp., (1983)
2) G. Neumann and G.M. Neumann, *Surface Self-Diffusion of Metals,* Diffusion Monograph Series, Ed. F.H. Wöhlbier, Solothurn, Switzerland (1972)
3) *External and Internal Surfaces in Metal Oxides*, Eds. L.-C. Dufour and J. Nowotny, Mater. Sci. Forum, Vol. **29**, Trans. Tech. Publ., Aedermannsdorf, Switzerland (1988)
4) G. Ehrlich and F.G. Hudda, J. Chem. Phys. **44** (1966) 1039
5) J.C. Fisher, J. Appl. Phys. **22** (1951)74
6) P.G. Shewmon, J. Appl. Phys. **34** (1963)755
7) R.T. Whipple, Phil. Mag. **45** (1954)1225
8) T. Suzuoka, Trans. Jpn. Inst. Met. **2** (1961) 25 and J. Phys. Soc. Japan **20** (1965) 1259
9) A.D. LeClaire, Brit. J. Appl. Phys. 14 (1963) 351
10) V.S. Stubican, G. Huzinec and D. Damjanovic, J. Amer. Ceram. Soc. **68** (1985)181
11) V.S. Stubican, C. Lin and E. Mazey, Advances in Ceramics **23** (1987) 97
12) M.O. Marlowe and A.I. Kaznoff, J. Nucl. Mater. **25** (1968)328
13) H. Furuya and M. Koizumi, Nucl. Technol. **28** (1976) 226
14) G.C. Kuczynski, Trans. AIME **185** (1949) 796
15) T.O. Sørensen, Risø (Denmark) Report Risø-I-372(1988)
16) G. Dhalenne, A. Revcolevschi and C. Monty, Phys. Stat. Sol. (a) **56** (1979) 623
17) C. Herring, J. Appl. Phys. **21** (1950) 301
18) W.W. Mullins, J. Appl. Phys. **28** (1957) 333; **30** (1959) 77, Acta Met. **7** (1959) 746; Trans AIME **213** (1960) 354
19) H.P. Bonzel, in ref. (1), p. 195 (1983)
20) I. Amato, R.L. Colombo and G.C. Grappiolo, Sol. State Comm. **4** (1966) 237
21) G.L. Reynolds, J. Nucl. Mater. **24** (1967) 69
22) Hj. Matzke, J. Nucl. Mater. **114 (1983)** 121
23) Y. Ikuma and W. Komatsu, in ref. (3), p. 199 (1988)
24) E. Preuss, N. Freyer and H.P. Bonzel, Appl. Phys. **A41** (1986) 137
25) H.P. Bonzel, Surf. Sci. **21** (1970) 45
26) G.E. Rhead, Surf. Sci. **15** (1969) 353
27) H.P. Bonzel and N.A. Gjostein, Surf. Sci. **22** (1970) 216
28) J.E. Griffith and J.A. Kubby, Defect Diff. Data **A59** (1988) 167
29) D.M. Duffy, J.P. Hoare and P.W. Tasker, J.Phys. C: Solid State Phys. **17** (1984) L195
30) Hj. Matzke and J.L. Whitton, Can. J. Phys. **44** (1966) 995
31) Hj. Matzke, Z. Naturforschg. **22a** (1967) 507

32) Hj. Matzke, in *Physics of Ionized Gases*, Ed. B. Navinsek, Inst. J. Stefan, Ljubljana (1971)354
33) J.L. Whitton and Hj. Matzke, Can. J. Phys. 44 (1966) 2905
34) Hj. Matzke, Diffusion in Nonstoichiometric Oxides, in *Nonstoichiometric Oxides*, Ed. T.O. Sørensen, Academic Press, New York (1981) 155
35) Hj. Matzke, J.Chem. Soc., Faraday Trans. **83** (1987) 1121
36) D.R. Olander, J. Nucl. Mater. **96** (1981) 243
37) S.Y. Zhou and D.R. Olander, Surface Science **136** (1984) 82
38) W.M. Robertson, J. Nucl. Mater. **30** (1969) 36
39) Hj. Matzke, in *Plutonium 1975 and Other Actinides*, North Holland, Amsterdam (1976) 801
40) J. Magill, J. Bloem and R.W. Ohse, J. Chem. Phys. **76** (1982) 6227
41) Hj. Matzke, Diffusion in Ceramic Oxide Systems, Advances in Ceramics **17** (1986) 1
42) V.S. Stubican and C. Lin, Proc. NATO ASI, Nonstoichiometric Compounds, Eds, J. Nowotny and W. Weppner (1988), in press
43) J. Henney and J.W.S. Jones, J. Mater. Sci. **3** (1968) 159
44) A.S. Lythe and V.S. Stubican, J. Amer. Ceram. Soc. **65** (1982) 210
45) Z. Adamczyk and J. Nowotny, J. Electrochem. Soc. **127** (1980) 1117
46) A.E. Paladino and W.D. Kingery, J. Chem. Phys. **37** (1962) 957
47) Y. Oishi, K. Ando and Y. Kubota, J. Chem. Phys. **73** (1980)1410
48) T.K. Gupta, J. Amer. Ceram. Soc. **61** (1979) 191 and Advances in Ceramics **10** (1984) 70
49) A.R. Gaddipati and W.D. Scott, J. Mater. Sci. **21** (1986) 419
50) C. Monty and J. LeDuigou, High Temp.-High Press. **14** (1982) 709
51) R. Dieckmann and H. Schmalzried, Ber. Bunsenges. Phys. Chem. **79** (1975) 1108 and **81** (1977) 344
52) H. Bay and Hj. Matzke, unpublished results
53) M.G. Nicholas and E.N. Hodkin, J. Nucl. Mater. **38** (1971) 234
54) P.S. Maiya and J.L. Routbort, J. Mater. Sci. **7** (1972) 609
55) Hj. Matzke, Science of Advanced LMFBR Fuels, A Monograph on Solid State Physics, Chemistry and Technology of Carbides, Nitrides and Carbonitrides of Uranium and Plutonium, North Holland Publ. Co., Amsterdam (1986)
56) Hj. Matzke, J.L. Routbort and H.A. Tasman, J. Appl. Phys. **45** (1974) 5187
57) J.L. Routbort and Hj. Matzke, J. Amer. Ceram. Soc. **58** (1975) 81
58) H. Matsui and Hj. Matzke, J. Nucl. Mater. **88** (1980) 317
59) M.A. DeCrescente and E.G. Selleck, US Report PWAC-476 (1965)
60) J. Nowotny, in *External and Internal Surfaces in Metal Oxides*, Eds. J.C. Dufour and J. Nowotny, Mater. Sci. Forum **29** (1988) 99
61) D.T. Livey and P. Murray, Proc. 2nd Plansee Seminar, Metallwerke Plansee, Reutte, Austria (1975) 375 and J. Amer. Ceram. Soc. **39** (1956) 363

62) R.J. Bratton and C.W. Beck, , J. Amer. Ceram. Soc. **54** (1971) 379
63) E.N. Hodkin and M.G. Nicholas, J. Nucl. Mater. **47** (1973) 23
64) P. Nikolopoulus, S. Nazare and F. Thümmler, J. Nucl. Mater. **71** (1977) 89
65) Hj. Matzke, R. Warren and T. Inoue, J. Nucl. Mater. **91** (1980) 205
66) Hj. Matzke, Ed., *Indentation Fracture and Mechanical Properties of Ceramic Fuels and of Waste Ceramics and Glasses*, Harwood Acad. Publishers, London (1987)
67) T. Inoue and Hj. Matzke, J. Amer. Ceram. Soc. **64** (1981) 355
68) Hj. Matzke and R. Warren, J. Mater. Sci. Letters **1** (1982) 441
69) R.O.A. Hall, M.J. Mortimer and D.A. Mortimer, J. Nucl. Mater. **148** (1987) 237
70) P.W. Tasker, Harwell, Report TP-812, UK (1979), and Surface Science **87** (1979) 315
71) T. Inoue and Hj. Matzke, J. Amer. Ceram. Soc. **64** (1981) 355
72) W. Van Lierde, J. Mater. Sci. **5** (1970) 527
73) Hj. Matzke and A. Turos, unpublished results
74) M. Rodriguez Perezza and M.V. Davis, Trans. ANS **16** (1973) 73
75) L.E. Willertz and P.G. Shewmon, Metall. Transact. 1 (1970) 2217
76) Hj. Matzke, Radiation Effects **53** (1980) 219
77) M.E. Gulden, J. Nucl. Mater. **23** (1967) 30
78) C. Baker, Eur. Appl. Res. Reports 1 (1979) 19 and J. Nucl. Mater. **71** (1977) 117
79) F.A. Nichols and C. Ronchi, Advanc. Ceramics **17** (1986) 85
80) H. Thiele, private communication
81) B.J. Buescher and R.O. Meyer, J. Nucl. Mater. 48 (1973) 143
82) R.M. Cornell and G.H. Bannister, Proc. Brit. Ceram. Soc. **7** (1967) 355
83) L.C. Michels and R.B.Poeppel, J. Appl. Phys. 44 (1973) 1003
84) R.W. Weeks, R.O. Scattergood and S.R. Pati, J. Nucl. Mater. **36** (1970) 223
85) R.S. Nelson, J. Nucl. Mater. **19** (1966) 149
86) Hj. Matzke, H. Blank, M. Coquerelle, K. Lassmann, I.L.F. Ray, C. Ronchi and C.T. Walker, J. Nucl. Mater., in print
87) Hj. Matzke, C. Ronchi and C. Baker, Eur. Appl. Res. Reports 5 (1984) 1101
88) R.A. Verrall, Hj. Matzke, T. Ogawa and B.J.F. Palmer, Proc. Int. Conf. on CANDU Fuel, Chalk River, ed. I.J. Hastings, Can. Nucl. Soc., Toronto, ISBN O-919784-13-5 (1987) 558 and Report AECL-9475 (1987) 558
89) Hj. Matzke, I.L.F. Ray and R.A. Verrall, Proc. Techn. Comm. Water Reactor Fuel Behavior and Fission Products Release in Off-Normal and Accident Conditions, IAEA Vienna, IWGFPT/27 (1987) 183

GRAIN BOUNDARY DIFFUSION IN CERAMICS

A. Atkinson
Materials Development Division
Harwell Laboratory
Didcot
Oxfordshire OX11 ORA
England

C. Monty
Laboratorie Physique des Materiaux
Bellevue
92190 Meudon
France

ABSTRACT. Diffusion at a grain boundary occurs at a different rate from that in the bulk crystalline lattice. Usually bulk lattice diffusion is relatively slow and boundary diffusion is much more rapid so that the boundary acts as an easy (or short circuit) path for transport in parallel with bulk diffusion.

The experimental approaches to direct measurement of tracer diffusion along grain boundaries are well-established and, in principle, can give the grain boundary diffusion coefficient and the grain boundary width (or, in the case of an impurity atom, the product of boundary width and the segregation coefficient).

Published data cover a range of materials in doped and undoped forms, and specimen types (single crystals, bicrystals, sintered powders and thin films), with host atoms or impurities as the diffusing species. In addition, atomistic simulations have been carried out using static lattice methods to help interpret the experimental observations. There is much disagreement in the detailed results in this field, but certain general conclusions are emerging.

Introduction

Diffusional transport at grain boundaries in ceramic materials plays an important role in many technologically important phenomena such as: the stability of thin film structures, corrosion, ionic conduction, catalysis, sintering and creep. In most of these examples the grain boundaries have higher diffusivity than the bulk crystals and thus they provide fast diffusivity (or short circuit) pathways acting in parallel with the slower bulk diffusion. This is, however, not always the case and in some materials (e.g. fast ion conductors) transport across boundaries, from grain to grain, can be slower than in the bulk. In such cases grain boundary diffusion acts as a high resistance in series with faster bulk diffusion. In this paper we will be concerned exclusively with cases in which boundary diffusion is much faster than bulk diffusion. It is not possible to state a priori whether grain boundary diffusion in a given ceramic will be faster or slower than bulk diffusion. In extreme examples when lattice diffusion is very slow (e.g. Al

L.-C. Dufour et al. (eds.), Surfaces and Interfaces of Ceramic Materials, 273–284.

and O in Al_2O_3 or Zr in ZrO_2) then the increased disorder at a grain boundary is expected to lead to much faster diffusivity than in the bulk. On the other hand, when the bulk structure is especially favourable for diffusion (e.g. O in ZrO_2) the increased boundary disorder would be expected to impede diffusion. In many cases (e.g. Fe in FeO) the situation is not so clear cut and the relative rates of bulk and grain boundary diffusion could vary widely with non-stoichiometry and temperature.

In metals grain boundary diffusion has been studied more widely than in ceramics and is reasonably well understood [1,2]. As far as the authors are aware grain boundary diffusion is always much faster than bulk diffusion in metals and is typified by the summary for fcc metals shown in Figure 1. Diffusion along dislocations is similar to diffusion along grain boundaries since, in the limit of low angle boundaries, the boundary structure is an array of lattice dislocations. This summary diagram provides a useful reference base from which to compare observed behaviour in ceramics. In particular it is worth emphasising the following points:

(i) boundary diffusion extrapolates to a value characteristic of a liquid ($\sim 10^{-9}$ m^2s^{-1}) at the melting temperature,Tm,

(ii) at T/Tm = 0.5, $D_{gb}/D_l \simeq 10^7$,

(iii) the ratio of activation energies, Q_{gb}/Q_l, is about 0.5, and

(iv) the pre-exponential factor, D_o, is about 10^{-5} m^2s^{-1} for both lattice and grain boundary diffusion.

In ceramics the situation is far less clear and the reader is referred to earlier reviews for some of the data [5,6]. In what follows we reiterate some of the main points and update them were appropriate.

Experiments and Phenomenology

Techniques for measuring diffusion at grain boundaries can be classified either as direct or indirect. The direct methods involve monitoring the motion of isotopic tracers or impurity atoms along the interfaces in question. Of the various ways in which this can be accomplished, the best is to measure the concentration of diffusant as a function of distance from a planar source. The boundary condition at the source can be one of constant concentration or a fixed total amount of diffusant (usually in the form of a 'thin' source). These direct methods can only be used when diffusion along the interface is much more rapid than in the lattice. Typically an enhancement of at least 10^3 in diffusivity is required for the experiment to be successful. In the early experiments the concentration profile of the tracer could only be determined by mechanical serial sectioning which is extremely time consuming. More modern techniques of

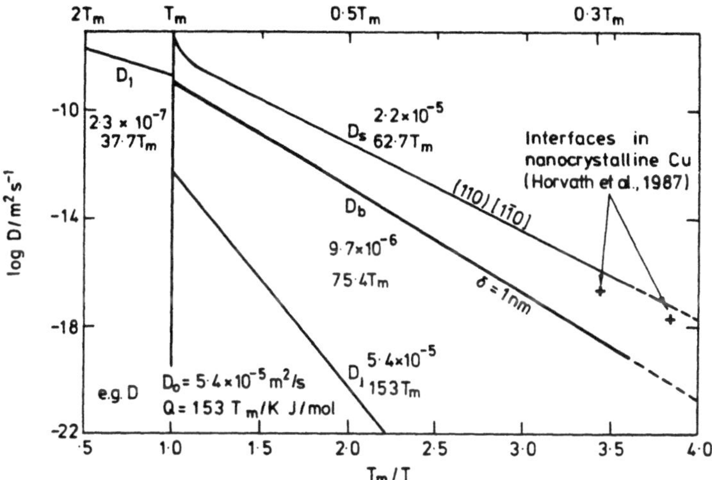

Figure 1 Generalised plot of self diffusion processes for fcc metals. D_s, D_b, and D_l refer to surface boundary and lattice diffusivities respectively. (After Gust et. al., 1985 [3]).

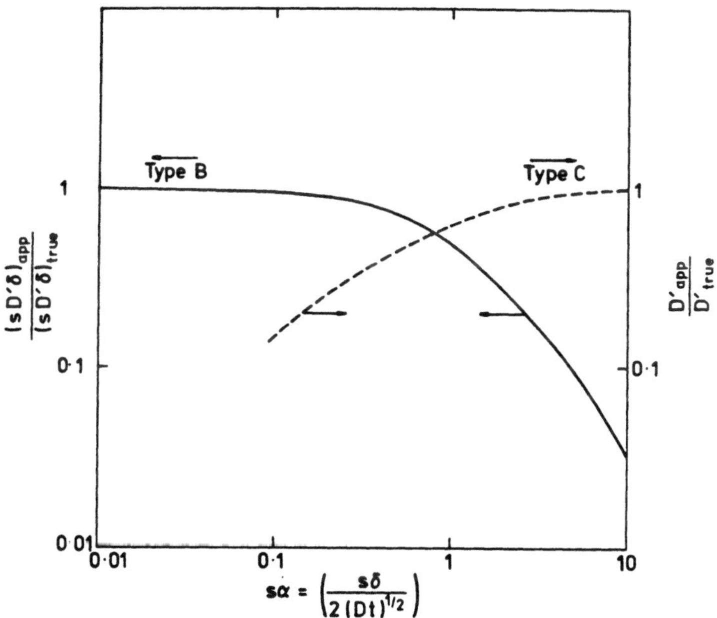

Figure 2 Calculated curves showing estimates of the errors which are generated by using the asymptotic solutions, for type B and C diffusion, in the region of transition between them. (s is the segregation coefficient).

microanalysis (e.g. electron probe microanalysis) have both improved depth resolution and speed. This is particularly so for secondary ion mass spectrometry (SIMS) which has high lateral and depth resolution, is fast and often has a sensitivity and dynamic range approaching those of radioactive tracers.

The phenomenological model that has been almost universally used to analyse fast transport along planar interfaces is to treat the interface as a slab of different material having a width, δ, and diffusion coefficient, D_{gb}, greater than the lattice diffusion coefficient D_{el}. The solutions of the diffusion equations and the analysis of experimental data have been discussed in detail by LeClaire [7]. Such a simple model is clearly very different from atomistic models of interfaces and there is a danger that it is too simple to be realistic. However, simulations using jump models of interfaces (but not true atomistic models) have shown that the slab model should be adequate for analysing experimental data [8,9].

It is important to appreciate that there are different regimes in the phenomenology which lead to different information being available from the experiment. These were classified by Harrison [10] as types A, B and C. Type A only applies in polycrystalline specimens and then only when the extent of lattice diffusion becomes greater than the grain size, g. In such circumstances one can only deduce an effective diffusion coefficient given as a weighted average of lattice and boundary diffusion:

$$D_{eff} = (1-f)D_l + fD_{gb}, \qquad 1(a)$$
$$\text{where, } f = 3\,\delta s/g \qquad 1(b)$$

The parameter s is the segregation coefficient of the diffusant to the boundary (assuming a linear isotherm) and is unity for self diffusion.

Type C diffusion occurs when lattice diffusion is negligible (short times, low temperatures). The penetration profiles of diffusant then give D_{gb} directly, since there is no simultaneous lattice diffusion to complicate the issue. In type B diffusion, lattice diffusion is appreciable and penetration profiles always show two regions. The near surface region is dominated by lattice diffusion and hence, in principle, gives D_l. The more penetrating 'tail' region should be linear on a plot of logarithm of concentration versus $y^{6/5}$ (where y is penetration depth) and the slope of this plot will give the parameter $sD_{gb}/(D_l)^{\frac{1}{2}}$ [7,11]. Thus, if D_l is known, type B diffusion gives $sD_{gb}\delta$. The 'strength' of the tail region of the penetration profile is inversely proportional to the grain size [5,12]. In attempting to satisfy the constraints of experiments and data analysis, it often turns out that the tail region is close to the detection limits of the experiments and this leads to large uncertainties in the final results for $sD_{gb}\delta$.

In principle, it is possible to differentiate between Type B and Type C diffusion by the different depth dependence in the different regions. In practice, however, this is not possible and the same data can give equally acceptable fits to both the ideal cases (see for example [11]). Nevertheless, it is possible, using the full solution to the diffusion equations, to compute the errors that are incurred by making the wrong assumption regarding the diffusion type. These are shown in Figure 2 where the ratio of the apparent diffusion parameter to the true parameter is plotted as a function of $\alpha = s\delta/(Dt)$. Making the wrong assumption about diffusion type leads to underestimating the diffusion parameter in both type B and type C cases. If diffusion measurements can be made to span both type B and type C at a single temperature then it is possible to determine D_{gb} (from type C) and $sD_{gb}\delta$ (from type B) and thus allow $s\delta$ to be estimated. Since $s\delta$ is not known a priori then a self-consistent description of the complete behaviour must be sought. Examples of this analysis have concluded that δ is about 1 nm for low angle and high angle boundaries in NiO [13,14] and have shown that s deduced from diffusion for Cr in high angle boundaries in NiO is reasonably consistent with s deduced from microanalysis [11]. However, it must always be borne in mind that the diffusion equations for impurities assume a linear segregation isotherm and diffusivity independent of concentration. These assumptions may not be valid for strongly segregating impurities even at low total concentrations.

Experiments can be made on single interfaces (e.g. bicrystals), or assemblies of interfaces (polycrystals). Only those on single controlled interfaces can ultimately lead to an appreciation of diffusion in terms of interface structure. Unfortunately they are the most difficult in practice. Experiments on polycrystals in principle measure diffusivity averaged over all the boundaries present. This is not always straightforward, however, since the high diffusivity interfaces tend to be overemphasised. Thus in metals Periantoni et al [15] observed different effects in bicrystals and polycrystals of Cu (Ni) alloys for the influence of S segregation on S diffusion.

Self Diffusion

NiO and Cu_2O are cation deficient oxides having a relatively small deviation from stoichiometry which is controllable through the ambient oxygen pressure. It is possible to obtain these materials in sufficiently pure form that the residual impurities have little effect on the populations of point defects responsible for bulk lattice diffusion. Such materials therefore offer the best opportunity to study 'intrinsic' (i.e. not influenced by impurities) grain boundary diffusion. (Nevertheless, strong segregation of impurities at grain boundaries could be significant for boundary diffusion even though their overall concentration is low enough to be negligible for lattice diffusion). Grain boundary diffusion studies have been reported for Ni in low angle boundaries in NiO single crystals [13] and for Ni and O diffusion in high angle boundaries in polycrystalline films formed by oxidation of Ni metal foil [14,16]. The results are summarised as an Arrhenius plot in Figure 3 and show that grain boundary diffusion is faster than

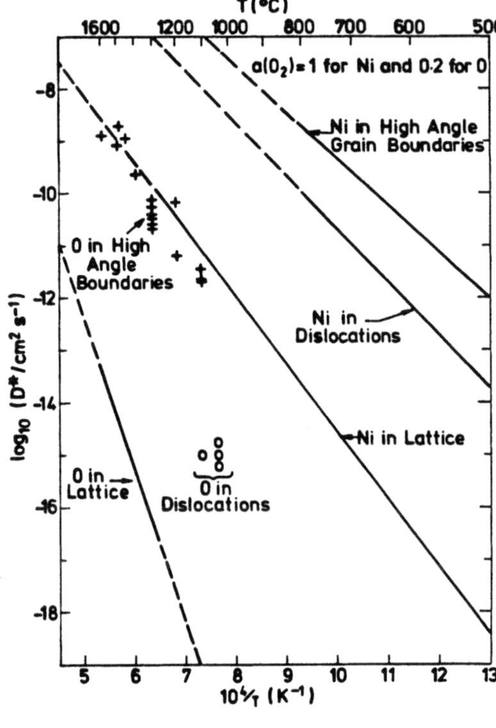

Figure 3 Diffusion of Ni and O in NiO. The grain boundary and dislocation diffusion coefficients are given assuming $\delta = 1$ nm. The data are from references [13, 14, 16, 17].

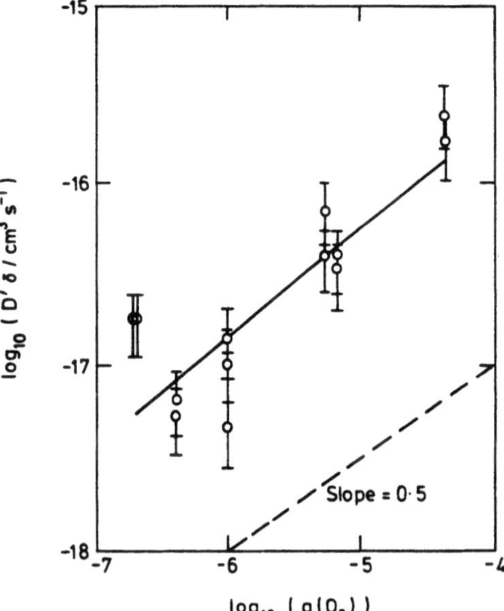

Figure 4 Diffusion of oxygen in Cu_2O grain boundaries at 712°C as a function of oxygen activity. The slope of 0.5 is expected for an uncharged interstitial oxygem atom being the diffusing species (After Perinet, 1987 [19]).

dislocation diffusion and that both Ni and O have greater diffusivity at grain boundaries than in the bulk; with Ni having greater diffusivity than O even in grain boundaries. Experiments at different oxygen pressure [13,14] have shown that Ni diffusivity along dislocations and grain boundaries increases as the concentration of Ni vacancies in the bulk crystal lattice increases. This implies that dislocation and grain boundary diffusion take place by a vacancy mechanism. This mechanism is supported by theoretical calculations of the formation and migration energies of Ni vacancies at symmetrical tilt boundaries in NiO as summarised in Table 1.

Table I

Activation energies, Q(eV), for lattice and grain boundary diffusion in NiO
(Duffy and Tasker [18])

	Experiment			Calculation		
	Q_{gb}	Q_l	Q_{gb}/Q_l	Q_{gb}	Q_l	Q_{gb}/Q_l
Ni diffusion (vacancy)	1.78	2.56	0.70	1.6-2.2	2.9	0.5-0.8
O diffusion	2.5	5.6	0.45	2.5	5.3	0.47

Recently, Perinet [19] reported measurements of lattice and grain boundary diffusion of oxygen in Cu_2O. Both D_l and $D_{gb}\delta$ were found to vary as $p(O_2)^n$ with n being close to 0.5 (Figure 4). This indicates that both lattice and grain boundary diffusion of oxygen in Cu_2O involves an uncharged oxygen interstitial defect (atom). It therefore appears that in both these materials grain boundary diffusion involves the same type of point defect as lattice diffusion.

Attempts have been made to study Ni diffusion in NiO bicrystals [20] and polycrystalline specimens produced by sintering [21] using autoradiography. In the bicrystals the penetration of diffusant along the grain boundary was inhomogeneous and sensitive to prior thermal treatment whereas in sintered polycrystals enhanced diffusivity at grain boundaries was not detected.

Impurity Diffusion

Diffusion of impurities is complicated by their segregation behaviour which leads to the parameter $sD_{gb}\delta$ being measured in type B diffusion. Impurity diffusion has been studied for Co, Cr and Ce (nominally in charge states 2+, 3+ and 4+ respectively) in NiO grain boundaries [11]. These offer good examples of distinctions between type B (Co) and type C (Ce) kinetic regimes. Cr behaves in an intermediate fashion since it has appreciable lattice solubility (like Co) and strong segregation (like Ce). In such a case it has proved possible to extract both the segregation coefficient, s, (assuming δ = 1 nm) and the diffusivity, D_{gb}, using a self-consistent analysis of the transition from type B to type C kinetics. The results (Figure 5) demonstrate that the ionic charge is the major factor controlling D_{gb} since the Ce^{4+} ion diffuses the slowest and Co^{2+} the fastest. In the case of Co diffusion, attempts have also been made to make measurements on bicrystal specimens. Chen and Peterson [22] report grain boundary diffusion to be higher than found by Atkinson and Taylor [11] and to have a greater activation energy (about the same as lattice diffusion). Barbier [25], on the other hand, studied a wide range of bicrystals and could detect no rapid diffusion of Co, thus implying boundary diffusivities which are several orders of magnitude lower than those reported by Atkinson and Taylor and Chen and Peterson.

There have been many studies of impurity diffusion in the refractory oxides MgO and Al_2O_3, but in these materials the concentration of intrinsic point defects is so low that all mass transport phenomena are believed to be dominated by impurities (either trace or deliberate dopants). In Figure 6 data for a variety of interface diffusion processes in Al_2O_3 are presented (the plotted parameter is $sD_{gb}\delta$).

Influence of Impurities on Diffusion

Impurities influence lattice diffusion in ceramics by entering solid solution in the host crystal lattice and thereby changing the concentration and mobility of point defects responsible for diffusion. Thus a trivalent cation (e.g. Al^{3+}) substituting for a divalent host (e.g. Ni^{2+} in NiO) will produce cation vacancies to maintain charge neutrality and hence increase host self-diffusion. This is indeed observed for Al-doped NiO [29], but this type of behaviour is additionally complicated by the attractive forces between oppositely charged defects and their aggregation into pairs or clusters [29].

The same phenomena could occur in principle at grain boundaries, but there are at least three significant differences. The first is that impurities will often segregate strongly to grain boundaries such that boundary concentrations can be high even when overall doping levels are small. The second is that fast boundary diffusion probably takes place by point defects hopping between a limited set of 'fast' sites [18]. The impurities may compete with host atoms and vacancies for occupancy of these sites. Thirdly, because the effective width of the boundary is only about 1 nm then boundary

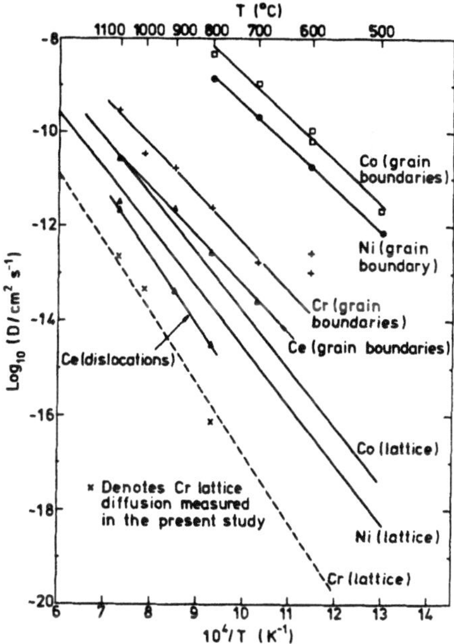

Figure 5 Arrhenius plot summarising the diffusion coefficients for cations in the NiO lattice and in grain boundaries (and dislocations in the case of Ce)[11].

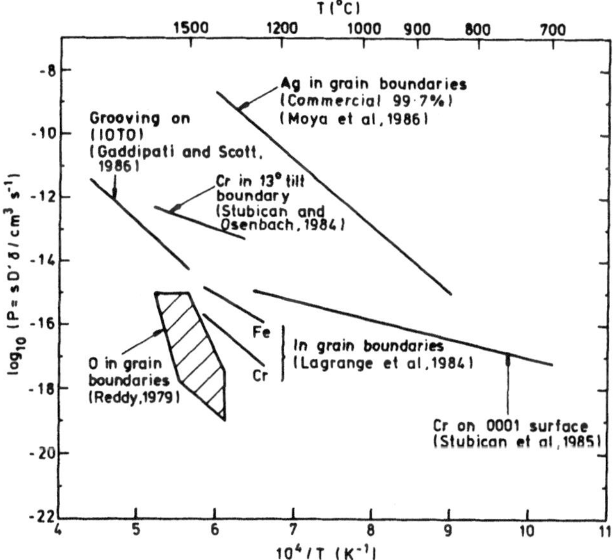

Figure 6 Diffusion parameters for grain boundary and surface diffusion in Al_2O_3 [24-28].

diffusion is almost two- dimensional. All these features tend to amplify the influence of impurities on boundary diffusion. In metals the tendency is for strongly segregating impurities to reduce grain boundary diffusivity probably by blocking the fast diffusivity sites [30]. In NiO, for example, the effect of Cr doping is similar, but is so strong that boundary diffusivity is almost entirely blocked despite the fact that Cr doping increases the concentration of lattice vacancies, and hence lattice diffusivity [31]. Atomistic modelling techniques [18] have revealed how this blocking might take place. The higher valent impurity ions are effectively positively charged and bind vacancies, which have an effective negative charge, to them. However, the higher valent impurity ion is relatively immobile and restricts the mobility of the bound vacancies not only by electrostatic attraction, but also by increasing their jump activation energy to a value greater than that for a jump in the lattice.

All oxides do not behave in a similar way as far as impurities are concerned, however. For example, Cr doping increases the diffusivity of Cr in MgO grain boundaries and in the lattice [24]. But this is impurity diffusion rather than self-diffusion and therefore may not necessarily be different from the effect of Cr doping on Cr grain boundary diffusion in NiO (which has not been measured). In many practical ceramics segregation at grain boundaries is so extensive that second phase, often amorphous, is formed at grain boundaries so that diffusivity in the boundary has no mechanistic relationship to that in the bulk (e.g. the fast diffusion of Ag in Al_2O_3 grain boundaries already referred to).

Influence of Boundary Type

Experiments in which diffusion in bicrystal boundaries has been studied as a function of misorientation have been carried out for Cr in MgO [24] and Co in NiO [23]. The experiments on low angle MgO boundaries gave results which are consistent with the dislocation model of such boundaries, just as in metals. In extensive measurements using NiO bicrystals, however, preferential diffusion of Co along the boundaries was not detected thus implying diffusivities lower by several orders of magnitude than those reported for polycrystalline NiO formed by oxidation of Ni. The reasons for this difference are not clear, but other experiments using bicrystals and sintered polycrystals of NiO have also given erratic results [20,21].

It therefore appears that boundaries fabricated in different ways display widely different diffusivity. This may be a secondary effect due to different impurity concentrations from different fabrication routes or it could be a primary effect due to different boundary structure. For example, boundaries formed by a relatively low temperature process (e.g. the NiO boundaries in NiO grown by oxidising Ni) may have a more 'open' or less well-bonded structure than those grown near the melting temperature. In metals, however, it would appear that this latter suggestion is unfounded since Au self-diffusion in evaporated Au thin films has the same Arrhenius behaviour as self-diffusion in melt cast polycrystalline Au [32]. Nevertheless, in nanocrystalline material, grain boundaries formed at yet lower temperatures display diffusivities more akin to surface diffusion [4].

It has also been suggested that when boundaries are moving (e.g. when there is grain growth during a diffusion experiment) they may exhibit enhanced diffusivity. However, a survey of such effects in metals [33] has concluded that there is no significant effect of boundary velocity on boundary diffusivity.

Conclusions

There is now a reasonable volume of data building up on grain boundary diffusion in ceramics, but it contains many discrepancies and inconsistencies.

The single most important parameter controlling boundary diffusivity would appear to be chemical composition and this may be at the bottom of many of the discrepancies. It would appear as a result that we can distinguish three broad types of boundary diffusion. The first could be termed 'intrinsic' and ideally is a property of the host material alone. In this class fall ceramics with small, but controllable deviations from stoichiometry such as NiO and Cu_2O. The boundary diffusion occurs by the segregation of point defects from the lattice and therefore is a fast version of lattice diffusion.

At the other extreme are boundaries to which impurities have segregated to such an extent that a recognisable second phase is formed. These may be described as "extrinsic-second phase" and their diffusion properties are essentially those of the second phase. An example of this type would be commercial alumina.

Finally, we have those with intermediate properties which may be described as "extrinsic-doped". In these materials the boundary diffusivity may be understood in terms of point defects similar to those in the host material, but whose behaviour is significantly perturbed by the dopant (e.g. Cr in MgO). As the dopant level increases, the defect clusters may generate diffusive properties similar to those of a second phase even though a discrete second phase is not formed. This type of behaviour is postulated for Cr-doped NiO where the reduction in diffusivity is consistent with that expected from $NiCr_2O_4$ (relative to NiO).

References

1. N.L. Peterson, Grain Boundary Structure and Kinetics (ASM, Metals Park, Ohio, 1980) p209.
2. R.W. Balluffi, Metall. Trans. A. 13A, 2069 (1982).
3. W. Gust, S. Mayer, A. Bögel and B. Predel, J. Phys. (Paris) 46, C4-537 (1985).
4. J. Horvath, R. Birringer and H. Gleiter, Solid State Comm. 62 319 (1987).

5. A. Atkinson, Solid State Ionics 12, 309 (1984).
6. A. Atkinson, J. Phys. (Paris) 46, C4-379 (1985).
7. A.D. LeClaire, Br. J. Appl. Phys. 14, 351 (1963).
8. P. Benoist and G. Martin, Thin Solid Films 25, 181 (1975).
9. G.E. Murch, Diffusion and Defect Data 32, 1 (1983).
10. L.G. Harrison, Trans. Faraday Soc., 57, 1191 (1961).
11. A. Atkinson and R.I. Taylor, J. Phys. Chem. Solids 47, 315 (1986).
12. T. Suzuoka, Trans. Jap. Inst. Metals 2, 25 (1961).
13. A. Atkinson and R.I. Taylor, Philos. Mag A39, 581 (1979).
14. A. Atkinson and R.I. Taylor, Philos. Mag A43, 979 (1981).
15. M. Pierantoni, B. Aufray and F. Cabané, J. Phys. (Paris) 46, C4-537 (1985).
16. A. Atkinson, F.C.W. Pummery and C. Monty in Transport in Nonstoichiometric Compounds edited by G. Simkovich and V.S. Stubican (Plenum, New York, 1985) p359.
17. C. Dubois, C. Monty and J. Philibert, Philos. Mag A46, 419 (1982).
18. D.M. Duffy and P.W. Tasker, Philos. Mag A54, 759 (1986).
19. F. Perinet, Thesé, Université de Paris-Sud, Orsay, 1987.
20. A. Atkinson, D.P. Moon, D.W. Smart and R.I. Taylor, J. Mater. Sci., 21, 1747 (1986).
21. L.B. Harris, R.I. Taylor and A. Atkinson, J. Mater. Sci., 22, 1993 (1987).
22. W.K. Chen and N.L. Peterson, J. Amer. Ceram. Soc., 63, 566 (1980).
23. F. Barbier, Thesé D'Etat, Université Paris-Sud, Orsay (1986).
24. J.W. Osenbach and V.S. Stubican, J. Amer. Ceram. Soc. 66, 191 (1983).
25. V.S. Stubican, G. Huzinec and D. Damjanovic, J. Amer. Ceram. Soc., 68, 181 (1985).
26. E.G. Moya, L. Badrour, J. Bernadini and F. Moya, to be published.
27. K.P.R. Reddy, Ph.D Thesis, Case Western Reserve University 1979.
28. M.H. Lagrange, A.M. Huntz and J.H. Davidson, Corros. Sci., 24, 613 (1984).
29. A. Atkinson, A.E. Hughes and A. Hammou, Philos. Mag. A43, 1071 (1981).
30. J. Bernadini, P. Gas, E.D. Hondros and M.P. Seah, Proc. R. Soc. Lond. A379, 159 (1982).
31. A. Atkinson, D.W. Smart and R.I. Taylor, Werks. u. Korros. 38, 704 (1987).
32. D. Gupta and K.W. Asai, Thin Solid Films 22, 121 (1974).
33. W. Gust, M.B. Hinz and B. Predel, J. Phys. (Paris) 46, C4-529 (1985).

SEGREGATION AT CERAMIC SURFACES AND EFFECTS ON MASS TRANSPORT

JACK M. BLAKELY and SHARMILA M. MUKHOPADHYAY
Dept of Materials Science and Engineering
Bard Hall,
Cornell University
Ithaca, N.Y., USA, 14853

ABSTRACT. A brief discussion is given of the factors that influence the concentration of impurities at ceramic surfaces and interfaces. The relationship between the experimentally measured temperature dependence of surface composition and the binding energy differences between surface and bulk is discussed. The simple Langmuir model of surface segregation describes the limiting case when no interactions occur among impurity ions and will generally be inadequate for real systems in which surface condensation and phase transition may occur. The use of segregation isosteres to determine heats of segregation is discussed.

Model calculations of the distribution of divalent impurities near alumina surfaces, using the mean field approach involving the Poisson equation, show that impurity concentrations may be enhanced both in the surface monolayer and in the near-surface space charge region.

Experimental studies of the segregation of Mg and Ca to the (0001) and (1010) surfaces of Al_2O_3 are described. These experiments demonstrate the importance of suppressing net evaporation in order to approach thermodynamic equilibrium. The application of the sine-way decay method to determine mass transport parameters for Al_2O_3 is described, with particular emphasis on the effect of segregated Mg.

1. Introduction

Segregation of one component of an alloy or solution to an interface is a process that is generally accompanied by a decrease in configurational entropy. In order that it should occur spontaneously, the excess free energy associated with segregation must be negative. If vibrational entropy effects can be neglected, the difference in 'binding energy' for a component between surface and bulk will be the factor that determines the surface to bulk partitioning. This binding energy difference will have a simple interpretation only in the limit of very low surface concentrations of the segregating element. In this case it is helpful to use a 'one ion' type of energy level diagram to represent the binding energies of the segregating component in various types of sites; an example of such a diagram is shown in fig(1). At low concentrations the positions of the energy levels are independent of the occupation of the corresponding sites. As the surface concentration of the segregating ions builds up, their mutual interactions will become important and new 2-dimensional phases or 3-dimensional precipitates may

L.-C. Dufour et al. (eds.), Surfaces and Interfaces of Ceramic Materials, 285–299.

form. These effects would have to be represented on the diagram of fig(1) by allowing the levels to shift with occupation. In ionic materials the interactions between segregated ions may be of very long range due to the Coulomb potential and the overlap of extended polarized regions.

The segregation of ions of different ionic charge from the matrix will interact with and contribute to dipole layers that may exist at the surface; such dipole layers may be electronic or ionic in origin. The segregating element may concentrate on the surface monolayer or in the near-surface diffuse space charge region. A brief discussion of ionic space charge regions will be given below with special reference to Al_2O_3 surfaces.

Changes in morphology of ionic crystal surfaces can be effected by a number of mechanisms including evaporation-condensation, volume diffusion, and surface diffusion. As the dimensions of the morphological features decrease, surface diffusion becomes an increasingly important mechanism. Since surface diffusion fluxes are defined as excess quantities, there will be contributions both from the first monolayer where the bonding is significantly altered and from the diffuse near-surface region where the densities of point defects may be different.

In a compound or alloy the fluxes of the different species will be coupled by a composition or volume constraint, so that the effective diffusivity describing the overall mass transport process will be a combination of the individual ionic diffusivities; this effective diffusivity will usually be determined by the value for the less mobile species.

1.1 THERMODYNAMICS OF SURFACE SEGREGATION

The analysis of segregation data in terms of thermodynamics is appropriate provided it can be demonstrated that equilibrium was in fact acheived in the experimental system(1). It is often difficult to obtain convincing evidence for the attainment of equilibrium states(2); tests might, for example, include measurements of the extent of reversibility of surface concentrations with temperature cycling, but such a test will not always distinguish equilibrium from a steady state condition. The thermodynamic treatment of segregation is really very similar to that for adsorption from a gas phase; in segregation we usually think of the chemical potentials as being controlled by bulk concentrations while in adsorption they are controlled through vapor pressures. With this modification, the extensively developed theory of adsorption can be carried over to the segregation case(3).

The simplest possible situation is that in which the surface presents a set of energetically equivalent sites at which impurity ions may attach with a binding energy different from that of the bulk lattice. Fig(1) is appropriate for this case when the binding energy in the surface sites is independent of their occupation. If X_s is the fraction of surface sites occupied by impurities, then it can readily be shown that

$$X_s/(1-X_s) = X_b/(1-X_b)\ \exp(-\Delta G_{seg}/kT) \tag{1}$$

where X_b is the fraction of available bulk sites occupied by the impurity, and $\Delta G_{seg}(=\Delta H_{seg}-T\Delta S_{seg})$ is the partial atomic free energy of segregation; ΔH_{seg} and ΔS_{seg} are the corresponding heat and entropy of segregation respectively. A so-called Langmuir plot, ie, $\ln[X_s/(1-X_s)]$ vs 1/T should give a straight line of slope equal to $-\Delta H_{seg}/k$. Equation (1) is frequently used in the analysis of experimental data even when there is no evidence that the Langmuir model is appropriate; the slope should then be referred to as the 'effective Langmuir heat' of segregation. The difficulty arises when one tries to interpret this effective heat in terms of differences in bonding between surface and bulk. If the interaction between segregated ions is repulsive the interpretation of the effective Langmuir heat may still be relatively straightforward(4), but the existence of even a small attractive interaction between segregating ions can result in abrupt condensation of impurities in the surface region and perhaps coexistence of surface phases(5).

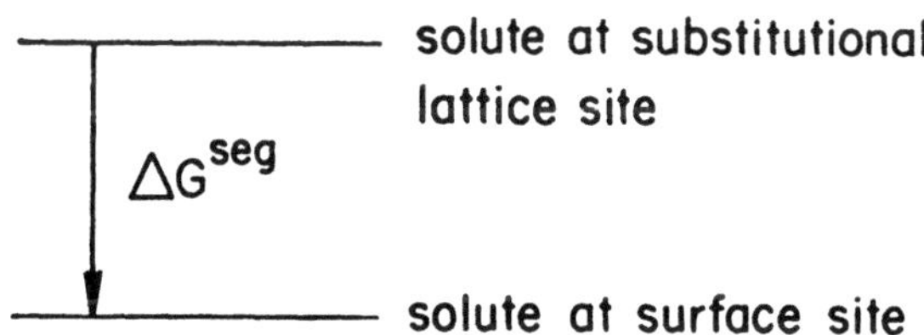

Figure(1)
Schematic energy diagram for a solute element which is more tightly bound at the surface than in the interior of the crystal.

The use of segregation isosteres leads to the determination of heats of segregation which are model independent. Segregation isosteres are plots of bulk concentration versus 1/T for constant surface concentration; they are more difficult to obtain since they require the measurement of surface concentration as a function of temperature for a set of bulk doping levels. Few determinations of segregation isosteres have been made for any kind of system. The isosteric heat of segregation for a crystal with a single impurity may be defined as

$$-\Delta H_{seg}/RT^2 = [\, d \ln X_b/dT \,]_{\Gamma_s} \tag{2}$$

where Γ_s is the surface excess(5) of the impurity. The authors are not aware of any determination of the isosteric heat of segregation for a ceramic system; for the purpose of testing recent calculations(4) of heats of segregation such measurements would clearly be desirable.

1.2 IONIC DOUBLE LAYERS AT CERAMIC SURFACES

Surfaces and interfaces in ionic materials will usually have a net charge at equilibrium due to (i)the existence of surface electronic states in the bulk band gap, (ii)differences between surface and bulk in the binding energies of the ions, or (iii) combinations of (i) and (ii). For many ionic solids,(including the material, Al_2O_3, used in the experiments described below), the density of mobile electronic charge is so low that electronic carrier contributions to the surface double layer can probably be neglected, so that the surface double layer characteristics depend primarily on the ionic binding. We will consider this case only.

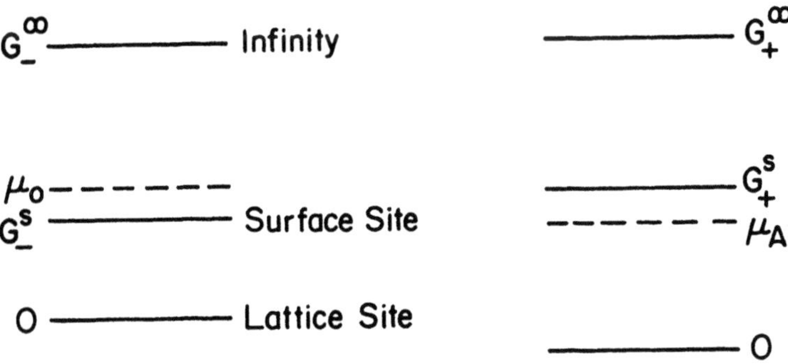

Figure(2a)
Schematic of the binding energies associated with ions occupying different sites in an intrinsic alumina crystal. From ref.(6) the free energies to remove ions to infinity, G_+^{∞} and G_-^{∞} are ~61eV and ~21.7eV respectively; the Schottky defect energy ($2G^s_+ + 3G^s_-$) ~25.2eV.

In order to derive the characteristics of the space charge layer we need to know the binding energies of the various ions at different sites in the crystal. Fig(2a) is a schematic representation of the binding energies of the Al and O ions at different sites in an *intrinsic* Al_2O_3 crystal. The reference zero of energy has been taken in each case as that corresponding to the normal lattice sites. Fig(2b) is intended to illustrate the physical meaning of the quantities in Fig(2a). The Schottky formation energy, ΔG_{sch}, (for 3 Al^{3+} vacancies + 2 O^{2-} vacancies) has been estimated(6) to be ~25.2eV; referring to fig(2a), $\Delta G_{sch} = 2G^s_+ + 3G^s_-$. The density of intrinsic defects in

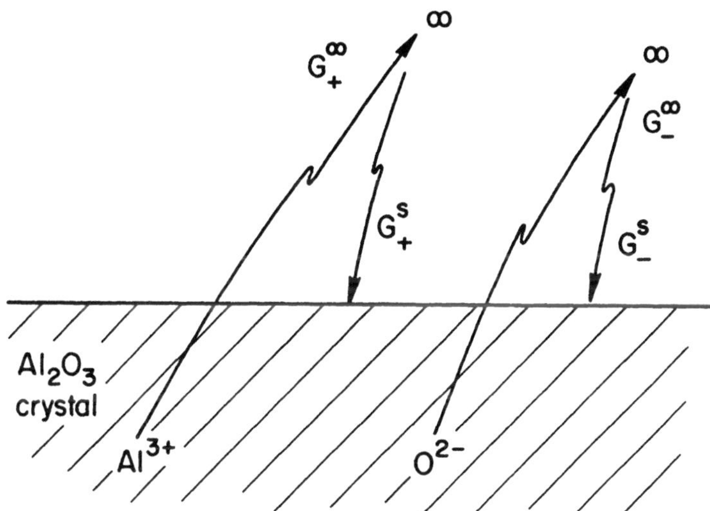

Figure(2b)
Diagram illustrating the meaning of the free energies involved in figure(2a). The free energy to form an Al vacancy $= G_+^\infty - (G_+^\infty - G^s{}_+) = G^s{}_+$.

alumina at all temperatures below the melting point should therefore be extremely low. (In fact, for this case it may not be reasonable to neglect the electron and hole contributions; we are however, discussing the intrinsic case only for the purpose of establishing the ionic defect model). The chemical potentials of the ions depend on the bulk properties of the crystal and are indicated schematically in fig.(2a). The fractional occupations of the various sites of the crystal are then determined by the energies of these sites relative to the chemical potentials using a Fermi-Dirac filling function[7]. For the intrinsic case and the situation depicted in fig.(2a) the surface would acquire a net negative charge and an electrostatic potential difference would exist between surface and bulk to bring the chemical potentials into coincidence with the corresponding surface site energies; the space charge region would have an excess of oxygen ion vacancies. The equilibrium electrostatic potential difference between surface and bulk and the charge distribution are determined from the solution of the simultaneous equations expressing overall charge neutrality and the Poisson relation.

The case when the ionic defect densities are dominated by the presence of impurities,ie the *extrinsic* case, is of more interest in connection with the interpretation of experiments. Fig.(2c) schematically represents the relative[6] binding energies of Mg^{2+} ions, when present as a substitutional impurity in alumina. Defect complexes can occur and the energies associated with these have been the subject

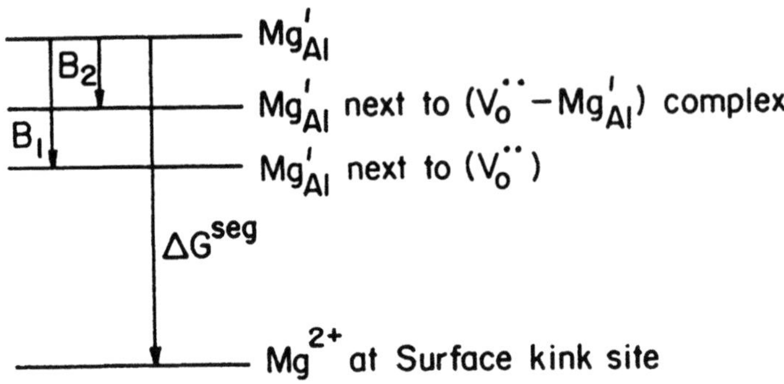

Figure(2c)
Schematic of the binding energies of Mg^{2+} in an alumina crystal; at lower temperatures Mg will predominantly be present in complexes with oxygen ion vacancies. In the figure Mg'_{Al} and $V_O^{\bullet\bullet}$ denote Mg on an Al site and an oxygen ion vacancy, respectively. As calculated in ref.(6), the binding energies associated with forming complexes, B_1 and B_2 are ~ -1.4 eV and ~ -1.2 eV respectively.

of recent theory[6]; where available the best estimates for these energies are indicated on the figure. In the presence of Mg the number of vacancies on the cation sublattice is negligible; the principal defects now consist of substitutional Mg^{2+} on the cation sublattice, isolated O^{2-} vacancies, Mg-Oxygen vacancy(positively charged) complexes, and oxygen vacancy-Mg-oxygen vacancy (neutral) complexes. Since the oxygen vacancies seem to form these complexes, and the neutral complex involves two Mg ions on the same sublattice, we cannot treat the sites as being independent of one another, and therefore cannot use the simple Fermi function to express the individual occupations of the various sites. However, as described in detail in another publication[8], the problem can be handled by finding directly the spatial distributions that minimize the Helmholtz free energy, (consistent with the Poisson equation), of this inhomogeneous system. Fig(3) gives the calculated depth dependence of the electrostatic potential for the parameters indicated in the caption, while fig(4) shows the corresponding concentration profile for Mg. It is of interest to note that most of the Mg occurs in neutral complexes. From these calculated Mg concentration profiles it is now possible to evaluate the temperature variation of the Mg signal that would be detected through an Auger spectroscopy study. Langmuir plots of the Mg 'coverage' as a function of temperature, computed in this way, are shown in fig(5) using the parameters indicated in the figure caption. It should be noted that these plots are

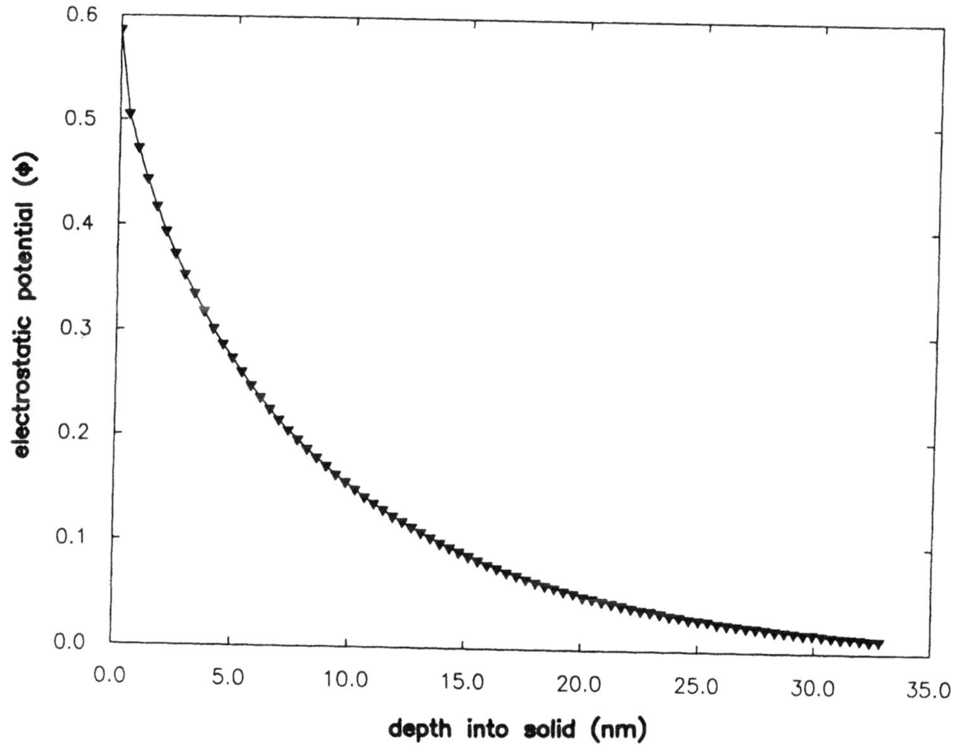

Figure(3)

Electrostatic potential (Φ) as a function of distance from the surface for an alumina crystal for a particular choice of the parameters involved. In computing the curve shown here we have used the following values of the parameters indicated in figures (1) and (2): B_1= -1.4 eV, B_2= -1.2 eV, (ref.6), $G^s{}_+$ =3.0 eV, Number of surface sites $=10^{18}/m^2$ (arbitrarily chosen), ΔG_{seg} = -1.0 eV, bulk Mg concentration = 10 parts per million, and temperature=1300°C.

reasonably linear but that because of the surface double layer effect the slopes do not correspond in any simple way to the assumed binding energy difference for Mg between surface and bulk. This should be kept in mind when interpreting segregation data of the type described in a later section.

In order to test experimentally the validity of the segregation profiles predicted by this analysis we require a technique with high sensitivity and depth resolution. We are currently investigating the use of x-ray fluorescence emission near external reflection conditions[9].

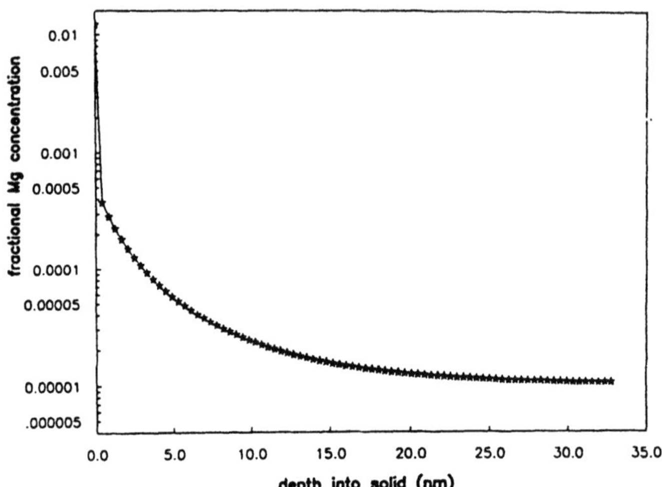

Figure(4)
Calculated depth profile of total Mg concentration near an alumina surface using the same parameters as for figure(3). It must be noted that this profile is very dependent on the choice of the parameters.

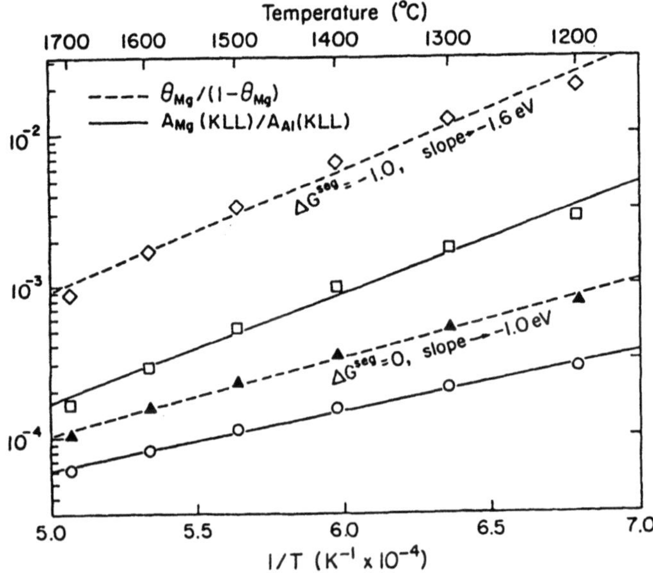

Figure(5)
The points on the solid lines are the computed values of the ratio of Mg (KLL)to Al (KLL) Auger signals, plotted vs 1/T, for profiles of the type shown in figure(4), for two assumed values of ΔG_{seg}. The dashed lines are the corresponding Langmuir plots calculated in the same way as was done for the experimental data shown below in figures(8) and (9). Note that the slopes of these lines do not give effective Langmuir heats of segregation equal to the assumed values of ΔG_{seg}.

1.3 MASS TRANSPORT PROCESSES

The atomic transport processes that are resposible for bringing about changes in surface morphology in crystals are surface diffusion, volume diffusion and evaporation-condensation. The relative contributions of these processes depends of course on the material, the temperature and the characteristic dimension of the surface features. For model studies of the important processes of mass transport it is highly desirable to deal with surface shapes whose evolution with time can be described by a relatively simple analysis. The sine wave decay method, introduced by Blakely and Mykura(10) using an analysis given by Mullins(11), satisfies this criterion.

Under the assumption of orientation independent surface tension for the range of surfaces exposed, it can be shown that a sinusoidal surface will decrease its amplitude with time according to an exponential decay law

$$y(x)= A(t)\sin \omega x = A(0) \exp (-Kt) \sin \omega x \tag{3}$$

where (neglecting evaporation-condensation) the decay constant is related to the characteristic wavelength and material rate parameters by : $K = C\omega^3 + B\omega^4$, with $C = D_V\gamma\Omega / kT$, $B = D_S\gamma\Omega^{4/3} / kT$. Here γ is the surface tension, D_s the effective molecular surface diffusivity, D_V the effective volume diffusivity, and Ω the molecular volume. It should be noted that surface diffusion will become the most important transport mechanism at small feature dimensions (large ω). The surface diffusivity that appears in this equation is to be thought of as an excess quantity(12) and thus will include effects due to extended defect regions of the type discussed in the preceding section, as well as surface monolayer contributions.

Experimentally, surface sine waves can be produced by a combination of lithography (or mechanical ruling) to fabricate surface gratings followed by annealing to allow higher harmonics of the initial shape to decay. Fig(6) below is an example of such a profile on a sapphire crystal formed by ion etching and annealing. We will describe in a later section measurements of the rate of decay of the amplitudes of such profiles with emphasis on the effect of surface Mg segregation.

2. Segregation to Al_2O_3 Surfaces

We have carried out a series of experiments on the segregation of Mg and Ca to the (0001) and the $(10\bar{1}0)$ surfaces of sapphire(13,14). We have focussed on the temperature dependence of the surface impurity concentrations and have attempted to make our measurements under near-equilibrium conditions. The Mg was introduced into the crystal either by adding MgO to the melt(15) or by ion implantation(14).

annealing, and sputtering to remove the damaged region; Ca was present as an impurity at the level of 10's of ppm in the sapphire wafers used in these experiments.

Initial attempts to detect Mg segregation in vacuum proved unsuccessful probably due to rapid evaporation from the free surface(14). We have found that reproducible results on the temperature dependence of the surface Mg concentration could be obtained by annealing in air at 1 atm using a 'quasi-closed' system; this

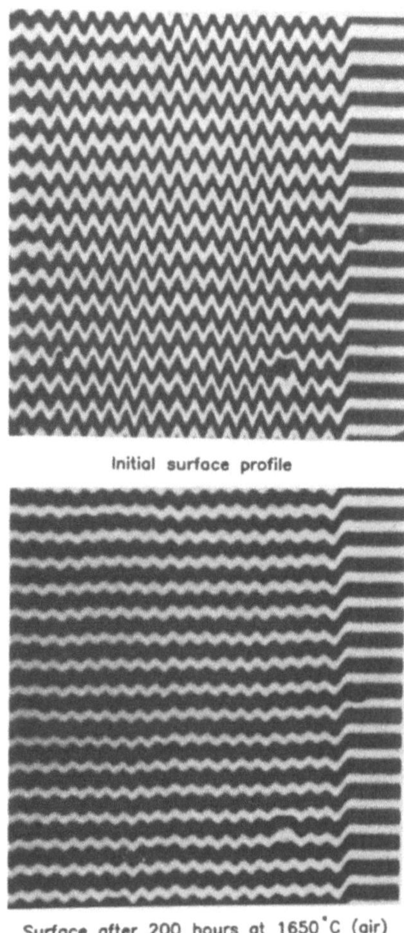

Figure(6)
Interference micrographs showing a sinusoidal surface grating of spacing 7.5μ before and after an anneal in air for 200 hr. at 1650°C. A significant decrease in amplitude has occurred.

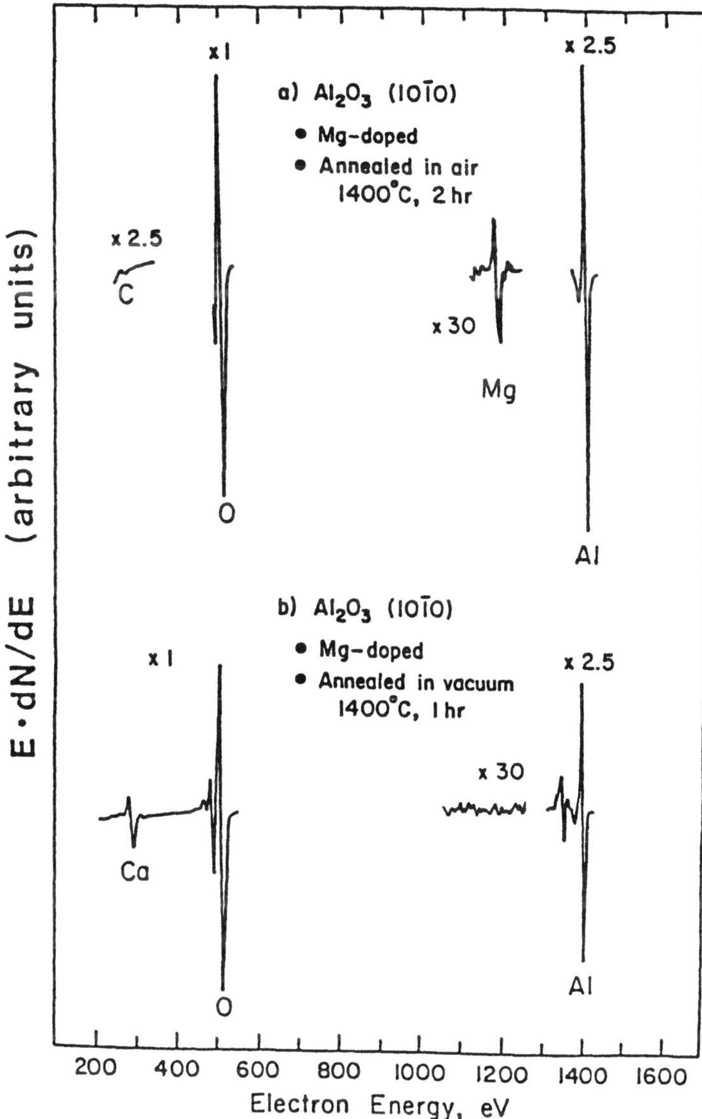

Figure(7)

Typical Auger spectra displayed in the E.(dN/dE) vs E mode where E is the electron energy. (a) indicates Mg enrichment in the surface region of the (10$\bar{1}$0) surface of alumina after annealing in air for 2 hr at 1400°C. (b) shows that no such enrichment can be detected after vacuum annealing; note the presence of a significant Ca peak in this case. [Incident beam energy 3 keV, beam current 20μA, modulation voltage 10 eV].

consisted of two identically doped crystals in close contact during the annealing treatments. Under these conditions Mg was indeed found through Auger spectroscopy studies to concentrate in the surface region. This experimental arrangement probably comes close to simulating the conditions at pore surfaces in a sintering compact.

Fig(7) shows portions of the Auger spectra from vacuum annealed and air annealed ($10\bar{1}0$) surfaces using the quasi-closed arrangement mentioned above. The air annealed sample shows clear evidence for Mg surface segregation while the vacuum annealed sample shows no Mg but a distinct Ca peak. In all of our studies of segregation at sapphire surfaces we have consistently observed this correlation in the surface Mg and Ca concentrations. It is interesting to speculate that the beneficial effect of Mg doping on the sintering characteristics of alumina may be due in part to its ability to keep the surface free from Ca, a larger ion which is a slow diffuser and which may promote faceting.

2.1 EFFECTIVE LANGMUIR HEATS OF SEGREGATION

Using the best available estimates for the parameters involved, the measured temperature dependences of the surface Mg concentration for (0001) and ($10\bar{1}0$) have been converted to Langmuir plots(14) as shown in fig(8). The corresponding plot for Ca segregation to ($10\bar{1}0$) in vacuum is shown in fig(9). The Mg data fall on reasonably good straight lines and can therefore be described by effective Langmuir heats of segregation as indicated on the figure. Our Ca data does not fit an Arrhenius relation at all.

The values of -1.4 eV and -1.9 eV for the heats of segregation of Mg to the ($10\bar{1}0$) and (0001) surfaces are certainly in the range predicted by recent theory by Tasker and coworkers(16). However in light of the discussion given above of the effect of ionic double layers the agreement may be fortuitous. We could, on the other hand, take the agreement between theory and experiment to indicate that in this system double layer effects are unimportant. Further experiments on surface double layer effects are clearly needed; measurements of segregation isosteres would greatly help in the interpretation of the data.

3. Effect of Segregation on Mass transport

Studies of the rate of decay of surface sinusoidal modulations on sapphire in air have been carried out under the same experimental conditions as the segregation experiments; details of these measurements are given in another pulication(17) An extensive series of measurements have been made using a range of sinusoidal wavelengths to establish the relative contributions of surface and volume diffusion to the observed mass transport. We have also investigated the effect of large surface Mg concentrations on the rate of mass transport. This was accomplished by annealing the surface in contact with a

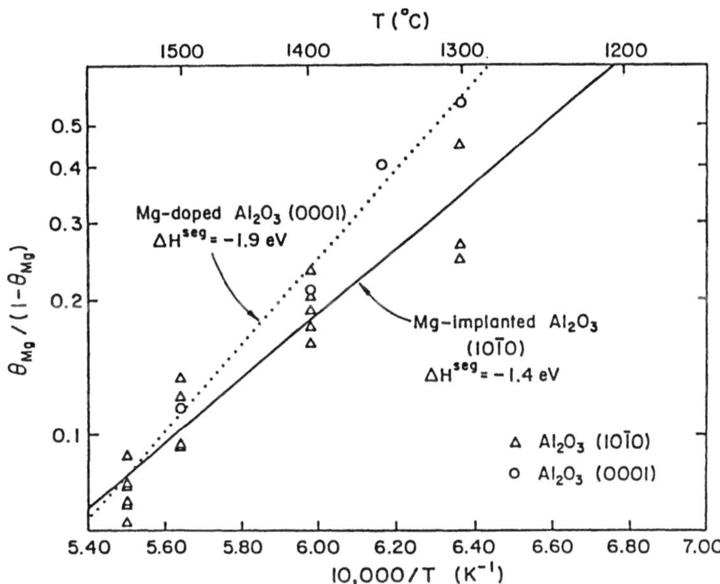

Figure(8)
Langmuir plots from experimental Auger data on the segregation of Mg to the (0001) and (1010) surfaces of alumina. The coverages were obtained from the Auger peak heights as described in the text and in reference(14).

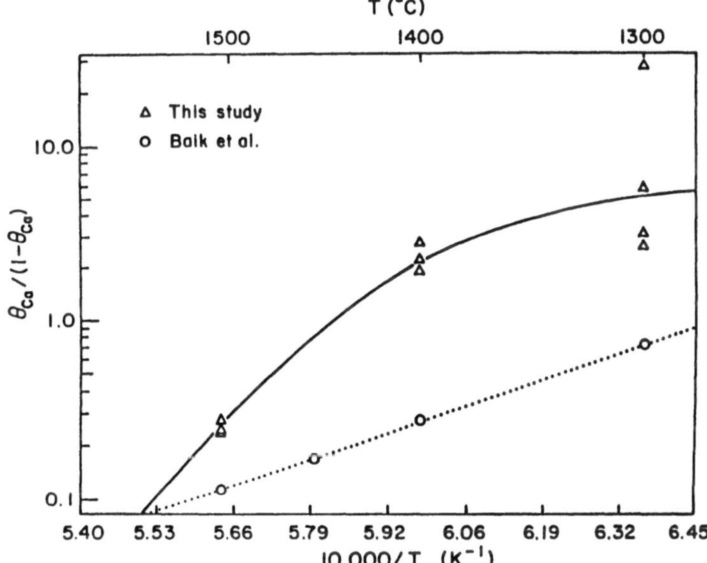

Figure(9)
Langmuir plots calculated from data on the variation of Ca Auger peak heights with temperature for(1010) alumina surfaces annealed in vacuum (see ref.14).

sintered Mg-Al spinel compact(17). An example of the mass transport data is shown in fig(10). This shows the rate of decay of sine wave amplitudes at a wavelength of 2.5 microns at 1500°C for the surface Mg concentration levels indicated on the figure. At low Mg levels the mass transport rate is unaffected to within experimental error. However under conditions where surface precipitates are optically visible and the Auger measurements indicate a surface Mg level of ~30 atomic % there is a large enhancemant in the rate of amplitude decay.

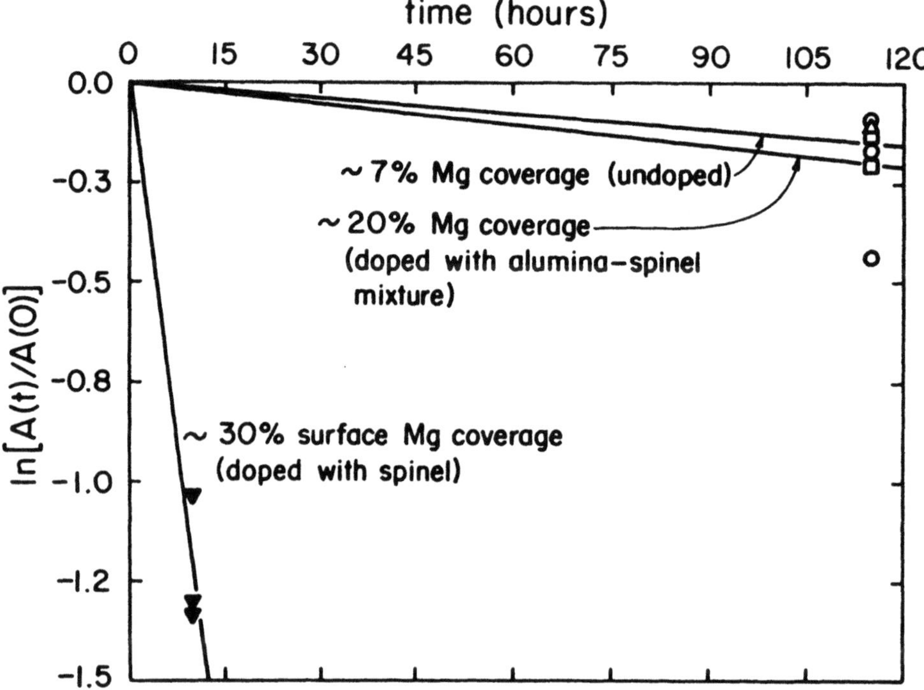

Figure(10)

Rate of decay of the amplitude of 2.5μ sine wave profiles on the (10$\bar{1}$0) surface of alumina with different surface Mg concentrations. Even at a coverage of ~20% no change in mass transport rate can be detected; however at a surface Mg coverage in the region of ~30% the decay constant is strongly enhanced.

The main conclusion from this work seems to be that small surface Mg levels have little or no effect, but that the presence of a new phase (probably spinel) at the surface has a major effect on surface mass transport. Further experiments, perhaps using electron microscopy to identify this surface phase, would clearly be desirable.

4. Acknowledgements

This work has been supported by the National Science Foundation through the Materials Science Center at Cornell and by the Semiconductor Research Corporation.

5. References

1. J.M. Blakely and J.C. Shelton, in Surface Physics of Materials, Vol.1, 189, Academic Press, New York, (1975)
2. P. Wynblatt and R.C. Ku, in Interfacial Segregation, ed. W. Johnson and J.M. Blakely, ASM, Metals Park, Ohio, (1979), p.115.
3. D. McLean, in Grain Boundaries, Oxford Univ. Press, London, (1957).
4. W,C. Mackrodt and P.W. Tasker, J. Amer. Cer. Soc., (to be published).
5. J.M. Blakely and H.V. Thapliyal, in Interfacial Segregation, ed W.C. Johnson and J.M. Blakely, ASM, Metals Park, Ohio, (1979).
6. W.C. Mackrodt, in Adv. in Ceramics, vol 10, ed. W.D. Kingery, Amer. Cer. Soc., Columbus, Ohio, (1984), p. 62.
7. R.B. Poeppel and J.M. Blakely, Surface Sci., 15, 507, (1969).
8. S.M. Mukhopadhyay and J.M. Blakely, (to be published).
9. J.M. Bloch et al., Phys. Rev Lett., 54, 1039, (1985).
10. J.M. Blakely and H. Mykura, Acta Metall., 9,23, (1961).
11. W.W. Mullins, J. Appl. Phys., 30, 77, (1959).
12. J.M. Blakely and C.Y. Li, Acta Metall., 14, 279, (1966).
13. S. Baik, D.E. Fowler, J. M. Blakely, and R.Raj, J. Amer. Cer. Soc., 68,281, (1985).
14. S.M. Mukhopadhyay, A.P. Jardine, J.M. Blakely, and S. Baik, J. Amer Cer. Soc., 71, 358, (1988).
15. J.S. Haggerty, NASA Report, CR-120948, May, (1982).
16. W.C. Mackrodt and P.W. Tasker, Mater. Res. Soc. Symposium Proceedings, 60, 291, (1986).
17. S.M. Mukhopadhyay and J.M. Blakely, Mater. Res. Soc. Symposium Proceedings, (to be published).

THE ROLE OF GRAIN BOUNDARIES AND INTERFACES ON SUPERCONDUCTIVITY

DUANE DIMOS and DAVID R. CLARKE
IBM Research Division, Thomas J. Watson Research Center
P.O. Box 218, Yorktown Heights, NY 10598

ABSTRACT. The influence of grain boundaries and interfaces on critical current density in both high-temperature, cuprate superconductors and conventional superconductors is discussed. Recent measurements of the critical current density of individual grain boundaries in yttrium barium cuprate are also reviewed.

I. Introduction

It was recognized early in the development of superconducting metals that processing to incorporate grain boundaries and interfaces led to an increase in the attainable transport critical current density without affecting other superconducting properties. Such approaches as extrusion and cold working continue to be exploited industrially despite the fact that the mechanisms for the increase in critical current are not completely understood. It was thus expected that the established metallurgical techniques for improving critical current density, such as grain size refinement, could be used to optimize the values for the new high-T_c superconductors. However, following the discovery of the first cuprate superconductor by Bednorz and Müller, it was quickly established that the critical current densities of polycrystalline ceramic samples were not only quite low, but were typically 10^3-10^4 times lower than those of single crystals or single crystal (epitaxially grown) films. Furthermore, the critical current densities of polycrystalline samples were significantly more sensitive to applied magnetic fields than were those of single crystals. These observations indicate that, in contrast to the conventional, metallurgical superconductors, grain boundaries in the cuprates superconductors adversely affect the attainable critical current density. Whether this behavior is unique to the cuprate ceramics remains to be established, since few measurements of the earlier ceramic superconductors, such as barium lead bismuth oxide and lithium titante, have been made. In large part this is probably due to the fact that these earlier ceramic superconductors, having lower superconducting transition temperatures than the metal superconductors available at the time, were regarded by many as curiosities rather than potentially important superconducting materials.

In this contribution we review the role of grain boundaries and interfaces on the critical current densities of superconducting materials. We restrict our discussion to the transport critical current density since, as distinct from a number of other superconducting properties, such as the transition temperature, it is not an intrinsic thermodynamic property but rather is sensitive to the microstructure and processing of the material. Until recently the effect of grain boundaries has been deduced from measurements on polycrystalline materials. However, the discovery of high temperature superconductivity has come at a time when many techniques of microelectronics can be applied to research. By using some of these techniques it has been possible to isolate individual grain boundaries and, thus, measure the transport critical current across grain boundaries having a well-defined crystallography. Some of the important results ob-

L.-C. Dufour et al. (eds.), Surfaces and Interfaces of Ceramic Materials, 301–318.

tained in this manner for single boundaries in $YBa_2Cu_3O_7$ will be described together with some of their implications. To begin we briefly review some of the concepts of superconductivity pertinent to a consideration of critical currents and interface dominated phenomena. More complete introductions to superconductivity can be found in a number of texts[1-3].

II. Introductory Concepts

Superconducting materials can be broadly classified as type I (elemental superconductors such as Nb and Pb), and type II (alloys and compounds, such as Nb-Ti, Nb_3Sn, and the CuO-based high-temperature superconductors) according to their magnetic properties. Type I superconductors exhibit perfect diamagnetism (i.e., the total flux density inside the superconductor is zero) upto the superconducting transition temperature, T_c and below a critical magnetic field, $H_c(T)$, that is a function of temperature. This perfect diamagnetism is due to the formation of surface screening currents that produce a flux density exactly cancelling the applied flux density. The presence of these surface currents increases the free energy of the superconducting state so that above $H_c(T)$ it exceeds the free energy of the normal state and the material becomes a normal conductor. In type II materials, superconductivity can persist to very large field strengths by entering a mixed state. In this state, between H_{c1} and the upper critical field strength, H_{c2} , the material can be thought of as consisting of a perfectly diamagnetic material threaded by an array of cylindrical flux lines (or fluxons) lying parallel to the applied field. Each of these flux lines consists of a normal core that contains a quantized amount of magnetic flux, $\phi_0 = hc/2e$. The large values of H_{c2} exhibited by some materials has led to their use in high field (> 10 Tesla) magnets. Complete flux exclusion in these materials is exhibited only below a much lower critical field strength, H_{c1} , which is typically a few hundred Oersted.

The free energy associated with the formation of the mixed state is determined by two characteristic length scales of the superconducting state, the coherence length, ξ, and the penetration depth, λ, both of which are functions of temperature. The coherence length is a measure of the spatial extent over which a significant change in the superconducting order parameter (Δ) will be averaged. The order parameter, which is equivalent to the density of superconducting electron pairs (n_s) or, equivalently, to the amplitude of the electron-pair wave function ($|\psi_s|^2$), cannot vary rapidly with position. The penetration depth is a measure of the distance that a magnetic field can penetrate into a superconductor, and so is equivalent to the distance in which the surface screening currents are confined. An excess free energy is associated with the transition region from the core of a fluxon to the bulk superconductor; it is a sum of two terms. One, the decrease in free energy due to an increase in the order parameter or n_s , occuring over a distance ξ, and two, the increase in free energy associated with the penetration of the magnetic field into the superconductor over a distance λ. Since the loss of order in the core of a fluxon produces an increase in free energy which is proportional to its area, the effective radius of the normal core will be roughly equal to ξ, which is the minimum value it can take. When $\xi < \lambda$, the excess free energy of the transition zone is negative and so the mixed state will be the lowest free energy state in the presence of a magnetic field above H_{c1} .

The distinction between type I and type II superconductors can be made in terms of the relative values of ξ and λ. It is usual to define the ratio κ ($= \lambda/\xi$) such that type II materials are materials with $\kappa > 1/\sqrt{2}$ and materials with $\kappa < 1/\sqrt{2}$ are type I superconductors. In typical type II materials, the coherence lengths are < 300 Å, whereas the penetration depths are $\simeq$ 1000-2000 Å. In contrast, the typical values for type I superconductors are $\xi \simeq 0.2$ - 2 μm and $\lambda \simeq$ 200-1000 Å. The shorter coherence lengths and longer penetration depths of the alloy and compound superconductors are directly related to the shorter mean free paths of the conduction electrons in the normal state as compared to the relatively pure elemental superconductors.

The magnetic flux contained within each flux line is shielded from the bulk of the superconductor by a circulating supercurrent extending to an approximate radius of λ from its center. The overlap of the magnetic fields of the individual fluxons produces interactive forces that are repulsive (attractive) if their magnetic fields are parallel (anti-parallel). As a result of their mutual interactions, and in the absence of strong pinning, the fluxons arrange themselves into a regular periodic array known as the flux line lattice. As the flux contained within each normal core has a fixed value of ϕ_o, the density of flux lines increases with increasing field up to H_{c2}. At this field, the entire volume of the superconductor is filled with a close-packed array of fluxons, and superconductivity is lost. Thus, there is a direct geometrical relationship between the upper critical field strength, $H_{c2}(T)$, and the temperature-dependent coherence length of a type II superconductor, given by $H_{c2}(T) \simeq \phi_o/\pi\xi^2$. According to this equation, the shorter the coherence length the larger is the value of H_{c2}.

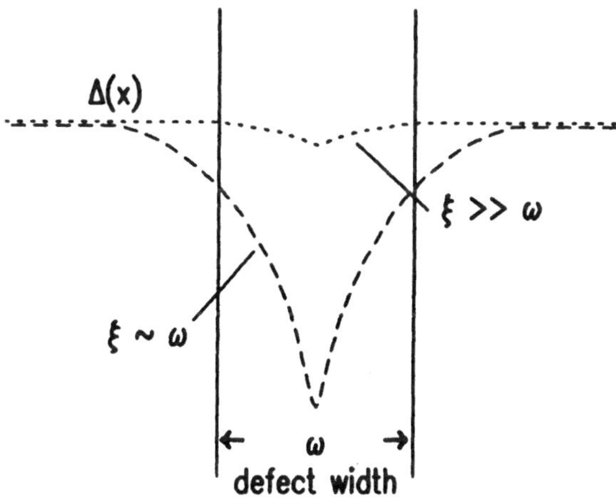

Figure 1. Schematic diagram showing the variation of the order parameter, Δ, near a defect for a superconductor with 1) $\xi >> \omega$ and 2) $\xi \simeq \omega$, where ω is the width of the defect.

The magnitude of the coherence length also determines the extent to which the properties of a type II superconductor are affected by microstructural heterogeneities. Both structural and compositional variations can be expected to cause changes in the strength of the electron pair potential. The length scale for which significant variations in the pair potential can occur is set by the coherence length, ξ; thus, Δ will be strongly depressed only at defects wider than about 2ξ and be weakly depressed (if at all) at defects significantly smaller than 2ξ, as shown in Fig. 1. Consequently, narrow defects, such as grain boundaries, would be expected to significantly perturb the local order parameter only in materials having very short coherence lengths.

III. Critical Currents in Conventional Type II Superconductors

A consequence of the requirement that magnetic flux be completely excluded from the interior of type I superconductors, is that supercurrents in these materials can only flow in a thin layer ($\simeq\lambda$ thick) near the surface. In contrast, since magnetic fields can penetrate type II superconductors, currents are no longer confined to a superficial layer and type II materials can support resistanceless current flow throughout their cross section. However, when a current per unit area, J, flows through a type II sample when it is in its mixed state, a Lorentz force ($\mathbf{F} = \mathbf{J} \times \boldsymbol{\phi}_0$) is exerted on each flux line in a direction perpendicular to both the current and the magnetic field. (In the absence of an applied magnetic field, the self magnetic field induced by the transport current will dominate.) If the Lorentz force causes motion of the flux line lattice, a voltage will be created by the motion of the magnetic flux contained within the cores. This voltage will be proportional to the number of moving fluxons and to their velocity, which is determined by the driving force (i.e., the current density). The minimum current density that causes a detectable or measureable voltage due to flux motion is the usual operational definition of the critical current density (J_c).

The critical current density of a type II sample in the mixed state is not, however, infinitesimally small because flux lines can be pinned at heterogeneities inhibiting their motion. Even if every flux line is not pinned by a defect, as is typically the case, the repulsive interaction between flux lines is sufficient to effectively pin the remaining flux lines that are not tied to a defect. The critical current density is then determined by the overall interaction between the entire flux line lattice and the various microstructural defects in the material. Although calculation of the pinning force is a complex many-body one (and currently unsolved except within certain statistical mechanical models), a first step is to consider the interaction between an individual defect and a fluxon.

Traditionally, it was assumed that pinning occurs because defects prohibit superconducting condensation in their locality. Thus, when a flux line overlaps a heterogeneity, in which the order parameter goes to zero, the free energy of the system will be reduced due to the recovery of the superconducting condensation energy ($\mu_o H_c^2/2$) within the overlapping volume (V_O). The pinning force due to this core interaction is then given by the derivative of the free energy with respect to the direction of the driving force, yielding

$$f_p \simeq \mu_o H_c^2 V_O / 2\xi \quad .$$

This force exists at any nonsuperconducting region of a sample, such as an insulating precipitate or a void[4].

Based on this line of reasoning it was anticipated that grain boundaries would be unable to pin fluxoids, since the grain boundary "width", which may extend only a few lattice spacings, would be significantly smaller than the fluxoid diameter of 2ξ. However,

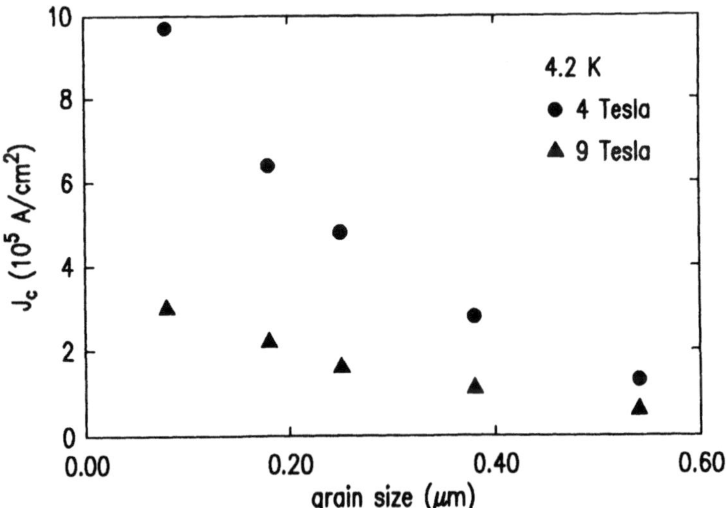

Figure 2. Critical current density as a function of grain size for a Nb_3Sn material at two different field strengths[5].

it has been well established that narrow defects, such as grain boundaries, not only pin fluxoids, but can actually be the dominant pinning centers in type II superconductors. This point is illustrated in Fig. 2, which is a plot of critical current density as a function of grain size for a Nb_3Sn material[5]. In this metal $a_o \simeq 5Å$ and $\xi \simeq 50Å$. It should be noted that the critical current density of very fine-grained Nb_3Sn can exceed 10^5 A/cm^2, even in the presence of fields exceeding 10 T. Similar correlations between critical current density and grain size have been demonstrated for a number of other technologically important type II superconductors, such as NbN[6] and V_3Ga.

A similar effective pinning is demonstrated by two-phase superconductors, such as Nb-Ti. This material consists of α-Ti precipitates, which remain normal, embedded in a superconducting Nb+Ti matrix phase. The critical current densities of these materials are usually maximized by extensive drawing which deforms the equiaxed α-Ti precipitates into long, thin ribbons parallel to the drawing direction, as shown in Fig. 3. It has been recently shown that these normal metal ribbons act as the dominant source of flux pinning even when the fluxoid diameter, which has a value of $2\xi \simeq 100Å$, is about an order of magnitude larger than the precipitate thickness ($\simeq 10Å$)[7]. The core interaction between a fluxon and one of these precipitates might be expected to be especially weak in comparison to insulating precipitates, since superconductivity will be induced in these thin ribbons by the proximity effect[8]. However, the experimental finding that the pinning strength is not reduced as the precipitate thickness is decreased suggests that it is the properties of the interface (i.e., the phase boundary) that are responsible for pinning.

Since both grain boundaries and phase boundaries are planar defects, they can only act as efficient flux pinning centers when their plane is 1) parallel to the length of the fluxoids (i.e., the direction of the magnetic field) and 2) perpendicular to the direction of the Lorentz force. The first point was convincingly demonstrated by Das Gupta et. al.[9], who showed that the pinning strength of an individual grain boundary in a niobium bicrystal had a sharp maximum as a function of angular orientation when the boundary plane was parallel to the applied field. These results are reproduced in Fig. 4. The second point was illustrated by the work of Gray et. al.[6] as part of a study of critical current densities in NbN thin films with columnar grains. The critical current density measured with the applied field perpendicular to the film surface was a strong function of grain size as expected. However, J_c with field parallel to the film was essentially independent of grain size, because the grain boundaries were approximately perpendicular to the flux lines and were, thus, ineffective as pinning centers. Gray et. al. also showed that as the grain diameter became less than the coherence length, critical current densities were reduced presumably because the fluxons experience a spatially more uniform pinning potential.

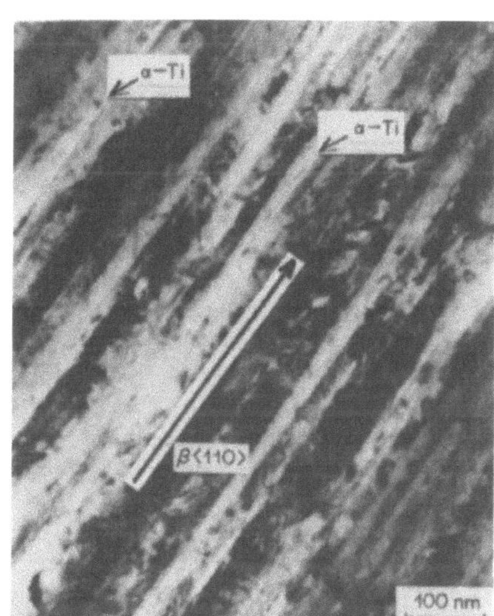

Figure 3. TEM image of a drawn Nb-Ti filament[7]. Two representative α-Ti precipitates are indicated.

Theoretical attempts to explain the strong pinning that occurs at interfaces have largely been based on the role of the interface as a scattering center. One approach[10,11], which is based on a Ginzburg-Landau perturbational formalism considers pinning to be the result of a local reduction of the coherence length in the vicinity of the boundary due to a reduction in the electron mean free path perpendicular to the interface. Due to the local reduction in ξ, a fluxon at the interface would have a smaller cross-sectional area and, thus, a lower free energy per unit length than a fluxon in the bulk. This reduction in energy, which is equivalent to the fluxon-defect binding energy, leads to pinning. An alternative approach developed by Thuneberg[12] uses the quasiclassical method to cal-

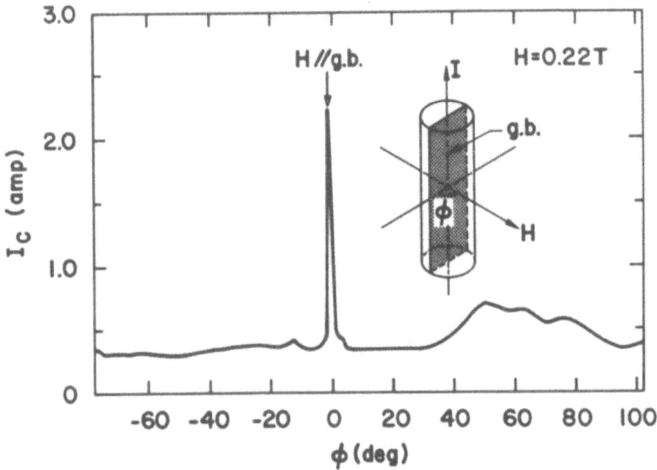

Figure 4. Critical current as a function of the angle between the applied magnetic field and the grain boundary in a Nb bicrystal[9]. As indicated in the schematic diagram (inset), the current flow is perpendicular to the field direction. A sharp maximum in the critical current, indicating a maximum in the pinning strength, is obtained when the field and the grain boundary are parallel.

culate pinning energies brought about by quasiparticle scattering off small defects. In this approach, the physical basis of the pinning mechanism is that a scattering center helps to sustain deformations of the order parameter up to distances of approximately ξ, which occur at the core of a vortex, even when little depression of Δ occurs at the defect. This model, which requires further verification, predicts that the binding energy per unit length between a fluxon and a grain boundary would be proportional to ξ^2, which is larger by the factor (coherence length)/(defect width) from the core overlap effect. This conclusion disagrees with the result of the perturbational approach that the pinning strength is reduced as the coherence length becomes so large that changes in the mean free path of approximately a factor of two are unimportant.

IV. Critical Currents In High-Temperature Superconductors

Since the initial discovery of superconductivity at ~ 30 K in $(La_{1-x}Ba_x)_2CuO_4$ by Bednorz and Müller[13], a number of cuprate-based ceramic superconductors with unprecedented transition temperatures have been found. The first compound for which T_c exceeded the boiling point of liquid nitrogen was $RBa_2Cu_3O_7$ (T_c = 90-94 K), where R = Y or any rare earth element other than Ce or La. More recently, transition temperatures above 100 K have been achieved in Bi-Sr-Ca-Cu-O and Tl-Ba-Ca-Cu-O compounds. These cuprate superconductors are structurally related compounds, since they all contain Cu-O "planes" in which each copper ion has four strongly bonded oxygen neighbors in a square-planar coordination. These CuO_2 planes are separated by layers that contain the other cation species; as illustrated in Fig. 5, which is a structural model of the unit cell for $YBa_2Cu_3O_7$. The stacking arrangement and, thus, the unit cell dimension parallel to the stacking direction, (the c axis [001]), is determined by the stoichiometry of the cuprate. However, the basal plane lattice spacing (i.e., parallel to the CuO_2 planes) is roughly 3.8-3.9 Å for all the compounds. (For a more complete discussion of the various structural features of these materials see Beyers and Shaw[14]). The charge carriers in

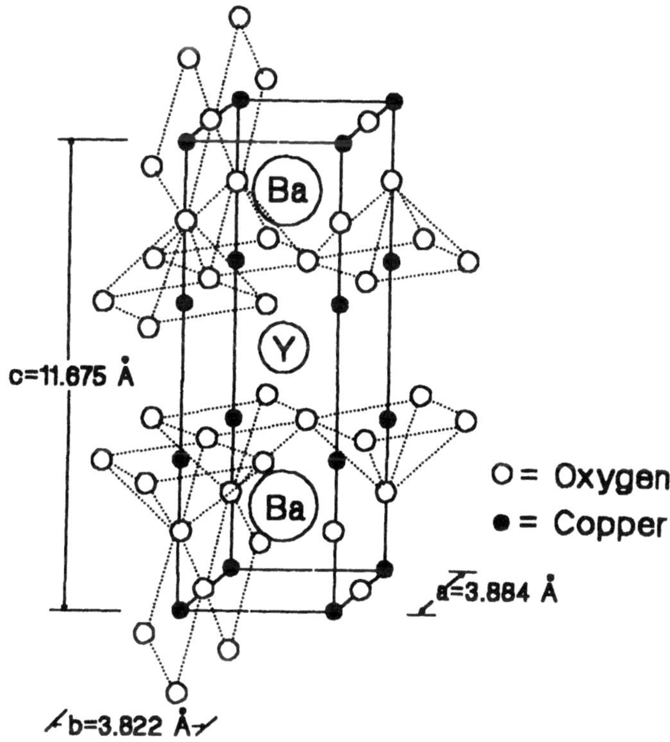

Figure 5. Structural model of the unit cell for $YBa_2Cu_3O_7$. The Cu-O planes are comprised of the bases of the square pyramids in the diagram.

these materials are now known from a wide variety of physical measurements to be mobile holes in the oxygen 2p orbitals which are largely confined to the CuO_2 planes. While it has been verified that these holes form pairs with charge $+2e$ in the superconducting state, the pairing mechanism has not been identified. In addition to their high transition temperatures, these materials display a number of remarkable properties. In particular, the upper critical field strengths of these compounds are believed to be in excess of 100 Tesla, which is significantly larger than for any conventional, low-temperature superconductors. The coherence lengths estimated from these magnetic measurements are extremely small (on the order of the unit cell dimensions) and highly anisotropic. The penetration depths of the cuprates are on the order of 1000 Å, so that these materials are type II superconductors with large κ values. In addition, the low carrier densities (less than one electron hole per unit cell) coupled with the very short coherence lengths imply that virtually all of the mobile charge carriers participate in the superconducting state.

One of the reasons that technological applications have not yet been found for these new superconductors in bulk form has been the fact that their transport critical current densities have been so low[15-17]. Since both single crystals and single crystal films of yttrium barium cuprate can be made that exhibit considerably larger critical current densities[18,19], it was quickly established that the low values were associated with the polycrystallinity of bulk samples. For comparison, the maximum critical current densities that have been measured on epitaxial films of $YBa_2Cu_3O_7$ at 77 K, currently exceed 10^6 A/cm^2 [20,21], some 10^3-10^4 times larger than for the bulk polycrystalline ceramics, and comparable to the values for commercial superconductors, NbTi and Nb_3Sn, at 4.2 K. More recently it has been shown that the critical current densities of single crystals of Bi-Sr-Ca-Cu-O compounds are also orders of magnitude larger than the values obtained for polycrystalline materials[22,23]. From experience with conventional superconductors, it seemed unusual that rather perfect single crystals were capable of supporting large critical current densities since large concentrations of defects are required for effective volume pinning of flux lines. However, one consequence of a short coherence length is that even point defects or point defect clusters may be able to suppress the order parameter and, thus, pin fluxons. The possibility that point defects are the dominant pinning centers is strongly suggested by vortex decoration experiments showing that flux lines tend to be pinned at sites that are rather uniformly distributed and are not associated with any crystal defect identifiable in the optical or scanning microscope[24].

Many of the measurements of transport critical current densities for polycrystalline materials, especially in the early studies of yttrium barium cuprate, have been carried out on poorly characterized samples, largely because of the initial lack of information about the effect of many important processing parameters. These samples were often very porous and contained appreciable amounts of second phases. In many samples of $YBa_2Cu_3O_7$, a liquid phase was formed during sintering due to the presence of a low melting peritectic in the CuO-BaO rich region of the phase diagram; this melt produced an insulating phase that coated grain boundaries. A number of investigators also detected carbonate-like bonding at the grain boundaries; since carbonates are also insulating phases, their presence would presumably reduce the attainable critical current density. Consequently, it was generally assumed that relatively high bulk critical current densities would be achieved with improved processing. However, even with a great deal of subsequent work directed towards achieving improved and reproducible processing, only modest increases in J_c have been achieved for bulk samples of yttrium barium cuprate. While the transport properties of the other cuprate superconductors have not been as extensively characterized, it appears that the critical current densities attainable for these polycrystalline samples are also quite low compared with those obtainable for single crystals.

Partially because of the difficulties involved in making direct transport measurements of J_c in bulk samples (very large currents are required) macroscopic magnetic measurements have been extensively used to evaluate critical current densities. The basis of this approach, the Bean critical state model[25], relates the critical current density of a homogeneous type II superconductor to the hysteretic magnetization in an M-H loop, a measure of the remnant magnetization or trapped flux. While the applicability of this technique to the measurement of critical current densities in single crystals has been verified[26], it has been found that a reliable estimate of J_c for a polycrystalline sample can not be obtained. In fact, the hysteresis loop size for a powdered sample, where the

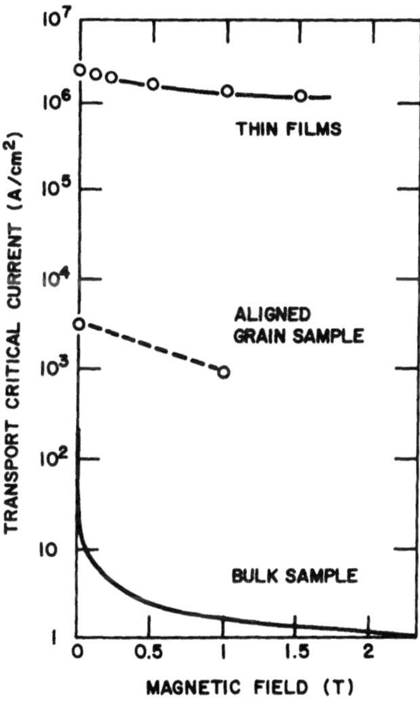

Figure 6. Critical current density as a function of applied magnetic field for a single crystal film and a typical sintered sample. Highly-textured polycrystalline materials have intermediate values.

transport $J_c = 0$, is almost as large as for a sintered pellet of comparable volume[27]. This result is strong evidence that these polycrystalline materials are not homogeneous but actually behave as granular superconductors with only weak coupling between strongly superconducting regions. By assuming that the samples are homogeneous on the scale of the grain size, an intergranular J_c can be calculated from the magnetization measurements. In general, these values are similar to the critical current densities obtained for single crystal films, strongly suggesting that the regions of weakened superconductivity are associated with the grain boundaries.

The detrimental effect of the grain boundaries is further evidenced by the fact that the transport critical current densities for polycrystalline materials are strongly suppressed by small magnetic fields ($\approx$ 10-100 Oe) [16,17], whereas such fields do not lower the J_c of single crystals significantly. These facts are summarized in Fig. 6, which is a plot of the critical current density as a function of magnetic field for a typical single crystal film and a typical sintered sample at 77 K. The strong field dependence exhibited by polycrystalline samples lead several authors to suggest that the grain boundaries behave as Josephson junctions[28,29].

IV.1. Experiments on Individual Grain Boundaries

While evidence from macroscopic measurements, such as those mentioned above, that grain boundaries were superconducting weak links was compelling, the first direct demonstration that the grain boundaries are responsible for the low critical current densities and the extreme field sensitivity was provided by measurements of the transport properties of single grain boundaries and their adjacent grains[30-32]. By isolating individual grain boundaries, it has proved possible to directly measure their critical current densities as a function of magnetic field, temperature, and misorientation angle and to compare these values to those of the grains on either side.

To systematically investigate the superconducting properties of single grain boundaries and of their adjacent grains it was necessary to 1) produce large-grained samples with well-defined grain boundary geometries and 2) develop a patterning technique for electrically isolating the regions of interest. To accomplish the first requirement, we took advantage of the fact that single crystal films of $YBa_2Cu_3O_7$ could be prepared by epitaxial growth on $SrTiO_3$ substrates[19]. Thin film bicrystals of yttrium barium cuprate were thus

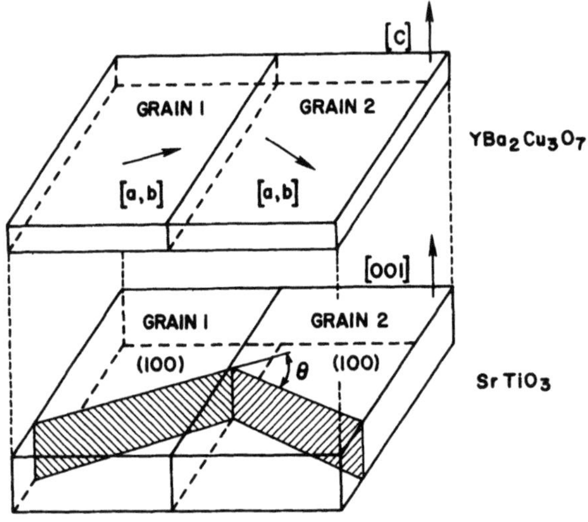

Figure 7. Schematic diagram showing the important crystallography of the bicrystal. The grain boundary geometry in the superconducting film is determined by the substrate.

fabricated by using oriented bicrystals of $SrTiO_3$ as substrates. The films were prepared by electron-beam evaporation using a heated substrate (≈ 400°C) with a background oxygen pressure of about 20 μtorr. The as-deposited films which are amorphous and nonsuperconducting, are then crystallized at about 900°C and oxygenated by slow cooling in flowing oxygen to produce superconducting films with T_c (midpoint) ≈ 85-88 K and ΔT_c ≈ 2-6 K. The films grow with an epitaxial relationship to the substrate so that the crystallography prescribed by the substrate is reproduced in the film. For this study, nominally symmetric [001] tilt boundaries were fabricated with the c-axis of the yttrium barium cuprate essentially perpendicular to the plane of the film, as shown in Fig. 7. The acute angle θ between {100} planes defines the misorientation (tilt) angle between the principle directions in the Cu-O planes. However, alignment errors occasionally caused the [001] axes to be misaligned by a small but non-negligible angle (ϕ) in a plane normal to the grain boundary plane.

The films were patterned as shown schematically in Fig. 8 by using an excimer laser to ablate the unwanted regions of the film. Narrow conducting lines for the transport measurements, approximately 10 μm wide, were formed within the two single grain regions and across the grain boundary. Electrical contacts to the large regions of the pattern were made directly to the film using aluminum wire bonds. The current-voltage characteristics of each line were measured using a four-point technique, where the current and voltage leads are independent. One of the advantages of using thin superconducting films (approximately 0.5 μm thick) is

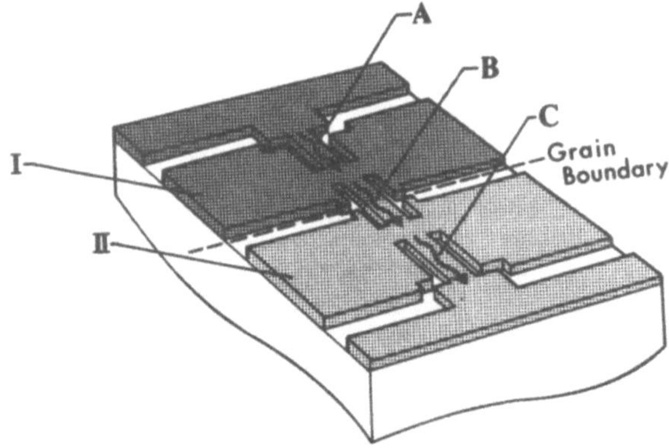

Figure 8. Schematic diagram of the patterned film showing the two grains (regions I and II) and the three narrow lines. Lines A and C are contained entirely within grains I and II, respectively, while line B straddles the grain boundary.

that large current densities are achievable with only modest measuring currents applied to the samples.

Typical I-V curves from conducting lines in a single grain and across a single grain boundary from the same sample (4.2 K) are shown in Fig. 9. In both cases, a zero-voltage current is exhibited up to a reasonably well-defined maximum value, the critical current, I_c, for the line. The critical current density is then obtained by dividing I_c by the measured cross-sectional area of the patterned line. In this case, the critical current density of the single grain line is more than an order of magnitude larger than for the grain boundary line. While the difference in critical current densities depended on the misorientation angle, it was always found that the critical current density of the line containing the grain boundary was lower than for either adjacent grain. Consequently, the lower critical current density is a property of the grain boundary, and one which clearly limits the maximum zero-voltage current that can flow between adjacent grains.

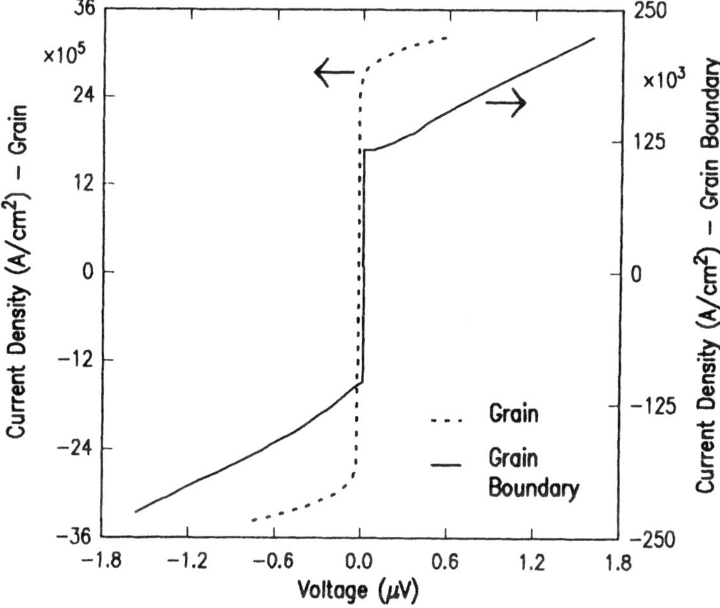

Figure 9. Typical current-voltage curves for a single grain and a single (high-angle) grain boundary line at 5 K.

The grain boundary critical critical current density, J_c^{gb}, was also quite sensitive to small magnetic fields. A typical example of the field dependence for a large-angle grain boundary is illustrated in Fig. 10. In this case, the grain boundary J_c is reduced by about a factor of 10 from its peak value due to a field increment of about 10 G. While the exact position of the peak depends on the magnetic history of the sample (i.e., there is some hysteresis), the peak is at zero field, $B_{applied}$ ($\equiv$ H) = 0, for samples initially cooled in the absence of an applied field. The amount by which J_c is suppressed due to a field varies from one sample to the next. However, the general shape of the J_c vs B curve for these single grain boundaries is qualitatively similar to the field dependence typically observed for polycrystalline samples. Over a similar field range, up to $\simeq$ 300 G, the intragranular J_c is independent of applied field. This observation demonstrates that the magnetic field preferrentially penetrates at the grain boundary and, thus, verifies that the grain boundary is a region of weakened superconductivity. However, there were no secondary maxima in the J_c^{gb} vs B curves as would be expected if the grain boundaries behaved as a classic Josephson junction.

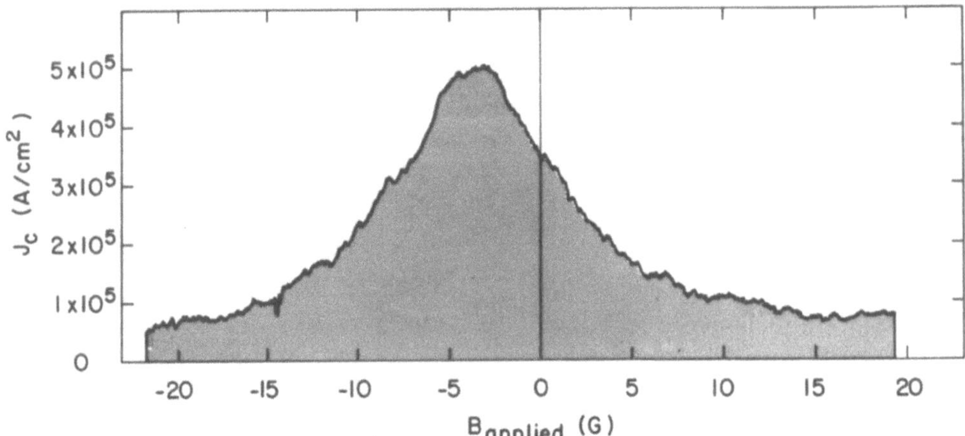

Figure 10. Critical current density vs applied magnetic field (perpendicular to the film) for a high-angle grain boundary at 5 K. In this case, the field was swept from left (-20 G) to right (20 G).

The relationship between the superconducting transport properties and the structural properties of the tilt boundaries was explored by measuring the critical current density as a function of misorientation angle; this data is summarized in Fig. 11. The ratio of the grain boundary J_c to the average value for the two grains emphasizes the role of the grain boundary in limiting transport critical currents and also allowed samples of different quality to be compared directly. The measured critical current density ratio is plotted on a logarithmic scale showing that the grain boundary J_c decreases very rapidly with increasing misorientation angle until it reaches a saturation value of about 1/50 for $\theta > 15°$-$20°$. While the data used in this plot were taken at 4.2-5 K, the critical current ratio is almost independent of temperature. For the relatively high angle boundaries, exhibiting large depressions in J_c with strongly field dependent behavior, the grain boundary critical current densities (in the best samples) were in excess of 10^5 A/cm^2 at 4.2 K and 10^4 A/cm^2 at 77 K. In contrast to the high-angle behavior, the critical current densities for low angle boundaries ($\theta < 5°$) were only slightly reduced and were independent of small applied fields; a behavior similar to that exhibited by the grains. The transition from strong coupling to weak-link behavior occurs over an angular range θ = 5° - 10°. The square data points in Fig. 11 are for samples where the out-of-plane misorientation angle, ϕ, is comparable to or larger than the in-plane misorientation angle, θ. The two data points at θ = 5°, which are for samples with ϕ = 0° (circle) and 9° (square), exhibit qualitatively different behavior. While the former exhibits little supression of J_c and no low field sensitivity, the latter behaves as a weak link with a J_c ratio of about 1/10. Such a comparison illustrates the importance of additional degrees of misorientation but at the time of writing this has not been quantified.

In general, a superconducting weak link is a section of a superconductor in which the order parameter, Δ, is strongly depressed, but which can nevertheless support a supercurrent. The most common types of weak links in conventional superconductors are formed by having a thin layer of a nonsuperconducting phase separating two bulk

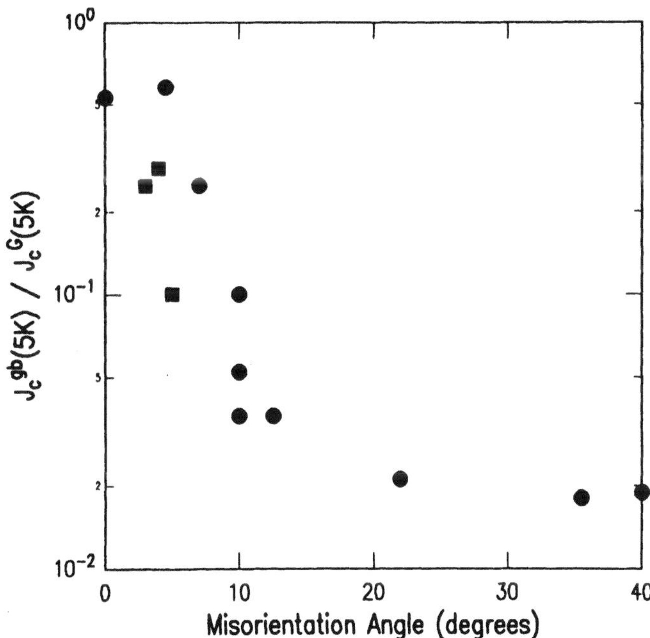

Figure 11. Ratio of the grain-boundary critical current density to the average value of the critical current density in the two adjacent grains as a function of the misorientation angle in the basal plane. It should be noted that 45° is the largest misorientation angle, even in this orthorhombic crystal, due to fine-scale twinning on (110) planes.

superconductors. For very thin (10-30 Å) insulating barriers, the supercurrent transport is due to tunneling of electron pairs; thus, these superconductor-insulator-superconductor (SIS) sandwiches are often referred to as tunnel junctions. In the case of a normal metal, a weak form of superconductivity is induced in the normal metal layer (due to the proximity effect) so that the order parameter is less severely depressed than in a tunnel junction. Consequently, the critical current densities of these SNS structures can be much higher than for SIS junctions. Weak links can also be formed by creating localized regions of damage (e.g., by ion implantation), which decreases the condensation energy, or by forming a small (essentially point) contact between bulk superconductors.

While the relationship between grain boundary structure and critical current density remains to be established, considerable insight to the origin of the weak-link behavior can be made be considering a grain boundary as a region of structural disorder in which a local depression of the pair potential can occur. As discussed previously, the order parameter will be strongly suppressed in the vicinity of such a structural defect only if the effective width of the defect, ω, is comparable to 2ξ. While the effective structural width of a grain boundary is not a well-defined concept, diffraction experiments on a variety of materials indicate that the lateral extent of the displacement perpendicular to the grain boundary is largely confined to within 1-2 lattice spacings on each side of the boundary plane for high-angle boundaries[33]. While 2ξ is generally at least an order of magnitude larger than typical lattice dimensions in conventional superconductors, this is not necessarily true in the high-T_c materials. For the [001] tilt boundaries in $YBa_2Cu_3O_7$, the appropriate lattice spacing is $\simeq$ 3.85 Å; thus, the effective grain boundary width for large angle boundaries is estimated to be in the range of $\simeq$ 10-15 Å. A reliable and generally accepted value for the coherence length has not yet been determined for any of the cuprate-based superconductors, since there is disagreement as to the value of the thermodynamic upper critical field. The magnetic anomaly commonly reported as H_{c2} (it is actually associated with an irreversibility line) can, however, be taken as a lower limit for the true value of H_{c2} [34,35]. From these measurements an upper limit for the coherence length can be obtained which for $YBa_2Cu_3O_7$ gives an upper limit for the basal plane coherence length (parallel to the copper-oxygen planes) of $\xi_{ab} \simeq$ 20-30 Å. However, even this rather small value of the coherence length appears somewhat too large to account for the observed weak-link behavior, which suggests that ξ may be smaller still.

While high-angle grain boundaries ($\theta \geq 10°$) always act as weak links, strong coupling between adjacent grains is maintained for low-angle grain boundaries ($\theta \leq 5°$). This result is not suprising since the degree of structural mismatch at the boundary decreases as the misorientation angle decreases. As $\theta \rightarrow 0$, the grain boundary J_c must approach the grain J_c, as observed. Describing low-angle boundaries as consisting of an array of parallel edge dislocations with a spacing, d, of $d = b/\theta$ with $b \simeq 3.85$ Å ($\mathbf{b}$ = [100] or [010])[36], the dislocation density decreases smoothly with decreasing θ. It is unclear whether the variation in grain boundary critical current density can be understood in terms of dislocation models. However, it is noted that if the size of a dislocation core is b, they would overlap when $d \simeq 2b$ and this would occur for $\theta \simeq 20°$, the value corresponding to the angle for which J_c^{gb}/J_c^G reaches its minimum value.

Although there is uncertainty about the absolute values of ξ, it has been established that the coherence lengths in all of the cuprate superconductors are highly anisotropic. In $YBa_2Cu_3O_7$, the value of ξ_c (i.e., parallel to the c axis) may be only 2-4 Å, which is as much as an order of magnitude shorter than ξ_{ab}. Consequently, the depression of Δ and, thus, the reduction of J_c should be most severe when the grain boundary normal is parallel to [001], but this has yet to be tested. Chemical segregation or deviations from the bulk stoichiometry in the vicinity of the grain boundary could also lead to an additional reduction in the local condensation energy although this too has yet to be established. This additional contribution may be particularly important for $(La_{1-x}Sr_x)_2CuO_4$, since the dopant species, Sr, preferrentially segregates to grain boundaries[37] Although such solid solution segregation of cations should not occur in the stoichiometric compound $YBa_2Cu_3O_7$, there is some evidence that local deviations in stochiometry can occur at grain boundaries[38-40]. It should, however, be noted that not all investigations are in agreement on this point and that these measurements are quite difficult.

IV.2. IMPLICATIONS

The results of the work to date have important implications for the attainment of high critical current densities in polycrystalline samples. For instance, a number of investigators have assumed that by texturing polycrystalline samples such that the c axes of the grains are aligned, with the CuO planes being parallel, very high critical current densities could be obtained. However, even for samples with almost perfect c-axis alignment, misorientations between the principal in-plane directions produce high-angle [001] tilt boundaries that on the basis of these experiments will limit the bulk J_c. Although this conclusion may seem to be rather restrictive, this data suggests that the critical current densities of such uniaxially textured samples should be at least as large as the values for the high-angle grain boundaries, which exceeded 10^4 A/cm^2 at 77 K (in the best samples). In fact, critical current densities in this range, which might be quite acceptable for some applications, have been achieved by producing highly textured microstructures [41,42]. However, to achieve critical current densities which approach the single crystal values, a high degree of alignment both normal to and within the basal plane must be achieved; this is an unprecedented texturing requirement for a polycrystalline material.

While these experiments directly show that grain boundaries in $YBa_2Cu_3O_7$ act as superconducting weak links, there is a variety of evidence which suggests that this behavior is generic to all the cuprate superconductors. For instance, the critical current densities of bulk polycrystalline samples of La-Sr(Ba)-Cu-O, Bi-Sr-Ca-Cu-O, and Tl-Ba-Ca-Cu-O compounds are typically both low and highly field dependent[17,43,44]. Measurements of the complex a.c. susceptibility in bismuth- and thallium-cuprate superconductors also exhibit loss peaks (peaks in the imaginary susceptibility) associated with weak intergrain coupling that are similar to those observed for polycrsytalline $YBa_2Cu_3O_7$[45,46]. However, quite high critical current densities have been reported for polycrystalline thin films of $Tl_2Ba_2CaCu_2O_y$[47] (J_c > 1x10^5 A/cm^2). But since these films have a strong tendency to grow with textured or even highly-oriented microstructures under a wide variety of conditions, the large values may be the result of their preferential platey growth morphology rather than being indicative of large grain boundary critical current densities.

If the intergranular weak-link behavior exhibited by the cuprate superconductors is simply due to the structural change at the grain boundaries, and is not related to the seemingly unique nature of the superconductivity mechanism in these materials, then other superconductors with very small coherence lengths should behave similarly. Consequently, the Chevrel phase superconductors, which are ternary molybdenum chalcogenides, should make an interesting comparison[48]. In particular, the coherence length and lattice parameter of $PbMo_6S_8$ are about 25 Å and 10Å, respectively. Even with such a short coherence length, the properties of polycrystalline $PbMo_6S_8$ appear quite similar to those of more conventional superconductors. For example, reasonably high critical current densities with conventional field dependences have been obtained for $PbMo_6S_8$[49]. An inverse relationship between critical current density and grain size has also been reported, which suggests that grain boundaries do act as centers for flux pinning in these ceramics. However, such a relationship does not rule out there being a depression of the order parameter across grain boundaries, since in the absence of strong intragranular pinning, weak-link behavior may well be masked by the dominant role of grain boundaries as pinning sites.

V. Conclusions

The critical current densities of superconductors are strongly influenced by their microstructures. In conventional superconductors, microstructural defects (e.g., grain boundaries, nonsuperconducting phases, and voids) are important as pinning centers for magnetic flux lines leading to increased critical current densities. The effectiveness of grain boundaries as pinning sites is determined by both its misorientation angle and by its orientation with respect to an applied magnetic field.

However, when the coherence length of the superconductor approaches a dimension characteristic of the structure of a grain boundary, as appears to be the case for the cuprate-based high-T_c superconductors, the boundary can act as a superconducting weak link. In the limit of strong intragranular flux pinning, these grain-boundary weak links control the critical current density of polycrystalline samples. This weak-link behavior, which may be an intrinsic property of grain boundaries in cuprate superconductors, has important consequences for the potential usefulness of these revolutionary new materials.

ACKNOWLEDGEMENTS. The authors wish to acknowledge P. Chaudhari and J. Mannhart, with whom most of the experiments on the single grain boundaries were done. We also thank T.M. Shaw, C.C. Tsuei, and R.F. Cook for advice and comments.

VI. References

1. A.C. Rose-Innes and E.H. Rhoderick, *Introduction to Superconductivity,* (Pergamon Press, New York, 1978).
2. P.G. DeGennes, *Superconductivity of Metals and Alloys,* (W.A. Benjamin, New York, 1966).
3. R.D. Parks, *Superconductivity Vol. I and II,* (M. Dekker, New York, 1969).
4. C.C. Koch, H.C. Freyhardt, and J.O. Scarbrough, 'Fluxoid Pinning in Bulk Niobium by Voids Produced During Neutron Irradiation', *IEEE Trans. Mag.*, **MAG-13,** 828 (1977).
5. R.M. Scanlan, W.A. Fietz, and E.F. Koch, 'Flux Pinning Centers in Superconducting Nb_3Sn', *J. Appl. Phys.*, **46,** 2244 (1975).
6. K.E. Gray, R.T. Kampwirth, J.M. Murduck, and D.W. Capone II, 'Experimental Study of the Ultimate Limit of Flux Pinning and Critical Currents in Superconductors', *Physica C,* **152,** 445 (1988).
7. P.J. Lee and D.C. Larbalestier, 'Development of Nanometer Scale Structures in Composites of Nb-Ti and Their Effect on the Superconducting Critical Current Density', *Acta. Metall.*, **35,** 2523 (1987).
8. E.J. Kramer and H.C. Freyhardt, 'The Proximity Effect in Flux Pinning', *J. Appl. Phys.*, **51,** 4930 (1980).
9. A. DasGupta, C.C. Koch, D.M. Kroeger, Y.T. Chou, 'Flux Pinning by Grain Boundaries in Niobium Bicrystals', *Phil. Mag. B,* **38,** 367 (1978).
10. G. Zerweck, 'On Pinning of Superconducting Flux Lines by Grain Boundaries', *J. Low Temp. Phys.*, **42,** 1 (1981).
11. W.E. Yetter, D.A. Thomas, and E.J. Kramer, 'Grain-Boundary Flux Pinning by the Electron-Scattering Mechanism', *Phil. Mag. B,* **46,** 523 (1982).
12. E.V. Thuneberg, J. Kurkijärvi, and D. Rainer, 'Elementary-Flux-Pinning Potential in Type II Superconductors', *Phys. Rev.* **B29,** 3913 (1984).
13. J.G. Bednorz and K.A. Müller, 'Possible, High-T_c Superconductivity in La-Ba-Cu-O System', *Z. Phys. B* **64,** 189 (1986).
14. R. Beyers and T.M. Shaw, 'The Structure of $Y_1Ba_2Cu_3O_{7-\delta}$ and its Derivatives', p. 135-212 in *Solid State Physics, Vol. 42,* (Academic Press, London, 1989).
15. D.C. Larbalestier, M. Daeumling, X. Cai, J. Seuntjens, J. McKinnell, D. Hampshire, P. Lee, C. Meingast, T. Willis, H. Muller, R.D. Ray, R.G. Dillenburg. E.E. Hellstrom, and R. Joynt, 'Experiments Concerning the Connective Nature of $YBa_2Cu_3O_7$', *J. Appl. Phys.*, **62,** 3308 (1987).

16. S. Jin, R.A. Fastnacht, T.H. Tiefel, and R.C. Sherwood, 'Transport Critical Current in Rare-Earth-Substituted Superconductors $RBa_2Cu_3O_{7-\delta}$ (R= Gd, Dy, Sm, Ho, Y)', *Phys. Rev.* **B37** 5828 (1988).

17. K. Togano, H. Kumakura, and D.R. Dietderich, 'Critical Currents and Magnetic Properties of Bi- and Tl-Based New High T_c Superconductors', *Cryogenics*, **29**, 286 (1989).

18. T.R. Dinger, T.K. Worthington, W.J. Gallagher, and R.L. Sandstrom, 'Direct Observation of Electronic Anisotropy in Single-Crystal $Y_1Ba_2Cu_3O_{7-x}$ ', *Phys. Rev. Lett.*, **58**, 2687 (1987).

19. P. Chaudhari, R.H. Koch, R.B. Laibowitz, T.R. McGuire, and R.J. Gambino, 'Critical Current Measurements in Epitaxial Films of $YBa_2Cu_3O_{7-x}$ Compound', *Phys. Rev. Lett.*, **58**, 2684 (1987).

20. G. Koren, A. Gupta, E.A. Giess, A. Segmüller, and R.B. Laibowitz, 'Epitaxial Films of $YBa_2Cu_3O_{7-\delta}$, on $NdGaO_3$, $LaGaO_3$, and $SrTiO_3$ Substrates Deposited by Laser Ablation', *Appl. Phys. Lett.*, **54**, 1054 (1989).

21. A. Inam, M.S. Hegde, X.D. Wu, T. Venkatesan, P. England, P.F. Miceli, E.W. Chase, C.C. Chang, J.M. Tarascon, and J.B. Wachtman, 'As-Deposited High T_c and J_c Superconducting Thin Films Made at Low Temperatures', *Appl. Phys. Lett.*, **53**, 908 (1988).

22. R.B. van Dover, L.F. Schneemeyer, E.M. Gyorgy, and J.V. Waszczak, 'Critical Current Densities in Single-Crystal $Bi_{2.2}Sr_2Ca_{0.8}Cu_2O_{8+\delta}$', *Appl. Phys. Lett.*, **52**, 1910 (1988).

23. T.R. McGuire, S.A. Shivashankar, S.J. LaPlaca, G.V. Chandrashekhar, R.F. Boehme, T.M. Shaw, D.S. Lee, M.W. Shafer, and J.J. Cuomo, 'Superconductivity in Bi-Sr-Ca-Cu-O', *J. Appl. Phys.*, **64**, 5792 (1988).

24. G.J. Dolan, G.V. Chandrashekhar, T.R. Dinger, C. Feild, and F. Holtzberg, 'Vortex Structure in $YBa_2Cu_3O_7$ and Evidence for Intrinsic Pinning', *Phys. Rev. Lett.*, **62**, 827 (1989).

25. C.P. Bean, 'Magnetization of Hard Superconductors', *Phys. Rev. Lett.*, **8**, 250 (1962).

26. J. Mannhart, unpublished results.

27. H. Küpfer, I. Apfelstedt, W. Schauer, R. Flükiger, R. Meier-Hirmer, and H. Wühl, 'Critical Current and Upper Critical Field of Sintered and Powdered Superconducting $YBa_2Cu_3O_7$', *Z. Phys. B* **69**, 159 (1987).

28. J.F. Kwak, E.L. Venturini, and D.S. Ginley, 'The Consequences of Intergrain Josephson Coupling in Ceramic High-Temperature Superconductors', *Physica B*, **148**, 426 (1987).

29. R.L. Peterson and J.W. Ekin, 'Josephson-Junction Model of Critical Current in Granular $Y_1Ba_2Cu_3O_{7-\delta}$ Superconductors', *Phys. Rev.* **B37**, 9848 (1988).

30. P. Chaudhari, J. Mannhart, D. Dimos, C.C. Tsuei, J. Chi, M.M. Oprysko, and M. Scheuermann, 'Direct Measurement of the Superconducting Properties of Single Grain Boundaries in $Y_1Ba_2Cu_3O_{7-\delta}$ ', *Phys. Rev. Lett.*, **60**, 1653 (1988).

31. D. Dimos, P. Chaudhari, J. Mannhart, and F.K. LeGoues, 'Orientation Dependence of Grain-Boundary Critical Currents in $YBa_2Cu_3O_{7-\delta}$ Bicrystals', *Phys. Rev. Lett.*, **61**, 1653 (1988).

32. J. Mannhart, P. Chaudhari, D. Dimos, C.C. Tsuei, and T.R. McGuire, 'Critical Currents in [001] Grains and Across Their Tilt Boundaries in $YBa_2Cu_3O_7$ Films', *Phys. Rev. Lett.*, **61**, 2476 (1988).

33. C.B. Carter and S.L. Sass, 'Electron Diffraction and Microscopy Techniques for Studying Grain Boundary Structure', *J. Am. Ceram. Soc.*, **64,** 335 (1981).

34. A.P. Malozemoff, T.K. Worthington, Y. Yeshurun, F. Holtzberg, and P.H. Kes, 'Frequency-Dependence of the A.C. Susceptibility in a Y-Ba-Cu-O Crystal - A Re-interpretation of H_{c2}', *Phys. Rev.* **B38,** 7203 (1988).

35. A.P. Malozemoff, 'Macroscopic Magnetic Properties of High Temperature Superconductors', pp. 71-150 in *Physical Properties of High Temperature Superconductors I.* edited by D.M. Ginsberg (World Scientific, Singapore, 1989).

36. M.F. Chisholm and D.A. Smith, 'Low-Angle Tilt Grain Boundaries in $YBa_2Cu_3O_7$ Superconductors', *Phil. Mag. A,* **59,** 181 (1989).

37. Y.-M. Chiang, D.A. Rudman, D.K. Leung, J.A.S. Ikeda, A. Roshko, and B.D. Fabes, 'Effects of Grain Size and Grain-Boundary Segregation on Superconducting Properties of Dense Polycrystalline $La_{1.85}Sr_{0.15}CuO_4$ ', *Physica C,* **152,** 77 (1988).

38. D.M. Kroeger, A. Choudhury, J. Brynestad, R.K. Williams, R.A. Padgett, and W.A. Coghlan, 'Grain-Boundary Compositions in $YBa_2Cu_3O_{7-x}$ from Auger Electron Spectroscopy of Fracture Surfaces', *J. Appl. Phys.*, **64,** 331 (1988).

39. S.E. Babcock, T.F. Kelly, P.J. Lee, J.M. Seuntjens, L.A. Lavanier, and D.C. Larbalestier, 'Investigations of Composition Variations near Grain Boundaries in High-Quality Sintered Samples of $YBa_2Cu_3O_{7-\delta}$ ', *Physica C,* **152,** 25 (1988).

40. Y.-M. Chiang, J.A.S. Ikeda, A. Roshko, 'Grain Boundary Segregation and Critical Current Density in $YBa_2Cu_3O_{7-x}$ Superconductors', pp. 607-618 in *Ceramic Superconductors II, Research Update 1988.* edited by M.F. Yan (American Ceramic Society, Westerville, OH, 1988).

41. S. Jin, T.H. Tiefel, R.C. Sherwood, R.B. van Dover, M.E. Davis, G.W. Kammlott, and R.A. Fastnacht, 'Melt-Textured Growth of Polycrystalline $YBa_2Cu_3O_{7-\delta}$ with High Transport J_c at 77 K', *Phys. Rev.* **B37,** 7850 (1988).

42. T. Hikata, K. Sato, H. Hitotsuyanagi, 'Ag-Sheathed Bi-Pb-Sr-Ca-Cu-O Superconducting Wires with High Critical Current Density', *Jap. Jour. Appl. Phys. Part 2-Letters,* **28,** L82 (1989).

43. G.R. Wagner, J. Talvacchio, and A.J. Panson, 'Critical Currents of 85-K Bi-Sr-Ca-Cu-O', *Materials Lett.*, **6,** 390 (1988).

44. H. Küpfer, S.M. Green, C. Jiang, Yu Mei, H.L. Luo, R. Meier-Hirmer, and C. Politis, 'Weak Link Problem and Intragrain Current Density in Polycrystalline $Bi_1Ca_1Sr_1Cu_2O_6$ and $Tl_2Ca_2Ba_2Cu_3O_{10}$', preprint - to be published in *Z. Phys. B - Condensed Matter.*

45. H. Kumakura, K. Togano, K. Takahashi, E. Yanagisawa, M. Nakao, and H. Maeda, 'A.C. Complex Susceptibility Measurement in Sintered Bi-Sr-Ca-Cu-O and Tl-Ba-Ca-Cu-O Superconductors', *Jap. Jour. Appl. Phys.*, **27,** L2059 (1988).

46. H. Kumakura, K. Togano, H. Maeda, E. Yanagisawa, T. Morimoto, 'Critical Current Density and Coupling Nature of Grains in Textured Bi(Pb)-Sr-Ca-Cu-O Superconductors', *Jap. Jour. Appl. Phys.*, **28,** L176 (1989).

47. J.F. Kwak, E.L. Venturini, R.J. Baughman, B. Morosin, and D.S. Ginley, 'High, Magnetic-Field Insensitive Transport Critical Currents in Unoriented Polycrystalline Thin Films of $Tl_2CaBa_2Cu_2O_y$ ', *Physica C,* **156,** 103 (1988).

48. O. Fischer, 'Chevrel Phases: Superconducting and Normal State Properties', *Appl. Phys.*, **16,** 1 (1978).

49. C. Rossel and O. Fischer, 'Critical Current Densities in Bulk Chevrel-Phase Samples', *J. Phys. F: Met. Phys.*, **14,** 455 (1984).

GROWTH AND MASS TRANSPORT IN CERAMIC TYPE PROTECTIVE SCALES ON METALS

W.W. Smeltzer
Institute for Materials Research
McMaster University
Hamilton, Ontario, L8S 4M1, Canada

ABSTRACT. A survey is presented of growth and mass transport in ceramic type protective scales on metals and alloys at high temperatures by lattice and short-circuit diffusion of metal and oxygen through low resistance diffusion paths in oxide boundaries. A methodology based on correlating phenomenological theory of oxidation with reactant short-circuit diffusion across reaction product layers is discussed in some detail for Cr_2O_3, Al_2O_3 and SiO_2 scales on the pure metals and on iron and nickel alloys because independent measurements of diffusivities by isotopes are available to confirm the oxidation models. Isotope studies using sequential oxidation in O^{16} and O^{18} have been definitive in describing oxide growth within protective Cr_2O_3 and Al_2O_3 scales. Mechanisms relating loss of scale adherence to the metal substrate with the formation of new oxide within these protective scales are discussed and they are shown to remain open to speculation.

INTRODUCTION

An understanding of the growth, mass transport and integrity of Al_2O_3, Cr_2O_3 and SiO_2 scales is of critical importance because these scales have proven to be most significant for development of high temperature corrosion resistance alloys and metal coatings on structural alloys.

The point defect structure of Cr_2O_3 which behaves as an intrinsic semiconductor has not been completely elaborated but probably consists of cation intersititials, cation and anion vacancies. At 1000°C chromium diffusion by low angle boundaries and dislocations is $\tilde{}10^4$ more rapid than lattice diffusion. The parabolic oxidation rate constants governing scale growth from different investigators vary as much as three orders of magnitude, a behavior believed to be associated with variations in scale adherence, oxide grain size and porosity. Metal diffusion along the oxide grain boundaries will be shown to play a pivotal role in the growth of the scales.

The point defect structure of α-Al_2O_3 which behaves intrinsically as an

L.-C. Dufour et al. (eds.), Surfaces and Interfaces of Ceramic Materials, 319–336.

ionic conductor is largely determined by the nature of impurities in the host lattice. Both Al and O are mobile, the diffusion coefficient of Al is approximately one order of magnitude greater than that of O in polycrystalline oxide. Diffusion of oxygen is structurally sensitive, its diffusivity in polycrystalline oxide being 10^2-10^4 larger than for single crystal oxide. In the case of Al_2O_3 layer growth, outward aluminum and inward oxygen transport occur predominantly by grain boundary diffusion to a degree dependent on the type of alloy but the prevailing species appears to be oxygen.

SiO_2 is a network type oxide often existing at high temperatures for long exposures in the vitreous state. Diffusion and permeation measurements indicate that interstitital oxygen is the prevailing point defect and that this species interacts and exchanges with network oxygen during growth of SiO_2 scales on pure Si at high temperatures. The problem with the formation of protective SiO_2 films on alloys is that the Si content required for its exclusive growth is very high and corresponds to alloys with poor mechanical properties. Silicon as a minor alloying element in Fe-Cr alloys does lead to growth of healing passive type barriers beneath protective Cr_2O_3 scales.

The purpose of this contribution , accordingly, is threefold: first, to demonstrate that a sound empirical knowledge base is under active development for describing the growth mechanisms of Cr_2O_3, Al_2O_3 and SiO_2 scales on the pure metals and iron-nickel alloys; second, to illustrate present findings on the "reactive rare earth metal effect" on Cr_2O_3 and Al_2O_3 scale growth and adhesion and, third, to document the need for a deeper understanding of factors influencing diffusion mechanisms, scale microstructures and adhesion resulting from segregation of alloying elements at oxide boundaries.

OXIDE GROWTH MODELS

Metals and alloys are frequently covered during early exposure by microcrystalline films resulting from the growth processes and the metal-film system minimizing strain. Ion transport across films at low temperatures proceeds under the influence of an internal electric field. Thermal energy, however, is sufficient to promote oxide growth by chemical diffusion at elevated temperatures because the ratio of boundary to lattice diffusion is as high as 10^4-10^6 with ratios of activation energies ranging from 0.4-0.7. An extensive body of research now demonstrates that short-circuit diffusion of reactants by oxide grain boundaries can determine the oxidation rates of metals and alloys at moderately elevated temperatures ranging up to 1000°C.

At sufficiently high temperatures, the oxide grain size may become sufficiently large for volume diffusion by interstitials or vacancies in the superficial oxide layer to determine the reaction kinetics. These oxidation kinetics for a pure metal are then amenable to analysis by theory advanced by Wagner [1,2] based upon ambipolar diffusion of the reactants which leads to parabolic kinetics for scale growth

$$X^2 = k_p t + B \qquad (1)$$

where X is the oxide layer thickness at time t, k_p is the parabolic oxidation rate constant and B is a constant. If the flux is regarded as divergence free and the scale is a metal conducting semiconductor (transport number of electrons ~1), k_p is proportional to the parabolic oxidation rational rate constant

$$k_r(eq/cm\ s) = \bar{C} \int_{a_o(o)}^{a_o(x)} D_M \, d\ell n a_o \qquad (2)$$

here, D_M represents the metal diffusivity of the oxide. $\bar{C}$ is the average oxide composition and a_o is the oxygen activity at the inner (o) and outer (x) scale interfaces. Interpretation of scaling by means of the Wagner theory is treated in depth in several texts and recent reviews [3].

An essential property to be considered is that oxide films and scales which are formed at elevated temperatures, but less than one-half the oxide melting point, usually contain grains of sufficiently small size such that boundaries and dislocations act to short-circuit lattice diffusion. The actual reaction kinetics are then several orders of magnitude larger than those predicted by lattice diffusion. Utilizing an analysis given by Hart [4] for short-circuit diffusion (random walk of species via lattice sites and internal boundaries), Smeltzer and co-workers [5] assigned a single effective volumetric diffusion coefficient D_{eff} to describe reactant transport across an oxide layer,

$$D_{eff} = D_L(1-f) + D_B f \qquad (3)$$

where D_L and D_B are diffusivities for the lattice and internal boundaries, respectively, and f is the fraction of the total available diffusion sites lying in the low resistance dislocation and boundary paths. If oxidation proceeds by chemical diffusion,

$$\frac{dX}{dt} = \Omega D_{eff} \frac{\Delta c}{X} \qquad (4)$$

which gives the following modified parabolic oxidation rate equation to describe growth of the oxide layer

$$X^2 = 2 \Omega D_L \Delta c \int_o^t (1 + \frac{D_B}{D_L} f)dt = k_p \int_o^t (1 + \frac{D_B}{D_L} f)dt = k_p(eff)t \qquad (5)$$

In these expressions, X is the layer thickness, Ω is the volume of oxide formed by a diffusing particle and $\Delta c/X$ is the concentration gradient across the oxide layer. Since the parabolic rate constant $k_p = 2\Omega D_L \Delta c$ represents ambipolar lattice diffusion of metal or oxygen by point defects, equation (5) describes an oxidation curve exhibiting parabolic relationships over two limiting intervals. If the oxide grain size remains practically unchanged during early exposure, the oxidation curve conforms to a parabolic relationship in which the effective rate constant contains terms for boundary and lattice diffusion. At long time, the classical parabolic oxidation relationship is obtained for lattice diffusion when the large density of boundaries is decreased by grain growth processes.

Protective Cr_2O_3, Al_2O_3 and SiO_2 scales grow on iron and nickel alloys as a result of the selective oxidation of the appropriate alloying element. For example, the standard free energies for formation of Al_2O_3, SiO_2 and Cr_2O_3 per mole of oxygen at 1273 K are in the order -202, -154 and -126 kcal which are to be compared to -86 and -60 kcal for formation of FeO and NiO, respectively. The scale-alloy diffusion model advanced by Wagner in 1952 [6] and depicted schematically in Fig.1 has been extensively applied by investigators to interpret growth of these protective scales at high temperatures. If parabolic growth of the

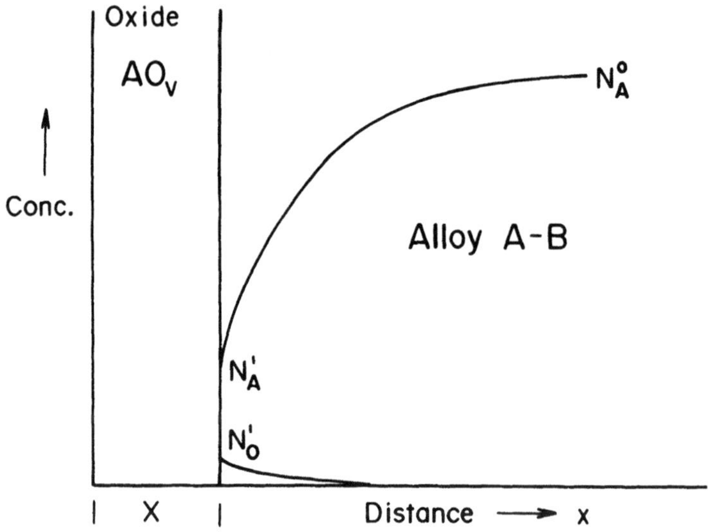

Figure 1. The oxide scale/alloy model with symbols corresponding to those in the text for selective oxidation of the alloying element A. After Wagner [6].

scale results from metal and oxygen diffusion in the alloy, the concentration profiles in the alloy (Fig.1) for concentration-independent diffusivities are given by the following expressions:

$$N_A = N_A{}^o - (N_A{}^o - N_A') \frac{\text{erfc } x/2\sqrt{D_{AB}t}}{\text{erfc} \sqrt{k/2D_{AB}}} \; ; \; x \geq X \tag{6}$$

$$N_o = N_0' \frac{\operatorname{erfc} x/2\sqrt{D_o t}}{\operatorname{erfc}\sqrt{k/2D_o}} \; ; \; x \geq X \qquad (7)$$

Here, N_A^o, N_A', and N_A are the atom fractions of A in the bulk alloy, at the alloy/scale interface, and locally at distance x and time t, respectively, where position x is measured from the original alloy surface. N_o' and N_o are the corresponding atom fractions of oxygen. D_{AB} and D_o are the interdiffusion coefficient and the oxygen diffusion coefficient in the alloy, respectively. The position of the actual alloy surface X at any time is related to the corrosion constant, k, by

$$X = \sqrt{2kt} \qquad (8)$$

Since pure oxide, AO_y is assumed to form, a mass balance at the alloy/scale interface leads to

$$N_A' = \frac{N_A^o - F(u)}{1 - F(u)} \qquad (9)$$

where u replaces $\sqrt{k/2D_{AB}}$ and the auxiliary function F(u) is given by

$$F(u) = \sqrt{\pi}\, u \exp u^2 \operatorname{erfc} u \qquad (10)$$

The following equilibrium defines the relationship between the oxide and alloy components at the alloy/oxide interface

$$AO_v = A(\text{alloy}) + vO\ (\text{alloy}) \qquad (11)$$

$$K = N_A'.N_o'^{v} \text{ at } x = X \qquad (12)$$

Accordingly, internal oxidation occurs when the product $N_A.N_o'^{v}$ exceeds the solubility constant at positions within the alloy, away from the interface.

Equation (6)-(12) can be utilized to define the limiting atom fraction of the alloying element above which internal oxidation would not occur (concentrations are only sufficient for satisfying the solubility product right at the scale/alloy interface):

$$N_A^o \geq \frac{F(u) + \frac{1}{v}\left(\frac{D_o}{D_{AB}}\right)^{1/2} u\pi^{1/2}}{1 + \frac{1}{v}\left(\frac{D_o}{D_{AB}}\right)^{1/2} u\pi^{1/2}} \qquad (13)$$

More recent studies have been carried out to demonstrate that this critical alloy concentration for prevention of internal oxidation can be decreased as a result of attractive non-ideal thermodynamic solution interaction of oxygen with alloying elements [7] and more rapid diffusion of the selectively oxidized alloying element to the alloy interface brought about by repulsive alloying interactions [8].

OXIDE GROWTH ON PURE METALS

Short-circuit diffusion of reactants in scales has led to a much more complete understanding of the oxidation mechanisms of metals at temperatures less than one-half the values for melting of the formed oxides. Mass transport remains relatively rapid across the scales because of the large retained density of boundaries even though lattice point defect concentrations are vanishingly small [9,10]. These considerations apply to the growth of Cr_2O_3 on Cr up to 1000°C. On the other hand, SiO_2 remains vitreous up to 1000°C, and its growth on Si is determined by volume diffusion within the SiO_2 network. Aluminum metals at 660°C; consequently, investigations on oxidation are confined to relatively moderate temperatures where growth of both amorphous and polycrystalline allotropes of Al_2O_3 occur.

Cr/Cr_2O_3: Recently presented reviews [9,11] summarize the many attempts from as early as 1965 to correlate the oxidation properties of chromium with diffusion of the reactants across the Cr_2O_3 scale. The measurements of chromium tracer diffusion coefficients for Cr_2O_3 single crystals and of isotopic oxygen diffusion in growing polycrystalline Cr_2O_3 scales support the conclusion that chromium diffusion accounts predominantly for oxide growth. The point defect structure of Cr_2O_3 which behaves as an intrinsic semiconductor, has not been completely elaborated but probably consists of cation interstitials, cation and anion vacancies. At 1000°C, chromium diffusion by low angle boundaries and dislocations is $\sim 10^4$ more rapid than lattice diffusion. The parabolic oxidation rate constants governing scale growth from different investigations vary by as much as three orders of magnitude, a behaviour believed to be associated with variations in scale adherence, oxide grain size and porosity. The temperature coefficient of these results over the range 500°-1200°C is given in Fig.2. An estimate by Atkinson and Taylor [12] of the parabolic rate constant using the Wagner expression for chromium lattice diffusion in Cr_2O_3, eqn. (2), is also included in this figure. A comparison of the experimental and calculated rate constants at 1100°C clearly demonstrates that metal lattice diffusion cannot account for the rapid growth of these polycrystalline oxide scales. One thus concludes that metal diffusion along the oxide grain boundaries determines growth of these scales. Upon prolonged exposure at high temperature, the Cr_2O_3 scale evolves to a highly convoluted multilayered microstructure as illustrated in Fig.3.

Recent investigations concerning the actual extent of oxygen migration through compact and convoluted Cr_2O_3 scales using SIMS [14] and NRA [15] techniques for profiling of oxygen isotopes, respectively, have clearly demonstrated that small quantities of new oxide are actually formed within the oxide scale and mostly in its outer regions. A model schematically illustrating the preferential growth of this new oxide at grain boundaries in the outer regions of the scale is shown in Fig. 4. These

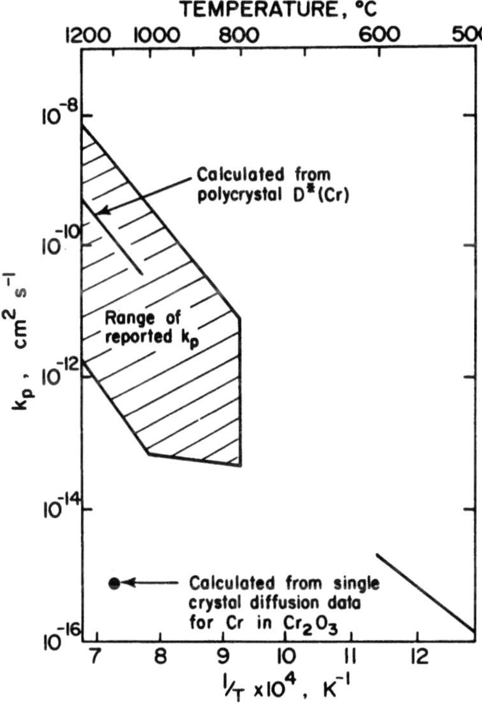

Figure 2. The parabolic oxidation rate constants for growth of Cr_2O_3 on Cr as determined experimentally and by estimates using chromium diffusivities for single crystal and polycrystalline Cr_2O_3. After Atkinson and Taylor [12].

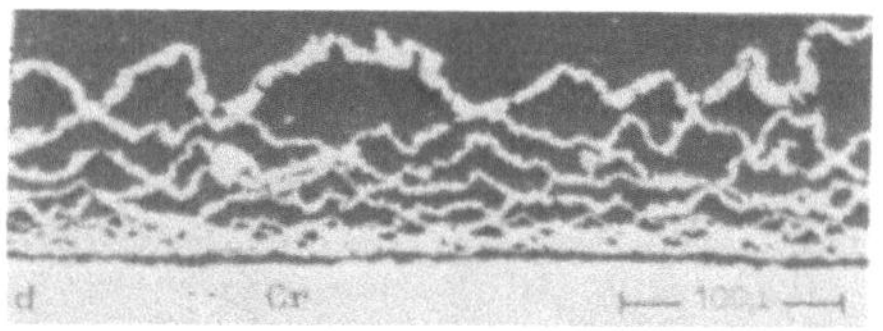

Figure 3. Convulated Cr_2O_3 scale on Cr after 21h exposure at 1090°C. After Caplan and coworkers [13].

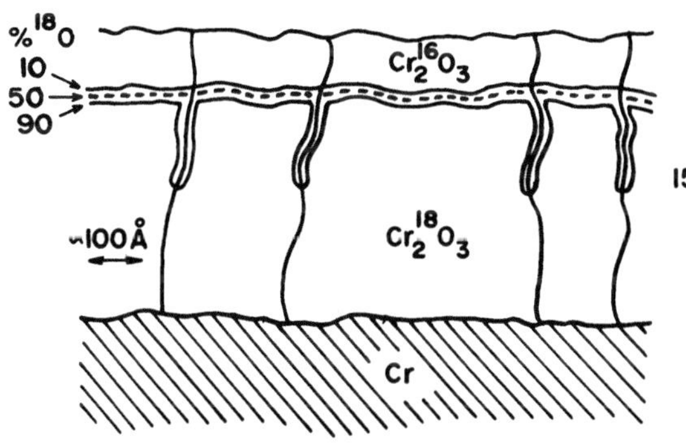

Figure 4. Schematic representation of early-stage Cr_2O_3 scale growth sequentially in O_2^{18} and O_2^{16} atmosphere. After Hussey et al. [14].

isotope techniques, however, do not distinguish how this oxide was produced, e.g. by point defect exchange reactions at grain boundaries or by healing of fissures. Consequently, these experimental findings do not resolve a prevailing consideration that lateral growth of this new oxide at grain boundaries leads to generation of stress and convoluted Cr_2O_3 scale by scale separation at the metal interface [16].

Al/Al_2O_3: The oxide formed on this solid metal above 450°C and to its melting point consists of a thin outer amorphous Al_2O_3 film, underneath which grows γ-Al_2O_3 island growths [17]. An oxidation model therefore must incorporate growth of the outer amorphous Al_2O_3 film, lateral and inward growth of the γ-Al_2O_3 islands. Elridge and coworkers [17] have recently completed an intensive investigation of oxide growth on (110), (111) and (100) Al crystals at 550°C, of oxygen transport using O^{18} SIMS combined with SEM and TEM observations. The protective Al_2O_3 film is very thin ~ 200Å with rates of growth and ultimate thickness dependent upon the Al crystal face orientation.

Different γ-Al_2O_3 island morphologies were associated with each metal substrate orientation; the O isotope results provided strong evidence for oxygen transport and exchange through local channels in the amorphous Al_2O_3 film with oxygen incorporation at the peripheries of the laterally growth γ-Al_2O_3 islands without much diffusion within the islands.

Si/SiO_2: Parabolic growth of the vitreous SiO_2 film and oxygen permeation through this film exhibit a direct dependence on O_2 pressure at temperatures near 1000°C. This behavior has been interpreted to arise from the transport of electrically neutral O_2 molecules through interstitial space of the SiO_2 network structure. Deal and Grove [18] have shown that the parabolic oxidation rate constant is then directly proportional to the diffusivity for interstitial molecular oxygen and the ratio of oxygen interstitial concentration to oxygen ion concentration in the SiO_2 network structure. Cawley and coworkers [19] have recently carried out an investigation on the sequential oxidation of Si using O^{16} and O^{18} isotopes near 1000°C which confirms that oxidation occurs by the transport of electrically neutral interstitial oxygen with a small degree of simultaneous atomic self diffusion in the SiO_2 network, resulting in characteristic isotopic fraction distributions near the gas/scale interface. The Arrhenius plots in Fig.5 demonstrate that the diffusivities determined for oxygen interstitial and network diffusion from oxidation measurements (current work) are in agreement with those obtained from independent diffusivity and permeation measurements.

OXIDE GROWTH ON ALLOYS

Scales formed from Cr_2O_3 and Al_2O_3 are slow growing and ensure useful lifetimes to heat resistant iron and nickel alloys. The primary mechanisms of oxide growth are believed to be short-circuit diffusion of metal and oxygen in Cr_2O_3 and Al_2O_3 scales, respectively, as a result of their inherently small point defect concentrations. Oxide growth kinetics, transport properties, microstructures and adhesion of these scales to the alloys are inherently complex and under certain circumstances, modified by oxide precipitation in the alloys. Hindam and

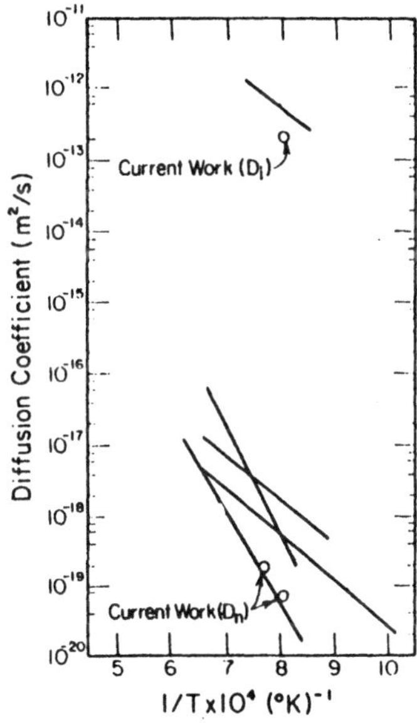

Figure 5. Arrhenius plots of the oxygen diffusivities in vitreous SiO_2. The large diffusivities (D_I) refer to molecular O_2 interstitial diffusion and the band of small values refer to network O self-diffusion. After Cawley and coworkers [19].

Whittle [20] concluded from a survey of the investigations on Cr_2O_3 and Al_2O_3 scale formation on iron and nickel alloys that the reaction kinetics can be represented by an effective parabolic oxidation rate constant as defined for short-circuit diffusion by eqn, (5). In turn, the previously discussed theory for selective oxidation defines the diffusional growth of these polycyrstalline fine-grained scales. This theory has been most intensively applied to growth of Cr_2O_3 scales on Fe-Cr alloys and then only under the condition that the parabolic oxidation rate constant for growth of the scale is known experimentally [21,22].

Cr_2O_3 and Al_2O_3 Scale Growth: The findings substantiating the viewpoint that short circuit diffusion of Cr through the Cr_2O_3 scales on alloys determined the parabolic oxidation kinetics correspond to those previously discussed for oxidation of pure Cr. That short-circuit diffusion of reactants determines the growth of Al_2O_3 scales on alloys is illustrated by the representation in Fig.6 where the parabolic oxidation rate constant was related to D_L and D_{eff} by k_p(eff), eqs. (3) and (5).

It is evident that the magnitude of the scaling rate ≤ 1200°C can be adequately appraised according to short-circuit by grain boundaries with oxygen as the more mobile species. Reddy and coworkers [23] completed direct measurements of the diffusion process in Al_2O_3 scales undergoing growth on Ni-Cr-Al alloys at 1100°C: alternate oxidation exposures in ^{16}O and ^{18}O containing atmospheres revealed that oxygen inward diffusion, presumably by grain boundaries, was the primary mechanism for oxide layer growth.

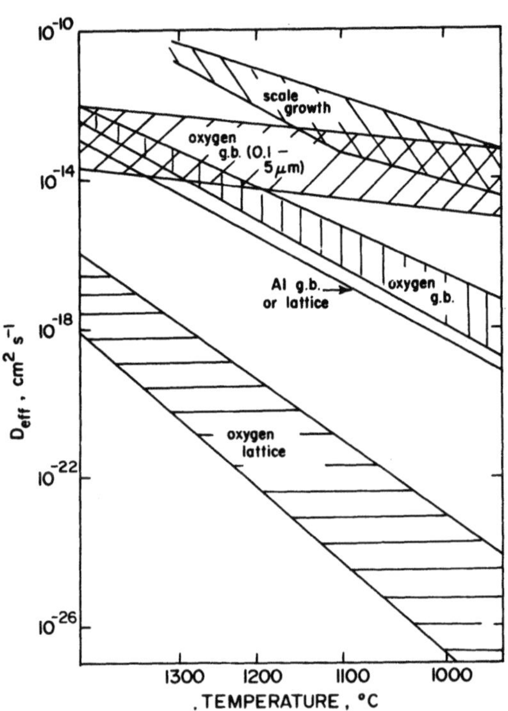

Figure 6. A comparison of lattice and grain boundary diffusivities for aluminum and oxygen diffusion in single-crystal and polycrystalline Al_2O_3 as a function of temperature from different investigations. The calculated values of D_{eff} from parabolic growth kinetics for Al_2O_3 growth on alloys are within or higher than the band of values for oxygen grain boundary diffusivities. After Hindam and Whittle [20].

Transition from Internal to External Oxidation: Criteria to explain this transition are presently based on relatively simple premises of satisfying either the oxide solubility product at the alloy interface (eqn. 13) or diffusional blocking of oxygen by internal oxide precipitates [24]. Interpretation, however, has required a knowledge experimentally of the parabolic scaling rate constant and the critical volume fraction of blocking oxide. The introduction of a percolation model to define this oxide volume using perturbation theory may lead to partial resolution of this problem [25]. Even so, the precipitation of internal oxide is not at random in many cases but in preferentially aligned geometries of lamellar rods or plates. In these cases, fast transport of the reactants at the interfaces can preserve internal oxidation to fairly high volume fractions of precipitates. An example of this behavior occurs upon oxidizing a Ni-6Al alloy at high temperature as illustrated in Fig.7. There are three noteworthy features exhibited here: growth of a duplex scale consisting of an outer columnar NiO layer and an inner NiO + $NiAl_2O_4$ layer, growth of an internal oxidation zone consisting of $NiAl_2O_4$-Al_2O_3 rods, and lastly the growth ultimately of a NiO/$NiAl_2O_4$/Al_2O_3 scale resulting from lateral growth of an Al_2O_3 film at the internal oxidation front. Growth of these types of fully developed scales by diffusion usually occur sufficiently close to local equilibrium such that the spatial distributions of oxide phases can be determined from diffusion paths (composition profiles) on appropriate phase diagrams.

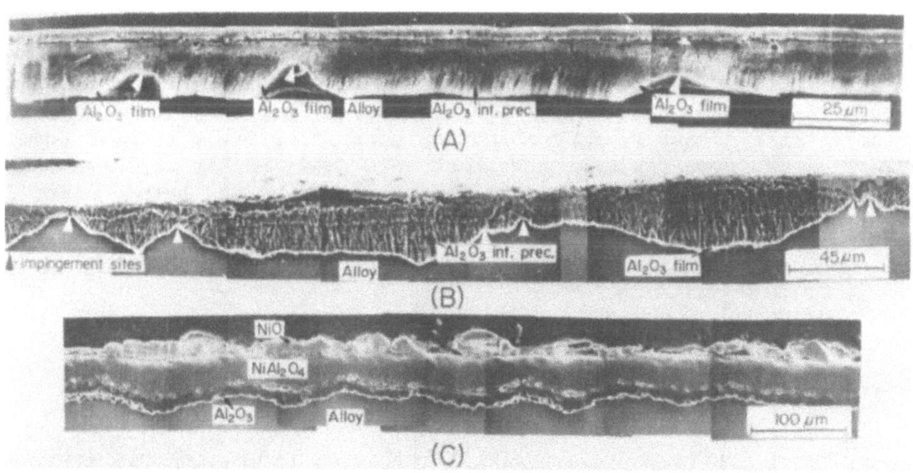

Figure 7. Morphologies of scale and subscale of Ni-6 Al alloy oxidized in 1 atmosphere oxygen at 1000°C: (a) internal oxide precipitation zone and discontinuous Al_2O_3 film at oxidation front, (b) Al_2O_3 continuous film formation at internal oxidation front, (c) fully developed $NiO/NiAl_2O_4/Al_2O_3$ layered scale. After Hindam and Smeltzer [26].

Synergistic Effects of Alloying Elements: In practice, there is a sort of synergistic effect between the alloying elements in forming protective scales on ternary and multicomponent alloys. For example, in the oxidation of Ni-Cr-Al alloys, both alloy elements may contribute to establish the alloy contents above which selective oxidation of the most active element aluminum leads to growth of an Al_2O_3 scale as a consequence possibly of multicomponent metal diffusion interactions in the alloy phase of a repulsive nature increasing the diffusion of aluminum to the alloy/scale interface [27]. Also, research on the oxidation properties of Fe-Cr steels containing silicon has demonstrated that the required amount of silicon for healing layer formation at the alloy/scale interface reduces with increasing chromium concentration [28].

Early studies on the oxidation of ternary Fe-Si-Al alloys indicate that in combination with silicon, the aluminum content required to form a protective Al_2O_3 scale is significantly lower than in binary Fe-Al alloys [29]. Oxidation of binary and ternary Fe-Al-Si alloys containing 0-6 at % Al and 0-5 at % Si in oxygen at 900°C can be classified into distinct modes based upon oxidation rates and scale morphologies as illustrated in Figs. 8 to 10. Iron containing 5-6 at % Al or Si form duplex $Al_2O_3/Fe(Al)_2O_3$ or SiO_2/Fe_2O_3 scales interspersed with thicker $Al_2O_3/FeAl_2O_4$ + Fe_3O_4/Fe_2O_3 or SiO_2/Fe_2SiO_4 + Fe_3O_4/Fe_2O_3 nodules (Fig.9). Iron containing greater than 16 at % Al or Si form highly protective scales containing a complete layer of Al_2O_3 or SiO_2 surmounted by island type film segments of Fe_2O_3. The Al_2O_3 layers tend to exhibit convolutions; the SiO_2 layers remain flat. In contrast, ternary Fe-6Al-1Si, Fe-4Al-4Si and Fe-1Al-5Si alloys exhibit parabolic oxidation rates comparable to the above high content binary alloys and the same duplex scale consisting of an Al_2O_3 layer covered by Fe_2O_3 segments (Figs. 8 and 10). These results are consistent with the hypothesis that Al and Si in combination more effectively prevent formation of iron type oxides and that SiO_2 only exists as a transient reaction product during the early oxidation stage before onset of parabolic kinetics governing growth of the protective Al_2O_3 scale layer.

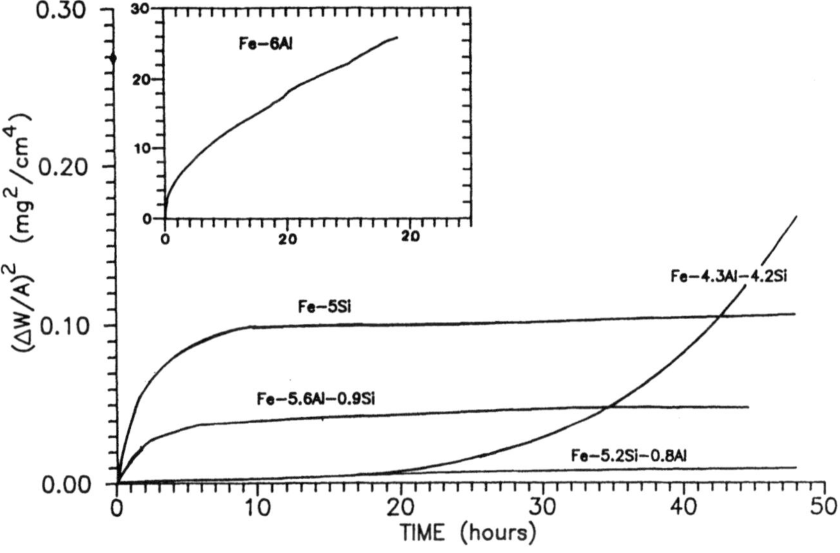

Figure 8. Oxidation kinetics plotted in parabolic form for Fe-Al-Si alloys oxidized at 900°C in 100 Torr of dry oxygen. (After Corkum and Smeltzer [29].

Here too, strong repulsive alloying interactions increasing the aluminum diffusion flux to the alloy interface and strong O-Al attractive interactions possibly lead to a decrease in the contents of alloying elements required to suppress internal oxidation [29]. Experimental results, however, are not available to further substantiate these suggestions of alloying synergistic effects.

Mechanism of Convoluted Scale Growth: Al_2O_3 and Cr_2O_3 scales on iron and nickel alloys often develop convoluted morphologies (see Figs.3 and 10) and there is general consensus that lateral growth in oxide scales is clearly indicated by scale buckling (convolutions) [16,30]. Another major feature is that lateral growth of oxide in a scale can occur by lattice diffusion of appropriate anion and cation point defects and/or

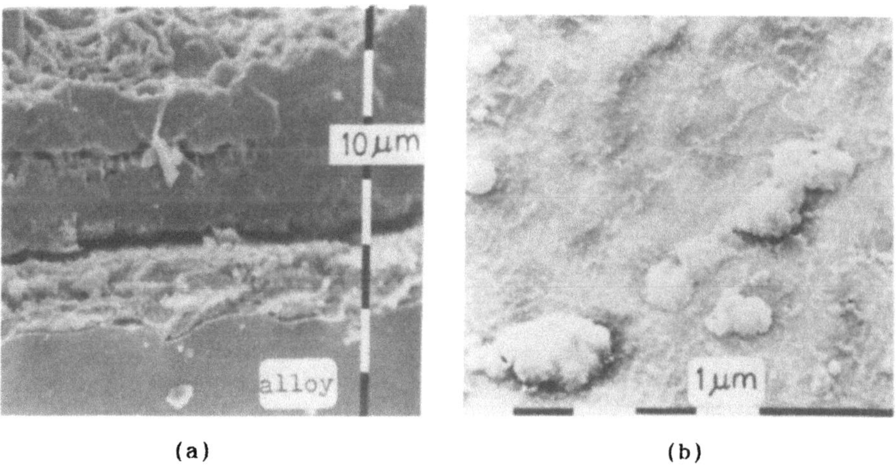

(a) (b)

Figure 9. (a) cross-section of $Fe_2O_3/(FeAl)_3O_4/Al_2O_3$ scale formed on Fe-6Al alloy; (b) SEM micrograph of the oxide film formed on Fe-5Si alloy. Oxidation curves given in Figure 8. After Corkum and Smeltzer [29].

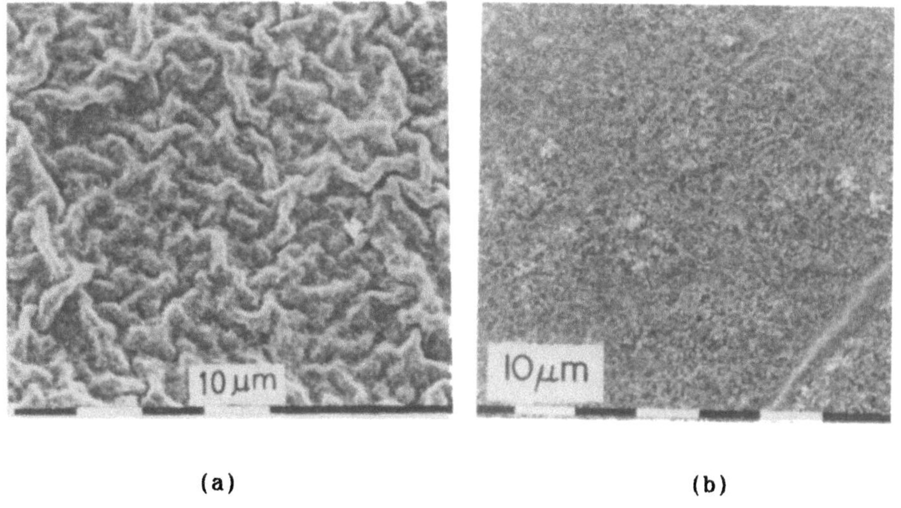

(a) (b)

Figure 10. SEM observations of the surface morphology of Al_2O_3 films formed on (a) the Fe-4.3 Al-4.2 Si alloy and (b) the Fe-5.2Si-0.8 Al alloy. Oxidation curves given in Figure 8. After Corkum and Smeltzer [29].

by grain boundary diffusion of one or both metal and oxygen as reactants [31]. That new oxide forms in growing Cr_2O_3 scales was discussed in the section dealing with chromium oxidation. Transport of reactants in α-Al_2O_3 scales, however, is not presently well understood; diffusion and marker measurements indicate that this layer grows inward by oxygen transport occurring predominantly at grain boundaries, and a lesser but significant flux of aluminum diffusing through the lattice [20,23]. Oxide formation at grain boundaries would give rise to compressive stress in the oxide and a residual tensile stress in the alloy interfacial region. If a pre-existent flaw occurs at the alloy/oxide interface, buckling of the oxide occurs when the compressive stress in the oxide exceeds a critical value which depends on the ability of the oxide to relieve stress and the energy required to produce a new surface. The oxide when separated from the alloy either as a result of void formation or massive separation by buckling is observed to thicken and grow laterally. This necessitates a mode of aluminum transport from the alloy surface to the underside of the detached scale but a mechanism based upon Al evaporation and/or interfacial diffusion often cannot be distinguished from presently available information.

Development of convolutions can lead to scale fracture; for example, compressive stress in a scale may cause interface sliding, resulting in plastic deformation of the alloy substrate and a wave type interface [32]. The convex curvature then leads to mechanical breakdown of the protective scale after a critical oxide thickness and radius of curvature is achieved. This effect of curvature could be a factor in the acceleration of the oxidation kinetics observed for the Fe-4.3Al-4.2 Si alloy after oxidation for 20h (Fig.8). Manning [33] has demonstrated that the critical scale thickness for oxide fracture is related to the elastic modulus of the oxide and the fracture surface energy per unit area expended generating new alloy/oxide surfaces. A first order calculation based on bulk values of these parameters and the estimated oxide thickness after 20h of oxidation yields estimate of the critical radius of 0.1mm which is consistent with observations on the Fe-4.3Al-4.2Si alloy interface [29].

The Reactive Rare Earth Element Effect: Several recent experimental investigations specifically designed to control the sulfur content of chromium and nickel alloys forming protective Cr_2O_3 and Al_2O_3 scales offer supporting evidence to the hypothesis that sulfur segregated to the scale/metal interface exhibits a deleterious effect on scale adhesion. Rare earth elements and dispersions of their oxides exert a beneficial effect because they prevent this interfacial sulfur segregation with a consequent promotion of scale adhesion [34-37]. The presence of sulfur is without doubt harmful but removing all sulfur from an alloy still allows scale spallation unless the microstructure of the scale is modified to prevent the strain energy in the scale from reaching a critical value that overcomes the energy of adhesion at the alloy/scale interface.

It is now documented experimentally, as illustrated in Fig.11, that addition of yttrium to alloys that form Cr_2O_3 scales does alter the growth rate favorably, as long as sufficient yttrium is added to the alloy [16,38-40]. Apparently, the rare earth element acts as a sulfur getter, modifies the scale microstructure and also influences the relative contribution of metal and oxygen to the overall material transport. As shown in Fig.12, the segregation of yttrium ions at Cr_2O_3 grain boundaries slows the transport of chromium cations along the grain boundaries to the extent that oxygen anion transport predominates [40].

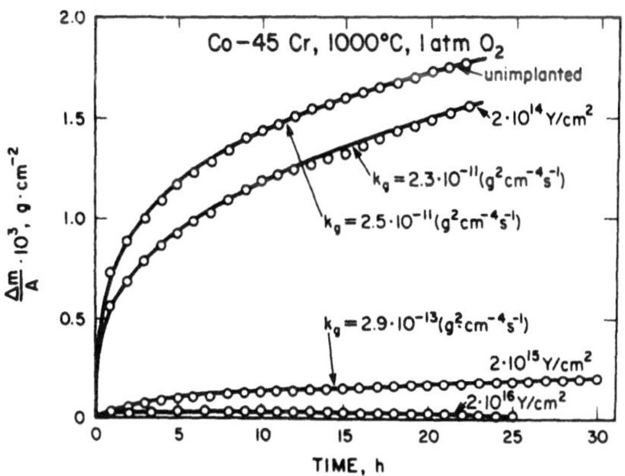

Figure 11. The growth of the Cr_2O_3 scale on a Co-45Cr alloy at 1000°C after Y implantation to various doses ringing up to 2 x 10^{16} Y ions/cm^2. After Przybylski and coworkers [38].

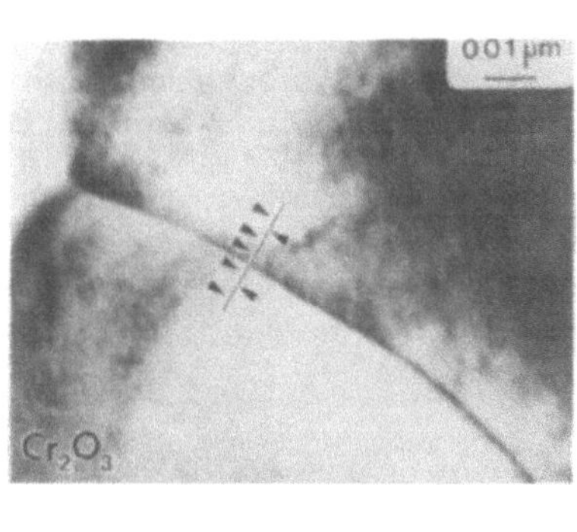

(a)

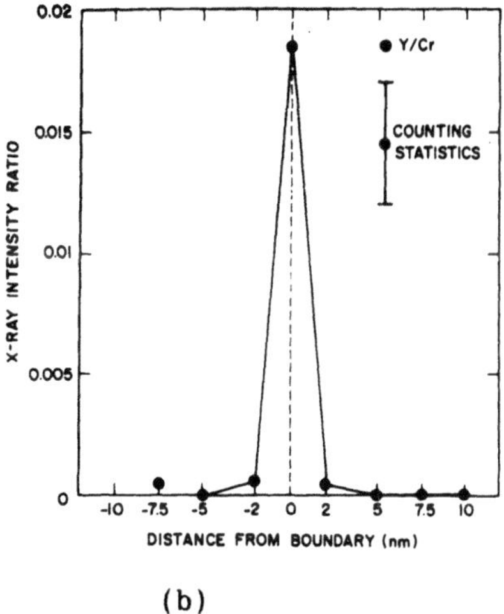

(b)

Figure 12. Segregation of Y at grain boundaries in Cr_2O_3 scale formed on Co-45Cr alloy implanted with 2 x 10^{16} Y ions/cm^2. (a) region of EDAX analysis at oxide grain boundary using STEM; (b) Y analysis at region of oxide grain boundary. (After Przybylski and coworkers [38].

One concludes from the results of recent work that the amorphous phase and metastable allotropes play a decisive role in determining the scale microstructure and relative proportions of aluminum and oxygen migration during stages of Al_2O_3 scale growth. As previously discussed, observations on the oxidation of pure aluminum near its melting point suggest oxide growth by oxygen transport via local pathways through an amorphous Al_2O_3 layer and oxygen diffusion and incorporation at the periphery of the underlying laterally and inward growing γ-Al_2O_3 islands [17]. The θ to α-Al_2O_3 transformation has a drastic effect on the oxidation behavior of the NiAl intermetallic compound phase over the temperature range 700-1200°C [41,42]. A scale is composed practically entirely of θ-Al_2O_3 up to 900°C and of α-Al_2O_3 above 1000°C for oxidation extending up to 100h; θ- and α-Al_2O_3 scales support predominant Al and O transport via grain boundaries, respectively [41]. Yttrium segregates to grain boundaries in NiAl and its concentration is enhanced in the oxide scale above these metal grain boundaries; since yttrium retards the θ to α transformation, the Al_2O_3 scale containing both allotropes exhibits growth by short-circuit diffusion of Al and O [42]. On the other hand, SIMS observations on Al_2O_3 scales formed at 1000°C on an Fe-20Cr-5Al and this alloy containing 0.5% Y_2O_3 support Al and O migration but, in this case, Al migration appeared to be strongly suppressed by the Y addition [43]. In light of these diverse findings, we are able only to conclude that the detailed growth mechanisms of Al_2O_3 scales depend on the allotropes present and on the base metal material.

CONCLUDING REMARKS

In general, a sound empirical knowledge base is under active development for describing growth and properties of Cr_2O_3, Al_2O_3 and SiO_2 scales. Predictive theory with respect to formation of these scales will probably focus on binary and ternary alloys and involve methodologies of perturbation theory to interpret transient oxidation kinetics before onset of parabolic kinetics, selective and internal oxidation of alloying elements. Experimental and theoretical advances are especially needed for correlating short-circuit diffusion of reactants with point defect parameters controlling mass transport, segregation of impurities and alloying elements at oxide boundaries.

ACKNOWLEDGEMENTS

The writer very much appreciated discussions with A. Atkinson, M.J. Graham, W.J. Quadakkers, J.H.W. de Wit, J.L. Smialek and G.J. Yurek and preprints of research with their colleagues.

REFERENCES

1. C. Wagner, Z. Physik. Chem. B21, 25 (1933).
2. C. Wagner, Atom Movements, ASM, Cleveland (1951), 153.
3. P. Kofstad, High Temperature Corrosion, Elsevier Appld. Sci. (1988).

4. E.W. Hart, Acta Metall. 5, 597 (1957).
5. W.W. Smeltzer, R.R. Haering and J.S. Kirkaldy, Acta Metall. 9, 880 (1961).
6. C. Wagner, J. Electrochem. Soc. 99, 369 (1952).
7. W.W. Smeltzer and D.P. Whittle, J. Electrochem. Soc. 125, 1116 (1978).
8. J.E. Morral, M.S. Thompson and O.F. Devereux, Scripta Metall. 20, 1355 (1986).
9. A.Atkinson, Rev. Mod. Phys. 57, 437 (1985).
10. W.W. Smeltzer, Materials Science Forum, 29, 151 (1988).
11. K. Kofstad, High Temperature Corrosion, ed. R.A. Rapp, NACE 6, 123 (1981).
12. A. Atkinson and R.I. Taylor, Transport in Nonstoichiometric Compounds, ed. G. Simkovich and V.S. Stubican, NATO AISI Series B: Physics 129, 285 (1985).
13. D. Caplan and G.I. Sproule, Oxid. Met. 12, 459 (1975).
14. R.J. Hussey, D.F. Mitchell and M.J. Graham, Werkstoffe und Korrosion, 38, 575 (1987).
15. M. Skeldon, J.M. Calvert and D.G. Lees, Oxid. Met. 28, 109 (1987).
16. International Workshop on Critical Issues Concerning the Mechanisms of High Temperature Corrosion, ed. A. Rahmel, G.C. Wood, P. Kofstad and D.L. Douglass, Oxid. Met. 23, 251 (1985).
17. J.I. Eldridge, D.F. Mitchell, R.J. Hussey and M.J. Graham, Oxid. Met. (in press).
18. B.E. Deal and A.S. Grove, J. Appl. Phys. 36, 3370 (1965).
19. J.D. Cawley, J.W. Halloran and A.R. Cooper, Oxid. Met. 28, 1(1987).
20. H.M. Hindam and D.P. Whittle, Oxid. Met. 28, 1 (1987).
21. M. Seo, G. Haltquist, F. Baba and N. Sato, Oxid. Met. 25, 163 (1986).
22. F. Gesmundo and F. Viani, Oxid. Met. 25, 269 (1986).
23. K.P.R. Reddy, J.L. Smialek and A.R. Cooper, Oxid. Met. 17, 429 (1982).
24. C. Wagner, Z. Elektrochem. 63, 772 (1959).
25. P.S. Hagan, R.S. Polizzotti and G. Luckman, SIAM, J. Appl. Math. 45, 956 (1985).
26. H.M. Hindam and W.W. Smeltzer, J. Electrochem. Soc. 127, 1622 (1980).
27. J.E. Morral, M.S. Thompson and O.F. Devereux, Proc. Norman L. Peterson Memorial Symp., Oxidation of Metals and Associated Mass Transport, ed. M.A. Dayananda, S.J. Rothman and W.E. King, Met. Soc. Inc. (1987) 315.
28. F.H. Stott, G.J. Gabriel, F.I. Wei and G.C. Wood, Werkstoffe und Korrosion, 38, 521 (1987).
29. J.L. Corkum and W.W. Smeltzer, Proc. NASA Workshop, Oxidation of High-Temperature Intermetallics, ed. J. Doychak, Met. Soc. Inc. (in press).
30. F.A. Golightly, F.H. Stott and G.C. Wood, Oxid. Met. 10, 193 (1983).
31. A. Atkinson, Corros. Sci. 22, 347 (1982).
32. A.G. Evans, G.B. Crumley and R.E. Demaray, Oxid. Met. 20, 193, (1983).

33. M.I. Manning, Aufbau von Oxidschicten auf Hochtemperaturverkstoffen un ihre technische Bedeutung, ed. A. Rahmel, Deutches Gessellschaft Metallkunde, 283, (1983).
34. J.L. Smialek, ref 27, 297.
35. D.G. Lees, Oxid. Met. 27, 75 (1987).
36. D.R. Sigler, Oxid. Met. 29, 23 (1988).
37. J.G. Smeggil and G.L. Peterson, Oxid. Met. 29, 103 (1988).
38. K. Przybylski, A.J. Garratt-Reed and G.J. Yurek, J. Electrochem. Soc. 135, 509, 517 (1988).
39. P.Y. Hou and J. Stringer, Oxid. Met. 29, 103 (1988).
40. C.M. Cotell and G.J. Yurek, R.J. Hussey, D.F. Mitchell and M.J. Graham, J. Electrochem. Soc. 134, 1871 (1987).
41. G.C. Rybicki and J.L. Smialek, Oxid. Met. (in press).
42. P.A. van Manen, E.W.A. Young, D. Schalkoord, C.J. van der Wekken and J.H.W. de Wit, Surface and Interface Analysis (in press).
43. W.J. Quadakkers, private communication.

DYNAMIC SEGREGATION IN MULTICOMPONENT OXIDES UNDER CHEMICAL POTENTIAL GRADIENTS

G. Petot-Ervas
CNRS, Laboratoire d'Ingénierie des Matériaux et des Hautes Pressions, Université Paris-Nord, Av. J.B. Clément, 93430 - VILLETANEUSE (France)

ABSTRACT

Reactivity and demixing tendency of multicomponent oxides are discussed on the basis of a formal treatment of matter transport under oxygen potential gradients. The present study shows how the dynamic segregation of cations near interfaces takes place in p-type semi-conducting oxides. This effect has been considered for different cases encountered depending on the range of stability of initially homogeneous multicomponent oxides. Experimental examples illustrate the importance of these phenomena.

I. INTRODUCTION

At sufficiently high temperature oxide solid solutions are often not in thermodynamical equilibrium with their surrounding atmosphere. The constituants are then mobile and a transport of matter occurs in the oxide solid solution which results in oxidation or reduction of the material. Depending on the mobilities of the different species this may lead to a demixing of the impurities or of the constituants of the solid solution. This demixing tendency modifies the reaction kinetics and can lead in some cases to the decomposition of compounds or to the formation of new phases, either dispersed or as a continuous layer near the interface.

This survey concerns especially the behavior of p-type semiconducting oxides under an oxygen potential gradient. Among these, transition metal oxides have received great attention not only due to their technological applications but also from a fundamental point of view. A number of results are thus available for these materials. In the first part of this paper general equations of oxidation or reduction kinetics are given. The demixing tendency near interfaces is

L.-C. Dufour et al. (eds.), Surfaces and Interfaces of Ceramic Materials, 337–349.

then discussed and its influence on the behavior of oxides is illustrated with experimental examples.

II. KINETICS DEMIXING OF OXIDE SOLIDE SOLUTIONS

Although the basic treatment of kinetics demixing under oxygen potential gradients is presented elsewhere [1], the essential points are summarized here.

II.1. Basic equations

In many p-type semiconducting oxides (e.g. CoO) the majority defects are cationic vacancies[2]. These defects are responsible for the transport of matter in the material while the oxygen ions are close packed and practically immobile. When these materials are exposed to an oxygen potential gradient a vacancy concentration gradient is set up in the sample (Eq. 4). Hence, an oxidation or reduction front moves in the crystal with a velocity, v, given by the relation [1] :

$$v = J_V \, V_M \tag{1}$$

where J_V is the flux of vacancies and V_M the molar volume of the oxide.

The flux of vacancies under the vacancy concentration gradient is related to the chemical diffusion coefficient, $\tilde{D}$, by Fick's law [2,3]:

$$J_V = - c_M \, \tilde{D} \, \frac{d \, [V^{\alpha'}]}{dx} \tag{2}$$

where $[V^{\alpha'}]$ is the molar concentration of vacancies and c_M the overall concentration (in mol. cm^{-3}) of cation sites in the host lattice AO of the oxide solid solution $AO\text{-}BO_Y$.

Furthermore, after a sudden change of the oxygen partial pressure in equilibrium with the material, the mean penetration depth of the oxidation or reduction front after a time t is given by :

$$\Delta x \simeq \sqrt{\tilde{D} \, t} \tag{3}$$

Let us consider an oxide in the form of a plate of thickness Δx whose two main surfaces are exposed to different oxygen partial pressures. One can assume a local chemical equilibrium on each surface which can be described by the expression :

$$1/2\ O_2 \rightleftarrows V^{\delta'} + \delta\ h^{\circ} + O_0 \qquad (4)$$

$$\text{with } K = \frac{[V^{\delta'}]\ [h^{\circ}]^{\delta}}{P_{O_2}^{1/2}} \qquad (5)$$

where h° is an electron hole, $V^{\delta'}$ a cationic vacancy and δ its degree of ionisation. The square brackets indicate molar concentration.
Since the two main sides of the sample are exposed to different oxygen partial pressures, the concentration of vacancies on these surfaces will be different (Eqs 4 and 5). These defects therefore diffuse in the direction of the lower oxygen partial pressures. This flux of vacancies is accompanied of course by a flux of cations in the opposite direction. Away from vacancy sources and sinks lattice sites are conserved and so these fluxes are coupled through the condition :

$$J_V + \Sigma\ J_c = 0 \qquad (6)$$

where $\Sigma\ J_c$ is the sum of the cationic fluxes in the material.

Under the influence of a chemical potential gradient, the fluxes of cations A and B in the laboratory reference frame, are given by the relation[3] :

$$j_{A^{2+}} = -c_M\ (D_A\ /RT)\ (1-m)\ (\partial\eta_{A^{2+}}\ /\ \partial x) + v\ c_M\ (1-m) \qquad (7)$$

$$j_{B^{\alpha+}} = -c_M\ (D_B\ /RT)\ m\ (\partial\eta_{B^{\alpha+}}\ /\ \partial x) + v\ c_M\ m \qquad (8)$$

where m is the molar concentration of the cations $B^{\alpha+}$ (with $\alpha = 2\gamma$ in the AO-BO_{γ} solid solution), D_A and D_B the diffusion coefficients of A and B, v the drift velocity of the crystal (Eq. 1) due to the drift of the point defects in the oxygen potential gradient, $\eta_{A^{2+}}$ and $\eta_{B^{\alpha+}}$ the electrochemical potential related to the chemical potential by the relation :

$$\eta_i = \mu_i + z_i\ F\ \varphi \qquad (9)$$

with z_i the electrical charge of species i, F the Faraday constant and φ the internal electrical potential generated by the difference of mobilities of the charged species.

In the oxide considered here the oxygen ions can be assumed immobile. Consequently, if the reference frame is fixed on the crystal (on one end for example) the second term in Eqs. 7 and 8 disappears and these equations can be written, for particles of type i, :

$$J_i = -(D_i c_i / RT)\ \partial\eta_i / \partial x \qquad (10)$$

II.2. Oxidation or reduction kinetics

From Eq. 1 it follows that the shift velocity of the oxidation or reduction front is proportional to the flux of vacancies in the oxides considered. This gives rise to migration of the cations and a motion of the crystal as a whole in the direction of the high oxygen potential side (relative to the laboratory reference frame) with a shift velocity given by Eq. 1. For a solid solution (A,B)O this equation can be written according to Eq. 6 :

$$v = -\left(J_{A^{2+}} + J_{B^{\alpha+}}\right) V_M \qquad (11)$$

Since the concentration of vacancies is generally small and the cation chemical potential not easy to determine we have expressed the shift velocity i.e. the flux of vacancies as a function of the oxygen potential gradient. The mean steps are recalled in the following.

During the migration process the total electrical current which passes through the sample is equal to zero (4). Then, it follows from Eq. 9 that :

$$\partial\mu_{h^\bullet}/\partial x = -F\ (\partial\varphi/\partial x) \qquad (12)$$

Furthermore, assuming a local chemical equilibrium between ions, electron holes and atoms[5] $\left(d\mu_A + 2d\mu_{h^\bullet} = d\mu_{A^{2+}}\right)$ and making use of Eqs. 9 and 12, one can write :

$$d\eta_{A^{2+}} = d\mu_A = d\mu_{A^{2+}} - 2\ d\mu_{h^\bullet} \qquad (13)$$

where μ_A is the chemical potential of the electrically neutral species related to the thermodynamic activity by the relation :

$$d\mu_A = RT\ d\ \ln a_A \qquad (14)$$

Equivalent equations hold for species B. Therefore, if we take into account the local chemical equilibrium between the different neutral species one obtains :

$$d\mu_A = d\mu_{AO} - d\mu_O \qquad d\mu_B = d\mu_{BO_\gamma} - \gamma d\mu_O \qquad (15)$$

Using the Gibbs-Duhem equation $(x_{AO}\ d\mu_{AO} + x_{BO_\gamma}\ d\mu_{BO_\gamma} = 0)$ and assuming that the solid solution AO - BO_γ forms an ideal solid solution Eq. 10 can be written, using Eq.13 and its equivalent for B,[1]:

$$J_{A^{2+}} = c_M\ D_A\ [\partial m/\partial x + (1-m)\ \partial \ln a_O / \partial x] \qquad (16)$$

$$J_{B^{\alpha+}} = - c_M D_B [\partial m/\partial x - m \gamma \partial \ln a_O / \partial x] \quad (17)$$

As we have seen previously (Eq.1) the flux of vacancies through the surface controls the oxidation or reduction kinetics of the material. Consequently, substituting Eqs. 16 and 17 in Eq. 11 yields for the shift velocity of the reaction front in the solid solution $AO\text{-}BO_\gamma$ the following relation :

$$v = - (\gamma D_B - D_A) [m \partial \ln a_O / \partial x] + (D_B - D_A) \partial m/\partial x - D_A \partial \ln a_O/\partial x \quad (18)$$

In the pure oxide, the drift velocity of the oxidation or reduction front is simply[6,8] :

$$v = - D_A \partial \ln a_O /\partial x \quad (19)$$

Consequently, comparison of Eqs. 18 and 19 shows that the effect of impurities on the reaction kinetics depends both on their mobilities and on their demixing tendency ($\partial m/\partial x$) in the material.

II.3. Kinetics of demixing

Let us again consider a sample in the form of a plate of thickness Δx. The application of an oxygen potential gradient leads to a flux of vacancies toward the direction of the lower oxygen potential accompanied by a flux of cations in the opposite direction (Eq. 11). The exchange frequency between cations and vacancies is lower for the less mobile cations. Consequently, an enrichment of the less mobile cations is observed in the material at the lower oxygen potential side. From Eqs. 11 and 18 it follows that this demixing tendency under an oxygen potential gradient is described by the general equation :

$$c_M (D_B - D_A) \partial m/\partial x = J_V + c_M [D_A + m (\gamma D_B - D_A)] \partial \ln a_O/\partial x \quad (20)$$

When the mobilities of the cations are different, Eq. 20 allows to relate the demixing tendency near surfaces and interfaces to the flux of vacancies and to the oxygen potential gradient across the sample.

III. BEHAVIOUR OF OXIDES UNDER OXYGEN POTENTIAL GRADIENTS

If the fluxes of cations A and B are independent (i.e. no correlation occurs between A and B jumps) the less mobile cations tend to accumulate at the side of the lower oxygen potential. In the following, we will consider the different cases encountered depending on the range of stability of the initially homogeneous solid solution (see also ref. 11).

III.1. AO and BO_γ form a complete solid solution

The isotherm phase diagram is schematized in Fig. 1a. When a

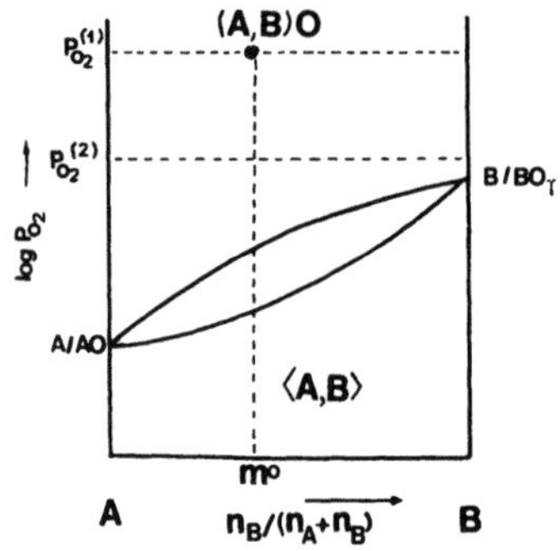

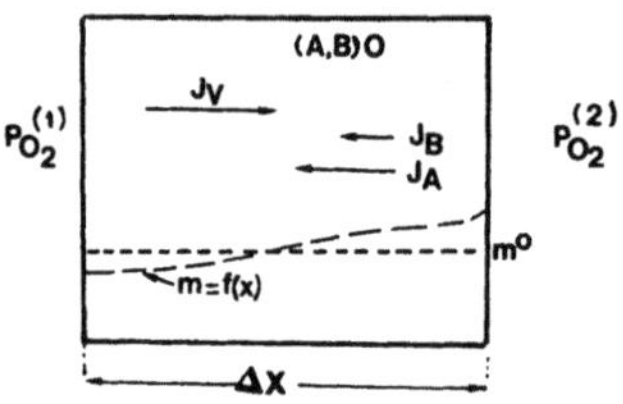

Figure 1a : Isotherm phase diagram when AO et BO_γ form a complete solid solution and concentration profile of BO_γ through the sample (in form of a plate of thickness Δx) after a time t. The two main surfaces of the sample have been exposed to different oxygen partial pressures with $P^{(1)}_{O_2} > P^{(1)}_{O_2}$. No correlation occurs between A and B jumps and $D_B < D_A$.

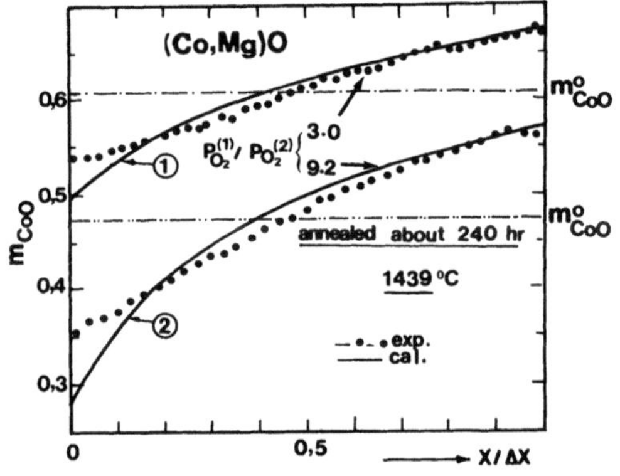

Figure 1b : Experimental and calculated concentration profiles when a steady state is reached for two (CO, Mg) O single crystalline solid solutions[9]. The two samples with homogeneous initial CoO concentration of 0.61 and 0.47, have been exposed to different oxygen potential gradients (Δx thickness of the oxide sample).

sample in form of a plate of thickness Δx and of initial composition $m^o_{BO_\gamma}$, is annealed at sufficiently high temperature in an oxygen potential gradient included between $P^{(1)}_{O_2}$ and $P^{(2)}_{O_2}$, B cations (if $D_B < D_A$) accumulate near the interface exposed to the lower oxygen partial pressure (fig. 1.a). This demixing tendency increases with time and leads to a change in the oxidation or reduction kinetics of the material (Eq. 18).

In the case considered (fig. 1a) a steady state can be reached in which all particles move with the same velocity[8] :

$$v = J_A \; / \; (1-m) \; c_M = J_B \; / \; m \; c_M \tag{21}$$

The concentration profiles of the cations do not change with time while the crystal as a whole moves with a constant velocity v with respect to the laboratory reference frame. According to Eqs 16,17 and 21 the concentration profile of the cations is then given by the relation :

$$\partial m/\partial x = [m \; (1-m) \; (\gamma D_B - D_A) \; / \; (m \; D_A + (1-m) \; D_B)] \; \partial \ln a_O/\partial x \tag{22}$$

This case has been treated by Schmalzried et al.[8,9,11]. In fig. 1b are shown the measured and calculated concentration profiles they obtained for the (Co, Mg)O solid solution after an annealing of 240 hr in two different oxygen potential gradients.

III.2. AO and BO_γ have a limited solubility range

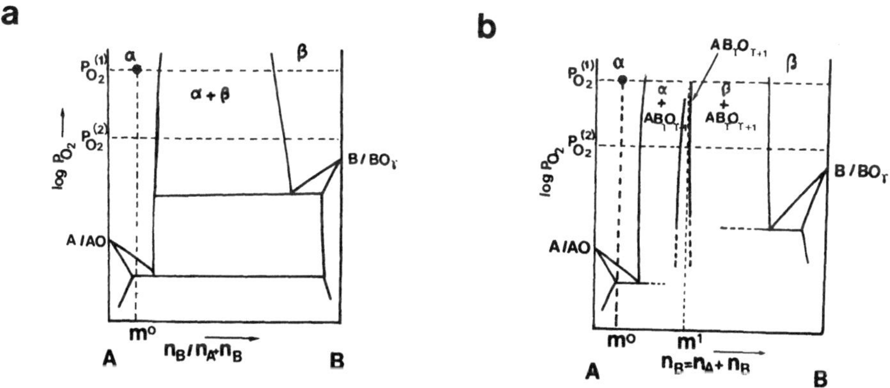

Figure 2(a) : Isotherm phase diagram when AO and BO_γ have a limited solubility range and form a miscibility gap.

(b) : Isotherm phase diagram when AO and BO_γ have a limited solubility range and form a ternary compound $AB_\gamma O_{\gamma+1}$.

The phase diagrams at constant temperature are shown schematically in Figs 2a and 2b. BO_γ dissolves only to a small extent in AO. These two oxides can form a miscibility gap (fig. 2a) or they can be in equilibrium with a ternary compound $AB_\gamma O_{\gamma+1}$ having a limited range of stability (fig. 2b).

As before, the initially homogeneous solid solution of composition $m^o_{BO_\gamma}$, stable in the oxygen potential range $P^1_{O_2} - P^2_{O_2}$, is exposed to a mean oxygen potential gradient $(P^1_{O_2} - P^2_{O_2})/\Delta x$. Due to the cationic vacancy flux the less mobile cations (B) again accumulate at the side of the lower oxygen potential $(P^{(2)}_{O_2})$. The chemical potential of BO_γ increases near this oxide / gas interface until it reaches the equilibrium value between the immiscible phases α and β (fig. 2a) or α and $AB_\gamma O_{\gamma+1}$ (fig. 2b). Therefore, if the oxygen potential is maintained, the solid solution β (or $AB_\gamma O_{\gamma+1}$) will precipitate in dispersed particles in the solid solution α. Eventually a continuous layer of β (or $AB_\gamma O_{\gamma+1}$) would be formed at the interface, as is observed in layers formed during alloy oxidation[13,14].

III.3. Ternary compounds

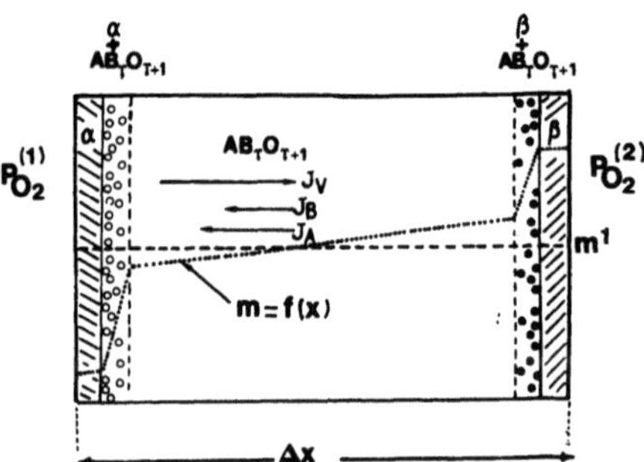

Figure 3 : Concentration profile of BO_γ when the ternary compound $AB_\gamma O_{\gamma+1}$ (in form of a plate of thickness Δx) is exposed to an oxygen potential gradient. After a time t the ternary compound starts to be decomposed near the surfaces in contact with high and low oxygen partial pressures.

Due to the narrow range of stability of the compound $AB_\gamma O_{\gamma+1}$ (fig. 2b) the chemical potential of the oxide corresponding to the lower moving cation (BO_γ in the present case) increases rapidly near the side exposed to the lower oxygen potential until it reaches the equilibrium value between the phases β and $AB_\gamma O_{\gamma+1}$. The same behaviour is observed for AO at the opposite side. Therefore, if the oxygen potential gradient is maintained the ternary compound will be decomposed (fig. 3). These reactions, which lead to the ageing of materials, have been investigated for different compounds such as Ni_2SiO_4[9], Fe_2SiO_4[8], $NiTiO_3$[10] etc ... As an example we have shown in fig. 4 backscattered scanning electron micrographs of a cross section through a Ni_2SiO_4 sample from an experiment of this type performed by Wolfenstine et al.[10]. Polycrystalline Ni_2SiO_4 samples were exposed to an oxygen potential gradient between 0.21 and 3.1 10^{-4} atm. at 1336°C for 96 hr. A layer of SiO_2 was formed near the low P_{O_2} side (fig. 4a) while NiO was found on the high oxygen potential side (fig. 4b). This distribution of NiO and SiO_2 is consistent with diffusion data for silicate olivines in which Ni diffuses faster than Si and oxygen.

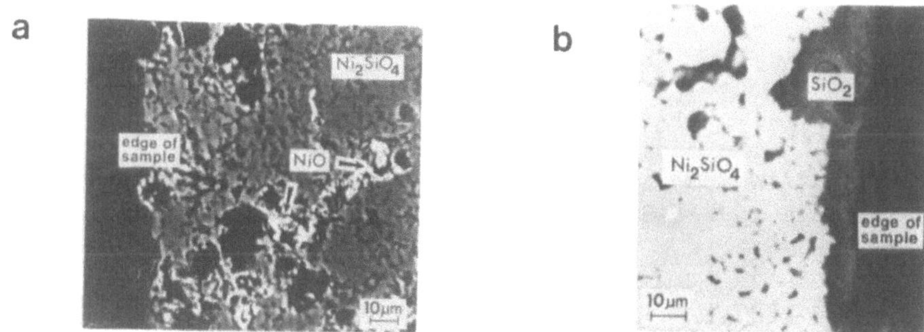

Figure 4 : Backscattered scanning electron micrograph of a cross section of a Ni_2SiO_4 sample[10]. (a) near the surface in contact with high oxygen partial pressures (b) near the surface in contact with low oxygen potentials.

IV. EXPERIMENTAL RESULTS AND APPLICATIONS

IV.1. Kinetics of formation of NiO scales on Ni-Cr alloys

During their formation these scales are exposed to an oxygen potential gradient included between the oxygen partial pressure of the Ni/NiO equilibrium and that of the oxide / gas interface (generally 0.21 atm). Ben Abderrazik[11] has observed in NiO an increased concentration of chromium near the metal / oxide interface after oxidation. Stott et al.[13,14] have also shown that a NiO layer develops at the surface of the alloy in the early stage of the high temperature oxidation of dilute chromium alloys (20 wt % Cr). The solute metal forms discrete Cr_2O_3 precipitates in the subjacent alloy. These precipitates are subsequently incorporated into the thickening

NiO scale. After a few hours they have observed that the oxidation rate considerably decreases due to the formation of a continuous layer of Cr_2O_3 at the alloy/oxide interface. This set of results is consistent with the analysis given above [1] .Indeed, during the corrosion process the nickel rich oxide scale grows outward due to the diffusion of cations from the alloy / oxide interface to the scale/gas interface. The diffusion coefficient of chromium is lower that of nickel [15]. Consequently, as we have shown previously this lead to an enrichment of chromium at the alloy/oxide interface. According to Eq. 18, this chromium demixing tendency should lead to a decrease of the rate of displacement of the oxidation front in the oxide scale ($\partial m/\partial x$ and $\partial \ln a_0/\partial x$ have opposite signs) as observed by Nowotny et al.[16]. Indeed, the simplest situation for analysis of these transport processes under oxygen potential gradients is the study of the exclusive oxidation of pure and doped oxide samples. These kinetic studies have been performed by Nowotny et al.[16] on pure and chromium doped NiO single crystals. In agreement with the previous analysis they have observed that chromium decreases the oxidation kinetics of NiO.

The oxygen potential is maintained continuously through the oxide scale during corrosion, the demixing phenomenon increases then with time leading eventually to the formation of a continuous layer of Cr_2O_3 as observed by Stott et al.[13,14] after few hours of oxidation at 800 - 1000°C.

IV.2. Solid state metal-ceramic bonding

A situation similar to that encountered in corrosion is found in some cases of solid state metal-ceramic bonding. When certain ceramics are held in intimate contact with metals such as Ni or Cu under air, argon or vacuum at sufficiently high temperatures a reaction bonding occurs between these materials. For example, the formation of a spinel phase is believed to be responsible for bonding when nickel is oxidized in contact with Al_2O_3.

During the nickel corrosion process the NiO layer grows outward due to the diffusion of cations through the oxide layer toward the oxidation front (at the NiO/Al_2O_3 interface). This oxide reacts then with Al_2O_3 with the formation of a NiO-Al_2O_3 solid solution. Consequently, an oxygen potential gradient is maintained continuously throughout the nickel rich oxide layer as in the corrosion scales. An increase concentration of the less mobile cations must then be expected near the Ni/NiO interface.

Experimental work has shown that the thickness of the NiO layer is critical for the strengh of the metal/oxide bonding[17]. Below a critical value, the thicker the NiO layer the stronger the bond and this effect is attributed to an increasing amount of the spinel phase. Above this critical thickness the nickel/oxide adherence decreases.

Although these solid state reactions appear simple, complementary studies are necessary to understand this type of bonding. At present the diffusion coefficient of Al in NiO for example is not known and the effect of different parameters such as temperature or oxygen partial pressure, have not been fully investigated.

IV.3. Influence of impurities on the reduction kinetics of oxides

The blast furnace reduction of iron oxides to iron provides the best known example of such a reaction and during the last few years the effect of impurities on the reduction kinetics of iron oxides has received considerable attention.

When an oxide such as wüstite is reduced by a CO/CO_2 mixture a metal scale is formed on its surface. The pores in the metal layer permit the reducing gas to reach the reaction interface directly. An oxygen potential gradient is established in the oxide which generate a flux of vacancies toward the reduction front accompanied by a flux of cations in the opposite direction. Consequently, an enrichment of the less mobile cations occurs near the metal/oxide interface (if the correlation effects can be neglected) which leads to a change of the reduction kinetics (Eq. 18). In agreement with this analysis it has been found for example that Cr or Mn decrease the reduction kinetics of wüstite[18,19], at least during the early stage of the reduction process, and this effect is accompanied by a segregation of Cr or Mn in FeO near the metal/oxide interface.

IV.4. Influence of cooling rate on the surface segregation of impurities

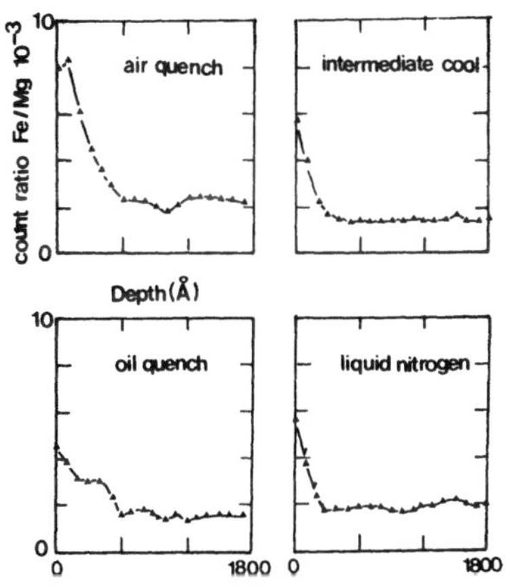

Figure 5 : Near surface iron concentration in MgO versus depth for different quenching rates[20].

The influence of cooling rate on the surface segregation of impurities such as chromium, iron or scandium in MgO has been shown by Black et al.[20]. The near-surface concentration of iron versus depth for different quenching rates is illustrated in fig. 5. These results show that the cooling rates has an influence both on the iron surface concentration and on the thickness of the near-surface layer enriched with iron. One can point out that these thicknesses are of the same order of magnitude as those found for example by Hirschwald et al.[21,22] for chromium in NiO and CoO.

These results may be attributed to non-equilibrium thermodynamical conditions. Indeed, during cooling one can assume a thermodynamical equilibrium between the surface of the crystal and the surrounding atmosphere. This leads in materials such as CoO, NiO or MgO to a decrease of the concentration of vacancies near the surface and consequently to an outward flux of vacancies. The formal treatment is then equivalent to that considered in the first part of this paper. In the present case it may be further complicated by the temperature gradient across the sample. Nevertheless, the thickness of the layer which is altered by the transport processes occurring during cooling may be estimated from the depth of penetration of the reduction front, assuming that the specimen is placed at a middle constant temperature for a lapse of time which decreases when the cooling rate increases. These thicknesses have been estimated for CoO and NiO from the depth of penetration of the reduction front (Eq. 3) . For example, at 400°C and for t = 200s these depths are equal to 1200 Å and 7300 Å for NiO and CoO, respectively. The calculated thicknesses are consistent with the observations of Hirschwald et al. [21,22] . Furthermore, in agreement with the experimental results, it has been shown that the thickness of the surface layer in which matter transport can occur during cooling is higher in CoO than in NiO.

Equilibrium segregation of cation impurities at the surface of oxides has been the subject of increasing scientific interest during the last decade. Nevertheless, the results reported here emphasize the difficulty in separating equilibrium segregation from the dynamic segregation when the observations are performed after cooling.

V. CONCLUSION

In this paper an analysis of the behavior of p-type semiconducting multicomponent oxides under chemical potential gradients has been given. The dynamic segregation of cations which can occur and in some cases lead to the ageing or even to the decomposition of the materials has been considered. The influence of these phenomena on the behaviour of these compounds and on industrial applications has been illustrated and discussed on the basis of the formal treatment of matter transport under oxygen potential gradients.

REFERENCES

1. G. PETOT-ERVAS, C. PETOT, to be published
2. P. KOFSTAD, Nonstoichiometry, Diffusion and Electrical Conductivity in binary metal oxides, J. Wiley, 1972.
3. J. PHILIBERT, Diffusion et Transport de matière dans les oxydes, Editions de Physique, 1985.
4. A.T. FROMHOLD, Theory of metal oxidation, North Holland, 163, 145, 1976.
5. C. WAGNER, Corrosion Science, 9, 91, 1969.
6. F. GESMUNDO, F. VIANI, J. Phys. Chem. Sol. 42, 777, 1981.
7. P. OCHIN, G. PETOT-ERVAS, C. PETOT, J. Phys. Chem. Sol. 46, 695, 1985.
8. H. SCHMALZRIED, Reactivity of solids 117, 1986.
9. H. SCHMALZRIED, W. LAQUA, P.L. LIN, Z. Naturforsch. 34a, 192, 1979.
10. J. WOLFENSTINE, D. DIMOS, D.L. WOHLSTEDT, J. Am. Cer. Soc. C 117, 1985.
11. H. SCHMALZRIED, W. LAQUA, Oxid. of metals, 15, 339, 1981.
12. G. BEN ABDERRAZIK, Thesis University Paris XI, 1986.
13. F.H. STOTT, G.C. WOOD, D.P. WHITTLE, B.D. BASTOW, Y. SHIDA, Solid State Ionic 12, 365, 1984.
14. F.H. STOTT, J.S. PUNNI, G.C. WOOD, G. DEARNALEY, Transport in Nonstoichiometric Compounds, Ed. G. Simkowitch, V. Stubican, Plenum Press, 463, 1985.
15. N.B. PETERSON, Solid State Ionics 12, 201, 1984.
16. J. NOWOTNY, J. OBLAKOWSKI, A. SADOWSKI, Bull. Polish. Acad of Sciences Chemistry 33, 100, 1985.
17. F.P. BAILEY, W.E. BORBIDGE, Surfaces, Interfaces in ceramic and ceramic metal systems, Ed. J. Pask, A. Evans, Plenum Press 525, 1981.
18. S. JASIENSKA, J. JANOWSKI, A. SADOWSKI, C. MONTY, G. PETOT-ERVAS Cryst. latt Def. and Amorph. Mat. 16, 145, 1987.
19. G. PETOT-ERVAS, H. KLIMCZYK, C. MONTY, C. PETOT, J. JANOWSKI. Low Dimensional Defects in Nonstoichiometric Compounds, Ed. J. Nowotny, W. Weppner. Kluver Academic Publishers (to be published)
20. J.R.H. BLACK, W.D. KINGERY, J. Am. Cer. Soc. 62, 176, 1979.
21. W. HIRSCHWALD, I. SIKORA, F. STOLZE, Surface and Interface Analysis 7, 155, 1985.
22. I. SIKORA, F. STOLZE, W. HIRSCHWALD, Surface and Interface Analysis 10, 424, 1987.

STUDY OF NON STOICHIOMETRIC PURE AND Zr-DOPED YTTRIA SURFACES BY X-RAY PHOTOELECTRON SPECTROSCOPY AND SCANNING ELECTRON MICROSCOPY

M. Gautier, J.P. Duraud, F. Jollet, N. Thromat,
Ph. Maire, C. Le Gressus
IRDI/DESICP/DPC/SPCM
Centre d'Etudes Nucléaires de Saclay
91191 Gif-sur-Yvette Cedex, France

ABSTRACT. Surfaces of oxygen-deficient yttrium oxide, pure or Zr-doped, have been studied by means of X-ray photoelectron spectroscopy and scanning electron microscopy.

The bulk local geometric structure of these non-stoichiometric compounds was previously determined around the Y atom by an EXAFS (Extended X-ray absorption fine structure) study.

The local electronic structure around both Y and O, at the surface, was investigated by X-ray photoelectron spectroscopy.

The partial transfer of the electronic distribution between the anion and the cation was probed using the Auger parameter.

Coupling of these experiments with microscopic observations show that :

- In the pure oxygen-deficient sample, the concentration of oxygen vacancies appears to be increased at the grain boundaries.
- The Auger parameter shows upon reduction an evolution of the Y-O bond towards a more covalent one, this evolution being modulated with the presence of ZrO_2.

I - INTRODUCTION

Point defects play an important rôle in the macroscopic properties of ceramics. These point defects can either be structural defects (vacancies, interstitials, lattice distortions), or they can be due to the chemical substitution of foreign cations with different ionic radius and coordination. However, the relationship between the structure of the defects and the resulting properties is not well understood, although it constitutes a promising way to set up high performance ceramics. Moreover, the structure of non-stoichiometric oxides is still not very well known / 1 /. Much work is needed to get a better insight into the electronic and atomic changes induced by the production of oxygen vacancies in oxides.

Yttrium oxide is a ceramic with high melting point, high resistivity, which crystallizes into the C-type rare earth sesquioxide structure. Only a few studies deal with the defect structure of this oxide,

L.-C. Dufour et al. (eds.), Surfaces and Interfaces of Ceramic Materials, 351–363.

which is widely used as a dopant in ZrO_2 for stabilization. Most of these studies are devoted to electrical conductivity measurements / 2, 3, 4, 5 /. In a previous paper / 6 /, we have shown that pure yttria exhibits a quite different fracture behavior when it is non-stoichiometric (oxygen deficient) compared to the stoichiometric case.

On the stoichiometric sample, a transgranular fracture path occurs, whereas in the oxygen deficient ones, the fracture features are intergranular.

This paper deals with the structure of oxygen-rich or deficient yttria ceramics produced by redox processes at high temperatures / 7 /. These ceramics are either pure or doped with zirconia. Indeed yttria is able to dissolve tetravalent cations such as Zr, without change in the crystallographic structure / 8/.

The local geometric structure around an yttrium atom located in the bulk was probed using EXAFS (Extended X-ray absorption fine structure) above the K-absorption edge of yttrium. The results are extensively reported in another paper / 9 / ; we only summarize them here.

The local electronic structure around yttrium and oxygen, at the surface, was investigated using X-ray photoemission to get insight into the nature of the Y-O bond and its modification with oxygen content.

Both methods give an averaged information on a large area (a few mm^2) of the sample, and do not allow determination of the structure (both geometric and electronic) at the grain boundaries, compared to the bulk of the grains. Therefore scanning electron microscopy was used to get informations on processes occuring at the grain boundaries.

The results show that large concentrations of oxygen vacancies may be produced in reduction condition, for pure yttria as well as ZrO_2-doped yttria.

When oxygen vacancies are introduced, there is no apparent change in the Y-O and Y-Y bond lengths, but there is a charge transfer from the anions to the cations, leading to a more covalent Y-O bond. For pure yttria, scanning electron microscopy shows that the surface potential at the grain boundaries is different from that of the bulk, indicating that oxygen vacancies are preferentially segregated at the grain boundaries.

II - MATERIALS AND METHOD

II.1 - Samples

Pure Y_2O_3 (5N) * and 14 wt % (23 mol %) ZrO_2-doped yttria powders were sintered under high isostatic pressure (HIP) : 1450°C, 150 MPa for 45 mn. Zr-doped powders were synthetized using a precipitation method. The ZrO_2 powder ** was dispersed into an aqueous solution of yttrium nitrate. Precipitation of yttrium hydroxide was then obtained by reaction with ammonium hydroxide, followed by calcination.

---

* Rhône-Poulenc, France

** Criceram, France

The green body was then cut into pellets of thickness 1 mm. The samples used for scanning electron microscopy were mechanically polished. All samples were first annealed for 2 hours at 1500°C (oxidation heat treatment) in the room atmosphere, in order to bleach the intrinsic oxygen vacancies and the residual stresses resulting from the sintering process and to produce oxygen rich samples (white color).

Some of them were heated at 1200°C in vacuum (10^{-6} torr) to obtain stoichiometric samples, according to the conductivity σ versus Log P_{O_2} plots measured by Tallan et West / 2 /.

Others were heated at 1700°C in vacuum (10^{-6} torr). This latter treatment corresponds to a dissociation condition of the oxide / 2, 10 / leading to its reduction. Therefore a rather large number of oxygen vacancies is expected, at least in the case of the pure Y_2O_3. This latter treatment (reduction heat treatment) led to black samples.

The Zr concentration was chosen according to the phase diagram of the yttria-zirconia system / 11 /. 14 wt % ZrO_2 is located at the boundary of the one solid solution (substituted Zr in Y_2O_3) field, and a two solid solution (Zr in Y_2O_3 and Y in ZrO_2) field. Therefore, only the doped sample annealed at 1700°C is constituted by a single phase, as verified by X-ray diffraction. The lattice parameter (10.56 Å) is found to be smaller than in the case of the pure Y_2O_3 (10.604 Å), as reported by Scott / 8 /.

II.2 - Methods : photoelectron spectroscopy

X-ray photoelectron spectroscopy is a very sensitive tool for measuring chemical shifts of electron binding energy, produced by modifications of the electronic environment. However, this technique is difficult to apply to the case of Y-O bonds in samples with different oxygen stoichiometry : this arises from the small chemical shifts of the Y3d peak in the photoelectron spectrum, and from a propensity for charge building up under the exciting beam.

To overcome such problems the Auger parameter introduced by Wagner / 12, 13 /, can be used. This parameter is defined as the difference between the kinetic energy of the Auger electron (ijk) and that of the photoelectron emitted from the inner shell i, as measured on the photoelectron spectrum :

$$A = E_{kin}\,(ijk) - E_{kin}\,(i)$$

It depends on the incident photon energy hν. For this reason the modified Auger parameter A' was introduced :

$$A' = E_{kin}\,(ijk) + E_B(i) = A + h\nu$$

where $E_B(i)$ is the binding energy of the photoelectron emitted from the shell i. Using Koopman's approximation, it can be shown that the Auger parameter variation ΔA for two different chemical environments is twice the variation of the extraatomic relaxation energy arising from the decrease of the screening charge for the electrons involved in the chemical bond when a positive hole is left on the inner shell i. Hence

ΔA is directly related to the electronic polarization energy variation ΔE_{pol} resulting from the ionization of the inner shell i / 14-17 / :

$$\Delta A = 2\ \Delta E^{ea}_{relax}(i) = -\ 2\ \Delta E_{pol}$$

Data obtained on several silicates / 18 / showed a direct correlation between this polarisation energy and the polarizability α_O of the oxygen ion in the chemical environment considered :

$$\Delta A = -\ K\alpha_O$$

where K is a constant.

II.2 - Experimental set-up

Photoemission experiments were carried out in a VG Escalab Mark II with a base pressure around 5.10^{-11} mbar. In the analysis chamber a monochromatized X-ray source equipped with an aluminum anode was available ($h\nu$ =1486.6 eV). On the other hand, a conventional twin anode source could be used (AlKα and ZrLα). In all cases, the electrons emitted from the sample were energy filtered in a hemispherical analyser operated in the constant analyser energy mode, with a pass energy of 50 eV. In order to overcome sample charging under the incident photon beam, an electron flood gun (5 eV, 200 µA) was used. The working conditions of this gun were chosen so as to get for the O 1s photoelectron line recorded on the stoichiometric sample the same value as that of the same line on Al_2O_3 . Detection was made by a counting method ; data could be stored and processed using a DAD system.

Scanning electron microscopy was performed using a high resolution scanning reflection electron microscope (JSM 880-JEOL). This microscope is equipped with an immersion lens which allows one to reach 1.5 nm resolution in the secondary electron imaging (SEI) mode or 5 nm in the backscattered electron imaging (BEI) mode. Primary currents as low as 10^{-12} A could be used , as well as low accelerating voltage such as 1.7 kV.

III - RESULTS

III.1 - Photoemission results

The modified Auger parameters of respectively oxygen (A'_O) and yttrium (A'_Y) were measured for pure Y_2O_3 samples submitted to either the oxidation (1500°C air) or the reduction (1700°C vacuum) heat treatments, and for the ZrO_2-doped samples (14 wt %) submitted to the reduction heat treatment (1700°C vacuum).

A'_O is associated to the O 1s photoelectron line, and the O KLL Auger peak. A'_Y is associated to the Y3d photoelectron line, and the Y $L_3M_5M_5$ Auger peak. The latter line was obtained using the Bremsstrahlung X-ray background produced by the non monochromatized source. Indeed M levels can be considered as core levels in yttria, and therefore a good assumption is that the chemical shifts of the 3d levels and the 2p levels differ by only tenths of eV.

Figure 1 shows the photoelectron spectrum including the part excited by the Bremsstrahlung background obtained on the pure oxidized Y_2O_3 sample.

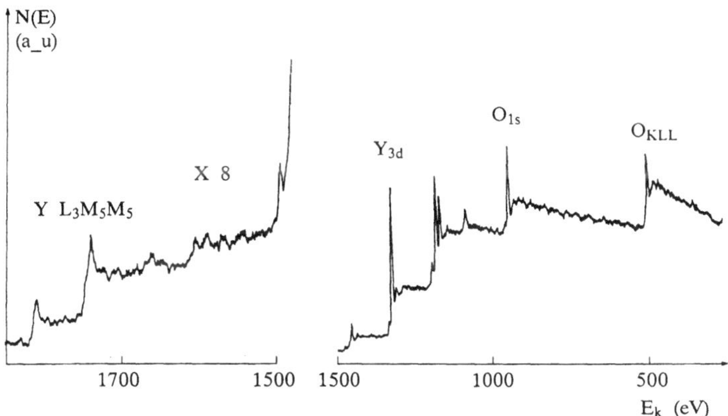

Figure 1. X-ray photoelectron spectrum of Y_2O_3 showing the $L_3M_5M_5$ Y-Auger peak and the Y-3d photoelectron line, used to calculate the Auger parameter (AlKα radiation).

For a given sample, A'_O and A'_Y do not vary with the experimental conditions : AlKα or ZrLα exciting line, charge compensation or not with the flood gun, within an uncertainty of 0.4 eV. In all cases, the pure reduced Y_2O_3 has the largest modified Auger parameters (A'_O = 1041.5 eV, A'_Y = 1895 eV), the pure oxidized Y_2O_3 the smallest (A'_O = 1037 eV, A'_Y = 1893.3 eV). The doped reduced Y_2O_3 has Auger parameters located inbetween , but closer to that of the pure reduced Y_2O_3 (A'_O = 1040 eV, A'_Y = 1894.6 eV).

The difference of the modified Auger parameters when going from the pure reduced Y_2O_3 to the pure oxidized Y_2O_3 are :

$$\Delta A'_O \simeq 4.5 \text{ eV} \quad \text{and} \quad \Delta A'_Y \simeq 1.7 \text{ eV}$$

On the other hand, considering the reduced samples, the difference of the modified Auger parameter when going from pure Y_2O_3 to doped Y_2O_3 are :

$$\Delta A'_O \simeq -1,2 \text{ eV} \quad \text{and} \quad \Delta A'_Y \simeq -0.4 \text{ eV}$$

III.2 - Scanning electron microscopy

The backscattered electron image (BEI) and the secondary electron image (SEI) of respectively the pure reduced Y_2O_3 and the doped reduced Y_2O_3 (14 wt % ZrO_2) are shown on fig. 2 (a,b) and fig. 3 (a,b).

For both samples, the BEI image reveals the grain structure at the surface (fig. 2.a and fig. 3.a). The contrast arises from crystalline effects / 19 /. Note that the grain size is slightly smaller (x 1/3) in the case of the doped reduced sample.

On the doped reduced Y_2O_3, the SEI image is similar to the BEI one, only the relative contrasts between the grains are different (fig. 3.b).

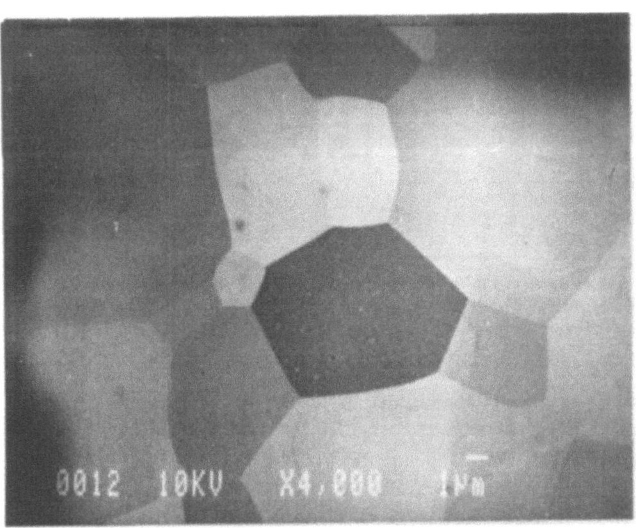

Figure 2. Scanning electron images of pure reduced Y_2O_3.
(a) BEI mode ; (b) SEI mode.

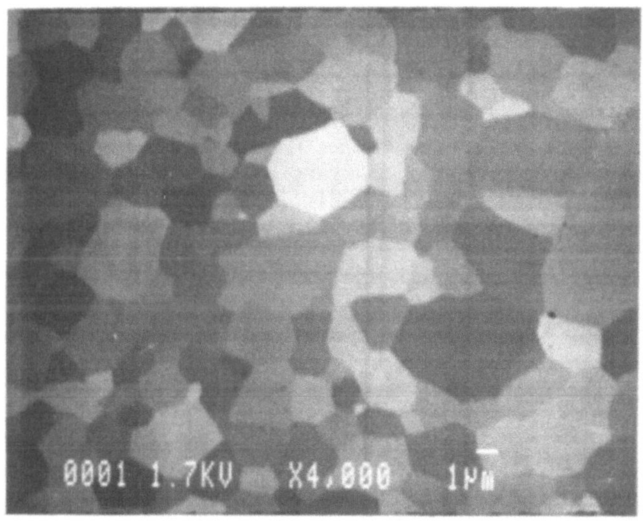

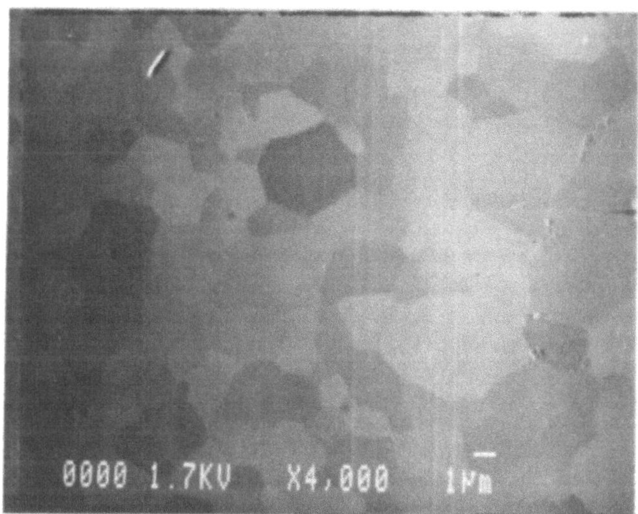

Figure 3. Scanning electron images of doped reduced Y_2O_3.
(a) BEI mode ; (b) SEI mode.

In the case of the pure reduced Y_2O_3, the SEI image is quite different : all grains have the same contrast, and along the grain boundaries a dark line can be observed (fig. 2.b).

These images were obtained without coating the samples. The experimental conditions : accelerating voltage and primary current ($I = 5.10^{-11}$ A) were chosen to avoid sample charging. Using the same conditions, it was possible to get a BEI image on the pure oxidized sample, but charging effects dit not allow us to obtain a SEI image.

IV - DISCUSSION

Yttrium oxide crystallizes into the C-type cubic rare earth sesquioxide structure. This structure is closely related to the fluorite structure. In the fluorite lattice, each cation is surrounded by a cube of eight anions. The C-type structure can be derived by removing one-quarter of the anions and slightly rearranging the remaining ions. For 25 % of the cations, the vacancies lie along a body diagonal and the anions are located at the vertices of a slightly distorted octahedron. For the remaining cations, the vacancies lie along a face diagonal resulting in distorted octahedra. In such a structure, yttrium is six-fold coordinated.

The yttrium oxide structure, in the stoichiometric composition, contains ordered empty anionic sites. Therefore excess oxygen atoms are expected to be incorporated into the lattice either by oxidation heat treatment, or by introducing foreign cations with a higher valency than yttrium. Y_2O_3 can dissolve different amounts of oxides with tetravalent cations such as ZrO_2 (20 mol % ZrO_2 at 1500°C). The solubility limit increases with temperature (24 mol % ZrO_2 at 2000°C) / 20 /. Addition of tetravalent Zr^{4+} cations induces interstitial anions, whose mobility is high / 21 /.

IV.1 - Bulk defect structure of pure and doped yttria

Most experimental studies on yttrium oxide defect structure are based on conductivity measurements / 2, 3, 4, 5, 22 /. They give a description of the defect structure in terms of isolated point defects without interaction, at the thermodynamic equilibrium. In fluorite-like oxides, the cation sublattice is considered very stable, and the cations are much less mobile than the anions. Therefore the predominant defects in Y_2O_3 are located on the oxygen sublattice. In pure yttria, the most probable point defects are anion Frenkel pairs O_i'', $V_O^{\cdot\cdot}$ / 22 /. It has been shown / 3, 4, 5 / that the conductivity strongly depends on water pressure, pointing out the role of $H^{\cdot}$ which compensates the positive O_i'' defects if the water pressure is important.

For ZrO_2-doped Y_2O_3, oxygen interstitials have been observed / 21, 23 / as expected from the C-type cubic crystallographic structure.

The detailed structure of defects in anion deficient fluorite oxides is not known, but the usual defect arrangement is based on the pairing of empty anion sites accross the diagonal of the cubic cation coordination polyhedron / 1 /.

The local geometric structure around an yttrium atom located in the bulk was probed using EXAFS (Extended X-ray Absorption Fine Structure) above the K absorption edge of yttrium, in the pure and doped reduced yttria samples of this work / 9 /.

Main results are the following :

Pure Y_2O_3 : In reduction conditions, an appreciable number of oxygen vacancies can be introduced (∿ 10 % in the first coordination shell around an yttrium atom). The production of such defects does not lead to any apparent reorganization of the lattice in the conditions

explored : the bond lengths remain constant within the accuracy of the method (2.10^{-2} Å) but static disorder in the first coordination shell increases in anion deficient samples.
Doped Y_2O_3 : For the same reduction treatment, the number of oxygen vacancies in the first coordination shell (∿ 30 %) is larger, and the static disorder much weaker than for pure Y_2O_3, while the bond lengths remain constant too. Then, there is a competing process between the introduction of oxygen-excess by doping with a tetravalent cation, and the production of anion vacancies upon reduction. In doped fluorite oxides, the oxygen diffusion coefficients are enhanced / 1, 20 /. This could account for the higher deficit of oxygen in Zr-doped yttria, compared to the case of pure yttria with the same reduction heat treatment.

IV.2 - Electronic structure of pure and doped yttria surfaces

In a scheme of isolated point defects, as anionic vacancies $V_O^{\cdot\cdot}$ (or $V_O^{\cdot}$) as well as dopant ions $Zr_Y^{\cdot}$ (in substitution) bear positive effective charges, the perturbation of charge equilibrium in the bulk must be restored by the introduction of negative species, such as delocalized electrons in the conduction band.

In other terms, an electron transfer must occur for charge compensation. This is shown by the XPS results, which concern the surface of the ceramic (20-30 Å). For the purpose of this part, we will assume that the conclusions on the local geometric structure obtained by EXAFS, which concern the bulk of the ceramic, are qualitatively valid at the surface. SEXAFS (surface EXAFS) measurements are currently in progress to make this point precise. However, oxygen vacancies exist at the surface, but may be in a different concentration to that in the bulk. The main difference between bulk and surface is that the surface need not necessarily be neutral, and the local electronic structure around either an yttrium or an oxygen atom may be quite different in the bulk or at the surface / 24 /.

As mentioned in II.2, in an ionic bond model, which can be applied to be case of an ionocovalent oxide such as Y_2O_3, the variation of Auger parameter can be related to the electronic polarization energy variation and to the polarizability variation of the oxygen ion.

In the case of the reduced samples (pure or doped) the oxygen Auger parameter is larger than for the oxidized pure sample. This result shows that the creation of oxygen vacancies leads to a charge redistribution which makes the Y-O bond more covalent on average, and which corresponds to a decrease in the mean oxygen polarizability.
A charge transfer occurs from the anion to the cation. The same conclusion can be obtained considering the yttrium Auger parameter, whose variation is weaker.

This conclusion concerning the decrease of the ionic character of the oxide when reducing was confirmed by the study of the oxygen KLL Auger lineshape, and by the energy of the final state of the ion in the O KLL Auger transition / 9, 25 /.

As far as the reduced doped Y_2O_3 sample is concerned, the Auger parameters of Y and O indicate a charge transfer at most as great as that observed in the pure sample having experienced the same reduction treatment, although the oxygen deficit around Y is larger. That would imply that the change of ionicity of the Y-O bond is also related to the static disorder in the first coordination shell.

IV.3 - Microscopic aspects of the surface modifications induced by a reduction heat treatment

The EXAFS investigations and the XPS experiments show differences in the local geometric structure around an yttrium atom, as well as in the local electronic structure around both oxygen and yttrium. This information is averaged over the part of the surface irradiated by the X-ray beams (i.e., for EXAFS : 2 cm x 5 mm and for XPS : 8 mm x 2 mm).

As these ceramics are polycrystalline it is necessary to investigate the spatial homogeneity of this information. These microscopic observations have been obtained by scanning electron microscopy.

The behavior of uncoated yttrium oxide under an electron beam is quite different when the sample is stoichiometric or oxygen deficient. This different behavior toward charging up under the beam has been observed for pure Y_2O_3 in a previous paper / 6 /.

On the doped reduced sample, the differences in the contrasts of the different grains observed on the backscattered electron image and in the secondary electron image arise from channeling of the primary electron beam / 26 /. This effect modulates the backscattering coefficient of the primary beam, and can be observed also in the secondary electron imaging mode (SEI), due to the contribution of backscattered electrons to the total intensity of the detected electrons.

The pure reduced sample exhibits a quite different type of contrast when observed in SEI. The crystalline contrast is no longer observed, but a black zone appears along the grain boundaries. This contrast, similar to a voltage contrast, can be attributed to a modification of the surface potential in the neighborghood of the grain boundaries. The grain boundaries are transition regions between crystals with the same nature but with different crystallographic orientation. Therefore defects may find favored sites to diffuse. Thus, in the case of the pure oxygen deficient sample, oxygen vacancies could diffuse and give rise because of their effective charge to a local potential, which could be at the origin of the contrast observed in the secondary electron image. This contrast was observed to disappear under oxidation conditions / 6 /, indicating that it is related to the presence of oxygen vacancies which are segregated at the grain boundaries of the pure reduced Y_2O_3 sample.

However, this electrical activity at the grain boundaries is not observed on the doped reduced sample, which is known to contain more anionic vacancies than the pure reduced one, as deduced from the EXAFS results.

This would imply that, in the case of the doped reduced samples the anionic vacancies are not so strongly segregated at the grain boundaries. This may be due to a competitive segregation of Zr at the grain boundaries. The difference in the grain growth in the doped sample supports this hypothesis. Indeed for the same H.I.P. processing and the same subsequent heat treatment, the mean grain size in the doped reduced sample is three times smaller on average than in the pure reduced one, indicating that Zr plays an important role in the grain boundary motion.

Furthermore, this result concerning the segregation of vacancies in the pure and doped Y_2O_3 has to be related to the differences observed in the local geometric structure around an yttrium atom in the pure and doped yttria. In the doped reduced sample, the number of oxygen vacancies is larger, and the static disorder much less important than in the pure reduced one, but no point defect association or ordering can be deduced from the EXAFS results, as outlined by Tang et al. / 27 /.

However it is known that, when the concentration of defects is high, the induced disorder may be stabilized either by defect clustering or aggregation, or by the elimination of point defects with the formation of extended planar defects. Defect clustering has been observed for example in oxygen deficient CeO_2 (doped with Y^{3+}), which crystallizes in the fluorite structure.

Elimination of point defects with formation of extended planar defects, such as shear planes, has been observed in oxygen deficient titanium oxide /1/.

The formation of such extended defects in the doped reduced Y_2O_3 ceramics could on the one hand explain the lack of static disorder in the first coordination shell around Y, compared to the case of the pure reduced samples, and on the other hand the difference in the defect segregation at the grain boundaries, which changes their electrical activity, and therefore their SEI contrast. Further experiments on grain boundaries must shed light on these hypothesis.

V - CONCLUSION

Surfaces of oxygen deficient yttrium oxide, pure or doped (14 wt % ZrO_2), were studied by means of X-ray photoelectron spectroscopy and scanning electron microscopy.

After a reduction heat treatment (1700°C under vacuum) an important concentration of oxygen vacancies is produced as shown for the bulk by the EXAFS measurements, the oxygen deficiency depending on the doping.

The presence of these oxygen vacancies leads to a modification of the local electronic distribution of the surface around both Y and O. A partial charge transfer from the anion to the cation occurs, leading to a more covalent Y-O bond.

However this information is averaged over several grains of the polycrystalline ceramics. Using scanning electron microscopy, an inhomogeneity in the anionic vacancies concentration was observed, depending on the presence of altervalent substituting cations (Zr).

An electrical activity of the grain boundaries is detected in the case of pure reduced Y_2O_3 : oxygen vacancies are strongly segregated at the grain boundaries. However, this electrical activity cannot be seen for doped reduced Y_2O_3 in which the anionic vacancies concentration is larger.

This result has to be related to some defect rearrangement in doped reduced Y_2O_3, like defect clustering or formation of extended defects (shear planes), in agreement with the differences in the static disorder obtained by the EXAFS experiments.

REFERENCES

/1/ C.R.A. Catlow, 'Defect clustering in nonstoichiometric oxides', in 'Non stoichiometric oxides', 61-98. ed. by O. Toft Sorensen. Academic Press, New York (1981).

/2/ N.M. Tallan, R.W. Vest, 'Electrical properties and defect structure of Y_2O_3', J. Am. Ceram. Soc., 49 (8), 401-404 (1966).

/3/ T. Norby, P. Kofstad,'Direct Current Conductivity of Y_2O_3 as a function of water vapor pressure', J. Am. Ceram. Soc., 69 (11) 780-783 (1986).

/4/ T. Norby, P. Kofstad, 'Electrical conductivity of Y_2O_3 as a function of oxygen partial pressure in wet and dry atmospheres', J. Am. Ceram. Soc., 69 (11), 784-789 (1986).

/5/ T. Norby, P. Kofstad, 'Proton and native-ion conductivities in Y_2O_3 at high temperatures', Sol. Stat. Ionics, 20, 169-184 (1986).

/6/ F. Jollet, Ph. Maire, M. Gautier, J.P. Duraud, C. Le Gressus, 'Influence of stoichiometry on the electrical and mechanical behavior of yttrium oxide ceramics', J. Am. Ceram. Soc., 71, (9), C396-C398 (1988).

/7/ H.L. Tuller, 'Thermodynamics and defect structure of non stoichiometric oxides', in 'Non-Stoichiometric oxides', 2-59, ed. by O. Toft Sorensen, Academic Press, New York (1981).

/8/ M.G.SCOTT, 'Phase relationships in the zirconia-yttria system', J. of Materials Science, 10, 1527-1535 (1975).

/9/ J.P. Duraud, F. Jollet, N. Thromat, M. Gautier, Ph. Maire, C. Le Gressus, E. Dartyge, 'Non-stoichiometry of pure and Zr-doped yttria ceramics : an EXAFS and XPS study', submitted to J. Am. Ceram. Soc.

/10/ R.C. Anderson, in 'High temperature oxides', ed. by M. Aper, Academic Press, New York, 1-40 (1970).

/11/ P. Duwez, F.H. Brown, F. Odell, 'The zirconia-yttria system', J. Electrochem. Soc., 98, 350-352 (1951).

/12/ C.D. Wagner, 'Auger parameter in electron spectroscopy for the identification of chemical species', Anal. Chem., 47, 1201-1203 (1975).

/13/ C.D. Wagner, D.E. Passoja, H.F. Hillery, T.G. Kinisky, H.A. Six, W.T. Jansen, J.A. Taylor, 'Auger and photoelectron line energy relationships in aluminium-oxygen and silicon-oxygen compounds', J. Vac. Sci. Technol., 21 (4), 933-944 (1982).

/14/ S. Kohiki, S. Ozaki, T. Hamada, 'Characterization of silicon compounds using the Auger parameter in X-ray photoelectron spectroscopy (XPS)', Appl. Surf. Science, 28, 103-110 (1987).
/15/ N.F. Mott, R.W. Geviney, '<u>Electronic Processes in Ionic Crystals</u>', Clarendon, Oxford, 1948.
/16/ C.D. Wagner, 'Chemical shifts of Auger lines, and the Auger parameter', Faraday Disc. Chem. Soc., 60, 291-301 (1975).
/17/ C.D. Wagner, A. Joshi, 'The Auger parameter, its utility and advantages : a review', J. of Electron Spectroscopy and Related Phenomena, 47, 283-313 (1988).
/18/ R.H. West, J.E. Castle, 'The correlation of the Auger parameter with refractive index : an XPS study of silicate using ZrLα radiation', Surf. Interf. Analysis, 4 (2), 68-75 (1982).
/19/ H. Niedrig, 'Electron backscattering from thin films', J. Appl. Phys., 53 (4), R15-R49 (1982).
/20/ T.H. Etsell, S.N. Flengas, 'The electrical properties of solid oxide electrolytes', Chem. Rev., 70, 339-376 (1970).
/21/ K. Ando, Y. Oishi, 'Oxygen self diffusion in Y_2O_3 and Y_2O_3-ZrO_2 solid solutions', NATO-ASI, ser. B, 129, 203-215 (1985).
/22/ Ph. Odier, J.P. Loup, 'An unsual technique for the study of non-stoichiometry : the thermal emission of electrons. Results for Y_2O_3 and TiO_2', J. of Solid State Chem., 34, 107-119 (1980).
/23/ R. Bratton, 'Defect structure of Y_2O_3-ZrO_2 solid solutions', J. Am. Ceram. Soc., 52, 213 (1969).
/24/ A.M. Stoneham, 'Ceramic Surfaces : theoretical studies', J. Am. Ceram. Soc., 64 (1), 54-60 (1981).
/25/ C.D. Wagner, D.A. Zatko, R.H. Raymond, 'Use of the oxygen KLL Auger lines in identification of surface chemical states by electron spectroscopy for chemical analysis', Anal. Chem., 52, 1445-1451 (1980).
/26/ D.C. Joy, D.E. Newbury, D.L. Davidson, 'Electron channeling patterns on the scanning electron microscope', J. Appl. Phys., 53 (8), R81-R122 (1982).
/27/ C. Tang, P. Georgopoulos, J.B. Cohen, 'Study of extended X-ray absorption fine structure for possible use in examining local atomic arrangements in oxides', J. Am. Ceram. Soc., 65 (12), 625-629 (1982).

SEGREGATION IN ZrO_2-Y_2O_3 CERAMICS

G.S.A.M. THEUNISSEN, A.J.A. WINNUBST, A.J. BURGGRAAF
University of Twente, Faculty of Chemical Technology, Lab. for Inorganic Chemistry, Materials Science and Catalysis,
P.O. Box 217, 7500 AE ENSCHEDE, The Netherlands.

ABSTRACT The composition of grain boundaries and surfaces often control phenomena like sintering, grain growth, wear resistance and corrosion behaviour. In the present study the surface of zirconia ceramics doped with various amounts of Yttria was examined in order to find an explanation for observed differences in grain growth. This was done by using AES (Auger Electron Spectroscopy) and XPS (X-ray Photoelectron Spectroscopy). Specimen heat treated at 873K had a surface composition similar to their bulk composition. Specimen heat treated at 1273K showed a surface composition containing 30-34 at% Yttria independent of the bulkcomposition. This results in a larger solute drag for zirconia with low amounts of yttria which probably causes a slower grain growth by means of an impurity drag mechanism. It could be shown that the relative increase of the 76eV Augerline was not due to Silicon segregation but solely to the segragation of Yttrium.

1. Introduction

In solid solutions a difference is often present in chemical composition between bulk and interface of the materials. This difference frequently occurs because of segregation of impurities or main components to the interfaces. Several types of driving forces can give rise to segregation in solids [1].

1. Surface tension (energy) differences and (surface) interaction effects.
2. Strain energy (relaxation).
3. Electrostatic potential and charge compensation.
4. Spread in Madelung potentials.

Group 1 and 3 have in common that they are based on differences in atomisation (ionisation) energies between the components in the mixture. The composition of grain boundaries and of surfaces of ceramic materials often control fabrication behaviour like sintering and grain growth [2]. Besides this, segregation of main components under the influence of water may also influence the thermal stability of tetragonal zirconia as was dicussed by Winnubst et al. [3].

In most zirconia ceramics grain growth behaviour varies with composition. Different mechanisms has been proposed to explain this difference in grain growth. Lange [4] attributed this difference to the

L.-C. Dufour et al. (eds.), Surfaces and Interfaces of Ceramic Materials, 365–372.

presence of different crystal phases in the ceramic (the cubic grains control the growth of the tetragonal ones). Jurado [5] assumes that the larger grains in tetragonal zirconia have grown at the expense of the smaller ones. In our zirconia materials however, with a bulk composition ranging from 4 to 8.9 mol% $YO_{1.5}$, no cubic phase or abnormal grain growth was observed. The mechanisms mentioned before therefore do not account for the observed differences [6].

Another phenomenon, which may explain the difference in grain growth rate is a decrease in grain boundary mobility through an "impurity" drag mechanism [2]. This implies the presence of a segregation layer at the surface, which consists of impurities or solute atoms.

In order to find an explanation for the difference in grain growth (and grain boundary conductivity) in Zirconia materials doped with various amounts of Yttria the chemical composition of pure ultrafine grained $(ZrO_2)_{100-x}(YO_{1.5})_x$ ceramic interfaces was studied by means of Auger Electron Spectroscopy (AES) and X-ray Photoelectron Spectroscopy (XPS).

2. Experimental procedure

As starting materials for the synthesis of the $(ZrO_2)_{100-x}(YO_{1.5})_x$ ceramics (sample code ZYx) ultra fine grained powders were used (8 nm). These powders were prepared by the alkoxide or chloride method [6,7], which yields a pure and sinterreactive powder. After isostatic pressing at 400 MPa and consequently sintering at temperatures ranging from 1100-1400 ^oC a relative density of more than 95% could be obtained. The resulting grain sizes range from 0.1 um with low yttria content to 0.4 um with high (> 13 mol% $YO_{1.5}$) yttria content.

Before the surface analysis experiments were carried out, the specimen were carefully polished with Al_2O_3 (0.05 um) and ultrasonically cleaned in ethanol. These specimen were submitted to a temperature treatment in an oxygen atmosphere for 5 hours at either 600, 700 or 1000^oC. As a standard, rods were prepared which fractured transgranular under ultra high vacuum. In this way surface contamination could be avoided and a fresh surface having the bulk composition is obtained.

The Auger analysis were performed with a cylindrical mirror analyser, type 10-155 (Physical Electronics). The analysis were performed using a primary beam current of 0.25 uA. No sample charging occured under these conditions when clean specimen were used.

The XPS measurements were carried out on a Kratos ES 200B spectrometer, using Al-K_α radiation (1487 eV), in a vacuum chamber with a background pressure lower than 10^{-8} Torr. Calibration of the XPS-spectra was performed using the C_{1s} peak.

In case of AES the spectra were examined in a semi-quantitative way using Palmberg's equation [8]:

$$C_x = (I_x/S_x)/\sum_\alpha(I_\alpha/S_\alpha) \qquad (1)$$

where C_x is the concentration (in at%) of element x. I_x and S_x are respectively the peak intensity and the relative elemental sensitivity factor of the Auger line of element x used for quantification. The Yttrium and Zirconium sensitivity factors are obtained from the standard spectra, the oxygen sensitivity was given a fixed value of 0.5 [8].

To quantify the XPS spectra an equation similar to equation (1) was used [9]. The method of analysis is described elsewhere [10].

To give a good interpretation of the Auger spectra, good knowledge of the correct position of the Augerpeaks is necessary.
Yttrium has an MNN Augerpeak at 76 eV, Zirkonium a MNN peak at 147 eV and Oxygen a KLL peak at 503 eV. In measuring the Yttrium peak one has to be careful because of the presence of a LMM Silicon peak on exactly the same place (76 eV). This is important because in some cases a small amount of Silicon (which segregates quite easily) is present in the ceramic. Therefore part of the peak intensity at 76 eV can be due to Silicon. In order to eliminate the Silicon part of the 76 eV peak a calculation has to be made. This can be done by deducing the Silicon intensity at 76 eV from the intensity of the free KLL Silicon peak at 1606 eV (oxidic form; in elemental form the peak is situated at 1619 eV [8]).

element	oxide form
Zr	.10
Y	.13
O	.50
Si (1619 eV)	.081
Si (76 eV)	.373

Table I: Auger sensitivity factors (E_p = 3 keV).

By using the sensitivity factors as stated in table I, the transformation is performed as follows:

$$I_{Si,76} = I_{Si,1606} \cdot S_{Si,76}/S_{Si,1606} \qquad (2)$$

The sensitivityfactors of Silicon at 76 and 1606 eV are obtained from a standard SiO_2 Augerspectrum [8]. In cases where Silicon is found to be present the Yttrium intensity is obtained by substracting the Silicium intensity from the total intensity at 76 eV. Because only the upper part of the Augerpeak at 76 eV has been taken into account, the calculated $I_{Si,76}$ has to be divided by a factor 2 to obtain the Yttrium intensity. In formula:

$$I_Y = I_{tot,76} - I_{Si,76}/2 \qquad (3)$$

in which $I_{tot,76}$ represents the measured peakheight at 76 eV.
In the XPS-spectra however a clear distinction can be made between the Y_{3d} and the Si_{2s} peak. The Y_{3d} peak has its maximum at 161.6 eV, Si_{2s} at 157.8 eV.

3. Results and Discussion

The Auger spectra of a polished surface of ZY17 heat treated at T = 873 K and 1273 K respectively are shown in fig. 1a and b. From this it

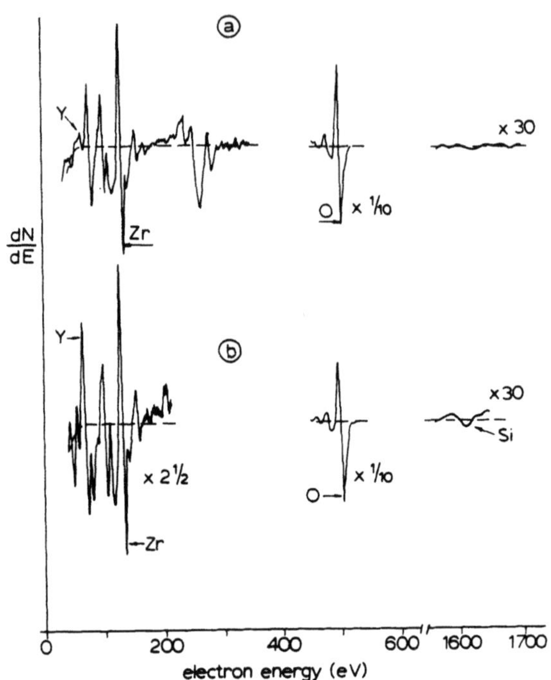

Figure 1: Augerspectra of a polished surface of ZY17 heat treated for 5 hr at 873 K (a) or 5 hr at 1273 K (b).

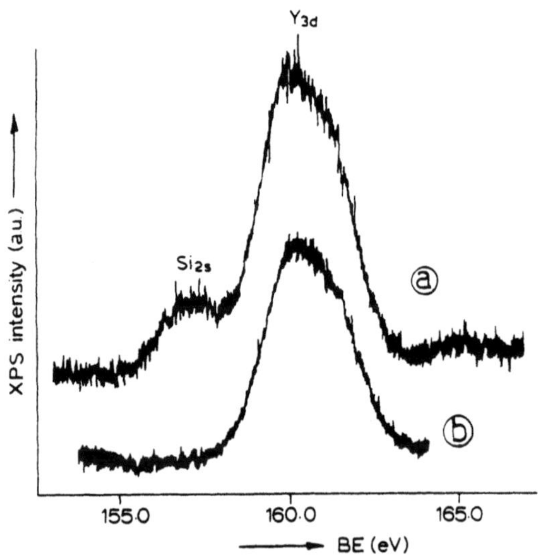

Figure 2: XPS spectra of ZY specimen treated at 1273 K during 5 hrs
a) ZY4 with Silicon contamination.
b) ZY6.1 without Silicon contamination.

can be seen that the Y/Zr peak intensity ratio of the high temperature treated specimen (fig. 1b) is higher than that of the bulk (fig. 1a). The inaccuracy in the choice of the base line is smaller than the observed differences in the peak intensity ratio between the two spectra. Therefore it can be concluded that Yttrium enrichment occurs at the surfaces of ZY17.

The Yttrium enrichment at the surface is also indicated by a decrease of the Yttrium signal as function of the Ar^+ ion sputtering time. After 30-40 seconds of ion sputtering the Y/Zr peak intensity ratio becomes constant (surface composition is then about ZY10). This quantitative result, obtained from the observed sputterprofile, is not very reliable because of some reduction of Zr^{4+} to Zr metal (The maximum of Zr metal doesn't occur at the same energy as Zr^{4+} in oxides). Therefore the thickness of the Yttrium-segregation layer cannot be determined with any precision but is estimated to be of the order of 2-4 nm [10].

Bulkcomposition sample	Temperature treatment (K)	Surfacecomposition x (from ZYx)
ZY4	873	5
ZY4	1273	31
ZY5*	873	6
ZY5*	1273	30
ZY6.1	873	6
ZY6.1	1273	30
ZY13	873	11
ZY13	1273	34
ZY17	873	17
ZY17	1273	35
ZY17*	1273	31
ZY26.4	1273	34
ZY32	873	33

* : single crystals

Table II : Chemical surface composition Y_2O_3-ZrO_2 ceramics heat treated during 5 hours as determined by AES. (Materials with compositions between 4 and 9 at % $YO_{1.5}$ have a tetragonal structure, the others have a cubic fluorite structure).

The quantitative results of the measurements are given in table II, using the sensitivity factors shown in table I. From this it can be seen that after temperature treatment a pronounced Yttrium enrichment takes place up to a composition of about ZY30-ZY34.

These results are already corrected for possible Silicon contamination if this is present. That the larger augerpeaks at 76 eV are not due to Silicon is shown in the XPS spectra of fig. 2. In this figure (2b) no significant presence of Silicon is shown (the detectionlimit is about 0.2 at%).

In AES (fig. 1), the peak to peak height of Silicon at 1606 eV (with a 30* peak enlargement) should be at least 5 mm to be visible above the backgroundlevel. In case of a peakheight of 5 mm (1606 eV) a Silicon intensity of less than 0.5 mm is calculated at 76 eV which, in general, causes a correction of only a few percent in the total peakheight at 76 eV (and therefore in the Yttrium intensity).

The large influence of a possible Silicon contamination is shown in the following example. A sample was purposely contaminated with Silicon. In this particular case the real peakintensity of Yttrium was about 20% lower compared to the case with ignorance of Silicon where the total 76 eV peakintensity was ascribed to Yttrium. The composition was calculated to be ZY42 (ignorance of Silicium) and ZY34 with Silicon taken into account. In this case the Silicium concentration at the surface was found to be 3 at%.

All the samples which received a temperature treatment of 1273K showed an Yttrium enrichment of about 30 at% to the surface (and probably also to the grain boundary [1,10,11]). This is independent of their crystal structure (both 100% tetragonal and 100% cubic materials show the same result) and bulk composition. So this concentration seems to be a saturation value. In the XPS spectrum also an Yttrium enrichment is observed. Using the sensitivityfactors found by Wagner et al. [9], a surface composition of ZY20 is calculated (bulkcomposition ZY6.1). To calculate the exact composition standards has to be made, which has not been done in this case. The AES results were confirmed by Steele [12] who found with XPS on the (001) surface of ZY15 a concentration of 33 at% Yttrium. These results mean that the Yttrium enrichment factor (defined as $C_{Y,surface}/C_{Y,bulk}$) for materials containing low amounts of Yttria (i.e. tetragonal materials) is much larger than the (cubic) materials containing high percentages of Yttria. A stronger segregation (higher enrichmentfactor) results in a steeper Yttrium gradient between the surface and the bulk of the material, and therefore in a stronger "solute" drag. Probably a stronger "solute" drag results in a stronger retardation of the grain growth (which means a smaller grain size after sintering). This is what has been observed while sintering Yttria stabilized zirconia materials doped with various amounts of Yttria. All the materials started with the same initial crystallite size [6]. This might also explain why in PSZ ceramics containing both tetragonal and cubic phases the larger grains are cubic.

At low temperatures no segregation of any component was found to be present. Measurements were also carried out on powdercompacts because it is believed that the segregationbehaviour is different from that of fully dense, coarse grained ceramics (in fine grained ceramics a large exhaustion of the bulk takes place if the same amount of segregation occurs as in coarse grained ceramics). These measurements (on compacted powders) seems to show that also in very fine grained compacts (grainsize below 20 nm) no segregation of Yttrium occurs. The impact of this observation on sintering will be discussed in a next paper [13].

Because the as prepared single crystals (by skull-melting) show the same segregation behaviour as the ceramical materials, a retardation in domaingrowth in single crystals may also be possible.

In conclusion:

1. Al low temperatures no segregation of any component has been observed.

2. After temperature treatment at 1273 K for 5 hr $(ZrO_2)_{100-x}(YO_{1.5})_x$ ceramics show a surface composition of about 30-34 at% Yttrium regardless the bulk composition of the ceramic. The segregation layer has a thickness of a few atom layers (≃ 2 nm)
3. A similar surface composition of about 30 at % Yttrium for ceramics containing low amounts of Yttria (in the bulk) results in a larger Yttrium enrichment factor compared to ceramics containing larger amounts of Yttria. So the Yttrium enrichment factor increases from 1.5 to 7 with a bulk concentration of Yttria decreasing from 26 to 4 at% $YO_{1.5}$.
4. Grain growth is dependant on the amount of Yttria in the bulk, which results for tetragonal Yttria stabilized Zirconia in a slower grain growth compared to the cubic ones.
5. To obtain reliable results, special attention has to be paid to the presence of Silicon.

ACKNOWLEDGEMENTS
We would like to thank Elcoma (Philips Electronic Industries Eindhoven) for use of their AES equipment. A. van de Berg is acknowledged for his assistance in performing the XPS measurements and discussion of the results. These investigations were partly supported by the Innovative Research Program on Technical Ceramics (IOP-TK) with financial aid of the Dutch Ministry of Economic Affairs.

REFERENCES

1. A.J. Burggraaf, A.J.A. Winnubst, "Segregation in oxide surfaces, solid electrolytes and mixed conductors" in: Surface and Near-Surface Chemistry of Oxide Materials, ed. by J. Nowotny and L.-C. Dufour, Elsevier Science Publishers B.V. Amsterdam, 1988 pp.449-477.
2. R.J. Brook, in Treatise of Materials Science and Technology, **9** edited by F.F.Y. Wang (Academic Press, New York, 1976) p. 331.
3. A.J.A. Winnubst, A.J. Burggraaf, Proceedings of the 1986 Zirconia III conference, Advances in Ceramics ed. by the American Ceramic Society, to be published.
4. F.F. Lange, J. Am. Ceram. Soc., **69** (1986) 240.
5. J.R. Jurado, C. Moure, P. Duran, J. Phys. Coll., **47** (1986) C1-789.
6. A.J.A. Winnubst, G.S.A.M. Theunissen, W.F.M. Groot Zevert, A.J. Burggraaf, "The sintering behaviour of fine grained ZrO_2-Y_2O_3 ceramics", in: Science of Ceramics **14**, ed. by D. Taylor, The Institute of Ceramics, Shelton, Stoke-on-Trent UK, 1988, pp. 309-314.
7. M.A.C.G. van de Graaf, J.H.H. ter Maat, A.J. Burggraaf, J. Mat. Science **20** (1985) 1407.
8. Palmberg P.W., Riack G.E., Weber R.E., McDonald N.C., Handbook of Auger spectroscopy, Physical Electronics, Edina, Minnesota (1972).
9. C.D. Wagner, W.M. Riggs, L.E. Davis, J.F. Moulder, G.E. Muilenberg, Handbook of X-ray Photoelectron Spectroscopy, Perkin-Elmer Corp., Minnesota (1979).

10. A.J.A. Winnubst, P.J.M. Kroot, A.J. Burggraaf, J. Phys. Chem. Solids **44** (1983) 955.
11. A.J. Burggraaf, M. van Hemert, D. Scholten, A.J.A. Winnubst, in Reactivity of Solids (Elsevier, Amsterdam, 1985) p. 797.
12. B.C.H. Steele, E.P. Butler, in Britisch Ceramic Proceedings, 1985, edited by (The institute of Ceramics, Stoke on Trent, 1985) p. 45.
13. G.S.A.M. Theunissen, A.J.A. Winnubst, W.F.M. Groot Zevert, A.J. Burggraaf, in preparation.

MICROSTRUCTURAL CHANGES AND DYNAMIC SEGREGATION NEAR AN INTERFACE FORMED DURING THE REDUCTION OF $(Fe_{1-x-y}Ca_y)O$

J. Kusinski *, S. Jasienska *, A. Rivière ** and C. Monty **
* Academy of Mining and Metallurgy, Krakow (Poland)
** Laboratoire de Physique des Matériaux, CNRS, Meudon (France)

Numerous studies have been done about the influence of foreign ions in solid solutions on the structure and reduction of wustite because some of these ions are present in the iron ores and wustire remains a reference material in many fundamental studies. The present study is one among the series of our investigations devoted to the role of Ca on the structure and the reduction process of $(Fe_{1-x-y}Ca_y)O$ wustite (1). Calcium has relatively high ionic size ($r_{Ca^{2+}}$ = 0.99 Å) as compared to Fe ($r_{Fe^{3+}}$ = 0.564 Å); this causes expansion of the crystal lattice and therefore may change the defect structure of wustite and furthermore influence the reduction process.

The samples of $(Fe_{1-x-y}Ca_y)O$ were prepared as follows: a) oxidation of the high purity iron plates for 6 days at 1150 °C (Fig. 1), b) thermochemical

Fig. 1. SEM image of $Fe_{1-y}O$ formed by oxidation of iron plates

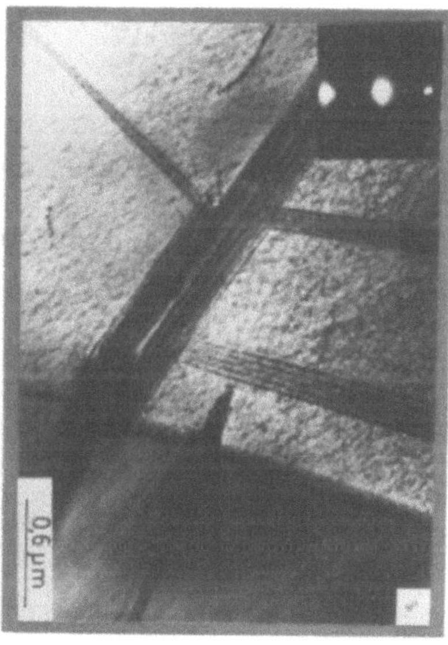

Fig. 2. Stacking faults interaction with the interface $Fe_{1-z}Ca_z/Fe_{1-x-y}Ca_yO$ in partially reduced samples

L.-C. Dufour et al. (eds.), Surfaces and Interfaces of Ceramic Materials, 373–374.

treatment in a CaO powder at 900 °C for 4 hrs, c) homogenization at 900 °C for 5 days. All these treatments were done at P = 1 atm. in the following gas mixture: 10% CO - 10% CO_2 - 80% N_2. Such a treatment enabled to get a uniformly distributed Ca concentration as controlled by Electron Microprobe Analysis (Cameca). The values of Ca concentration ranged from 0.05 to 0.82 at%. TEM images of Ca doped wustite showed a high density of stacking faults present in the structure with some dislocations associated. (Fig. 2). Such a phenomenon was already observed in Na doped FeO (2).

After reduction at 900 °C in CO, the samples (bulk and thin films) were again analysed by TEM and EMA. During reduction, nucleation of Fe crystals occurred in the areas containing a large number of stacking faults (Fig. 2). An increase of dislocation density was observed in the vicinity of Fe nuclei. These Fe crystals were very regular and oriented parallel to the <110> directions in wustite. Differences in the Ca distribution were evidenced by EMA near the reaction interface as a function of time of reduction (Fig. 3). Increased Ca amount (higher than 1.5 At% in a 30 μm thick layer) leads to precipitation of $CaFe_5O_7$ at stacking faults or in their vicinity. These results illustrate a phenomenon of dynamic segregation controlled by the diffusion of Fe and Ca (3) and by the difference of Ca solubility between FeO and Fe.

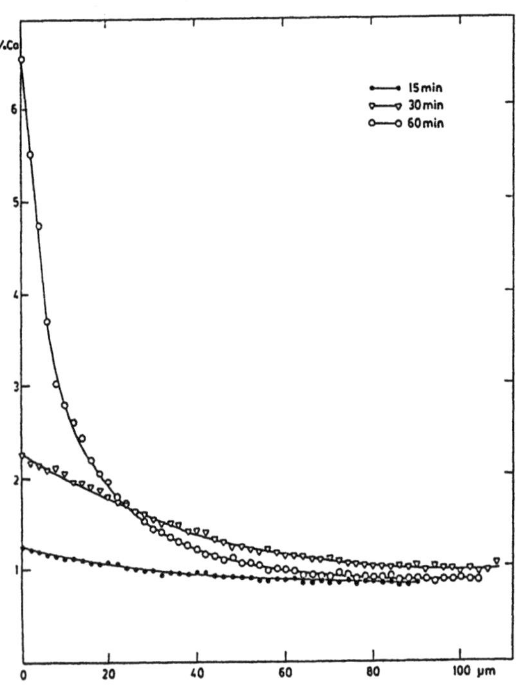

Fig. 3 Dynamic segregation of Ca near an interface $Fe_{1-z}Ca_z/Fe_{1-x-y}Ca_yO$. In the nucleated Fe, the Ca concentration is around 0.01 - 0.03 at%.

References

1 S. Jasienska *et al.*, Crystal Latt. Def. and Amorph. Mat., 16 (1987) 145
2 A.M. El Balkhi, M. Zanne and C. Gleitzer, J. Solid State Chem., 18 (1976) 293
3. G. Petot-Ervas, this book

THERMODYNAMICS AND CHEMISTRY OF CERAMIC-METAL INTERFACES

J.T. KLOMP
Centre for Technical Ceramics
Eindhoven University of Technology
P.O. Box 513, 5600 MB Eindhoven, The Netherlands

Abstract
In this tutorial on metal-ceramic interaction the fundamentals of metal-ceramic bonding will be considered from the equilibrium thermodynamics point of view. This includes the wetting of a solid ceramic by a liquid or solid metal or vice versa, the effect of the gas atmosphere on the wetting behaviour and on the chemical reactions that can occur at a metal-ceramic interface. From this approach the conditions for a metal-ceramic reaction can be predicted and it also gives indications to the chemical stability of the metal-ceramic bond. Additional to this the effect of impurities or second phases present in the ceramic, such as segregation of impurities or melt formation, on the formation of a metal-ceramic bond is discussed. Finally the formation of interfacial phases and the effect of the interface morphology is briefly considered.

1. Introduction

The use of ceramics is ever increasing in the electronic field and structural or engineering ceramics are now on the verge to be applied in power sources. Ceramic research does and will continue to result in new types of ceramics such as superconducting ceramic and bioceramics, to mention a few. Most ceramics have to be used in combination with metals to take full advantage of the properties of the materials and to provide the required freedom in the design of the structure. There also is a continuing interest in structural composites that are built up from ceramic fibers in a metal matrix. This makes the field of metal-ceramic interface research important because it will provide the required knowledge about controlled metal-ceramic interactions to proceed for bonding applications and about the stability of a bond in operational conditions. This includes the kinetics of metal-ceramic interactions which information is, with a few exceptions (Backhaus-Ricoult ref. 1), almost lacking. This knowledge is essential to give the technology a background based on materials science rather than based on receipes obtained by trial and error. As in many technological areas the development of successful techniques for metal-ceramic bonding was far ahead the development of interface science. The first development in

L.-C. Dufour et al. (eds.), Surfaces and Interfaces of Ceramic Materials, 375–392.

metal-ceramic bonding started in 1936 by sintering refractory metal powder on steatite ceramic at high temperature. This technology, although modified, is still the most widely applied for oxide ceramic-metal systems used in the electronic field and in vacuum systems. The background of this process is the migration of the intergranular phase of the ceramic into the porous metal layer where it wets the metal particles. This is known as the glass migration theory (Cole and Sommer ref. 2). Thus an oxide-metal composite is formed that adheres to the ceramic.

Wetting phenomena also play an important role in later developments, the so called ceramic frit process (Klomp and Botden ref. 3) and the active metal brazing process (Nicholas ref. 4) where at elevated temperatures a liquid oxide mixture or a liquid metal alloy respectively fills a capillary space between a metal and a ceramic body. In the solid-state bonding of metal to ceramic and in the sintering of composites wetting in the sense of interface formation is of equal importance (ref. 5). Having emphasized in the foregoing the importance of wetting we have to give this macroscopic phenomenon a thermodynamic meaning because wetting can occur in metal-ceramic systems with and without chemical reaction at the metal-ceramic interface. In the next sections the fundamentals of wetting, capillary action and chemical reaction of metal-ceramic systems are considered from an equilibrium thermodynamics point of view. Attention is given to the effects of segregation of impurities or purposely added alloying elements on the metal-ceramic interactions.

2. Interactions at metal-ceramic interfaces: adsorption and segregation effects.

The first stage in every interfacial interaction process between materials is the creation of intimate contact between the surfaces so that the atomic species are brought within interatomic distances.

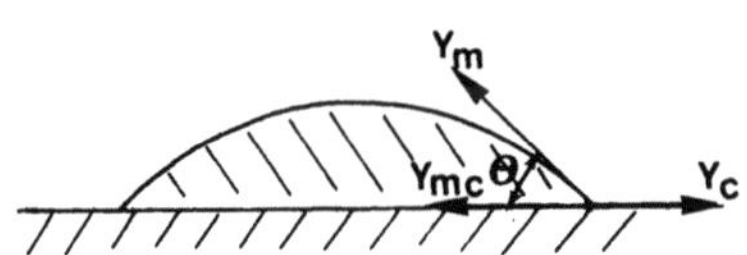

Fig. 1. Contact angle (Θ) of a liquid wetting a solid substrate.

This is independent of the physical state (solid, liquid or vapour) of the materials, but we consider here the ceramic being always in the solid state. The driving force for interaction between the materials is the decrease of the free energy of the system. If between a metal and a ceramic with surface energies in a given environment of γ_m and γ_c respectively an interface is formed with energy γ_{mc} then the energy change per unit area (ΔG) of the system is given by the Dupré eguation:

$$\Delta G = \gamma_m + \gamma_c - \gamma_{mc} \qquad 1)$$

This energy change is identical to the work of adhesion (W_{ad}), that is the work required to separate a unit area of interface into the two original surfaces. The consequence of this definition is that no changes of the surface, neither structurally nor chemically have occurred. It is common practice to derive W_{ad} or γ_{mc} from the measured contact or wetting angle (Θ), which a solid or liquid material establishes in equilibrium with a solid surface in a given atmosphere (ref. 6) see Fig. 1.

The Young-Dupré equation states for the equilibrium conditions:

$$\gamma_c - \gamma_{mc} - \gamma_m \cos\Theta = 0 \qquad 2)$$

From 1) and 2) it follows

$$W_{ad} = \gamma_m (1 + \cos\Theta) \qquad 3)$$

Wetting occurs if $\gamma \leq 90°$.
Although it is common practice to determine W_{ad} or γ_{mc} experimentally according to eq. 3, the validity of its application to an actual system has to be justified since one has to be sure that no changes at the materials surfaces have occurred during the measurements. Otherwise corrections on the values of γ_m and γ_c have to be made. Such changes can be caused by a number of effects such as: charge transfer to the outermost layers of the materials, chemisorption of gas at the surfaces, segregation of impurities to the interface and by interfacial chemical reactions.
Since interfaces are not easily accessible for analytical work the influence of most of the above mentioned effects on the original surface energy of the materials can hardly be made quantitative. The effect of chemisorption of oxygen on γ_m of a metal can be large, e.g. $\gamma_{Al_2O_3} = 892$ and $\gamma_{Al} = 1160$ mJm^{-2}.
If chemical reactions occur the situation becomes much more difficult. Not only the surface energy of the reaction products that form the interface has to be known but the in situ measurement of the contact angle (Θ) is impossible (fig. 2).

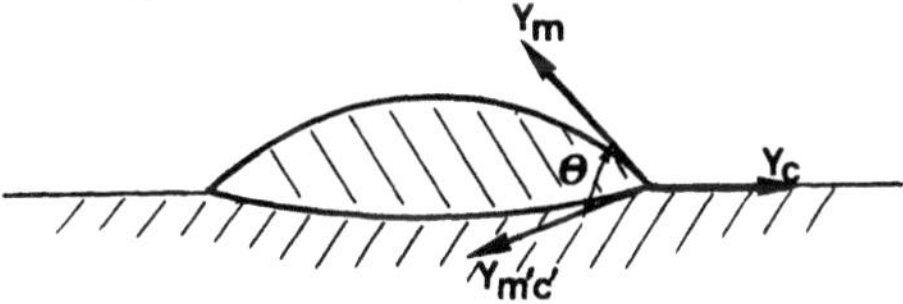

Fig. 2. Contact angle (Θ) of a liquid by reaction with a solid substrate.

Moreover, the requirement that equilibrium of the system is established has still to be fulfilled. Segregation of impurities to free metal surfaces can be well predicted (Niessen and Miedema ref. 8) but it is much more difficult to rationalize this for ceramic systems and for metal-ceramic systems hardly any prediction can be made (Hofmann ref. 9). This is because the nature of the bonds in metal-ceramic systems is not well understood. The consequence of impurity segregation is that the surface energy of the material changes according to the Gibb's adsorption isotherm:

$$\Gamma \approx d\gamma / RT \ln Ci$$

in which Γ is the excess surface concentration per unit area, $d\gamma$ is the change in surface energy and C_i is the impurity concentration of the material. A review of models to predict segregation parameters is in ref. 9). Here we will determine the energy involved in the surface segregation of a metal by using the Miedema model (ref. 8). This relates the heat of segregation of A in B ($\Delta H^{segr}_{|A| \text{ in } |B|}$) to the surface enthalpy of A and B (ΔH^{surf}_{A}) and (ΔH^{surf}_{B}) respectively and the mixing enthalpy of A in B ($\Delta H^{bulk}_{|A| \text{ in } |B|}$) according to:

$$\Delta H^{segr}_{|A| \text{ in } |B|} = \Delta H^{surf}_{A} - \Delta H^{surf}_{B} - (1 - f^{A}_{B})\Delta H^{bulk}_{|A| \text{ in } |B|} \qquad 5)$$

where f^{A}_{B} is the degree to which impurity atoms in the surface layer are surrounded by neighbours of the bulk material.

The atomic composition of the surface layer (C^{1}_{A}) can be predicted if the bulk composition (C_{A}) is known, from:

$$\frac{C^{1}_{A}}{C_{A}} = \exp \frac{(-\Delta H^{segr}_{|A| \text{ in } |B|})}{RT} \qquad 6)$$

However, if segregation of impurities to a free metal surface does not occur because it is energetically unfavourable, this does not mean that it will not occur if the metal forms an interface with a ceramic. It will occur if the impurities can react with the ceramic. Then it is to be considered as reaction induced wetting which also occurs in the process of active metal brazing (ref. 4).
Reactions occurring at metal-ceramic interfaces are discussed later on.
Theoretical work on the predictability of W_{ad} or γ_{mc} of a non-reactive metal-ceramic system is in a rather elementary stage.

Nevertheless, qualitative prediction on the wettability of a ceramic by a non-reactive liquid metal can be made as is shown (Stoneham ref. 10) from the calculation of the interaction of image charges. This showed that wetting occurs if the refractive index of the ceramic is larger than about 2.3.

An approximation of W_{ad} has been made by calculating the interaction energy (μ_p) of a metal atom-oxygen pair (dispersion interaction, London) from the expression:

$$-\mu_p = \frac{\varepsilon}{R^n} - \frac{\lambda}{R^m} \qquad 7)$$

where ε/R^n is the attractive potential and λ/R^m is the repulsive potential, ε and λ are interaction constants and R is the distance between the center of the interacting species. If $n = 6$, ε/R^6 represents the energy of dispersion interaction. Then the pair interaction energy is given by the London equation, ref. 11):

$$-\mu_p = \frac{3}{2} \frac{\alpha_m \cdot \alpha_{O^{2-}}}{R^6} \cdot \frac{I_m \cdot I_{O^{2-}}}{I_m + I_{O^{2-}}} \qquad 8)$$

in which α is the polarizability and I the ionization potentials of the metal(m) and the negative ion of the ceramic. Integration of μ_p over the available pair interaction sites per unit area yields W_{ad}. Calculated values of W_{ad} for some fcc metal-sapphire interfaces taken from the literature (ref. 12,13,14), are represented in Fig. 3, in relation to the surface energy of the metals at the reduced

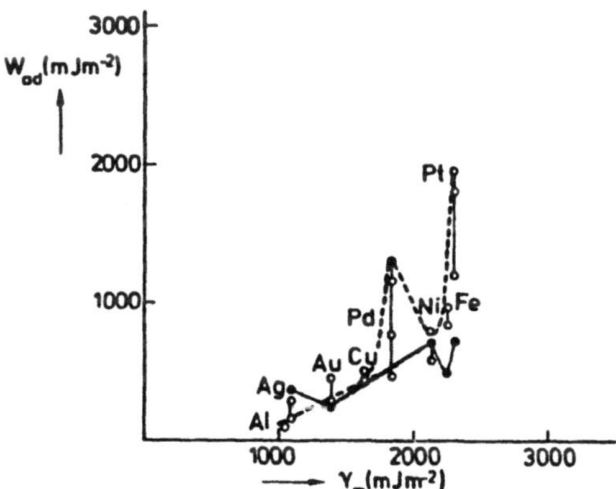

Fig. 3. Experimental (----) and calculated(——) work of adhesion at 0.9Tm of some metal-sapphire (0001) interfaces (ref. 14).

temperature T/T_m = 0.9. In Fig. 3 also values of W_{ad} are represented that are recalculated from experimental data from literature by introducing a temperature coefficient (dW_{ad}/dT) (ref. 14), to obtain W_{ad} at the reduced temperature, according to:

$$dW_{ad}/dT = \tfrac{1}{2} \left(\frac{d\gamma_m}{dT} + \frac{d\gamma_c}{dT}\right)$$

The curves in Fig. 3 show that the calculated and the experimental values are only for a few metals in reasonable agreement. This can be due to a number of uncertainties in the various parameters. In the calculations there are uncertainties in the distance R(eq. 8) which can have a large effect and also in the data of α and I. In the experimental data there is the uncertainty whether the bond is of pure physical nature and if the surface energy of the materials used in the calculations has changed during the experiments, e.g. by segregation or adsorption effects.

3. Wetting phenomena in metal-ceramic systems

As is mentioned in sect. 1, wetting effects play a decisive role in every bonding process. To illustrate this we briefly consider a few metal-ceramic bonding processes that are based on wetting of a liquid on a solid material and on capillary action.

In the processes of metallization of an oxide ceramic where a metal powder - whether or not a refractory metal - is applied to the ceramic surface, use is made of a liquid phase that is either supplied by melting of an intergranular phase in the ceramic, e.g. debased alumina ceramics, or by melting of an oxide phase that is purposely added to the metal powder. The latter applies to high temperature metallization of ceramic that does not contain an intergranular phase or only a very small amount. It also applies to low temperature metallization with non-refractory metals as is common for low temperature applications such as electronic circuitry (ref. 15).

During the high temperature metallization of a debased alumina ceramic with Mo powder in wet H_2/N_2, with dewpoint of 25°C, the intergranular glassy phase becomes less viscous and penetrates the porous metal layer. The glass wets the molybdenum so that a composite is formed that adheres well to the ceramic. The gas atmosphere, although it is wet, does not oxidize the molybdenum (ref. 16) and prevents the SiO_2 in the glassy phase to become reduced. The same phenomenon occurs if Mn or Ti is added to the molybdenum powder, only the metallizing temperature can be lower than if pure molybdenum powder is used because Mn or Ti oxidizes in the gas atmosphere and lowers the melting temperature and thus the viscosity of the glass by a dissolution reaction. The theory of glass-migration (ref. 2) has been made more quantitative by Twentyman (ref. 17) by developing the capillary flow model that relates the capillary pressures on the glassy phase in the ceramic to those on the glassy phase that penetrates the porous metal layer. Fig. 4.

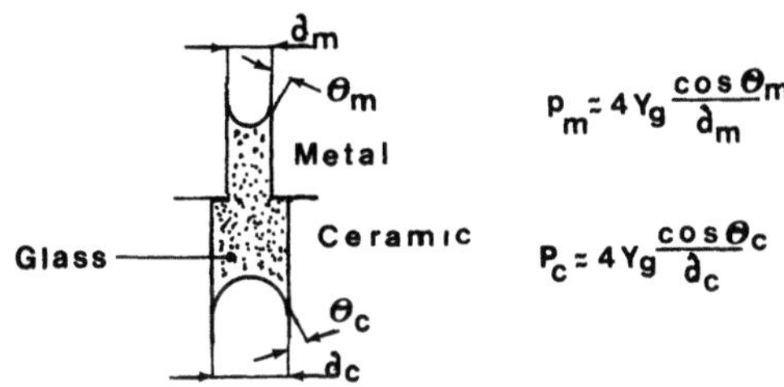

Fig. 4. Capillary flow model: P is the capillary pressure in the metal (m) and the ceramic (c) (ref. 17).

The glass penetrates the metal layer if the capillary pressure (p_m) in the layer is greater than (p_c) in the ceramic so if $p_m > p_c$. Since under the given atmospheric conditions $\Theta_m \approx \Theta_c$ (Klomp ref. 18) the metal layer will be penetrated by capillairy flow of the glass if $\delta_m \leq \delta_c$. With this the size of the metal powder particles (d_m) which determine the size of the pores in the metal layer, and the layer thickness (t_m) can be determined (ref. 18) from the geometrical approach:

$$d_m = \frac{\delta_c}{1-\Delta} \quad \text{and} \quad t_m = \frac{l_c}{2(1-\Delta)}$$

in which Δ is the average shrinkage of the layer during sintering, which is about 35% and l_c is the average length of the ceramic grains which has to be determined metallographically.
If the thickness of the intergranular phase is small the capillary pressure rises to extreme high values which means that the powder particles in the metal layer have to be of nanometer size dimensions to form yet a smaller capillaries. Then a glass powder is added to the metal powder and during metallization the glass flow is from the layer to the ceramic. The capillary that has to be filled is the space between the layer and the ceramic surface and of the internal pores of the layer. This is also done in thick film metallizing processes for hybrid electronic applications.
In the ceramic frit joining process use is made of oxide mixtures that melt at high temperatures and fill a capillary space between a metal and a ceramic part. For the case that the oxide melt does not react, neither with the ceramic nor with the metal, the capillary action is given by the following relations (ref. 19). For a horizontal capillary space with spacing (d) between the metal and the ceramic the capillary pressure (p_h) across the liquid surface is given by:

$$p_h = \frac{\gamma_m \cos\Theta_m + \gamma_c \cos\Theta_c}{d}$$

In the filling of a vertical capillary space the capillary pressure is opposed by the hydrostatic pressure of the melt column which is ρ.g.l. (ρ is the density of the melt, l is the length of the column, g is the gravitational constant) so the capillary pressure (p_V) in the vertical space is

$$p_v = \frac{\gamma_m \cos\theta_m + \gamma_c \cos\theta_c}{d} - \rho.g.l.$$

In many metal-ceramic systems wetting is accompanied by interfacial reaction either with the metal or the ceramic or with both. This changes the properties of the melt and with that also θ changes with the melt advancing in the capillary. Reactions at metal-ceramic interfaces are discussed in the next section.

4. Thermodynamics of metal-ceramic chemical reactions

The second stage in metal-ceramic interaction at the interface is connected with chemical reactions occurring. Thus redox and dissolution reactions with and without the assistance of a reducing gas occur. Such reactions are generally considered to be the key factor in adherence since they further decrease the free energy of the metal-ceramic system once physical interaction has occurred. However, this is only true if other conditions are fulfilled as will be shown. The driving force for the reactions is the chemical potential of the species involved. So charge transport across the interface occurs and mostly also mass transport. The latter can lead to the formation of interfacial phases which properties can markedly differ from the starting materials. Whether or not new phases form depends on the nature of the materials and on the reaction conditions such as temperature, time and atmosphere. The chemical reactions are irreversible and, to be of value for technological applications, they should not proceed until equilibrium is established. So the reaction kinetics are important. Therefore, a description of the reactions and their kinetics should be given by using the thermodynamics of irreversible processes to determine the entropy production of the reactions. Unfortunately, the data relating the flow of matter and heat and of the thermodynamic potentials of the species involved are, with a few exceptions (ref. 1), still lacking.
Therefore, we consider the metal-ceramic reactions by using the equilibrium thermodynamics from which the conditions for a metal-ceramic reaction to proceed can be determined. These conditions are: the temperature and the atmosphere in relation to the concentration of the ceramic species already present in the metal or dissolved in the metal by the reaction with the ceramic.
In many systems a reaction is not expected if only the affinity of the metal to the non-metallic species of the ceramic is considered or if a reducing atmosphere is involved which reduction potential is not sufficient to reduce the ceramic. But taking into account all possible reactions the free energy change of the reactions can be negative (ref. 20).

To illustrate this we describe the reactions of metals with different types of ceramics and their thermodynamic relationships from which the reaction conditions are determined.

4.1. The Al_2O_3-Pt-H_2 system

In the Al_2O_3-Pt-H_2 the Pt-O_2 reaction has a positive value of the Gibbs free energy (ΔG^o) but the ΔG^o of the Pt-Al reaction is negative. If we neglect the heat of solution of H_2 in Pt, the reactions in the system, including their thermodynamic relations in the temperature range 1000-2000K, are under standard conditions (ΔG^o in cal/mole):

$1/3Al_2O_3 \rightarrow 2/3Al + 1/2O_2$	$\Delta G^o = 134330 - 25.89T$	(ref. 21)	13)
$H_2 + 1/2O_2 \rightarrow H_2O$	$\Delta G^o = -59630 + 13.64T$	(ref. 21)	14)
$2Pt + 2/3\ Al \rightarrow 2/3\ Pt_3Al$	$\Delta H^o = \Delta G^o = -32520$	(ref. 8)	15)
$2/3Pt_3Al \rightarrow 2/3|Al|Pt + 2Pt$	$\Delta G^o = 3464 - 1.2T$		16)

$$1/3Al_2O_3 + H_2 \rightarrow 2/3|Al|Pt + H_2O \qquad \Delta G^o_R = 45644 - 13.45T \qquad 17)$$

A few remarks on the thermodynamic relations of eq. 15) and 16) have to be made. Regarding eq. 15), ΔG^o of Pt_3Al is not known. Therefore, ΔH^o is taken to be equal to ΔG^o. This assumes that ΔS^o is negligibly small which is generally true for this type of reactions (ref. 21). Regarding eq. 16, ΔG^o is calculated from the equilibrium between Pt_3Al and the saturated solid solution of Al in Pt by taking the sum of reaction eq. 15) and 16):

$$2/3\ Al \leftrightarrow 2/3|Al|Pt \text{ with } \Delta G^o_R = 0 \qquad 18)$$

Using the given value ΔH^o of Pt_3Al (eq. 15) we can calculate ΔG^o of the dissolution of Al in Pt, using the van 't Hoff's isochore:

$$\Delta H^o = -R\,\frac{d(\ln x)}{d(1/T)} \qquad 19)$$

$$\text{and } \Delta S^o = \frac{d\Delta G^o}{dT} \qquad 20)$$

In the further descriptions we simplify the calculations by assuming that the solid-solutions obey Raoult's law which implies that the concentration of the solute (x) equals the activity (a) of the solute in the solvent, so $x = |Al|_{Pt} = a$. It has to be noted that this assumption is not generally allowed because the activity of a solute can be high even for small concentrations.
In order to calculate ΔH^o of $|Al|_{Pt}$ we take the numerical values of the solid-solution in equilibrium with Pt_3Al from the Pt-Al phase diagram (ref. 22) in the temperature interval e.g. 1000-1100K. Then, with eq. 19 and ΔG^o from eq. 15) we find ΔH^o of $2/3|Al|_{Pt}$. With eq. 20) and ΔG^o from eq. 15) we find ΔS^o. This yields for reaction

16): ΔG^o = 3464 - 1.2T. So, for the sum of reactions 15) and 16) we get, ΔG^o = -29056 - 1.2T. If T = 0, then $\Delta G^o = \Delta H^o$ = -29056 cal which is close to -26826 cal, that is predicted by the model of ref. 8. It is therefore more useful to use the referred Miedema model (ref. 8) direct for metal alloys, the more so if it regards transition metals, than doing the laborous calculations using van 't Hoff's isochore. However, the introduction of the latter in considering metal-ceramic reactions can be a necessity as will be shown later in this paper.
As the free energy of the total reaction eq. 17) shows, the reaction will not proceed under standard conditions in the given temperature range. So, for reaction to proceed the H_2O activity has to be lower than 1, which means that the conditions change to non-standard conditions. Then the activity of Al in Pt changes also to values < 1. The energy relation of eq. 17 then becomes:

$$\Delta G^1 = \Delta G^o_R + RT\ln \{x^{2/3} . (\frac{pH_2O}{pH_2})\} \qquad 21)$$

For given concentrations of Al dissolved in the Pt the ratio pH_2O/pH_2 can be calculated for equilibrium conditions, that is for $\Delta G^1 = 0$, according to:

$$\frac{pH_2O}{pH_2} = \exp-(\frac{\Delta G^o_R}{RT} + 2/3 \ln x) \qquad 22)$$

In Fig. 5 the calculated ratio pH_2/pH_2O, for pH_2 = 1 atm. is given with 1/T and the concentration (x) of Al in Pt as the parameter.

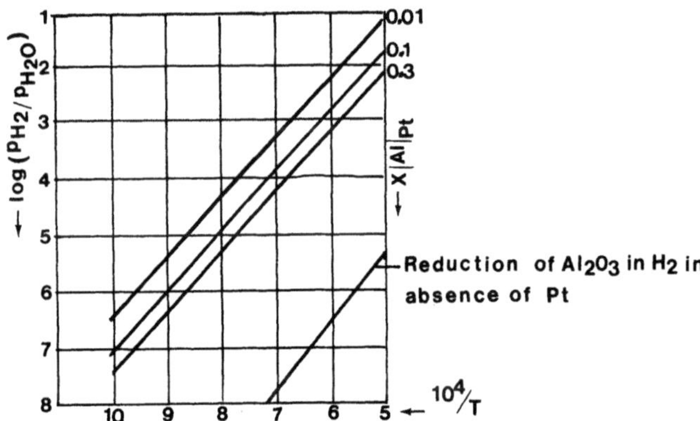

Fig. 5. Reaction conditions of the Al_2O_3-Pt-H_2 system. x = atomic concentration of Al in Pt.

In Fig. 5 a line is present that indicates the reduction of Al_2O_3 by H_2 in absence of Pt. The position of this line relative to the other lines indicates the strong energy effect of the dissolution of Al in Pt.

4.2. Ni-Si_3N_4 reaction

The following equations and energy relations are considered for the temperature region 1000-1500K under standard conditions. The data for ΔG^o in the thermodynamic relations are in cal/mole from ref. 16 and the ΔH^o from ref. 8.

$$1/3\ Si_3N_4 \rightarrow Si + 2/3\ N_2 \qquad \Delta G^o = 62933 - 36.8T \qquad 23)$$

$$3\ Ni + Si \rightarrow Ni_3Si \qquad \Delta G^o = \Delta H^o = -25400 \qquad 24)$$

$$Ni_3Si \rightarrow |Si|Ni + 3\ Ni \qquad \Delta G^o = 14412 + 6.8T \qquad 25)$$

$$1/3\ Si_3N_4 \rightarrow |Si|Ni + 2/3\ N_2 \qquad \Delta G_R^o = 51945 - 30T \qquad 26)$$

Here, ΔG^o of eq. 25) is calculated for the equilibrium reaction:

$$Ni_3Si \leftrightarrow |Si|Ni + 3\ Ni \qquad 27)$$

using ΔG^o from eq 24) and the van 't Hoff's isochore according to eq. 19) and ΔS^o according to eq. 20). This procedure is more accurate than calculating the heat of alloy formation in a transition-non-transition metal system using the model in ref. 8.

Looking to eq. 26) we see ΔG_R^o is positive within the given temperature range. Reaction can only occur if the N_2 activity is changed by lowering the N_2 pressure. Then the free energy changes to:

$$\Delta G^1 = \Delta G_R^o + RT\ln(a_{|Si|Ni} \cdot P_{N_2}^{2/3}) \qquad 28)$$

Identical with the foregoing system we assume: $a|Si|Ni$ = concentration (x). For equilibrium, ΔG^1 of eq. 28) = 0 and from this follows the equilibrium N_2 pressure:

$$P_{N_2} = \exp{-3/2}\left(\frac{\Delta G_R^o}{RT} + \ln x\right) \qquad 29)$$

For given values of x the N_2 pressure can be calculated. In Fig. 6 the result of such calculations is represented as log $PN_2 = f(1/T)$ with x as the parameter. The equilibrium N_2 pressure for the reaction 1/3 $Si_3N_4 \leftrightarrow Si + 2/3\ N_2$ is also indicated. The shaded area in Fig. 6 indicates the maximum solubility of Si in Ni and this solid solution is in equilibrium with Ni_3Si.

As Fig. 6 shows reaction between Ni and Si_3N_4 depends on both the temperature and the N_2 pressure in the environment. The boundary of Ni_3Si formation does not mean that this phase forms only if the Si concentration in Ni reaches the maximum value. Ni_3Si can form directly when the reaction of Ni with Si_3N_4 occurs. Then at the metal-ceramic interface a layer of Ni_3Si is present and the Ni-Si solid solution is further away from the interface.

So far no experimental evidence for the above descriptions is available.
Like in the foregoing system the effect of the Ni-Si alloy formation on the destabilisation of the Si_3N_4 ceramic is large.

4.3. The SiC-Ni system

In the system Si-C-Ni the Ni can alloy with both the Si and the C according to the respective phasediagrams (ref. 22). Here we assume that the Ni-Si and Ni-C compounds do not have a solid solution range. The relevant reactions in the temperature range 800-1200K are:

$SiC \rightarrow Si + C$	$\Delta G^o = 12770 - 1.66T$	30)
$3\ Ni + Si \rightarrow Ni_3Si$	$\Delta G^o = \Delta H^o = -25400$	31)
$Ni_3Si \rightarrow \lvert Si \rvert Ni + 3\ Ni$	$\Delta G^o = 14412 + 6.8T$	32)
$3\ Ni + C \rightarrow Ni_3C$	$\Delta G^o = 9402 + 3.93T$	33)
$Ni_3C \rightarrow \lvert C \rvert Ni + 3\ Ni$	$\Delta G^o = 2013 - 1.37T$	34)
$SiC \rightarrow \lvert Si \rvert Ni + \lvert C \rvert Ni$	$\Delta G^oR = 13197 - 7.7T$	35)

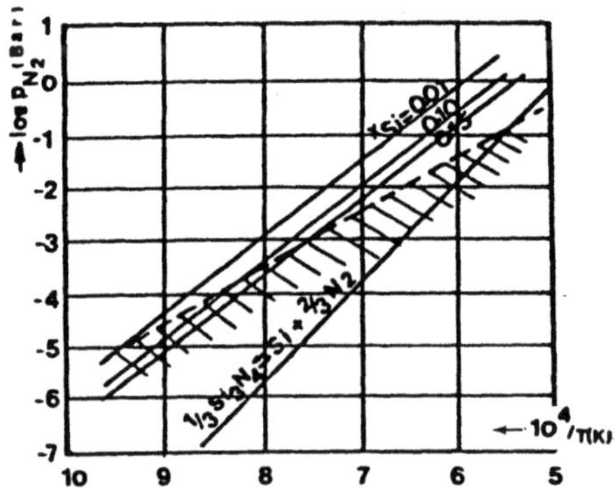

Fig. 6. Equilibrium N_2-pressure of Si_3N_4 and for the system Si_3N_4-Ni with x = atomic concentration of Si in Ni.

Also in this set of reactions the Gibb's free energy of the total reaction (eq. 35) is positive. Looking at eq. 30) and 31) and taking the sum of these reactions we get:

$$SiC + 3\ Ni \rightarrow Ni_3Si + C \qquad \Delta G^o = -12630 - 1.66T \qquad 36)$$

The sum of eq. 31) and 32) gives:

$$Si \rightarrow \lvert Si \rvert Ni \qquad \Delta G^o = -10988 + 6.8T \qquad 37)$$

ΔG^o of both eq. 36) and 37) is negative, so these reactions can proceed. In this system the activities of the species cannot be changed by outside influences. This means that Ni reacts with SiC forming silicide (Ni_3Si or more Si-rich Ni compounds), solid

solutions of Si and C in Ni and free carbon. The Ni-C solid solution can contain up to 0.48 at % C at 1018 K (ref. 22). Given the fact that ΔG^o of Ni_3C is highly positive, the Ni-C solid solution will be in equilibrium with free carbon and Ni_3Si.
Experimental evidence for SiC-Ni reaction is available Fig. 7, (from ref. 23).

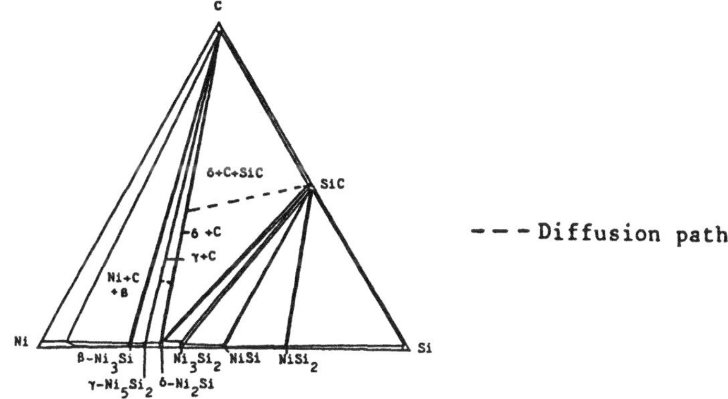

Fig. 7. Ni-Si-C system at 1123 K (ref. 23)

It appeared that at the HIP-SiC-Ni interface at 1123 K a reaction layer was formed consisting of alternating layers of Ni_2Si and Ni_2Si with C, alternating layers of Ni_5Si_2 and Ni_5Si_2 with C and a layer of Ni_3Si in contact with Ni. A Ni-C solid solution was not reported in ref. 23. Although eq. 35) indicates a positive free energy for the solution of Si and C in Ni the solid-solution can appear if Ni is in excess and if transport of matter is not hampered by other reasons. Then the free energy relation gets the form:

$$\Delta G = \Delta G_R^o + RT\ln(a|Si|_{Ni} . a|C|_{Ni}) \qquad 38)$$

From eq. 38 it follows that e.g. at 1000 K, $a|Si|.a|C| = 4.10^{-2}$, which means that the Ni is not saturated by either one or both ceramic species.
Identical calculations are carried for other metal-ceramic systems. In Table 1 the total reaction between the metal and the ceramic, its Gibb's energy and one relation for the equilibrium gas pressure required in the ambient are given. From this and the foregoing illustrations a few conclusions can be drawn.

First, that in the metal-ceramic reaction not only the free energy (ΔG^o) of oxide, nitride or carbide formation of the metal is important but also, and in some cases the most important, the energy of metal alloy formation. Such cases are metal-ceramic systems where the affinity of the metal to the non-metallic species of the ceramic is small e.g. Pt, Cu, Au etc.
The second conclusion is that in metal-ceramic systems where the ceramic contains a gaseous species, a reaction can be controlled by changing the activity of the species in the ambient.

Table 1 Condition for metal-ceramic interface reaction.
(metal in excess, ΔG in cal/mole)

	System	ΔG
1.	$1/4Si_3N_4 + Ti \rightarrow 3/4|Si|_{Ti} + |N|_{Ti}$	$-41953 + 0.57T + RT\ln(a_{Si}^{3/4}.a_N)$
2.	$SiC + Ti \rightarrow |Si|_{Ti} + |C|_{Ti}$	$-106535 + 0.44T + RT\ln(a_{Si}.a_C)$
3.	$SiC + Cu \rightarrow |Si|_{Cu} + C$	$3246 - 1.66T + RT\ln(a_{Si})$
4.	$SiC + W \rightarrow |Si|_W + |C|_W$	$13367 - 9.40T + RT\ln(a_{Si}.a_C)$
5.	$SiC + W \rightarrow W_5Si_3 + 3WC$ (W not in excess)	

Equil N2-pressure of system 1:

$$PN_2 = \exp - \{(\frac{67517}{T} - 23.5 + 2\ln(a_N)\}$$

The third conclusion is that when the activity of the non-metallic species cannot be controlled by outside means such as for SiC, and if the Gibb's energy for the metal-SiC reaction is negative the reaction cannot be controlled. There will always be a chemical potential gradient until the ceramic has disappeared.

From the last two conclusions criteria for the interface stability or materials selection can be derived.

As is mentioned in sect. 2, the presence of impurities in the metal or purposely added alloying elements can have large effects not only on the surface energy of the matrix metal but also on the interface phenomena. A well known example is the use of an alloying element (Ti) that has high affinity to the non-metallic species of the ceramic Ag-braze alloy (Nicholas ref. 4). Using the eq. 5) and 6) and the data given in ref. 8 it appears that Ti will not segregate to the free surface of the alloy. But if the alloy is in contact with a ceramic (SiC, Si_3N_4, Al_2O_3 etc.) it diffuses to the interface where it reacts with the ceramic. The reduced metallic component of the ceramic then dissolves in the brazing alloy. The thermodynamics predict this behaviour. The Ti forms hypo-stoichiometric TiO, TiN or TiC with oxide, nitride and carbide ceramics respectively (ref. 4).

The use of a reactive element in a non-reactive matrix metal and whether or not in the liquid state is a means to control the extend of the reaction, because in diluted alloys the gradient of the thermodynamic potential caused by the reaction of the active element with the ceramic is very small in comparison to that of the active metal as a bulk. So, limiting both the amount and the activity of the active element gives the opportunity to limit the extent of the interfacial reaction layer. This widens also the range of materials that can be selected because the reactivity can be separated from physical and mechanical requirements.

The question may arise what the contribution is of an interfacial chemical reaction between the metal and the ceramic to the adherence or the work of adhesion. Kurkjian and Kingery (ref. 24) concluded that the strength (S) of a metal-ceramic interface is a function of the free energy of oxide formation of the metal, so: $S = f(\Delta G^o)$. McDonald and Eberhart (ref. 13) proposed for metal-oxide ceramics the relation $W_{ad} = f(\Delta G^o + b)$ where b is the contribution of the energy of dispersion interaction between the metal and the cations in the ceramic. From the foregoing thermodynamics it appears that the heat of alloy formation has to be included in the energy the system gives up by chemical reaction. This leads to the expression for the total energy change:

$$\Delta G^1 = \Delta G^o_{MeX} - (\Delta G^o_{meX} + \Delta H_{i-k}) \qquad 39)$$

where ΔG^o_{MeX} and ΔG^o_{meX} refer to the free energy of the ceramic and of the reaction product of the metal with the non-metallic species of the ceramic and ΔH_{1-k} refers to the heat of mixing of the ceramic species in the metal. Then the expression for the work of adhesion will have the form of:

$$W_{ad} = f\{\Delta G^o_{MeX} - (\Delta G^o_{meX} + \Delta H_{i-k})\} + \Delta G_s. \qquad 40)$$

This and any other description for the work of adhesion requires to define the actual interface because if reaction products are produced more interfaces are formed. This area is rather unexplored.
The developments in surface and interface studies, both theoretically and experimentally will certainly contribute to the understanding of metal-ceramic interfaces.

5. Effect of interface morphology on the interfacial strength

The morphology of a metal-ceramic interface depends on the type of interaction that has occurred. If only physical interaction has occurred the structure of the materials is not changed. The only change that is observed is that the close packed crystallographic planes of the metal orient itself to those of the ceramic e.g. $(111)_{Pt}//(0001)_{Al_2O_3}$, $(111)_{Cu}//(0001)_{Al_2O_3}$ (ref. 25) and $(011)_{Nb}//(0001)_{Al_2O_3}$ (ref. 26). If chemical reactions have occurred the structure of the metal at the interface depends on whether only (solid)-solutions or also new interfacial phases are formed. The structure of the metal does not change by the formation of solid-solutions (ref. 27) but mostly the lattice parameters change. However, the formation of satured solutions at the reaction temperature can lead to internal reaction of the dissolved species in

the metal as is shown by Rühle et al. (ref. 27) in the Nb-Al_2O_3 system. The dissolved Al and O react during cooling forming θ-Al_2O_3 in the Nb close to the Nb-Al_2O_3 interface. If new interfacial phases are formed not only the structure but also the physical and mechanical properties can differ markedly from the parent metal. The thickness of the interfacial layer depends on the reaction conditions, in particular on the temperature and the time. Another example of a reaction, not mentioned in the previous section but still a type of reaction to be included in the thermodynamic description, is the ceramic compound formation. This occurs in the bond formation of Cu to Al_2O_3 using a Cu/CuO eutectic composition as a liquid bonding agent. The CuO reacts with Al_2O_3 to $CuAlO_2$ and $CuAl_2O_4$ and solidifies without preferential orientation to the ceramic (ref. 29).
For some metal-ceramic systems the strength of interfaces with different morphologies is measured and can be compared. In Table 2 these systems and the strength values (from ref. 20) are given. Unfortunately mutual comparison is not possible due to the differences in the methods of measuring.

Table 2 Morphology of metal-ceramic interfaces and their strength at room temperature.

System.	Interfacial phases.	Strength*(MNm^{-2}).	Layer thickness.	Ref.
Pt-Al_2O_3	none	σ_t, 250		20
Pt-Al_2O_3	Pt_3Al	170	2μm	20
Nb-Al_2O_3	NbO_x	120	2μm	28
Nb-Al_2O_3	NbO_x	10	20μm	28
Nb-Al_2O_3	none(ss)	τ_b,365		31
Cu-Al_2O_3	none	τ_{max}.50		30
CuO/Cu-Al_2O_3	$CuAlO_2$	τ_{max}30-40	0.2	30
Al-SiO_2	not observed	σ_t,20		20
Al-SiO_2	α-Al_2O_3	σ_t,0	20	20

* σ_t = tensile, τ_b = bending, τ = shear
ss = solid solution

From Table 2 it is to be seen that the interface morphology plays an important role in the strength of interfaces. Thick reaction layers affect the interfacial strength negatively. This seems to contradict the role ascribed to chemical reactions to further lower the interface energy once physical interaction has occurred. In fact the latter is still true because a chemical bond is stronger than a physical bond whether it leads to an ionic or a covalent bond. The relevant conclusion is therefore that the reaction should be limited so that only an extremely thin reaction layer can be formed and effects such as lattice and thermal expansion mismatch between the phases, and non-coherency of reaction products are eliminated as far as possible. This points to a preference for reactions and reaction conditions that will dominantly result in (solid)-solutions.

Conclusion

Equilibrium thermodynamics are useful in the description of reactions between metals and ceramics and to determine the conditions of a reaction to proceed. In the same way the effect of impurities or purposedly added alloying elements to the metal on the metal-ceramic reaction can be well described. On that basis criteria for the materials selection for a particular application can be derived and indications on the chemical stability can be obtained. The kinetics of metal-ceramic interfacial reactions still have to be determined experimentally. This is urgent in view of the use of structural ceramics at high temperatures because the extent of interfacial reactions and the interfacial strength are closely related.

References

1. M. Backhaus-Ricoult, Ber. Bunsenges. Phys. Chem. 90 (1986) 684
2. S.S. Cole and G. Sommer, J.Am.Ceram.Soc. 44 (1961) 1961
3. J.T. Klomp and T.P.J. Botden, Am.Ceram.Soc.Bull. 49 (1970) 204
4. M.G. Nicholas, Fundamentals of Diffusion Bonding, Ed. Y. Ishida, Elsevier Amsterdam (1987) 25
5. J.T. Klomp, Electronic Packaging Materials Science, MRS, vol. 40, ed. E.A. Giess, K.N. Tu and D.R. Uhlman (1985) 381
6. J.A. Pask and A.P. Tomsia. Surfaces and Interfaces in Ceramic and Ceramic-Metal Systems, Ed. J.A. Pask and A. Evans Plenum, New York (1981) 411
7. S.K. Rhee, J.Am.Ceram.Soc. 55 (1972) 300
8. A.K. Niessen and A.R. Miedema, Cohesion in Metals, to be published by North Holland Publ.Comp. Amsterdam
9. S. Hofmann, J. de Chimie Physique 84 (1987) 141
10. A.M. Stoneham and P.W. Tasker, Ceramic Microstructures '86, Role of Interfaces, Ed. J.A. Pask and A. Evans. Plenum. New York (1987) 155
11. L. Pauling, General Chemistry, 3rd ed. Freeman, San Francisco (1970) 118
12. Y.V. Naidich, Progress in Surface and Membrane Science, Ed. D.A. Cadenhead and J.F. Danielli, Acad. Press New York (1981) 353
13. J.E. McDonald and J.G. Eberhart, Trans.Metall.Soc. AIME 233 (1965) 512
14. J.T. Klomp, Fundamentals of Diffusion Bonding, Ed. Y. Ishida, Elsevier Amsterdam (1987) 1
15. R.W. Vest, Ceram. Bull, 65 (1986) 631
16. I. Barin and O. Knacke, Thermochemical Properties of Inorganic Substances. Springer New York 1973
17. M.E. Twentyman, J.Mat.Sci. 10 (1975) 765
18. J.T. Klomp, Powder Metall. Intern. 3 (1971) 142
19. D.R. Millner, British Welding Journal. March (1958) 90
20. J.T. Klomp, Ceramic Microstructure '86 Role of Interfaces, Ed. J.A. Pask and A. Evans. Plenum. New York (1987) 307
21. J.T. Klomp, Proceedings E.C. Colloquim Petten April 1988, to be published
22. M. Hansen, Constitution of Binary Alloys, Mc Graw Hill New York 1958

23. R.J.C. Schiepers, F.J.J. v. Loo and G. de With, J.Am.Ceram.Soc., 6 (1988) C-284
24. C.R. Kurkjian and W.D. Kingery, J. Phys. Chem. 60 (1956) 961
25. C.A.M. Mulder and J.T. Klomp, J. de Physique 46 C4 (1985) 111
26. M. Florjanic, W. Mader, M. Rühle and M. Turwitt, J. Physique 46 C4 (1985) 129
27. M. Rühle, K. Burger and W. Mader, J. Microsc. Electron 11 (1986) 163
28. S. Morozumi, M. Endo, M. Kikuchi and K. Hamajuna, J. Mater. Sci. 20 (1985) 3976
29. C. Beraud, Thesis 86 ISAL 0017, Institute National des Sciences Appliquées Lyon 1986
30. M. Courbiere Thesis ECL 86.13 L'Ecole Centrale de Lyon. 1986
31. G. Elssner and U. Krohn, Z. Metallkde 70 (1979) 71

CERAMIC-METAL INTERFACES

M.G. Nicholas
Harwell Laboratory
Oxfordshire
United Kingdom

ABSTRACT. High integrity ceramic-metal interfaces are required for several areas of advancing technology but are difficult to produce because of the markedly differing lattice bonding characteristics of ceramics and metals. If the required properties are to be achieved, the chemistries of interfaces must be changed and this paper considers to what extent simple chemical concepts can account for the behaviour of brazed and diffusion bonded ceramic-metal systems. Particular attention is paid to the role of Ti promoting the wetting of ceramics through the formation of metal rich, hypostoichrometric, reaction product and to the beneficial effects of incorporating such metals as Sn and In in Ti containing brazes. The need to control the extent as well as the nature of the reactions if strong bonds are to be produced is high lighted, and comparisons are made between the fabrication processes and properties of brazed and diffusion bonded interfaces.

1. INTRODUCTION

A wide range of ceramic-metal interfaces occur naturally and are applied in technology. Thus the surface of almost every metal is covered by an oxide film, and the properties of the oxide-metal interfaces can exert a dominant influence on film characeristics. Further the interfaces formed by carbide precipitates with their matrices control important properties of steels and other metal alloys. This presentation, however, is concerned with the synthetic interfaces created in response to demands for ceramic-metal components and composite materials. In particular, attention is devoted to the interfaces formed with ceramics by liquid metals during brazing and by solid metals during diffusion bonding.

Compared to metals, ceramics are usually refractory and brittle electrical and thermal insulators that have low thermal expansivities, and these differences reflect the bonding characteristics of their lattices. While the bonding electrons in metals are delocalised, the movement of electrons in ceramics is restricted by the ionicity or covalency of their lattices. Thus the nature of the lattice bonding

L.-C. Dufour et al. (eds.), Surfaces and Interfaces of Ceramic Materials, 393–417.

changes abruptly at ceramic-metal interfaces, and it is important to understand this behaviour and in particular how this is affected by the relative ionicity or covalency of the ceramics.

The best established approach to predicting the lattice bonding characteristics of ceramics is that of Pauling [1], who developed a semi-empirical relationship between ionicity and the electronegativities of the component elements. This predicts for example that SiO_2 is 51% ionic, Si_3N_4 is 30% ionic, and SiC is only 12% ionic. Further, Pauling regarded covalent bonding as intermediate between ionic and metallic bonding [2] and hence these variations in ionicity could provide guidance about interface effects. Thus the discontinuities of the interfaces formed with srongly ionic oxides should be severe while those of some carbide-metal interfaces should be negligible. Such discontinuities are energy barriers, and are reflected in interfacial energy values and hence influence processes such as wetting and sintering. In fact, knowledge and understanding of interfacial energy values is an essential component in the interpretation of many technologically important ceramic-metal phenomena.

2. CERAMIC-METAL INTERFACIAL ENERGIES

2.1. Values

Interfacial energy values can be derived from the characteristic geometries assumed at interphase boundaries when the total energies of ceramic-metal systems are minimised [3]. Thus Figure 1 illustrates the equilibrium configurations adopted by a small liquid drop resting on horizontal solid surface, and grain boundaries emerging at surfaces in contact with an inert gaseous atmosphere or a liquid metal. The angles θ, ϕ and α can be related to the surface and interfacial energies of the systems by the equations:

$$\gamma_S = \gamma_L \cos\theta + \gamma_{SL} \tag{1}$$

$$\gamma_{GB} = 2\gamma_S \cos\phi/2 \tag{2}$$

$$\gamma_{GB} = 2\gamma_{SL} \cos\alpha/2 \tag{3}$$

where γ_S and γ_L are the energies of the solid ceramic and liquid metal, γ_{SL} is the solid-liquid interfacial energy, and γ_{GB} is the energy of a grain boundary in the solid. Thus if one energy is known, usually γ_L, which can be derived from careful measurements of the profile of a sessile drop, the others can be calculated using various combinations of angular measurements, for example

$$\gamma_{SL} = \gamma_L \frac{\cos\theta \cos\phi/2}{(\cos\alpha/2 - \cos\phi/2} \tag{4}$$

Using such multi-phase equilibration techniques several groups of workers have generated energy data. In particular, Kingery et al. [4] in the USA and both Eremenko and Naidich and their co-workers [5] in the USSR have derived solid-liquid interfacial energies, while Piliar and

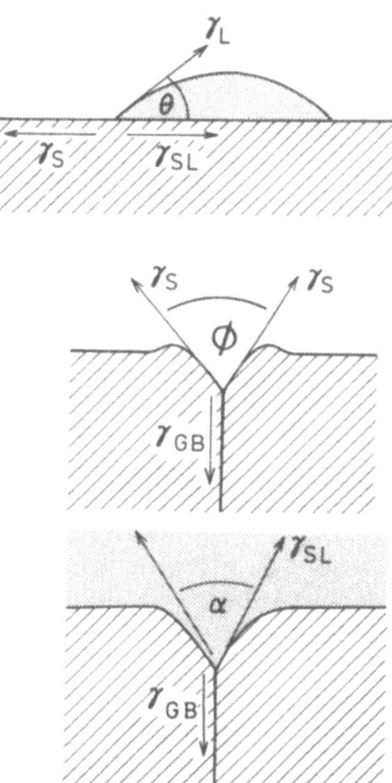

Figure 1. Equilibrium configurations assumed at surfaces and interfaces; top - a sessile drop resting on a horizontal substrate, middle - grain boundaries emerging at a surface exposed only to its own vapour, bottom - grain boundaries emerging at a surface exposed to a liquid environment.

Nutting [6] and, more recently, Murr [7] and Klomp [8] have used similar techniques to derive solid-solid energies from shape measurements for minute metal particles resting on oxide substrates. By far the commonest ceramic employed in such work has been Al_2O_3 as can be seen by consulting compilations such as those of Naidich [5], but a significant body of data exist also for carbides and carbon - some of which is cited in Table I.

For the better documented oxides, certain general observations can be made. For systems in which chemical interactions are limited to electron exchange, the interfacial energies are generally large - values of 1.3 to 2.4 Jm^{-2} being measured for interfaces formed in vacuum or inert gas atmospheres between Al_2O_3 and molten Ag, Co Cu, Fe or Ni. In contrast, if the environment or the melt contains sufficient oxygen to permit the formation of new phaser such as the complex oxides $CuAlO_2$ or $NiAl_2O_4$, the restructuring of the interfaces can be associated with smaller energy values - only 1.0 and 1.6 Jm^{-2} for the systems cited. (In contrast, alloying melts with carbon decreases their ability to dissolve material from graphite substrates and interfacial energy and contact angle values are increased [14].) Redox reactions also can produce changes that decrease the interfacial energy values. For some systems, the restructuring associated with an energy decrease requires a high temperature, thus the energies of Al-Al_2O_3 interfaces decrease abruptly from approximately 1.5 to 0.3 Jm^{-2} when the formation temperature is raised above about 950°C and the metal is no longer coated by a tenacious oxide skin. The few data available for solid-solid interfacial energies are similarly high for high inert systems as can be seen by inspection of Table I but it is unclear whether they are generally larger or smaller than those for liquid-solid interfaces.

Apart from the authors who quote directly derived interfacial energies, many others report their findings as the work of adhesion, W_{ad}, which can be defined as the reversible work needed to separate an interface into its component surfaces. Thus

$$W_{ad} = \gamma_S + \gamma_L - \gamma_{SL} \tag{5}$$

TABLE I - Interfacial energies for some metal-ceramic systems

System	Liquid Metal-Solid Ceramic [5,10-12]		Solid-Metal-Solid Ceramic [9]	
	γ, $J.m^{-2}$	T, K	γ, $J.m^{-2}$	T, K
Al_2O_3-Ag	1.62	1273	1.57	973
-Au	1.75	1373	1.8	1273
-Cu	2.08	1373	2.21	1123
-Fe	2.37	1823	2.20	1273
-Ni	2.44	1773	2.73	1275
SiO_2-Ni	1.32	1773		
ZrO_2-Ni	1.75	1773		
ThO_2-Ni			2.0	1473
Al_2O_3-Cu(O)	1.0	1503		
Al_2O_3-Ni(O)	1.6	1873		
TiC-Cu	1.48	1423		
VC-Cu	0.46	1423		
WC-Cu	1.56	1423		
ZrC-Cu	1.17	1423		
C-Cu	1.72	1373		
-Ag	1.45	1253		
-Sn	1.24	1273		

and by substituting equation (1)

$$W_{ad} = \gamma_L (1 + \cos \theta) \tag{6}$$

2.2 Models

The difficulties encountered in creating technologically useful synthetic interfaces have provoked many speculations about their nature and several attempts have been made to develop predictive models based on thermodynamic, electron or electrostatic bonding concepts. The thermodynamic approach received an early impetus from the conclusion of Humenik and Kinergy [4] that the wetting behaviour of oxides by molten metals, that is, the ability to form low energy interfaces, could be related to the oxygen affinity of the metals,

$$W_{ad} = f\,(\Delta G^{\circ}_{MO}) \tag{7}$$

where ΔG°_{MO} is the free energy of formation of the oxide of the liquid metal. For Al_2O_3, McDonald and Eberhart [15] developed a model based on the ceramic surface structure shown in Figure 2. The metal atoms were considered to locate on two types of site. Metal atoms directly above locations that would extend the $A\ell^{+++}$ cation lattice were regarded as being ionically or chemically bonded while the others were considered to be attached only by Van der Waals forces. The work of adhesion, W_{ad}, of the interfaces was related to these bonding forces by the expression

$$W_{ad} = -\frac{n}{2a} \cdot \Delta G^{\circ}_{MO} + n \cdot \frac{3}{2} \frac{\alpha_M \alpha_o}{R^6} \cdot \frac{I_M I_o}{I_M + I_o} \tag{8}$$

where n is the density of ionically bonded sites, $\frac{1}{2}\Delta G^{\circ}_{MO}$ is the free energy of ionic bonds in $J.mole^{-1}$, a is the number of bonds in a gram mole of MO, α_M and α_o the polarizability of the metal atom and oxygen ion, I_M and I_o the ionisation potentials and R the separation distance of a Van der Waals bonded metal-oxygen pair. For most of the data sets reviewed by McDonald and Eberhart, the ionic bonding term was by far the the larger than that for Vander Waals interactions and an excellent correlation between ΔG_F and experimental work of adhesion data was observed, Figure 3. This combination of Van der Waals and ionic bonding has also been used by Naidich [5]. Although his mathematical treatments were similar he assumed ionic bonding would be insignificant unless the oxygen affinity of the liquid metal was more than that of the oxide ceramic cation.

The McDonald and Eberhart model accounts reasonably for the experimental data cited by its authors. The model is noteworthy because it is equally applicable to solid-solid as slid-liquid interfaces, but but it can be criticised on several grounds:

- It does not take account of macroscopic chemical interactions that can occur at interfaces to change their phase structures.
- It suggests that W_{ad} should decrease with temperature as does ΔG°_{MO}, whereas in practice increases are usually observed.

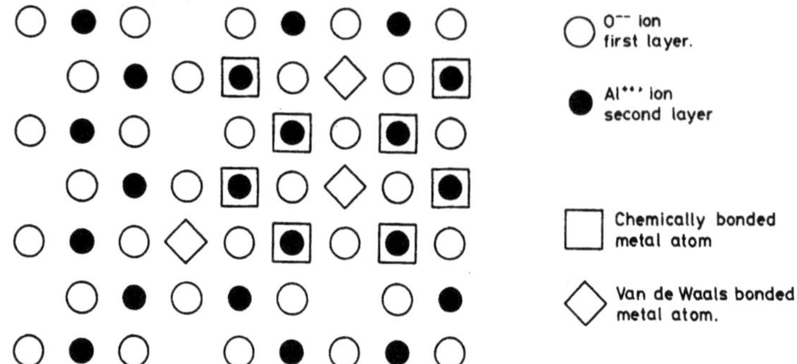

Figure 2. Al_2O_3 surface used in McDonald and Eberhart model.

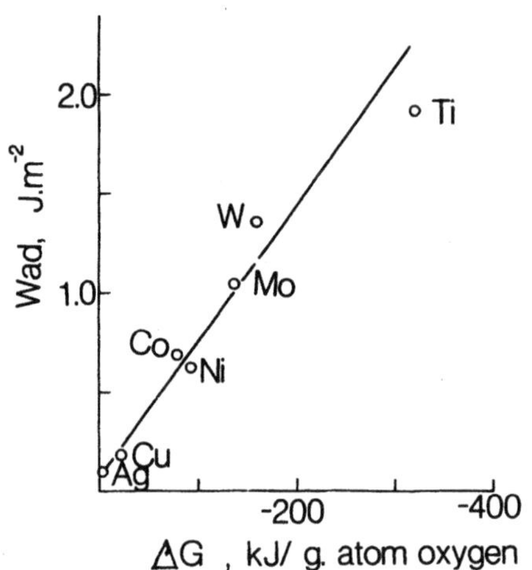

Figure 3. Comparison of experimental data for the wettability of Al_2O_3 by various metals with the predictions of the McDonald and Eberhart model.

- It ignores the possible contribution of metal-metal interactions.

In contrast, Klomp [8] has paid particular attention to this last shortcoming and has shown that interactions with the Al^{+++} ions can have a decisive influence on Pt-Al_2O_3 interfaces and that the contribution of the heat of solution of Al in Pt increases the W_{ad} value far above that predicted by the McDonald and Eberhart model. Similarly, Chatain et al. [16] have recently analysed the behaviour of inert oxide-metal systems in terms solely of the energy changes due to solution of both the cations and the oxygen ions in the metal.

Attention has been paid also to electron bonding models. Johnson and Pepper [17] used a molecular orbital model to analyse the interfaces of Al_2O_3 with Ag, Cu, Fe and Ni and concluded that a chemical, electron exchange, bond was formed with the oxygen anions. Very recently, Tasker and Stoneham [18] have discussed the significant changes in the polarisation energy of charges near an interface between materials with differing dielectric constants, (∞ for a metal, 1 for a vacuum and > 1 for a ceramic). The model regards the ceramic-metal interface as a reflecting plane and calculates the effect of the resultant image charges on the energy of the interface. Although still being developed, the model predicts realistic W_{ad} values (0.5-1.5 J m^{-2}) for non-reactive oxide-metal systems and can account qualitatively for a number of other interfacial effects such as the radiation enhancement of bonding.

For non-oxide ceramics, interfacial modelling work is much less developed although in principle some of the models for oxide systems could be adapted and approaches such as the quantum mechanical calculations of Briant et al. [19] for metal/non-metal systems could be relevant. One approach that has been pursued already is that of calculating Van der Waals interaction energies for chemically inert metal-carbon systems. Naidich and Kolesnichenko [20] derived W_{ad} values from the expression

$$W_{ad} = \frac{3}{2}\,\frac{n.\alpha_M\alpha_c}{R^6}\cdot\frac{I_M I_c}{(I_M + I_c)} \qquad (9)$$

where α_c and I_c refer to the polarisability and ionisation potential of carbon, that were in reasonable accord with the experimental data to which they were compared, as illustrated in Figure 4.

Naidich and Kolesnickenko assumed the interfacial structures to be close packed with the separation distances equal to the sum of the metal and carbon atomic radii, but it could be argued that the metal atoms occupy projected ceramic lattice sites, as in the McDonald and Eberhart oxide model and hence the separation distances are equal to the interlayer spacing of graphite, 335 pm. The effect of this second option is generally to decrease the calculated W_{ad} values, as shown in Figure 4, but the correlation with experimental values remains reasonable. Uncertainty about interfacial separations must be felt also for the oxide systems referred to earlier, but these are not of such numerical significance for the McDonald and Eberhart model because Van der Waals interactions represent only a minor part of the total calculated W_{ad} values.

3. INTERFACE CREATION

Knowledge and understanding of interfacial behaviour finds application in joining and other technological processes in which iterfaces with solid ceramics are created by the flow of liquid or solid metals to produce contact. The phenomena of interface creation by the flow of liquid metals, wetting, will be discussed first.

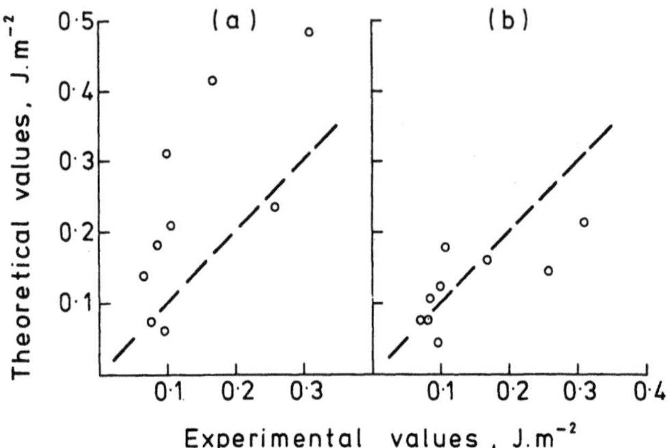

Figure 4. Comparison of experimental data for the wettability of graphite by chemically inert metals with predictions of the Naidich and Dolesnichenko metal. In the left hand plot the separation distance is equated to the sum of the carbon and metal atom radii. In the right hand plot the separation distance is equated to the graphite interlayer spacing.

3.1 Liquid-Solid Interfaces

3.1.1 **Wettability**

A liquid is said to wet when it spreads over a solid surface or penetrates small channels and pores within the solid. This process of interface formation is a necessary condition for many technologically important processes such as brazing, pressureless infiltration of fibre reinforced composites, and some coating techniques. For brazing and coating, excellent spreading and penetration is desirable but a marginal ability to penetrate can suffice for successful infiltration processes.

A way to describe the excellence or otherwise of wetting was formulated in 1805 by Thomas Young [21] and is still the basis of current practice. He argued from observation of the rise of wetting liquids in glass tubes and against partially immersed plates that wetting behaviour was determined entirely by the "superficial cohesion", the surface and interfacial energy, of the liquid and solid and that the angle at which the liquid front contacted the solid, θ, could be related to the "superficial cohesion" by the expression

$$\gamma_S = \gamma_L \cos\theta + \gamma_{SL} \tag{1}$$

The contact angle θ is unique for every combination of materials according to Young, with the liquid remaining balled up on the surface with a smaller contact than equatorial area if θ is greater than 90°, but spreading out and penetrating any emergent porosity if θ is less than 90°. Clearly, θ must be less than 90° for successful brazing, infiltration or coating, and practice suggests θ should be less than 20° for easy coating or brazing.

The experimental technique used by most workers to derive contact angles for ceramic-metal systems is measurement of the profile of a sessile metal drop resting on a horizontal ceramic surface viewed in a vacuum or controlled atmosphere furnace. The drop volume is usually small, and its surface is approximately spherical. If care is taken to ensure the solid surface is horizontal and the system is not noticeably vibrating, contact angles can be measured with an accuracy of about ± 1 or ± 2°.

The importance of reliable wetting data has long been recognised as necessary for both technological developments and theoretical modelling and several collections have been prepared. Unfortunately, the data quoted are not reliable enough to fall into the category of reference material or to permit careful and precise comparisons to be made with the predictions of the models cited in the previous section. The basic reason is not that the experimental work was careless but that the ceramics, metals or environments were not or could not be characterised sufficiently carefully to permit other, later workers to reproduce the work. Further, surprisingly few attempts have been made to measure the wettability of complete groups or families of ceramics. There are of course some notable exceptions, such as the work some years ago with carbides reported by Ramqvist [12] and very recently the painstaking work of Chatain et al [16] with Al_2O_3 and chemically inert metals. Generally, guidance to trends in wetting behaviour still must be sought from rather uncritical compilations such as that of Naidich [5], some of whose data are summarised in Table II.

In the introduction, reference was made to the variable ionicity of ceramics and the conclusion by Pauling that covalent bonding was intermediate between metallic and ionic bonding. It is of interest therefore that there is some evidence of a correlation between wettability and ionicity, as illustrated in Figure 5 for carbide-copper systems [1,12]. The least wetted ceramics tend to be the most ionic, and it should be noted that the very ionic oxides are also the most chemically stable. The effect of this on wetting behaviour has been commented on earlier, but simple correlations of unselected wetting data and oxide thermodynamic stabilities such as that suggested by Naidich [5] exhibit considerable scatter, Figure 6. Some improvement can be made by selecting only wetting data that satisfy certain experimental criteria, such as the use of smooth substrates and confirmation by other workers, as can be seen in Figure 6 for the Al_2O_3 wetting data preferred by Chatain et al [16]. However, the chemical stability of a specific system depends on more factors than merely the free energy of formation of the stoichiometric compound. For example, if the ceramic is not stoichiometric, its chemical stability and surface properties will be affected and wetting behaviour can be changed. Thus increasing the O/U ratio of UO_2 from 2.001 to 2.084 causes the contact angle of copper at 1150°C to change from 116 to 84° [21]. Similar enhancement of wetting by hypostoichiometry has been noted for carbides, as illustrated in Figure 7 for TiC [22].

TABLE II - Contact angle data reported for some ceramic-metal systems [5,64,65]

	Al_2O_3	BeO	MgO	SiO_2	ThO_2	TiO_2	UO_2	SiC	TiC	WC	C	BN
Ag	144	137	136	139				128	138		136	140
Al	118			140			145	34	118		157	
Au	138							138				145
Co	114					109			39	0	48	35
Cr	65	100	160									
Cu	138	140	123	148	92		113		112	30	140	137
Fe	141	127	116		112	72			36	0	37	
In	124		155					130			141	
Ni	128	128	152	125	134	104		65	32	0	45	75
Pt		125			82						87	
Rh					92							
Si	133	76	95			107	90	36	0		15	110
Sn	123	158	121	127	147		140	135		120	149	
V		35										

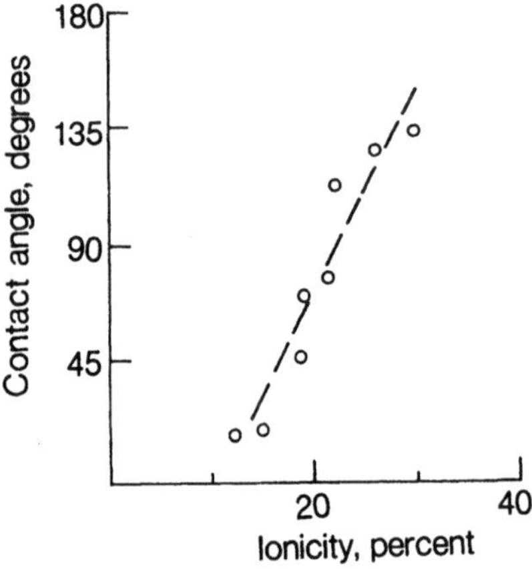

Figure 5. The influence of Pauling ionicity on the wetting of some carbides by copper.

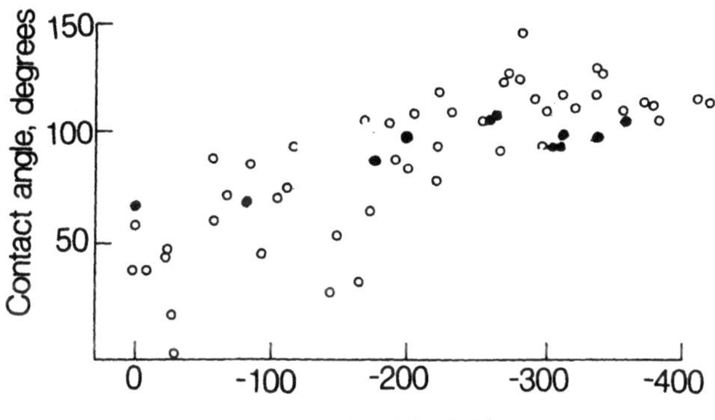

Figure 6. The contact angles of various metal-oxide systems plotted as a function of the relative oxygen affinfity of the ceramic cation and the metal. The data are derived mainly from reference 5, but the full symbols refer to the reference 16 preferred data for Al_2O_3.

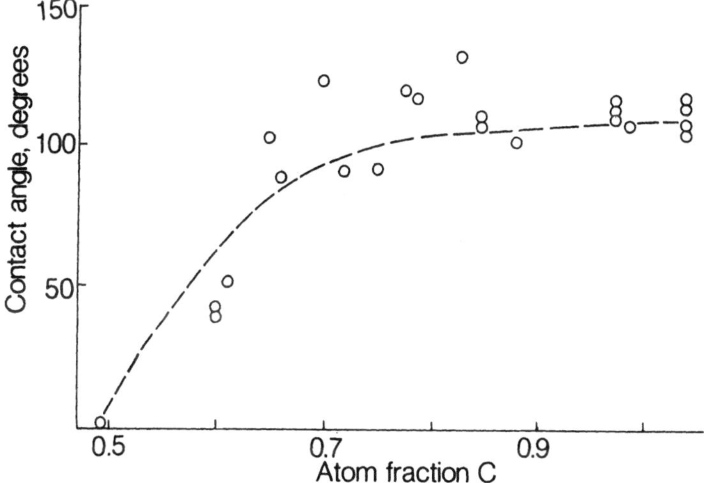

Figure 7. The wettability of titanium carbide by copper plotted as a function of the Ti/C ratio.

If the oxygen activity in the melt is significant, complex reaction products can form at oxide interfaces and wetting can be promoted as indicated by the data for Cu-Al_2O_3 shown in Figure 8 [10]. But it is not always true that the introduction of substrate material is beneficial. Thus carbon diminishes the wettability of graphite by Co, Fe and Ni [14] by suppressing the dissolution of substrate material, and Pask and his co-workers [26] view such dissolution as being beneficial by providing an extra, chemical, driving force and modifying the Young equation to:

$$\gamma_S = \gamma_L \cos\theta + \left(-\frac{\partial \Delta G}{dt} + \gamma_{SL}\right) \qquad (10)$$

Certainly beneficial effects can be caused by reactions as well as dissolution and these are the basis of active metal brazing technology.

3.1.2 **Wetting by Chemically Active Alloys**

The energy of an interface is a function of its chemistry and hence reactions at the interface should affect wetting behaviour. However the detailed nature of the reaction is important because formation of a more stable ceramic compound should in general diminish wettability. Nevertheless, technical and patent literature contains many references to brazes containing reactive components, the most common of which is Ti. The chemical changes caused at ceramic surfaces by Ti can be complex, but analyses can be attempted for a few ceramics such as Al_2O_3 by making simplifying assumptions.

Ti has a high affinity for O, taking up to 29 atom% into solution at 1127°C (1400K) and forming a family of oxides such as TiO_2, Ti_2O_3 and TiO - which latter can have a stoichoimetry ranging from $TiO_{1.2}$ to $TiO_{0.88}$ at 1127°C, a typical brazing temperature of ceramics [23]. Thus many possible reactions might occur between Ti and oxides.

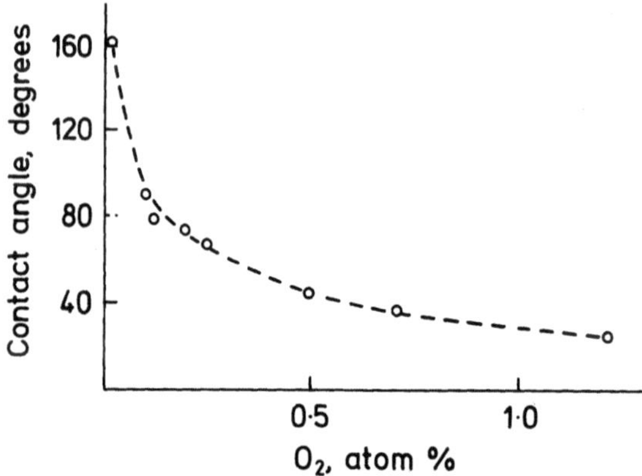

Figure 8. The wettability of Al_2O_3 by Cu-O alloys in vacuum at 1230°C [10].

Consider first the reaction

$$3[Ti] + Al_2O_3 \rightarrow 3TiO + 2[Al] \tag{11}$$

in which square brackets identify solutes. Each reaction is associated with an energy change ΔG°_R which can be related to the free energies of formation of the oxides and the metal activities by equations such as

$$\Delta G^\circ_R = 3\Delta G^\circ_{TiO} - \Delta G^\circ_{Al_2O_3} \tag{12}$$

$$= -RT\ln \left[\frac{a^2_{Al} \;.\; a^3_{TiO}}{a_{Al_2O_3} .\; a^3_{Ti}} \right] = -RT\ln \frac{a^2_{Al}}{a^3_{Ti}} \tag{13}$$

in which the activities of the solid oxides are equated to unity. Thermodynamic data [24] indicate that this reaction will continue until the activity ratio reaches an equilibrium value of 0.9679 at 1127°C, 1400°K, or until the Ti activity level decreases below 4.09×10^{-4} when Ti_2O_3 wil be formed by the degradation reaction

$$3TiO \rightarrow [Ti] + Ti_2O_3 \tag{15}$$

The nature of the solvent does not affect these activity ratios or values but it can affect the Ti concentration needed to form a particular oxide. The activity coefficients of Ti and particularly of Al are low [25] in most braze solvents so that the activity of Al builds up very slowly. Hence dilute Ti alloys can be depleted to such an extent that Ti_2O_3 is formed by degrading TiO. This change in product is important because TiO is metallically bonded [26] and associated with wetting whereas Ti_2O_3 is not [5,27].

While in principle it would seem sensible to saturate the brazes with Ti and thereby maximise the titanium activity and the chance of achieving wetting, high concentrations of Ti are undesirable because they can cause excessive stiffening of the solidified brazes. However, the concentration of Ti needed to cause wetting sometimes can be decreased by introducing a third element. Thus about 8 atom% of Ti is needed to induce Cu to wet Al_2O_3 at 1150°C, but this is decreased to 3.5 atom% if 11 atom% of Sn is added, Figure 9 [28]. The solubility of Ti in Cu at 1150°C is about 67%, but is only 7.6% in Sn [23] so the effect of adding Sn was to cause the melt to be more easily saturated by Ti and hence to increase the activity of the Ti present. Similar effects can be observed with In and to a lesser extent with Ag and Au. In reviewing these data it was suggested that Sn and In were particularly beneficial because they decreased the surface energy of Cu as well as enhancing the Ti activity [28].

Commercially exploited reactive brazes exemplify some of these laboratory derived principles. Thus considerable effort has been devoted recently to decreasing the Ti concentrations of Ag-Cu alloys to about 2% Ti [29], but considerable care has to be taken in controlling the Ti level because as Figure 10 shows [30] the Ag-Cu-Ti system has a liquid miscibility gap. If the Ag-22Cu eutectic is saturated, with

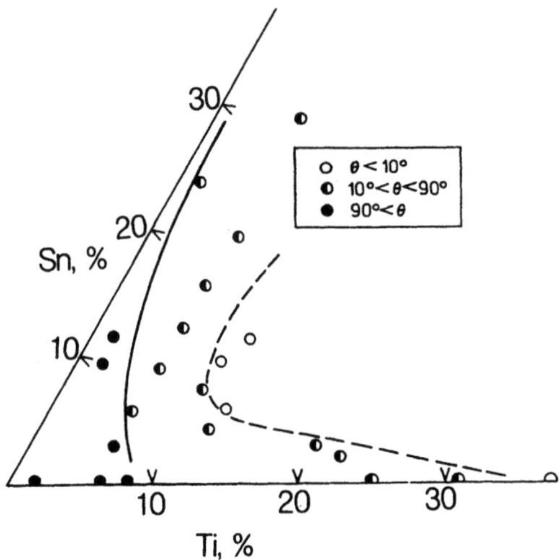

Figure 9. The wettability of Al_2O_3 by Cu Sn Ti alloys [28].

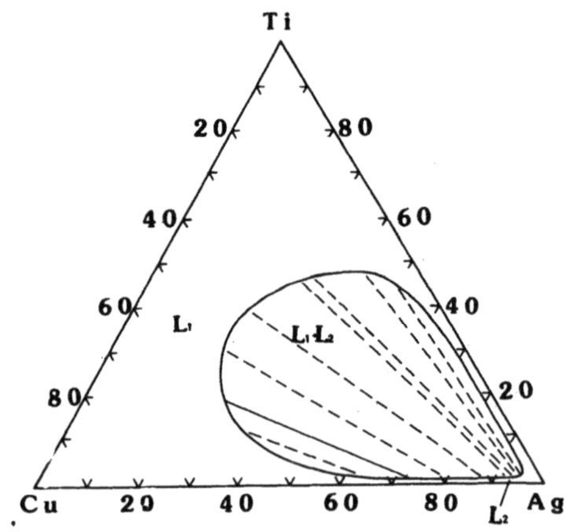

Figure 10. The equilibrium phase diagram for the Ag Cu Ti system at 1300°C [30].

2% Ti, the activity will be high and equal to that of a Cu-12Ag-22Ti alloy as indicated by the tie-lines, but if the Ti concentration increases further an inhomogeneous two phase melt will form and if the Ti concentration decreases by even a fraction of a percent the activity will be diminished sharply. Similarly, Ag-Cu-In-Ti alloys have been developed in which 10-15% of In has been added to the Ag-Cu eutectic and the Ti level decreased to about 1% [31].

Active metal brazes can be used also with non-oxide ceramics and the range of possible reactions is very large. Nevertheless, in practice at this time the materials favoured as brazes are again either alloys containing Ti or Al and its alloys. The simplest case to consider is the wetting of carbon by chemically inert metals alloyed with Ti. Only one reaction product can be formed, titanium carbide, but its stoichiometry is of crucial importance because of its effect on wetting behaviour as illustrated in Figure 6. In practice about 10% Ti has to be added to Cu to produce wetting of graphite at 1100°C, as compared to the thermodynamic prediction that less than 1% is needed to form stoichiometric TiC and that about 20% is needed to form $TiC_{0.60}$ [32], which is wetted by Cu. The comments made previously about large Ti additions causing unacceptable stiffening of brazes are equally true for non-oxide ceramics, but fortunately the beneficial effects of ternary additions such as In and Sn are also equally true for non-oxide ceramics.

Rather more complicated reactions have to be considered in attempting to understand the active metal brazing of nitrides and carbides because interactions with the ceramic cations can be of importance. Thus multiphase layers 10-30 microns thick can be formed by Cu alloyed with 10% Ti at 1150°C in contact with BN and Si_3N_4 for 900s [33]. Substantial reaction product layers are formed by both wetting and non-wetting alloys suggesting that the details of the product chemistries may be important. Thermodynamic calculations suggest that less than 1% of Ti in Cu is needed to form stoichiometric TiN by reactions such as

$$Si_3N_4 + 4[Ti] \rightarrow 4TiN + 3[Si]$$

with the Si liberated then forming silicides with any excess Ti in the melts. However, experimental work with Cu-Ti alloys suggests that the formation of stoichoimetric TiN will not be beneficial since it is not wetted as a bulk ceramic by Cu(5), and in practice several percent of Ti can be needed to produce wetting, Figure 11 [33]. However, TiN is not the only product of interaction with the nitrides. Reaction with with BN or Si_3N_4 can cause B or Si to go into solution or, as indicated by multilayered products and EPMA studies [33], to form Ti compounds adjacent to the brazes. It may be expected that the formation of Ti silicides should promote wetting because of their metallic nature, but literature data indicate that at least TiB_2 is not wettable by Cu [5] and this could account for the poor wettability of BN relative to Si_3N_4.

Al can also be used as a braze for nitride ceramics [33] and several careful studies of the interfacial reaction products have been made for Si_3N_4. Unlike the reactions with Ti containing brazes, the product layers are very thin and Japanese workers have had to use TEM techniques to resolve them as Si-Al-O-N and AℓN compounds [34,35].

The relationship between reactivity and wettability for SiC is similar to that for Si_3N_4. Thus Iseki et al [36] found pressureless sintered SiC to be wetted when at least 1.5% of Ti was added to Ag-Cu eutectic melts, and this was accompanied by the formation of layers of TiC and Ti_5Si_3 with thicknesses of a micron or more. Interestingly, wetting was less readily achieved for reaction bonded SiC containing free Si.

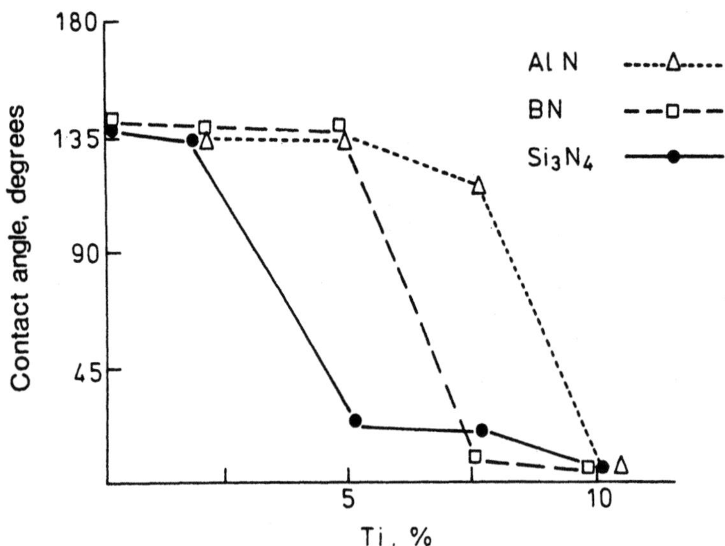

Figure 11. Wettability of some nitrides by Cu Ti alloys [33].

Ag-Cu-Ti alloys are also efficient brazes for nitride ceramics. Thus the addition to Ag-Cu eutectic melts of a saturating 2% Ti, Figure 11, enables Si_3N_4 to be wetted at less than 850°C with the formation of thin reaction product layers. However BN is less readily wettable [33] with an initiating temperature greater than 900°C.

3.2 Solid-Solid Interfaces

The creation of solid-solid interfaces by diffusion bonding, applying pressure at temperatures below the melting point of either workpiece, can be attractive to the fabricator not only because of the modest temperatures but also because the materials that can be bonded are not restricted to those that wet. Detailed quantitative models for metal-metal bonding have been developed for predicting the influence of fabrication parameters on the bond area of metal-metal interfaces [37]. However, the application of diffusion bonding to ceramic-metal systems is more recent. Theoretical modelling has yet to be developed, but Derby [39] has suggested Al grain boundary diffusion to control Al_2O_3-Al bonding and some work in the author's laboratory has revealed the growth of Al_2O_3 - Aℓ/Mg bonding to have an activation energy equal to that for the volume diffusion of Mg in Al [40].

Bond area measurement by cross-sectioning ceramic-metal couples is difficult, but Klomp has derived some data from examination of fracture surfaces [41] which demonstrate that the fabrication temperature can have a dramatic effect, Figure 12. Other parameters such as the bonding pressure and time also have some influence strength but their quantitative effect on the bonded area is unclear. If the fabrication conditions are chosen correctly, near perfect interfaces can be produced with the metal and ceramic lattices extending to the bonding plane [42].

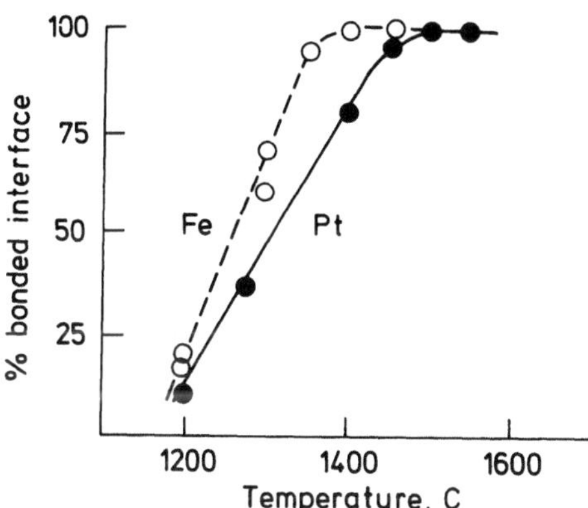

Figure 12. Influence of fabrication temperature on the diffusion bonded area of Al_2O_3-Fe and Al_2O_3-Pt interfaces.

4. INTERFACE PROPERTIES

The primary property required by the users of interfaces is that they are adequately strong, although the aspect of strength that is important varies with the application. Further, commentary on interfacial strength is complicated by that fact that while national and international standards for measuring the strength of monolithic materials have been established those for characterising interfaces are still being developed. The ASTM method for measuring the tensile strength of brazed joints between metallised ceramics [43], has been adapted by some workers for active metal brazed and diffusion bonded joints, but many authors report tensile data derived from other experimentally convenient measurements such as the push-off loads of solidified sessile drops or the bursting pressures of capped tubes, Figure 13. These tests are by their nature very specific and at best yield only tensile data rather than the toughness parameters now required by designers.

In response to this need, notched or pre-cracked 3- and 4-point bend tests are being adopted [62,63]. Nevertheless, difficulties can arise in their use because of uncertainties in the selection of elastic moduli and in ensuring that fracture occurs along the plane of the interface. The avoidance of interface damage during the preparation of the samples calls for great care and has stimulated an interest in developing the use of crack inducing microhardness indentations as a technique for assessing interface as well as bulk ceramic strength characteristics [44].

The absolute strength values suggested by these various tests differ considerably - the bend test tensile strength values being typically 4 or 5 times those derived from ASTM tests - but the conclusions that derive from the tests about the influence of material and fabrication parameters are generally consistent.

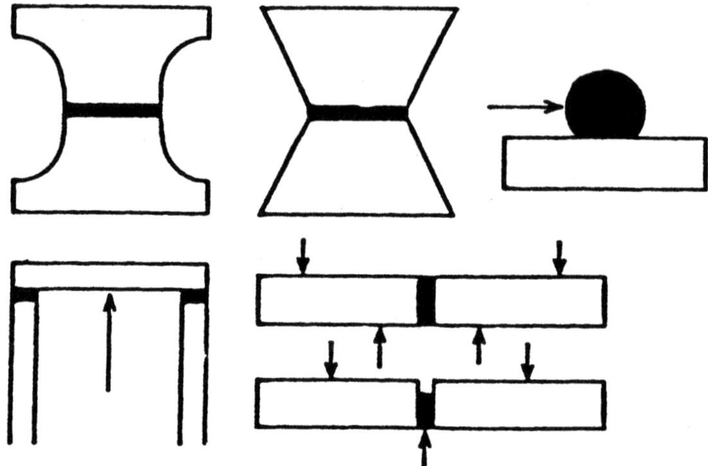

Figure 13. Some techniques used to measure bond strengths; clockwise from top left; ASTM test, truncated cone tensile test, push off test, burst test, bend tests.

4.1 Brazed Interfaces

Relatively little information is available about the strength of joints formed by bringing pure liquid metals into contact with solid ceramics because these systems are often non-wetting and hence brazing is impractical. However, strength data derived from push-off tests show that solidified non-wetting sessile drops can be strongly bonded to oxide substrates [45,46]. For chemically unreactive systems the tensile strengths are not affected by wettability changes caused by varying the contact temperature or time, suggesting that the strengths reflect material rather than fabrication parameters.

Some models of oxide-metal interfaces such as that of McDonald and Eberhart relate W_{ad} and room temperature strengths. In fact, that particular correlation is poor, but strengths can be related to W_{ad} values for solid- solid interfaces, Figure 14 [45], when the metals forming the interfaces were oxide-free and hence both the strength and W_{ad} values reflected direct interactions between the metals and the oxygen ions dominating the Al_2O_3 surface.

The reactions needed to cause wetting of ceramics also affect interfacial microstructures and hence strengths. Thus a reaction initiates by nucleating a product, a new microstructure, at discrete sites and progress of the reaction is associated with lateral spreading of the product and further nucleation so that there is a continuing increase in the extent of the interface affected. When the product covers the interface, the reaction can continue only by forming thicker product layers. The effect of these changes on strengths is complex; interfaces can initially strengthen but ultimately weaken. It can be argued that the reaction products formed at the interfaces create chemical bridges between the ceramics and the brazes, and hence the

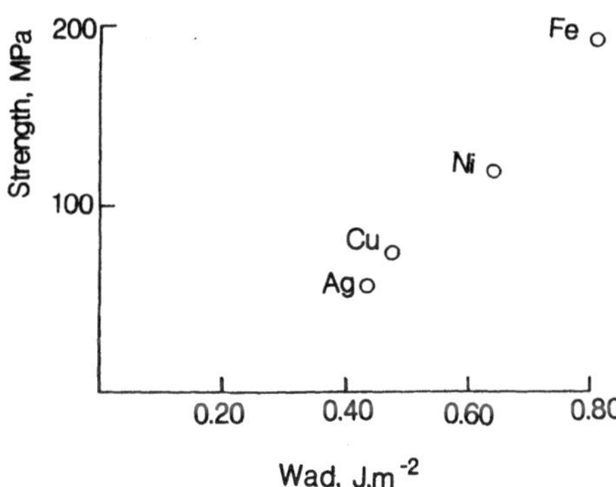

Figure 14. Comparison of the tensile strengths of some Al_2O_3-molten metal systems with W_{ad} values for solid-solid interfaces [45].

initial lateral spread of the products is associated with strengthening. Ultimately the entire interface is covered, and the growth of thick reaction product layers weakens the interface due to excessive growth stresses to cause a strength peak as illustrated in Figure 15. Thus for transparent diamonds brazed with copper alloys, observations through the ceramic and post-fracture examination linked peak strengths to the first achievement of complete interfacial coverage by the reaction products [47].

There is, therefore, an optimum degree of reaction for bonding and this is reflected in the effects of reactive additive concentration, brazing temperature and brazing time. Peak strength values for some sets of systems with possibly similar interfacial microstructures can be related to the chemical affinities of the additives for the ceramic non-metal components, as illustrated in Figure 16 for interfaces formed between Al_2O_3 and binary Ni alloys [48]. This relationship is purely empirical since complex reaction products will be formed at many interfaces and hence the relevance of simple M-O interactions has to be proven, but chemical affinities are noteworthy anyway as one of the few guides for bonding behaviour available at present.

The additive concentrations at which the peaks shown in Figure 15 occurred were lower than those needed to induce wetting. As commented earlier, most reactive additions are unlikely to promote wetting and even for those that do such as Ti, wetting follows rather than precedes complete coverage of the ceramic-braze interfaces by reaction products. For these binary alloys, therefore, the choice is between a strongly bonded but non-wetting composition or one that promotes wetting but causes excessive reaction and bond weakening. Happily this behaviour is not characteristic of all compositions, and in particular the ternary and quaternary brazes referred to earlier, in which the activity of Ti is promoted by decreasing its solubility, exhibit composition regimes in which both excellent wetting and strong bonding can be achieved.

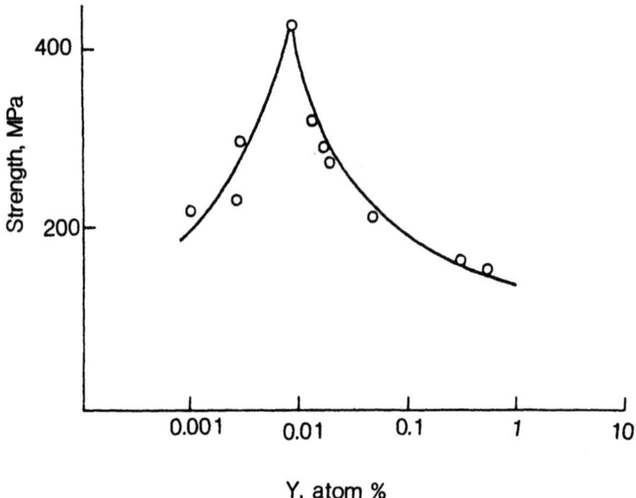

Figure 15. Tensile strengths at interfaces formed by contacting Al_2O_3 with molten Ni-Y alloys at 1500°C for 20 minutes in vacuum.

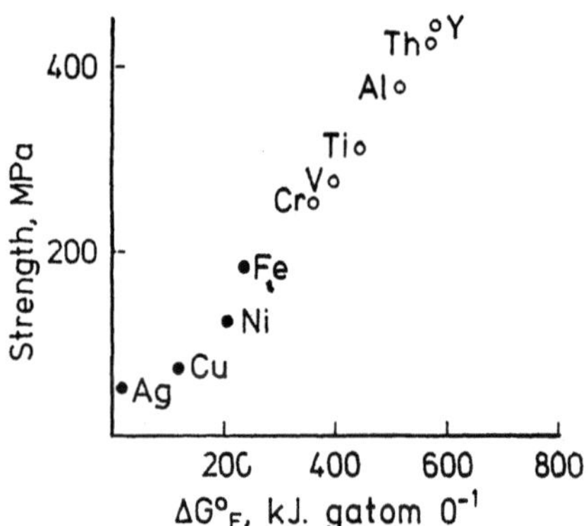

Figure 16. Maximum tensile strengths for interfaces formed by Al_2O_3 with, •, some molten pure metals and, o, Ni alloys plotted as a function of the oxygen affinity of the metals or alloying elements [48].

The actual strengths produced in practice by these brazes depend on many factors besides the nature and extent of the interfacial reactions. Thus the roughness of the ceramic surface and the presence of micro-cracks in ground workpieces can affect strengths, as can bulk properties such as coefficients of thermal expansion and component size and design. These technically important factors, however, lie outside the scope of this paper.

4.2 Diffusion Bonded Interfaces

Strong diffusion bonds have been achieved with many ceramic-metal systems and particularly with oxide-metal systems. Thus Al_2O_3 has been bonded to Ag, Al, Cr, Cu, Fe, Nb, Ni, Pb, Pt, Zn [49,55] as a number of alloys. Provided care is taken to ensure the ceramic surfaces do not contain high crack populations and the mating surfaces are clean, bonding can be achieved using a wide range of fabrication conditions: the literature quotes pressures of 1-100 MPa, temperatures of 50 to 98% of the metal melting point, and times ranging from 1 second to several hours. Some of the conditions employed in bonding Al_2O_3 are summarised in Table III. Usually the larger pressures or longer times produce stronger bonding, but the variables having the greater effects are the environment and temperature.

TABLE III - Some conditions reported for the successful vacuum diffusion bonding of metals to sapphire or polycrystalline alumina

Substrate	Metal	Temperature °C	Time h	Pressure MPa	Bond Strength MPa	Ref.
Sapphire	Cr	1100	2	32	79	55
	Ni	1100	2	64	64	53
Polycrystalline alumina	Al	600	0.3	8.5	90	54
	Cu	980	0.4	8.5	46	54
	Nb	1575	1	15	120	51
	Ta	1400	4	10	27	54

Using a technical vacuum rather than a reducing environment has been found to promote the bonding of some oxide ceramics as illustrated in Figure 17 for the Cu-Al_2O_3 system. While the oxygen activity of a technical vacuum of about 1 mPa is modest, perhaps 10^{-9}, it is still sufficient to permit formation of spinel oxides such as $CuAlO_2$ and $NiAl_2O_4$ which act as bridging compounds. In air, de Bruin and others [56,57] achieved satisfactory bonding of noble and base metals to oxides, but transmission electron microscopy [58] suggests that a liquid phase may have been involved in the interface creation. Interface bridging can be induced also by chemical reactions with components in metal alloys as in active metal brazing, but the process of strength peaking is much slower [40], being controlled by the diffusional supply of the reacting metal component.

Earlier, the effect of temperature on bond area has been commented upon, and similar dramatic effects have been observed for bond strengths. Thus for Al_2O_3-Al ASTM test pieces bonded in a technical vacuum, strengths can be measured once a fabrication threshold of 450°C

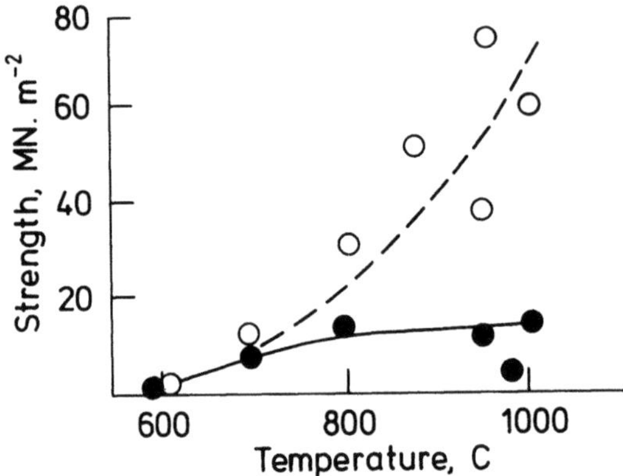

Figure 17. Strengths of Al_2O_3-Cu interfaces diffusion bonded at various temperatures in oxidising (hollow symbols) and reducing (solid symbols) environments [66].

has been exceeded, with the bond strength values having an apparent activation energy of 83 $kJ.mol^{-1}$. Other metals similarly bonded and tested have higher threshold temperatures, as shown in Figure 18 [59], that can be related to the melting points. However, neither the thresholds nor the strength levels indicated in the figure can be regarded as material parameters. Pepper [60] for example bonded the metals to sapphire at room temperature by using ion bombardment and UHV, while Fischmeister et al [63] achieved one hundredfold improvements in the fracture energy of Al_2O_3-Cu samples diffusion bonded at high temperatures by changing from a technical vacua to UHV as the pressing environment. Similarly, other work has shown that bond strength levels for Al_2O_3-Cu [61] can be doubled by low temperature anneals to relieve stresses generated by mismatched post-fabrication oxide and metal thermal contractions.

5. CONCLUDING REMARKS: INTERFACES IN JOINTS

The paper has not yet discussed components and, therefore, important effects of design and post fabrication processing have been ignored. For the present purpose, a joint can be defined as a junction between macroscopic workpieces achieved using an intermediate metal layer, as in brazing or diffusion bonding using ductile interlayers.

The generally smaller thermal contractions of ceramics than metals can exert a dominant influence on joint characteristics since the mismatch causes stresses to be generated as bonds cool after fabrication. Attention has been given by many workers to developing ways of avoidance or mitigation. Thus whenever possible materials with closely matched thermal characteristics are used, and joints with outer metal sleeves are preferred because ceramics are generally better able

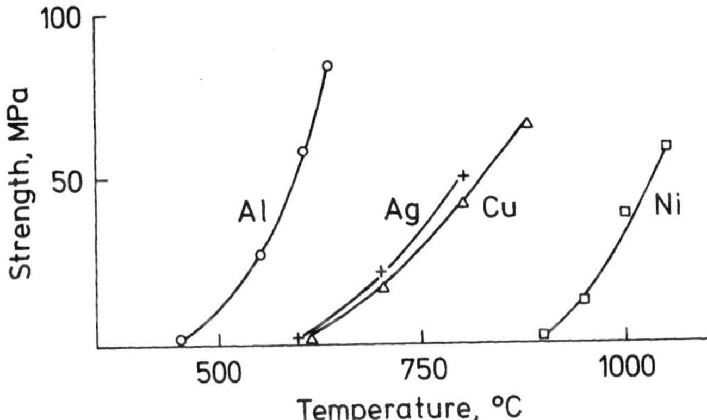

Figure 18. Bond strengths of some Al_2O_3-metal systems: the interfaces were formed using a pressure of 50 MPa applied for 0.5 h in a technical vacuum [59].

to resist compressive than tensile stresses. Unfortunately close matching is seldom possible. Again, active metal brazes can be too sluggish to flow into the capillary annuli of sleeve joints and this configuration also can make it difficult to achieve the flow needed for diffusion bonding. Thus in many cases butt joint configurations must be used, but some mitigation of the mismatch stressing can be achieved by using transition pieces to produce joints with graduated contraction characteristics or by post-fabrication stress relief anneals.

However, even when the design of the joint and selection of materials has been optimised, there remains a need to create well bonded interfaces. This requirement is not easy to satisfy but reference has been made to some of the progress being made in developing techniques and gain understanding of the controlling factors and mechanisms. Rapid advances in future can be anticipated as a result of scientific curiosity and, perhaps more importantly, the demands arising from the widening use of ceramics in engineering and advanced technology.

REFERENCES

1. L. Pauling, 'The nature of the chemical bond', 2nd ed., Cornell Univ. Press, 1948, p. 69.
2. Ibid., p. 421.
3. C.S. Smith, Trans AIME, 1948, **15**, 175.
4. M. Humenik and W.D. Kingery, J. Amer. Ceram, Soc., 1954, **37**, 18.
5. Yu Naidich, Progress in Surface and Membrane Science, 1981, **14**, 353.
6. R.M. Piliar and J. Nutting, Phil. Mag., 1967, **16**, 181.
7. L.E. Murr, Mat. Sci. and Eng., 1973, **12**, 277.
8. J.T. Klomp, Mat. Res. Soc. Proc., 1985, **40**, 381.

9. E.D. Hondros, Inst. Phys. Conf. Ser. #75, 'Science of hard metals', ed. E.A. Almond, C.A. Brooks and R. Warren, pub. A. Hilger, Bristol, 1986, p.121.
10. A.C.D. Chaklader, A.M. Armstrong and S.K. Misra, J. Am. Ceram. Soc., 1968, **51**, 630.
11. K. Ogino and H. Taimatsu, J. Jap. Inst. Met., 1979, p. 871.
12. L. Ramqvist, Int. J. Powder Met., 1965, **1**(4), 2.
13. D.A. Mortimer and M.G. Nicholas, J. Mat. Sci., 1985, **1**, 657.
14. M. Humenik and T.J. Whalen, in "Cermets" ed. J.R. Tinklepaugh and W.B. Crandell, Reinhold Publ, (1960), 37.
15. J.E. McDonald and J.G. Eberhart, TAIME, 1965, **233**, 512.
16. D. Chatain, I. Rivollet and N. Eustathopoulos, J. Chem., Phys., (1986), **83**, p.561.
17. K.H. Johnson and S. Pepper, J. App. Phys., 1982, **53**, 6634.
18. A.M. Stoneham and P.W. Tasker, Mat. Sci. Res., 1986, **21**, 155.
19. C.L. Briant and R.P. Messmer, Studies in Physics and Theoretical Chemistry, (1987), **48**, 261.
20. Y. Naidich and G.A. Kolesnichenko, Russ. Metally, 1968, (4), 141.
21. E.N. Hodkin and M.G. Nicholas, J. Nucl. Mat., 1977, **67**, 171.
22. D.A. Mortimer, "The effects of composition on some properties of titanium oxycarbide", Harwell Laboratory report R-7942 (1975).
23. T.B. Massalski, "Binary Alloy Phase Diagrams", ASM (1986).
24. M.W. Chase et al, J. Phys. Chem. Ref. Data, 1975, 4 (1), 1 and Ibid, 1978, **7**(3), 793.
25. R. Hultgren, P.D. Desai, D.T. Hawkins, M. Gleiser and K.K. Kelley, Selected Values of the Thermodynamic Properties of Binary Alloys (ASM, Metals Park, 1973).
26. W. Rostoker, Trans. AIME, **194**, 981 (1952).
27. A.J. Moorhead and M.L. Santella, Presented at Am. Ceram. Soc. Meetintg, Pittsburgh, April 1987, Paper 27-F11-87.
28. T.M. Valentine, M.G. Nicholas and M.J. Waite, J. Mat. Sci., **15**, 2197 (1980).
29. H. Mizuhara and K. Mally, Welding J., **64**, 27 (1985).
30. Y. Chang, D. Gldberg and J.P. Neuman, J. Phys. Chem. Ref. Data, **6**, 621 (1977).
31. H. Mizuhara and E. Huebel, Welding J., **65**, 43 (1986).
32. E.K. Storms, 'Refractory carbides', 1967 Academic Press, 16.
33 M.G. Nicholas, D.A. Mortimer, L.M. Jones and R.M. Crispin, "Some observations on the wetting and bonding of nitride ceramics", Harwell report R-12940 (1988).
34. M.Naka, H. Mori, M. Kubo, I. Okamoto and H. Fujita, J. Mat. Sci. Lett, **5**, 696 (1986).
35. X.D. Ning, K. Suganuma, M. Morita and T. Okamoto, Phil. Mag. Lett., **55**, 93 (1987).
36. T. Iseki, H. Matsuzaki and J.K. Boadi, Am. Ceram. Soc. Bull., **64**, 322 (1985).
37. B. Derby and E.R. Wallach, Met. Sci., 1982, **16**, 49.
38. B. Derby and E.R. Wallach, ibid, 1984, **18**, 427.
39. B. Derby, Proc. 'Ceramic microstructures '86: role of interfaces', Conf., UC Berkeley, USA, 1986, to be published.
40. M.G. Nicholas and R.M. Crispin, Sci. Ceram., 1987, **14**, 539.

41. J.T. Klomp, Bull. Amer. Ceram. Soc., 1972, **51**, 683.
42. A.G. Evans, M.C. Lu, S. Schmauder and M. Rühle, Acta. Met., 1986, **34**, 1643.
43. 'Standard method for tension and vacuum testing metallised ceramic seals', 1964, ASTM F19-64.
44. C. Hsueh and A.G. Evans, J. Amer. Ceram. Soc., 1985, **68**, 120.
45. M.G. Nicholas, J. Mat. Sci., 1968, **3**, 571.
46. J.E. Ritter and M.S. Burton, Trans. AIME, 1967, **239**, 21.
47. P.M. Scott, M.G. Nicholas and B. Dewar, J. Mat. Sci., **10**, 1833 (1975).
48. R.M. Crispin and M.G. Nicholas, J. Mat. Sci., **11**, 17 (1976).
49. D.A. Mortimer, 'Solid phase bonding method for joining ceramics to metals' AERE report R9533, 1979.
50. J.T. Klomp, Bull. Amer. Ceram. Soc., 1972, **51**, 683.
51. S. Morozumi, J. Mat. Sci., 1981, **16**, 2137.
52. G. Elssner, H. Jehn and E. Fromm, High Temp.-High Pressure, 1978, **10**, 487.
53. C.A. Calow and I.T. Porter, J. Mat. Sci., 1971, **6**, 156.
54. G. Heidt and G. Heimke, Ber. Dt. Keram, Ges., 1973, **50**, 303.
55. C.A. Calow, P.D. Bayer and I.T. Porter, J. Mat. Sci., 1971, **6**, 150.
56. F.P. Bailey and W.E. Borridge, Mat. Sci. Res., 1981, **14**, 249.
57. H.J. de Bruin, A.F. Moodie and C.E. Warble, J. Mat. Sci., 1972, **7**, 909.
58. A.F. Moodie and C.E. Warble, Phil. Mag., 1977, **35**, 201.
59. R.M. Crispin and M.G. Nicholas, Forts. Dt. Keram. Ges., 1985, **1**, 212.
60. S.V. Pepper, J. App. Phys., 1976, **47**, 801.
61. M.G. Nicholas, Proc. 'Ceramic microstructures '86: role of interfaces' Conf., UC Berkeley, USA, 1986, to be published.
62. A.J. Moorhead and P.F. Becher, J. Mat. Sci., 1987, **22**, 3297.
63. H.F. Fischmeister, G. Elssner, B. Gibbesch and W. Mader, M.R.S., International Meeting on Advanced Materials, Tokyo, June, 1988, paper F5.5.
64. G. Samonov, 'The oxide handbook', Plenum Press, New York, 1982.
65. G. Geirnaert, Bull. Soc. Fr. Ceram., 1975, **41**, 40.
66. M.G. Nicholas, Mat. Sci. Forum, (1988), **29**, 127.

SMALL PARTICLES AND THIN FILMS OF METALS ON CERAMIC OXIDES

Louis-C. DUFOUR and Marcel PERDEREAU
Réactivité des Solides, UA 23 CNRS,
Université de Bourgogne
BP 138, 21004 Dijon Cedex
France

ABSTRACT. This paper is aimed at presenting some recent ideas and experimental results on characterization and properties of small particles and thin films of metals on oxide surfaces. Analysis is restricted to metallic films on well-defined surfaces, mainly monocrystalline surfaces and to ceramic oxides having a marked ionic character. The properties of metallic films obtained by chemical reduction of oxides and those of deposits are compared. Nucleation and growth modes are described. Interfacial properties are analysed in terms of size or thickness effect and in the general framework of the ceramic-metal bonding.

1. Introduction.

Due to the fast-expanding development of materials science and to important need in advanced composite materials, investigations on thermodynamic and structural properties of ceramic-metal bonding and interfaces have strongly developed in the last years. Recent progress in this field is analysed by J.T. Klomp and M.G. Nicholas in this book (Klomp, 1989; Nicholas, 1989). Among works devoted to ceramic-metal interfaces, studies dealing with characterization and properties of small particles and thin solid films of metals on oxide surfaces are of fundamental importance in metallurgy, electronics (miniaturization), energetics, optics (optical coatings), ceramic-ceramic bonding, decorative applications, heterogeneous catalysis, etc... This is particularly crucial in catalysis since the properties of supported metals, determining activity and selectivity of this type of catalysts, are often directly dependent on specific interaction between particles of metallic compounds and ceramic support.

This paper is aimed at reporting recent development in the field. It does not pretend to present an exhaustive survey of the subject but rather to remind some sbasic ideas through some experimental results selected in the latest literature. Most examples here reported are dealing with well-defined surfaces of ceramic oxides, preferably pure and monocrystalline, which are generally well-adapted to modern experimental techniques. Nevertheless, we will also take into account some of the numerous results published on preparation and characterization of small metallic particles deposited on very finely divided ceramic oxides to be used as catalysts. The analysis is also

L.-C. Dufour et al. (eds.), Surfaces and Interfaces of Ceramic Materials, 419–448.

restricted to some representative oxides having rock-salt (MgO and NiO), corundum (Al_2O_3) or rutile (TiO_2) structure. The present paper continues and completes a previous one recently published and containing a more exhaustive list of references to which the reader is invited to refer (Dufour and Perdereau, 1988).

2. Small Particles and Thin Films on Oxide Surfaces

2.1 GENERAL

When they are considered from a basic point of view, the properties of metals (or alloys) are generally studied according the following "hierarchic" order:

ATOMS --> CLUSTERS --> SMALL PARTICLES --> THIN FILMS --> THICK FILMS --> MASSIVE METALS

For practical applications, interest is rather directed to small particles and thin films in so far as these latter have size or thickness low enough to give unusual properties when compared to those of massive metals. From the point of view of electronic properties, the dividing line between clusters and small particles is set at the size of 2 nm (~ 150 atoms for an octahedral nickel aggregate) above which usual bulk electronic properties are measured (Poppa, 1984). Nevertheless, unusual chemical or structural properties can be observed for sizes up to 5-10 nm (~ 1900-15000 nickel atoms in the above aggregate) and even 50 nm (~ 1.9 10^6 nickel atoms).

Ceramic metal oxides are interesting support materials for small metal particles. Mostly, they have specific interaction with supported metals and the overall properties are specific and often the result of synergetic effects between the two types of materials.

2.2 CHEMICALLY REDUCED AND DEPOSITED FILMS

There are three main ways to prepare small particles and thin films of metals on ceramic oxides. Each of them lead to specific properties but nucleation and growth of these films correspond to similar processes (Fig. 1).

Films deposited by direct metal condensation. They are usually formed by condensation of a metallic flux in ultra high vacuum conditions. Ion sputtering is also possible. Structural quality and morphology of very thin films are controlled both by the density of the metallic flux and the temperature and properties of the substrate (crystal structure, defective state,..). In strictly controlled deposition conditions, nearly-monocrystalline and well-oriented films can be grown. Surface contamination by gaseous impurities or thermally segregated species often plays an important role.

Deposits by chemical vapour processes. Here metallic particles or films are formed at ceramic surfaces or deposited after chemical reaction of species present in the vapour phase. As generally observed in chemical vapour deposition (Teyssandier, 1989), the numerous steps and factors in this process

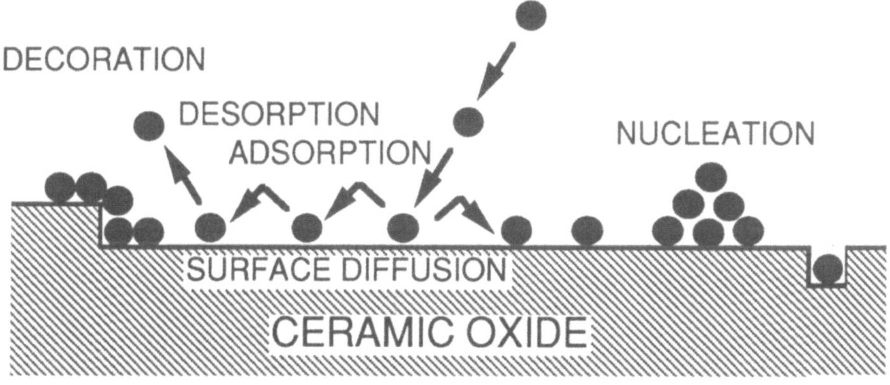

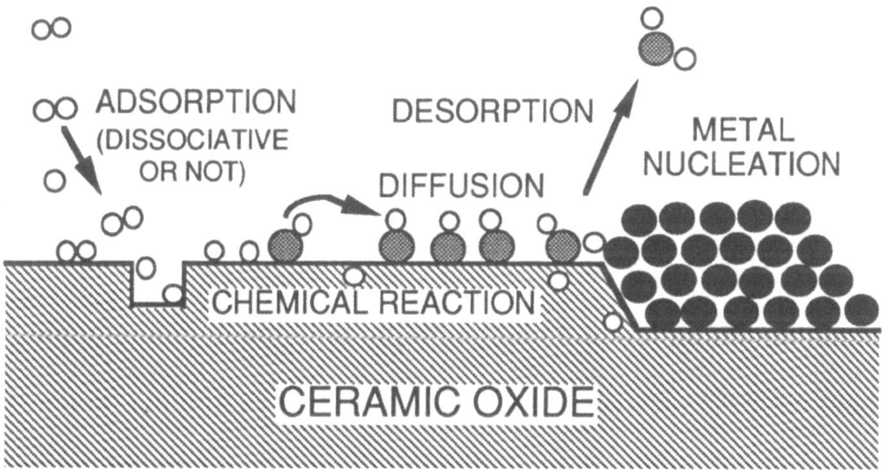

Fig. 1. Main steps in the formation of small particles and thin films on oxides surfaces.

are difficult to be entirely controlled when homogeneous and very thin films are requested. Similar ways have been developed to produce small particles by chemical or electrochemical reduction of species deposited from a liquid phase (for instance, by impregnation...)

Chemically reduced films. Ultra thin metallic films can be grown by chemical reduction of the oxide surface layers. Reducing agents are generally gaseous, as atoms or molecules, and the process is controlled by both thechemical potential of reducing species and temperature. When appropriate prerequisite for both quality of initial surface (defective state, purity...) and experimental growth conditions are selected, this process leads to high grade films having excellent orientation and epitactic relationships towards the substrate. It is obviously limited to metals on their own oxides.

2.3 NUCLEATION AND GROWTH

Most steps involved in the formation of these three types of thin metallic films are similar (Fig. 1). They are mainly constituted by:

2.3.1. Adsorption-desorption processes.

For chemical films, adsorption of reducing species (with or not dissociation) and desorption of reaction products can be intricate multi-step processes. In all cases, these steps preferentially occur at surface defects or irregularities or low coordinated sites which may be classified according to a hierarchical order towards adsorption when their energy level is considered (*e.g.* Dufour, 1982; Boudriss and Dufour, 1989, and ref. therein). Dissociation of molecules on oxide surfaces is always needed for chemical reduction. Mostly, it requires the presence of surface defective sites able to lead to a binding energy between chemisorbed species and site higher than that for dissociating molecules to be adsorbed (*e.g.* Colbourn and Mackrodt, 1982; Surratt and Kunz, 1979; Wepfer *et al.*, 1980). The situation is different when metallic atoms or clusters are present at the oxide surface. These metallic entities often enable direct dissociation of the molecules striking the surface or favours adsorption of species formed after specific chemical reaction. High coverage in reducing species leads to high density of metallic nuclei. For instance, it is well-established that hydroxylation of clean oxide surfaces is very active (*e.g.* Bourgeois and Perdereau, 1982). Then, by using hydrogen atoms easy to be adsorbed, very thin nickel films, few monolayers thick, can be grown by chemical reduction. Such films were obtained from NiO{001} surfaces (Rickard *et al.*, 1979; Dufour and Floquet).

The same is true with metallic deposits. Preferential deposition of metals at steps or extended defects of ionic alkali halide surfaces is used for decorative applications or to reveal these surface irregularities (*e.g.* Métois *et al.*, 1978 and refs therein). With ionic oxide surfaces such as MgO{001} surface, the adsorption energy on normal sites is around 0.5 eV for noble metals such as Pd (Chapon *et al.*, 1985), value to be compared with those found for the noble metal-alkali halide systems [0.5-0.8 eV(Venables *et al.*, 1984)].

2.3.2. Surface and near-surface diffusion or migration.

Surface diffusion may lead to surface rearrangement either for adsorbed species or for metallic entities formed by chemical reduction. As these species preferentially adsorb or form at sites having high energy level, it is difficult to know whether the presence of metallic particles at irregularities is due to adsorption or is the consequence of possible migration. The experimental approach of the surface diffusion for oxide ceramics is difficult and the data are scarce (Matzke, 1989). Generally, surface diffusion is considered at relatively high temperatures for atoms belonging to crystal lattice or more or less strongly bound. Nevertheless, as for noble metals on alkali halide surfaces, diffusivity of metallic heteroatoms or migration of metallic particles at surfaces of ionic ceramic oxides can be considered as possible at moderate and low temperatures (Masson *et al.* 1971, Poppa, 1984). It probably depends on concentration and distribution of imperfections such as ledges, kinks, point and extended defects. Little is known on these phenomena but it can be assumed that their activation energy is low, although higher than those measured for metals on alkali halides [0.1-0.4 eV; Venables *et al.*, 1984]. Theoretical investigation has also been done on these problems (Milchev and Markov, 1985). Moreover, it clearly exists a relationship between surface mobility of atoms and growth mode of particles. Surface mobility and growth mode are, as illustrated further, strongly dependent on stoichiometry, defective state, ordering and crystalline state of the substrate. Lastly migration of small metallic particles could occur if chemical or thermal gradients and, therefore, surface energy gradients exist at the oxide surface. This processes are more and more efficient when particles become more and more viscous at high temperatures and can bend out of their shape. Such a problem has been recently discussed for liquid droplets lying on a solid surface (Brochard, 1989).

2.3.3. Nucleation and growth phenomena.

A lot of works have been devoted to nucleation and growth processes of small particles and thin solid films on ionic substrates (*e.g.* reviews and general articles from : Kaishev, 1951 in: Kern, 1978; Baucr, 1958; Chopra, 1966; Masson *et al.* 1971; Kern,. 1978; Kern *et al.*, 1979; Rhead *et al.*, 1981; Poppa, 1984; Venables *et al.*, 1984; Gibson and Dawson, 1985; Pashley, 1985; Aita and SreeHarscha, 1986; Marks, 1986; Masson, 1986; Grabow and Gilmer, 1988; and refs therein). Most of reported works deal with metallic particles deposited on alkali halide surfaces which are easier to be prepared than ceramic oxide surfaces and can present nearly perfect features, suitable for theoretical

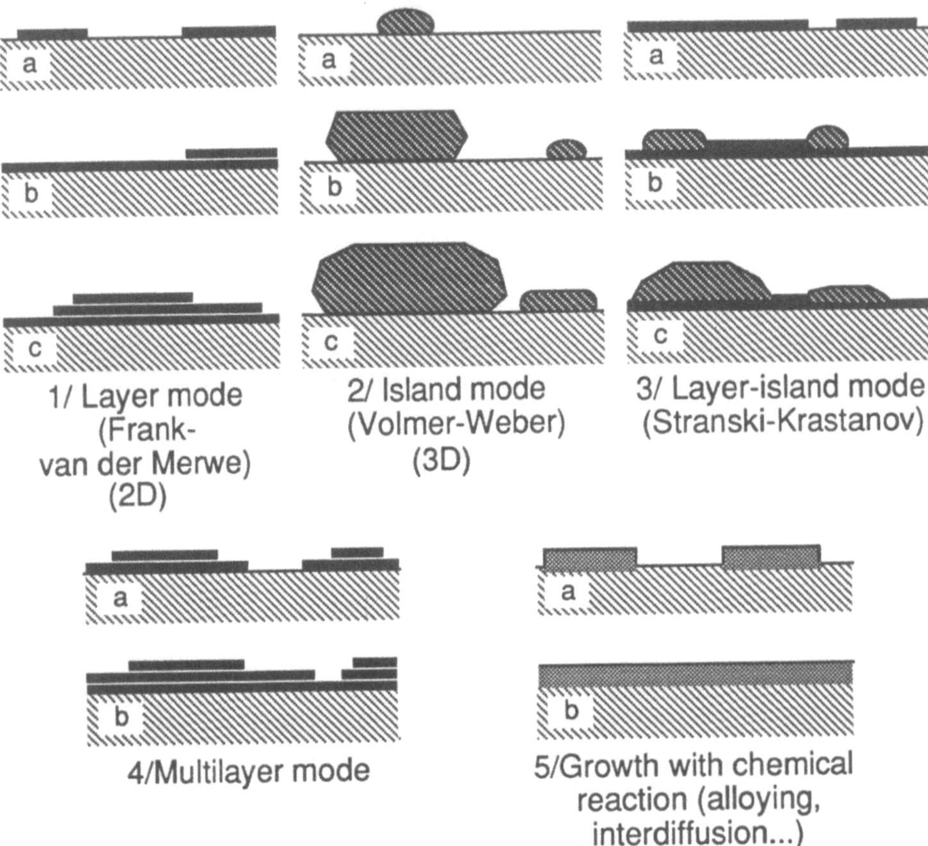

Fig. 2. Possible growth modes of thin metallic films on oxides (see the text)

calculations. Conversely, little work is devoted to nucleation and growth of chemical metallic films on monocrystalline oxide surfaces (Dufour, 1982, 1984; Dufour *et al.*, 1985; and refs therein).

Figure 2 schematically illustrates the possible growth modes of metallic thin solid films on solid oxides according to the classical classification (Kaishev, 1951 in: Kern, 1978; Bauer,.1958; Kern, 1978; Venables *et al.*, 1984 and refs therein).

1/ the layer-by-layér or Frank-Van der Merwe mode (bidimensional or 2D mode) corresponding to a strong metal-oxide adhesion strength compared with cohesion between metallic atoms.

2/ the island or Volmer-Weber mode (threedimensional or 3D mode) is the reverse case where the cohesion energy of the metal is strong if compared with the adhesion between particle and substrate.

3/ the layer-island or Stranski-Krastanov mode is the most usual one in the present case. It corresponds to formation of 3D nuclei from a metallic film grown layer by layer.

Two specific growth modes have been added in Figure 2. The multilayer growth mode can occur for chemical films. It locally corresponds to layer-by-layer mode but the number of layers can vary along the surface. It is difficult to be differentiated from the layer-island mode. Also depicted is a growth mode with concomitant formation of a chemical compound between metal and oxide by alloying, interdiffusion or chemical reaction. This case becomes important with temperature increasing.

Most of the problems related to growth mechanism have been and are still discussed theoretically in relation with those of crystal growth in general but several remarks can be made in the case presently discussed:

a/ Nucleation and surface ordering

Some theoretical aspects of the nucleation and growth of clusters and small particles are currently discussed in terms of fractal geometry (Mandelbrot, 1977, 1982) and computer simulation can be performed from simple models for surface diffusion, deposition or aggregation (see *e.g.* , Witten and Sander, 1981, 1983; Meakin, 1983; Bottet and Jullien, 1988 and refs therein). For instance, a model was very recently proposed for the two-dimensional growth through surface diffusion and through direct deposition (Marner and Schmickler, 1989). It leads to a growing 2D-cluster, not fractal, having a radial density depending on time.The shape of this cluster is roughly circular and similar to that of the disk-shaped nucleus as hypothesized by other authors (*e.g.*, Avrami, 1939, 1940, 1941).

These models have not been essentially conceived to be directly applicable to metallic atoms or clusters on oxide surfaces. In particular, an important point not systematically considered is concerning the interaction between particles and ionic oxide support. As briefly discussed further through some examples, minimization of the interfacial energy often requires ordering at the interface and crystallographic relationships to exist between the two materials. Then crystal lattice symmetry of the initial oxide surface as well as crystallographic orientation and symmetry of the metallic nuclei with possible periodical misfit defects at the interface make rather unlikely the development of these nuclei as fractal objects. Nevertheless, recent analysis of small metallic particles by use of electron microscopy techniques has been interpreted by the fractal character of their surface roughness (Romeu *et al.*, 1986). More generally, growth and coalescence of small particles could obey fractal geometry as far as experimental conditions exist where Brownian motion of metallic atoms and random diffusion at the oxide surface would be actual. Similarities can be found in the growth of a new solid phase in the solid state reactions. This growth can occur according to well-defined crystallographic relationships between nucleus and matrix. But, in specific conditions, the nuclei can also propagate by "branching" of chains inside the initial solid (see *e.g.*: Garner, 1955; Young, 1966; Delmon, 1969). It would be interesting to know whether the boundary region of these nuclei has a fractal character.

b/ Can the growth mode be predicted?

It can be observed (Table 1) that the mechanism of nucleation and growth of particles is directly depends on the quantity: $W_{ad} - 2\,\gamma_M$ also called S, the spreading coefficient, by analogy with liquids.

Table 1 Thermodynamical criteria for the growth mode of metals (M) on oxides (O). $W_{ad} = \gamma_M + \gamma_O - \gamma_{MO}$ = work of adhesion (Dupré law). (θ = contact angle)

0	γ_M	$2\gamma_M$	<-- Relative value of W_{ad}
		X	complete wetting /2D growth mode
	XXXXXXXXX		partial wetting /3D ($0 < \theta < \pi/2$) growth mode
XXXXXXXXX			partial wetting /3D ($0 < \theta < \pi$) growth mode
X			non wetting /3D growth mode

W_{ad} denotes the work of adhesion [= $\gamma_M + \gamma_O - \gamma_{MO}$] or reversible work required to separate a unit interfacial area of the two solid phases. Here the system is considered without adsorption (inert medium) and without interdiffusion at the metal-oxide interface. γ_M and γ_O refer to the surface tension (also called surface energy but, for solids, confusion is possible with surface free energy) of the metallic "adsorbate" and oxide substrate respectively. Let us recall that the surface tension of a solid is defined from its surface free energy (= F^s per unit area) as:

$\gamma = d(F^s.\,A)/dA. = F^s + A.(\partial F^s/\partial A)_{T,V}$ where A is the surface area.

The latter term corresponds to the energy required to have the surface stretched (mainly by elastic deformation for small particles) without change of the total number of its ions or atoms The interfacial tension γ_{MO} is the reversible work of formation of the unit area of this interface at constant volume, temperature and chemical potentials. Amongst the terms not strictly independent which can contribute to this interfacial energy, several are important: changes in electrostatic energy, dispersion and short-range repulsion forces, spreading of metal conduction electrons into the oxide and epitaxial relationships (Stoneham and Tasker, 1988).

Despite the existence of thermodynamical criteria (Table 1), it is rather difficult to predict *a priori* the mode of growth of small metallic particles for each system. The surface tension of pure metals is generally higher than that of ceramic oxides (*e.g.* Nicholas,1989), but can decrease more strongly with temperature increasing. Little is known on W_{ad} and γ_{MO} but the growth of metals with high cohesion energy will probably occur in the island mode. In fact, it can be expected that the growth mode will strongly depend on experimental conditions. Nevertheless, as pointed out further, many of the published results for metals on clean crystallized oxide surfaces put forward a Stranski-Krastanov growth mode.

Recent papers have been devoted to theoretically discuss and to perform simulation about these problems (*e.g.* Nieminen and Kaski, 1987; Stoyanov, 1988; Grabow and Gilmer, 1988 and refs therein). In particular, a model has been proposed making possible to estimate the thickness of the uniform film from which 3D-nucleation occurs in the layer-island mode (Grabow and Gilmer, 1988). The contact angle (θ) between the metallic three dimensional nucleus and the film obeys the classical Young-Dupré equation (Fig. 3) :

$$\gamma_{\text{film}} = \gamma_{\text{film-nucleus}} + \gamma_{\text{3D nucleus}} \cdot \cos\theta$$

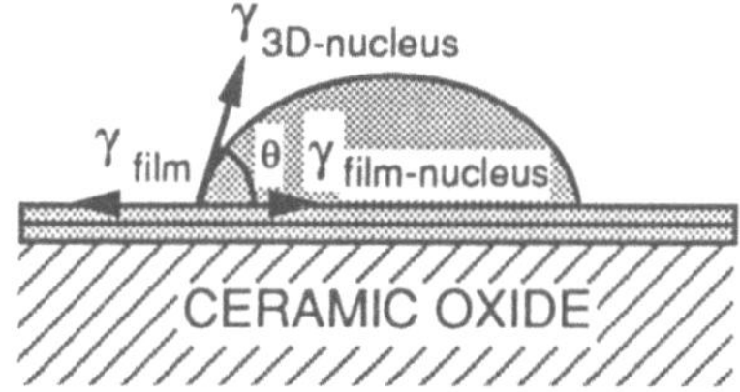

Fig.3. Schematic representation of a three-dimensional nucleus in the layer-island growth mode (S-K growth mode)(after Grabow and Gilmer, 1988)

θ is different from zero because of epitaxial strains present in the film. It is shown that the uniform film never is the equilibrium state. The driving force for 3D nucleation can be calculated as a function of the number of monolayers and the degree of misfit between metallic film and oxide substrate. It is predicted that, for most systems, this 3D nucleation becomes kinetically favoured when the number of layers in the film is in the range 2-5 corresponding to the maximum value of the driving force. Nevertheless the question of how the layers form remains open.

c/ Growth mode and oxide-metal bonding

Some of the factors taken into account in wetting of ceramic oxides by liquid metals or in the ceramic-metal bonding (*e.g.* Klomp and Van de Ven,

1980; Stoneham, 1982; Chatain *et al.*, 1986,1987; Pask, 1987; Tasker and Stoneham, 1987; Courbière *et al.*, 1987; Dalgleish *et al.*, 1988; Hicter *et al.*, 1988; Nicholas, 1988; Loehman, 1989; and refs therein) can be expected to play an important role in the growth mode and strongly oriented the morphological features of metallic particles. Table 2 schematically illustrates the possible influence of these factors. The subject constitutes a formidable task and unfortunately only a limited number of experimental results in the field are

---

Table 2. Some of the factors involved in bonding of metals to ceramic oxides and possibly operating to orientate the growth mode of small metallic particles on oxide surfaces.

+ ⇐	Wettability of ceramic oxides	⇐-
2D (layer by layer mode)	2-3D (layer island mode)	3D (island mode)

⇒ oxide ionicity increasing ⇒
⇒ free energy of formation of the oxide increasing ⇒
⇒ oxide stoichiometry increasing ⇒
⇒ cohesive energy of the metal increasing ⇒
⇐ oxygen activity in metal increasing ⇐
⇐ formation of a more stable compound at the interface ⇐
impurities or defects preexisting at the oxide surfaces or segregating after heat treatment ⇐(according to their affinity increasing for oxygen) ⇐
mismatch of physical properties at the interface (crystal structure, dielectric properties,...)

published with complete characterization of the bulk and surface properties of both materials before and after bonding. As reported further, many difficulties exist particularly with very thin metallic films.

d/ Equilibrium shape of small particles and growth mode

If the crystallographic surface fluctuations are excluded (see Section 4), small crystallized particles could have their surface constituted by flat facets or by facetted regions together with curved regions or by totally curved regions depending on the anisotropy of the Wulff plot. It should be noted that the curved regions can be considered as consisting in a succession of elementary facets. As pointed out by several authors (*e.g.*: Herring, 1952; Pilliar and Nutting, 1967; Winterbottom, 1967; see also Wang *et al.*, 1985),

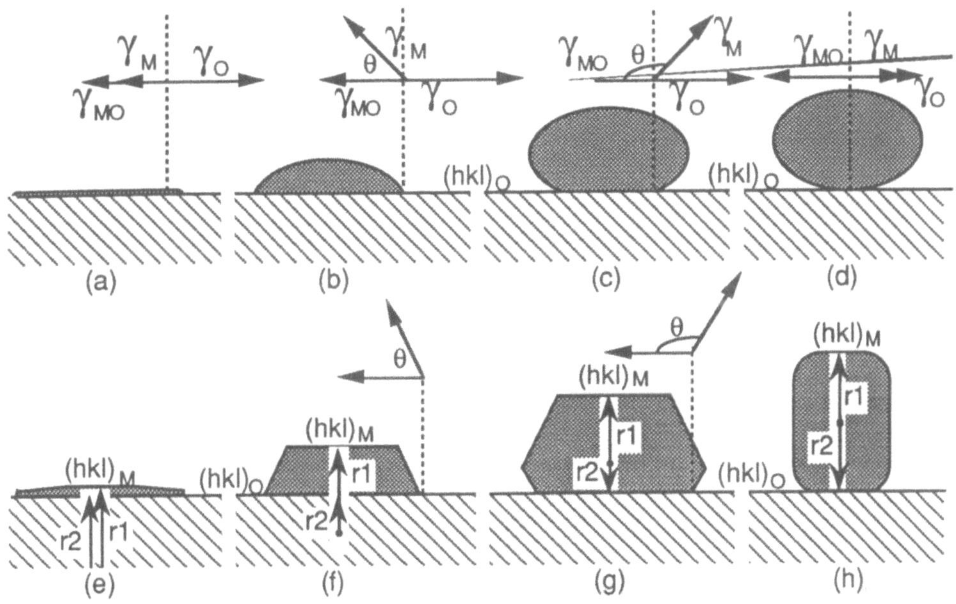

Fig. 4. Possible shapes of small metal particles on oxide surfaces.

(a)-(d): liquid drops or curved solid particles having a surface tension quasi independent of orientation (spherical γ-plot).

(a) $\theta = 0$; $\gamma_O - \gamma_{MO} \geq \gamma_M$; $W_{ad} \geq 2\gamma_M$; complete wetting.

(b) $0 < \theta < \pi/2$; $0 < \gamma_O - \gamma_{MO} < \gamma_M$; and (c) $\pi/2 < \theta < \pi$; $0 < \gamma_{MO} - \gamma_O < \gamma_M$; partial wetting.

(d) $\theta \to \pi$; $\gamma_{MO} - \gamma_O = \gamma_M$; $W_{ad} = 0$; non-wetting;

(e)-(h): solid particles with facetted and curved regions. The same criteria may be used for single crystals under specific conditions (see the text). One may also use r_1 and r_2, the vectors on the γ plot of the particle to the (hkl) interface and (hkl) surface planes, respectively. Then (e), (f), (g) and (h) correspond to $r_1/r_2 \leq -1$, $-1 \to 0$, $0 \to 1$, $= 1$ respectively.

(After Kaishev in Kern, 1978; Pilliar and Nutting, 1967; Winterbottom, 1967)

---

there is often indeterminacy in the configuration of intersecting interfaces when the various interfaces of the metal particle adopt orientations at equilibrium. The Young-Dupré equation:

$$\gamma_{MO} = \gamma_O + \gamma_M \cos\theta \qquad \text{(see Fig. 4)}$$

used for liquid particles having their surface tension independent of orientation is not theoretically valid but under specific conditions. For instance, consider monocrystalline particles, 10-1000 nm in size, with a very limited fraction of edge and corner sites at their surfaces and let us assume that the metal-oxide interface plane is not coherent but crystallographically well-defined [$(hkl)_M$] Then a treatment parallel to that used for liquid metal particles may be applied to curved and facetted particles leading to also define a contact angle, θ (Fig. 4). A similar way consists to use a modified Young-Dupre equation:

$$\gamma_{MO} = \gamma_O + \gamma_{(hkl)M}\, r_1/r_2$$

where r_1 and r_2 are the vectors on the particle γ-plot (or Wulff-plot) corresponding to external (hkl) surface and (hkl) interface respectively. (see Winterbottom, 1967; Pilliar and Nutting, 1967).

In reality, with particle size decreasing, new features may make the problem more intricate. Due to packing effects, the relative fraction of surface planes can be strongly modified and large deviation from the Wulff construction are identified (*e.g.*,Marks, 1985)(see Section 4 below). The relative fraction of corner or edge sites having a low coordinence number becomes more and more important according to particle structure and shape (Van Hardefeld and Hartog, 1969). Moreover, the particles, even very small, often are multiply twinned to increase their stability by the existence of new symmetry at the interface and in the bulk. The presence of multiply twinned particles (MTP) and the applicability of the Wulff construction have been discussed (Ino, 1969; Marks, 1983-1986; Howie and Marks, 1984).

3.. Characterization of Small Particles or Very Thin Films

Characterization of small particles and very thin solid films of metals on their oxide substrate is still a very difficult task despite development of very sensitive experimental techniques enabling to get information below the nanometric scale. Figure 5 schematically illustrates some of the most important properties to be characterized. The main modern techniques used for such investigations are gathered in Table 3.

The first information relates to coverage or mean thickness, *i.e.* the mean number of metallic atoms per unit surface area of oxide substrate. It is achieved from the kinetic measurement of the amount of metallic atoms fixed or reduced at the oxide surface. Ultragravimetry is often used for this purpose. AES and SIMS are also used if careful and reliable calibration can be performed. Additional information on mean size and size dispersion of particles is necessary when the oxide surface is not totally covered, the most usual case. Specific chemisorption (*e.g.* of hydrogen for group VIII metals) is commonly used to determine the degree of dispersion of the metallic film. Electron microscopy techniques, magnetic measurements and EXAFS for

CHARACTERIZATION OF SMALL PARTICLES
(or VERY THIN FILMS) OF METALS ON OXIDES
= A DIFFICULT CHALLENGE

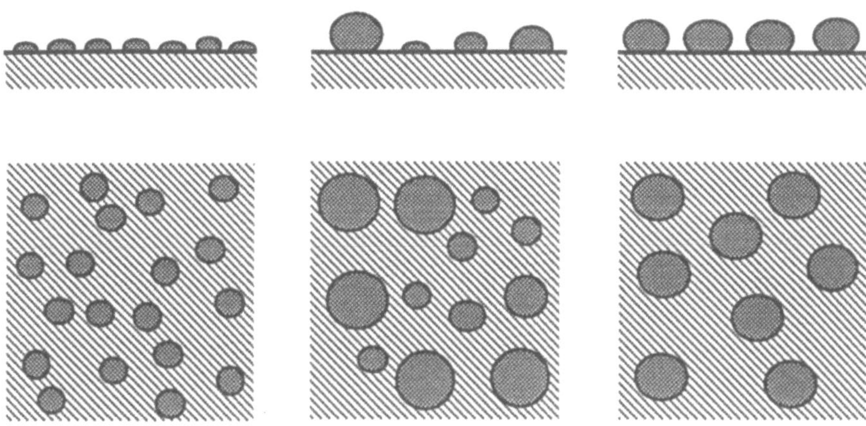

- coverage or mean thickness (in number of metallic atoms per unit surface area of the oxide)
- particle size and size distribution
- crystal structure and shape
- structural relationships at metal-oxide interface
- chemical composition
- electronic structure ......

Fig. 5. Schematic presentation of some of the most important elements needed to characterize small particles and thin films of metals on oxides. The mean coverage (or equivalent mean thickness) is the same in the three cases but particle size and size distribution are different.

highly dispersed or embedded metals are operative for such a purpose. Determining crystal structure and shape, crystallographic orientation and chemical composition is a difficult task on the nanometric scale, but of basic importance for understanding many structure or size sensitive properties or their modifications due to specific metal-oxide interaction. HREM, STEM, STED, REM, EDX, RHEED and STM are powerful tools for such a study. Amongst the physical properties of clusters and small particles and their interfaces with ceramic oxides, electronic structure is important particularly in catalysis to explain size effects. Lastly AES and SIMS are usually used to

Table 3
Main experimental methods to characterize the chemical and physical properties of small particles and thin films of metals on oxides

LARGE AREA TECHNIQUES	HIGH SPATIAL RESOLUTION TECHNIQUES (*on the nanometric scale)
Auger Electron Spectroscopy (AES)	Scanning Auger Microprobe (SAM)
(Static) Secondary Ion Mass Spectrometry [(S)SIMS]	Secondary Ion Microprobe (SIM)
Ion Scattering Spectroscopy (ISS) Rutherford BackScattering (RBS)	* Scanning Transmission Electron Microscopy (STEM)
Extended X-ray Absorption Structure (EXAFS)	* Transmission Electron Microscopy (HREM,TEM)
X-ray and Ultraviolet Photoelectron Spectrometry (XPS and UPS)	Scanning Reflection Electron Microscopy (SREM, REM,SEM)
X-ray Absorption Near Edge Stucture (XANES)	Energy Dispersive X-ray Microanalysis (EDX)
Reflection High Energy Electron Diffraction (RHEED)	* Scanning Tunneling Microscopy (STM)
Low Energy Electron Diffraction (LEED)	*Scanning Transmission Electron Diffraction (STED)
Electron Energy Loss Spectroscopy (EELS)	
Surface Extended Energy Loss Fine Structure (SEELFS)	
Photoemission Yield Spectrometry (PYS)	
Magnetic Measurements	
Ultragravimetry	
Chemical Methods	
-> HIGH SENSITIVITY ->MORPHOLOGY BY INDIRECT WAY	->DIRECT VISUALIZATION ->POSSIBLE LIMITATION BY OXIDE THICKNESS

->RADIATION (ELECTRON, ION...) INDUCED DEFECTS IN OXIDE
->CHARGING EFFECT ON INSULATING OXIDES
(DEPENDING ON STOICHIOMETRY)

determine the growth mode of these very thin metallic films although measurements do not lead always to conclusive answer.

The experimental methods of characterization have recently been reviewed for thin films (Werner and Garten, 1984) and oxide surfaces (Hirschwald, 1988). Some of the techniques of characterization are listed in Table 3. Their application to characterize bulk and interfacial properties of small metallic particles have been discussed particularly by:

- Le Lay and Kern, 1978; Kern, 1978; Rhead *et al.*, 1981; Werckmann *et al.*, 1981; [Growth mode]
- DiCenzo and Wertheim, 1985; Wertheim, 1987; Wertheim and DiCenzo, 1988; [AES, XPS, UPS]
- Haller, 1985; Evans, 1986; Kip *et al.*, 1987; Wijnen *et al.*, 1988; [EXAFS]
- Brown and Packwood, 1986; [EDX]
- Poppa, 1984; Gallezot *et al.*, 1985; Smith, 1985; Brown L.M., 1986; Freund *et al.*, 1986; Gillet, 1986; Heineman and Soria, 1986; Sanders, 1986; Venables *et al.*, 1986,1987; Smith *et al.*, 1987; [TEM, SEM, HREM, STEM]
- Morita *et al.*, 1987; Venables *et al.*, 1987; Chidsey *et al.*, 1988; [STM]

It should be noticed that, when nanometer-size metal aggregates are analysed, extreme care must be used in the interpretation of results obtained by any technique. Definitive arguments cannot be put forward as far as two different techniques do not lead to same conclusions. Limitation by contamination before or during the observation is particularly important to consider for these very reactive particles as well as the oxide support.

TEM is still the main technique to get atomic-level information on the shape of very small metal particles (*e.g.*, Marks and Smith, 1983; Wallenberg *et al.*, 1985; Bovin *et al.*, 1985; Harris, 1987; Datye and Long, 1988; and refs therein). Nevertheless, among new techniques, scanning tunneling microscopy (STM) promises to be a very useful tool for the study of deposition processes, particularly on thick substrates when the TEM imaging is no more operative. This technique is generally used together with complementary techniques making possible the characterization of structure and orientation of particles on their substrate. A recent example concerns the surface morphology of gold particles on mica (Chidsey *et al.*, 1988). STM enabled to directly evidence the relation between mobility of metal atoms and fusion of nuclei into large islands with temperature increasing.

4. Some Properties of Small Particles

Two series of questions are raised about the properties of small metal particles and very thin metal films. The first one is: how do these properties vary as a function of mean thickness or particle size? The second one: how are these properties modified through interaction with an oxide support? In this section a brief survey is given about some of these properties.

4.1. INDIVIDUAL SMALL METAL PARTICLES.

Size effect and unusual properties of free small particles are firstly due to strong changes in the relative fraction of low coordinated surface sites with size decreasing. As recalled above, the relative fraction of these different types of surface sites depend not only on the size but also the shape and crystal structure of the particle (Van Hartefeld and Hartog, 1969). Packing effect in the very small solid particles ($\phi < 10$ nm) of noble and transition metals has been experimentally evidenced and theoretically discussed (see *e.g.* : Khanna *et al.*, 1983; Borel and Chatelain, 1985; Solliard and Flueli, 1985; Marks, 1985; Buttet, 1986; Solliard *et al.*, 1988) This crystal lattice contraction is directly related to the importance of the surface stress and can be described by a Laplace-type law:

$$\Delta a/a = -2/3\, g\, K/r$$

where g = mean surface stress coefficient defined as: $g = \gamma + A\, d\gamma/dA$
($g < \gamma$ if contraction)(Solliard and Flueli, 1985)
K = compressibility coefficient
and r = particle radius

Change in the lattice parameter up to 1% has been measured for various shapes of particles and the values obtained are in good agreement with the calculated values. Nevertheless, for very small particles, conclusive results through electron patterns can be obtained only after careful calculations of the possible effects of electron diffraction leading to apparent size effects (Gallezot *et al.*, 1985).

Another important question concerns the structural rearrangements of ultrafine metal particles (smaller than 50 Å) recently evidenced by HREM (Iijima and Ichihashi, 1986; Smith *et al.*, 1986). This structural instability leads to continual changes in the shape of particles and can be accompanied by motion of the particle on its substrate. The rate of fluctuation depends on the degree of contact between particle and substrate, the electrical conductivity of the substrate and the current density of the incident beam. Nevertheless, this phenomenon, similar to quasi-melting (Marks *et al.*, 1986; Ajayan and Marks, 1988) would not be due to thermal heating effect but is induced by other electronic phenomena (Dundurs *et al.*, 1988). It becomes sluggish with particle size increasing as a consequence of controlling diffusion kinetics.

Results and calculations on physics of microclusters have been recently reviewed (Halicioglu and Bauschlicher Jr, 1988). Despite large progress in the characterization of these very fine particles and the study of structural and electronic properties as well as chemical reactivity, the question 'when does a cluster become a bulk?' still remains unanswered. In particular, changes of electronic properties from the isolated atoms to the bulk metal have been largely studied. This investigation is of great interest not only theoretically but also for understanding the size effects observed with ultrafine metal particles in catalysis (see refs in the paper of the authors, 1988 and some recent works or reviews in the field *e.g.* : Mason, 1983; Poppa, 1984; Dicenzo and Wertheim, 1985; Zdetsis and Kunz, 1985; Kohiki, 1986; Masson *et al.* 1986; Fritsch and Legaré, 1987; Wertheim, 1987; Cordes, 1988; Wertheim and Dicenzo, 1988). Individual and collective optical properties of small metallic particles can be inspected for studying the electronic size effects (Kreibig and Genzel, 1985). Lastly, the mechanical properties of thin films have recently been reviewed (Hardwick, 1987).

4.2. SMALL PARTICLES SUPPORTED ON CERAMIC OXIDES.

The two main questions concern the change of properties due to the existence of a metal particle-oxide interface and that related to more or less strong interaction between the two materials or with environment and bringing about reversible or irreversible changes in the physical and chemical properties of the particle.

For an abrupt interface (without atomic interdiffusion or alloying) the work of adhesion at the metal particle-oxide substrate depends on various factors, including:

- ionicity and stoichiometry of the oxide,
- cohesive strength of the metal,
- electronic properties of the two materials in the interfacial region
- structural or geometrical factors determining the epitaxial relationship

The misfit between the two crystal lattices in epitaxial relationship (see *e.g.*: Matthews, 1979 and refs therein) is usually defined as:

$$f = (p.a_O - q.a_M)/q.a_M$$

where a_O and a_M are the stress-free lattice parameters of oxide and metal particle respectively. The integers p and q are defined from: $a_O / a_M = p/q$. Then it can be assumed that a fraction of the interfacial atoms lie in atomic sites common for the two lattices. For small particles on oxides, it is convenient to define the misfit as:

$$f = (a_O - .a_M)/.a_M$$

For very fine particles, no dislocation exists at the interface and mostly the metal lattice will be strained elastically with an elastic strain proportional to f. In these conditions, if the effect is tensile (f often positive) for the metal lattice, the above size contraction effect for free ultrafine particles will be partially or totally compensated by the epitaxial strains created at the metal-oxide interface. The resulting effect will be contraction or expansion. For instance, expansion (1-3%) was observed for very fine palladium particles (1.5-4 nm) on mica (Heinemann and Poppa, 1985) and MgO (Chapon *et al.*, 1985). When the particle is bigger, the creation of misfit dislocations at the interface is an efficient process to minimize the overall free energy. Then the elastic strain in the particle decreases discontinuously for each new misfit dislocation appearing. Periodic and non periodic arrangement of misfit dislocations are possible but the latter is often more efficient to release the epitaxial strains.

Chemical and thermal stability. A lot of problems, rather difficult to solve for very small particles, follow from chemical interaction during thermal treatment or gas exposure. The lack of knowledge on the dynamic properties and the thermal stability of small particles is still important despite the growing number of works in this field. The situation is made very complicated because the chemical, structural and electronic characterization of very small particles on their ceramic oxide support remains still a difficult task. Amongst the thermal and chemical processes involved to modify the properties of small metal particles on their oxide support, there are:

- adsorption of gases and segregation of additives or impurities present in the oxide. These processes can drastically modify the interfacial tension as well as surface tension and shape of metallic particles according to the chemical affinity of the metal for these gases, additives or impurities. They can lead to spreading of the particle.
- surface migration as presented above and leading to the growth of the particle size by coalescence or Ostwald ripening
- encapsulation or covering of the particle by a thin layer of the oxide substrate or a product of the reaction between metal and oxide.
- chemical transformation of the particle by interdiffusion, alloying, dissolution...

Some of these processes can lead to strong changes in the properties of small particles as a result of the formation of new chemical bonds. That is why their study was and is still the subject of many investigations by EXAFS, electron and ion spectroscopies or TEM. Particular attention is paid to the phenomenon so called "SMSI" (Strong Metal-Support Interaction), often observed when small particles of a group VIII metal are deposited onto a multivalent reducible oxide and thermally treated at moderate temperature in specific environment. This phenomenon is important in heterogeneous catalysis where large surface area ceramic oxides are used as support to prevent sintering of small metallic particles and preserve their state of high dispersion. When metal particles are dispersed on certain transition metal oxides (TiO_2, Ta_2O_5..), the SMSI state arises when the catalysts is reduced above ≈ 650 -773K. In this state, the capacity of the catalysts to adsorb gases like hydrogen or CO becomes very low whereas the particle size remains unchanged. This state can be removed by oxidation at high temperature. A lot of papers have been published about this phenomenon since its discovery (Tauster *et al.*, 1978). Several explanations were proposed for this effect: localized or delocalized electron transfer between metal and oxide, change in the metal structure, encapsulation or decoration effect by migration of oxide species onto the metal, creation of vacancies or special sites at the metal-oxide boundary, spillover effect.... In reality, no clear evidence exists in favour of any of the models proposed. Nevertheless the model proposing a modification of the valence levels of the metal as a result of the presence of a suboxide encapsulating the metallic particle seems to be accepted most widely in 1988. It is difficult to give an exhaustive list of the interesting papers in this field. We propose some recent papers, but surely many other works should be also quoted (see *e.g.*: Bond and Burch, 1983; Sadeghi and Henrich, 1984, 1988; Broekhoven and Ponec, 1985; Haberlandt and Ritschl, 1986; Sun *et al.*, 1986; Ruckenstein, 1987; Sanchez and Gazquez, 1987; Martens *et al.*, 1988; Shpiro *et al.*, 1988 and refs therein).

5. Some recent examples of results on growth and properties of very thin metallic films on oxides

In conclusion of this paper, some examples of recent experimental results on "chemical" and "physical" films are briefly commented. It is difficult to quote all papers published in the field of small particles and we have extracted some of them in a prolific literature often dealing with supported metals for catalysis.

5.1 CHEMICALLY REDUCED FILMS ON WELL-DEFINED CERAMIC OXIDE SURFACES

Experimental results on the first stages of nucleation and growth of metals by chemical reduction of well-defined oxide surfaces are rather scarce. Amongst the oxide-metal systems, the model system NiO{001}-Ni has been studied by the present authors (Rickard *et al*, 1979, Floquet *et al.*, 1981; Floquet and Dufour, 1983; Dufour *et al.*, 1985; Boudriss and Dufour, 1989) mainly because it enables to perform a relatively comprehensive investigation of the structural features of the metal formation. Figures 6 and 7 illustrate the main experimental results of the structural study. Roughly speaking, nickel nuclei form in the three-dimensional growth mode when the NiO{001} surface, prepared "in situ" by cleavage, is exposed to hydrogen or carbon monoxide. Nevertheless two very different growth modes can be evidenced by RHEED as a function of temperature.

The first one is observed at low temperature when nickel forms in a pseudo-hexagonal stacking. This unusual stacking corresponds to the best fitting between the two crystal lattices and it directly results from the interfacial arrangement between the NiO{001} fcc planes and the Ni{11.0} hexagonal planes. Nevertheless, the diffraction patterns show, besides spots due to three-dimensional Ni crystallites, specific features interpreted by the existence of Ni atom rows ordered along only one direction (see Fig. 7) and a structural desordering in the stacking of the {00.1} hexagonal planes perpendicular to the initial oxide surface. It may therefore be hypothesized that the nickel nucleus forms from atomic rows in the NiO <010> directions

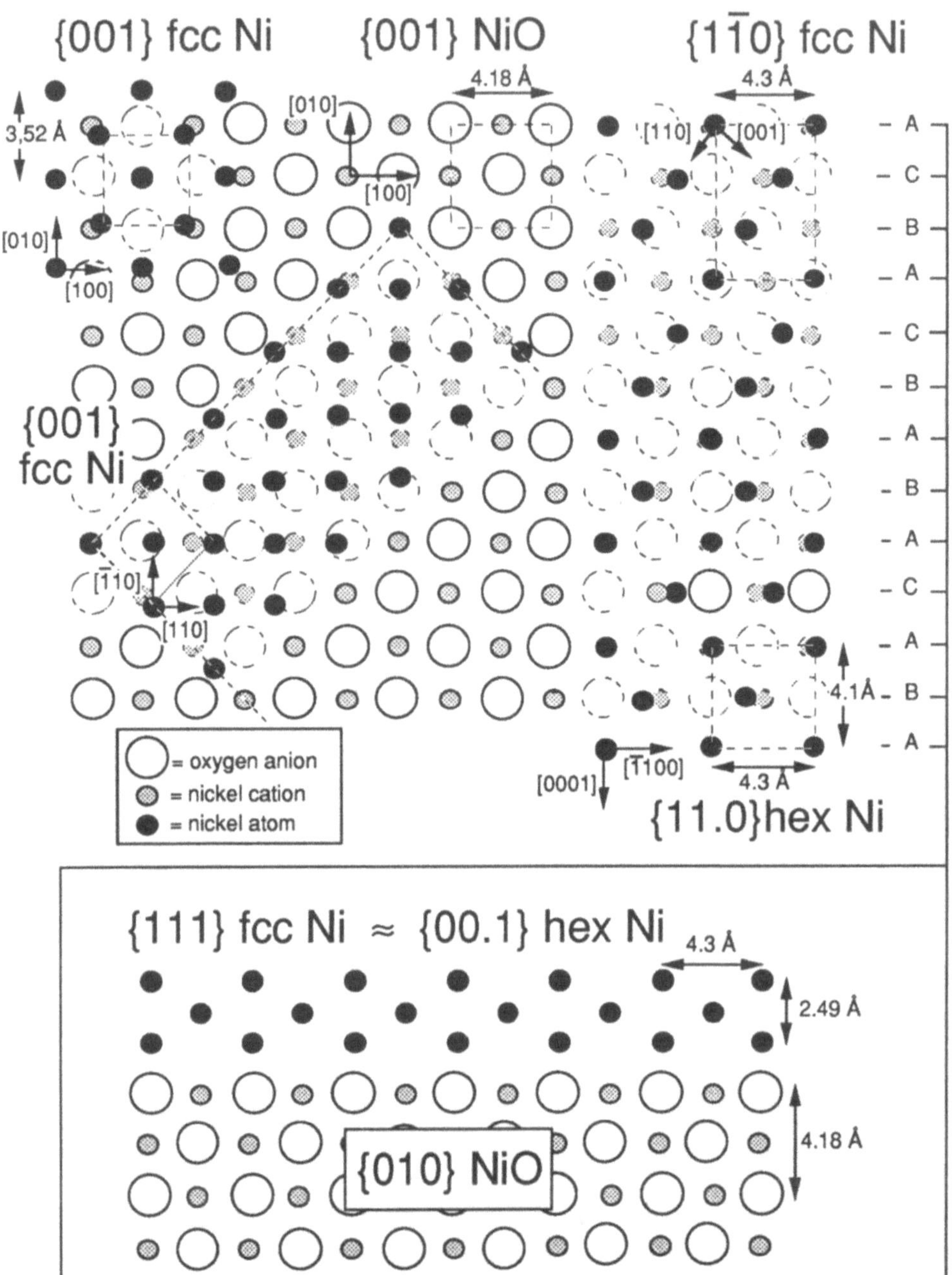
{001} fcc Ni
{001} NiO
{1$\bar{1}$0} fcc Ni
4.18 Å
4.3 Å
3,52 Å
[010]
[100]
[110]
[001]
{001}
fcc Ni
[$\bar{1}$10]
[110]
A
C
B
A
C
B
A
B
A
C
A
B
A
4.1Å
= oxygen anion
= nickel cation
= nickel atom
[0001]
[$\bar{1}$100]
4.3 Å
{11.0}hex Ni
{111} fcc Ni ≈ {00.1} hex Ni
4.3 Å
2.49 Å
4.18 Å
{010} NiO

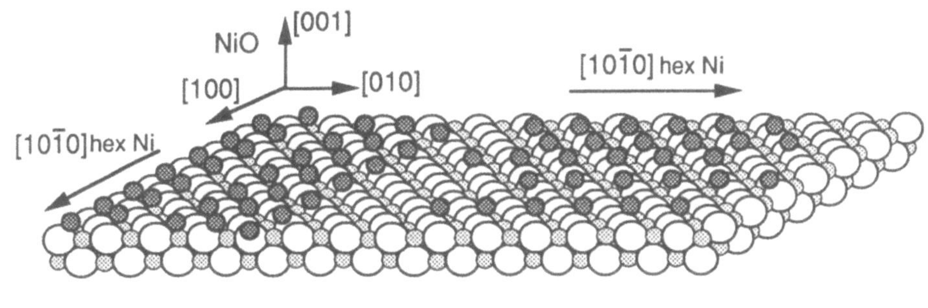

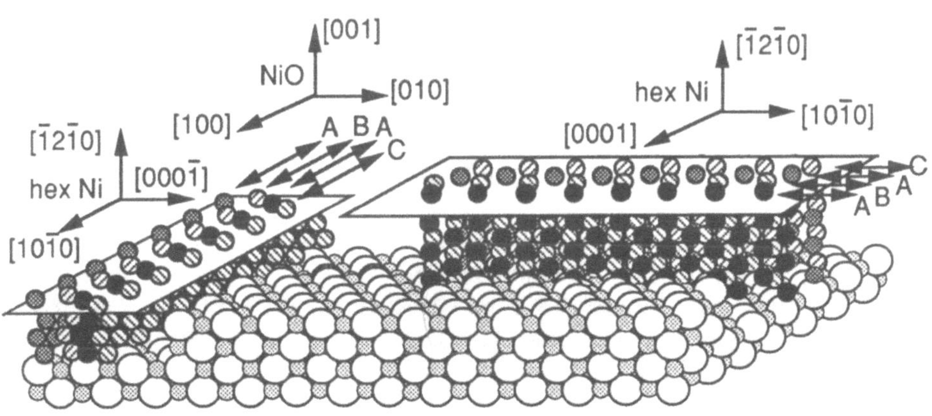

Fig. 7. Schematic view of the low temperature growth mode of Ni-nuclei from the NiO{001} surface as deduced from RHEED-AES investigations (see Fig. 6 and text). Ni-rows and Ni-nuclei are represented outside the oxide lattice but, in fact, they are embedded inside matrix. Coalescence of nuclei having one or the other equivalent orientation possibly leads to twinning.

<---- Fig.6. Relative position of the crystallographic planes at the oxide-metal interface as deduced from the RHEED analysis of the first stages of the reduction of a nearly perfect, "in situ" cleaved NiO {001} surface.

Nickel nucleates in the unusual hexagonal structure in hydrogen or CO at low temperatures (low temperature mode illustrated at bottom right). The ABACAB... sequence of the (0001) hexagonal planes was often observed but stacking faults were also evidenced. The ABCABC... sequence in the face centered cubic stacking was formed from the hexagonal stacking during the nucleus growth (top right).

At top left, the structural orientation commonly observed in hydrogen and CO in the high T growth mode of the metal nucleation (fcc {001} nickel). At bottom left, the fcc {001} nickel orientation, at 45° from the previous one, was only detected in CO. This latter orientation corresponds to an excellent coincidence between the two lattices in the NiO <110> directions.

The position of the metallic atoms with respect to the oxide {001} plane is hypothetized. The atomic arrangements are represented with the same parameters as in the bulk but the lattices actually are slightly distorted at the interface. (after Boudriss and Dufour, 1989)

where the interatomic distances are nearly preserved ($d_{<010>NiO} / d_{<10.0>Ni}$ ≈ 0.973). Nevertheless a bidimensional ordered arrangement of these Ni rows is not energetically favoured and a three-dimensional hexagonal lattice is preferably built up by spatial arrangement of these rows. The periodicity of the initial NiO {100} planes is preserved and the stacking of the Ni {00.1} planes always leads to the sequence A∗A∗A∗.. As a consequence of the initial arrangement of the Ni rows, the distribution of the other Ni {00.1} planes (∗ = B or C) leads either to random stacking or to the sequence ABACABAC.(both observed) but never to hcp structure (ABAB)(non observed). Fcc Ni structure with the ABCABC stacking of the {111} planes can appear afterwards particularly with temperature increasing.

The second growth mode is observed at higher temperature and its three-dimensional character is undoubted since, in the most general case, the crystallographic axes of the two fcc lattices of the oxide and metal remain parallel during the growth of the nucleus. From TEM results, a mechanism for the nickel island growth was proposed in which the lattice misfit dislocations, at the metal-oxide interface, are used as diffusion paths (Floquet *et al.*, 1981).

From the results obtained in this experimental case, it is clear that both morphology and properties of these small metal particles basically depend on the experimental conditions of reduction. At low temperature, with hydrogen atoms, very thin metallic films , few layers thick, can be grown.

5.2. DEPOSITS ON WELL-DEFINED CERAMIC OXIDE SURFACES

Table 4 summarizes some selected results recently published. Two conclusive remarks can be deduced about growth mode:

- as expected, the Stranski-Krastanov and island growth modes are often observed when metal particles grow on crystalline, stable and clean oxide surfaces. The metal atoms have sufficient mobility, even at relatively low temperatures, to form three-dimensional nuclei.
- the influence of stoichiometry or crystalline state of the oxide on the growth mode of metal particles is clearly evidenced. Non stoichiometric surface defects or amorphous state favour the layer-by-layer growth mode.

6. Conclusion

As a short conclusion, three points are noteworthy:

a/ Basic research on the properties of metals on oxides is still needed, particularly on structure and chemical interaction at interfaces

b/ Many data are still lacking on the atomic-level properties of ceramic metal oxide surfaces when compared with those of metals. This strongly

Table 4
Some experimental results obtained on the metallic films deposited on MgO, Al_2O_3 and TiO_2

INFLUENCE OF THE DEFECTIVE STATE				
MgO	<	Al_2O_3	<	TiO_2
one oxidation state (Mg^{2+})		------------->		various oxidation states (Ti^{4+}/Ti^{3+})

SYSTEMS	REFERENCES	TECHNIQUES	RESULTS

MgO {001}

SYSTEMS	REFERENCES	TECHNIQUES	RESULTS
Au/-	Cowley *et al.* ,84	TEM	3D gth mode, epitaxy, decoration
Cu/-	He *et al.* ,86	AES,LEED,ELS	SK growth mode, epitaxy
Cu/-	Alstrup *et al.*, 88	XPS	SK g. mode, non-metallic state for Cu
Fe/-	Hubert *et al.* ,83-86	AES,TEM,XPS	3D g.mode, epitaxy, Fe dissolution
Fe/-	Lee *et al.*, 87	UPS,EELS,SEELFS	Fe-O bonding for low coverage
Mg/-	Janssen *et al.*, 75	LEED,AES	island growth mode,epitaxy
Pd/-	Chapon *et al.*,85	TEM	3D growth mode, epitaxy
Pd/-	Renou *et al.*,85	TEM	3D g. mode,morphol.,epit.,infl.of T

Al_2O_3

SYSTEMS	REFERENCES	TECHNIQUES	RESULTS
Cu/(0001)S	Lin *et al.*,88	SEM	high T morphology,epitaxy
Fe/(1102)-	Poppa *et al.*,86	TEM, AES, STED	3D mode (?) on various substrates
Mo/(0001)-	Katrib *et al.*,88	XPS	non-met. Mo,Mo-Al_2O_3 interaction
Pd/-	Kohiki,86	XPS	change of electronic state
Pd/(1012)-	Gillet *et al.*,87	TEM,AES	3D mode, epitaxy,diffusion at RT
Pt/amorph.	Masson *et al.*,86	RHEED,XPS,TEM	VW mode,2D-3D trans.,size effect
Pt/-	Fritsch *et al.*, 87	XPS	electronic structure
Pt/(0001)-	Altman *et al.*,88	AES,T Prog.Des.	3D mode
Pt/amorph.-	Altman *et al.*,88	AES,T Prog.Des.	2D mode

TiO_2

a lot of papers devoted to catalysis

SYSTEMS	REFERENCES	TECHNIQUES	RESULTS
Fe/(001)-	Brugniau *et al.*,84	AES,LEED,Work F	2D layer by layer mode
Ni/(110)-	Kao *et al.*,80	UPS,AES,LEED	electronic properties
Ni/-	Bourgeois *et al.*,88	SIMS, AES,PYS	2D -> 3D infl. of stoich., Ni diffusion
Ni/-	Munuera et al., 89	XPS	phase mobility
Pt/(001)-	Brugniau *et al.*,84	AES,LEED,Work F	2D layer by layer mode
Pt/(110)-	Sun *et al.*,86	SIMS, AES	SMSI, no encapsulation
Pt/100-	Tamura *et al.*, 88	XPS	SMSI,
Rh/-	Belton *et al.*,84	SIMS,TDS,AES	SMSI
Rh/(110)-	Sadeghi *et al.*, 84-88	XPS,UPS,AES	Rh-Ti bonding,Rh ordering, SMSI
Rh/-	Fuentes *et al.*, 88	TEM	1D->2D->3D with T increasing
Ti/(110)-	Rocker et al., 87	XPS,AES,ELS	electronic structure

inhibits the studies defined in (a).

c/ Practical applications and experimental works about small metal particles supported on oxides are now essentially directed to improve selectivity and efficiency of catalysts. Nevertheless, development in other industrial areas (microelectronics, energetics, welding ..) is in progress.

7. References

Aita C.R. and SreeHarscha K.S. (Eds), 'Thin Films: the Relationship of Structure and Properties', Mat. Res. Soc. Symp. Proc., Vol. 47, Mat. Res. Soc., Pittsburgh, 1986

Altman E.I. and Gorte R.J., *A study of small Pt particles on amorphous* Al_2O_3 *and* α-Al_2O_3 *{0001} substrates using TPD of CO and* H_2, J. Catalysis, **110** (1988) 191-196

Alstrup Ib. and Møller P.J., *X-ray photoelectron spectroscopy study of the vapor deposition of copper onto a MgO(100) surface, Appl. Surface Sci.*, **33/34** (1988) 143-151

Ajayan P.M. and Marks L.D., *Quasimelting and phase of small particles*, Phys. Rev. Lett., **60** (1988) 585-587

Avrami M., *Kinetics of phase change.* (a) *I. General theory*, J. Chem. Phys., **7** (1939) 1103-112. (b) *II Transformation time relations for random: distribution of nuclei*, ib., **8** (1940) 212-224. (c). *III. Granulation, phase change and microstructure*, ib., 9 (1941) 177-184

Bauer E., *Phänomenologische Theorie der Kristallabscheidung an Oberflächen*, Z. Kristall., **110** (1958) 372-394 (I) and 395-431 (II)

Belton D.N., Sun Y.M. and White J.M., (a)*Thin-film models of strong metal-support interaction catalysts. Platinum on oxidized titanium*, J. Phys. Chem., **88** (1984) 1690-1695; (b) *Metal-support interactions on Rh and Pt/*TiO_2 *model catalysts*, J. Phys. Chem., **88** (1984) 5172-5176

Bond G.C. and Burch R., *Strong metal-support interactions*, Specialist Periodical Repts Royal Soc. Chem. (London), Catalysis, **6** (1983) 27-60

Borel J.P. and Châtelain A., *Surface stress and surface tension: equilibrium and pressure in small particles*, Surface Sci., **156** (1985) 572-579

Botet R. and Jullien R., *A theory of aggregating systems of particles: the clustering of clusters process*, Ann. Phys. Fr., **13** (1988) 153-221

Boudriss A. and Dufour L-C., *Defects and reactivity at oxide surfaces: experimental aspexts of the interaction of hydrogen, CO and* CO_2 *with the NiO{001} surface*, in 'Non Stoichiometric Oxides', Nowotny J. and Weppner W. (Eds), Kluwer, Dordrecht, 1989, in print

Bourgeois S.and Perdereau M., *SIMS study of the interactions of Ni-O surfaces with* H_2 *and* H_2O, Surface Sci., **117** (1982) 165-168

Bourgeois S., Diakité D., Jomard F., Perdereau M., Poirault R., *A SIMS and AES study of nickel deposition on* TiO_2*(100). Influence of the stoichiometry of the support*, Proc. 1st Int. Conf. Structure and Reactivity of Surfaces, Trieste, 13-16 Sept. 1988

Bovin J.O., Wallenberg R. and Smith D.J., *Imaging of atomic clouds outside the surfaces of gold crystals by electron microscopy*, Nature, **317** (1985) 47-49

Brochard F., *Motions of droplets on solid surfaces induced by chemical or thermal gradients*, Langmuir, **5** (1989) 432-438

Brown L.M., *Recent applications of STEM to small particles*, Ultramicroscopy, **20** (1986) 39-42

Brown J.D. and Packwood R.H., *Methods for quantitative electron probe microanalysis of small particles and thin films*, Appl. Surface Sci., **26** (1986) 294-305

Brugniau D., Parker S.D. and Rhead G.E., *Monolayers of platinum and iron on $TiO_2(001)$: characterization by Auger electron spectroscopy, secondary electron emission and work function changes*, Thin Solid Films, **121** (1984) 247-257

Buttet J., *Geometrical structure of metal clusters*, Z. Phys. D, Atoms, Molecules and Clusters, **3** (1986) 155-157

Chapon C., Henry C.R. and Chemam A., *Formation and characterization of small Pd particles on MgO as a model catalyst*, Surface Sci., **162** (1985) 747-754

Chatain D., Rivollet I and Eustathopoulos N., (a) *Adhesion thermodyna-mique dans les systèmes non-réactifs métal liquide-alumine*, J. Chim. Phys., **83** (1986) 561-567; (b)*Estimation du travail d'adhésion et des angles de contact dans les systèmes non réactifs métal-oxyde ionocovalent*, J. Chim. Phys., **84** (1987) 201-203

Chidsey C.E.D., Loiacono D.N., *Sleator T. and Nakahara S., STM study of the surface morphology of gold on mica*, Surface Sci., **200** (1988) 45-66

Chopra K.L., 'Thin Film Phenomena', Mc Graw-Hill Book Company, New York,

Colbourn E.A. and Mackrodt W.C., *Theoretical aspects of H_2 and CO chemisorption on MgO surfaces*, Surface Sci., **117** (1982) 571-580

Cordes O. and Harsdorff M., *Investigation of gold clusters with photoelectron spectroscopy*, Appl. Surface Sci., **33/34** (1988) 152-159

Courbière M., Tréheux D., Béraud C. and Esnouf C., *Liaison métal-céramique: techniques et aspects physicochimiques*, Ann. Chim. Fr., **12** (1987) 295-312

Cowley J.M. and Neumann K.D., *The alignment of gold particles on MgO crystal faces*, Surface Sci., **145** (1984) 301-312

Dalgleish B.J., Lu M.C. and Evans A.G., *The strength of ceramics bonded with metals*, Acta Met., **36** (1988) 2029-2035

Datye A.K and Long N.J., *The use of nonporous oxide particles for imaging the shape and structure of small metal crystallites in heterogeneous catalysis*, Ultramicroscopy, **25** (1988) 203-208

Delmon B., 'Introduction à la Cinétique Hétérogène', Technip, Paris, 1969

DiCenzo S.B. and Wertheim G.K., *Photoelectron spectroscopy of supported metal clusters*, Comments on Solid State Phys., **11** (1985) 203-219

Dufour L.C., *Réactivité des surfaces monocristallines d'oxydes métalliques dans l'hydrogène: formation et déplacement des interfaces solide-solide*, J. Microsc. Spectrosc. Electron., **7** (1982) 341-357

Dufour L.C., *Surface and bulk chemical reactivity of binary metal oxides. Relation with some physical properties of the solids involved*, in'Basic Properties of Binary Oxides', Dominguez-Rodriguez A., Castaing J.and Marquez R. (Eds), Publ. Univ. Sevilla, 1984, pp. 259-277

Dufour L.C and Floquet N., unpublished

Dufour L.C., Floquet N. and de Rosa B., *Progress in the understanding of the mechanisms of nucleation and growth of metals from oxides. A study from the NiO-Ni model system*, in 'Reactivity of Solids' Barret P. and Dufour L.C. (Eds), Elsevier, Amsterdam, 1985, pp.47-52

Dufour L.C. and Perdereau M., *Metals on oxides: formation, characterization and properties*, in 'Surface and Near-Surface Chemistry of Oxide Materials', Nowotny J. and Dufour L.C. (Eds), Elsevier, Amsterdam, 1988, pp. 577-610

Dundurs J., Marks L.D. and Arayan P.M., *Structural fluctuations in small particles*, Phil. Mag. A, **57** (1988) 605-620

Evans J., *Characterization of small particles by EXAFS*, Chem. in Britain, 1986, pp. 813-818

Floquet N., Dufour P. and Dufour L.C., *Caractérisation par résolution de plans de cristallites de nickel CFC obtenus par réduction de l'oxyde NiO*, J. Microsc. Spectrosc. Electron., **6** (1981) 473-481

Floquet N. and Dufour L.C., *Stability and reactivity of NiO(001) and (111): a RHEED -AES investigation of Si surface segregation and Ni formation by gas reduction*, Surface Sci., **126** (1983) 543-549

Freund E., Lynch J. and Szymanski R., *Small particles in the field of industrial supported catalysts*, Ultramicroscopy, **20** (1986) 107-118

Fritsch A. and Légaré P., *XPS study of the Pt/Al_2O_3 interface*, Surface Sci., **185** (1987) L355-L360

Fuentes S., Vazquez A., Silva R., Perez-Ramirez J.G. and Yacaman M.J., *Characterization by TEM and image processing of rhodium supported on TiO_2*, J. Catal., **111** (1988) 353-359

Gallezot P., Avalos-Borja M., Poppa H. and Heinemann K., *Structure and morphology characterization of nanometer-size metal aggregates by electron scattering patterns*, Langmuir, **1** (1985) 342-347

Garner W.E. (Ed.),'Chemistry of the Solid State', Butterworths, London, 1955

Gibson J.M. and Dawson L.R.(Eds), 'Layered Structures, Epitaxy and Interfaces', Mat. Res. Soc. Symp. Proc., Vol. 37, Mat. Res. Soc., Pittsburgh, 1985

Gillet M., *Etude des petites particules par microscopie électronique*, J. Microsc. Spectrosc. Electron., **11** (1986) 449-470

Gillet E., Legressus C. and Gillet M., *Etude de l'interface Pd/Al_2O_3. Application aux phénomènes de mouillage et d'adhésion*, J. Chim. Phys., **84** (1987) 167-174

Grabow M.H. and Gilmer G.H., *Thin film growth modes, wetting and cluster nucleation*, Surface Sci., **194** (1988) 333-346

Haberlandt H. and Ritschl F., *Quantum chemical investigation of support-metal interactions and their influence on chemisorption. 2. Strong metal-support interaction in H...Ni-MOx (M = Ti,Si)*, J. Phys. Chem., **90** (1986) 4322-4330

Halicioglu T. and Bauschlicher Jr C.W., *Physics of microclusters*, Rep. Prog. Phys., **51** (1988) 883-921

Haller G.L., *Spectroscopic analysis of local structure and small particles of catalysts*, Appl. Surface Sci., **20** (1985) 351-381

Hardwick D.A, *The mechanical properties of thin films: A review*, Thin Solid Films, **154** (1987) 109-124

Harris P.J.F., *The morphology of platinum catalyst particles studied by transmission electron microscopy*, Surface Sci., **185** (1987) L459-L466

He J.W. and Möller P.J., *Epitaxial and electronic structures of ultra-thin copper films on MgO crystal surfaces*, Surface Sci., **178** (1986) 934-942

Heinemann K. and Poppa H. *In-situ TEM evidence of lattice expansion of very small supported palladium particles*, Surface Sci., **156** (1985) 265-274

Heinemann K. and Soria F. , *On the detection and size classification of nanometer-size metal particles on amorphous substrates*, Ultramicroscopy, **20** (1986) 1-14

Herring C., 'Structure and Properties of Solid Surfaces', Gomer R. and Smith C.S. (Eds), Chap. I, 1952, University of Chigago Press

Hicter P., Chatain D., Pasturel A. and Eustathopoulos N, *Approche électronique de l'adhésion thermodynamique dans les sytèmes non-réactifs métal-SP-alumine*, J. Chim. Phys., **85** (1988) 941-945

Hirchwald W., *Selected experimental methods in the characterization of oxide surfaces*, in 'Surface and Near-Surface Chemistry of Oxide Materials', Nowotny J. and Dufour L.C. (Eds), Elsevier, Amsterdam, 1988, pp. 61-187

Howie A. and Marks L.D., *Elastic strain and the energy balance for multiply twinned particles*, Phil. Mag. A, **49** (1984) 95-109

Hubert R.A., Darville J. and Gilles J.M., *Characterization of the Fe/MgO interface*, Physica Scripta, **T4** (1983) 179-182

Hubert R.A. and Gilles J.M., *Fe/MgO(001) model catalyst: electron microscopy and spectroscopic studies*, Appl. Surface Sci., **22/23** (1985) 631-637

Hubert R.A. and Gilles J.M.,*The iron-MgO(001) system studied by AES, XPS and TEM*, Microbeam Anal., **21** (1986) 599-601

Iijima S. and Ichihashi T., *Structural instability of ultrafine particles of metals*, Phys. Rev. Lett., **56** (1986) 616-619

Ino S., *Stability of multiply-twinned particles*, J. Phys. Soc. Jap., **27** (1969) 941-953

Janssen A.P., Schoonmaker R.C. and Chambers A., *A study of the epitaxial growth of magnesium on magnesium oxide {001} using reflexion diffraction, LEED and Auger spectroscopy*, Surface Sci., **49** (1975) 143-160

Kao C.C., Tsai S.C., Bahl M.K. and Chung Y.W., *Electronic properties, structure and temperature-dependent composition of nickel deposited on rutile titanium dioxide (110) surfaces*, Surface Sci., **95** (1980) 1-14

Kaishev R., 1951, quoted by Kern, 1978

Katrib A., Petit C., Légaré P., Hilaire L. and Maire G., *Metal-support interaction and interatomic Auger transitions in vapor-deposited molybdenum on silica and α-Al_2O_3*, J. Phys. Chem., **92** (1988) 3527-3531

Kern R., *Croissance épitactique (aspects topologiques et structuraux)*, Bull. Minéral., **101** (1978) 202-233

Kern R., Le Lay G. and Métois J.J., *Basic mechanisms in the early stages of epitaxy*, in 'Current Topics in Materials Science, Vol. 3', Kaldis E. (Ed.), North Holland, Amsterdam, 1979, pp. 139-419

Khanna S.N., Bucher J.P., Buttet J. and Cyrot-Lackmann F., *Stability and lattice contraction of small platinum particles*, Surface Sci., **127** (1983) 165-174

Kip B.J., Duivenvoorden F.B.M., Koningsberger D.C. and Prins R., *Determination of metal particle size of highly dispersed Rh, Ir and Pt catalysts by hydrogen chemisorption and EXAFS*, J. Catal., **105** (1987) 26-38

Klomp J.T. and Van de Ven, *Parameters in solid-state bonding of metals to oxide materials and the adherence of bonds*, J. Mat. Sci., **15** (1980) 2483-2488

Klomp J.T., *Thermodynamics and chemistry of ceramic-metal interfaces*, this book, 1989

Kohiki S., *Photoemission from small Pd clusters on Al_2O_3 and SiO_2 substrates*, Appl. Surface Sci., **25** (1986) 81-94

Kreibig U. and Genzel L., *Optical absorption of small metallic particles*, Surface Sci., **156** (1985) 678-700

Lee Y.C., Tong P. and Montano P.A. , *Deposition of iron on MgO(100): reaction of CO and H_2*, Surface Sci., 181 (1987) 559-572

Le Lay G. and Kern R., *Physical methos used for the characterization of modes of epitaxial growth from the vapor phase*, J. Crystal Growth, **44** (1978) 197-222

Lin S.T. and Ownby P.D., *A microscopic study of preferred orientation and adhesion of solid copper particles on (0001) sapphire substrates*, J. Mat. Sci., **23** (1988) 3187-3194

Loehman R.E., *Interfacial reactions in ceramic-metal systems*, Cer. Bull., **68** (1989) 891-896

Mandelbrot B., 'Fractals, Form, Chance and Dimension', Freeman, San Francisco, 1977

Mandelbrot B., 'The Fractal Geometry of Nature', Freeman, San Francisco, 1982

Marks L.D., *Modified Wulff constructions for twinned particles*, J. Crystal Growth, **61** (1983) 556-566

Marks L.D., *Surface structure and energetics of multiply twinned particles*, Phil. Mag. A, **49** (1984) 81-93

Marks L.D., *Particle size effects on Wulff constructions*, Surface Sci., **150** (1985) 358-366

Marks L.D., *Solid-like growth*, Thin Solid Films, **136** (1986) 309-315

Marks L.D. and Smith D.J., *Direct imaging in small metal particles*, Nature, **303** (1983) 316-317

Marks L.D., Ajayan P.M. and Dundurs J., *Quasi-melting of small particles*, Ultramicroscopy, **20** (1986) 78

Marner B. and Schmickler W., *Model for the growth of a two-dimensional cluster*, J. Phys. Chem., **93** (1989) 3186-3188

Martens J.H.A., Prins R., Zandbergen H. and Koningsberger D.C., *Structure of Rh/TiO2 in the normal and the SMSI state as determined by extended X-ray absorption fine structure and high-resolution transmission electron microscopy*, J. Phys. Chem., **92** (1988) 1903-1916

Mason M.G., *Electronic structure of supported small metal clusters*, Phys. Rev. B, **27** (1983) 748-762

Masson A., Métois J.J. and Kern R., *The development of epitaxy by surface migration of crystallites*, in 'Advances in Epitaxy and Endotaxy', Schneider H.G. and Ruth V., VEB Deut. Verlag für Grundstoffindustrie, Leipzig, 1971

Masson A., *Peculiar aspects of heterogeneous nucleation and growth processes related to metal supported catalyst*, in 'Contributions of Cluster Physics to Materials Science and Technology', Davenas J. and Rabette P.M. (Eds), Martinus Nijhoff Publishers, Dordrecht, 1986, pp. 295-306

Masson A., Bellamy B., Hadj Romdhane Y., Che M., Roulet H. and Dufour G., *Intrinsic size effect of platinum particles supported on plasma-grown amorphous alumina in the hydrogenation of ethylene*, Surface Sci., **173** (1986) 479-497

Matthews J.W., *Misfit dislocations*, in: F.R.N. Nabarro (Ed.), 'Dislocations in Solids', Vol. 2, North-Holland, Amsterdam, 1979, pp. 461-545

Matzke Hj., *Surface diffusion and surface energies of ceramics*, this book

Meakin P., (a) *Diffusion-controlled cluster formation in two, three and four dimensions*, Phys. Rev. A, **27** (1983) 604-607; (b) *Diffusion-controlled cluster formation in 2-6 dimensional space*, Phys. Rev. A, 27 (1983) 1495-1507; (c)*Diffusion-controlled deposition on fibers and surfaces*, Phys. Rev. A, **27** (1983) 2616-2623

Métois J.J.,Heyraud J.C. and Kern R., *Surface decoration: localization of crystallites along the steps*, Surface Sci., **78** (1978) 191-208

Milchev A. and Markov I., *The effect of realistic forces in finite epitaxial islands: equilibrium structure, stability limits and substrate-induced dissociation of migrating clusters*, Surface Sci., **156** (1985) 392-403

Morita S., Okada T., Ishigame Y. and Mikoshiba N., *Scanning tunneling microscopy of metal surfaces in air*, Surface Sci., **181** (1987) 119-125

Munuera G., Gonzalezelipe A.R. and Espinos J.P., *XPS study of phase mobility in Ni/TiO2 systems*, Surface Sci., **211** (1989) 1113-1122

Nicholas M.G., *Interactions at oxide-metal interfaces*, in 'External and Internal Surfaces in Metal Oxides', Dufour L.C. and Nowotny J., Eds, Materials Science Forum, **29** (1988) 127-150

Nicholas M.G., *Ceramic-metal interfaces*, this book, 1989

Nieminen J.A. and Kaski K., *Criteria for different growth modes of thin films*, Surface Sci., **185** (1987) L489-L496

Pashley D.W., *The growth and structure of epitaxial layers*, in'Layered structures, epitaxy and interfaces', Gibson J.M. and Dawson L.R. (Eds), Mat. Res. Soc. Symp. Proc., Vol. 37, Mat. Res. Soc., Pittsburgh, 1985, pp 67-76

Pask J.A., *From technology to the science of glass/metal and ceramic/metal sealing*, Cer. Bull., **66** (1987) 1587-1592

Pilliar R.M. and Nutting J., *Solid-solid interfacial energy determinations in metal-ceramic systems*, Phil. Mag., **16** (1967) 181-188

Poppa H., *Model studies in catalysis with uhv-deposited metal particles and clusters*, Vacuum, **34** (1984) 1081-1095

Poppa H., Papageorgopoulos C.A., Marks F.and Bauer E., *Growth and properties of particulate Fe films vapor deposited in UHV on planar alumina substrates*, Z. Phys. D - Atoms, Molecules and Clusters, **3** (1986) 279-289

Renou A. and Rudra A., *Epitaxial growth of thin monocrystalline MgO substrates: transmission electron microscope characterization of palladium deposits,* Surface Sci., **156** (1985) 69-84

Rhead G.E.,Barthès M.G. and Argile C., *Determination of growth modes of ultrathin films from Auger electron spectroscopy: an assessment and commentary,* Thin Solid Films, **82** (1981) 201-211

Rickard J.M., Perdereau M. and Dufour L.C., *Croissance épitaxique de films de nickel par réduction d'une face NiO(100) dans l'hydrogène à basse pression,* J. Microsc. Spectrosc. Electron., **4** (1979) 95-110

Rocker G. and Göpel W., *Titanium overlayers on* TiO_2 *(110),* Surface Sci., **181** (1987) 530-558

Romeu D., Gomez A., Perez-Ramirez J.G., Silva R., Perez O.L., Gonzalez A.E. and Yacaman M.J., *Surface fractal dimension of small metallic particles,* Phys. Rev. Lett., **57** (1986) 2552-2555

Ruckenstein E., *Effect of the strong metal-support interactions on the behavior of model nickel/titania catalyst,* J. Catal., **104** (1987) 259-278

Sadeghi H.R. and Henrich V.E., (a)*Rh on* TiO_2*: model catalyst studies of the strong metal-support interaction,* Appl. Surface Sci., **19** (1984) 330-340; (b) *SMSI in Rh/*TiO_2 *model catalysts: evidence for oxide migration,* J. Catal., **87** (1984) 279-282

Sadeghi H.R., Resasco D.E., Henrich V.E. and Haller G.L., *The implications of one-dimensional ordering of Rh on the (110) surface of rutile,* J. Catal., **104** (1987) 252-255

Sadeghi H.R. and Henrich V.E., *Electronic interactions in the rhodium/*TiO_2 *system,* J. Catal., **109** (1988) 1-11

Sanders J.V., *Structure of catalytic particles,* Ultramicroscopy, **20** (1986) 33-38

Sanchez M.G. and Gasquez J.L., *Oxygen vacancy model in strong metal-support interaction,* J. Catal., **104** (1987) 120-135

Shpiro E.S., Dysenbina B.B., Tkatchenko O.P., Antoshin G.V. and Minachev Kh.M., *Strong metal-support interaction: the role of electronic and geometric factors in real Me/*TiO_2 *catalysts,* J. Catal., **110** (1988) 262-274

Smith D.J., *Atomic-resolution studies of surface dynamics by electron microscopy,* J. Vac. Sci. Technol. B, **3** (1985) 1563-1567

Smith D.J., Petford-Long A., Wallenberg R. and Bovin J.O., *Dynamic atomic-level rearrangements in small gold particles,* Science, **233** (1986) 872-875

Smith D.J., Bovin J.O., Bursill L.A., Petford-Long A.K. and Ye H.Q., *High resolution electron microscopy of surfaces and surface reactions,* Surf. Interf. Anal., **10** (1987) 135-141

Solliard C. and Flüeli M., *Surface stress and size effects on the lattice parameter in small particles of gold and platinum,* Surface Sci., **156** (1985) 487-494

Solliard C. and Borel J.P., *Diffraction et microscopie électroniques sur des petits grains métalliques: Effets de taille. I Défauts et structures anormales multimaclées dans l'or et le platine,* Helv. Phys. Acta, **61** (1988) 712-729

Solliard C., Flüeli M. and Borel J.P., *Diffraction et microscopie électroniques sur des petits grains métalliques: Effets de taille. II. Facteur de Debye-waller de l'or,* Helv. Phys. Acta, **61** (1988) 730-739

Stoneham A.M., *Systematics of metal-insulator interfacial energies: a new rule for wetting and strong catalyst-support interactions,* Appl. Surface Sci., **14** (1982) 249-259

Stoneham A.M., *Ceramic surfaces: theoretical studies,* J. Am. Ceram. Soc., **64** (1981) 54-60

Stoneham A.M. and Tasker P.W., *The theory of ceramic surfaces,* in 'Surface and Near-Surface Chemistry of Oxide Materials', Nowotny J. and Dufour L.C. (Eds), Elsevier, Amsterdam, 1988, pp. 1-22

Stoyanov S., *Layer growth of epitaxial films and superlattices,* Surface Sci., **199** (1988) 226-242

Sun Y.M., Belton D.N. and White J.M., *Characteristics of Pt thin films on* TiO_2*(110),* J. Phys. Chem., **90** (1986) 5178-5182

Surratt G.T. and Kunz A.B., *Theoretical study of H chemisorption on NiO. Perfect surfaces and cation vacancies*, Phys. Rev. B, **19** (1979) 2352-2358

Tamura K., Bardi U. and Nihei Y., *An investigation by angular resolved X-ray photoelectron spectroscopy of strong metal-support interaction (SMSI) in the Pt/TiO_2 system*, Surface Sci., **197** (1988) L281-L286

Tanaka N. and Mihama K., *High resolution electron microscopy and microdiffraction of metal-MgO composite films*, Appl. Surface Sci., **33/34** (1988) 472-483

Tasker P.W. and Stoneham A.M., *The stabilization of oxide and oxide-metal interfaces by defects and impurities*, J. Chim. Phys., **84** (1987) 149-155

Tauster S.J., Fung S.C. and Garten R.L., *Strong metal-support interactions. Group 8 noble metals supported on TiO_2*, J. Am. Chem. Soc., **100** (1978) 170-175

Teyssandier F., *Growth of ceramic layers from vapor phase*, this book, 1989

Van Broekhoven E.H. and Ponec V., *Surface chemistry of small particles*, Surface Sci., **162** (1985) 731-741

Van Hardeveld R. and Hartog F., *The statistics of surface atoms and surface sites on metal crystals*, Surface Sci., **15** (1969) 189-230

Venables J.A., Spiller G.D.T. and Hanbücken M., *Nucleation and growth of thin films*, Rep. Prog. Phys., **47** (1984) 399-459

Venables J.A., Batchelor D.R., Hanbücken M., Harland C.J. and Jones G.W., *Surface microscopy with scanned electron beams*, Phil. Trans. R. Soc. London, **A 318** (1986) 243-257

Venables J.A., Smith D.J. and Cowley J.M., *HREM, STEM, REM, SEM and STM*, Surface Sci., **181** (1987) 235-249

Wallenberg L.R., Bovin J.O , Petford-Long A.K. and Smith D.J., *Atomic-resolution study of structural rearrangements in small platinum crystals*, Ultramicroscopy, **20** (1986) 71-76

Wang T., Lee C. and Schmidt L.D., *Shape and orientation of supported Pt particles*, Surface Sci., **163** (1985) 181-197

Wepfer G.G., Surratt G.T., Weidmann R.S. and Kunz A.B., *Theoretical study of H chemisorption on NiO.II. Surface and second-layer defects*, Phys. Rev. B, **21** (1980) 2596-2601

Werckmann J.,Mosser A. and Deville J.P., *Détermination du mécanisme de croissance d'une nouvelle phase par spectrométrie d'électrons Auger*, Analusis, **9** (1981) 213-219

Werner H.W. and Garten R.P.H., *A comparative study of methods for thin-film and surface analysis*, Rep. Prog. Phys., **47** (1984) 231-344

Wertheim G.K., *Auger shifts in metal clusters*, Phys. Rev. B, **36** (1987) 9559-9562

Wertheim G.K. and DiCenzo S.B., *Cluster growth and core-electron binding energies in supported metal clusters*, Phys. Rev. B, **37** (1988) 844-847

Wijnen P.W.J.G., Van Zon F.B.M. and D.C. Koningsberger, *Determination of metal particle size in partly reduced Ni catalysts by hydrogen/oxygen chemisorption and EXAFS*, J. Catal., **114** (1988) 463-468

Winterbottom W.L., *Equilibrium shape of a small particle in contact with a foreign substrate*, Acta Met., **15** (1967) 303-310

Witten T.A. and Sander L.M., (a) *Diffusion-limited aggregation, a kinetic critical phenomenon*, Phys. Rev. Lett., **47** (1981) 1400-1403; (b)*Diffusion-limited aggregation*, Phys. Rev. B, **27** (1983) 5686-5697

Young D.A., 'Decomposition of Solids', Pergamon Press, Oxford, 1966

Zdetsis A.D. and Kunz A.B., *Ab initio cluster study of Ni adsorption on alumina*, Phys. Rev. B, **32** (1985) 6358-6362

MODELLING OF METAL-OXIDE INTERFACE BEHAVIOUR DURING OXIDE SCALE GROWTH CONTROLLED BY CATION DIFFUSION

B. Pieraggi
Lab. Physical Metallurgy - URA CNRS 445 - ENSCT
118, Route de Narbonne - 31077 Toulouse Cédex
France

ABSTRACT - The growth of cation-diffusing scales on pure metals is described from a modelling of metal-scale interface in terms of intrinsic dislocations for an epitaxial scale ; such a model is consistent with the experimental observations. It is proposed that the annihilation of cationic vacancies occurs at the metal-scale interface by the climb into the metal of some fraction of the intrinsic misfit interface dislocations, a process which generates tensile stress in the metal and compression in the scale. Above a critical interfacial strain, the glide of dislocations in the metal, in combination with dislocation glide in the scale, recreates the interface dislocations. These processes provide plastic deformation in both phases near the interface and maintain metal-scale epitaxy during oxide growth. The model may explain the origin of stresses arising during the growth of cation-diffusing scales on an extensive flat surface.

1. INTRODUCTION

At high temperature, metallic materials must be protected against the usually corrosive action of their gaseous environment. For the simplest case of oxidation, this protection is dependent on the growth of a thin scale separating the metallic substrate from the gas phase. Thus, good resistance to oxidation or corrosion requires the growth of a continuous, pore-free and adherent scale of oxide or corrosion products so that subsequent growth would involve mass transport of either anions, leading to growth at the metal-scale interface, or cations, leading to growth at the scale-gas interface. Usual growth and mass transport mechanisms for protective oxide scales on metals and alloys are reviewed by Smeltzer in the present volume (1) ; details on the growth mechanisms of oxide scales can also be found in references (2-5). However, many questions concerning scale growth remain unanswered. Most of these questions are related to the behaviour of the scale-metal interface during growth, i.e. the interface dynamics in relation with its structure and the diffusion fluxes associated to the scale growth. In

L.-C. Dufour et al. (eds.), Surfaces and Interfaces of Ceramic Materials, 449–454.

the following only the simple case of cation-diffusing scales would be considered, the main features associated to the growth of cation-diffusing scale were recently reviewed by Rapp (6).

2. EXPERIMENTAL EVIDENCES

The usual mechanisms (Fig.1) for the growth of a cation-diffusing scale on an extensive flat surface of a metal conclude that stress-free scale growth occurs at the external scale surface such that the oxidation of a metal with thickness W_m leads to a scale thickness $W_{ox} = \phi W_m$ at the external scale/gas interface, where ϕ is the Pitting-Bedworth ratio for the metal/scale system ($\phi > 1$). However, tensile stresses in the metal (compression in the scale) are known to arise (7-10) ; the suggestion (11) that compressive stresses in the scale result from oxide growth at grain boundaries within the scale because of grain-boundary diffusion of both metal and oxidant species has not found general experimental support or acceptance (12-13).

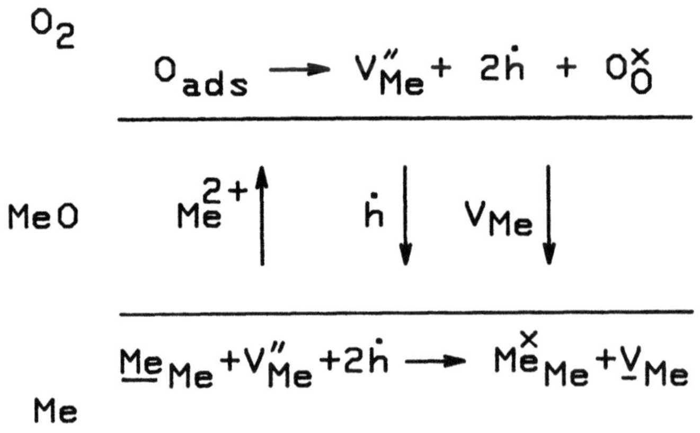

O_2

$$O_{ads} \rightarrow V''_{Me} + 2\dot{h} + O_O^x$$

MeO $\quad Me^{2+}\uparrow \quad \dot{h}\downarrow \quad V_{Me}\downarrow$

$$\underline{Me}_{Me} + V''_{Me} + 2\dot{h} \rightarrow Me^x_{Me} + \underline{V}_{Me}$$

Me

Figure 1 - General mechanism for scale growth by cation-vacancy difusion

As schematically shown by Fig. 1, the growth of a cation-diffusing scale implies the continuous arrival at the scale metal interface of a significant vacancy flux which must be annihilated at the scale-metal interface or otherwise injected into the metal where their eventual annihilation occurs. In that case, the injected vacancies are assumed to be annihilated at internal defects such as dislocations, grains boundaries ... so that the porosity sometimes observed in the metal adjacent to the scale, or the growth of an inner hollow tube for cylindrical specimens (14-17), is explained by the diffusion into the metal of vacancies associated with scale growth according to the interfacial reaction at the scale-metal interface :

$$\underline{Me}_{Me} + V''_{Me} + 2\dot{h} \rightarrow Me^x_{Me} + \underline{V}''_{Me} \qquad (1)$$

where $\underline{Me}_{Me}$ and $\underline{V}_{Me}$ represent a metal atom and a vacancy in the metal phase.

In addition, recent TEM observations of cross-sections of the Ni-NiO interface for an oxide scale grown on pure Ni show an epitaxial layer of NiO in excellent contact with the Ni substrate, even for relatively thick scales (18-21). No voids or porosity were observed in the metal but these observations show that plastic deformation, as evidenced by high dislocation densities, is localized in both the scale and the metal near the epitaxial Ni-NiO interface.

3. MODEL FOR INTERFACIAL DYNAMICS

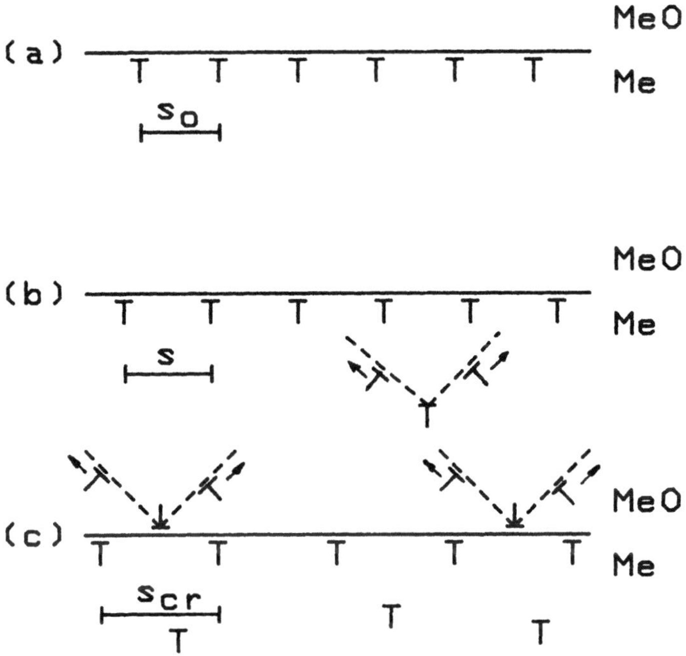

Figure 2 - Schematic illustration of combined climb and return glide of interface dislocations to effect vacancy annihilation.
- -a- MeO-Me interface at equilibrium, ie no growth and minimum interfacial strain
- -b- Climb and return glide in the metal for an intermediate situation
- -c- Dislocation emission at the MeO-Me interface and glide to MeO-gas interface.

From these observations, Pieraggi and Rapp (22) proposed that the annihilation of cation vacancies occur by the climb of a fraction of the intrinsic misfit dislocation of the epitaxial metal-oxide interface. Likewise, the associated interfacial dynamics result in stress-generation in the interface, while accounting for interface recession absence of voids or porosity and plastic deformation of metallic substrate.

Figure 2(a) shows schematically an epitaxial metal-scale interface at equilibrium, oriented to reveal one set of parallel edge dislocations which accounts for the difference in lattice parameters between the two phases. The present model assumes that the climb of some fraction of the intrinsic interface dislocations constitutes the means of vacancy annihilation. Accordingly, as shown in Fig. 2(b), the remaining interfacial dislocations adjust their spacing to homogenize and minimize the strain accross the interface. In Fig. 2(c), the climb and glide of dislocations in the metal has increased the spacing s of the residual intrinsic interface past some critical spacing s_{cr} such that dislocations in the scale glide towards the gas-scale interface to reduce the compressive stress, and dislocations climbed into the metal from the interface return via glide to reduce the tensile stress. The operation of cyclic dislocation climb and return glide in the metal with simultaneous glide in the oxide would locally provide a mechanism to insure good epitaxial metal-scale contact with vacancy annihilation. This combination of the adjustement of interfacial dislocations and the deformation by glide in the metal and in the scale constitutes the cooperative movement between the metal and the scale sought by Harris (23). At this point, the model is also consistent with the conclusion of Harris (23) that the interface between a metal and its oxide is not a significant source or sink for vacancies when oxidation does not occur.

4. CONCLUSION

The proposed model for vacancy annihilation and stress generation during scale growth limited by cation-vacancy diffusion is consistent with the Wagnerian mechanism for lattice diffusion, or for short-circuit diffusion. The model is consistent with experimental evidence for epitaxial relationships between the metal and the innermost grains of a thick scale, for the generation of growth stresses and local plastic deformation, for the absence of porosity in the underlying metal, for the effect of surface orientation on oxidation, etc. The proposed model may be applied to other kinds of diffusional solid-solid transformations involving a change in phase composition where an interface between solid phases needs to serve as a source or sink of dislocations or/and vacancies (24).

5. REFERENCES

1) W.W. Smeltzer, 'Growth and Mass Transport in Ceramic Type Protective Scales on Metals', this volume.

2) High Temperature Corrosion, R.A. Rapp, ed., NACE, (1983).

3) High Temperature Corrosion, R. Streiff, J. Stringer, R.C. Krutenat and M. Caillet, eds., Elsevier (1987).

4) Oxidation of metals and associated mass transport, M.A. Dayananda, S.J. Rothman and W.E. King, eds., TMS Publications (1987).

5) R. Kofstad, High Temperature Corrosion, Elsevier Applied Science (1988).

6) R.A. Rapp, 'The High Teperature Oxidation of Metals forming Cation-Diffusing Scales', Met. Trans, **15**, (1984), 765-782.

7) J. Stringer, 'Stress Generation and Relief in Growing Oxide Films', Corr. Sci., **10**, (1970), 513-543.

8) P. Hancock and R.C. Hurst, 'The Mechanical Properties and Breakdown of Surface Oxide Films at Elevated Temperatures', Adv. Corr. Sci. and Techn., **4**, M.G. Fontana and R.W. Staehle, eds., Plenum Press, (1974), 1-74.

9) Stress Effects in the Oxidation of Metals, J.V. Cathcart, ed., AIME, (1974).

10) G. Beranger, A.M. Huntz and B. Pieraggi, 'Stress Effects in High Temperature Corrosion : Internal Stresses, Applied Stresses', Corrosion des Matériaux à Haute Température, G. Beranger, J.C. Colson and F. Dabosi, eds., Editions de Physique (1987) 227-269.

11) F.N. Rhines and J.S. Wolf, 'The role of Oxide Microstructure and Growth Stresses in the High Temperature Scaling of Nickel', Metall. Trans., **1** (1970), 1701-1710.

12) M.V. Speight and J.E. Harris, 'The Generation of Stresses in Oxide Growing by Cation Diffusion', Acta Met., **26**, (1978), 1043-1045.

13) A. Atkinson, 'Conditions for the Formation of New Oxide within Oxide Films Growing on Metals', Corr. Sci., **22** (1982), 347-357.

14) B.W. Dunnington, F.H. Beck and M.G. Fontana, 'Scale Formation on Iron at High Temperature', Corrosion, **8** (1952), 2-13.

15) R.F. Tylecote and T.E. Mitchell, 'Marker Movement in the Oxidation of Iron and some other Metals', J. Iron Steel Inst., **196** (1960), 445-453.

16) P. Hancock and R. Fletcher, 'The Oxidation of Nickel in the Temperature Range 700-1000°C', Metallurgie, **6** (1966), 1-9.

17) M. Cagnet and J. Moreau, 'The Readsorption of Gaps of Ferrous Oxide during Oxidation of Iron at High Temperature", CR Acad. Sciences Fr., **244** (1957), 2925-2928.

18) H.T. Sawhill, L.W. Hobbs and M.T. Tinker, 'Interface Structure of Nickel Oxide on Nickel Substrates', Adv. In. Ceramics, **6**, M.F Yan and A.H. Heuer, eds., Amer. Ceram. Soc., (1983), 128-138.

19) M. Leseur and B. Pieraggi, 'Structure of Metal-Oxide Interfaces', J. de Physique, **C4** (1985), 135-140.

20) J.M. Perrow, W.W. Smeltzer and J.O. Embury, 'Role of Structural Defects in the Growth of Nickel Oxide Films', Acta Met, 16, '1968), 1209-1218.

21) R. Herchl, N.N. Khoi, T. Homma and W.W. Smeltzer, 'Short-Circuit Diffusion in the Growth of Nickel Oxide Scales on Nickel Crystal Faces', Oxid. Metals, **4** (1972), 35-49.

22) B. Pieraggi and R.A. Rapp, 'Stress Generation and Vacancy Annihilation during Scale Growth Limited by Cation-Vacancy Diffusion', Acta Metall., **36**, (1988), 1281-1289.

23) J.E. Harris, 'Vacancy Injection during Oxidation - A Re-examination of the Evidence', Acta Metall., **26**, (1978), 1043-1045.

24) B. Pieraggi, F.J.J. van Loo and R.A. Rapp, 'Fluxes and interface Migration in Diffusion-Driven Phase Transformations', Diffusion Analysis and Applications, A.D. Romig, M.S. Dayananda and R.H. Heckel, Eds, TMS Proceedings, (1988), to be published.

INTERFACIAL TENSION AND CONTACT ANGLE IN IMMISCIBLE SYSTEMS BY CAPILLARY PRESSURE MEASUREMENTS.

L. LIGGIERI, E. RICCI, N. RANDO and A. PASSERONE
Istituto di Chimica Fisica Applicata dei Materiali del C.N.R.
Lungobisagno Istria 34
16141 Genova
Italy

The recent development of experimentation in space, under microgravity conditions, makes it possible to measure the interfacial tension (σ) between immiscible fluids (organic compounds, liquid alloys...) by a straigthforward application of the Laplace equation in its simplest form.

The method presented here has been selected to fly on the Sounding Rocket MASER-4 of ESA. It consists essentially in measuring the pressure difference across a liquid-liquid (immiscible) or liquid-vapour interface as a function of its radius of curvature, using the simplified boundary conditions

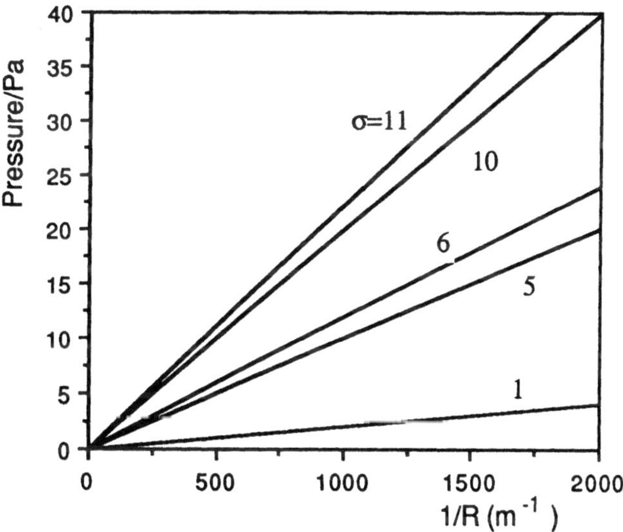

Fig. 1. Theoretical dependence of capillary pressure on the radius of a spherical drop for various values of the interfacial tension

L.-C. Dufour et al. (eds.), Surfaces and Interfaces of Ceramic Materials, 455–458.

provided by the microgravity environment. In fact, if the product $\Delta\rho \cdot g = 0$ (where $\Delta\rho$ is the density difference between the two fluids and g the acceleration of gravity), the drop's shape is exactly spherical so that the Laplace equation can be written[1]:

$$\rho = 2\sigma / R \quad (1)$$

with R = radius of the drop.

The capillary pressure is then a linear function of 1/R with a slope equal to 2σ as illustrated in Fig. 1 for different values of surface tension.

Our method consists in calculating this slope by a linear fit of the experimental P and 1/R data.

In spite of this simple theoretical approach, this method requires very high instrumental sensitivity and experimental care due to the fact that we are dealing with measurements of capillary pressures not greater than 100 Pa.

Experimental layout and methodology [2]

The apparatus consists of (Fig. 2): 1/ a differential pressure transducer; 2/ a system of plug and microsyringe; 3/ a plug drive motor; 4/ the experimental cell; 5/ a capillary tube with a particularly shaped tip. The whole apparatus is contained in a thermostatic chamber.

A drop is formed on the tip of the capillary tube by displacing continuously and uniformly the plug. During the formation of the drop, the pressure signal is sampled by an automatic digital acquisition system. Fig. 3 shows a typical output signal. The maximum in the signal is reached when the drop is exactly hemispherical. From the time corresponding to this maximum and since the plug speed is constant, a radius of the drop can be associated to time.

In order to obtain the needed precision on the choice of the maximum and on the pressure data, firstly, the signal sampling must be optimized with respect to signal noise, then, filtering methods like gaussian filters and time averaging have to be applied[3].

Experimental Results

To test the method in a gravitational environment, we used couples of immiscible liquids having the same or nearly the same density. In fact, drops having small volume can be considered as spherical if the density difference is not vanishing.

Preliminary tests show a good reproducibility of measurements. Table 1 shows some results obtained with the couple water-benzene in saturation conditions.

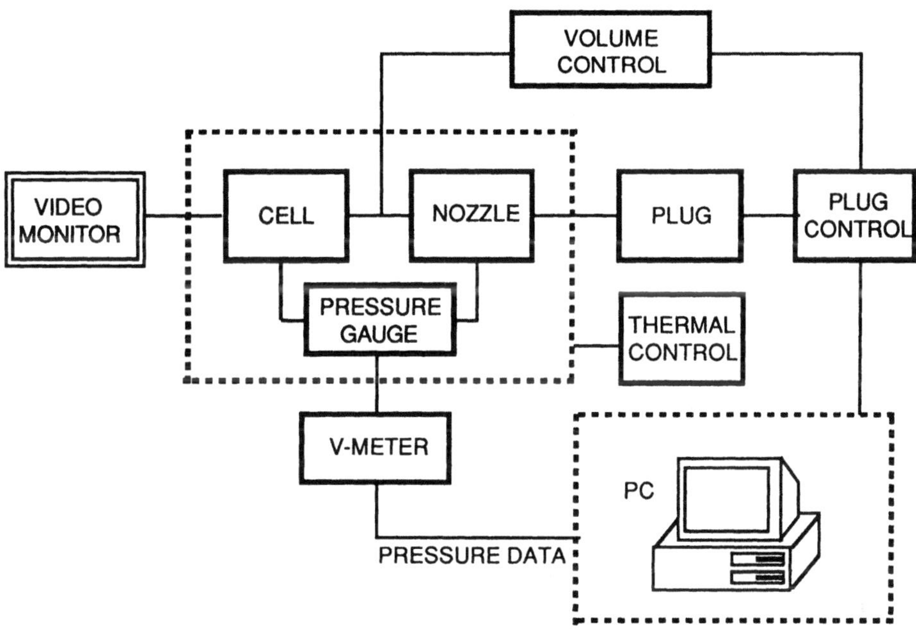

Fig. 2. Block diagram of the experimental layout

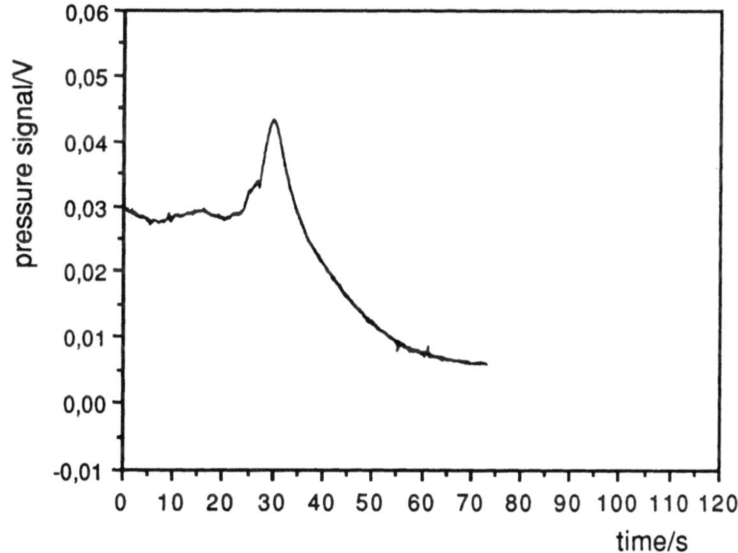

Fig. 3. A typical pressure signal output

The absolute values obtained are quite different from the literature values: that is due to the fact that, in this run, we were interested only in the reproducibility of the measurements without specific attention to the

Table 1	Run	σ (mN/m)	Corr. Coeff.	Std. Dev.
	1	27.84	.9998	4.2 E-2
	2	28.40	.9996	4.2 E-2
	3	27.77	.9998	1.5 E-2
	4	27.92	.9998	1.2 E-2
	5	28.02	.9999	1.0 E-2
	6	27.62	.9999	1.8 E-2

chemical purity of liquids. However, results centered on the best literature values (about 35 nM/m) have been obtained in a few dedicated runs where a particular care has been paid to preserve solutions from pollution.

This method, apart from measurements in equilibrium conditions, is particularly useful in stationary conditions when adsorption phenomena or some mass transfer across the interface take place.

Contact Angle Measurements

The method presented here can be used to measure, nearly simultaneously, the interfacial tension and the contact angle of the liquid-liquid interface with a solid substrate[4]. In fact, if a drop of the liquid 1 is formed inside the liquid 2, then, maintaining constant the drop volume V, it is possible to withdraw the capillary tube and to bring the drop into contact with a solid substrate.

At this point, a new configuration sets up with a definite value of the contact angle and a new value for the radius of the drop. Now, as the volume is already known and a new pressure value can be read, the contact angle can be calculated using the following relation:

$$V = 8\pi R^3 (2 - \cos\theta + \cos 3\theta) / (24 \sin^3\theta) \qquad (2)$$

References

1) R. Defay, I. Prigogine, I. Bellemans and D.H. Everett, *Surface Tension and Adsorption*, Longmans, London, 1966
2) Patent Applied for
3) A.W. Oppenheimer, R.W. Schafer, *Digital Signal Processing*, Prentice Hall Inc New York, 1975
4) J.F. Padday, 'Surface Tension' in E. Matijevic, Ed., *Surface and Colloid Science* Vol. 1, Wiley-Interscience, New-York, 1969

CHARACTERIZATION, PROPERTIES AND PROCESSING OF CERAMIC POWDERS

Terry A. Ring
Laboratoire de technologie des poudres
Ecole Polytechnique Federale de Lausanne
CH-1015 Lausanne, Switzerland

ABSTRACT. A review of the fundamentals required to 1) produce narrow size distribution ceramic powders 2) make suspensions of ceramic powders and 3) make green bodies with a uniform packing of particles is presented. In all cases, the interfaces that ceramic powders present to their environment is very important in controlling the properties of the powders during processing. For this reason, methods of surface characterization are presented.

1. INTRODUCTION

Processing ceramics often involves the dispersion of powders as slips and slurries in aqueous and nonaqueous solvents. These slips and slurries are convienient for several shaping/forming technologies - e.g., slip casting, tape casting, filter pressing and electrophoretic deposition - used to produce ceramic green bodies. As the demand to improve the properties of ceramic materials increases, more emphasis is being placed on controling the microstructure of a ceramic piece because the microstructure influences its properties. The final sintered microstuctures are highly dependant on 1) the initial powders; degree of agglomeration, and powder size distribution, 2) packing of the powder in the green body and 3) the subsequent thermal processing: drying, binder burn-out, and sintering. Uniformity of powder packing within a ceramic green body, is needed to control differential shrinkage in single parts (e.g., capacitors) and in complex parts (e.g., multilayer silicon-chip substrates). Uniformity of powder packing is controlled by the ceramic powder used and the process used to make the ceramic green body. The ability to control these parameters and properties reliably and reproducibly is needed to maintain quality control, (i.e., keep rejection

L.-C. Dufour et al. (eds.), Surfaces and Interfaces of Ceramic Materials, 459–484.

rates low, and produce ceramic parts with consistent mechanical, electrical, and thermal properties).

This paper reviews the pertinent fundamentals that are required to 1) produce narrow size distribution ceramic powders 2) make suspensions of ceramic powders and 3) make green bodies with a uniform packing of particles. With precipitation, the size distribution produced must be narrow. For ceramic suspensions, agglomeration must be prevented for single componet slips. For multicomponet suspensions, flocculation is often used to prevent segregation of the various componets. The floccs are then broken during green body pressing. For green body production segregation due to size, density and shape must be prevented when large quantities of solvent are removed.

2. PRODUCTION OF NARROW SIZE DISTRIBUTION SOL-GEL POWDERS

To precipitate a narrow size distribution, a solution must be slowly supersaturated until nucleation takes place. During nucleation a sufficient surface area of particles must be produced so that any subseuqent supersaturation can be relieved by growth onto the existing particles and not by another nucleation event. The particles will continue to grow until the supersaturation is relieved. Since nucleation is not instantaneous, the nucleii produced first will grow until the last nucleii is generated giving a distribution of particle sizes at the end of the nucleation event. To narrow this nucleii size distribution, growth must be limited by diffusion in the boundary layer surrounding the particle. The diffusing species or growth unit can be an ion, a complex, a polymer or an aggregate. To maintain a narrow particle size distribution, coagulation of the nucleating and growing particles must be prevented by either electrostatic or steric means.

Jean and Ring[1] have produced narrow size distribution TiO_2 powders by the following reaction sequence

$$Ti(OC_2H_5)_4 + 4H_2O \xrightarrow{C_2H_5OH} Ti(OH)_4 + 4C_2H_5OH$$

to produce $Ti(OH)_4$ which undergoes condensation, eliminating water

$$Ti(OH)_4 \xrightarrow{C_2H_5OH} TiO_2 \cdot XH_2O(solid) + (2-X)H_2O.$$

TiO2 particles are precipitated when the ratio of the initial concentrations of H_2O to $Ti(OC_2H_5)_4$ is greater than 2.5 The time after mixing the solution until nucleation takes place is referred to as the induction time. The induction time is a function of the initial concentrations of $Ti(OC_2H_5)_4$ and H_2O. Once nucleated the particles grow by a diffusion limited mechanism without further nucleation, as shown in Figure 1. In Figure 1 the yield and means size reach steady state values at ~50 min while the number density decreases from $3x10^{10}$ #/cm³ to $2x10^{10}$ #/cm³ in the first 20 minutes. Separate electron diffraction experiments showed that the particles formed where composed of growth units ~40Å in size.

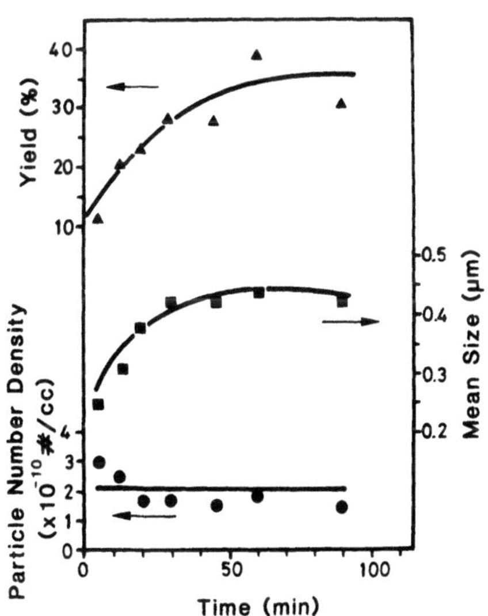

Figure 1. Yield, TEM mean size, and particle number density as a function of reaction time for 0.075 M $Ti(OC_2H_5)_4$ and 0.5 M H_2O and $0.427x10^{-3}$ gm/cc HPC.

In order to go to higher particle concentrations, coagulation of the large agglomerates must be prevented. Mates and Ring [2] found that hydroxypropyl cellulose (HPC) could be used as a sterically stabilizing surfactant during the nucleation and growth of TiO_2 particles, allowing the production of suspensions 8% by weight in TiO_2 with less than 1% multiplets of the largest particles.

To continuously produce these narrow size distribution sol gel agglomerates a reactor design must prevent the backmixing of fluid elements as they flow through the reactor. This type of chemical reactor is referred to as a plug flow reactor. In real reactor designs a small amount of backmixing cannot be prevented. Four types of chemical reactors minimize backmixing. They are 1) packed bed reactor, 2) motionless mixer reactor, 3) turbulent flow reactor and 4) constant curvature laminar flow reactor. The degree of backmixing in each of these reactors is measured by the dispersion number, D/(uL). Ring[3] has developed a theory shown in Figure 2 which predicts the geometric standard deviation, $\sigma_g = \exp(\sigma_z)$, of the particle size distribution precipitated in a reactor as a function of the dispersion number, D/(uL), for both single and multipoint nucleation models. As the degree of back mixing increases, the dispersion number increases. As the dispersion number increases, the geometric standard deviation of the particle size distribution increases.

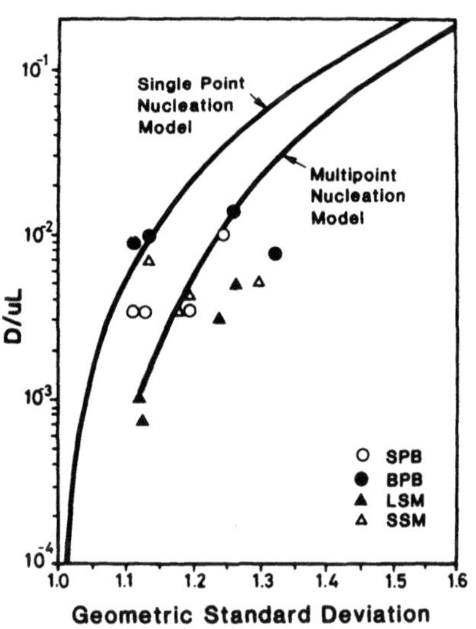

Figure 2. Dispersion number versus geometric standard deviation. SPB-Short packed bed, BPB-Long Packed Bed, LSM-Long Static Mixer, SSM-Short Static Mixer.

For the $Ti(OC_2H_5)_4$ reaction system with HPC discussed above Ring[3] and Jean, et al.[4] have shown reasonable agreement with the multipoint nucleation model using both packed bed and motionless mixer reactors. An example of the powder produced by a motionless mixer reactor is given in

Figure 3. It has a geometric mean size of 0.27μm and a geometric standard deviation 1.12. It is amorphous in crystal structure (by x-ray diffraction). The solids density was 8% exiting the reactor at a flow rate of 0.319 l/min.

Figure 3. TiO_2 particles produced with long 198 cm long static mixer.

Another approach to continuously generate narrow size distribution powders is to aggregate the particles in a backmixed reactor (also called a constant stirred tank reactor, CSTR). With aggregation the population balance for a backmixed reactor (CSTR) is

$$G(d\eta(L)/dL) = -\eta(L)/\tau + B(L) - D(L) \qquad [1]$$

where $\eta(L)$ is the population of particles of size L, G is the growth rate, and τ is the mean residence time for the CSTR. The birth function, B(L), and the death function, D(L), both due to agglomeration are

$$B(L) = \int_o^L K\eta(L-x)\ \eta(x)\ dx = KL\eta^2(L) \qquad [2]$$

$$D(L) = \int_o^\infty K/2\ \eta(L)\ \eta(x)\ dx = K/2\ \eta(L)N_T \qquad [3]$$

where K is the agglomeration rate constant and N_T is the particle number density. An exact solution to the population balance equation is given by Delpech[5]:

$$\eta(L) = \eta^o[\gamma(L/L_o + 1) + (1 - \gamma)\exp(L/L_o)]^{-1} \quad [4]$$

where

$$L_o = G\tau/[1 + KN_T\tau/2] \quad [5]$$

and

$$\gamma = KG\eta^o\tau^2/[1 + KN_T\tau/2]^2 \quad [6]$$

The exact solution is plotted in Figure 4 for various values of γ between zero and 10. When the nucleii agglomerate quickly (γ > 1.0), a narrow size distribution of agglomerates is predicted by this theory. When little or no agglomeration occurs (γ < 1.0), a broad particle size distribution is predicted.

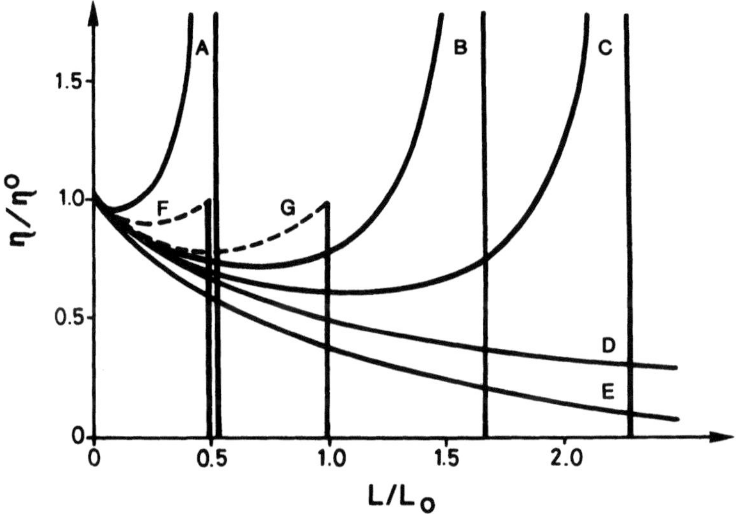

Figure 4. Population size distribution A) γ = 10.0, B) γ = 2.0, C) γ = 1.5, D) γ = 1.0 (Approximate solution F) γ = 4.0, G) γ = 3.0).

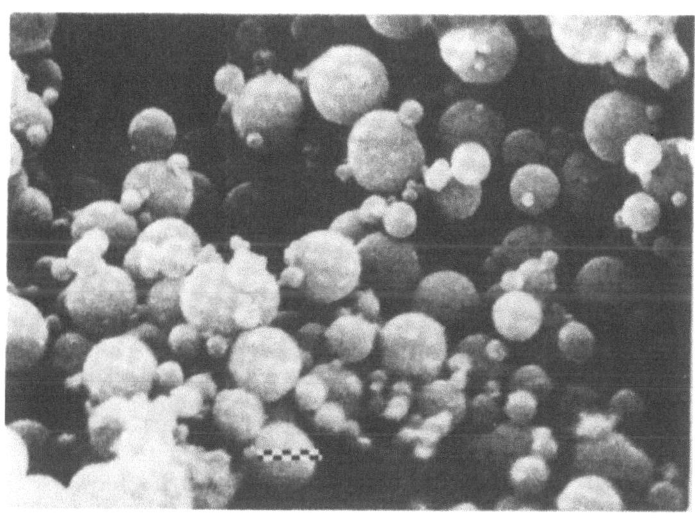

Figure 5. Particles from backmixed reactor without agglomeration 0.2 M $Ti(OC_2H_5)_4$, 1.0 M H_2O τ = 4.65 min.

For the $Ti(OC_2H_5)_4$ reaction system without HPC discussed above, Lamey[6] showed that without agglomeration smooth particles with a broad particle size distribution were produced (Figure 5) and that with agglomeration rough particles with a narrow particle size distribution were produced (Figure 6) in a backmixed reactor.

Primarily three methods are available for the precipitation of narrow size-distribution powders. Each requires a different technical approach which is summarized in Table 1. A plug flow reactor requires a single burst of nucleation. All the subsequent supersaturation must be relieved by the high growth rate which should be diffusion controlled. Coagulation must be prevented during growth by steric or electrostatic means. A backmixed reactor requires a continuous, high nucleation rate. Particle growth should be dominated by the coagulation of individual nucleii with larger particles and not by molecular growth.

Table 1. Generation of Narrow Size Distribution Powders

Reactor Type	Nucleation Rate	Growth Rate	Coag ulation Rate
Batch	Low	High	Low
Plug Flow	Low	High	Low
Back Mixed	High	Low	High

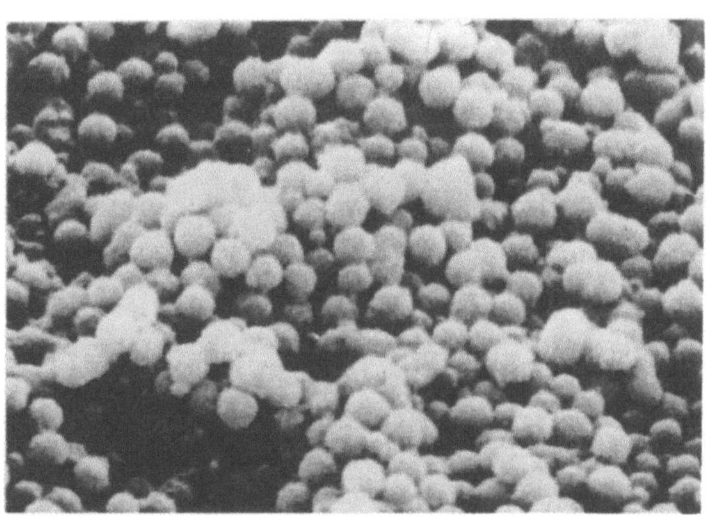

Figure 6. Particles from backmixed reactor with agglomeration 0.2 M $Ti(OC_2H_5)_4$, 1.0 M H_2O τ = 13.3 min.

3. CERAMIC POWDER SUSPENSIONS GENERATION

To generate a ceramic powder suspension the following sequence must be followed: 1) ceramic powder must be wetted by the solvent, 2) the powder must be de-agglomerated, 3) the particles once de-agglomerated must be prevented from further agglomeration until a uniformly mixed suspension is prepared. With single componet slips the suspension must remain colloidally stable throughout the process of dewatering the slip to form a green body. With multicomponet slips, flocculation is usually performed on the well mixed suspension to prevent segregation due to density, size or shape differences between the various componets of the slurry. The resulting green body after flocculation will have a low density and will require pressing to break up the floccs and give a high density well mixed green body.

3.1 Wetting of a Ceramic Powder

To obtain wetting the contact angle between the ceramic powder and the solvent must be less than 90°. With a zero contact angle the liquid spreads over the surface easily. When the contact angle is >90°, the liquid does not wet the surface. Young's equation[7] governs the contact angle, θ,

$$\cos\theta = (\gamma_{SV} - \gamma_{SL})/\gamma_{LV} \qquad [7]$$

where γ is the specific surface energy and the subscripts correspond to Surface, Vapor, and Liquid respectively. For wetting to occur, γ_{SL} and γ_{LV} should be made as small as possible. From the practical view point this is usually done by the addition of a surfactant to the liquid phase, which adsorbs on and lowers the surface energy of the solid-liquid and liquid-vapor interfaces. The surfactants used for this purpose in ceramics processing are called dispersants and often have both ionic and hydrophilic parts to the molecule.

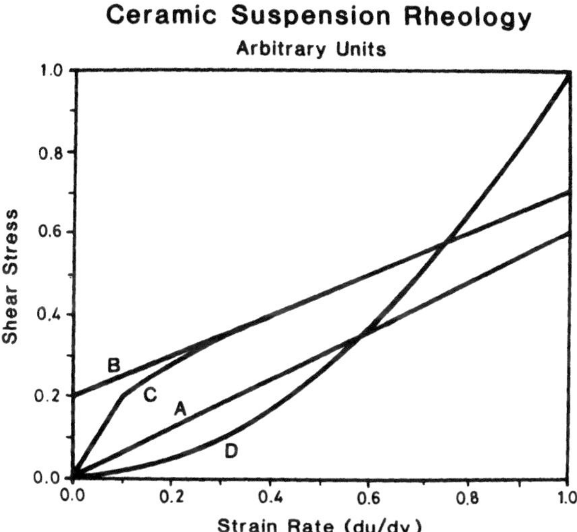

Figure 7. Rheological Behavior of Ceramic Powder suspensions, A) Newtonian, B) Bingham Plastic, C) Pseudoplastic, D) Shear Thickening.

3.2 Suspension Rheology

There are four basic types of rheological behavior for dispersions. These are shown in Figure 7. One type is a Newtonian fluid where the shear stress is proporational to the strain rate ($\partial u/\partial y$). The slope of the Newtonian curve shown in Figure 1 is the viscosity. The second type is a Bingham plastic where there is a yield stress given by τ_0 at zero strain rate followed by a linear stress strain relationship. The third type is pseudoplastic which has a curving stress strain relationship that has a shape roughly similar to that of the Bingham plastic. The fourth type is shear thickening in which the shear stress increases with strain rate with an increasing slope.

Both Newtonian and Bingham plastic types of rheological behavior are commonly observed with suspensions of ceramic powders. Saks[1] has shown both Newtonian and Bingham plastic characteristics are observed for silicon powder dispersions at 30% by volume in aqueous solution. For pH's of 9 and 10, Newtonian behavior was observed with a viscosity that was dependent on pH, while for pH 4 Bingham plastic behavior was observed with a yield stress of 2 pa.

4. COAGULATION IN POWDER PROCESSING

The coagulation time for a concentrated ceramic suspension undergoing shear due to differential settling is too complex for calculation. For systems where the particle size is less than 1 μm, perikinetic (Brownian) coagulation dominates over orthokinetic (shear induced) coagulation. Increasing the volume fraction decreases the settling velocities, which reduces the differential settling. The reduction in shear rate by an increase in volume fraction to one percent is sufficient to make perikinetic coagulation dominate for particle sizes up to 3 μm.

The following analysis of the coagualtion time is for a concentrated polydisperse suspension under going perikinetic coagulation. The theory developed uses the colloid stability factor calculated for dilute, one componet suspensions (i.e. two body interaction) and extrapolates to a concentrated suspension (i.e. multi body interaction).

4.1 Coagulation Time

The time, τ_c, for a fraction, f, of the particles to coagulate is[8]

$$\tau_c = \frac{2\pi\eta \; W_T \; \langle a^3 \rangle}{kT\phi} \left(\frac{f}{1-f}\right) \qquad [8]$$

where η is the viscosity, $\langle a^3 \rangle$ is the third moment of the particle size distribution, ϕ is the volume fraction, k is the Boltzmann constant, and T is the temperature. Particle size distribution and interactions between particles affect the coagulation time through the stability factor, W_T:

$$W_T = \left\{ \int_0^\infty \int_0^\infty \frac{(a_i+a_j)^2}{a_i a_j} \frac{P(a_i)P(a_j)}{W(a_i,a_j)} da_i da_j \right\}^{-1} \qquad [9]$$

where P(a) is the distribution of the particle sizes, $W(a_i,a_j)$ is the pair stability factor for the particles of radii a_i and a_j, and is of the form:

$$W(a_i,a_j) = (a_i+a_j)^2 \int_{a_i+a_j}^\infty \frac{\exp[V_{ij}(R)/kT]}{[a_i/\beta(u_j)+a_j\beta(u_i)]} \frac{dR}{R^2} \, . \qquad [10]$$

Where R is the center to center separation of the two particles, β is the hydrodynamic interaction parameter and u is twice the distance from the surface of a particle to the plane of symmetry between the particles divided by the radius. For ease and speed of calculation, the approximation for $\beta(u)$ derived by Honig et al.[9] is used. It is valid to within a few percent over a large range of values. The stability factor is very sensitive to the interaction energy, $V_{ij}(R)$, through its exponential dependance. The interaction energy in turn is strongly dependant on particle sizes a_i and a_j:

$$V_{ij}(R) = V_r^{ij}(R) + V_a^{ij}(R). \qquad [11]$$

Where V_r is the coulombic repulsion obtained from Hogg, Heally and Fuerstenau (HHF)[10] which states

$$V_r^{ij}(R) = \epsilon\Psi_0^2\left(\frac{a_i a_j}{a_i+a_j}\right) \ln\{1+\exp(-\kappa[R-(a_i+a_j)])\} \quad . \quad [12]$$

Where ϵ is the permitivity of the medium, Ψ_0 is the surface potential, and κ^{-1} is the double layer thickness, which is a function of the electrolyte concentration(Note! The higher the electrolyte concentration the smaller the double layer thickness as shown in Table 2). V_a is the van der Waals attraction[11] given by

$$V_a^{ij}(R) = -\frac{A}{6}\left\{\frac{2a_i a_j}{R^2-(a_i+a_j)^2} + \frac{2a_i a_j}{R^2-(a_i-a_j)^2} + \ln\left[\frac{R^2-(a_i+a_j)^2}{R^2-(a_i-a_j)^2}\right]\right\} \quad [13]$$

where A is the Hamaker constant for the particle-solvent-particle system.

Table 2. Double Layer Thickness

Concentration (Molar)	Electrolyte 1:1	2:2
10^{-5}	0.1µm	0.05µm
10^{-3}	0.01µm	0.005µm
10^{-1}	10Å	5Å

The HHF electrostatic interaction energy has some implicit assumptions: 1) the surface potential of the particles should be less than 25 millivolts (however the electrostatic energy has been used for surface potentials up to 50 or 60 millivolts), 2) the particle size must be at least 5 times the characteristic length of the electrical double layer, 3) the particles must be spherical, 4) the dielectric constant of the fluid is constant throughout, and 5) the electrolyte ions act as point charges. For oxides, the surface potential is a function of solution pH as shown in Figure 8. At pH values below the isoelectric point (IEP), a positive surface potential is observed. For pH values above the isoelectric point, a negative surface potential is observed. Table 3 gives the isoelectric points for common ceramic materials[12].

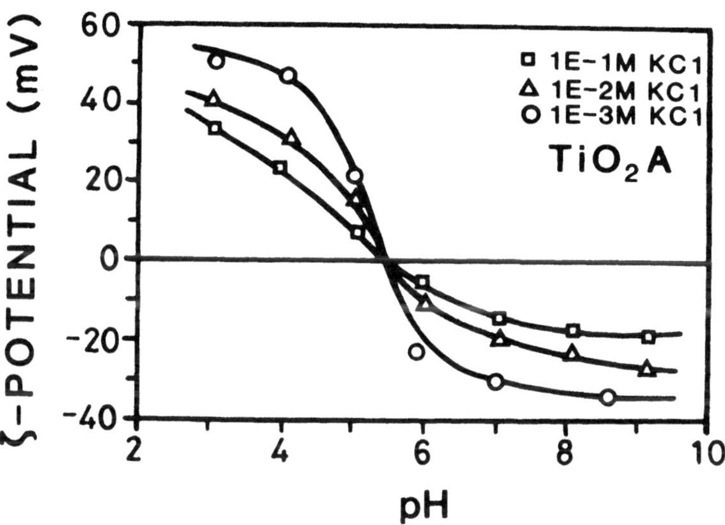

Figure 8. Zeta-potential of TiO_2 as a function of pH for the KCl electrolyte. From Barringer[13]

Table 3 Nominal Isoelectric Points (IEP) of Ceramic Materials

Material	IEP(pH)
Quartz	2
Potassium Feldspar	3-5
Zirconia	4-6
Tin oxide	4-5
Titania	5-6
Kaolin	5-7
Mullite	6-8
Chromium oxide	6-7
Hematite	8-9
Zinc oxide	9
Alumina	8-10
Calcium carbonate	9-10
Magnesia	12

Figure 9 illustrates the effect of particle size on the interaction energy. Two large particles must overcome a large energy barrier to coagulate, while two smaller particles must overcome a much smaller barrier. A large and a small particle together have a greatly reduced energy barrier to overcome. This fact coupled with the increased probability of an interaction between small and large particles, explains why small particles preferentially attach to large particles in polydisperse system.

A lognormal distribution is used to describe the particle sizes,

$$P(a)da = \frac{1}{\sigma\sqrt{2\pi}} \exp\left(\frac{-[\ln(a/a_g)]^2}{2\sigma_z^2} \right) \frac{da}{a} \quad . \qquad [14]$$

The lognormal width parameter, σ_z, is related to the geometric standard deviation by $\sigma_z = \ln(\sigma_g)$. The geometric mean particle-size, a_g, differs from the average particle size by $\langle a \rangle = a_g \exp(\sigma_z^2/2)$. In general, the nth moment of a lognormal distribution is $\langle a^n \rangle = a_g^n \exp(n^2\sigma_z^2/2)$.

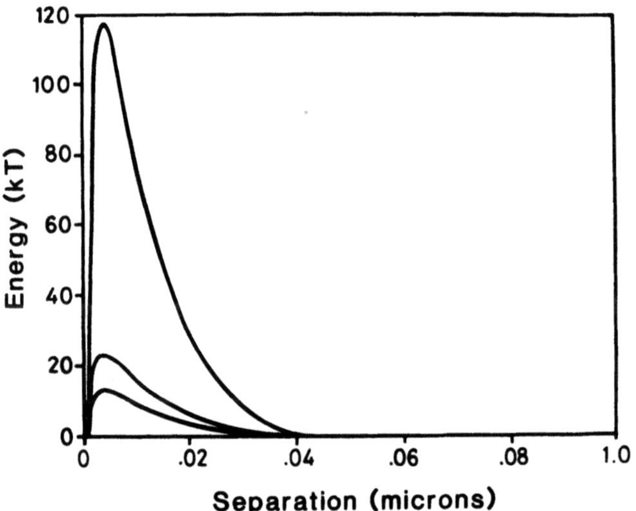

Figure 9. Interaction energy versus separation for different particle size combinations. The electrolyte concentration is 0.001 M, ψ_0 = 25 mV, A = 2.1×10^{-20} joules, a) $a_1 = a_2$ =0.5μm, b) a_1 = 0.5μm a_2 =0.05μm, c) $a_1 = a_2$ =0.05μm.

For ceramic powders subjected to grinding, a wide distribution corresponds to $\sigma_z = 0.75$. For 1 μm average size particles, 95% of the particles, by number, are between 0.17 and 3.28 μm. Ceramic powders made by chemical precipitation methods generally have much smaller width parameters. Silica powders have been produced this way with $\sigma_z = 0.05$. Titania powders have been produced this way with $\sigma_z = 0.113$. For a distribution with $\sigma_z = 0.1$ and $\langle a \rangle = 1.0$ μm, 95% of the particles are between 0.82 and 1.1 μm.

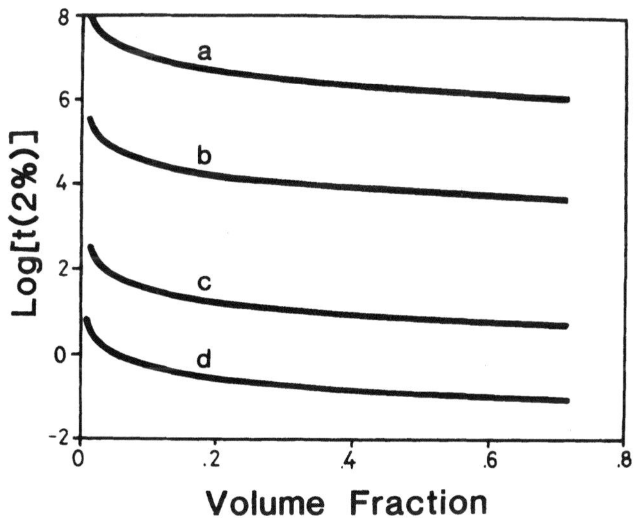

Figure 10. Coagulation time in seconds for 2% of particles coagulated versus volume fraction at different distribution width parameters. Electrolyte concentration is 0.01 M, ψ_0 = 25 mV, A = 2.1×10^{-20} joules, a) $\sigma_z = 0.0$, b) $\sigma_z = 0.25$, c) $\sigma_z = 0.5$, d) $\sigma_z = 0.75$.

The coagulation time calculations described are for titania (TiO_2) powders suspended in aqueous salt solutions at 25°C. The density of the titania powder is 3.81 gm/cm³. The Hamaker constant for titania suspended in water is 2.1×10^{-20} joules and water has a dielectric constant of 78.1. In all cases, the coagulation time should be larger than the time required for processing of the ceramic suspension. The logarithm of the coagulation time in seconds is plotted versus volume fraction for different degrees of polydispersity in Figure 10. For most of the range, the volume fraction has just a small effect on the coagulation time. Significant increases in the coagulation time occur at very low volume fractions where the particles are far apart, and therefore the diffusion distances and times are very large.

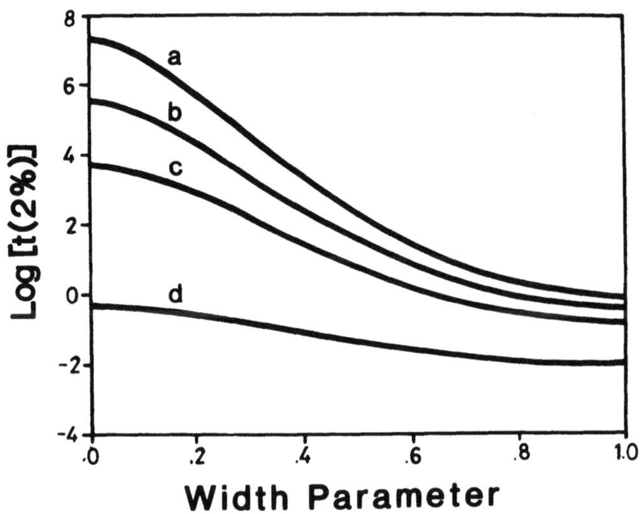

Figure 11. Coagulation time in seconds for 2% of particles coagulated versus distribution width parameter for different average particle sizes. Electrolyte concentration is 0.01 M, ψ_0 = 25 mV, A = 2.1×10^{-20} joules, a) $\langle a \rangle$ = 0.3μm, b) $\langle a \rangle$ = 0.25μm, c) $\langle a \rangle$ = 0.2μm, d) $\langle a \rangle$ = 0.1μm.

Narrowing the width of the particle size distribution has a much larger effect on increasing the coagulation time. This is seen more directly in Figure 11. Along each curve $\langle a \rangle$ is kept constant and the logarithm of the coagulation time is plotted versus σ_z. For example, monodisperse 0.6 μm particles take roughly a year to coagulate. While those with a σ_z of 0.4, coagulate in about an hour. Larger particles are inherently more stable because of their high energy barriers, therefore, the rapid coagulation in polydisperse systems is due to the small particles in the distribution. The total stability factor for a polydisperse system is found by summing the reciprocals of the individual pairs of stability factors, much like resistances in parallel. In this way, pairs of particles which have large stability factors contribute little toward the total, while pairs which have small stability factors contribute greatly.

The energy barrier between particles is enlarged by increasing the surface potential or decreasing the electrolyte concentration, which in turn increases the coagulation time. Even though the energy barrier for the average particle size may be high enough to hinder coagulation, polydispersity vill cause some of the energy barriers to be small or nonexistant, thereby reducing the coagulation time.

4.2 Floccuation in Ceramic Processing

With mixed ceramic suspensions, flocculation is used to prevent segregation of the two or more ceramic powder componets. To induce flocculation polymers are often used. The polymer must 1) adsorb on the surfaces of the componet ceramic powders, and 2) destabilize the suspension so that all of the particles stick together when they come into contact by Brownian diffusion.

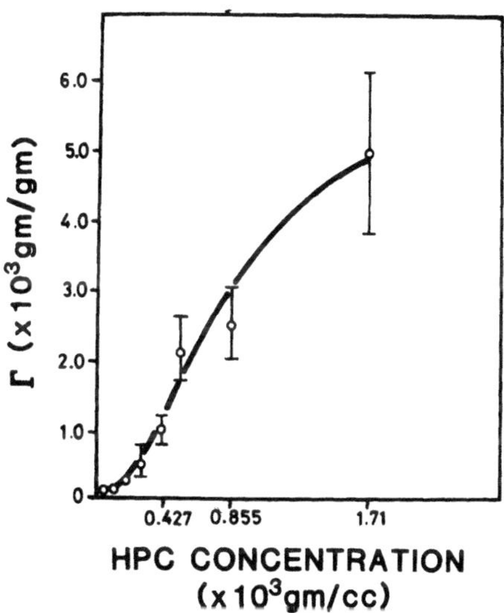

Figure 12. Adsorption density of HPC versus initial concentration.

Polymer adsorption takes place when the polymer has an

affinity to the powder surface or when it is pushed out of solution onto the powder surface by the approach of a solubility limit or a cloud point. A polymer's surface affinity can be manipulated by its constituent functional groups. A polymer's solubility can be manipulated by its molecular weight, conformation and functionality. The adsorption of polymers by ceramic powders often follows the Langmuir adsorption isotherm given by

$$\theta = ba_2/(1+ba_2) \qquad [15]$$

where b is the distribution coefficient and a_2 is the activity of the polymer. An example of the Langmuir adsorption isotherm for a polymer is shown in Figure 12 for the adsorption of Hydroxy-propyl-cellulose onto TiO_2 in ethanol. Higher molecular weights achieve monolayer coverage at lower concentrations.

Once adsorbed a polymer can either stabilize or destabilize a ceramic suspension. When the polymer likes itself better than it likes the solvent, flocculation occurs. Conversely, when the polymer likes the solvent more than it likes itself, steric stabilization occurs. In both cases, the same equation governs the interaction energy, $V_T(H)$, for two particles with radius, r, coated with a layer of polymer of thickness, δ, at a separation, H,[14]:

$$\begin{aligned} V_T(H) = {} & 4/3\cdot\pi RTBC_i^2(\delta-H/2)^2(3r+2\delta+H/2) \\ & -1/12\cdot\{[A_{22}^{1/2}-A_{33}^{1/2}]^2[(r+\delta)/H] \\ & +[A_{33}^{1/2}-A_{11}^{1/2}]^2[r/(H+2\delta)] \\ & +4r[A_{22}^{1/2}-A_{33}^{1/2}][A_{33}^{1/2}- \\ & A_{11}^{1/2}][r+\delta]/[(H+\delta)(2r+\delta)]\} \end{aligned} \qquad [16]$$

The first term is the steric interaction energy derived by Fischer[15] and the other terms are the van der Waals interaction energy derived by Vold[16]. In the first term, B is the second virial coefficient for the polymer and C_i is the polymer concentration in the adsorbed layer. The other terms, A_{11}, A_{22} and A_{33} are the Hamaker constants of the particle, solvent and polymer, respectively.

4.3 Flocculation Kinetics

The collision rate between particles consisting of i units and particles consisting of j units is given by[17]

$$dN_{ij}/dt = 4\ \pi\ R_{ij}\ D_{ij}\ N_i\ N_j \qquad [17]$$

where N is the number density of par[illegible]les of size r_i and r_j,

respectively. D_{ij} is the mutual diffusion coefficient of the two particles, and R_{ij} is the sum of the radii of the ith and j^{th} particle. The product $R_{ij}*D_{ij}$ is given by

$$R_{ij}*D_{ij} = [r_i + r_j]*[1/r_i + 1/r_j]*kT/(6\pi\eta) \qquad [18]$$

where k is the Boltzmann constant, T is the absolute temperature and η is the viscosity of the solvent.

If we let k = i + j, so $N_k = N_{ij}$, and assume that initially all particles are identical i = 1 (monodisperse), and that the aggregates are spherical, then the rate of formation of K particles form an initially monodisperse system:

$$dN_k/dt = 2\pi \sum_{\substack{i=1 \\ j=k-i}}^{i=k-1} R_{ij}\ D_{ij}\ N_i\ N_j \ - 4\pi\ N_k \sum_{i=1}^{\infty} R_{ik}\ D_{ik}\ N_i \qquad [19]$$

The first term gives the rate at which the k-type particles appearand the second term gives the rate of disappearance of k type particles. If the the size of the ith and j^{th} particles are the same, then

$$R_{ij}*D_{ij} = 4*kT/(6\pi\eta). \qquad [20]$$

The above equation holds for $1<(i/j)<2$, for larger ratios the contsant 4 becomes significantly larger. For unequal spheres the collision rate is larger than that for equal spheres. If the total number of particles, N_T, is accounted for in a differential equation we find

$$dN_T/dt = d\sum_{k=1}^{\infty} N_k/dt = -8\pi\ N_T^2\ kT/(6\pi\eta) \qquad [21]$$

which can be integrated to give

$$N_T = N_o/(1 + 8\pi\ N_o\ t\ kT/(6\pi\eta)). \qquad [22]$$

Where the flocculation half life, τ, is given by

$$\tau = 1/[8\pi\ N_o\ kT/(6\pi\eta)] \qquad [23]$$

and the time dependance of any size aggregate, N_k, is given by

$$N_k = N_o\ (t/\tau)^{k-1}/[1 + (t/\tau)]^{k-1}. \qquad [24]$$

where k is any integer. The results of this analysis are presented in Figure 13. In this figure we see the number of singlets, doublets, etc. as a function of time. For short time scales this theory fits well with experimental flocculation data. These agglomerates are used to lockin the well mixed structure of the suspension preventing segregation of particles due to differences in size, shape or density. Unfortunately the agglomerates give a inhomogeneous, low density packing and therefore must be broken up by pressing.

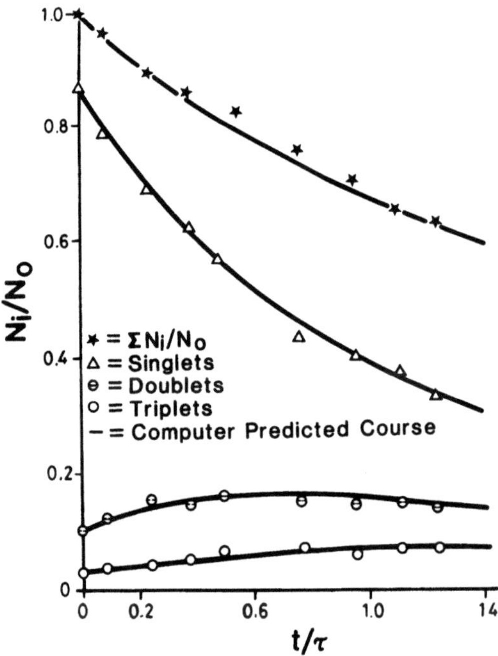

Figure 13. Distribution of singlets, doublets and triplets versus time for diffusion controlled agglomeration[18].

5. GREEN BODY FORMATION

The formation of comeramic green body requires one or more forces to remove the excess solvent from a ceramic suspension. With slip casting a combination of hydrostatic pressure and mold capillarity provide the necessary dewatering forces. With filter pressing or colloid presssing, only hydrostatic pressure is used. For tape casting, a polymeric binder is used which shrinks upon drying pulling the particles closer together. The fundamentals of each of these forming technologies will be discussed.

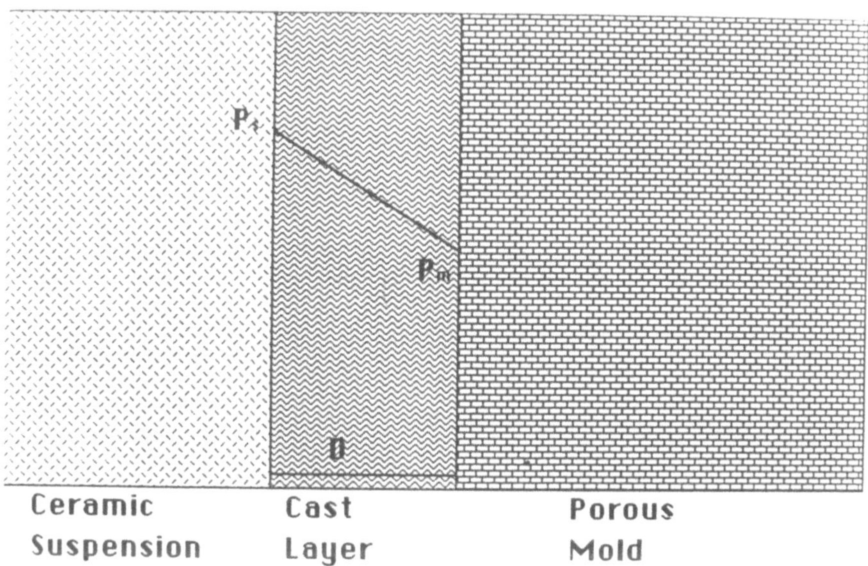

Figure 14. Schematic representation of the formation of a slip-cast layer formed by the extration of solvent by the capillary action and hydrostatic pressure of a porous mold.

5.1 Slip Casting

With slip casting a porous mold is used to give shape to the green body. The mold is filled with a ceramic suspension and a combination of hydrostatic pressure and capillary suction dewaters a thin section of the ceramic suspension adjacent to the mold as seen in Figure 14. The thickness of the slip cast, D, is given by Darcy's Law[19]:

$$D^2 = 2\ K\ \rho\ \kappa\ (\Delta P + 2\gamma/r)\ t \qquad [25]$$

where K is the permeability of the cast layer, ρ is the density of the cast layer, κ is the factor converting the volume of water removed to the volume of ceramic powder deposited, ΔP is the hydrostatic pressure head, γ is the surface tension of the solvent with all of its additives including dispersants and wetting agents and binders, r is the mean pore radius of the mold and t is time.

The packing that is obtained during slip casting is essentially that of the collapsed slurry. When the dewatering rate is slow the particles have ample time to diffuse laterally to find a packing location which minimizes the defects in the structure before they are locked into location by the next layer of particles. Conversely, when the dewatering rate is high the particles do not have enough time to rearrange themselves before they are fixed into a packing location by the arrival of the next layer. For this reason, the rate of dewatering controls the packing of the particles in a slip cast.

5.2 Pressure Filtration

In pressure filtration, a piston is used with a frit bottom to allow the excess liquid to leave the chamber and catch the ceramic powder on the frit. The consolidated layer of ceramic powder grows according to a parabolic rate law, derived from Darcy's Law. The piston displacement, d, is given by

$$d = [2\ K\ (v_1/v_0 - 1)\ P\ t/u]^{1/2} \qquad [26]$$

where K is the permeability of the consolidated layer, P the applied pressure, v_1 is the volume fraction of solids in the consolidated layer, v_0 is the volume fraction of the solids in the suspension, u is the piston velocity and t is time. Using this equation, a plot of the piston displacement, d, versus $t^{1/2}$ gives a straight line from which the permeability of the powder compact can be determined.

The permeability decreases as the applied pressure increases resulting in increased packing density. Flocculation of the ceramic suspension before filtraton gives much higher permeabilities, often by a factor of ten. Increasing the applied pressure decreases the permeability and increases the density to a much larger degree with flocced suspensions ,than with stable suspension. To break floccs and aggregates applied pressures in excess of 80 MPa are commonly needed.

Like slip casting, the packing that is obtained is essentially that of the collapsed slurry. When the dewatering rate is slow the particles have ample time to diffuse laterally to find a packing location which minimizes the defects in the structure before they are locked into location by the next layer of arriving particles. Conversely, when the dewatering rate is high the particles do not have enough time to rearrange themselves before they are fixed into a packing location by the arrival of the next layer. For this reason, the rate of filtering controls the packing of the particles in the cast layer. But unlike slip casting the high pressures applied used in filtration can cause further consolidation of the cast layer, increasing the packing density.

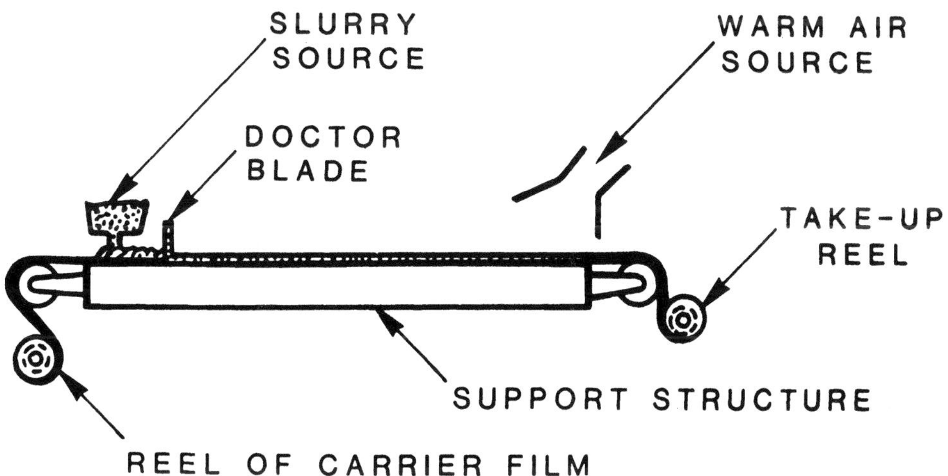

Figure 15. Schematic of Doctor Blade Tape Casting Process.

5.3 Tape Casting

Tape casting is a common way to produce thick film ceramic green bodies. A schematic of a tape casting machine is shown in Figure 15. It consists of a moving belt onto which a ceramic powder slurry is spread by a stationary doctor blade. Control of the thickness of the cast is of critical importance in tape casting. To determine the cast thickness it is necessary to know the velocity profile in the gap between the doctor blade and the moving belt. Once the velocity profile is known the integrated average velocity can be calculated and the cast thickness determined. For a Bingham plastic fluid, the thickness, δ, is given by[20]

$$\delta/h = (1/2 + P/6)(1 - [\tau_0^* - 1 + P]/[2P]) \quad [27]$$

where h is the thickness of the gap between the bottom of the doctor blade and the moving belt, P is the dimensionless pressure gradient across the doctor blade ($P = h^2(-dp/dx)/(2\eta u)$) and τ_0^* is the dimensionless Bingham yield stress ($T_0^* = \tau_0 h/(\eta u)$). The pressure gradient across the doctor blade is dp/dx, u is the belt velocity, τ_0 is the Bingham yield stress of the suspension and η is the viscosity of the suspension. When the pressure gradient is increased the thickness of the cast increases for all Bingham yield stresses as seen in Figure 16. At a constant pressure, the thickness of the cast decreases as the yield stress increases.

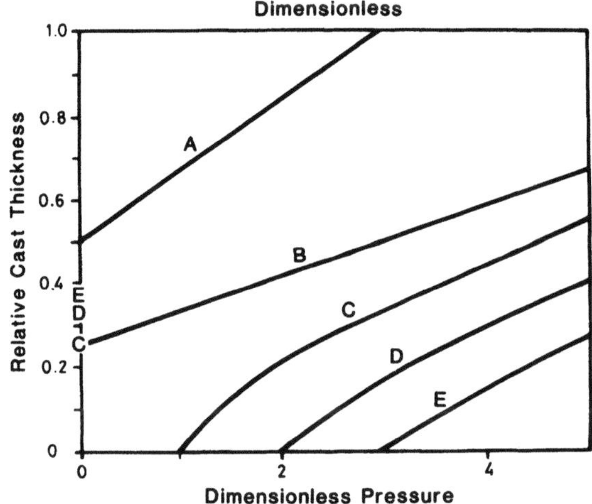

Figure 16. Cast thickness for a Bingham Fluid as a Function of Dimensionless Pressure Gradient and Dimensionless Bingham Yield Stress; A) $\tau_0^*=0$, B) $\tau_0^*=1$, C) $\tau_0^*=2$, D) $\tau_0^*=3$, E) $\tau_0^*=4$.

The packing of the particles in tape casting is highly influenced by shear. High shear rates tend to orient non-spherical particles, leading to non-uniform packing. The forces which act to pull the particles together are essentially those of polymer shrinkage upon drying. A polymeric binder is used to give the dried tape structural integrity. When this polymer dries it shrinks pulling the particles closer together giving a higher green-body density.

6. CONCLUSION

This paper has reviewed the generation of ceramic powders by chemical precipitation methods, the generation of ceramic powder suspensions, and the processing of ceramic powders to make ceramic green bodies. These precipitation methods are used to produce chemically pure ceramic powders with narrow particle size distributions. Agglomeration must be prevented by electrostatic or steric means with single componet suspensions. With multicomponet suspensions, flocculation is often performed to lockin the well mixed nature of the suspension preventing segregation due to differences in particle size, shape and density. During green body formation, large amounts of solvent must be removed while packing the particles uniformly. The fundamentals of green body formation are discussed with special attention to the details of particle packing in the green body since green body microstructure has such a large influence on the final sintered microstructure and the mechanical and electrical properties of the piece.

REFERENCES

1. Jean, J.H. and Ring T.A., Langmuir 2,251(1986)

2. Mates, T. E. and Ring, T. A., Colloids and Surfaces 24, 299 (1987)

3. Ring, T. A., Chem Eng. Sci. 39, 1731 (1984)

4. Jean, J. H., Goy, D. M. and Ring, T. A., Am. Ceramic Soc. Bulletin 66,1517(1987)

5. Delpech de Saint Guilhem X., and Ring. T. A., Chem Eng. Sci. 42,1247(1987)

6. Lamey, M. D. and Ring, T. A., Chem Eng., Sci. 41, 1213 (1986)

7. Adamson, A.W. "Physical Chemistry of Surfaces", 3rd edition, Wiley, New York, 1976

8. Strauss, M., Ring, T.A., Bleier, A. and Bowen, H.K. J. Appl. Phys. 58,3871(1985)

9. Honig, E.P., Roebersen, G.J., Wiersema, P.H., J. Colloid Interface Sci.33,97(1971)

10. Hogg, R., Healy, T. and Fuerstenau, D., Trans. Faraday. Soc. 62,1638(1966)

11. Hamaker, H. Physica 4,1058(1937)

12. Reed, J.S., "Principles of Ceramic Processing", Wiley, New York, 1988.

13. Barringer, E.A., PhD thesis MIT, 1983

14. Vincent, B., J. Colloid Interface Sci. 42,262(1973)

15. Fischer, E.W., Kolloid-Z 160,120(1958)

16. Vold, M.J., J.Colloid Sci. 16,1(1961)

17. Smoluchowski, M. von, Versuch einer Mathematischen Theorie der Koagulationskinetik Kolloide Losungen, Zeitschr. Phys. Chem., 92,129(1917)

18. Kruyt, H.J., "Colloid Science", Elsevier, New Your, N.Y., 1952, vol.I., p.278.

19. Kingery, W.D., Bowen, H.K. and Uhlmann, D.R., "Introduction to Ceramics", 2nd Edition, Wiley, New York, 1976,p385

20. Ring, T.A. submitted J. Am. Ceramic Soc. 1988

SOME CURRENT ISSUES ON POLYCRYSTALLINE STRUCTURES AND GRAIN GROWTH

A. Mocellin
Laboratoire de céramique
Ecole Polytechnique Fédérale de Lausanne
Chemin de Bellerive, 34
CH - 1007 LAUSANNE
Switzerland

KEYWORDS / ABSTRACT : polycrystals / geometry / grain growth / computer simulations / ceramics

A non exhaustive selection of open questions arising when characterizing the structures of polycrystals and modelling their changes during grain growth is presented. Geometrical parameters describing size, shape and their distributions, both static and as they evolve with time are first discussed theoretically. Then an account is given of those computer simulation studies on two-dimensional structures undergoing grain growth which have been published recently. Finally, recently observed complexities of grain boundary migration in ceramic systems are mentioned briefly.

1. INTRODUCTION

In polycrystalline ceramics, grain growth which is driven by the reduction of the solid-solid interfacial energy content of the system, is an as yet insufficiently understood phenomenon with very important practical implications. Most of the mechanical, chemical, physical or thermal service properties of sintered ceramics are directly related to the structure and composition of grain boundaries and, in general, when theoretical models are developed which aim at relating the properties of materials to their microstructures, the so-called average grain size is very often explicitely taken as a key structural parameter. In practice, a rather limited control can be exerted on grain size in ceramics via the duration of high temperature heat treatments, and to a lesser extent, via well-chosen impurity doping and/or adjustments to the gas atmosphere in the firing furnace. In contrast to metals, ceramics so far exhibit but limited intragranular ductility up to rather high temperatures, which prevents the recourse of primary recrystallization as a grain refining procedure. Such a limitation has had an important effect during the past 30 to 40 years in motivating a significant part of ceramic research programs, due to the fact that in a majority of cases

L.-C. Dufour et al. (eds.), Surfaces and Interfaces of Ceramic Materials, 485–505.

property improvements could only be obtained throughgrain size reductions. The race toward finer and finer grain sizes (at zero or minimal residual porosities) has intensified in recent years, in connection with and as a consequence of progress in powder processing in general. Rather spectacular results have thus been obtained in a number of cases and it is now becoming customary to produce fully dense sintered bodies with average grain sizes well in the submicron range. Interestingly however, several cases have arisen recently where an ultra fine equiaxed isotropic grain structure is no more synonymous of superior material properties. For example, the mechanical behavior of sintered silicon nitride (and for that matter of other ceramics and ceramic composites as well) has been shown to be improved when a fibrous-like structure is produced with individual grains having well defined geometries. Also, it is now argued that the achievement of practically acceptable current densities in the modern high T_c ceramic superconductors necessitates the development of strong textures with probably as large an individual grain size as can be tolerated. Challenges such as these undoubtedly raise problems that are new to many professional ceramists and for the solution of which, very little directly useful information is yet available.

This presentation which in no respect should be expected to provide a comprehensive overview of the subject, will consist of two parts. First, the geometry of ideal single phase isotropic polycrystals will be discussed, from both metric and topological points of views. Particular attention will be given to recent research describing local changes in such structures and their overall stability toward grain growth. A brief account will also be given of the modelling of the grain growth kinetic law.

The second part will present and discuss results obtained in recent years on the computer simulation of the normal grain growth process. For obvious reasons, most studies actually producing synthetic structures have been performed in two dimensions. The discussion will be centered on comparing simulations based on deterministic versus stochastic approaches respectively, and on discrete versus quasi continuous dynamics.

To conclude, some important physico-chemical aspects of grain growth in real ceramic systems for which still little experimental and theoretical information is available, will be brought to attention briefly.

2. THE IDEAL POLYCRYSTAL AND ITS NORMAL EVOLUTION

Most polycrystalline materials are optically opaque and their microstructures therefore are observed in reflexion from polished sections (Figure 1). As a result, by far the majority of studies aimed at describing the geometry of such structures and their changes has been restricted to truly 2-D cellular networks. And it still is not clear whether the latter may be considered as fully adequate representations of planar cuts through the physical 3-D reality of polycrystals, even when ideal. Since the ideal character of polycrystals and their normal grain growth is often implicitly assumed or referred to in the literature, it may be useful here to suggest some definitions for the sake of clarifying terms and concepts.

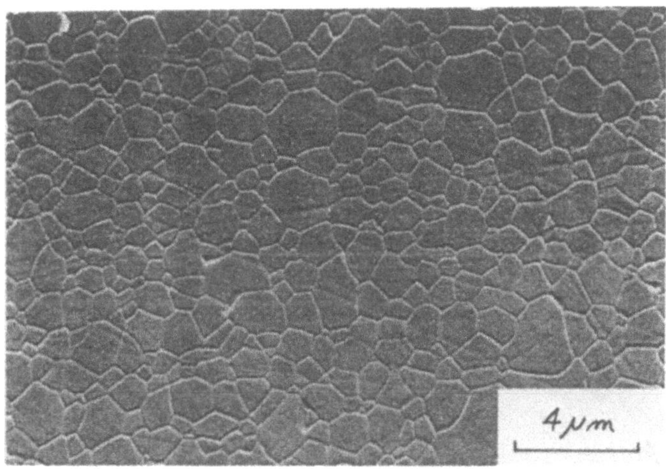

Figure 1. Typical microstructure of polycrystals (sintered Al_2O_3).

An "ideal" polycrystal is proposed to be one which fulfills the following requirements: it is single phased, ignores crystalline anisotropy effects. Its grain boundaries all have the same energy and - when undergoing displacements - the same mobility per unit area. Obviously then in an isothermal environment and in absence of any applied field (e.g. stress), an ideal polycrystal should remain equiaxed and statistically homogeneous, i.e. exhibit the same geometrical characteristics for every representative volume element, regardless of its position within the actual piece of material. Similarly, it is proposed that an ideal polycrystal undergoes "normal" grain growth when isothermally annealed at suitable temperatures, such that its microstructure remains "self-similar", i.e. the microstructures of samples cut from the same piece of material and subjected to annealings under different temperature and time conditions may be made to coincide statistically by mere adjustments in the magnifications of the microscopes that are used to make observations. More specifically, self-similarity or scaling is thought to imply that the reduced grain size distribution $f(D/\bar{D})$ is time-invariant. More recently, it has also been proposed by various authors that the distributions $\{f_n\}$ of polygonal grain shapes, as represented by their respective coordination numbers "n" also should be invariant during so-called normal grain growth. Although rather appealing per se, this latter property remains open to questions, both experimental and theoretical. And at the moment also, it is not known whether the self-similarity criterion has other direct quantitative implications on the architecture of polycrystals.

Since good well documented reviews have recently appeared in the literature (e.g. ref. 1), we shall restrict the following discussion to a few points of particular significance or interest.

2.1. Static description

It is convenient to organize our current understanding of an ideal polycrystalline geometry in terms of sizes, shapes and spatial arrangements respectively. Such an order of presentation also roughly corresponds to decreasing degrees of knowledge or amounts of available information.

2.1.1. Size distributions. It is usually assumed by theoreticians and experimentalists as well that the "equivalent spherical diameter", i.e. the diameter of the sphere having identical volume, adequately represents the true size of a grain in the polycrystal. This equivalent diameter is statistically distributed and a number of theoretical models have been derived to account for such distributions, including:

- the lognormal behaviour (2;3) which is most frequently invoked but rests on rather aproximate thus not quite convincing arguments. An alternative to the lognormal distibution, namely the gamma function:

  $$f(x) \equiv \frac{\alpha^{\alpha}}{\Gamma(\alpha)} \cdot x^{\alpha-1} e^{-\alpha x} \; ; \; \alpha > 0 \; ; \; x \equiv D/\bar{D}$$

  has been shown recently (4) capable of providing equivalently good (or approximate) statistical fit to experimental observations.
- the Hillert model (5) which derives an ad-hoc distribution by assimilating normal grain growth to an Ostwald ripening process. Experimentally observed distributions however do not correspond to the predicted function. But Hillert (6) and others have rightly argued that no satisfactory proof has yet been provided that steady state had indeed been achieved in grain growth experiments. In order for this to be checked, a suitable statistical test or criterion should be agreed upon. And it remains unclear whether polycrystalline systems undergoing grain growth can effectively reach steady state within reasonable experimental times.
- The Louat model (7) which leads to a Rayleigh distribution:

  $$f(x) = \frac{\pi}{2} \; x \exp\left(- \frac{\pi x^2}{4} \right) \; ; \; x \equiv D/\bar{D}$$

  This distribution for the moment, although probably not universal, rests on reasonable theoretical assumptions and has met (8) with some experimental justification (Figure 2). It also has the remarkable property of retaining its analytical expression regardless of whether x refers to 2- or 3-dimensions.

2.1.2. Topological parameters and distributions. The subject encompasses grain shapes and arrangements. The main constraints in this respect are:

- the space filling requirement
- the near equilibration of surface forces (at grain boundary junctions).

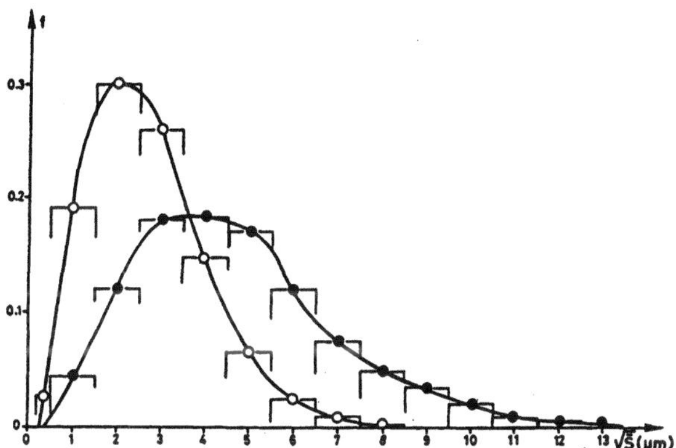

Figure 2. Examples of observed grain size distributions (Cr-doped Al_2O_3) apparently fit by the Louat-Rayleigh function (o: 0.4% Cr , $\sqrt{\bar{s}}$ = 3.1 μm; •: 0.7% Cr, $\sqrt{\bar{s}}$ = 5.3 μm).

The importance of such requirements was recognized early (9;10). But so far, no or largely unsuccessful attempts have been made at explicitly linking metric and topological features in theoretical models of polycrystalline structures and their changes. A notable exception is the recent work of N. Rivier (11), which awaits experimental confirmation. And by far, most of the present knowledge on the topology of polycrystals pertains to their two dimensional sections.

Two concepts are now generally accepted as basic, when discussing polycrystalline structures:

* that the polygonal grain shape is well represented by "n", i.e. its coordination number and the corresponding distribution $\{f_n\}$ such that: $n \geq 3$; $\sum_n f_n = 1$; $\langle n \rangle = 6$ from Euler's theorem (Figure 3)
* that any change of shapes within the system, results from either of two "elementary topological transformations" (ETT) and their repetitions (12) (Figure 4).

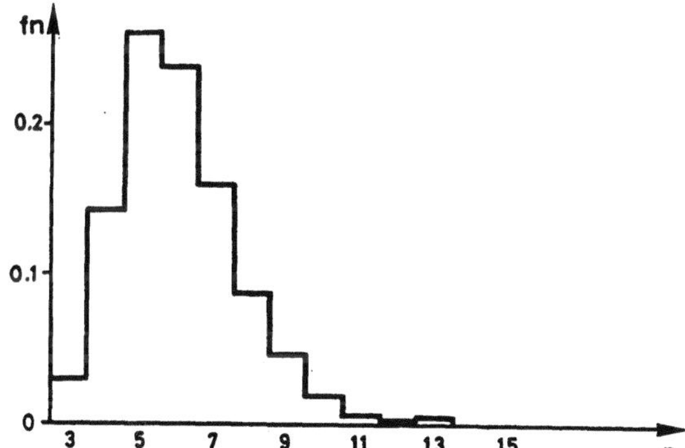

Figure 3. Example of 2-d grain shape distribution.

Figure 4. Schematic of elementary topological transformations through which overall structure changes are channeled.

It follows therefrom that:

- the shape n of a grain and that of its average nearest neighbour m_n are strongly correlated (13):

  $$m_n = 5 + \frac{6 + \mu_2}{n} \quad , \quad \text{where } \mu_2 \equiv \sum_n (n-6)^2 f_n$$

  as was first observed on sintered MgO by Aboav (14) and is illustrated on Figure 5.

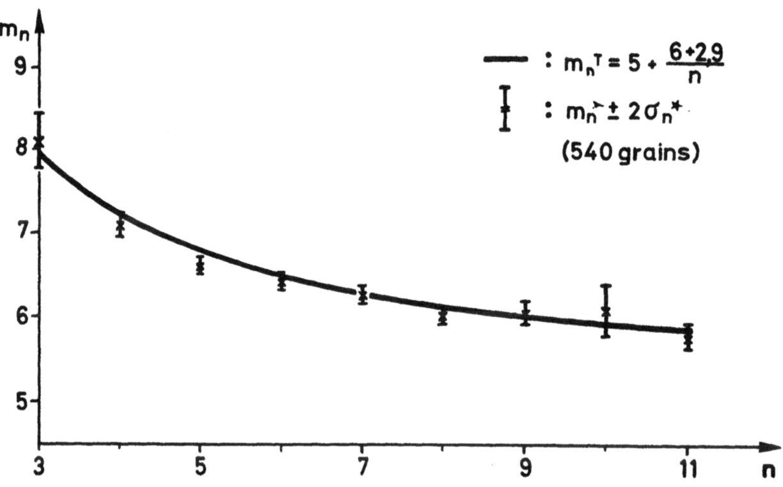

Figure 5. Aboav's law for shape correlation between neighbouring grains.

- the shape n of a grain and its size are also strongly correlated. This is known as Lewis' law for truly 2D structures (15):

$$\frac{D_n^2}{\langle D^2 \rangle} = 1 + \beta(n-6).$$

But in sections of polycrystals the correlation is experimentally found to be better represented by (Figure 6):

$$\frac{D_n}{\langle D \rangle} = 1 + \alpha(n-6)$$

with: D_n the average linear size of all n sided grains and $\langle D \rangle$ the population average; α,β are parameters which can be further specified (16) via straightforward geometrical arguments.

It is interesting to note in closing that both Aboav's and Lewis' laws have been incorporated in Rivier's attempt at describing ideal normal polycrystals as those random structures which maximize an ad-hoc configurational entropy suggested to be (11):

$$S_{conf} = - \sum_{n;i;\pi} p(n;i;\pi) \ln p(n;i;\pi)$$

where $n;i;\pi$ refer to individual grain shape, area and perimeter respectively. Rivier's is a very elegant and novel attempt at giving a unified treatment of the whole subject, and undoubtedly should trigger systematic experimental checks.

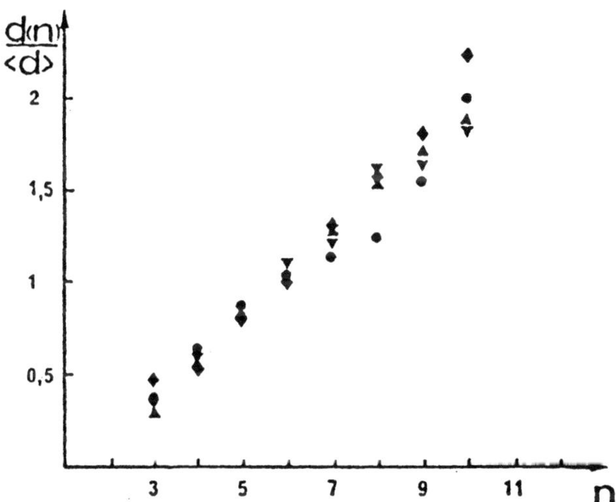

Figure 6. Linear relation between grain sizes and shapes, as observed in various heat-treated aluminas (ref. 8).

2.2. Normal grain growth theory

The modelling of polycrystal dynamics under conditions of normal grain growth raises two major issues:

- choosing a local or differential expression for the equation of grain boundary motion having physical significance
- integrating and averaging such an expression in order to obtain an overall kinetic law that can be checked experimentally.

Historically, the first theoretical treatment of grain growth is due to Burke and Turnbull (17) and their main argument goes as follows:

i) the velocity $\vec{v}$ of a non singular element of grain boundary surface is proportional to the local driving force $\vec{F}$: $\vec{v}=M\vec{F}$ where M is a mobility and $|\vec{F}|$ is proportional to the local curvature $1/R_1 + 1/R_2$. $\vec{v}$;$\vec{F}$ and the normal to the surface element are assumed to be colinear.

ii) taking first "shape averages", i.e. $v \sim \frac{dD}{dt}$ the rate of size change for the corresponding presumably spherical grain $\frac{1}{R_1} + \frac{1}{R_2} \approx 4/D$ yields $\frac{dD}{dt}$ proportional to $\frac{1}{D}$.

iii) assuming then that grain mobilities and sizes are unrelated and taking a "population average", the well-known parabolic law is obtained: $\bar{D}^2 - \bar{D}_0{}^2 = Kt$ where K is a rate constant incorporating the local mobility M.

In ceramics this law has rarely if ever been observed. Experimental time exponents more commonly fall in the 1/3 to 1/5 range. Two non exclusive explanations may be offered for such discrepancies:

- the mobility perturbating effects of impurites segregated toward grain boundaries or of second phase intergranular films or dispersoïds (including residual porosity) have been well recognized. They are discussed at length in the ceramics literature (33) and will thus not be dealt with here.
- the above quoted averaging assumptions leave much to be desired, particularly as they ignore the influence of singularities in the assumed equation of motion introduced by the presence of grain edges or corners. Considering also the nature and extent of correlations between metric and topological features just mentioned, it is not clear to which extent the overall grain growth process should be approached by purely stochastic versus deterministic arguments.

Problems therefore remain at both the local level and the ensemble behaviour.

2.2.1. Local equation of motion in 2 dimensions. Starting from the conventional definition of a driving force: $F = -\nabla E$ where $E = \gamma s$ is the total interfacial energy of the polycrystal, it may be shown (18) that $\vec{F}$ which is defined locally will have two different expressions:

$$\vec{F} \quad \vec{F}_s = \gamma \sum_{i=1}^{3} \vec{t}_i = \gamma \vec{T}$$

$$\vec{F} \quad \vec{F}_L = \gamma \frac{L}{\rho} \vec{n}$$

The meanings of t_i, $\vec{T}$, $\vec{n}$, L and ρ are made clear on Figure 7.

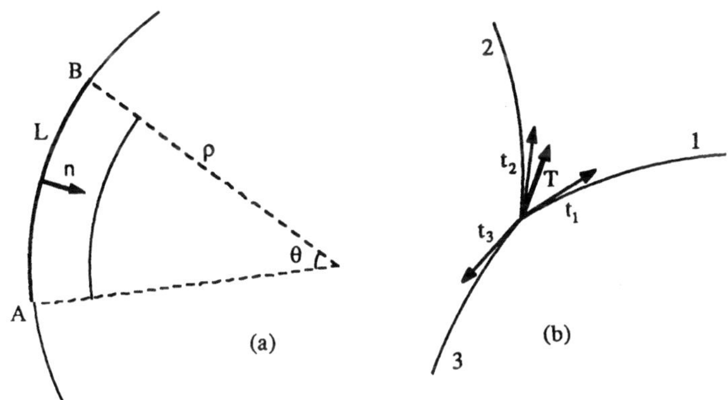

Figure 7. Local geometrical descriptions of the grain boundary network.

Equilibrium obviously requires; $F_S = F_L = 0$. But in general grain growth takes place when $F_S \neq 0$, $F_L \neq 0$ and observation suggests that deviations from equilibrium remain small such that the system may be viewed to evolve under quasi equilibrium conditions with two possibly interesting approximate situations:

a) $F_S = 0$; $F_L \neq 0$ corresponding to curved grain boundaries with equilibrated corners

b) $F_S \neq 0$; $F_L = 0$ corresponding to straight grain boundaries with non equilibrated corners.

Telley (18) has shown (Figure 8) that:

* the size changes of <u>regular</u> polygons under either aproximation are quite close to each other which justifies that to first order, either simplifying model is equally tenable.

* a general expression for the local equation of motion can be written (in 2D):

$$\dot{x} = M_\gamma \frac{-\nabla_x L}{1/2 \; L_\rho \; (\nabla_x L^{\perp})}$$

where x = characteristic geometrical feature (e.g. grain corner or grain boundary element)

L = length of boundary network related to x

$L_\rho \; (\nabla_x L^{\perp})$ = total projected length of network related to x, on the normal to $\nabla_x L$

M_γ = mobility x specific energy

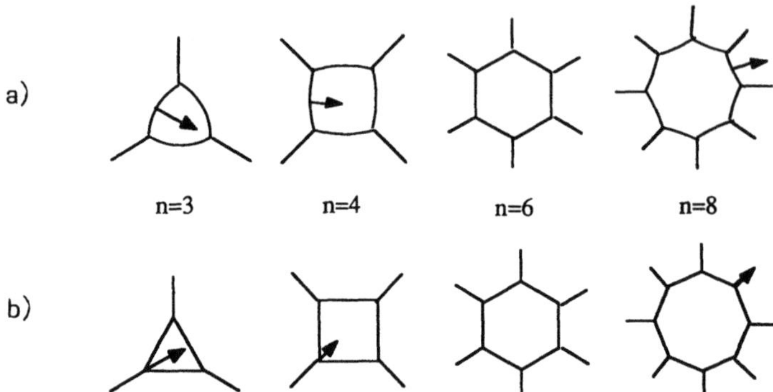

Figure 8. Schematic of regular n-sided polygons with a) $F_S = 0$; $F_L \neq 0$ and b) $F_S \neq 0$, $F_L = 0$, according to Telley (18).

It therefore appears a viable geometrical approximation, particularly in numerical simulations of grain growth, to restrict the polycrystalline network to a set of moving points: grain corners are an obvious but not compulsory choice in that respect.

2.2.2. Overal kinetics. A good synopsis of subsequent refinements brought to the original Burke and Turnbull treatment, is to be found in Atkinson's review (1). With the exception of Rhines and Craig's model (19), the geometry of which has also been questioned (20), these use a mean field approach to the grain size distribution f(R;t). In the most general case, the latter may be considered to be modified by:

* a deterministic (interface curvature related) driving force yielding a drift velocity v proportional to the force.
* a stochastic process, the physical origin of which is or need not be specified, yielding a Brownian diffusion-like motion (with coefficient D) of individual grains in phase space. Hence:

$$\frac{\partial f}{\partial t} = -\frac{\partial}{\partial R}(vf) + \frac{\partial}{\partial R}\left(D\frac{\partial f}{\partial R}\right)$$

drift term cf. Hillert (5) — diffusion term cf. Louat (7)

To this writer's knowledge, no general solution to the above equation has been offered yet. And partial solutions yield a $t^{1/2}$ dependences of grain size ($t^{1/3}$ in ref, 19), which is also a consequence of the sole self-similarity hypothesis, as rigorously established by Mullins (21) in reference to metric properties.

In conclusion then, major issues which remain open at this stage include the following. First, the stability of the grain shape distribution $\{f_n\}$ which has been conjectured (22) to be a consequence of the overall self-similarity assumption may not always be warranted by experiment and appears to deserve further scrutiny. Also deserving closer consideration is a recent suggestion (23) that deviations from the $t^{1/2}$ kinetic law may be indicative of a larger stochastic than deterministic nature of the grain growth process, and/or deviations from self-similarity. Finally, the approach to steady state and conditions for its achievement may be considerably more complex and time consuming than has so far been implicitly assumed, as may be argued from recent experimental work (24) on the dynamics of 2-D soap froths. This last question a priori seems to be well suited to be investigated with the help of computer simulations.

3. NUMERICAL SIMULATIONS OF GRAIN GROWTH

Modern large computers indeed allow complex processes to be numerically simulated with some degree of realism. In recent years some attention has been devoted to grain growth in two dimensions. Although encouraging results have been obtained with various simulation procedures, the topic is still in a quite exploratory stage, particularly because of purely numerical or statistical problems on the one hand, and modelling difficulties on the other.

A number of studies (e.g. 25) which limit themselves to only generating size distributions may provide useful insights but will not be discussed here. Another more informative approach consists of explicitly simulating 2-D networks in which structural changes can be monitored at all times. Two different levels of representation may be thought of, i.e.:

- the pseudo-atomic model in which the material is assumed to consist of a set of points on a periodic lattice. These points are given "colours" meant to represent crystalline orientations and can thus be clustered into connected cells or would be grains
- the microscopic model in which planar graphs directly mimicking the grain boundary network are constructed.

Similarly, configurational changes leading to observable grain growth can be either:

- random: given an initial state of the system, its next configuration is prepared according to some random process and effectively accepted depending upon some energy criterion, possibly also involving some randomness (e.g. Metropolis algorithm)
- deterministic: that is following equations of motion a priori set for an a priori chosen geometry.

The amplitude or extent of an iteration or elementary transition step should obviously be taken as small as is compatible with computation requirements, in order to deviate the least from continuity in the overall evolution. Obviously also, periodic boundary conditions are imposed to structures being simulated, the possible effects of which are still unknown.

In recent years all possible realistic simulation schemes have been explored as summarized in Table I and briefly discussed next.

TABLE I: Major grain growth simulation schemes.

type of motion \ representation	pseudo-atomic	microscopic
deterministic	(unrealistic)	Soares (1985) ref. 27 Frost (1988) ref.28;29
stochastic	Anderson et al. (1984) ref. 30; 31	Telley et al. (1986) ref. 32

3.1. Deterministic treatments

With the exception of Weaire and Kermode's attempt (26) who first published results obtained with a not clearly stated procedure on statistically small samples, two groups have undertaken simulations of the dynamics of 2-D polycrystals under either of the previously described idealizations of the local driving force.

3.1.1. The moving corner model. Soares et al. (27) have tackled the microscopic case under the basic assumption that grain boundaries remain straight at all times and that individual grain corners are moving with velocity v proportional to: $\vec{F}_s \sim \sum_{i=1}^{3} \vec{t}_i$ (Figure 7b) for some time interval Δt. Right after the structure is allowed to relax via an appropriate ETT (Figure 4) at those locations where two distinct grain corners have come closer to each other than some small (arbitrary) distance l_0. Figure 9 shows an example of the resulting network evolution.

This approach certainly is interesting in its simple mindedness. So far however it does not seem to have been extensively pursued. The statistical significance of the results (obtained with an initial sample of ~ 50 to 100 cells) has not been assessed. Some internal features of the model also would deserve more consideration, e.g. the choice that $l_0 = 100\ v\ \Delta t$.

3.1.2. The curved boundary model. Conversely to the previous authors, Frost et al. (28,29) have assumed local force equilibration at every grain corner and have considered curved grain boundaries which they idealized as polygonal lines (Figure 10). Such lines are moved parallel to themselves with velocity proportional to the estimated curvature. And after some given time increment the corner positions are readjusted in order to restore local equilibrium of forces. An example of the resulting structures is shown on Figure 11.

It is interesting to note here that much freedom is allowed in choosing the initial structure since it has to be "synthesized" first via some nucleation and growth mechanism before actual grain growth can get started. Since a wide range of nucleation and growth assumptions can be made a priori, then many initial structures can be devised possibly

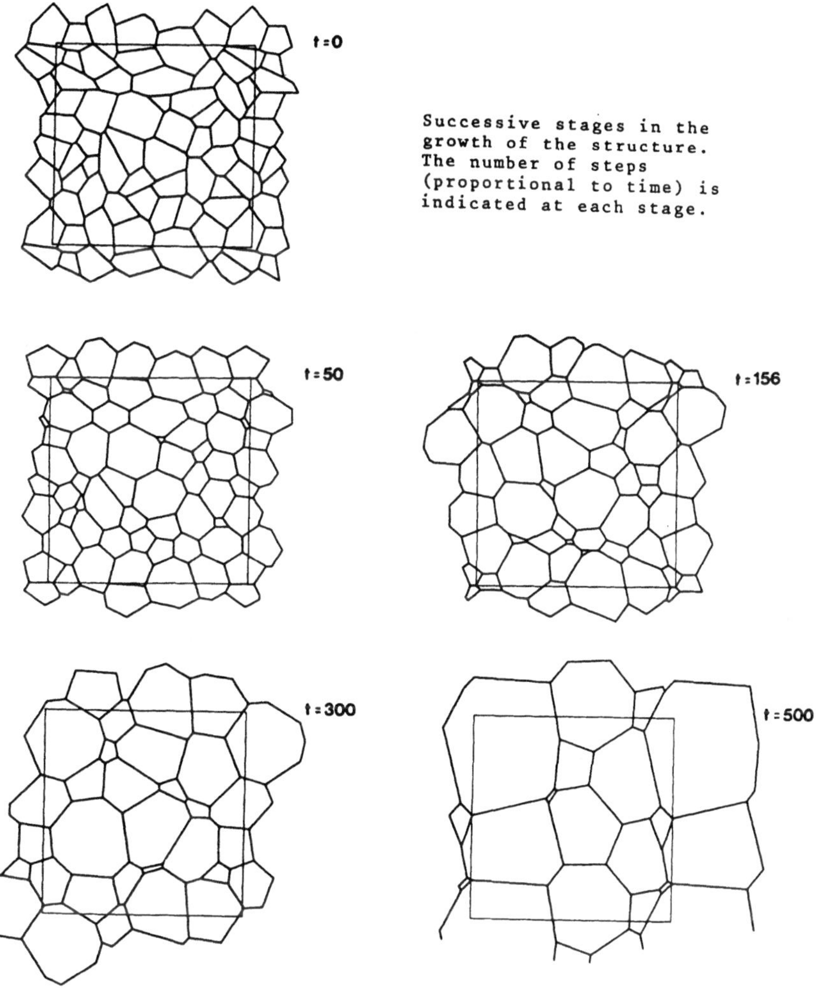

Figure 9. Growth stages in the moving corner model (ref. 27 - time in arbitrary units).

leading to rather different states after grain growth. In fact, the grain size distributions that have been obtained so far with this model do not seem to correspond to a known theoretical law. And it may also be pointed out that difficulties inherent to the present state of the model (e.g. in the handling of ETT or configurational readjustments at grain corners) may influence the overall process to an as yet unknown extent.

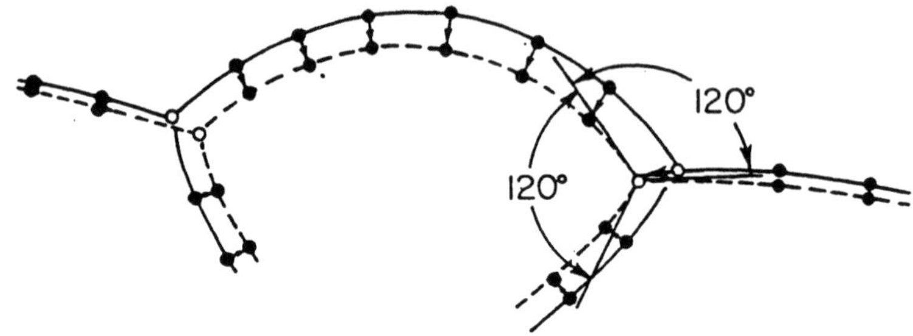

Figure 10. Schematic illustration of simulation technique for modelling boundary migration and triple point motion (ref. 28).

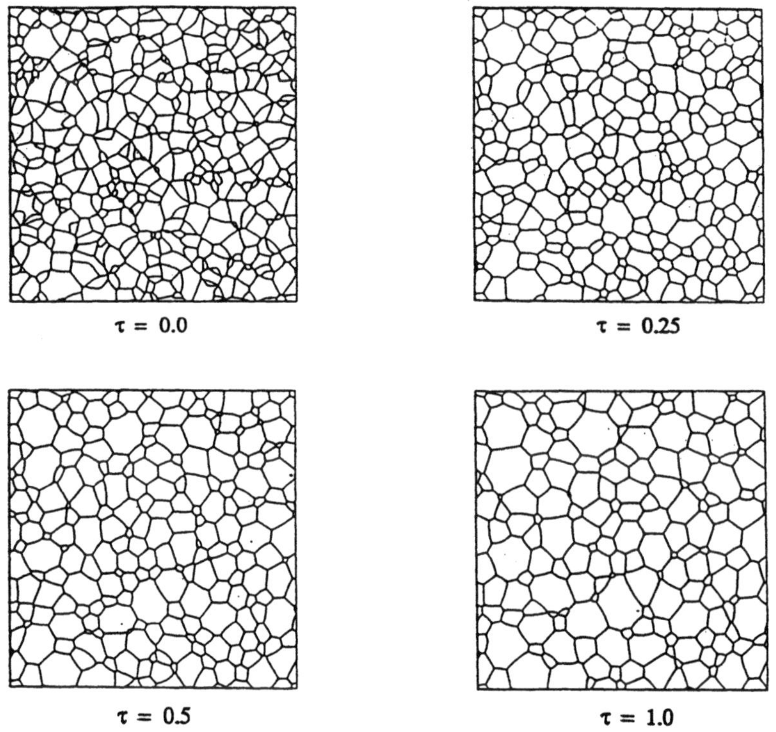

Figure 11. Evolution of the continuous nucleation structure at various normalized times. The initial structure contains approximately 400 grains (ref. 28).

3.2. Stochastic models

In this second group of simulations use is made of the Monte-Carlo procedure or rather the Metropolis algorithm, which consists in the following three essential steps:

- define a configurational energy E for the system as a whole, and set a geometrical procedure for performing structural changes
- randomly select sites for attempting an elementary transition $i \rightarrow j$ from configuration i toward j.
- accept and effect such a transition with probability 1 if $\Delta E_{ij} \leq 0$ or with probability $\exp\left[-\frac{\Delta E_{ij}}{T}\right]$ if $\Delta E_{ij} > 0$; where T is an ad-hoc temperature-like normalizing parameter.

In principle, other model-specific conditions may also be imposed in actual simulations. And two rather different cases have been considered so far.

3.2.1. The pseudo-atomic model. Anderson et al. (30) in a series of papers have carried out the most comprehensive 2-D study to date. They also provided a crude 3-D version of their model (31). These authors have adopted the pseudo-atomic level of representation in which point-like entities are interacting with their nearest neighbours only. This is an adjustment of the Potts model which itself generalizes the Ising model initially developed for describing magnetic systems. Here, elementary structure changes consist of "colour" (i.e. crystalline orientation) modifications of individual atom-like entities which affect the energies of interaction with their nearest neighbours. Figure 12 illustrates both how the geometry of polycrystalline structures is built from the pseudo-atomic model (Figure 12a) together with a sample of simulations at different stages (Figure 12b).

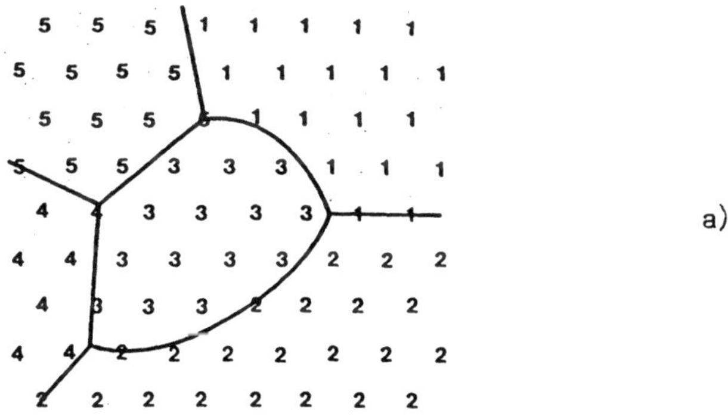

Figure 12. Sample microstructures of pseudo atomic model (ref. 30).
a) triangular lattice of "coloured" points represented by integers. Lines represent grain boundaries.

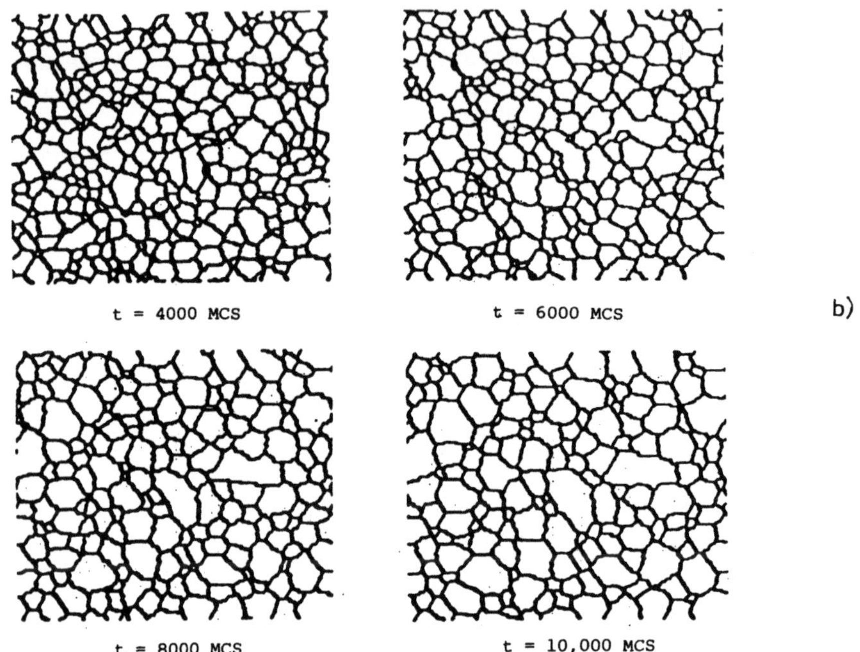

Figure 12. Sample microstructures of pseudo atomic model (ref. 30).
b) simulation results versus time (as amount of Monte Carlo steps MCS).

It is fair to say that in this fashion, a good deal of the familiar or essential geometrical features of normal grain growth could be reproduced, at least to first order. The simulated time dependence of the average cell size also was found close to being parabolic which generated interesting comments from the authors themselves. But it may also be argued that some shortcomings remain which may prove difficult to overcome. The level of representation is obviously not truly atomic: the reported simulations have considered some 40000 lattice points distributed among ~ 30 different orientation classes. The statistical significance of such limitations on the overall behaviour of the system again may deserve further consideration. Moreover, the possible bias which could result from the built-in anisotropies introduced by the discrete underlying lattices appears difficult to estimate precisely.

3.2.2. <u>The "mixed model</u>. More recently, Telley et al. (2) have attempted to devise a stochastic-microscopic model which hopefully would combine major advantages of each of the previous ones, yet retaining enough simple-mindedness in its conception. The reference system is a cellular network with straight edges but with overall energy:

$$E \sim \alpha L + \mu \sum_{\text{corners}} |\vec{T}| \qquad \text{where:}$$

L is the sum of all grain edge lengths, $\vec{T}$ is the previously defined sum of the three unit vectors along grain boundaries that meet at a given corner, α and μ are adjustable parameters.

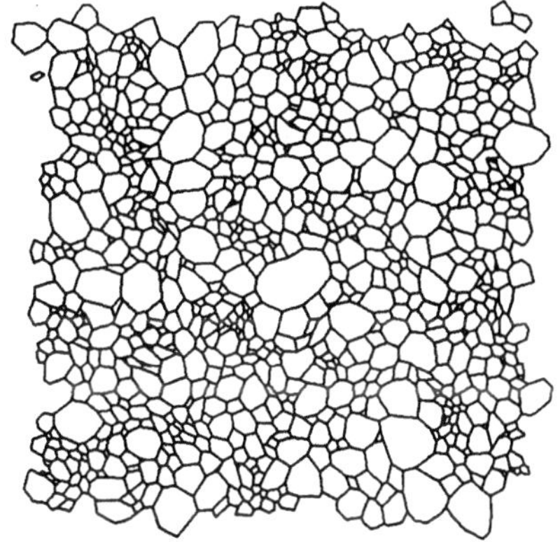

a) structure obtained
with T = 0.06
a/μ = 0.2

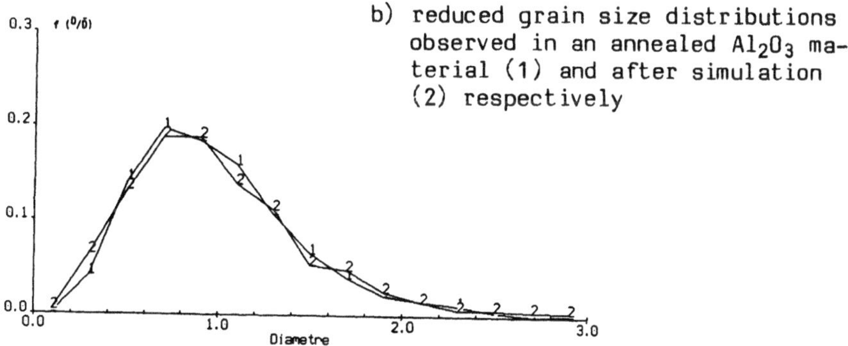

b) reduced grain size distributions observed in an annealed Al_2O_3 material (1) and after simulation (2) respectively

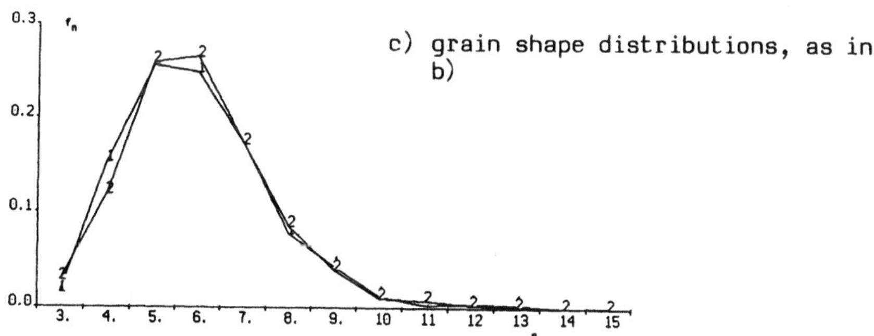

c) grain shape distributions, as in b)

Figure 13. Example of grain growth simulation with the "mixed" model (ref. 32).

The structure changes are effected via elementary topological transformations (Figure 4) subject to the Metropolis algorithm, and with a special refinement procedure to determine new positions of grain corners after an ETT has been carried out.

Initial simulation trials showed that there seems to exist an optimum α/μ ratio the physical significance of which is not readily apparent. It was also noted that relaxing the a priori constraint that every grain should remain convex throughout time also led to simulated structures looking closer to reality. As may be noted from Figures 13 and 14:

- satisfactory distributions $f(^D/\bar{D})$ and $\{f_n\}$ for grain sizes and shapes respectively can be obtained at high enough Metropolis temperatures.
- abnormal grain growth may also be observed when the simulation temperature turns out being too low, which would suggest that the abnormal growth phenomenon may be intrinsic under some circumstances and not necessarily related to anisotropy or second phase effects.

Telley's work however as well as others' is rather exploratory and major issues which remain open include the rationalization of the relationship between simulation temperature and the loss of self-similarity (abnormal grain growth). Also lacking so far is the definition of an adequate time scale so that growth kinetics may be assessed. And more generally the extent of transients which appear quite pronounced before the overall structure shows signs of having reached steady state would deserve more documentation.

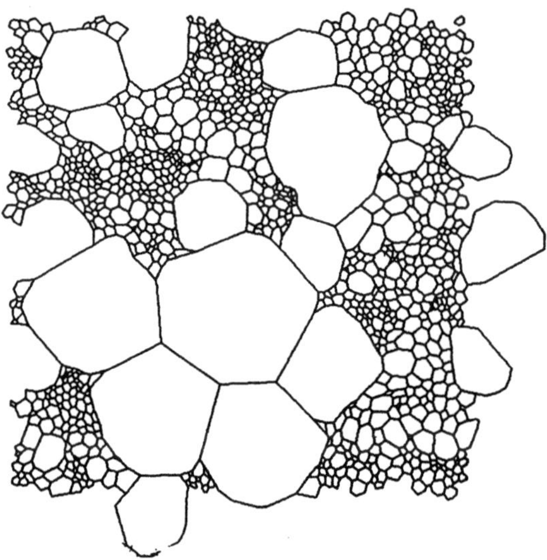

Figure 14. Example of simulated abnormal grain growth when T = 0.03; α/μ = 0.2

4. CONCLUSION: BOUNDARY MIGRATION IN CERAMICS

In contrast to more or less predictable behaviour of ideal polycrystals, grain boundary migration and grain growth in actual ceramic systems, even when presumed to be single phased, may be considerably more complex. It is not possible in a few words to give an overview of either theoretical models or experiments that have been recorded so far. It may nonetheless be pointed out that complications usually arise due to chemical effects (either residual insoluble impurities or solute elements), to crystalline anisotropies or even to externally applied fields such as stress.

Some years ago, good reviews on grain boundary migration and grain growth in ceramics were published (e.g. 33;34) focussing on effects due to residual impurities and minor amounts of dispersed phases. Such reviews are still useful as they collect experimental data which they rationalize in simple terms. A specially important case arises when impurities are condensed into intergranular, more or less connected liquid or glassy films, which is rather common in technical ceramics. Then, it has been argued (35) that such films should have an equilibrium thickness in the nm range. Therefore, it seems fair to assume that they are in equilibrium with the matrix and their motion during grain growth is effected through a solution/precipitation mechanism of the matrix species from one side of the film to the other without modification of the matrix composition.

In some other cases however, boundary films may only be transient, i.e. gradually dissolving into the matrix in more or less long periods of time. In such circumstances their migration characteristics may turn out to be rather peculiar and be equivalent to those taking place in phenomena known as Diffusion induced grain boundary migration (DIGM) or Diffusion induced recrystallization (DIR). A good comprehensive review on such processes has most recently been published (36) in which only a few ceramic systems are quoted as exhibiting this type of behaviour which in fact might have been overlooked in the past (37;38) and could be more frequent than currently realized.

Finally, with respect to crystalline anisotropies and variations of the local grain boundary mobility upon the relative misorientations of adjoining lattices, it seems worth drawing attention to a few recent observations. Local texture changes have been found to occur on a small scale during normal grain growth in some metals (39). They were claimed to be related to the proportions of boundaries that lie close to a CSL geometry, thus exhibiting different specific energies and mobilities. Similarly, different proportions of special versus general grain boundaries (40) together with some texturing (41) have been reported in some fine-grained Al_2O_3 depending upon their thermomechanical histories. And it is now known that ceramics which exhibit superplastic behaviour at small enough grain sizes (42) may also undergo concurrent grain growth at a somewhat faster rate under stress than during straight annealing. Reasons for and mechanisms of such phenomena remain largely unclear.

REFERENCES

(1) H.V. Atkinson, Acta Metall. 36 (1988) 469.
(2) P. Feltham, Acta Metall. 5 (1957) 97.
(3) S.K. Kurtz and F.M.A. Carpay, J. Appl. Phys. 51 (1980) 5745.
(4) M.F. Vaz and M.A. Fortes, Scripta Metall. 22 (1988) 35.
(5) M. Hillert, Acta Metall. 13 (1965) 227.
(6) M. Hillert, Scripta Metall. 22 (1988) 1035.
(7) N.P. Louat, Acta Metall. 22 (1974) 721.
(8) M. Blanc, Thesis; Ecole des Mines, Paris (1979).
(9) C.S. Smith, Metal Interfaces, p. 65; ASM, Cleveland, OH (1952).
(10) J. von Neumann, ibid, p. 108.
(11) N. Rivier, Phil. Mag. B52 (1985) 795.
(12) J.E. Morral and M.F. Ashby, Acta Metall. 22 (1974) 567.
(13) M. Blanc and A. Mocellin, Acta Metall. 27 (1979) 1231.
(14) A. Aboav, Metallography 3 (1970) 383.
(15) D. Lewis, Anat. Rec. 38 (1928) 351.
(16) M. Guillopé, Thesis Université de Paris VI (1981).
(17) J.E. Burke and D. Turnbull, Progr. Metal. Phys. 3 (1952) 220.
(18) H. Telley, Thesis EPFL (1989).
(19) F.N. Rhines and K.R. Craig, Metall. Trans. 5A (1974) 413.
(20) O. Hunderi, Acta Metall. 27 (1979) 167.
(21) W.W. Mullins, J. Appl. Phys. 59 (1986) 1341.
(22) E. Carnal and A. Mocellin, Acta Metall. 29 (1981) 135.
(23) K.J. Kurzydlowski and K. Tangri, Scripta Metall. 22 (1988) 785.
(24) J.A. Glazier, S.P. Gross and J. Stavans, Phys.Rev. A36 (1987) 306.
(25) V.Y. Novikov, Acta Metall. 26 (1978) 1739.
(26) D. Weaire and J.P. Kermode, Phil. Mag. B48 (1983) 245, ibid. B50 (1984) 379.
(27) A. Soares, A.C. Ferro and M.A. Fortes, Scripta Metall. 19 (1985) 1491.
(28) H.J. Frost, C.V. Thompson, C.L. Howe and J. Whang, Scripta Metall. 22 (1988) 65.
(29) H.J. Frost and C.V. Thompson, Symp. on "Microstructural science for thin film metallization in electronics applications" TMS-Annual meeting, Phoenix, Arizona, January 1988.
(30) M.P. Anderson, D.J. Srolovitz, G.S. Grest and P.S. Sahni
a) Acta Metall. 32 (1984) 783; 793; 1429
b) Acta Metall. 33 (1985) 509; 2233.
(31) M.P. Anderson, G.S. Grest and D.J. Srolovitz, Scripta Metall. 19 (1985) 225.
(32) H. Telley, Th.M. Liebling and A. Mocellin,
a) Proc. 7th Risø Int. Symp. on Metallurgy and Mat. Sci., Roskilde, Denmark (1986) pp. 573-578.
b) Acta Stereol. (1987); 6/III, pp. 337-342, Proc. ICS VII, Caen, France (1987).
(33) M.F. Yan, R.M. Cánnon and H.K. Bowen, "Ceramic Microstructures '76". R.M. Fulrath and J.A. Pask Edrs. Westview, Boulder, CO (1977) pp. 276-305.
(34) R.J. Brook, Treatise on Mater. Sci. and Tech., Vol. 9 (1976) 331.
(35) D.R. Clarke, J. Amer. Ceram. Soc. 70 (1987) 15.

(36) C. Handwerker, Diffusion induced grain boundary migration in thin films, "Diffusion phenomena in thin films". D. Gupta Edr. Noyes Publ. (1988).
(37) E. Roeder and M. Klerk, Z. Metallkunde 54 (1963) 462.
(38) a) J. Sautereau, Thesis Université de Paris VI (1974),
b) J. Sautereau and A. Mocellin, J. Mater. Sci. 9 (1974) 761.
(39) V. Randle et al., a) Acta Metall. 36 (1988) 267,
b) Proc. Roy. Soc. London, A415 (1988) 239.
(40) S. Lartigue and L. Priester, J. Amer. Ceram. Soc. 71 (1988) 430.
(41) J.D. Fridez, C. Carry and A. Mocellin, Advances in Ceramics 10 (1985) p. 720. W.D. Kingery Edr. The Amer. Ceram. Soc.
(42) C. Carry and A. Mocellin, Ceramics International 13 (1987) 89.

THE ROLE OF SURFACES IN CERAMIC PROCESSES

H. SCHUBERT
Max-Planck-Institut für Metallforschung,
Institut fur Werkstoffwissenschaften,
Pulvermetallurgisches Laboratorium
Heisenbergstr.5, 7000 Stuttgart 80, FRG

ABSTRACT. The preparation of ceramic materials involves a large number of independent steps. The path from the ores to the final product is influenced by the presence of surfaces and by their chemical and physical constitution. Since ceramic manufacturing is based on the consolidation of powder particles which contact each other via their surfaces, the control of processing requires control of the surfaces. The importance of surfaces for technical manufacturing and some associated difficulties will be reported. Powder preparation (ZrO_2), shaping processes (Si_3N_4) and thermo- electrical properties (ZrO_2) will be used as examples.

1. Introduction.

Ceramic materials have received attention because of their unique combination of chemical, thermal and mechanical properties. For 40 years, increasing technical use of electro ceramics has been observed. The dielectric properties have in particular led to application in electronics industry. The production of capacitors based on $BaTiO_3$ is now a mature commercial enterprise. Attractive mechanical properties are observed as a consequence of strong covalent bonding, which are dominating in materials such as Si_3N_4, SiC, BC_4 and SiB_6. High hardness, high Young's modulus, good wear resistance, corrosion resistance and outstanding high temperature properties make ceramics candidates for a wide range of structural applications. Great effort has been expended to develop ceramic parts for different mechanical applications and several areas have been sucessfully developed such as Al_2O_3-ZrO_2 cutting tools, Al_2O_3 for biomedical applications (hip joints), Al_2O_3 ball valves, corrosion resistant SiC in chemical apparatus and Si_3N_4 for high temperature turbocharge rotors and valve rings etc. (Fig. 1).

In spite of these promising results, engine designers and constructors have often been reluctant to use ceramics, an attitude caused by the sensitivity of these materials to stress concentrations around any manufacturing flaws. Unfortunately, there is no way to reduce such stress concentrations in a ceramic component other than by the formation of cracks. In metals, the plasticity brought about by the movement of dislocations, allows an adjustment of the material to the applied stress and leads to energy dissipation. In ceramics, such a dislocation movement is difficult and

L.-C. Dufour et al. (eds.), Surfaces and Interfaces of Ceramic Materials, 507–519.

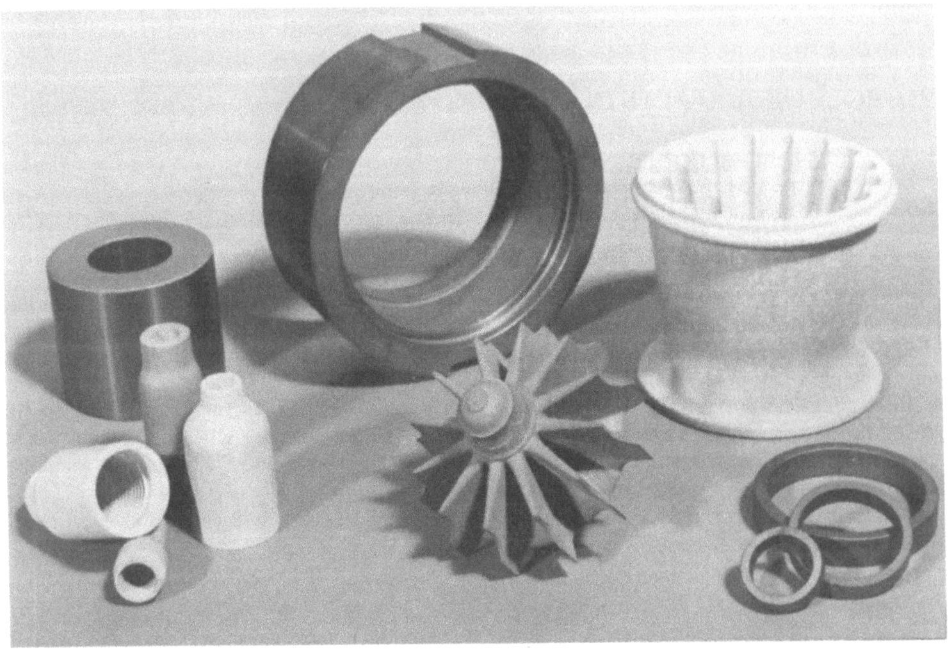

Fig. 1. Various ceramic parts for use under mechanical, thermal and corrosive load.

Fig. 2 High damage tolerance due to transformation toughening. A flat TPZ ceramic plate (Tetragonal Zirconia Polycrystals), ZrO_2 - 3 mole% Y_2O_3) can be surface ground to a thickness of only 80 μm without the formation of strength reducing surface cracks. This ceramic stripe can be strongly elastically bent without cracking.

PROCESSING FLOW DIAGRAM	CONSTITUTION	TYPE OF INTERFACE
$ZrOCl_2$ + YCl_3 solution	solution	
↓		
precipitation in ammonia	colloid/gel	solid-liquid
↓		
washing with H_20	colloid/gel	solid-liquid
↓		
washing with alcohol	colloid/gel	
↓		
drying	dry gel	solid-liquid-gas
↓		
calcination	powder	solid-gas
↓		
milling	suspension	solid-gas-liquid
↓		
spray drying	powder	solid-gas
------	------	------
milling	suspension	solid-liquid
(aggregate removal)		
↓		
drying	powder	solid-gas
↓		
milling/dispersion	suspension	solid-liquid
↓		
casting	compact	solid-liquid
↓		
high humidity drying	compact	solid-liquid-gas
↓		
sintering	solid state	solid-gas

Fig. 3: Schematic flow diagram for the preparation of ZrO_2 ceramics

therefore energy dissipation due to plasticity does not occur. As a consequence, ceramists have tried to develop transformation toughening /1,2,3,4/ and whisker/fibre reinforcement /4,5/ by which the critical stress intensity factor (frac- ture toughness) has been raised from 2-3 MPa $(m)^{1/2}$ up to values of 10 MPa $(m)^{1/2}$ and higher (Fig. 2). Nevertheless superalloys show

K_{IC} values of over 100 MPa $(m)^{1/2}$ and remain much more damage tolerant. The full use of ceramics is therefore dependent on the development of production routes which are capable of leading to microstructures with extremely small processing flaws and on the translation of such processes into commercial practice.

The processing of ceramics basically involves the manufacturing of a powder with suitable composition and morphology followed by a subsequent compaction or shaping process in which the powder particles are brought into close contact. During the following sintering process, atom diffusion in the contact areas between the particles results in bonding and shrinkage with the centres of the particles approaching each other. The result is a dense body with only small porosity.

During this process, the nature of the particle interfaces changes dramatically. Fig. 3 shows a flow diagram for the manufacturing of ZrO_2-Y_2O_3 ceramics by the coprecipitation of a powder and by subsequent compaction of the powder particles by either isostatic pressing or slip casting. The first part of this process is the manufacturing route for ZrO_2, which is representative of the methods used for the majority of commercially available powders. For slip casting, any agglomerates have to be destroyed primarily by an attrition process (step 7); a ball milling treatment ensures that the deflocculent is distributed (step 11). This particular process requires four drying processes, i.e. four changes from a gaseous to a liquid environment. This illustrates the difficulties associated with control of the surface composition during processing.

2. Powder Preparation

Conventionally, solid solution or compound powders have been prepared by mixing powders of the required components and by reacting them in a high temperature calcination. Moderately homogeneous powders have been achieved, but have tended to coarsen throughout the calcination. Wet chemical processing routes for such compound powders have become common place in industry within the last ten years. The chemical homogeneity of the product is already ensured by the perfect distribution of the cation within the precursor solutions. For oxide preparations, the precursors are converted into hydroxides by ammonia addition (co-precipitation) or by hydrolysis (alkoxides). These hydroxides form colloidal or polymeric gels, respectively. In order to remove the precursor anions, the precipitated gel has to be washed intensively. The removal of the anions which are bonded to the hydroxide surface is a key process because the presence of residual anions changes the surface energy of the product and leads to agglomeration. Different washing techniques such as acetone-toluene-acetone cycles or distillation with highly concentrated alcohols have been developed.

The decomposition of the hydroxides occurs during calcination at temperatures typically over 400°C, leading to oxides with small crystallites. Fig. 4 shows a TEM micrograph of a ZrO_2-3 mol.% Y_2O_3 material calcined at 450°C. The crystallite size is only 5 nm. With the unit cell of ZrO_2 being approximately 0.5 nm, it has to be recognized that such a crystallite has just 10 unit cells along its diameter! Assuming that the structure of a material is distorted to a depth of some 3 monolayers and that this region may have a different composition, then 90 % of the crystallites belongs to the direct near surface region. The behavior of a powder composed of such fine crystallites is expected to be quite different from a conventional coarse powder. Enhanced diffusion in the surface region and the high surface area can, for example, result in a very high sinter activity. While coarser conventional ZrO_2 powders require temperatures in excess of 1400°C for densification, the sinter temperature can be reduced to 1100-1200°C when fine crystallite sized powders are used /6/. However, it is necessary to keep the crystallites separate throughout the preparation i.e. to prevent any aggregation of these fine crystallites; otherwise, fast interagglomerate sintering takes place and coarsens the powder resulting in reduced sinter activity and in inhomogeneous microstructures.

3. Drying of Powders

The drying procedure is another key step in the processing route because it can result in the formation of agglomerates, which then play a central role in the subsequent ceramic processing. Such agglomerates are a form of morphological superstructure and powders with even nm sized crystallites behave like coarse powders when the individual crystallites are interconnected as agglomerates. The agglomerates do, however, make the powders easier to handle, which is beneficial to technical manufacturing. The required compromise is to achieve an agglomerate structure which can be destroyed by further processing so that the sinter activity of the powder is not diminished. Unfortunately the bondings between the particles may be very strong and relics of agglomerates are often left in the green (unfired) bodies. Such geometric inhomogeneities cause inhomogeneous sintering with resulting flaw formation* /7,8/.

In the final stage of drying, the liquid is located between the particles, in the form of bridges. Such a configuration is described by the two radii of curvature of the liquid surface (Fig. 4), by the contact angle and by the surface tension. In analogy to liquid phase sintering, there will be a capillary force

$$F = 2\pi\gamma R_l \sin\Psi_1 \sin(\Psi_1 + \theta) + \pi\gamma R^2_1 \sin^2\Psi \; [\, 1/\rho_2 - 1/\rho_1 \,]$$

---

* It should be noted that the word agglomerate is used in the ceramic literature also to describe aggregates, which have similar structures to agglomerates but which are not redispersible

attracting the particles to one another /9/ (the terminology is given in Fig. 5). The attractive effect is obviously strongly influenced by the nature of theliquid (Fig. 5). Nonpolar liquids (n-hexane, toluene, 2-butanol) have low dielectric constants and low surface tension and would be preferred for

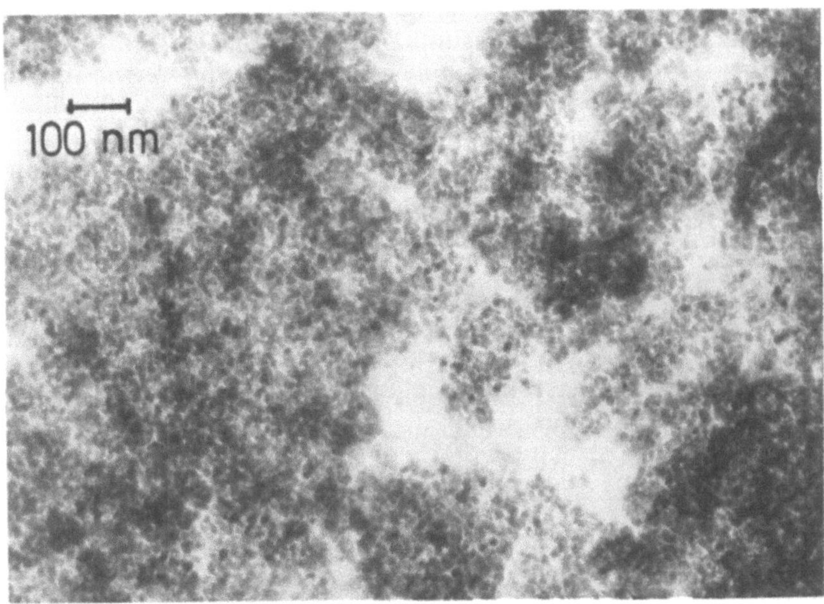

Fig. 4 TEM micrograph of precipitated ZrO2 powder. The crystallite diameter of 5nm is only 10 times the unit cell. Assuming that a layer of 3 unit cell depth under a free surface is distorted demonstrates that approximately 94% of the crystallite volume belongs to the surface near region.

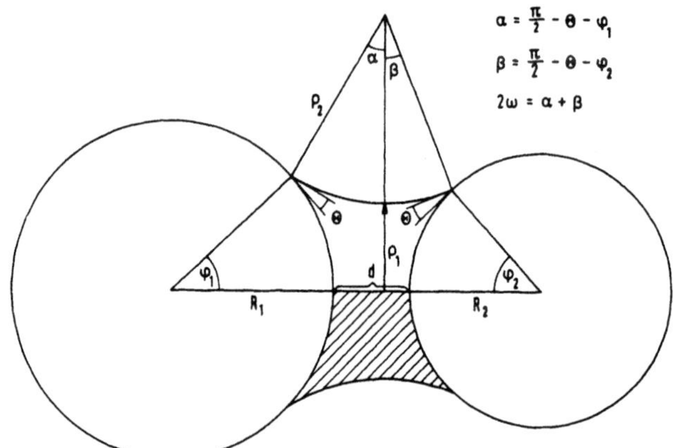

Fig. 5 Cross section through the contact between two particles by a liquid bridge after Somleij and Peovnik

ceramic processing but they are expensive and polluting. On the other hand, water is cheap but it has the highest surface tension and the highest dielectric constant.

Figure 6a shows a ZrO_2 powder dried with n-hexane at 20°C under vacuum, with very loose agglomerates. Increasing the temperature to 70°C creates denser and harder agglomerates (Fig. 6b). Hence, the use of a polar liquid with high surface tension like water (Fig. 6c) leads to very hard agglomerates which are unsuitable for ceramic process.

4. Shaping

The powders are formed into green bodies by either isostatic pressing*, injection molding or by wet forming techniques such as slip casting, pressure slip casting or sedimentation. The wetting techniques which involve a wet environment require knowledge of the powder particle/liquid interfaces, and control and manipulation of the surface chemistry.

Before the actual consolidation takes place, the particles must remain separate, which is realized by the use of repulsive interparticle forces. This can easily be established by means of electrostatic charges.

The surfaces of a ceramic powder exposed to a polar liquid or an electrolyte will be covered by a layer of adherent dipoles forming a double layer /10/. The charges of the outermost ion layer are compensated by countercharges on the dipole; at larger distan- ces from the surface, the dipoles are more and more statistically oriented, i.e. the potential falls off from the high value at the surface. The potential at a specific distance from the surface shear plane is used for characterization and is called the zeta potential /11/. In practical terms, the zeta potential depends on the chemical composition of the surface and on the pH value.

Oxygen ions typically form the outer surface of an oxide and the zeta potential/pH-curve is a relatively reliable form of characterization for oxides. In nitrides or carbides, which are subject to time dependent surface oxidation, the zeta potential curves are strongly dependent on the history of the samples. Figure 7 shows a number of zeta potential/pH curves for a Si_3N_4 powder as received and as subsequently oxidized. The virgin surface of Si_3N_4 is poor in oxygen and the zero point of charge is found at pH 8,3. With increasing oxidation, the zeta potential curves are shifted and the zero point of charge is found in the range of pH 4. The characterization of commercial Si_3N_4 powders is difficult using such a method because the surface oxygen content is a function of the actual manufacturing process and of the thermal

---

* In commercial production, pressing or dry bag isopressing techniques (non ideal stress distribution) are commonly used because these processes are faster and cheaper.

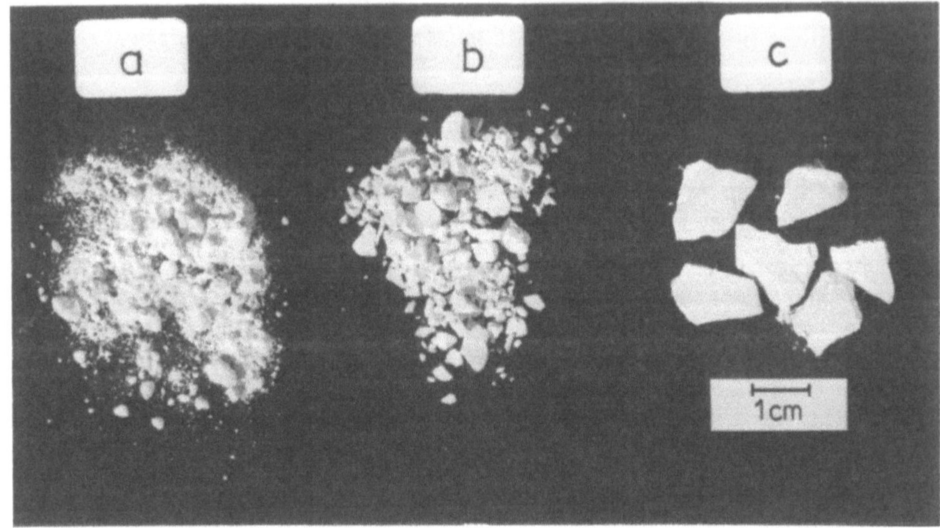

Fig. 6 The comparison of the effect of the dispersing liquid on the agglomerate formation of ZrO_2 powders. (a) n-hexane, vacuum drying at 20 °C; (b) n-hexane, dry box at 70 °C; (c) water, dry box at 70 °C

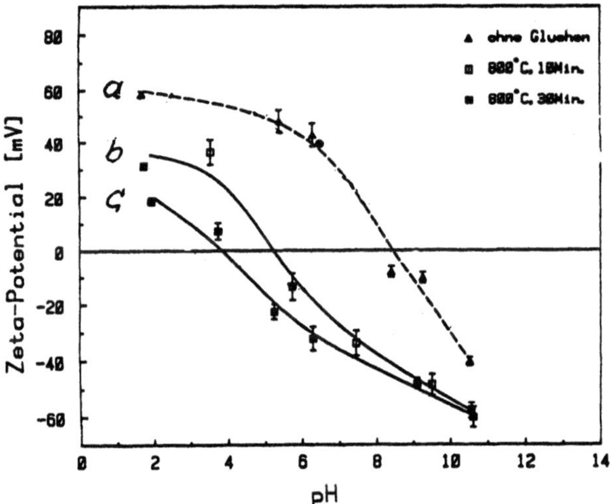

Fig. 7 Zeta potential curve for Si_3N_4 powder (Denka Corporation, Japan). (a) as received; (b) oxidized in air at 800 °C for 10 min; (c) oxidized in air at 800 °C for 30 min.

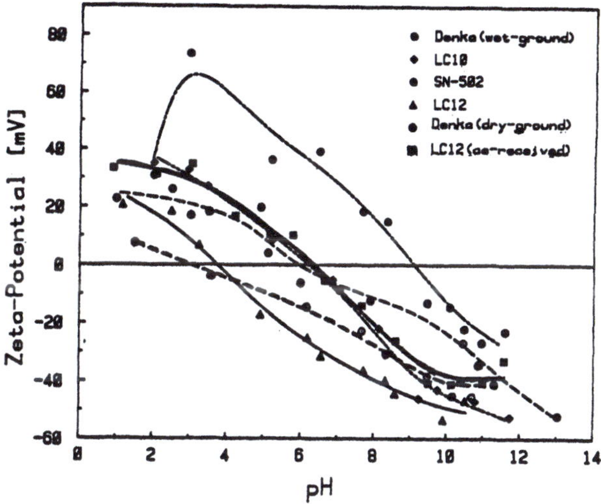

Fig. 8. Zeta potential curve for various Si_3N_4 powders in their as received state

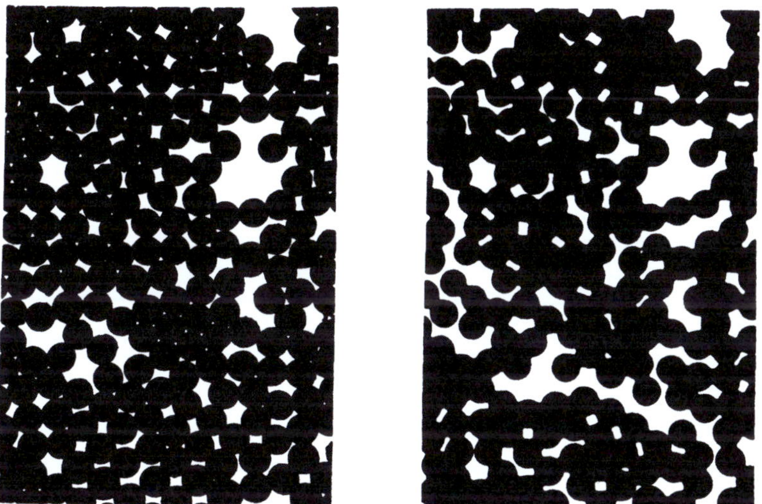

Fig. 9 Sintering experiment with a two dimensional array of Cu covered W spheres after ref. /12/. Packaging inhomogeneities increase in size during sintering

history of the material. Figure 8 shows a selection of data for several commercial Si_3N_4 powders.

The stabilisation of suspensions can be alternatively achieved by the use of steric molecules attached to the ceramic surface. These long chain molecules prevent the direct contact of the powder surfaces. A combination of steric and electrostatic stabilisation is thought about when the chain molecules are also charged on their outer ends (polyelectrolytes).

5. Densification

During the sinter process, bonding between the particles results in a body of almost full density. The role of the grain boundaries has been the subject of 20 years work from various groups. In this volume, the role of boundaries in sintering is addressed by R.J. Brook.

A model experiment by Huppmann /12/ underlines an important problem in densification (Fig. 9). It illustrates that even small variations in the density within a green compact can lead to the formation of larger voids in the final body; inhomogeneities are intensified during densification. Consequently, the aim of any ceramic processing route is to achieve a green

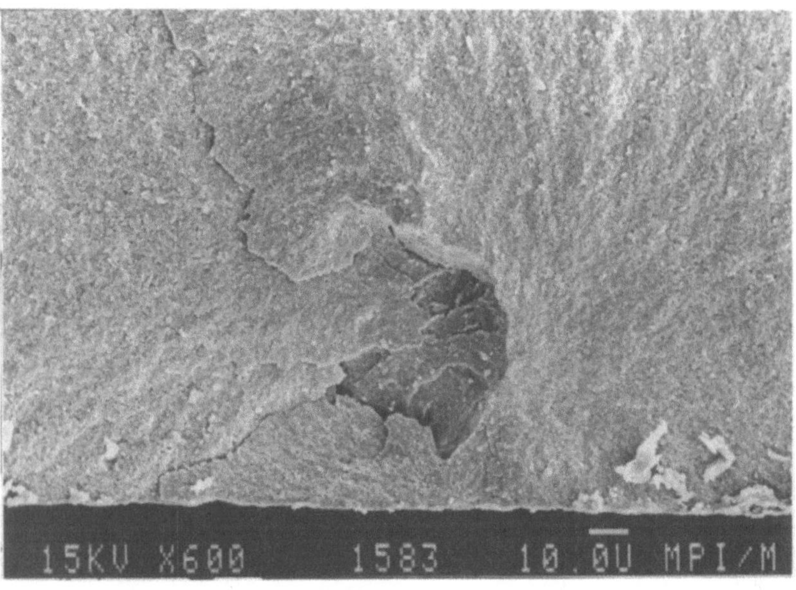

Fig. 10 The sintering flaw was fracture controlling for a TZP sample

body with as homogeneous a structure as possible. Figure 10 shows a typical flaw in a ZrO_2-Y_2O_3 ceramic; this is the strength controlling inhomogeneity in this ceramic. The limitation of the flaw size is a prerequisite for the manufacture of reliable ceramics.

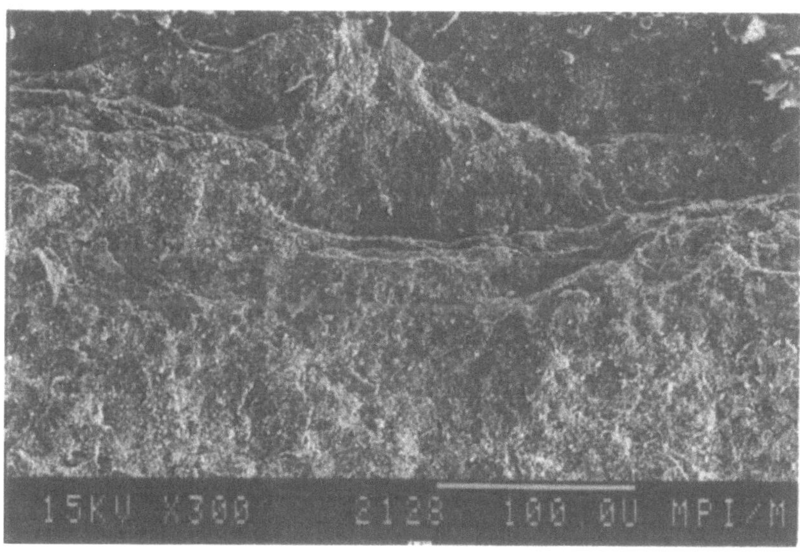

Fig. 11 A surface crack strength controlling in the Si_3N_4 samplesensitivity of ceramics to stress concentrations, cracks orthogonal to the sample surface can then occur (Fig. 11).

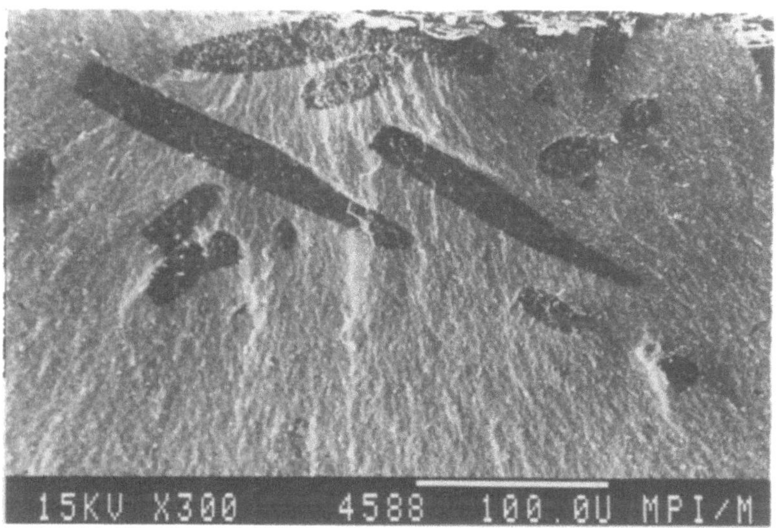

Fig. 12 Al_2O_3 fibre inclusions are strength controlling in ZrO_2-10 vol% Al_2O_3 fibre composites

6. Sample surfaces

For mechanical applications, the term 'surface region' should be defined as an area at least large enough to include the size of the critical flaw (~100 μm). During service, the surfaces of ceramics have to bear a higher stress than the bulk of the part; hence, sinter flaws, inclusions or surface cracks may be the origin of the fracture. Surface treatments such as grinding, cutting, lapping or polishing have to be seen in the context of the formation of surface cracks. Final surface treatment is accompanied with a stressing of the sample surface; because of the sensitivity of ceramics to stress concentrations, cracks orthogonal to the sample surface can then occur (Fig. 11)

When such a crack-containing surface is subject to stresses in service around the crack tip, stress concentrations develop and the part will finally fracture from one of these surface cracks. Especially when large second phase particles with a different thermal expansion coefficient from the matrix are present in the surface region, the formation of cracks occurs preferentially around the second phase. The superposition of the misfit stress- field with the applied grinding stress triggers surface cracking. Fig. 13 shows a surface crack around Al_20_3 fibres (low thermal expansion coefficient $\alpha = 8.6\ MK^{-1}$ in a ZrO_2 matrix ($\alpha = 10.7$). This ZrO_2 matrix is identical to that of Fig. 12. Without the Al_20_3 inclusions, sintering flaws would be strength controlling.

7. Summary

The preparation of ceramics is a process involving many steps. It must be recognized that a 1 cm^3 ceramic sample is prepared by contacting typically 10^{15} particle interfaces. The nature of the interface between the material being prepared and the environment changes several times.

During processing, the surface chemistry must be controlled and manipulated. An understanding and characterization of ceramic powder particle surfaces is necessary for a successful manufacturing of reliable ceramic components. The characterization of the various ceramic-liquid interfaces is, in particular, a field where many open questions remain to be answered by research in the future.

8. References

/1/ A.H. Heuer, W.M. Krieven, M. Rühle, and N. Claussen, "On the Stability of the Tetragonal Phase in Ceramic Matrices", J. Am. Ceram. Soc., **65** (1982) 642-50.

/2/ N. Claussen, "Microstructural Design of Zirconia Toughened Ceramics in Science and Technology of Zirconia II, Advances in Ceramics 1, ed. by

/3/ M. Rühle and A.H. Heuer, "Phase Transformations in ZrO2 Containing Ceramics II, The Martensitic Reaction in t-ZrO2", ibid /2/.

/4/ N. Claussen, K.-L. Weisskopf and M. Rühle, "Tetragonal Zirconia Polycrystals Reinforced with SiC Whiskers", J. Am. Ceram. Soc., **69** (1986)195-197.

/5/ G.C. Wei and P.F. Becher, "Development of SiC Whisker Reinforced Ceramics" Am. Ceram. Soc. Bull., **64** (1985) 298-304.

/6/ M.A.C.G. van de Graaf and A.J. Burggraaf, "Wet-Chemical Preparation of Zirconia Powders: Their Microstructure and Behaviour" ibid /2/.

/7/ F.F. Lange, "Sinterability of Agglomerated Powders" J. Am. Ceram. Soc., **67** (1984) 83-89.

/8/ M.A.C.G. van de Graaf, K. Keizer, and A.J. Burggraaf, "Influence of Agglomerate Structures in Ultra-Fine Substituted Zirconia Powders on Compaction and Sintering Behaviour", Science of Ceramics, **10** (1980) 83-92.

/9/ V. Smoleij and S. Pejovnik, "Some Remarks on the Driving Force for Liquid-Phase Sintering", Zeitschrift fur Metallkunde, **67** (1976) 603-605.

/10/ P.C. Hiemenz, "Principles of Colloid and Surface Chemistry", Marcel Dekker, Inc., New York, 1977.

/11/ R.J. Hunter, "Zeta Potential in Colloid Science, Principles and Application", Academic Press, London, 1981.

/12/ G. Petzow and W.J. Huppmann, "Flussigphasensintern", Zeitschrift für Metallkunde, **67** (1976) 579-590.

SOME ASPECTS OF THE INFLUENCE OF PARTICLE SIZE ON PROPERTIES AND BEHAVIOUR OF A DIELECTRIC MATERIAL : EXAMPLE OF BARIUM TITANATE

J.C. NIEPCE

(*) *Réactivité des Solides, UA 23 CNRS, Université de Bourgogne, BP 138, 21004 Dijon Cédex, France*

(+) *THOMSON-CSF, Avenue du Colonel Prat, Saint-Apollinaire, France*

The presented results are due to the contributions of : F. Batllo, A. Beauger+, B. Clerval*, S. Malbe*, J.C. Mutin* and J.C. Niepce*+.*

ABSTRACT. Anybody agrees with the idea that particle size plays an important role in the behaviour of a polycrystalline material but very few studies are giving precise information about this role. In this paper some experimental results are presented succintly to illustrate the importance of this physico-chemical factor. They are related to barium titanate, the main component of the dielectric material in the type II ceramic capacitors.

Particle size is playing a physical role. For example, the crystalline structure of the $BaTiO_3$ grains and, therefore, the dielectric properties of this material, are directly dependent on the grain size distribution.

However, particle size is also playing a chemical role. The reactivity of $BaTiO_3$ and, consequently, the dielectric properties of the resulting material can be apparently modified according to the grain size distribution.

These results prove a careful control of $BaTiO_3$ granulometry is needed. The chemical routes enable such a control. This control can be also obtained by the more traditional solid way synthesis. As an example, it is showed how granulometry of $BaTiO_3$ can be controlled by the granulometry of TiO_2, one of the two raw materials in this solid state synthesis.

INTRODUCTION

Barium titanate, $BaTiO_3$, is the main component of the dielectric material of the type II ceramic capacitors. Properties and behaviour of $BaTiO_3$ in the dielectric material are strongly related to the grain sizes of the initial $BaTiO_3$ powders. The present paper is

L.-C. Dufour et al. (eds.), Surfaces and Interfaces of Ceramic Materials, 521–533.

dedicated to this important problem.

1. CRYSTALLINE STRUCTURE OF $BaTiO_3$ IN RELATION WITH GRAIN SIZE

1.1. $BaTiO_3$ crystalline structure

The basic crystalline structure is the perovskite structure which is that of $BaTiO_3$ above 130°C (figure 1a). Between 0°C and 130°C $BaTiO_3$ has a distorted perovskite structure with an eccentricity of the titanium ions (figure 1b). Below the transition temperature, the atoms are displaced from their original symmetric positions, along one of the fourthfold axis. The distortion of the cell is due to the displacement of the Ti and O ions in opposite directions and, thus, two kinds of oxygen ions, O_1 and O_2, can be distinguished.

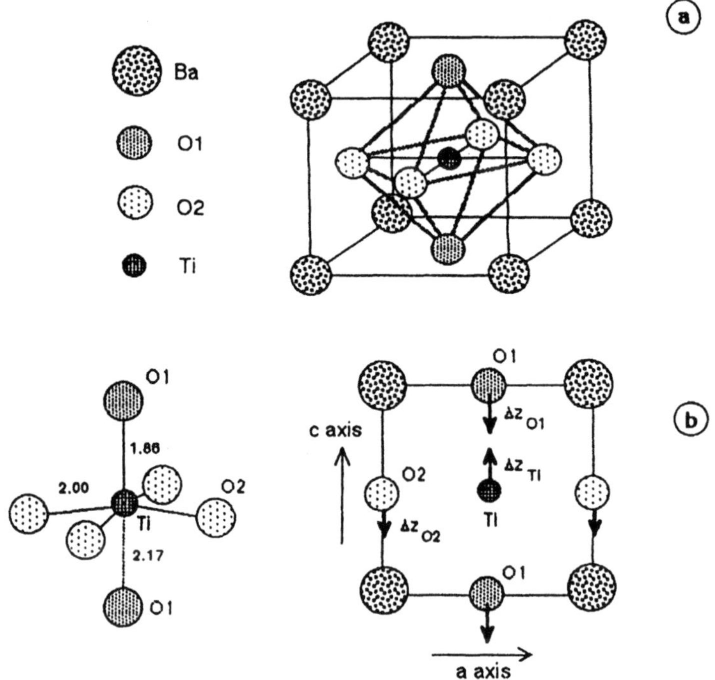

Figure 1. $BaTiO_3$ crystalline structure.
a) Cubic cell of ideal perovskite structure of $BaTiO_3$ above 130°C.
b) Tetragonal $BaTiO_3$ with the distortion of the TiO_6 octahedron and a schematic projection on (010). The oxygen O_2, superimposed upon Ti, is omitted.

The change in relative ion positions is associated with a slight deformation of the unit cell ; the cell becomes tetragonal with $a_T = b_T < a_c$ and $c_T > a_c$ (figures 1 and 2). The space group, P4 mm, is a noncentrosymmetrical group.

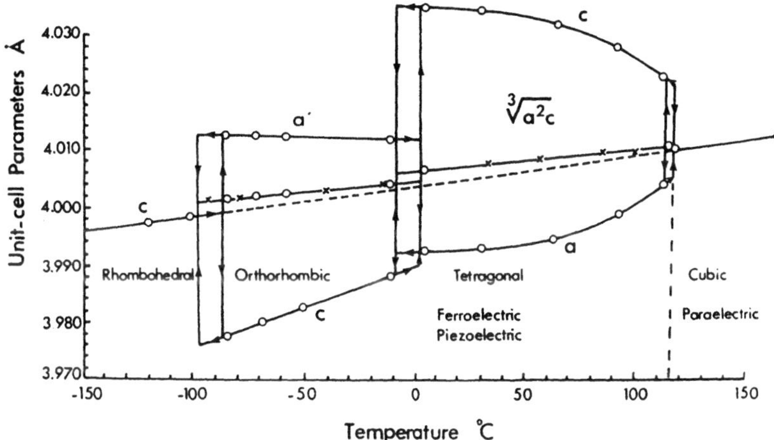

Figure 2. Lattice parameters, of $BaTiO_3$, as a function of temperature (1).

In addition, figure 2 exhibits a particular crystalline property of the tetragonal phase : the linear expansion coefficient along the c axis is negative although, usually, an expansion coefficient is positive. This reveals another particularity of the tetragonal-cubic phase transition of $BaTiO_3$ which becomes less and less tetragonal as the temperature increases up to the transition temperature.

These structural changes of $BaTiO_3$ associated with this phase transition originate the main physical uses and, particularly, the use of $BaTiO_3$ as dielectric material.

1.2. $BaTiO_3$ cell parameters in relation with grain sizes

The above information are given for large single crystalline grains of $BaTiO_3$ in which the approximation of infinite crystal is quite true.

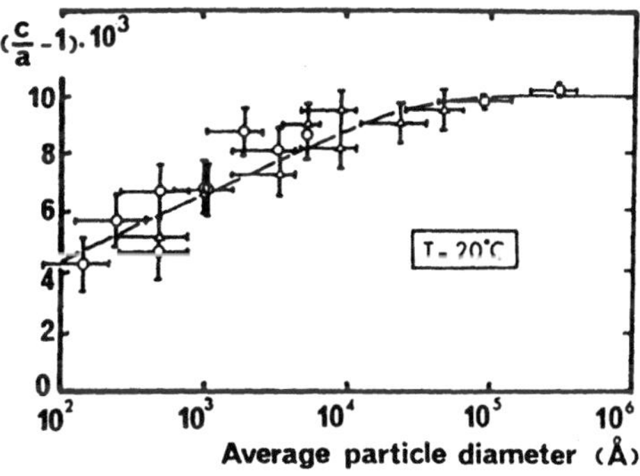

Figure 3. Tetragonal deformation of $BaTiO_3$ as a function of particle size (2).

When $BaTiO_3$ particle sizes decrease below few micrometers, the cell parameters of the tetragonal phase become a function of the powder granulometry : **the cell becomes less and less tetragonal as the particle size of the powders decreases** (figure 3)(2,3,4).

A recent and careful study shows a tight relation between the grain size distribution of the powder and the distributions of a and c parameters in two $BaTiO_3$ samples (5)(6)(Figure 4).

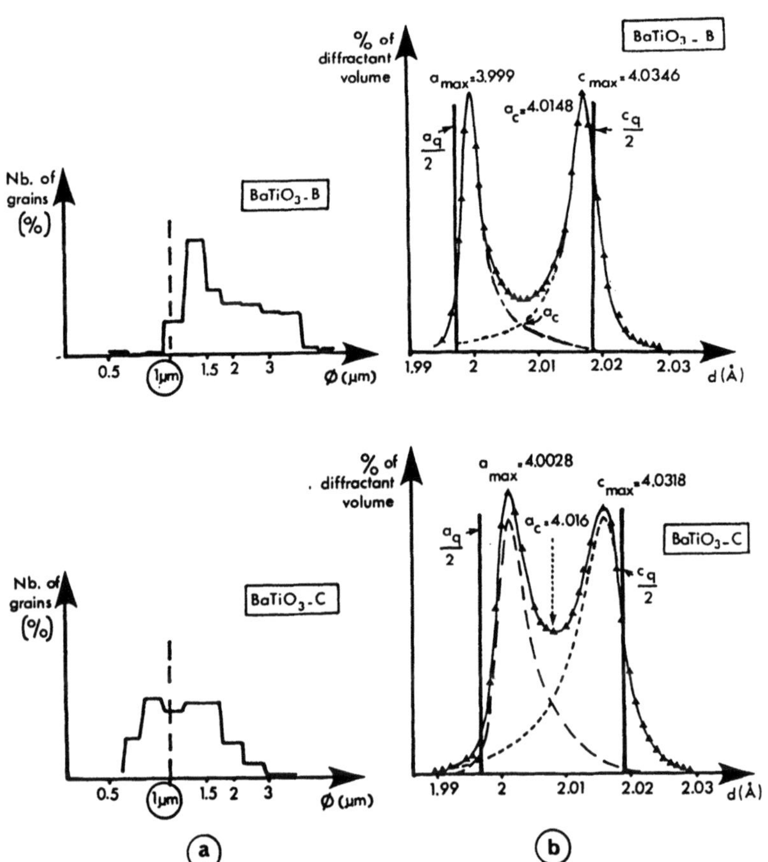

Figure 4. Relation between grain size distributions (a) and cell parameters distributions (b) for two $BaTiO_3$ samples.

For b) — — — — — distribution of a

- - - - - - - - - distribution of c

—▲—▲—▲— distribution of d_{hkl} around d_{200} and d_{002}

The two bold vertical lines give the true parameters of the $BaTiO_3$ tetragonal cell at room temperature, a_q = 3.992Å and c_q = 4.035 Å (cf. Figure 2).

1.3. Cell parameters inside a $BaTiO_3$ single crystalline grain

The grain size effect seems to be due to a surface effect : close to the surface of a grain, $BaTiO_3$ is apparently cubic at room temperature and tetragonal only in the bulk. Between the bulk and the surface there would be a **"gradient of tetragonality"** which has been called a **"tetragonal anisotropy defect"**(6)(figure 5).

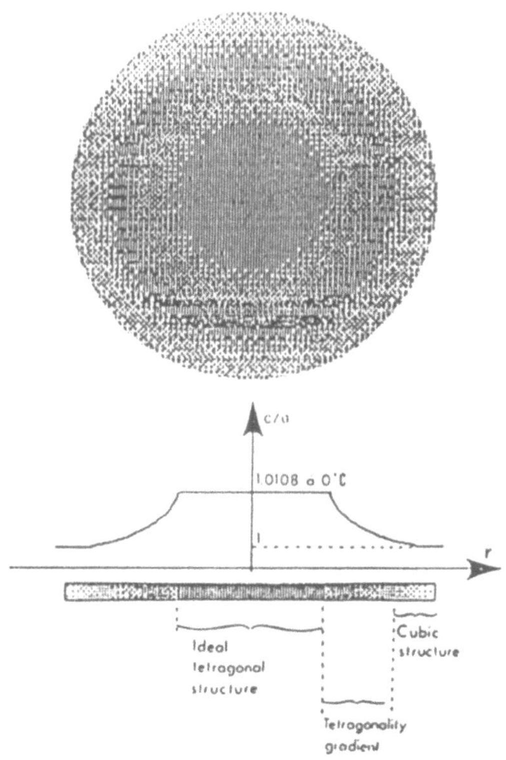

Figure 5. The tetragonal anisotropy defect in $BaTiO_3$ grain according to (6).

1.4. Tetragonal-cubic phase transition in $BaTiO_3$ as a function of grain size

The above information seems to evidence less and less difference between the two considered crystalline structures of $BaTiO_3$ as the grain sizes are decreasing. Thus, the phase transition between these two crystalline states of $BaTiO_3$ would be affected by the grain sizes. Effectively, the smaller the mean grain diameter, the lower the heat of transition (figure 6)(7)(8) although the transition temperature seems to be independent of the particle size (9)(figure 8).

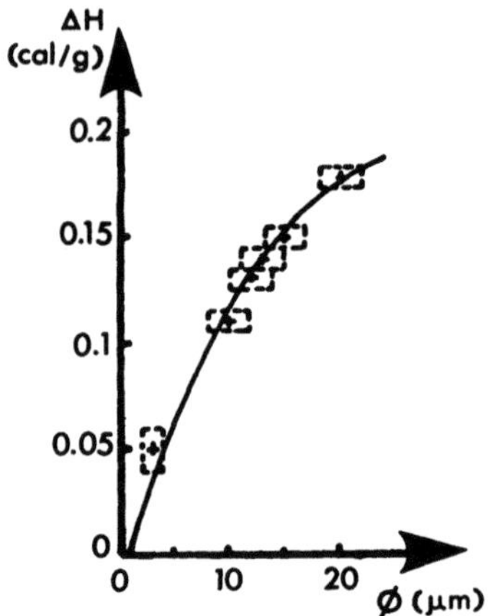

Figure 6. Heat of the tetragonal-cubic transition of $BaTiO_3$ versus mean grain diameter (8).

In barium titanate, **the phase transition,** between the tetragonal distorted perovskite structure and the true perovskite structure, **seems to vanish when the particle sizes become sufficiently small.**

2. DIELECTRIC PROPERTIES OF $BaTiO_3$ IN RELATION WITH GRAIN SIZE

2.1. The main dielectric property of $BaTiO_3$

Cubic $BaTiO_3$ is paraelectric and tetragonal $BaTiO_3$ is ferroelectric. The use of $BaTiO_3$ as dielectric material in ceramic capacitors is

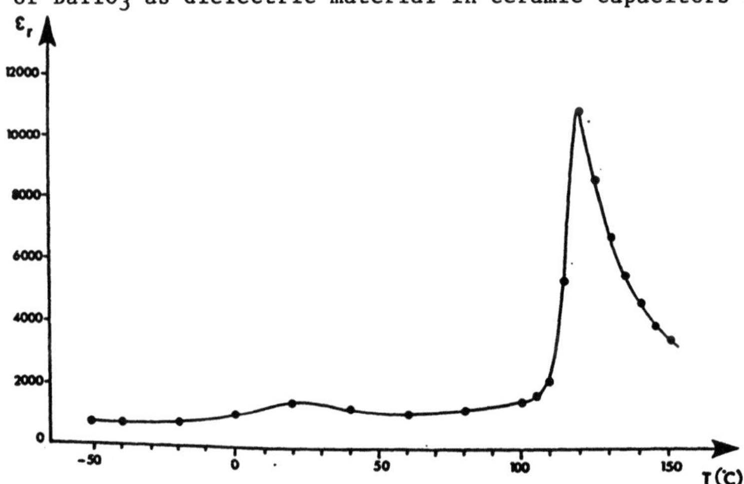

Figure 7. Relative dielectric constant ε_r of a pure polycristalline $BaTiO_3$ ceramic.

due to the very high value of the dielectric constant during the ferroelectric-paraelectric transition (figure 7). In the vicinity of this transition, in the tetragonal state, the amplitude of the displacement of the ions inside the oxygen octahedra becomes very large. This induces a strong variation of the polarizability and, consequently, a very important increase of the dielectric constant.

2.2. Dielectric properties of pure $BaTiO_3$ in relation with grain sizes

The cell polarizability is related to the tetragonal deformation. Thus it is obvious, from all the previous results, that the dielectric constant is a function of the particle size (9)(10)(figure 8) : the smaller the grain size, the higher the dielectric constant at room temperature but the smaller the dielectric constant during the ferro-electric - paraelectric transition. The evolution of the ε_r v.s.T curve

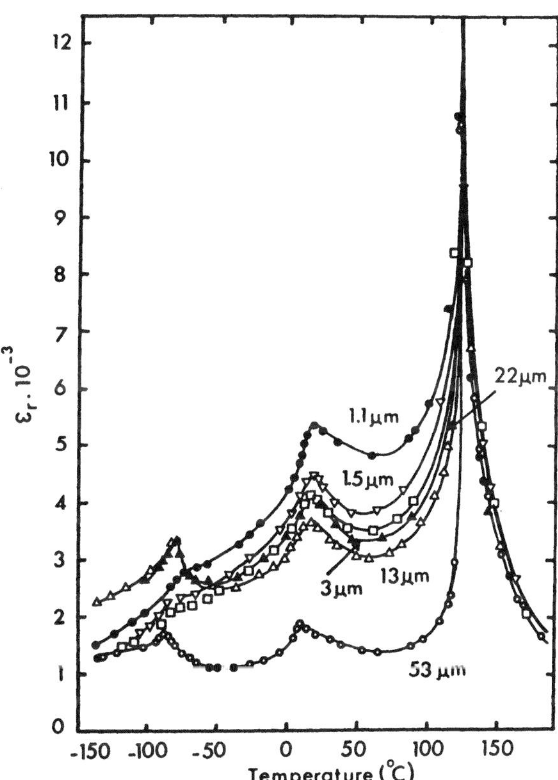

Figure 8. Dielectric constant ε_r versus temperature for different mean grain diameters in pure $BaTiO_3$ ceramics (10).

as a function of grain size, with size lower than 1 μm, would not be monotonous (figure 9)(11).

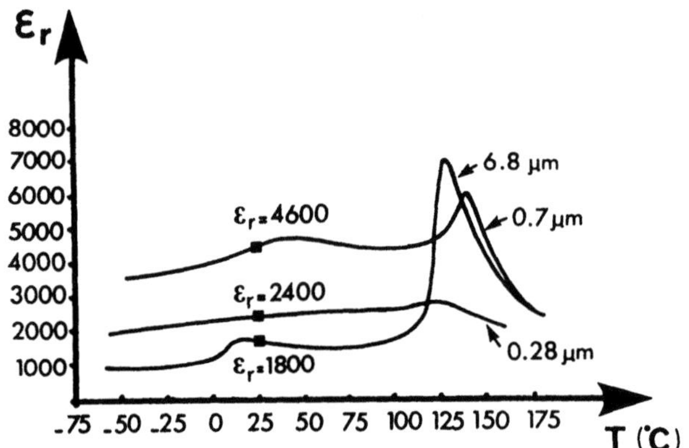

Figure 9. Dielectric constant ε_r v.s. temperature of $BaTiO_3$ for various grain sizes (11).

According to Arlt and Hennings (11) the increase of the dielectric constant at room temperature with decreasing the grain size would be due to an increase of the surface area of the walls of the 90° ferroelectric domains. When the size is lower than 0.5 μm, these domains disappear and, consequently, the dielectric constant is decreasing. The ferroelectric domains seem to vanish. This is confirmed by the decrease of the ferroelectric hysteresis loop (12) (figure 10) when decreasing the mean particle size in $BaTiO_3$ ceramics.

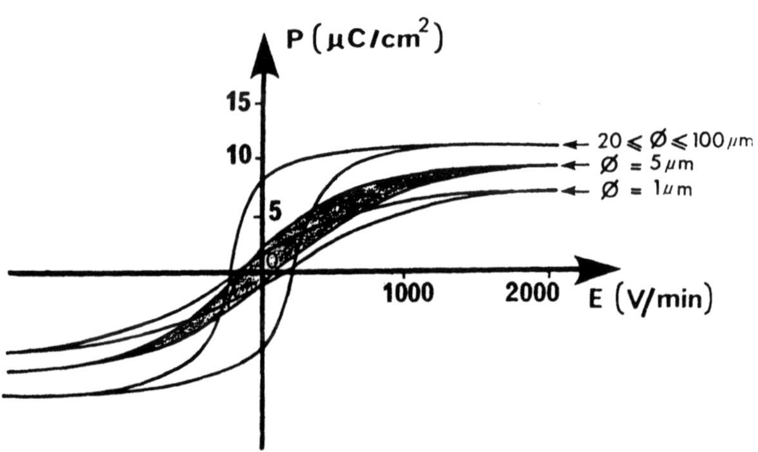

Figure 10. Ferroelectric hysteresis loop of $BaTiO_3$ as a function of the grain size in the ceramic (12).

2.3. Dielectric properties of Z5U ceramics in relation with the grain size of the initial $BaTiO_3$ powder, the basic component of the dielectric material

Pure barium titanate is not usable but almost all the type II dielectric compositions are made with barium titanate as main component. The characteristic curve for a Z5U dielectric is given (figure 11).

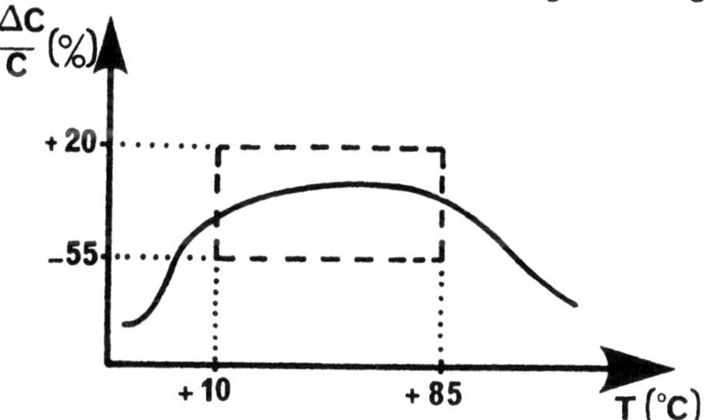

Figure 11. Relative capacitance variation as a function of temperature for a Z5U dielectric composition.

Such a composition is obtained with a cosintering of several oxides for example $BaTiO_3$, $CaZrO_3$, $MgZrO_3$, ... Figure 12 shows the differences between the characteristic curves of two Z5U compositions

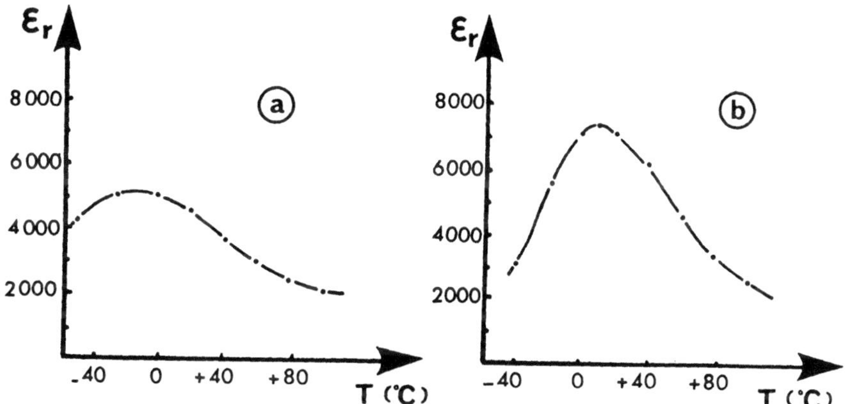

Figure 12. ε_r v.s. T curves of two Z5U materials which only differ by the grain size distribution curves of the two initial $BaTiO_3$ powders (figure 4 a)(13): **a)** Z5U ex. $BaTiO_3$ B **b)** Z5U ex.$BaTiO_3$ C.

which only differ by the grain size distribution curves of the two $BaTiO_3$ initial powders (figure 4a) ; smaller the mean crystallite size of the initial $BaTiO_3$ powder, greater the dielectric constant at the maximum of the ε_r v.s. T curve of the Z5U resulting material (figure 13)(13).

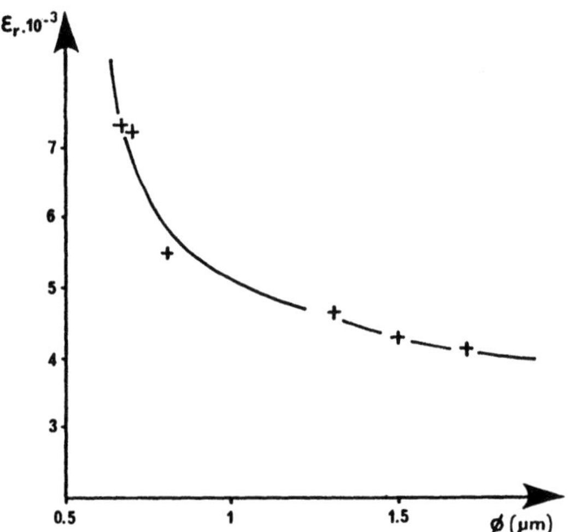

Figure 13. Maximum dielectric constant of Z5U materials, which only differ by the initial barium titanate, as a function of mean crystallite size of the initial $BaTiO_3$ powder (13).

In the case of these materials, $BaTiO_3$ is no more present; it reacted with the other oxides. The dielectric properties of the resulting material are a consequence of the chemical reaction ; during this reaction **the behaviour of the initial $BaTiO_3$ powder was related to its granulometry.**

2.4. $BaTiO_3$ reactivity during the elaboration of the Z5U ceramics as a function of the grain sizes of the initial $BaTiO_3$ powders

It must be said again that the elaboration process of the Z5U ceramics has been repeated as faithfully as possible ; only the initial $BaTiO_3$ powder was different from a ceramic to another one. Figure 14 shows the differences in the resulting ceramics : with only changing the grain size distribution curve of the initial $BaTiO_3$ powder, the phase composition of the resulting material is changed. That proves the reactivity of $BaTiO_3$ during the sintering is closely related to its grain size.

From all the above results it is clear that $BaTiO_3$ granulometry is a very important parameter for the dielectric properties. Thus it is obvious that this parameter has to be carefully controlled during the synthesis.

3. PREPARATION OF $BaTiO_3$ POWDERS WITH THE ADEQUATE GRAIN SIZES

It is well known that chemical routes enable to produce $BaTiO_3$ powders with the suitable granulometry. A recent thesis gives a good

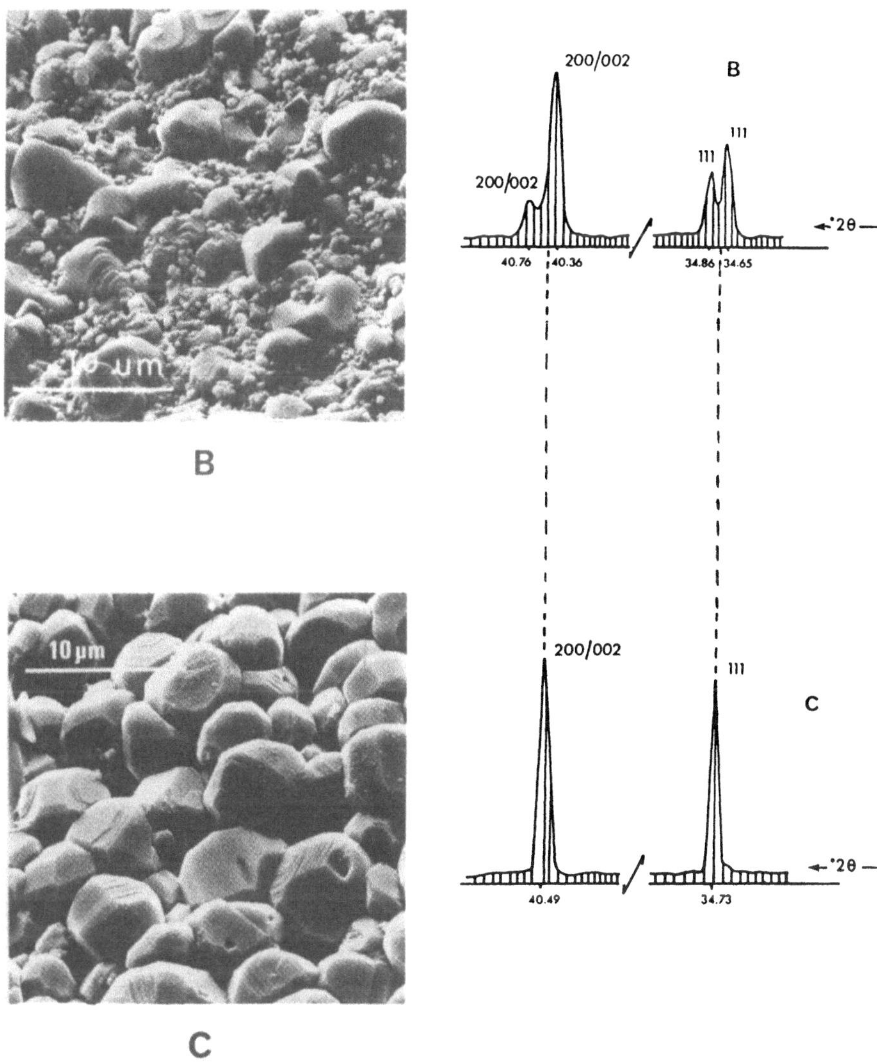

Figure 14. SEM photographs and X ray diffraction patterns of two Z5U ceramics made with two $BaTiO_3$ powders which differ only by their grain size distributions (13)

(a) From $BaTiO_3$ B (figure 4 a) : Resulting dielectric ceramic constituted of two phases.

(b) From $BaTiO_3$ C (figure 4 a) : Resulting dielectric ceramic constituted of only one phase.

overall view of these methods (14). However the solid way synthesis, which is still largely used by ceramic capacitor manufacturers, is also able to give a control mean of $BaTiO_3$ granulometry.

Figure 15 gives a diagrammatic representation of the reaction progress inside a TiO_2 grain during the $BaTiO_3$ solid way synthesis (15). **A resulting grain of $BaTiO_3$ derives from a TiO_2 grain.** Thus the

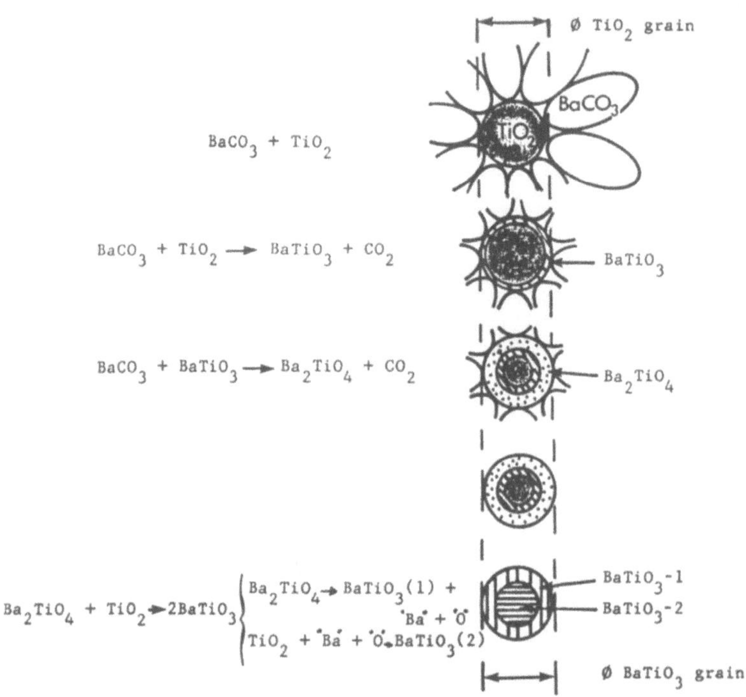

Figure 15. Schematic transformation of a TiO_2 grain surrounded by $BaCO_3$ grains during the $BaTiO_3$ solid state reaction synthesis (15).

choice of a titanium oxide, with a well adaptated granulometry, is necessary to produce a $BaTiO_3$ powder having the suitable particle size distribution. Of course, it is also necessary to control carefully the heating cycle to avoid any sintering during the reaction of the raw materials.

The grain size distribution of a powder material is able to modify drastically the properties and the behaviour of this material. The case of barium titanate described in this paper has to be considered only as an example. This phenomenon is rather general and, consequently, granulometry is a parameter which has to be controlled carefully for powder material applications.

REFERENCES

(1) H.F. KAY and P. VOUSTEN, Phil.Mag **40** (1949) 1019-1040.

(2) M. ANLIKER, H.R. BRUGGER and W. KANZIG, Helv. Phys. Acta **27** (1954) 99-124.

(3) M. ANLIKER, W. KANZIG and M. PETER, Helv. Phys. Acta **25** (1952) 474-475.

(4) W. KANZIG, Phys. Rev. **98** (1955) 549-550.

(5) S. MALBE, Défaut d'anisotropie quadratique dans $BaTiO_3$, Thesis (Dijon)(France), 1984.

(6) S. MALBE, J.C. MUTIN and J.C. NIEPCE, Phase Transitions (to be published).

(7) N.C. SHARMA and E.R. Mc CARTNEY, J. Austra. Cer. Soc. **10** (1974) 16-20.

(8) L. EYRAUD, Diélectriques Solides Anisotropes et Ferroélectriques Gauthier-Villars (Paris), 1967.

(9) G.H. JONKER and W. NOORLANDER, Science of Ceramics **1** (1962) 255-264.

(10) K. KINOSHITA and A. YAMAJI, J. Appl. Phys. **47** (1976) 371-373.

(11) G. ARLT, D. HENNINGS and G. de WITH, J. Appl. Phys. **58** (1985) 1619-1625.

(12) P. GAUCHER and J.C. KURFISS, Internal note Thomson (E.L. n° 41. 76.715), January 1988.

(13) F. BATLLO, A. BEAUGER, S. MALBE, J.C. MUTIN and J.C. NIEPCE J. Mat. Sc. (in publication).

(14) F. CHAPUT, Précurseurs sol-gel pour l'élaboration de céramiques diélectriques, Thesis (Limoges) (France), 1988.

(15) F. BATLLO, A. BEAUGER, J.C. MUTIN and J.C. NIEPCE, J. Mat. Sc. (in print publication).

VARIATION WITH PROCESSING CONDITIONS OF BULK AND GRAIN BOUNDARY PTCR PHENOMENA IN DOPED $BaTiO_3$

D.C. Sinclair and A.R. West,
University of Aberdeen,
Department of Chemistry,
Meston Walk,
Aberdeen AB9 2UE, UK

ABSTRACT. The use of combined impedance and modulus spectroscopy to analyse a.c. impedance data for doped polycrystalline $BaTiO_3$ shows the presence of two component impedances which are attributed to bulk and grain boundary effects. This assignment is based on the temperature dependence of their associated capacitances: the grain boundary capacitance does not change significantly with temperature, whereas the bulk capacitance passes through a maximum at the Curie point and shows Curie-Weiss behaviour at higher temperatures. The PTCR effect, characterised by a large increase in resistance as the temperature is raised through the Curie point, is shown by both components. The influence of processing conditions on the PTCR effect, especially the rate at which the sample is cooled to room temperature after sintering, is discussed. At the shorter rates, the PTCR response is dominated by the bulk impedance, whereas at longer times, the grain boundary impedance dominates.

INTRODUCTION The PTCR effect is a well-known but poorly understood phenomenon in which the electrical resistivity of certain materials undergoes a dramatic increase, often by several orders of magnitude, with increasing temperature. In doped, semiconducting barium titanate, the PTCR effect occurs as the temperature is raised through the ferroelectric Curie point, $\sim 120^{\circ}C$ and in which the resistance typically increases from $\sim$100 ohm to 100 Mohm over an interval of about $100^{\circ}C$. The effect is fully reversible with temperature cycling and leads to applications in thermal fuses and current overload protection devices.

The origins of the PTCR effect are not well-understood. It is usually regarded as being associated with grain

L.-C. Dufour et al. (eds.), Surfaces and Interfaces of Ceramic Materials, 535–543.

boundaries in barium titanate [1] and it is thought that additives which influence the PTCR effect, such as manganese, are segregated at the grain boundaries [2]. Various theories [2-5] have been proposed to explain the effect, but many doubts remain. The method of processing ceramics such as doped barium titanate appears to be critical to obtaining a large PTCR effect; factors such as sintering temperature, time, furnace atmosphere and cooling rate all appear to be important.

In polycrystalline ceramic materials which exhibit modest levels of electronic conductivity, the a.c. impedance technique provides a powerful, non-destructive method for probing electrical inhomogeneities associated with, for instance, crystal interiors, grain boundaries and surfaces. This is particularly the case when the results of different methods of data analysis are compared since often, one method alone, such as the use of complex impedance, admittance or electric modulus plane plots, may give information that is incomplete or even misleading. We have shown recently [6] that by use of combined impedance and modulus spectroscopy to analyse a.c. impedance data, it is possible to characterise both grain boundary and bulk phenomena in PTCR barium titanate.

The purpose of this paper is to show how one aspect of the ceramic processing, involving the nature of the final annealing treatment after sintering, can be investigated by a.c. impedance methods.

EXPERIMENTAL. Samples of barium titanate doped with a rare earth element to make them semiconducting and with manganese to enhance the PTCR effect were kindly provided by STC Components, Taunton, UK. The samples were in the form of cylindrical pellets with sputtered ohmic contact electrodes attached to opposite pellet faces. For impedance measurements, the electrodes on each sample were attached to the Pt leads of a conductivity jig and the arrangement heated inside a tube furnace, in air, with temperatures accurate to within 3°. Samples were allowed to equilibrate at the furnace temperature for about 45 mins prior to impedance measurements. A range of samples were measured in which their only difference was the rate at which the sample was cooled to room temperature after sintering.

A Solartron 1250/1286 Frequency Response Analyser and Electrochemical Interface was used for a.c. impedance measurements over the frequency range 30mHz to 65 kHz. This system was linked to a BBC microcomputer for programming the measurements, storing and handling the data. Blank measurements were made periodically to check that there were no stray leak resistances or high capacitances associated with the jig and measuring system. Impedance measurements were made over the range 25 to 450°C, on a heating cycle, with spot checks on the cooling cycle to confirm the reversibility of the results.

RESULTS AND DISCUSSION. Data are shown in Fig. 1 for the total resistance of three samples as a function of temperature. They illustrate well the PTCR effect: below

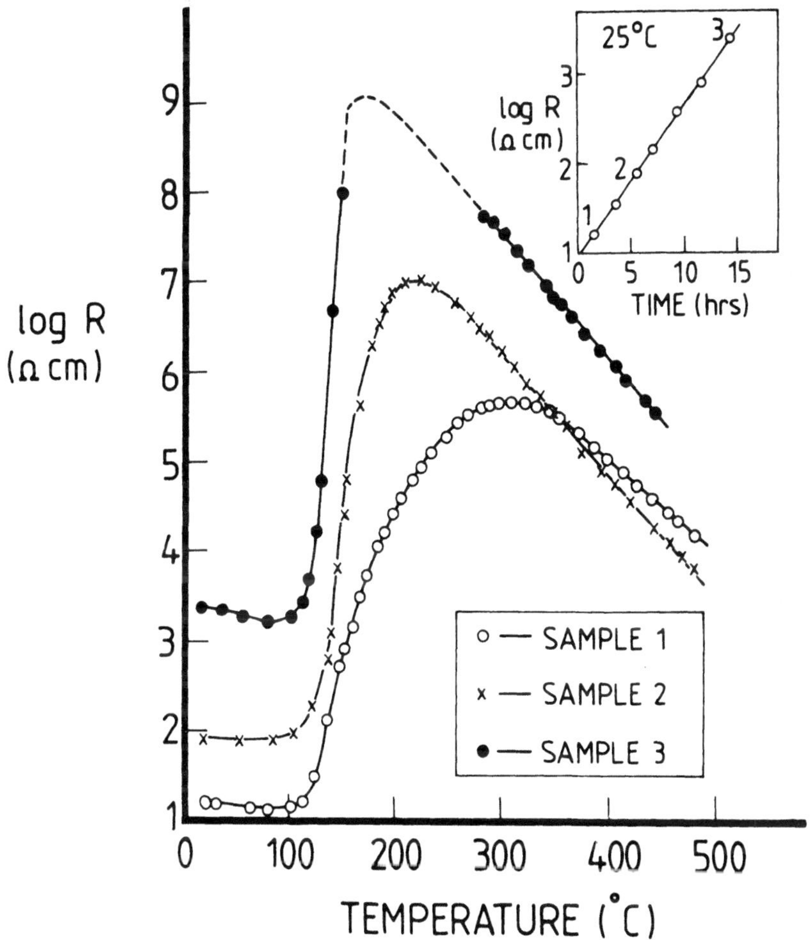

Figure 1. Resistivity against temperature for three samples of doped $BaTiO_3$, all showing a PTCR effect. Inset shows the room temperature resistivity of ceramics as a function of the time to cool the sample to ambient.

100°C, the resistance is low, in the range 15 to 2500 ohm cm, and decreases slightly with increasing temperature. Above ~100°C, the resistance undergoes a very large increase and passes through a maximum at some temperature between about 180 and 320°C; a fairly rapid decrease in resistance occurs at higher temperatures.

Several trends may be seen on comparing the behaviour of samples 1-3, which were characterised by progressively slower cooling rates. The room temperature resistance increases in this sequence, as shown further in the inset to Fig. 1 which includes data for other samples also. Thus, a linear correlation exists between the room temperature resistance and cooling rate. At the same time, both the steepness of the resistance rise and the magnitude of the resistance maximum increase with decreased cooling rate also, the temperature of the resistance maximum is displaced to lower temperatures.

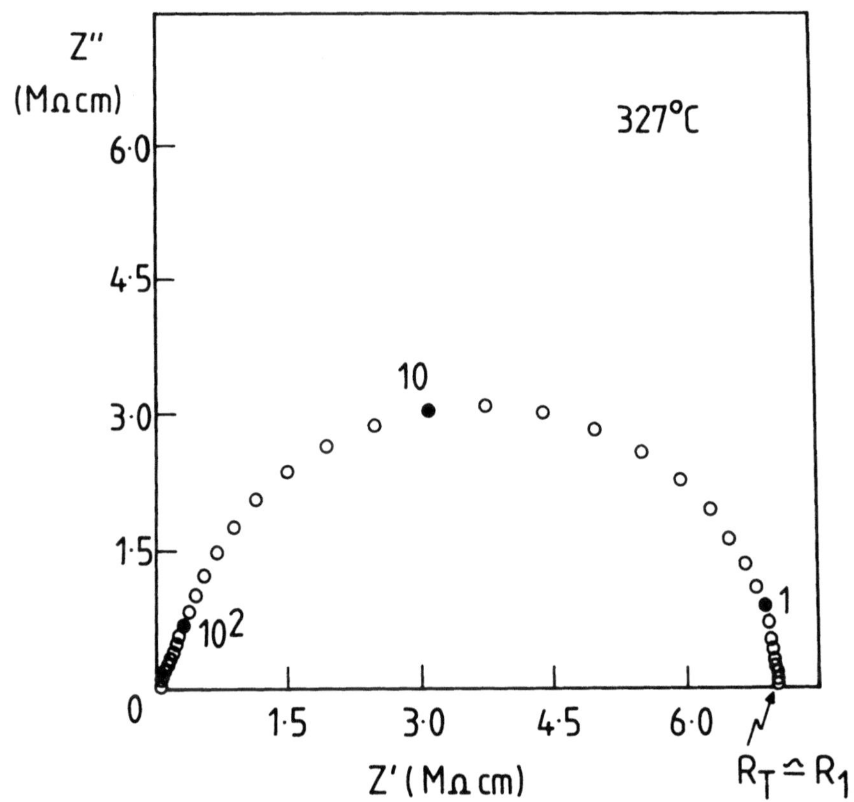

Figure 2. Typical complex impedance plot

The data and trends shown in Fig. 1 all pertain to the total resistance of the ceramics; in order to gain further understanding, it is imperative to separate the contributions from the bulk and grain boundary regions to the overall resistance. Use of conventional complex impedance plots alone are not particularly helpful, as shown in Fig. 2. Essentially, a single arc is seen with some indications of a second, small, poorly-resolved arc at high frequencies close to the origin. This plot was typical of all samples studied and permitted the extraction of total resistance values only.

A much more powerful way of analysing these particular data uses the complex electric modulus, M*, formalism. This is related to the complex impedance, Z*, by [7]:

$$M^* = j\omega C_o Z^* \qquad (1),$$

where C_o is the vacuum capacitance of the jig, without the sample. The data may be plotted in the complex modulus plane, M" vs M', or as spectroscopic plots of M" and M' against frequency.

A particularly useful plot for visual impact is to present combined plots of the imaginary components of the modulus, M" and the impedance, Z" against frequency, as shown in Fig. 3. Data are shown for each of the samples 1-3. In each case, the Z" plot, solid circles, shows a single peak which is almost Debye-like in shape; this peak becomes the arc approximating to a semicircle when plotted in the complex impedance plane, Fig. 2.

In striking contrast to the single peaks in the Z" plots, the M" plots based on the same raw data, show much more complexity with two, or in some cases three, overlapping peaks. In addition, there are considerable differences between the M" plots for the three samples.

An interpretation of the M" plots requires use of an equivalent circuit to simulate the a.c. response of the sample. This, and the circuit equations are treated more fully elsewhere [8], the main points and conclusions of which are as follows:

- the equivalent circuit comprises a series array of parallel RC elements; each parallel RC element represents an electrical region of the sample, such as the grain boundaries or the crystal interiors

- each parallel RC element gives rise to a Debye peak in the plots of Z", M" against frequency

- the size of the M" Debye peaks is inversely proportional to capacitance, ie $M''(max) = C_o/2C$. Consequently, spectroscopic M" plots are dominated by those elements with the smallest capacitance

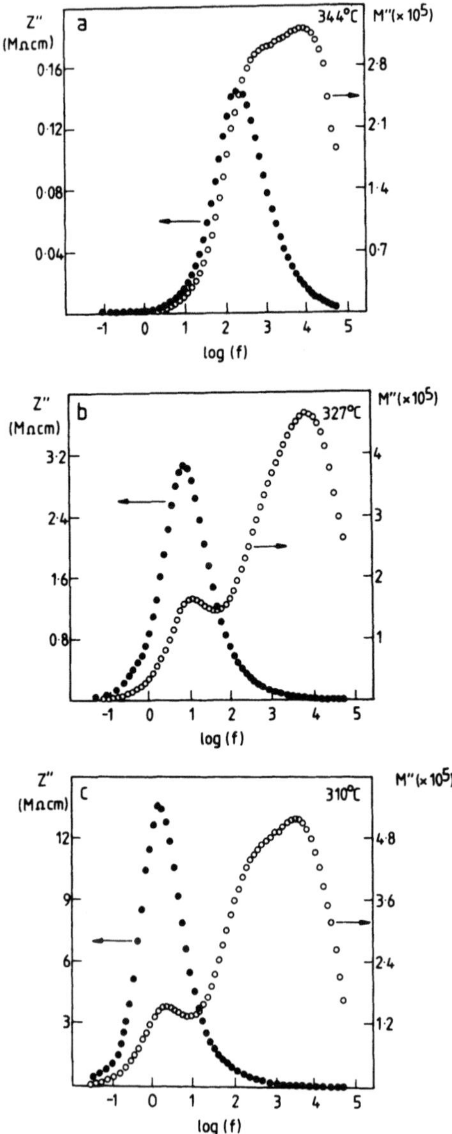

Figure 3. Impedance and modulus spectroscopic plots, Z" and M" vs log frequency, in Hz (where $\omega = 2\pi f$), for (a) to (c) samples 1 to 3, respectively.

- the size of the Z" Debye peaks is proportional to resistance, ie Z"(max) = R/2. Consequently, spectroscopic Z" plots are dominated by those elements with the largest resistance

- grain boundary capacitances can be distinguished from bulk capacitances in ferroelectric materials since, whereas grain boundary capacitances show little temperature dependence, bulk capacitances show a maximum at the Curie point, with Curie-Weiss behaviour in the higher temperature paraelectric region.

These considerations have been applied to the above data and yield the following conclusions.

Sample 1 contains two electrical components with similar capacitances, both of which show Curie-Weiss behaviour, Fig. 4a, with a Curie temperature of 124°C. The sample resistance is dominated by these same two components; hence any grain boundary resistances are too small, by comparison with the crystal resistances, to show up in the impedance data.

Sample 2 contains two distinct electrical components, one of which has a temperature independent capacitance, C_1 and the other a temperature dependent capacitance, C_2 with Curie-Weiss response, Fig. 4b and with a Curie temperature of 130°C. Of the two resistances, the grain boundary one, R_1 is larger by about two orders of magnitude, Fig. 5b, at least for temperatures >240°C. Nevertheless, both the crystal and grain boundary resistances show the PTCR effect.

Sample 3 is similar to sample 2 but now there is evidence for two crystal components, C_2 and C_3, in addition to the grain boundary one, C_1, Figs. 4c, 5c. One of the crystal resistances, R_2, is more resistive than the other, R_3, by about an order of magnitude, Fig. 5c, at least for temperatures >300°C. Nevertheless, the grain boundary resistance, R_1, remains the most resistive element in this sample.

CONCLUSIONS.

1. The use of combined impedance and modulus spectroscopy is more powerful than the complex impedance plane method alone for separating the contributions of crystal and grain boundary regions to the overall impedance of doped barium titanate ceramics.

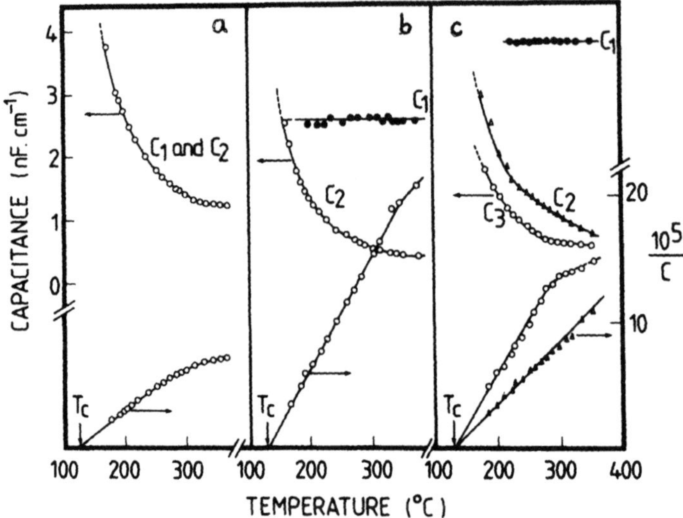

Figure 4. Capacitance data for samples 1 to 3 in (a) to (c), respectively. Upper graphs, capacitance against temperature; lower graphs, Curie-Weiss plots.

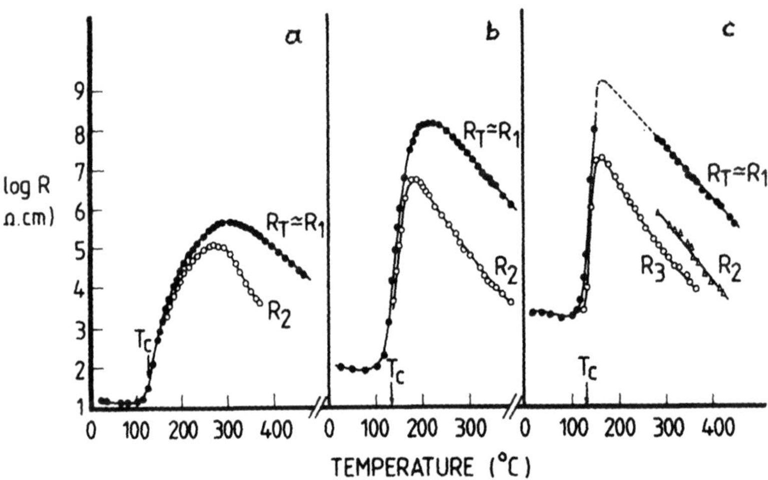

Figure 5. Component resistances as a function of temperature for samples 1 to 3. R_T is the total sample resistance.

2. It is possible to observe a PTCR effect in samples where the grain boundary resistance is negligible.

3. The relative magnitudes of crystal and grain boundary resistances are a sensitive function of the ceramic processing treatment.

4. Further work to understand the origins of the PTCR effect is required. The long-held view that it is only a grain boundary controlled effect appears to be no longer valid.

ACKNOWLEDGEMENTS. We thank the SERC for a grant to DCS. ARW thanks the Royal Society of Edinburgh for a Support Fellowship. The advice of REW Casselton is much appreciated.

REFERENCES

1. W. HEYWANG, J. Amer. Ceram. Soc., 47 (1964) 484
2. G.V. LEWIS and C.R.A. CATLOW, Proc. Brit. Ceram. Soc., 36 (1985) 187
3. W. HEYWANG, J. Mater. Sci., 6 (1971) 1214
4. J. DANIELS and R. WERNICKE, Phillips Research Report, 31 (1976) 544
5. T.R.N. KUTTY, P. MURUGARAJ and N.S. GAJBHIYE, Mat. Res. Bull., 20 (1985) 565
6. D.C. SINCLAIR and A.R. WEST, J. Mat. Sci. Letters, 7 (1988) 823
7. I.M. HODGE, M.D. INGRAM and A.R. WEST, J. Electroanal. Chem., 74 (1976) 125
8. D.C. SINCLAIR and A.R. WEST, in preparation

COPPER-CORDIERITE COSINTERING

V. OLIVER, J. GUILLE, J. C. BERNIER
I.P.C.M.S., Groupe de Chimie des Matériaux Inorganiques
E.H.I.C.S. 1 rue Blaise Pascal F-67008 Strasbourg Cedex

B. S. HAN, J. WERCKMANN, J, FAERBER, P. HUMBERT, B, CARRIERE*
I.P.C.M.S., Groupe Surfaces-Interfaces
Université Louis Pasteur, 4 rue Blaise Pascal F-67000 Strasbourg

ABSTRACT- The use of a sol-gel processed cordierite precursor sinterable about 900°C allows cosintering of the copper and the ceramic. A strong bonding between the copper film and the cordierite substrate can be achieved through an eutectic bonding technique. Three types of interfaces have been studied by EPMA, ESCA and SEM: (a) cosintered in hydrogen, leading to poor metal adhesion, (b) cosintered in wet argon with slow heating rate (500°C/h) leading to strong copper diffusion, (c) cosintered in wet argon with fast heating rate (800°C/h) leading to the best macroscopic properties, i.e. no deformation, limited diffusion and strong adhesion.

1. INTRODUCTION

The advantages of multilayer ceramic (M.L.C.) in microelectronics industry have been known for some time.

New ways for ceramic synthesis, such as "sol-gel" methods, were investigated in our laboratory /1/, since they allow a drastic decrease of the sintering temperature. Ceramics can thus be cosintered with low melting point metals (Au, Ag, Cu,...). This sol-gel method has been applied to cordierite (2MgO, $2Al_2O_3$, $5SiO_2$) synthesis. This ceramic has indeed many interesting properties for microelectronic packaging, particularly a low dielectric constant and a small thermal expansion. The metal used in this study was copper, its main advantages being its high conductivity and low cost.

Our purpose was to investigate the possibility of achieving both cosintering and metal bonding through an cutcctic bonding technique /2/. This paper is an extended abstract of our results on this subject.

---

* present address : Department of Electrical Engineering, Chonbuk National University, Chonju, South Korea

L.-C. Dufour et al. (eds.), Surfaces and Interfaces of Ceramic Materials, 545–551.

2. MATERIALS AND METHOD

2.1. Cordierite Synthesis and Sample Preparation

Figure 1 presents schematically the cordierite synthesis and the sample preparation. Since screen printed interfaces are not suited for surface analysis, due to the coating thicknesses (>10 μm), thin copper layers (10-100 nm) were sputter-deposited on green ceramic samples and cosintered in equivalent conditions.

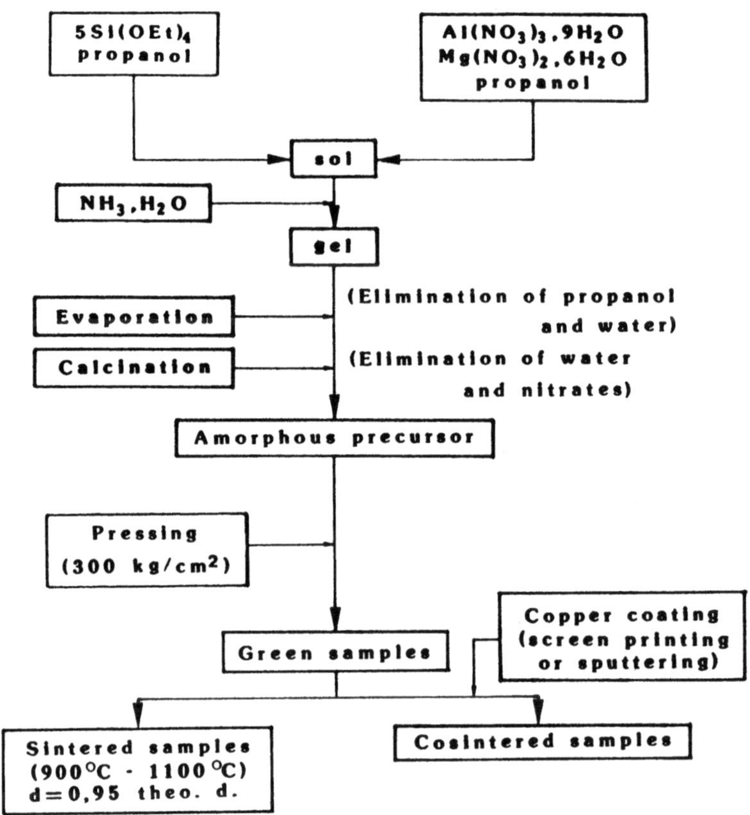

Figure 1 Diagram of cordierite synthesis and sample preparation

2.2. Cosintering and Sealing

The physical properties of these interfaces strongly depend on the cosintering conditions /3/. Three types of interfaces were studied (fig.2): (a) cosintered in hydrogen, (b) cosintered in wet argon with slow heating rate (500 °C/h), (c) cosintered in wet argon with fast heating rate (800 °C/h).

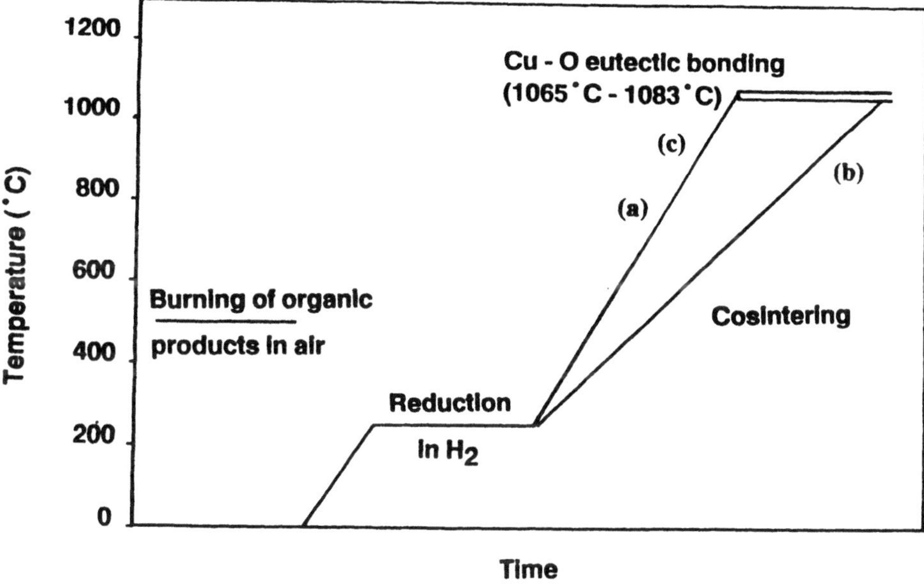

Figure 2 **Firing schedule**

2.3. Experimental

To explain the macroscopic observations and give a microscopic description of each type of interface, the following techniques were used :

- X-Ray diffraction,
- Scanning electron microscopy and microanalysis (SEM, EDS-XRF)
- Depth profiling by X-Ray photoelectron spectroscopy (XPS-ESCA)

We mainly investigated the copper diffusion in the substrate, the cristallographic structure of the ceramic and the chemical composition near the copper-cordierite interface.

3. RESULTS AND DISCUSSION

3.1. Green and sintered cordierite

- Bulk : composition near the theoretical value, presence of β, μ-cordierite and $MgAl_2O_4$ after sintering, compositional heterogeneities on a microscopic scale (ϕ 1μm).

- Surface : magnesium depletion and silicon enrichment down to 10 nm from the surface.

3.2 Sample (a)

EPMA : no copper diffusion
XPS : compositional measurements give values similar to green and sintered cordierite (fig.3).
No chemical interaction is obtained. This explains the poor adhesion of the metal to the substrate.

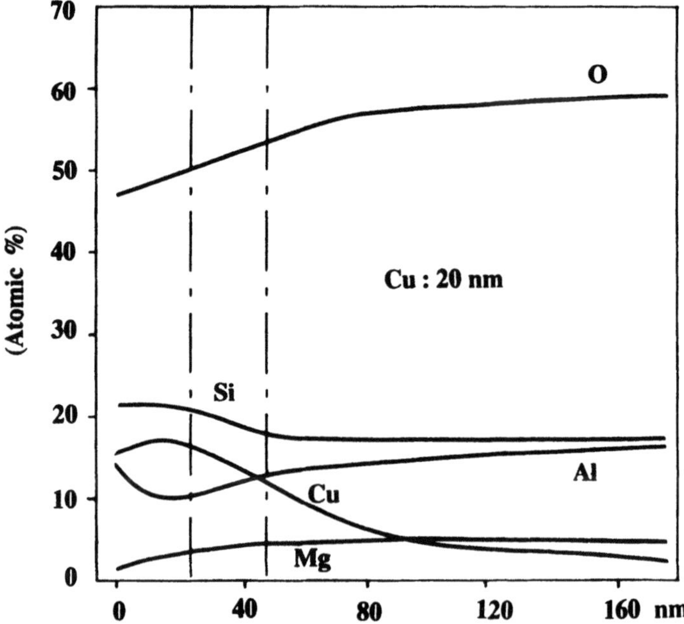

Figure 3

Elemental composition during depth profiling by XPS after firing in hydrogen

3.3 Sample (b)

EPMA : copper diffuses strongly into the bulk (fig.4) leading to a poor densification and thus to the camber of the ceramic /4/.

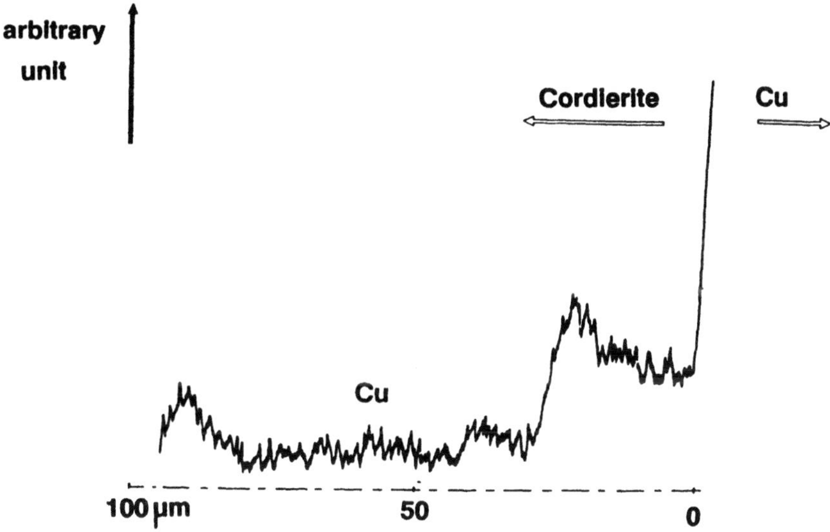

Figure 4 **Copper diffusion after firing (500°C/h)**

XPS : for 20 nm initial copper deposition, copper is not detected, the surface underwent important morphological and chemical modifications (fig.5), mainly characterized by silicon enrichment over about 10 nm.

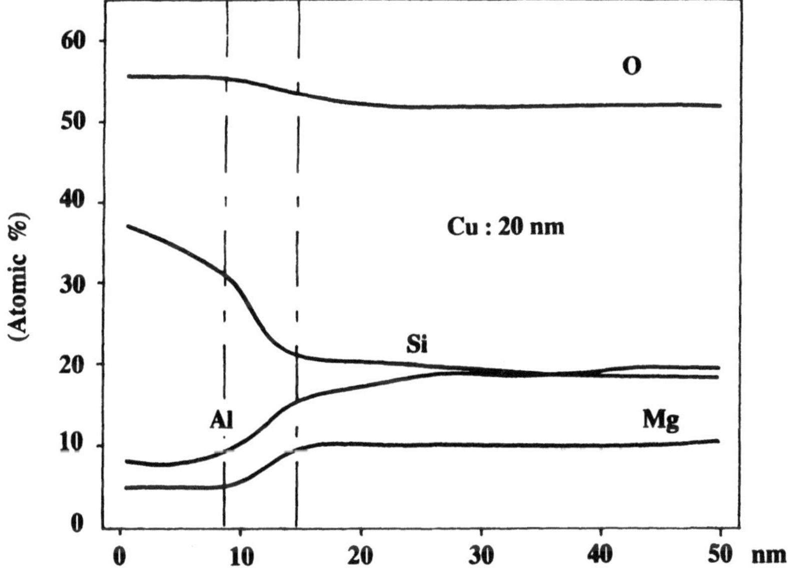

Figure 5 **Elemental composition during depth profiling by XPS after firing (500°C/h)**

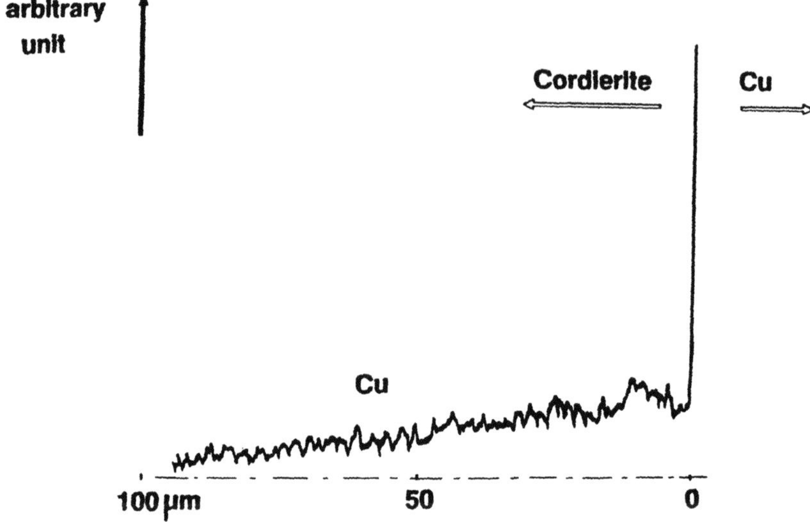

Figure 6 Copper diffusion after firing (500°C/h)

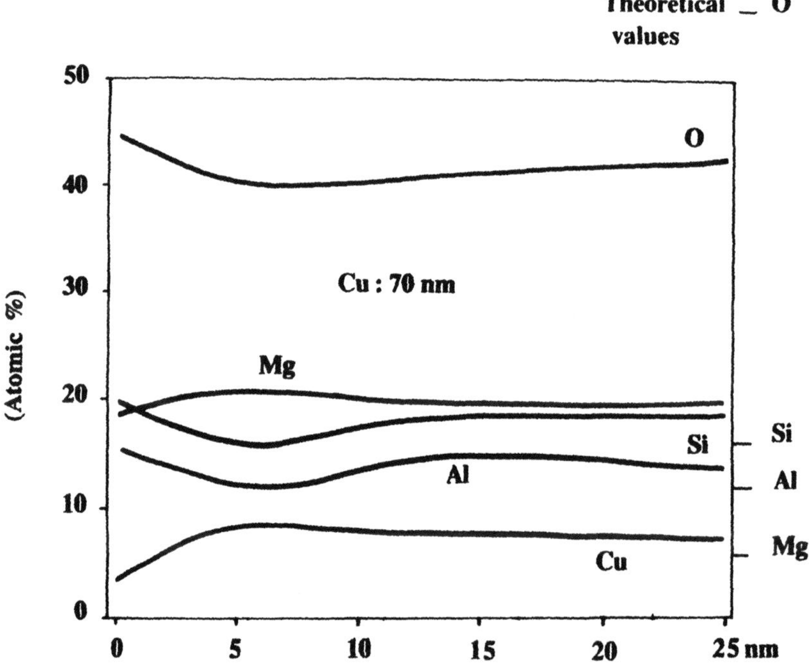

Figure 7 Elemental composition during depht profiling by XPS after firing (800°C/h)

3.4 Sample (c)

EPMA : copper diffuses slightly into the bulk (fig.6).

XPS : chemical composition is disturbed over a depth of a few ten nm (fig.7). Different phases coexist in this region : one enriched in magnesium and silicon, the others resulting from a preferential reaction between copper and aluminium /5/. These interfaces present in the same time the best macroscopic properties and the most severe chemical modifications.

4. CONCLUSION

In this work we investigated the possibility to achieve copper-cordierite cosintering using an eutectic bonding technique.
We optimized the cosintering conditions, in particular the firing atmosphere and heating schedule.

Work by ESCA and TEM is in progress to characterize more precisely the different interface phases formed during the cosintering process.

5. REFERENCES

[1] J.C. BERNIER, P. POIX, J.L. REHSPRINGER, G. VILMIN and S. VILMINOT, French patent n° 85-10873
[2] J.F. BURGESS et C.A. NEWGEBAUER, US patent 3.911.553
[3] V. OLIVER-BROUDIC, Thesis, Strasbourg I, France (1988)
[4] B.S. HAN, Thesis, Strasbourg I, France (1988)
[5] OLIVER-BROUDIC, J. GUILLE, J.C. BERNIER, B.S. HAN, J. WERCKMANN, J. FAERBER, P. HUMBERT and B. CARRIERE, J. Mater. Sc. Eng., in press

GRAIN BOUNDARY PHENOMENA IN THE EARLY STAGES OF SINTERING OF MO OXIDES

A.M.R. SENOS, M.R.SANTOS, A.P. MOREIRA, J.M. VIEIRA
Departamento de Engenharia Ceramica e do Vidro
Universidade de Aveiro 3800 Aveiro
Portugal

ABSTRACT. The linear Frenkel law of sintering is briefly reviewed. Sintering of pure and Cr^{3+} doped ZnO and NiO was studied by using computer assisted dilatometry. The linear shrinkage law was observed during sintering of both oxides for low values of shrinkage. Its extent is changed by doping with Cr_2O_3 . The sintering kinetics is discussed on the grounds of the MO oxides grain boundary .

1. Introduction

Grain boundary phenomena play a major role in high temperature processing of ceramics. Sintering being the usual production route of dense ceramic bodies is not an exception to that rule. Hardly measurable differences in chemical composition of the initial powders or of the sintering atmosphere may induce large microstructural differences during the sintering process, when chemical impurities segregate at the free surfaces and grain boundaries. These microstructural differences are easily observed in the final stage of sintering, but they can often be followed since the early stages of the process. Interface kinetics and anisotropy of the solid interfaces determine the morphology and the reactivity of the powder particles.

The investigation of ceramic interface phenomena via the quantitative study of sintering[1] is made difficult by the limitations of the known sintering models. Depending on the postulates, simplifications and approximations introduced in each of the models of a given sintering mechanism, different kinetic laws can be obtained. Discrepancies between the sintering models and the experimental results have led to frequent reviewing of the models. Nevertheless it is generally accepted that particle shape accommodation in solid state sintering of simple ceramic oxides is governed by diffusion controlled mechanisms of matter transport [2-4]. The diffusion mechanisms are driven by chemical potential gradients at the curved interfaces. Increased particle-particle contact area, or necking, is so produced, and densification is commonly observed. The theoretical densification and linear shrinkage laws of sintering have been observed in model experiments using powders of spherical, monodispersed particles [3,5].

L.-C. Dufour et al. (eds.), Surfaces and Interfaces of Ceramic Materials, 553–563.

Sintering kinetics of non-ideal powders is perturbed by particle coarsening, grain growth and differential sintering of particle domains and agglomerates [6-8]. The distribution of particle coordination numbers is changed during sintering of non-uniformly packed particles. Beside particle center-to-center approximation, grain boundary sliding, allowing for particle rearrangement under the effect of the sintering stress, must occur during sintering of all but the ideal, uniformly packed system [9,10]. Particle shape accommodation and grain boundary sliding are mutually dependent mechanisms in sintering as they are in creep [11]. The contribution of this mechanism has only recently been acknowledged in the study of solid state sintering [9,10,12,13].

Particle rearrangement has been identified with a nearly linear shrinkage kinetics as for instance in the first stage of liquid phase sintering [3]. The fundamental equation leading to linear shrinkage kinetics is the Frenkel's equation for neck growth between amorphous particles by viscous flow:

$$x^2 / R = 3\gamma t / 2\eta \tag{1}$$

where: x = neck size; R= particle size; γ = specific surface energy of free surfaces;

η = viscosity; t = time of isothermal sintering.

By assuming particle center-to-center approximation the linear shrinkage kinetics was derived [3] from equation (1):

$$\delta l / l_0 = 3\gamma t / 4\eta R \tag{2}$$

Equation (2) has been observed in the sintering of glass spheres [14,15]. It has also been reported in solid state sintering of single phase oxides such as TiO_2 [7,12,16], Al_2O_3 [17], MgO [18,19] and ZnO [20], although TEM investigation of the TiO_2 and Al_2O_3 particles could not reveal the presence of liquid films.

Two sintering models have been proposed as alternative to Frenkel's model of the linear shrinkage kinetics, for solid state sintering of ultrafine powders. Following M. Astier and P. Vergnon [7,12,16], the shrinkage rate for spherical particles which are sintering without formation of a concave surface at the neck, must be constant and proportional to $1/R^3$. The second model is due to G. C. Kuczynski [21]. The relative linear shrinkage corresponding to the removal of nanometer surface asperities at the particle:particle contacts can be as high as a few parts percent if the ultrafine particles are just one or two orders of magnitude larger than the asperities themselves. The shrinkage rate as deducted by G. C. Kuczynski is also constant and proportional to $1/R^2$.

A new approach, that goes back to particle rearrangement and grain boundary sliding processes, is here pursued in order to explain the observation of linear shrinkage kinetics in the initial stage of solid state sintering.

The effective viscosity of a polycrystal for creep controlled by grain boundary sliding is given by the Lifchits equation [22]:

$$\eta = k\,T\,R \;/\; D\,\Omega^{2/3} \tag{3}$$

where: Ω = atomic volume; k = Boltzmann constant; T = temperature. D, the effective diffusion coefficient is either the lattice diffusion coefficient D_1, or the grain boundary diffusion coefficient D_{gb}, or the diffusion coefficient in a grain boundary liquid film D^*, according to the nature of the mechanism of grain boundary sliding. Following R. C. Gifkins [22], grain boundary sliding can be hindered by the slow movement of grain boundary ledges. Lifchits equation (3) will then become:

$$\eta = 4\,L\,k\,T\,R \;/\; 3\,a\,f\,D_{gb}\,\Omega \tag{4}$$

where: a = numeric constant; f = constant form factor ($\approx$ 2); L = average grain boundary ledge length.
The average grain boundary ledge length, L, is dependent on grain boundary structure, but independent on the particle size [11,19,21-23]. By substituting the effective viscosity given by equation (4), equation (2) becomes:

$$\delta l \;/\; l_0 = 9\,a\,f\,\gamma\,D_{gb}\,\Omega\;t \;/\; 16\,k\,T\,L\,R^2 \tag{5}$$

This substitution is equivalent to assume that the sintering stress for grain boundary sliding of submicron size particles [7,12] is proportional to $2\,\gamma\,/R$.

As the exponent of the particle size factor in equation (5) is lower than for the shape accommodation sintering mechanisms, the grain boundary sliding rate may become the limiting factor in sintering of fine powders. Grain boundary structure may turn an important feature in the analysis of sintering kinetics. These relationships are further investigated in the present study.

2. Experimental

Pure light ZnO powders, O.4 µm particle size, were used (Analar grade from BDH), the main impurities being SO_4 0.01 w% and Pb 0.005 w%. NiO, O.7 µm particle size, was prepared by air calcination of pure $Ni(NO_3)_2.6H_2O$ at 800 °C during 8 hours. Cr^{3+} doping of both ZnO and NiO was done by preparing a thick slurry of the powder and $(NH_4)_2Cr_2O_7$ with ethanol. The slurry was mechanically agitated, then dried at 110°C and calcined at 350°C during 2 hours to decompose the chromium salt.

Cylindrical pellets, 25 mm long and 5 mm in diameter were isostatically pressed at 20 MPa. The pellets were sintered in air both under isothermal conditions and by constant heating rate (CHR) sintering. The linear shrinkage, as measured by a Chevenard-like dilatometer, was continuously recorded by a microcomputer. The shrinkage rate was calculated by numerical methods. The density of green samples and sintered samples was measured by the Archimede's mercury displacement method.

Porous sintered samples were impregnated with either epoxy resin or low melting point, low viscosity glass. Polished sections were prepared for scanning electron microscopy. Grain boundaries were developed by thermal etching (epoxy impregnated samples), or by controlled diffusive HF vapour etching trough a protective polymeric film (glass impregnated samples).

3. Results and Discussion

Figure 1 shows a linear relationship between the linear shrinkage, $Y = \delta l/l_0$, and time, t, in isothermal sintering of pure ZnO at 700°C. The same kinetics can also be observed in Moryioshi's results on the sintering of ZnO [20]. The corresponding shrinkage rate, dY/dt, is independent of the value of shrinkage. If CHR sintering rates are assumed to be identical to isothermal sintering rates [24], then, the shrinkage rate, that is measured in CHR sintering and corresponds to the linear dependence of shrinkage on time in isothermal kinetics, must also be independent of shrinkage

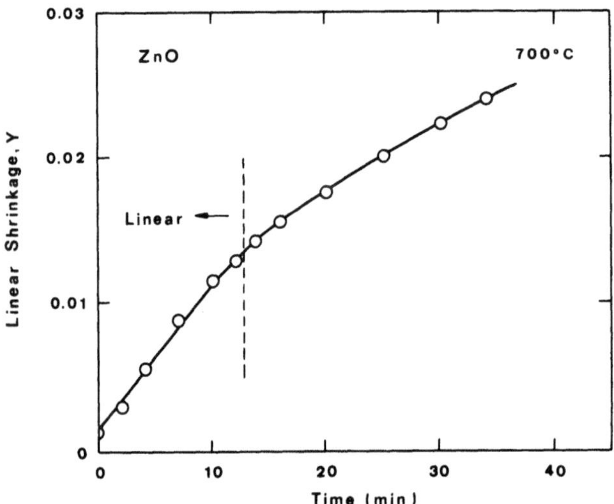

Fig. 1. Linear dependence of shrinkage on time in the isothernal sintering of pure ZnO.

(Fig. 2a). It is also independent of the value of c, the heating rate. The horizontal lines in Fig. 2a are equivalent to the linear shrinkage kinetics in Fig. 1.

With increasing temperature, the limits of the linear region move towards higher Y values . The general trend of the curves in Fig. 2a is maintained after doping with Cr^{3+}, but Cr^{3+} doped ZnO is more refractory than pure ZnO, the linear region being moved to quite low values of shrinkage.

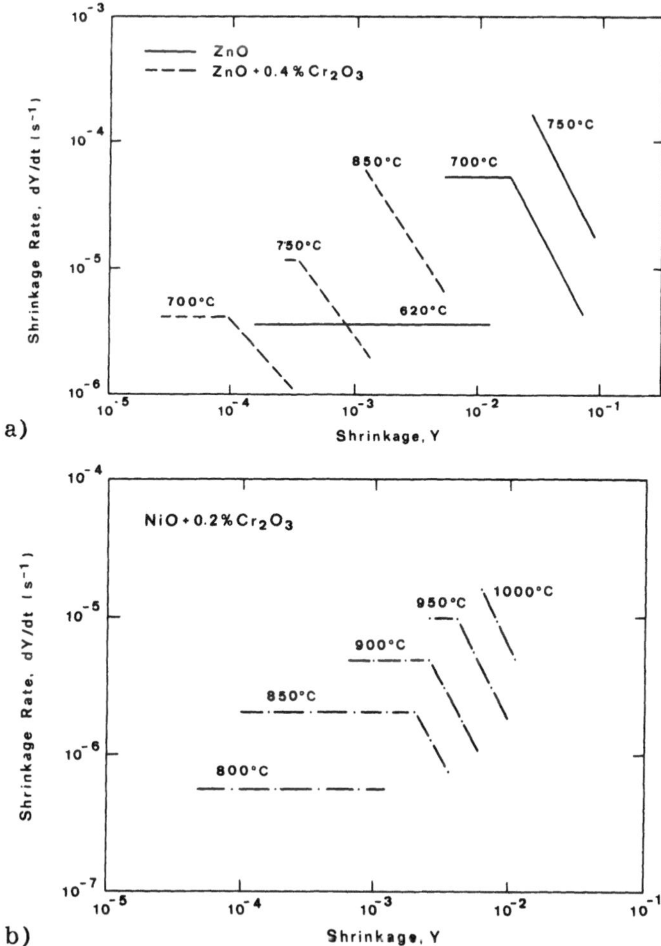

Fig. 2 Shrinkage rate versus shrinkage at constant temperature value in CHR sintering. (a) Pure and doped ZnO. (b) Doped NiO

The linear region is also observed in the CHR sintering of pure and doped NiO (Fig. 2b). As in the case of ZnO, doping NiO by Cr^{3+} leaves the general trend of its sintering curves unchanged, but increases its refractoriness.

The values of the activation energy of shrinkage rate in the linear region for both oxides are very similar, being 210 $kJ.mol^{-1}$ for ZnO and 206-214 $kJ.mol^{-1}$ for NiO. The low values of the activation energies may *a priori* be identified with grain boundary processes. Grain boundary films have not been observed in the study of ZnO and NiO with purity grade close to the one of the present investigation. Thinking of the linear region as being the result of shrinkage hindered by grain boundary sliding - equation (5) - the possibility of observing grain boundary ledges in the ZnO and NiO and the nature of the ledges need to be further clarified.

Grain boundary ledges can be formed as slip bands in plastically deformed crystals, or they may just be the result of the anisotropy of grain boundary specific surface energy. Hard second phase particles can also act as pinning points in the grain boundaries, restraining grain boundary slip and reducing grain boundary diffusion [25]. Grain boundary specific surface energy is minimized whenever grain boundary ledges are formed either as a result of surface tension anisotropy or by increased density of coincidental points, which reduces dislocation density (CSL theory) [26-29]

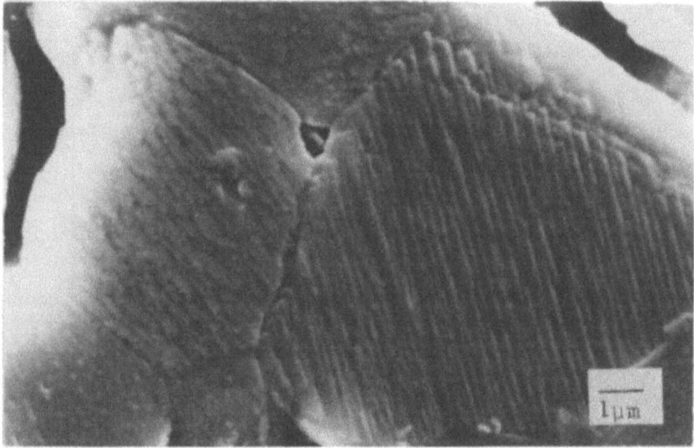

Fig. 3. Faceting in thermally etched chromium doped NiO

Although both ZnO and NiO crystalline structures are closed packed, they belong to different crystallographic systems. ZnO has the wurtzite-type hexagonal crystalline structure, with the c/a ratio equal to 1.63 [30]. The oxygen ions have hexagonal close packing, while the zinc ions occupy half of the tetrahedrally coordinated interstitial sites.

The NiO crystal structure is face centered cubic of the NaCl-type [30]. {001} and {011} tilt boundaries are the most commonly observed grain boundaries [29,31] and the less energetic ones, as they have null dipolar electrical moments and, therefore, bond relaxation at the interface is small [29]. The specific surface energy of the {011} tilt boundaries is estimated to be two to three times higher than that of the {001} grain boundaries [28,31]. Microfaceting of the {011} tilt

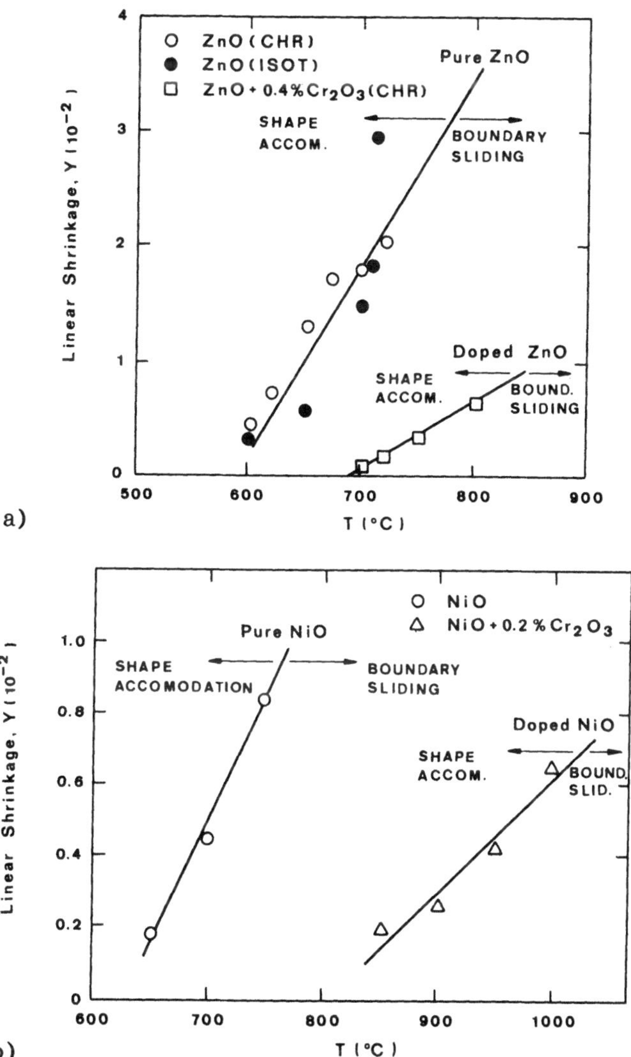

Fig. 4. Lines separating grain boundary mechanisms in the sintering of a) ZnO and b) NiO.

boundaries leading to {001} planes at high temperature is favoured. Faceting of NiO grain boundaries of various orientations has been reported [27]. The microfacets are 7- 90 nm wide and ≈ 1 nm thick [27]. Microfaceting of the free surfaces and of the grain boundaries of Cr^{3+}doped NiO is observed in the thermally etched sample in Fig. 3. The dimensions of the microfacets in Fig. 3 are one order of magnitude larger than those reported above, the microfacets representing multiple faceting.

Al_2O_3 is the most studied hexagonal closed packed crystalline oxide. The cation sublattice is different from the ZnO cation sublattice. Cations in Al_2O_3 occupy 2/3 of the octahedral interstitial sites. The calculated values of the specific surface energy of Al_2O_3 show an absolute minimum at rhombohedral $\{10\bar{1}2\}$ grain boundary planes [29]. Yet, the basal {0001} grain boundaries are the most frequently observed [32]. The predominance of these facets in the Al_2O_3 grain boundaries may be due to higher density of coincidental points or to kinetics [27,32]. From analogies between the crystal structures of ZnO and Al_2O_3, grain boundary energy anisotropy and grain boundary faceting would be expected in ZnO.

Bond relaxation at the interface of ionic crystals such as NiO and ZnO leads to relative movements of the anions and cations in opposite directions and may yield instable interfaces with irregular shapes. Bond relaxation of the {001} interfaces of MgO is very light, but it can be more severe for interfaces with non-zero dipolar electric moments [29].

The limits of the linear regions are shown in Fig. 4a and 4b for NiO and ZnO respectively. The linear region is identified with a grain boundary sliding mechanism. The sintering kinetics in the next region can be described by densifying sintering models of diffusion controlled neck growth [33], the shrinkage rate being dependent on shrinkage: $dY/dt \propto Y^{-p}$. The exponent p is characteristic of the sintering mechanism. The sintering kinetics of the second region for ZnO was discussed in a previous paper [33], the sintering kinetics being better described by the intermediate stage sintering models of grain boundary diffusion with simultaneous grain growth. If it is assumed that the first region corresponds to the intermediate stage of sintering by volume diffusion at constant grain size, as described by Coble's model[4], then the following points would be contradictory :

i) the volume diffusion mechanism would appear as the controlling mechanism before the grain boundary diffusion, at lower temperatures and when the system is finer. This is conflicting with Herring's scaling laws and with the expected dependence of the diffusion coefficients on temperature.

ii) the morphology of the sintering compact at the beginning of sintering is far different from that in the intermediate stage.

iii) the transition between sintering mechanisms in Fig. 2a is not coincidental with the onset of grain growth [33].

The beginning of shrinkage in Cr^{3+} doped samples is delayed for about 100 °C in ZnO and 200°C in NiO. In Figs 4a and 4b, Cr^{3+} doping decreases the extent of the grain boundary sliding region and changes the slope of the separation lines for both materials. The shrinkage rates of the second region are more reduced by Cr^{3+} doping than those of the grain boundary sliding region (Figs 2a and 2b). The position of the shrinkage rate lines of each region in Figs 2a and 2b relative to the extrapolated section of the lines belonging to the other region is in accordance with a sequence of mutually dependent mechanisms, the sintering rate being controlled by the slower mechanism.

ZnO is a non-stoichiometric n-type oxide semiconductor with excess metal. NiO is a non-stoichiometric valence p-type semiconductor with metal defect [34]. Both Frenkel disorder and Schottky disorder of the cation sublattice have been proposed as the dominant crystal lattice defects of ZnO [35,36]. Schottky disorder of the cation sublattice is dominant in NiO [34].

Lattice diffusion [37,38] and grain boundary diffusion of interstitial zinc, Zn_i [33,39,40], have been reported as the controlling mechanisms of sintering of ZnO in air. Densification of NiO appears to be controlled by oxygen lattice diffusion[41].

The ionic radii of Cr^{3+}, Zn^{2+} and Ni^{2+} are all close together. Cr_2O_3 forms substitutional solid solutions with ZnO and NiO [42,43]. At 1170°C and below, Auger electron microscopy shows that chromium is heavily segregated to the dislocations and grain boundaries of NiO, the width of the segregation region being lower than 50 nm [44,45].

Due to the electroneutrality balance, the addition of substitutional Cr^{3+} to ZnO will decrease the concentration of interstitial zinc, Zn_i, and the corresponding diffusion coefficient both in the lattice and in the grain boundary depletion region. Substitutional chromium in NiO will increase the cation lattice diffusion coefficient and it will lower the oxygen lattice diffusion coefficient. The blocking potential of chromium ion on fast diffusion sites can result in reduced grain boundary diffusion of the nickel ion [46]. Cr^{3+} doping would have similar effects on the effective diffusion coefficient of sintering of both oxides for volume diffusion or grain boundary controlled kinetics.

By postulating grain boundary controlled kinetics with two mutually dependent mechanisms for both regions in Fig. 4, one obtains a more coherent picture with respect to:

i) the low values of the homologous temperatures of the sintering experiments;

ii) the measured values of the activation energies;

iii) the analysis of the sintering kinetics of ZnO in the second region with the intermediate stage models[33];

iv) the net effects of Cr^{3+} in decreasing the shrinkage rates irrespectively of the region and of the oxide (given the reported values of chromium grain boundary segregation in these oxides at sintering temperatures, grain boundary processes are expected to be more sensitive to chromium doping than lattice kinetics [44,45]).

Particle coalescence was faster in Cr^{3+} doped samples than in the pure oxides, surface transport being promoted by the additions. Particle size differences between pure and doped materials cannot account for the differences of sintering rate at constant temperature and density.

Cr^{3+}in NiO, ZnO and CoO modifies the structure of free surfaces and of the grain boundaries [45,47], thus promoting important changes in the sintering behaviour of these oxides. The effects of chromium on the interfaces of ZnO and NiO will be further investigated by Auger electron microscopy in relation to the sintering behaviour.

4. Conclusions

The present results show that the diffusivities but also the structure of the interfaces must be investigated in order to explain the sintering kinetics of ceramics.

The linear shrinkage kinetics in solid state sintering of fine powders is explained as due to a grain boundary sliding mechanism.

Cr^{3+} additions change in similar ways the sintering kinetics of ZnO and NiO, that being inferred to be the result of grain boundary processes.

5. Acknowledgments

The authors gratefully acknowledge the financial support from INIC through the research project 2B85 of the Centro de Ceramica e Vidro do INIC (Universidade de Aveiro)

6. References

1. Su, M. *et al*.., Transport in Nonstoichiometric Compounds, Simkovich et al. Eds, NATO ASI Series B: **129**, 385, 1985
2. Johnson, D . L ., Mat . Sci . Res ., **11** ,137, 1978
3. Kingery, W. D., Berg, M., J. Appl. Phys., **26**, 1205, 1955
4. Coble, R. L., Gupta, T. K., Sintering and Related Phenomena, Kuczynski, G. C. Ed., Gordon and Breach, New York, 581, 1967
5. German, R. M., Munir, Z. A., Mat. Sci. Res., **10**, 259, 1975
6. Barringer, E. A.*et al*., Mat. Sci. Res.. **16**, 1, 1984
7. Astier, M. *et al*., Mat. Sci. Monographs, **4**, 150, 1979
8. Lange, F. F., J. Am. Ceram. Soc., **67**, 83, 1984
9. Hsueh, C. H., De Jonghe, L. C., J. Am. Ceram. Soc.. **67**, C-215, 1984
10. Weiser, M. W., De Jonghe, L. C., J. Am. Ceram. Soc., **69**, 822, 1986
11. Langdon, T ., Mat. Sci. Res . , **5**, 279, 1971

12. Vergnon, P. G. *et al.*, Sci. Ceram., **5**, 47, 1970
13. Uskokovic, D. *et al.*, Sci. Sintering., **9**, 265, 1977
14. Exner, H. E. *et al.*, Mat. Sci. Res., **10**, 279, 1975
15. Cutler, I . B ., J . Am . Ceram. Soc., **52**, 14, 1969
16. Vergnon, P. *et al.*, Rev. Int. Hautes Temper. et Refract., **9**, 275, 1972
17. Vergnon, P. *et al.*, Rev. Int. Hautes Temper. et Refract., **1**, 27, 1964
18. Wermuth, F. R. *et al.*, J . Am. Ceram. Soc., **56**, 401, 1973
19. Varela, J. A., J. Am. Ceram. Soc., **66**, 77, 1983
20. Moriyoshi, Y., J. Am. Ceram. Soc., **53**, 671, 1970
21. Kuczynski, G. C., Mat. Sci. Monographs, **4**, 245, 1979
22. Gifkins, R. C. *et al.*, Trans. Met. Soc. AIME, **239**, 910, 1967
23. Clarke, D. R., Mat. Sci. Res., **14**, 307, 1981
24. Young, W. S., Cutler, I. B., J. Am. Ceram. Soc., **53**, 660, 1970
25. Ashby, M. F., Scripta Met., **3**, 837, 1969
26. Bollman, W ., Surf. Sci., **31**, 1, 1972
27. Eastman, J. *et al.*, Adv. Ceram., **10**, 324, 1984
28. Dhalenne, G. *et al.*, Adv. Ceram., **6**, 139, 1983
29. Tasker, P. W., Adv. Ceram., **10**, 176, 1984
30. Wyckoff, R. W. G., Crystal Structures, John Wiley and Sons, New York, 528, 1968
31. Duffy, D. M.,Tasker, P. W., Adv. Ceram., **10**, 275, 1984
32. Carter, C. B., Morrissey, K. J., Adv. Ceram., **10**, 303, 1984
33. Senos, A. M. R. *et al.*, Sci. Ceram., **14**, 265, 1988
34. Hoshino, K. ,Peterson, N. L., J. Phys. Chem. Solids., **45**, 963, 1984
35. Kröger, F. A., The Chemistry of Imperfect Crystals, **2**, 2d ed., North-Holland, 743, 1974
36. Hagemark, K. I., J . Solid. State Chem., **16**, 293, 1976
37. Norris, L. F., Allsop, M. J., Trans. Farad. Soc., **55**, 1986, 1959
38. Gupta, T. K., Coble, R. L., J. Am. Ceram. Soc., **51**, 521, 1968
39. Whittmore, 0. J., Varela, J. A., J. Am. Ceram. Soc., **64**, C154, 1981
40. Whittmore, 0. J., Powell, S. L., Mat. Sci. Res., **16**, 137, 1984
41. Jimenez-Melendo *et al.*., Sci. Ceram., **13**, 385, 1986
42. Newman, G., Z. Phys. Chem., **5**, 825, 1980
43. Cimino, A. *et al.*, .Z. Phys., Neue Folge, **45**, 153, 1965
44. Bender, B. *et al.*, J. Am. Ceram. Soc., **63**, 542, 1980
45. Nowotny, J., Mat. Sci. Forum, **29**, 99, 1988
46. Atkinson, A., 'Surface and Interface Mass Transport in Ionic Materials', to be published
47. Nowotny, J. *et al.*, J. Am. Ceram. Soc., **65**, 192, 1982

SINTERING OF Nd_2O_3 AND CERAMIC STABILITY TO HYDRATION

J.M. HEINTZ[1], P. POIX[2] and J.C. BERNIER[2]

1) Laboratoire de Chimie du Solide. LP 8661 du CNRS.
351, Cours de la Libération. 33405 TALENCE Cedex (FRANCE)
2) Institut de Physique et Chimie des Matériaux de Strasbourg.
UM 390046 CNRS/EHICS/ULP.
1, rue Blaise Pascal. 67008 STRASBOURG Cedex (FRANCE)

ABSTRACT. Study of the sintering of Nd_2O_3 made possible to determine the best conditions for its high densification. Unfortunately, the compacts show a very poor resistance to humidity. This is mainly due to the A hexagonal structure. Stabilization against hydration was attempted by coating the Nd_2O_3 grains, to avoid the existence of an interface oxide-OH^-. Sintering of samples doped with TiO_2 leads to the hexagonal oxide which appears as Nd_2O_3 particles surrounded by an intergranular phase. The later was identified as Nd_2TiO_5 by use of microprobe analysis. The resistance to hydration of A-Nd_2O_3 was then four times improved.

1. Introduction

It is known that the first sesquioxides of the lanthanide series are extremely sensitive to hydration [1-3]. This phenomenon is an hindrance to their use as bulk ceramics, and explain that the sintering of these oxides has not been studied extensively [4-6]. The present paper deals with a first study of a commercial neodynium oxide powder, taking into account the two sides of the problem (sintering/hydration).

In a first time, we shall study the sintering behavior of Nd_2O_3 using classical parameters such as temperature, time, atmosphere composition. Unfortunately, the moisture influence on this oxide is not only noticeable on the

L.-C. Dufour et al. (eds.), Surfaces and Interfaces of Ceramic Materials, 565–574.

starting powder and during the sintering process but also on the aging of the sintered compacts. As far as the interface gas-oxide is first involved, we propose to realize a chemical coating of the grains to avoid the hydroxide formation and finally the destruction of the pellets.

2. Materials and Methods

2.1. EXPERIMENTAL METHODS

X-ray powder diffraction was used to identify the phases studied. The investigations were carried out using a Philips PW 1130/00 and Cu Kα radiation.

The granulometric distribution was determined by use of a laser diffusion type analyser with a lower detection limit of 1.2 µm (Malvern particle sizer 3600E).

The powders were compacted under unidirectional pressure of 300 MPa in order to obtain cubic samples (7 mm width). For sintering studies, we used an Adamel Lhomargy DI 24 dilatometer, under different atmospheres. The final temperature, 1440°C, was maintained for 10 mn . After cooling down to room temperature, the pellets were weighted and measured for density calculations. The apparent specific mass was determined independently by the water displacement method, according to the french standard NF A95-311.

Images of the sintered samples were obtained by use of a Jeol JSM 840 S.E.M.

Microprobe analysis were performed on sintered compacts using a Siemens apparatus.

2.2. POWDER DESCRIPTION

The neodynium sesquioxide (Nd_2O_3), used as starting material in this investigation, was obtained from Rhône Poulenc Recherches. Its purity was 99.95 % and its structure hexagonal. The average grain size was 4 µm with a gaussian distribution. The surface area was 1.8 m^2/g.

X-ray and TGA measurements revealed that the powder was hydrated : the diffraction pattern of $Nd(OH)_3$ was observed and a 2 % weight loss was noticed at 700°C. DTA and IR investigation confirmed that this was mainly due to neodynium hydroxide and in a lower proportion hydroxicarbonate. It was therefore necessary to calcine the starting powder at 700°C to obtain pure Nd_2O_3.

Titania used in order to realize Nd_2O_3 doping was a commercial rutile powder (Merck, 99 %). Another way has been proposed for doping which uses a sol-gel route with titanium ethoxide (Alfa products).

3. Results

3.1. SINTERING STUDY

The hydration phenomenon is critical for all the steps of the sintering. Obviously, the use of an organic binder in aqueous solution is impossible. Indeed, a green prepared without binder still remains sensitive to hydration : for example the green density falls from 4.25 down to 3.8 in two days and the pellets collapses a few days later (Fig. 1). It is then necessary to sinter the pellets just after pressing.

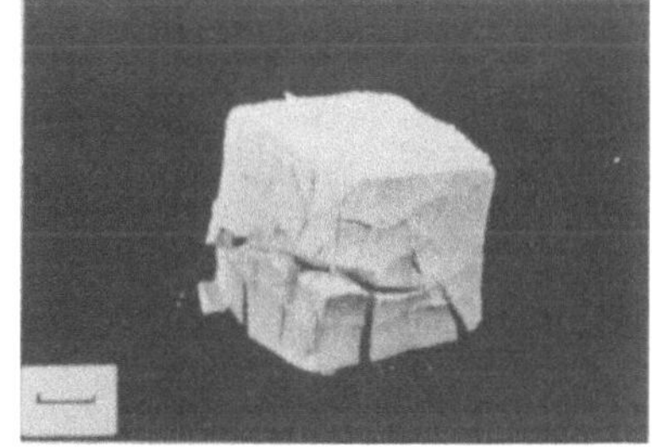

Fig. 1 : green pellet decomposed. bar=5 mm

Sintering of the crude powder (containing hydroxide) or of an hydrated green leads to ceramics with densities in the range 86-89 % of the theoretical. Moreover they show a lot of large cracks. On the other hand, using the calcined powder give final densities of about 91 % (Table I). Furthermore, the sintering is favoured by use of Ar or N_2 atmospheres (lack of H_2O or CO_2).

In order to improve densification of the oxide powder, its grinding was investigated. The results are presented in Table II and the best value is obtained when the average grain size is reduced by half (d_{50}=2.05 µm). Grinding implies one more calcination before pressing if not performed away from air.

TABLE I : influence of hydration and atmosphere on sintering

sample	calcination	atmosphere	green density	sintered density*
1	no	air	4.2	87.6
2	no	N_2	4.2	89.2
3	green hydrated	Ar	4.0	86
4	yes	air	4.4	90.7
5	yes	Ar	4.4	91.2

* exprimed in % of theoretical density (7.33).

TABLE II : influence of grinding (sintering in air).

sample	calci-nation	grinding		grain size (µm)	post calci-nation	green density	sintered density*
		dry	alcool				
1	no	-	-	4	-	4.2	87.6
4	yes	-	-	4	-	4.4	90.7
6	no	20'	60'	3	no	4.0	90.1
7	yes	20'	60'	3	yes	4.35	95.1
8	no	20'	-	2.05	yes	4.5	97.9

* exprimed in % of theoretical density.

3.2. RESISTANCE OF THE COMPACTS TO HYDRATION

Even when they have high final densities, Nd_2O_3 ceramics are sensitive to hydration. The evaluation of their resistance to hydration is achieved, in air, by measuring the time for appearance of $Nd(OH)_3$ and beginning of decomposition of the compact. Another information is given by the time necessary to obtain an opaque suspension of Nd_2O_3 in boiling water. As a general rule, the ceramic stability in air is poor : the decomposition starts after 3-4 days and the compact is entirely turned into hydroxide powder after two weeks. Nevertheless, a better densification increases the resistance to hydration (Table III). The air moisture attack versus time is illustrated by Fig. 2A through 2C. Growth of the neodynium hydroxide crystallites occurs preferentially in the macroscopic defects of the sintered sample, producing then its destruction.

TABLE III : resistance to hydration of the ceramics

Sample	% TiO_2	Sintered density	Decomposition time in air (days)	Decomposition time in H_2O (minutes)
4	0	6.64	4	4
8	0	7.18	-	8
9	10	6.98	13	-
10	10	6.57	11	-
11	20	5.65	14	16
12	Nd_2TiO_5	4.47	22	22

Fig. 2 : evolution of a ceramic hydrated versus time. bar = 1 mm.

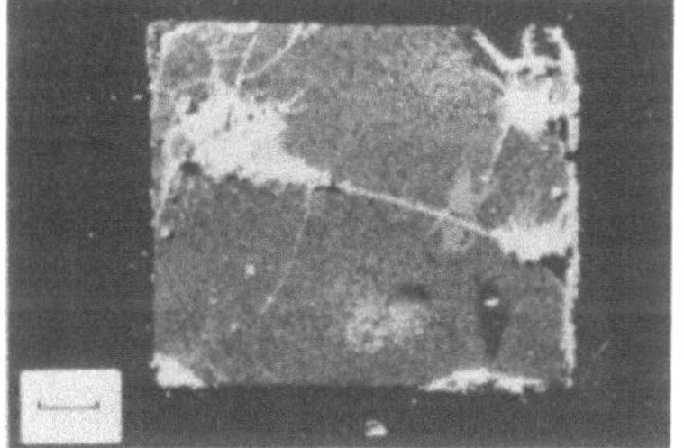

2 A : time = t_0

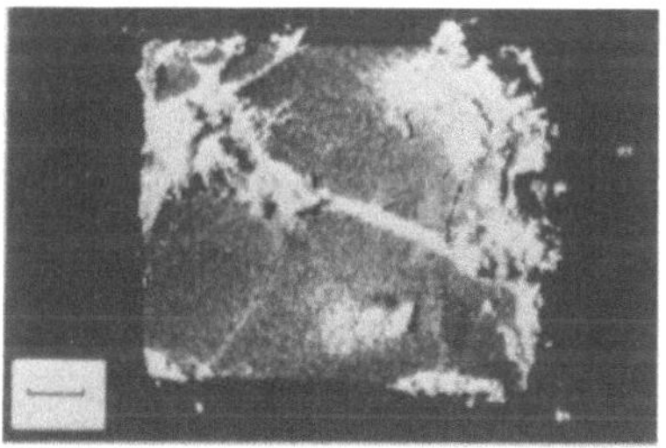

2 B : time = t_0 + 10 days

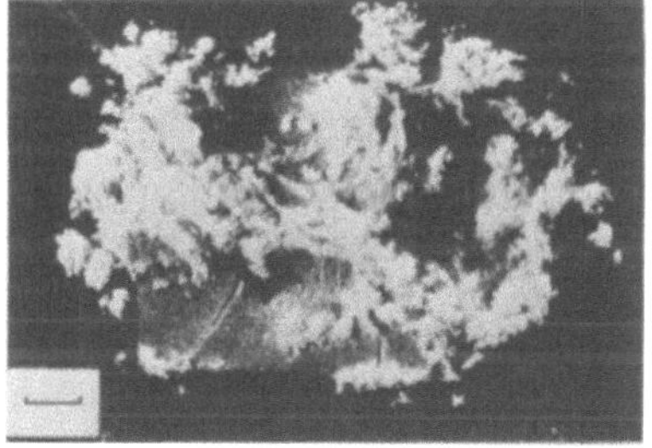

2 C : time = t_0 + 14 days

3.3. STABILIZATION OF THE CERAMICS

Stabilization of neodynium oxide ceramics is attempted by "protecting" the oxide grains with another compound which produces partial solid solutions on their outer core. For that purpose, titania doped samples can be prepared by solid state reaction. The following compositions have been used :

$(Nd_2O_3)_{0.9}(TiO_2)_{0.1}$	sample 9
$(Nd_2O_3)_{0.8}(TiO_2)_{0.2}$	sample 11
$(Nd_2O_3)_{0.5}(TiO_2)_{0.5}$	sample 12

An extra 10 % doped sample has been obtained using an alkoxide route (sample 10). The oxide powder was dispersed in an alcoholic solution (absolute ethanol with $Ti(OC_2H_5)_4$). Hydrolysis of the solution causes the precipitation of the titanium hydroxide. The resulting powder was calcined at 700°C for a night.

Sample 12 was identified by X-ray measurements as having the structure of Nd_2TiO_5. All other samples show the A-structure of Nd_2O_3 and the spectra exhibit very low intensity pattern which could be attributed to Nd_2TiO_5.

Addition of TiO_2 in low quantities does not modify strongly the sintering behavior of Nd_2O_3. For a given treatment, the shrinkage kinetics and the final densities are the same. All results of measurements of the resistance to hydration are presented in Table III. For a given density, they show, in all cases, an improvement by a factor close to four of the stability to moisture. Finally, one can notice that Nd_2TiO_5 is also hydration sensitive.

4. Discussion

4.1. STRUCTURAL DESCRIPTION

In the present work, Nd_2O_3 has the A-structure (hexagonal). As reported by CARO [7] this phase is anisotropic. It is constituted by a stacking of layers of $(LnO)_n^{n+}$ separated by planes of oxygen (Fig. 3). SEM photomicrographs (Fig. 4 and 5) of ceramic surfaces are characteristic of this sheet structure, showing the cleavage planes.

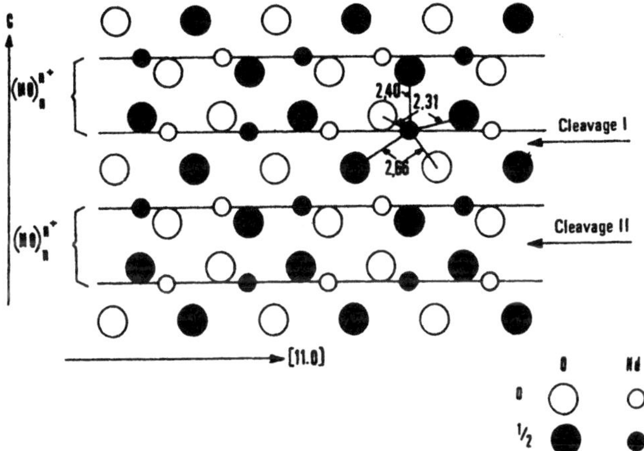

Fig. 3 : projection of the A-structure of Nd_2O_3 on the (110) plane (from [8]).

Fig. 4 : surface of a Nd_2O_3 ceramic. bar = 3 μm.

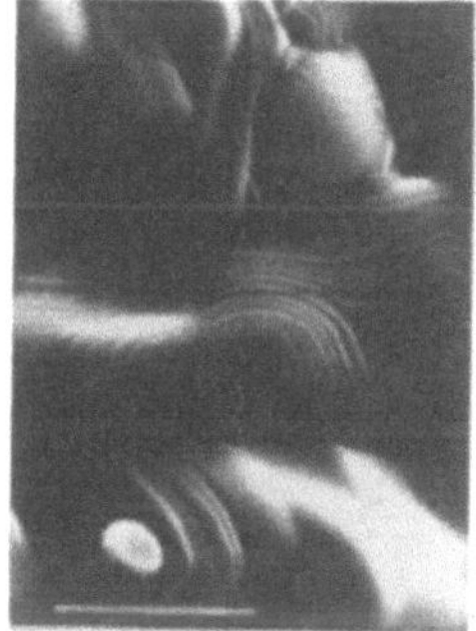

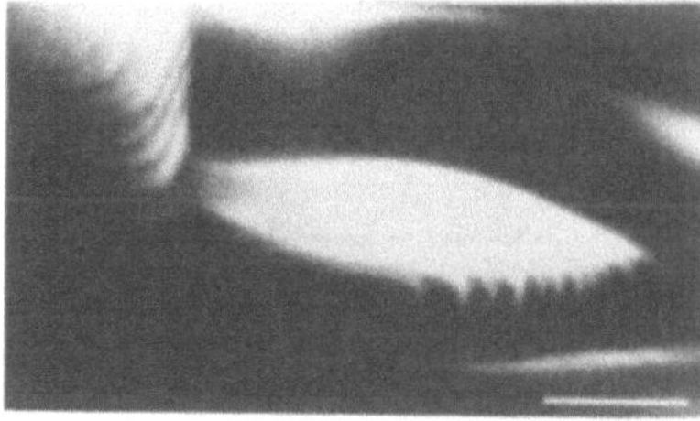

Fig. 5 : surface of a Nd_2O_3 ceramic at higher magnification. bar = 1 μm.

The mechanism of hydration can be understood from this structural description. An ionic character prevails in the metal-oxygen bonds between the $(LnO)_n^{n+}$ layers. Cleavage at this point (I of Fig. 3) implies then large residual charges (resulting from the fracture) which can be neutralized by a polar reagent.

Fig. 6 : sheet structure of a ceramic hydrated. bar = 10 μm.

DEXPERT [9] has shown that growth of hydroxyde on the oxide lattice occurs without any extensive reconstruction of the structure. This explains the easy hydration of Nd_2O_3 by water penetration between the $(NdO)_n^{n+}$ layers and the oxygen planes, and further decomposition of the ceramics. A look at a decomposed ceramic confirms this mechanism (Fig. 6) : the sheets that can be observed correspond to the original stacking of the layers in the oxide.

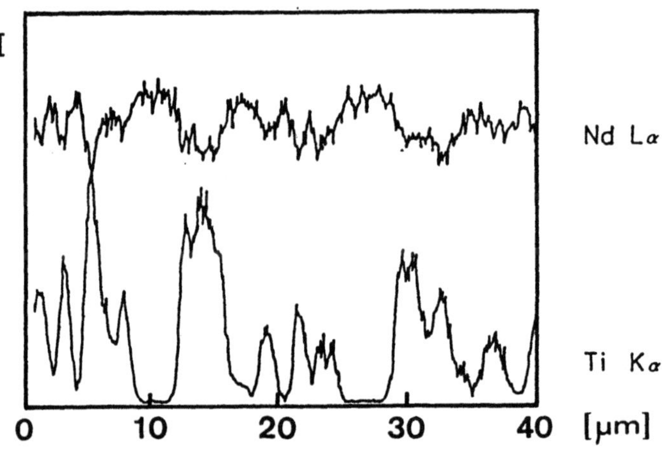

Fig. 7 : microprobe analysis of a $(Nd_2O_3)_{0.8}(TiO_2)_{0.2}$ ceramic.

4.2. DISCUSSION

It is necessary to obtain an adhesive and protective layer around the grains in order to avoid an interaction Nd_2O_3-OH^-. Consequently, the stabilization of the ceramic depends on the quality of the coating of the grains. Microprobe analysis of sample 11, treated at high temperature (Fig. 7), reveals actually Nd_2O_3 grains surrounded by an intergranular phase. Scanning allowed us to determine the size of the oxide grains, which lies between 2 and 10 μm when the intergranular compound has a thickness of 1 to 4 μm. Sample 12 presents a flat microprobe spectrum which is consistent with the X-ray pattern and allows to assert without any doubt that the coating phase of the other samples is Nd_2TiO_5. The later compound has an orthorhombic structure with distances neodymium-oxygen lower than the 2.66 Å bond which facilitates cleavage of the hexagonal structure. But in fact, addition of TiO_2 corresponds to TiO_4 tetrahedron inlaying in a Nd_2O_3 matrix [10]. It explains that this phase itself is sensitive to hydration.

When a microstructure with coated grains is well established, as for example in sample 11, the ceramic stability is greatly improved (Table III). This is even true in the case of a poor densification. Optimization of the sintering (sample 11) or of the microstructure (samples 9-10) would certainly increase the resistance to water, up to that of Nd_2TiO_5. But, as far as the protective layer Nd_2TiO_5 is also moisture sensitive, one cannot expect to exceed an improvement by a factor 5 of the time of resistance to hydration.

5. Conclusion

Nd_2O_3 ceramics with a high density (98 % d_{th}) can be realised by sintering at 1400°C. However, hydration remains a major problem for their use as bulk ceramics. This phenomenon, which occurs at the interface between the oxide grains and the atmosphere, may be slackened by a chemical coating of the grains. Doping with TiO_2 leads to a microstructure of A-Nd_2O_3 grains surrounded by Nd_2TiO_5. The sensitivity of the material to hydration in air is then decreased by a factor between four and five. A best choice of the intergranular phase and a more complete study of the hydroxide diffusion through this interface could lead to a real stabilization of the ceramics.

Acknowlegdements. This work was supported by 'Rhône Poulenc Recherches' under contract 1983/08/31.

References

[1] D. TOURET and F. QUEYROUX. 'Sur la réactivité vis-à-vis de l'eau de quelques matériaux réfractaires à base d'oxyde de lanthane' Rev. Chim. Min., 9, 883, (1972)

[2] S. BERNAL, F.J. BOTANA, R. GARCIA and J.M. RODRIGUEZ-IZQUIERDO. 'Thermal evolution of a sample of La_2O_3 exposed to the atmosphere' Thermochim. Acta, 66, 139, (1983)

[3] S. BERNAL, J.A. DIAZ, R. GARCIA and J.M. RODRIGUEZ-IZQUIERDO. 'Study of some aspects of the reactivity of La_2O_3 with CO_2 and H_2O' J. Mater. Sci., 20, 537, (1985)

[4] G.L. PLOETZ, C.W. KRYSTYNIAK and H.E. DUMAS. 'Sintering characteristics of rare-earth oxides' J. Am. Ceram. Soc., 41(12), 551, (1958)

[5] E. KARNALL and D. PEARLMAN. 'Transparent Gd_2O_3 ceramics and phosphors' Mat. Res. Bull., 7, 647, (1972)

[6] L.S. ALEKSEENKO, E.L. ZOZ, N.T. DYRA and N.V. GULKO. 'The structure and stability to hydration of ceramics based on Nd_2O_3' Russ. J. Inorg. Chem., 27(3), 825, (1982)

[7] P.E. CARO. 'OM_4 tetrahedra linkages and the cationic group $(MO)_n^{n+}$ in rare earth oxides and oxysalts' J. Less Common Metals, 16, 367-377, (1968)

[8] A.M. LEJUS and R. COLLONGUES. 'Lanthanide oxides' Current Topics in Materials Science, ed. E. KALDISS (North Holland Publishing Company), 4, chap 9, 481 (1980)

[9] H. DEXPERT and P.E. CARO. 'Processus réactionnels en phase solide pour certains composés des terres rares' Phys. Stat. Solidi., 45(a), 297, (1978)

[10] H. MULLER-BUSCHBAUM and K. SCHEUNEMANN. 'Zur kenntnis von Nd_2TiO_5' J. Inorg. Nucl. Chem., 35, 1091, (1973)

CONTROL OF CARBON-SILICATE INTERFACE AREA IN THE PREPARATION OF SiMON CERAMIC PRECURSORS FROM CLAYS.

F. KOOLI AND F. BERGAYA
C.N.R.S.-C.R.S.O.C.I.
F-45071 ORLEANS CEDEX 2
FRANCE

ABSTRACT. We describe the preparation and the microtexture of three classes of precursors for the synthesis of ceramic powders or films of the SiMON family (M=Al, Mg, Y). (i) simple mechanical mixtures of a non-swelling clay of the kaolinite family with charcoal, in which the two phases are separated at micronic or slightly submicronic level. The interface area, S, is of the order of a few m2/g. (ii) colloidal microcomposites obtained by a sol-gel method from mixed colloidal suspensions of a swelling clay (montmorillonite or hectorite) and graphitic oxide. This method, which permits the casting of films a few microns thick, leads to a material in which the two phases are dispersed as lamellar stacks a few nanometers thick. S is of the order of 100 m2/g. (iii) layered clay-carbon nanocomposites. The method involves the intercalation of acrylonitrile vapor in a swelling clay, followed by polymerisation and carbonisation of the organic molecules in the interlayer space. It leads to a material in which the carbon and silicate phases are organized as alternate layers of subnanometric thickness. S is larger than 800 m2/g.

1. Introduction

It is quite obvious that a high degree of dispersion favors the reaction between solid phases and often opens a way to low reaction temperatures and to small particle sizes . The purpose of this paper is to report about several ways to improve the degree of dispersion of carbon and clays, with the final purpose of synthesizing nitrogen ceramic powders and films.

Natural silicious materials have already been used for the synthesis of silicon nitride and sialons. Kaolinite clays have been nitrided by thermal reduction with carbon or aluminium powders [1,2], whereas montmorillonite clays have been nitrided by thermal reduction with intercalated organic molecules as reducing agents [3]. The first method is quite trivial since it involves merely the mechanical mixing of

L.-C. Dufour et al. (eds.), Surfaces and Interfaces of Ceramic Materials, 575–585.

kaolinite with a carbon source such as charcoal for instance. The kaolinite particles are rigid platelets which are typically a few tens of nanometers thick and a few hundreds of nanometers wide. Their surface area is in the range from 25 to 50 m2/g. Powdered carbons may develop a very high BET surface area, but they form often aggregates or agglomerates which are several hundreds or thouzands of nanometers large. In fact, the relevant surface area is the *external* surface area of the agglomerates, which is only a few square meters per gram, as estimated from the agglomerate size of charcoal. As a consequence, the clay-carbon interface area is quite low, even in pressed pellets. This is sketched in figure 1.

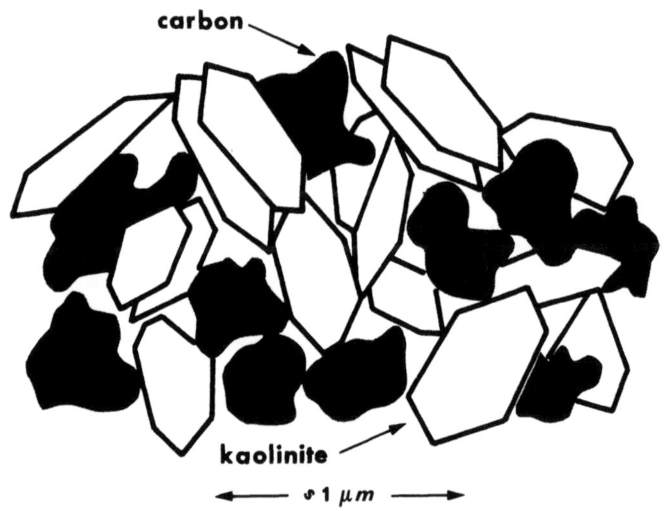

Figure 1. Sketch of a kaolinite - charcoal mixture.

We will concentrate on two other preparation methods, which allow for a more intimate contact between the carbon and the clay phases. The first one is the intercalation method already mentionned. The second one is a sol-gel method, which is carried out in aqueous medium with a swelling (or expanding) clay and a swelling form of carbon: graphitic oxide.

2. Materials

Swelling clays, or smectites. The "backbone" of clays is a continuous two-dimensional layer of tetrahedra sharing three corners. The apical oxygen at the fourth corner, usually directed normal to the sheet, forms part of an immediately adjacent octahedral sheet in which octahedra are linked by sharing edges. The unit formed by linking one octahedral sheet to one tetrahedral sheet is named a 1:1 layer, whereas the unit formed by sandwiching one octahedral layer between two tetrahedral sheets is named a 2:1 layer. 1:1 and 2:1 layers are not always neutral. Very often, as a result of ionic substitutions in their lattice, the layers bear a net negative charge which is balanced by interlayer cations. The highest charges are found in 2:1 layers. The layer type, the charge on the layer and the nature of the interlayer material provide the basis for the classification of clay minerals.

Smectites are 2:1 clays with properties almost opposite to those of kaolins. The layer charge is large, typically 0.5 negative charge per formula unit layer (it is zero in kaolinite, which has a 1:1 structure). This corresponds to a cation exchange capacity of the order of 1 meq/g. Statistically, each lattice charge covers a surface area of the order of 1.3 nm^2. In natural smectites, this charge is compensated by hydrated interlayer sodium or calcium cations. Most important, the hydration energy of the interlayer cations allow water to enter the interlayer space. This makes the original interlayer cations easily exchangeable by almost any cationic species. The interlayer space is also accessible to a wealth of neutral molecules [4]. Morphologically, smectites appear as large (~500 nm) and thin (~1 nm) flexible sheets. The most common members of the family are *montmorillonite* and *hectorite* (figure 2). In montmorillonite, the octahedral cations are Al ions and the interlayer charge originates from the substitution of some Al^{3+} by Mg^{2+}. In hectorite, the octahedral cations are Mg^{2+} and the lattice charge originates from the substitutions of some Mg^{2+} by Li^+.

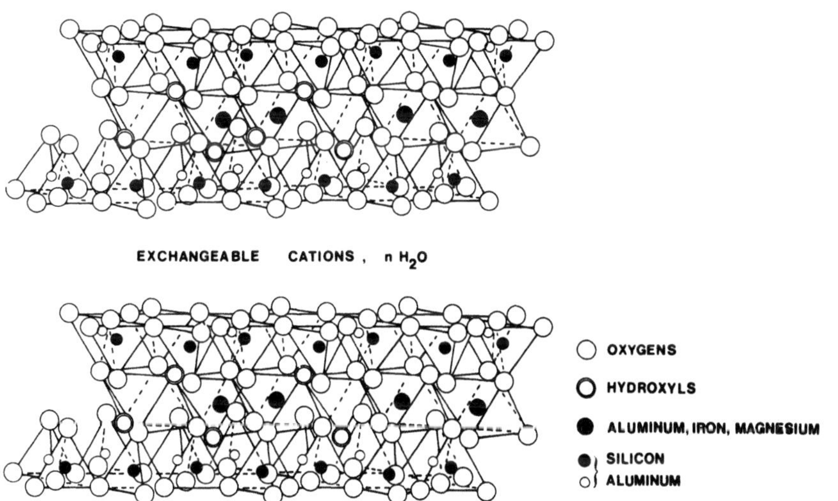

Figure 2. Stereoscopic view of the montmorillonite structure.

Among the best known properties of smectites is their ability to expand (to "swell") and to shrink upon addition or removal of water. They are also able to form stable colloidal suspensions and gels, when the solid/liquid ratio exceeds some critical value. Swelling is clearly associated with a statistical increase of the average distance between smectites layers. However, even in very dilute colloidal suspensions, the layers are not totally dissociated [5]. The basic microstructural units in smectite suspensions are aggregates of a few layers more or less parallel to each other. The size of the aggregates is strongly dependent upon the ionic strength of the medium, the nature of the counterions, and the presence of adsorbed (intercalated) molecules. High ionic strengths lead to the growth of the aggregates in all directions and flocculate the clay. Couterions with a large charge increase the thickness of the stacks.

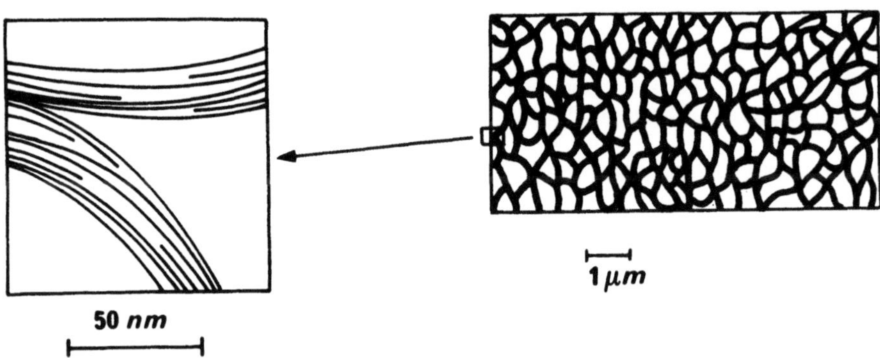

Figure 3. Microtexture of a montmorillonite gel in water, according to TESSIER et al. [6].

Smectite gels (figure 3) are formed by increasing the concentration of solid in water. Their structure is a continuous tridimensional network of delaminated and entangled aggregates, with a large porosity. The bending of the layers may be dramatic, but most layers keep in touch with their neighbours by face-to-face contacts (figure 4). Gels are most easily obtained with low-valent couterions which permit a high dispersion rate of the clay layers. The existence of stacked aggregates considerably modifies the solid-water interface area in suspensions and gels (and also the interface area betwwen the clay and other solid particles). A smectite suspension in which all the layers would be separated from each other would have a surface area close to 750 m^2/g. In addition, all this area would be in direct contact with the aqueous phase and could be used as an interfacial area with other solid particles. As soon as aggregation occurs, a fraction of the total area becomes part of the interlayer space of the clay and becomes "internal". The "external" area decreases accordingly. For instance, in a Na^+-hectorite/water suspension containing 10% by weight of solid, the external surface area drops to ~ 70 m^2/g [5].

Thus, if one is willing to increase the contact area between a smectite clay and another colloidal solid in aqueous medium, one has to use dispersion conditions which prevents the stacking of the clay layers. Quite obviously, using a colloidal solid with lamellar morphology would further improve the contact. This can be achieved with graphitic oxide.

Figure 4. Delaminated aggregate of smectite layers in water, with a large distribution of interlayer distances.

Graphitic oxide (GO). GO is a colloid which is remarkably similar to smectite clays. It has been known for more than a century [7]. It is obtained by oxidizing graphite with potassium chlorate in concentrated nitric acid [7] or in nitric-sulfuric acid mixtures [8]. Like the parent material, GO has a lamellar structure, but it is no longer strictly planar due to the loss of the sp_2 character of most of the carbon atoms. Its chemical composition is subject to some variability, according to the preparation method, the graphite used to prepare it, the number of oxidation treatments. The carbon/oxygen ratio ranges from ~ 1.7 to ~4. GO contains C-OH groups, and C=O as well as C=C bonds (figure 5). Most interesting, GO has an important ion exchange capacity (pH dependent) and swells in water, like smectites.

3. The Sol-Gel Route to Colloidal Composites

Thanks to the close similarity of morphological, colloidal , as well as surface chemical properties of smectites and graphitic oxide, one can easily prepare mixed aqueous suspensions of both materials. In order to decrease the length scale of the heterogeneity, dilute suspensions - typically below 1% of solid in water- have to be used. This prevents the growth of thick stacks of clay sheets on one hand, and thick stacks of GO on the other hand. Starting from a mixed dilute suspensions, several treatment can lead to a solid material. Freeze drying is the first one. It leads to a very low density material which has to be compacted before being used further. Spray drying is another possibility.

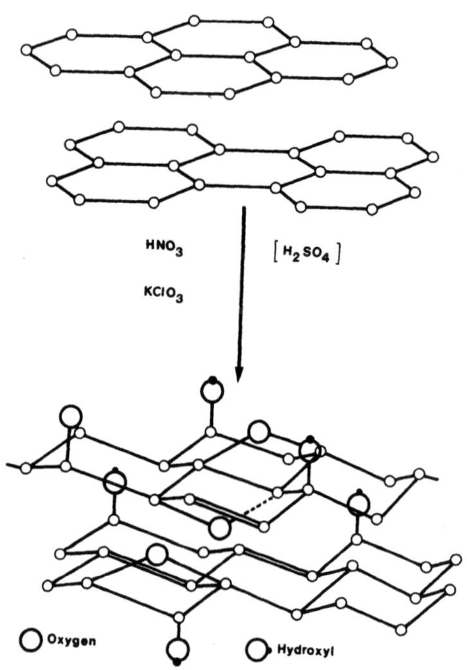

Figure 5. Model for the transformation of graphite into graphitic oxide (from CLAUSS, PLASS, BOEHM and HOFMANN [9]).

Simple evaporation of the water from the suspension is particularly interesting. As the water evaporates, the system passes through a gel state and, finally, leads to oriented and mechanically coherent deposits which can eventually be peeled from the support on which the suspension was poured. Thin films from a few microns to several hundreds of microns without craks can be prepared in this way. X-ray diffraction on such a film (~ 4mg/cm2; 30 microns thick) shows, by using SHERRER's formula [10], that the coherence length of the stacks is of the order of 5 nm, both for the clay and the GO. This is sketched in figure 6. Taking into account the period of the stacks determined from the 001 reflection (a period is the sum of the thickness of a sheet and that of the interlayer water), this corresponds to about four clay sheets and six GO sheets. A simple calculation based on a ordered model of stacked clay and GO aggregates shows that this leads to an interface area of the order of 200 square meters per gram. In fact, due to the delamination of the structure and the resulting microporosity (figure 6), the actual figure is probably smaller.

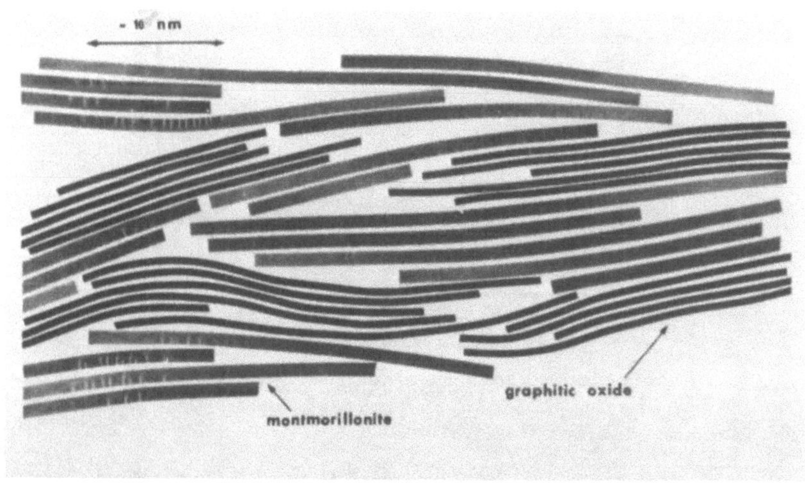

Figure 6. Microtexture of the montmorillonite - graphitic oxide colloidal composite obtained by the sol-gel method.

4. The Intercalation Route to Layered Nanocomposites

In 1984, SUGAHARA et al. [3] showed that Sialon powders could be obtained from a montmorillonite-polyacrylonitrile intercalation compound by carbothermal reduction. Acrylonitrile was first intercalated into an akylammonium - montmorillonite by immersing the clay powder in liquid acrylonitrile monomer. The intercalation compound was heat-treated at 50°C to polymerize the acrylonitrile, and further at 220°C for cyclisation of the polyacrilonitrile (remind that polyacrylonitrile, PAN, is one of the possible precursors for carbon fibers). Nitridation was achieved above 1150°C in a nitrogen atmosphere. On the other hand, KYOTANI et al. [11] showed recently that highly oriented graphite can be prepared from montmorillonite-PAN intercalates by first decomposing the intercalated polymer at 700°C, dissolving the silicate with HF and HCl, and finally heat treating the free carbonaceous sheets up to 2800°C.

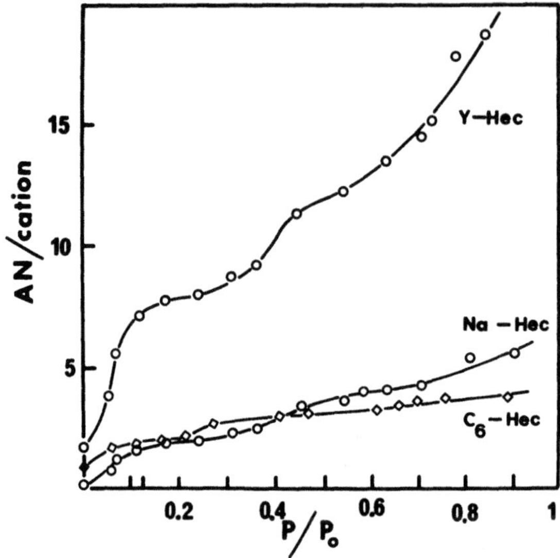

Figure 7. Adsorption isotherms of acrylonitrile vapor at room temperature on hectorite clays pre-exchanged with various cations: yttrium, sodium, and hexylammonium.

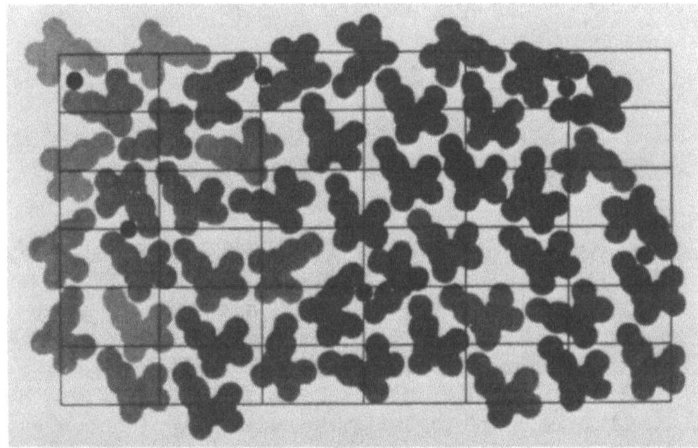

Figure 8. Arrangement of acrylonitrile molecules in the interlayer space of a montmorillonite or hectorite clay pre-exchanged with a trivalent ion such as Y^{3+}. The unit cell dimensions are a=0.52 nm and b=0.902 nm. The packing density shown corresponds to about seven molecules per cation.

In order to have a better control of the amount of intercalated monomer, we intercalated the acrylonitrile from the gas phase. The adsorption isotherms on three hectorite clays, ion-exchanged with yttrium, sodium and hexylammonium ions respectively, are shown in figure 7. A simple calculation from the molecular dimensions shows that, close to saturating vapour pressure (P_o), the interlayer gallery is filled with acrylonitrile molecules in close contact. The shift of the infrared-active C-N vibration around 2220 wavenumbers shows that the molecules are perturbed by the interlayer ions but, surprisingly, one is not able to separate the molecules close to the ions from those which are at a larger distance (figure 8).

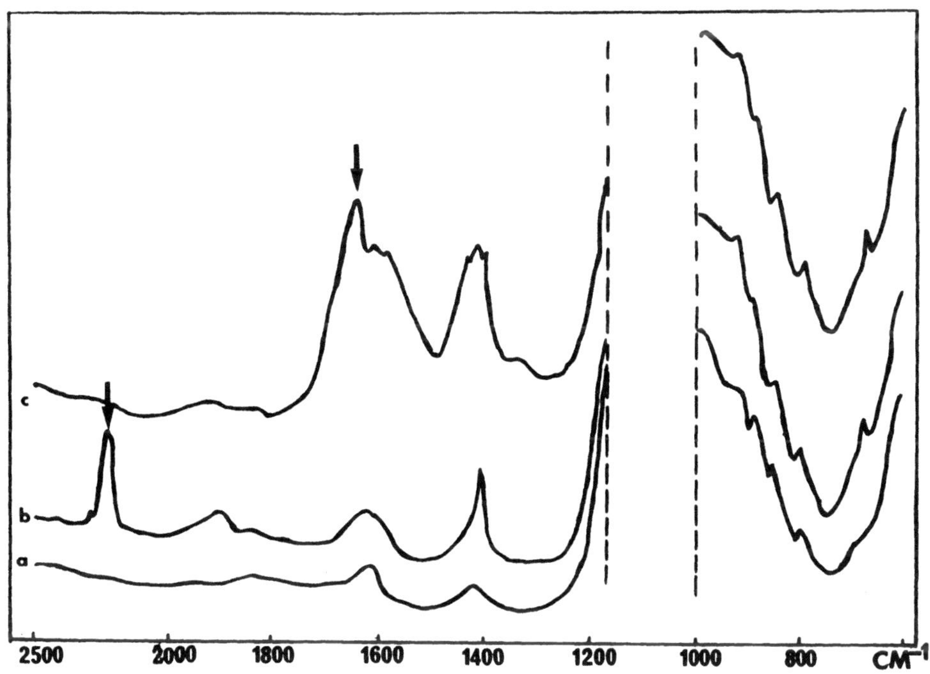

Figure 9. Infrared absorption spectra of a montmorillonite film (a), after intercalation of acrylonitrile (b), and polymerization-cyclization of the intercalated molecules (c). The arrows indicate the C-N stretching vibration band in (b), and the conjugated C=N obsorption band in (c).

We observed that polymerization could not be separated from cyclization. Both occur concomitantly at 220°C, as shown by the development of the IR band at 1660 wavenumbers, assigned to conjugated C=N bonds (figure 9). Heating further under nitrogen decomposes the PAN polymer into a layered carbonaceous compound with featurless spectra. At this stage, the material is a layered composite made of alternating carbon and silicate layers. The length scale of the heterogeneity is in the nanometer range (figure 10).

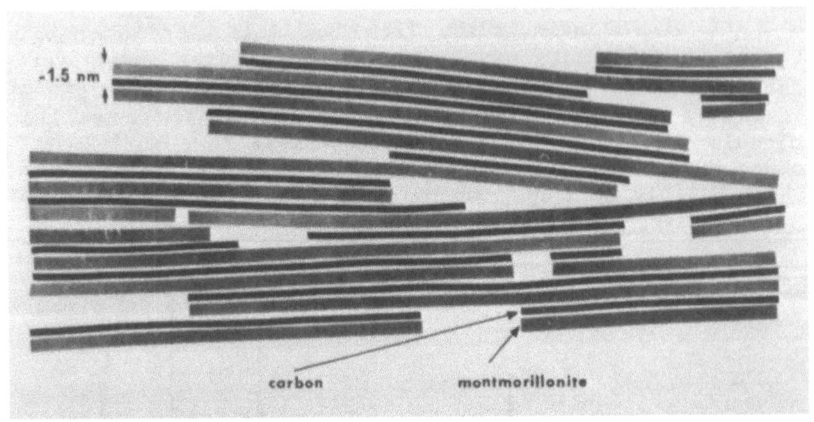

Figure 10. Microtexture of the carbon - silicate layered nanocomposite obtained by intercalating, polymerizing and decomposing a montmorillonite - acrylonitrile complex.

5. Conclusions

Unlike kaolinite, which has ben used so far for the synthesis of silicon nitride and Sialon from cheap materials, swelling clays, in combination with colloidal or polymeric carbon, offer the opportunity to construct very fine structures. The very short scale of their internal heterogeneity should allow for mild reaction conditions. In addition, their colloidal properties makes them able to be processed as matrix materials in composites or as thin coatings.

References

[1] J.G.LEE and I.B. CUTLER, *Am. Ceram. Soc. Bull.*, **58**, 869-871 (1979)

[2] H.YOSHIMATU, M.MITOMO, H.MIHASHI, S.OHMORI and T.YABUKI, *Yogyo Kyokaishi*, **91**, 442-449 (1983)

[3] Y.SUGAHARA, K.KURODA and C.KATO, *J. Am. Ceram. Soc.*, **67**, C247-C248 (1984)

[4] B.K.G.THENG, *'The Chemistry of Clay Organic Reactions'*, Adam Hilger, London (1974)

[5] J.J.FRIPIAT, J.CASES, M.FRANCOIS and M.LETELLIER, *J. Coll. Interf. Sc.*, **89**, 378-389 (1982)

[6] **D.TESSIER**, *'Etude Expérimentale de l'Organisation des Matériaux Argileux'*, INRA Ed., Versailles (1984); **H.BEN RHAIEM, C.H.PONS** and **D.TESSIER**, *Proc. Int. Clay Conf.*, Denver 1985, L.G.SCHULZ, H.VAN OLPHEN and F.A.MUMPTON Eds., 292-297 (1987)

[7] **B.C.BRODIE**, *Ann. Chim. Phys.*, **45** ,351 (1855) and **59**, 466 (1860); *Trans. Roy. Soc.*, **149**, 249 (1859)

[8] **L.STAUDENMAIER**, *Ber.*, **31**, 1481 (1898)

[9] **A.CLAUSS, R.PLASS, H.P.BOEHM** and **U.HOFMANN**, *Z. Anorg. Allgem. Chem.*, **291**, 205 (1957)

[10] **T.KYOTANI, N.SONOBE** and **A.TOMITA**, *Nature*, 331-333 (1988)

PREPARATION OF AN ALUMINA-ZIRCONIA SOL FOR PRODUCING MICROSPHERES

L. Montanaro
Politecnico - Dip. Scienza dei Materiali ed Ing. Chimica
Corso Duca degli Abruzzi 24
10129 - TORINO
Italy

ABSTRACT. A sol-gel process for producing microspheres of alumina with 10% by weight of zirconia has been developed. A fluid and concentrated sol has been obtained by peptization, in suitable conditions, of a gel of Al and Zr hydroxides. From this sol a powder of homogeneous chemical composition and characterized by well-defined geometric parameters (spherical shape, controllable size) has been produced in a pilot plant. The microspheres calcined at 1200°C were not agglomerated, were dense and uncracked, their average size was 20 microns and the particle size distribution was close to the average value. All the steps of this process (gel precipitation, washings, peptization, sol concentration, microspheres formation, heat treatment) have analitically been followed , above all by means of: X-Ray diffraction, thermogravimetric analysis, scanning electron microscopy, laser granulometry and micrograпulometry, rotational viscosimetry, B.E.T. technique.

1. INTRODUCTION

The properties of a ceramic mainly depend on the nature of the powders subjected to sintering (1): in fact all the powder characteristics, as chemical and phase composition, particle shape, size and size distribution, presence of agglomeration, determine, to a large extent, the microstructures developed during sintering (2).

Barringer and coll.(3) assert that it is possible to control the microstructure of a ceramic material; by closely controlling the chemical homogeneity of the powder, by using particles of spherical shape and sized between 0.1 and 1 micron, by having a particle size distribution close to the average value ($d_{max}/ d_{ave} < 3$) and by avoiding agglomeration.

Matijevic (4) and Kato (5) have also confirmed the importance of controlling particle size and shape.

Some researchers (6-7) have asserted that an ideal compact is obtainable by using monosized spheres: in fact, although a packing of different size and shape particles can give a high green density, its local densities are variable from point to point, and these cannot be determined by the compaction technique.

L.-C. Dufour et al. (eds.), Surfaces and Interfaces of Ceramic Materials, 587–598.

On the contrary, a packing of monosized spheres produces a uniform three-dimensional structure that will sinter uniformly and give rise to a uniform and reduced porosity (8).
An existing and successful method for preparing a ceramic powder with the described characteristics is the sol-gel process.
This process has become an area of particular interest and intense research because of its inherent advantages compared to more conventional processing, above all as regards the high purity and homogeneity of the obtained product (9).
Such technique is being used to produce many different items, including ceramic composites, ultrafilters, dental implants, catalysts, ferrites, solid electrolytes and optical gates (10).
Those applications that have reached the production stage have been very successful and the technology is on the threshold of a very rapid expansion. The market potential of sol-gel technology to produce ceramics and glasses from chemical precursors should approach $ 500 million by 1993 (10).
The process begins with a reactive precursor material that is converted into a final product by a chemical and/or thermal means. Schematically, a typical sol-gel process can be represented (11) as:
Starting material ---→ sol ---→ gel ---→ oxide.
Many methods for preparing a sol have been described in literature (11): in this work the peptization of well-washed metal hydroxides has been chosen.
This sol has been used for feeding a pilot plant for producing microspheres of controlled chemical composition and particle size distribution, as suggested by Fletcher and Hardy (12).
For the setting up of the process, an Al_2O_3-ZrO_2 gel has been used to obtain ceramic powders for producing toughened alumina whose properties can be compared to those of toughened aluminas produced in a traditional way (13-15).

2. MATERIALS AND METHODS

2.1 . Gel precipitation

To produce the mixture of aluminium and zirconium hydroxides, suitable quantities of $AlCl_3 \cdot 6H_2O$ and $ZrCl_4$ have been dissolved in distilled water to obtain an alumina doped with 10% by weight (7% by volume) of zirconia.
This ZrO_2 percentage has been chosen from the literature (16) as it falls in the 'optimum' area of the zirconia toughening effect on the matrix.
The inorganic salts of the two metals have been used as precursors, even if it is possible to obtain the aluminium and zirconium hydroxides from organo-metallic precursors. This choice has been motivated by the fact that the alkoxides, even if they produce a high purity gel, are generally expensive and extremely sensitive to moisture, heat and light (17).
The precipitation of the hydroxides has been carried out at 25°C by the addition of NH_4OH 4N (20 cc/min) with continuous stirring.
The pH variation during precipitation has been continually monitored by an electrode; using ζ potential measurement, a pH of 9 has been chosen

as the end of precipitation.

2.2 . Gel washing

The gel has been washed several times on a sintered filter and then centrifuged to remove the precipitation by-products,particularly the excess of Cl^- ions so that the level of Cl^- ions retained in the gel complied with the peptization conditions.
The washings have been carried out suspending everytime the gel in a quantity of distilled water whose weight was twice that of the gel and stirring the suspension for six hours.
During the washing the variation of the pH of the gel suspension and the Cl^- content in the gel have been followed.
The dimensional changes of the gel micelles have been checked by laser granulometry.
Any hydrated and crystal phases formed during the washings have been identified by using X-Ray diffraction and thermogravimetric analysis on the gel dried at 105°C.

2.3 . Peptization and sol concentration

The peptization (stabilization of a colloidal sol by the adsorption of electrolytes on the particle surfaces (9)) of the gel has been carried out to produce a stable sol that, at a suitable concentration, was fed into a pilot plant for the production of microspheres.
The peptization requires that a critical amount of certain acids be introduced into the gel and that the slurry be kept at a suitable temperature for a sufficient time (18,19).
In our case the gel is peptized by HCl addition in a sealed vessel at 80°C, for 48 hours.
This acid has been chosen since it is recommended to use as a peptizing agent an electrolyte having a common ion with the starting salts (20).
On the gel suspension, after washings, the molar concentration of residual Cl^- ions and of Al^{3+} ions have also been evaluated in order to know exactly the Cl^- contents in the gel, before and after the HCl additions for peptization, defined as Cl^-/Al^{3+} molar ratio.
The disaggregation of the micelles under the acid effect has been verified by laser granulometry and microgranulometry. Eventual crystallization has been checked by X-Ray diffraction using $CoK\alpha$ radiation.
The sol has then been concentrated at 80°C under stirring until an Al^{3+} ions content of about 4 mol/l was reached.
Again, the granulometry, to verify that agglomeration had not taken place, and X-Ray analysis, to verify if crystallization had occurred, have been carried out on the concentrated sol.
The rheological characteristics of the sol have been also determined using a rotational viscosimeter.

2.4 . Production of microspheres

The sol has been then introduced through a small nozzle and broken up into droplets which have been slowly dehydrated by contact with an organic liquid (1-octanol) and converted into easy to handle, solid gel microspheres.

The production of microspheres occurs in a pilot plant, previously described (21): a similar process was developed at the Oak Ridge National Laboratory for the production of large microspheres of Th and U oxides (22).
The microspheres have been subjected to a suitable heat treatment to convert the hydroxides into oxides.
The crystal phases developed during the heat treatment have been checked by X-Ray diffraction in a high-temperature chamber.
The specific surface areas of the powder during and at the end of the heat treatment have been determined by the B.E.T. technique.
The powder treated at 1200°C for 30 min. have been hot pressed (1550 - 1600°C, 40 MPa under vacuum). On the samples (4x3x18 mm^3) the flexural strenght σ_F (three point method) and the fracture toughness K_{Ic} (SENB method) have been evaluated as weel as the density of the sintered body.

3. RESULTS AND DISCUSSION

3.1 . Gel precipitation

The gel has been precipitated upto a neutralization pH of 9; in fact, the ζ potential at pH = 9 has been verified (23) to be close to the point of zero charge, that corresponds to maximum agglomeration.

3.2 . Gel washing

Figures 1 and 2 show that the pH of the aqueous gel suspension and the Cl^- content in the gel descrease in a different manner as a function of the number of washings.
After the first two washings, the Cl^- level decreases sharply to reach a practically constant value after the fifth washing: at this point the Cl^-/Al^{3+} molar ratio is 0.35.
The Cl^- ions level does not change after further washings: which presupposes that a close bond between gel and foreign ions is formed.
In Figure 3 the variation in the gel micelle size distribution as a function of the number of washings has been shown: their average dimension decreases from 35 to 7.5 microns.
The sharp decrease between the forth and the fifth washing is probably due to the fact that, from the fifth washing on, water has been extracted by centrifuging and no longer by filtering. After the fifth washing dimensional variations of the micelles are practically negligible.
The comparison of Figures 1,2,3 leads to think that the dimensional changes are due to the Cl^- removal from the gel rather than to the pH modifications.
The X-Ray analysis (Figure 4) showed that before washing the gel, contamined with NH_4Cl, contained small quantities of Gibbsite and Bayerite (Figure 4a); these latter two tended to crystallize in slightly increasing quantities as a consequence of the removal of foreign ions during washings (Figure 4b) (24).
The TGA confirmed the presence of trihydrate and also traces of Boehmite and pseudo-Boehmite (Figure 5) (25). The weight ratios between these three phases remain practically constant as washing proceeds.

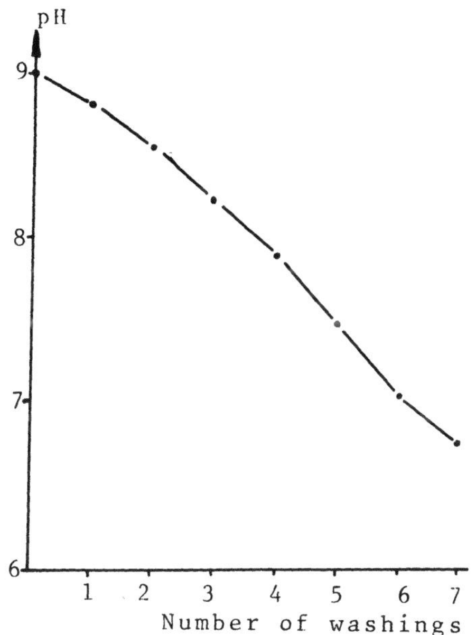

Figure 1

Variation of the pH of the gel suspension as a function of the number of washings.

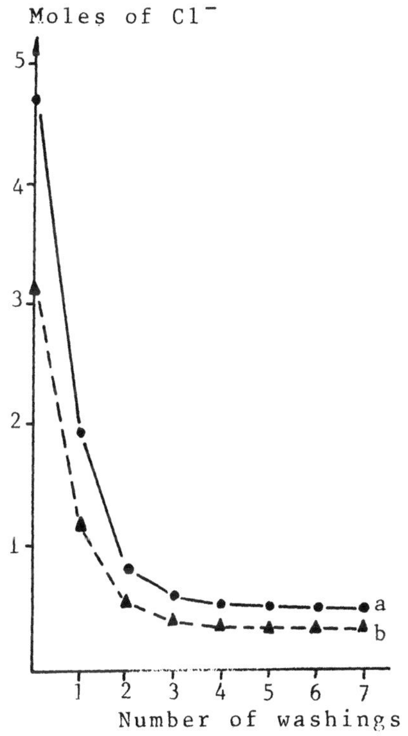

Figure 2

Influence of the number of washings on the Cl^- content:
(a) moles of total Cl^- in the gel;
(b) moles of Cl^- for one mole of Al^{3+}

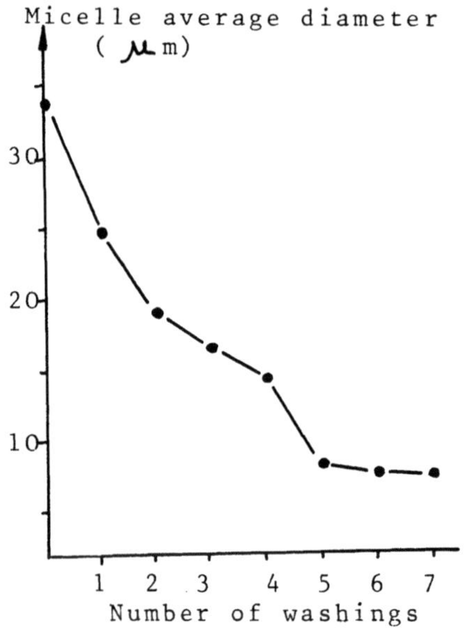

Figure 3

Variation of gel micelle average diameter as a function of the number of washings.

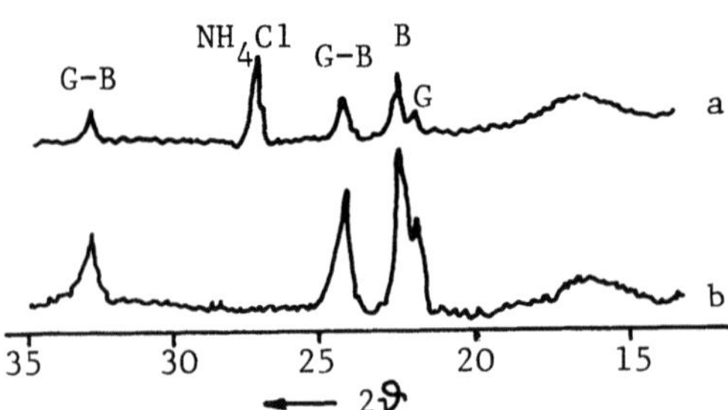

Figure 4

X-Ray diffraction pattern of gel : (a) after precipitation; (b) after seven washings.
G = Gibbsite
B = Bayerite

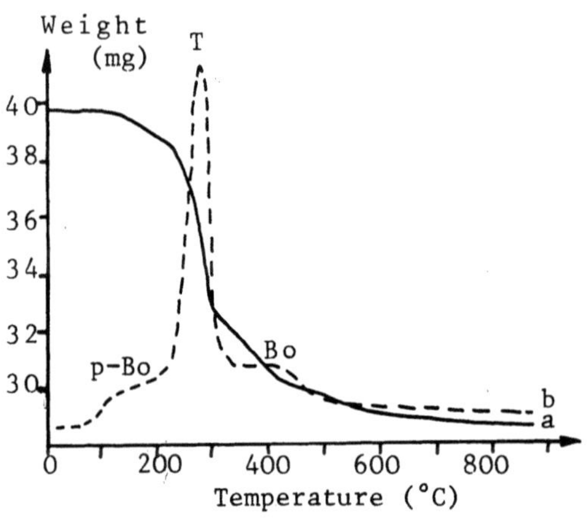

Figure 5

TGA (a) and DTG (b) of gel after seven washings
T = Trihydrate
Bo = Boehmite
p-Bo = Pseudo-Boehmite

3.3 . Peptization and sol concentration

First it was necessary to carry out some peptization tests adding HCl at different concentration to prepare a range of total Cl^-/Al^{3+} molar ratios and from these find the optimum ratio. On the ground of these tests an optimum final total Cl^-/Al^{3+} molar ratio of 0.4 has been found; this ratio allows to obtain a particularly stable sol that keeps its characteristics unchanged even at rather high Al^{3+} concentration.

The sol has been stabilized at 80°C for 48 hours in hermetically sealed vessels; in fact, preliminary tests at different temperatures (20-40-60-80°C) have shown that the peptization rate decreases drastically below 80°C, in agreement with Yoldas (18).

The gel-sol transformation was followed by laser granulometry and microgranulometry: the acid broke up the gel micelles and allowed to obtain a sol, that, according to the definition of Woodhead (11), is a dispersion of solid particles in a liquid medium in which particle dimension is between 1 nm and 1 micron.

The sol has then been concentrated at 80°C under stirring till an Al^{3+} concentration about 4 mol/l, without starting gelification, was reached. The high sol concentration and its fluidity, evaluated by viscosimetry (about 7 P), are fundamental for producing dense and uncracked microspheres.

X-Ray analysis and TGA have shown no substantial differences between the starting gel and the sol before and after concentration.

3.4 . Production of microspheres

The concentrated sol has been used to feed a capillary tube (Ø = 0.2 mm) in a pilot plant. The sol droplets are broken up by a 1-octanol stream, containing 1% by volume of a surfactant, the flow rate of which determines the microspheres diameter.

The 1-octanol also produces a slow dehydration of the sol preserving the microspheres whole. The microspheres do not agglomerate, and can thus be collected and subjected to heat treatment for converting the hydroxides into oxides.

Laser granulometry of the heat treated product shows that the microspheres average diameter is about 20 microns and that the particle size distribution is close to the average value (d_{max} / d_{ave} = 2.7) (3) (Figure 6); SEM observations confirm the granulometry results, show that the particle are spherical, dense and uncracked, that no agglomeration phenomena are present (Figure 7).

The X-Ray analysis in a high-temperature chamber has allowed to follow the development of the crystal phases during the heat treatment of the microspheres (Figure 8).

At 120°C Gibbsite and Bayerite are present (26); between 400°C and 1000°C only the appearence of a transition alumina (η -like structure) has been seen. At 1100°C a ϑ -alumina is present and a weak peak of tetragonal ZrO_2 becomes evident; it gets stronger at higher temperatures (1200-1300°C). The ϑ metastable alumina phase transforms to α beginning at 1300°C.

The specific surface area of the microspheres (Figure 9) develops according to the formed crystal phases. The area progressively decreases increasing the temperature of heat treatment up to the value of 30 m^2/g of

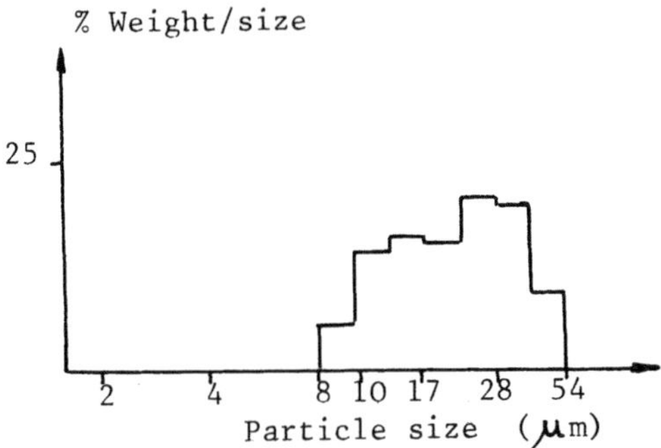

Figure 6

Laser granulometry of the microspheres treated at 1200°C for 30 minutes

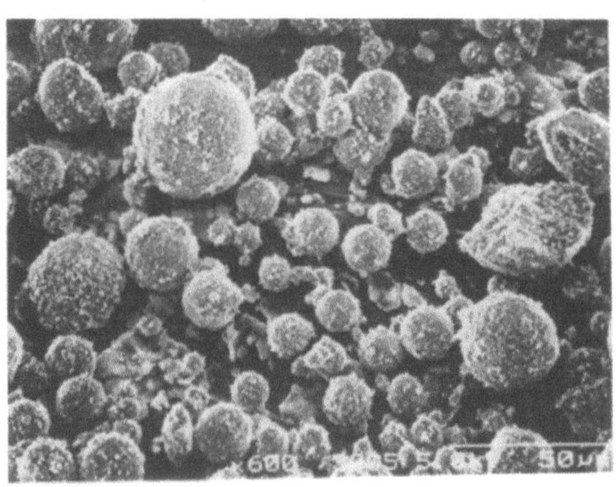

Figure 7

SEM micrograph of the microspheres treated at 1200°C for 30 minutes

the microspheres treated at 1200°C for 30 min.
This powder has then been hot pressed and on the sintered body some preliminary tests of mechanical properties (σ_F , K_{Ic}) have been carried out.
The microspheres easily sinter and the density of the sintered body is equal to the theoretical value.
The flexural strength is 400-600 MPa and the fracture toughness is 4-5

MPa$\sqrt{m}$.
Through chemical contrast in SEM observations, it is possible to check the good dispersion of ZrO_2 grains (0.2 μm in average diameter) in the alumina matrix

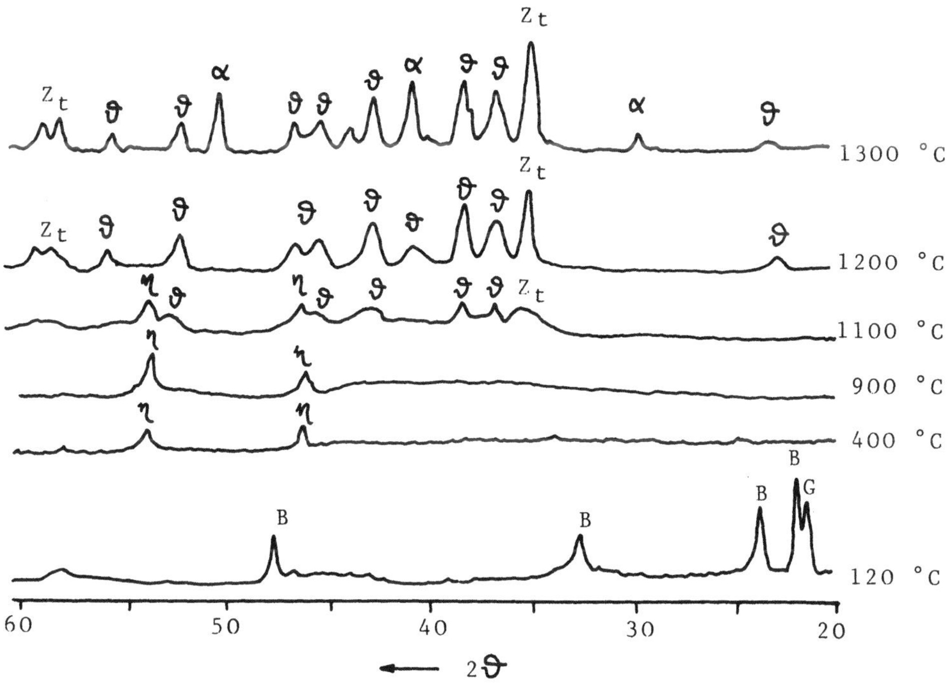

Figure 8

X-Ray analysis in a high-temperature chamber of the microspheres during heat treatment.

B = Bayerite
G = Gibbsite
Z_t = tetragonal zirconia
η = η-alumina
ϑ = ϑ-alumina
α = α-alumina

4. CONCLUSIONS

From the above results the following conclusions can be drawn:

- it is possible to obtain a fluid and concentrated sol by peptization, in suitable conditions, of a gel of Al and Zr hydroxides;

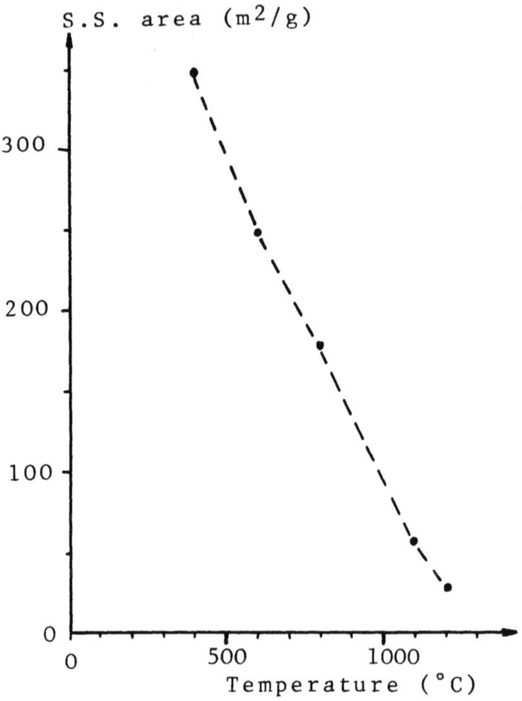

Figure 9

Variation of the specific surface area of the microspheres as a function of the temperature of heat treatment.

- from this sol one can produce a powder of homogeneous chemical composition and characterized by well-defined geometric parameters: spherical shape, controllable size;
- the powder is not agglomerated;
- the microspheres average size is 20 microns and the particle size distribution is close to the average value;
- the microspheres are dense and uncracked;
- preliminary tests show that the sintered body presents theoretical density and good mechanical properties;
- in our opinion, there exist the possibility of an industrial development of this process; this method, even if more complicated than others set up for obtaining ceramic powders from gel (for example, the spray-drying) allows to obtain powders with improved characteristics.

REFERENCES

1 . I.B.Cutler 'Active powders' in G.Y.Onoda,L.L.Hench, Ceramic Processing before firing, p.21, J.Wiley and Sons, New York (1978)

2 . E.A.Barringer, H.K.Bowen 'Formation,packing and sintering of monodisperse TiO_2 powders', J.Amer.Ceram.Soc. C-199,201 (Dec.1982)

3 . E.A.Barringer, N.Jubb, B.Fegley, R.L.Pober, H.K.Bowen 'Processing monosized powders' in L.L.Hench, D.R.Ulrich, Ultrastructure processing ceramics,glasses and composites, p.315, J.Wiley and Sons, New York (1984)

4 . E.Matijevic 'Colloid science in ceramic powder preparation' in P. Vincenzini, High Tech Ceramics, p.441, Elsevier Sci.Publ., Amsterdam (1987)

5 . A.Kato 'Recent production methods of ultrafine ceramic powders', Industrial Ceramics, 7 (2), 105-108 (1987)

6 . M.D.Sacks, T.Y.Tseng 'Preparation of SiO_2 glass from model powder compact: I, formation and characterization of powders,suspension and green compacts', J.Amer.Ceram.Soc. 67 (8),526-532 (1984)

7 . J.E.Blendell, H.K.Bowen, R.L.Coble 'High purity alumina by controlled precipitation from aluminium sulfate solutions',Ceram.Bull. 63 (6), 797-802 (1986)

8 . R.K.Iler (Inorganic colloids for forming ultrastructures' in L.L. Hench and D.R.Ulrich, Science of ceramic chemical processing,p.3, J.Wiley and Sons, New York (1986)

9 . D.W.Johnson 'Sol-gel processing of ceramics and glass',Amer.Ceram. Soc. Bull. 64 (12), 1597-1602 (1985)

10. Gorham Advanced Materials Institute ,Sol-gel production of high-performance ceramics and glasses, Gorham, Maine (1988)

11. J.L.Woodhead 'Gel processing', Journal de Physique in Science of Ceramics, Tome 47, C1, p.3-12 (Feb.1986)

12. J.M.Fletcher, C.J.Hardy 'Application of sol-gel processes to industrial oxides', Chem. & Ind.,48-51 (Jan.1968)

13. N.Claussen 'Microstructural design of zirconia-toughened ceramics (ZTC)', in Advances in Ceramics, 12 , p.325,The Amer.Ceram.Soc. (1984)

14. N.Claussen 'Strengthening strategies of ZrO_2 toughened ceramics at high temperatures' Mater.Science and Eng. 71,23-38 (1985)

15. F.Cambier 'Les composites à dispersoïdes.Céramique-zircone.Perspectives d'application' Silicates Ind. 11, 263-274 (1982)

16. N.Claussen 'Fracture toughness of Al_2O_3 with an unstabilized ZrO_2 dispersed phase' J.Amer.Ceram.Soc. 59, 49-51 (1976)

17. M.S.Mazdiyasmi 'Powder synthesis from metal-organic precursors' Ceram.Intern. 8 (2), 42-56 (1982)

18. B.E.Yoldas 'Alumina sol preparation from alkoxides' Ceram.Bull. 54 (3), 289-290 (1975)

19. A.C.Pierre, D.R.Uhlmann 'Gelation of aluminum hydroxide sols' J. Amer.Ceram.Soc. 70 (1), 28-32 (1987)

20. J.Duclaux Colloïdes et gels p.208, Ed. Gauthier-Villars,Paris (1953)

21. L.Montanaro, P.Orlans, J.P.Lecompte, B.Guilhot, V.Specchia, A.Negro, A.Gianetto 'Sol-gel process for preparing ceramic powders: application to Al_2O_3-ZrO_2 system', Conf. on emerging technologies in materials, CE 1.20,Minneapolis, Aug.18-20, 1987

22. P.A.Haas, S.D.Clinton 'Preparation of Thoria and mixed-oxide microspheres' I&EC Product Res. and Devel. 5 (3), 236-244 (1966)

23. P.Orlans, Thése 'Etude physicochimique des oxides de zirconium et du dispersoïde alumine-zircone' St.Etienne (France 1987)

24. M.V.Mathieu, Thése 'Contribution à l'etude des gels d'alumine désorganisée' Lyon (France 1956)

25. A.B.Carel, D.K.Cabbiness 'Analysis of Alumina by combined TG/X-Ray diffraction' Ceram.Bull. 64 (5), 716-719 (1985)

26. E.A.Pugar, P.E.D.Morgan 'Coupled grain growth effect in Al_2O_3/ 10% vol. ZrO_2' Comm.J.Amer.Ceram.Soc. 69 (6), C-120 (1986)

SOLID-GAS AND SOLID-SOLID INTERACTIONS OF CERAMIC OXIDES AT HIGH TEMPERATURES

H.J. Grabke
Max-Planck-Institut für Eisenforschung GmbH
Department of Physical Chemistry
P.O. Box 140260
D-4000 Düsseldorf 1, Fed. Rep. Germany

Abstract

Thermodynamics, kinetics and mechanisms of gas-solid interactions are exemplified for oxides. The thermodynamics of oxide reduction and of oxide evaporation are demonstrated. The solubility of extraneous nonmetal atoms in oxides is discussed. Concerning kinetics and mechanisms there are not many studies with ceramic oxides, their treatment has to be based mainly on results for more reactive oxides such as FeO, Fe_3O_4, Cr_2O_3. General rate equations are presented for the oxygen transfer to and from oxides in CO_2-CO and H_2-H_2O mixtures. Examples are given for the kinetics and mechanisms of oxide reduction and of oxide conversion to carbides and oxide reduction to intermetallic compounds. It is stated and exemplified that in many solid-solid interactions gaseous intermediates play a decisive role.

1. Thermodynamic Aspects

The oxide ceramics are generally very stable compounds and not easily attacked by gases and other phases. This can be demonstrated for example from the formation pressures of the oxides, i.e. the oxygen pressure at which the metal and its oxide are in equilibrium. The value of the formation resp. decomposition pressure of an oxide is characteristic for its stability and can be calculated from its Gibbs' free enthalpy of formation according to

$$p_{O_2} = \exp\left(-\Delta G^{o}_{oxide}/RT\right) \qquad (1)$$

Figure 1 shows a plot of log p_{O_2} vs. 1/T for such equilibria, it can be seen that the equilibrium lines for the less stable oxides such as FeO, PbO, NiO are in a range of higher oxygen pressures and that the

L.-C. Dufour et al. (eds.), Surfaces and Interfaces of Ceramic Materials, 599–624.

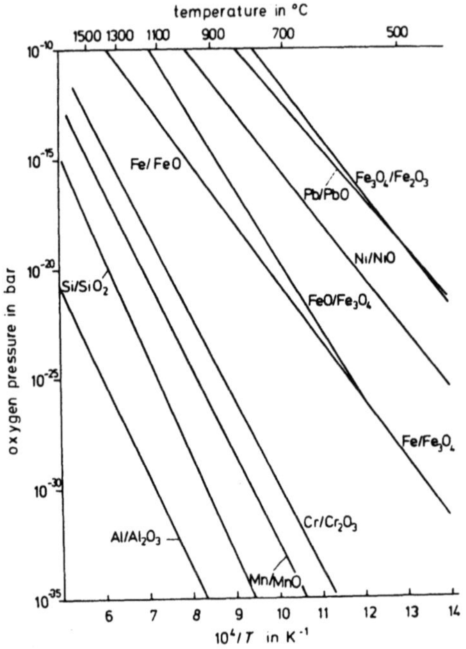

Fig. 1 Plot of the formation and decomposition pressure of several oxides in dependence on temperature, log pO_2 vs 1/T

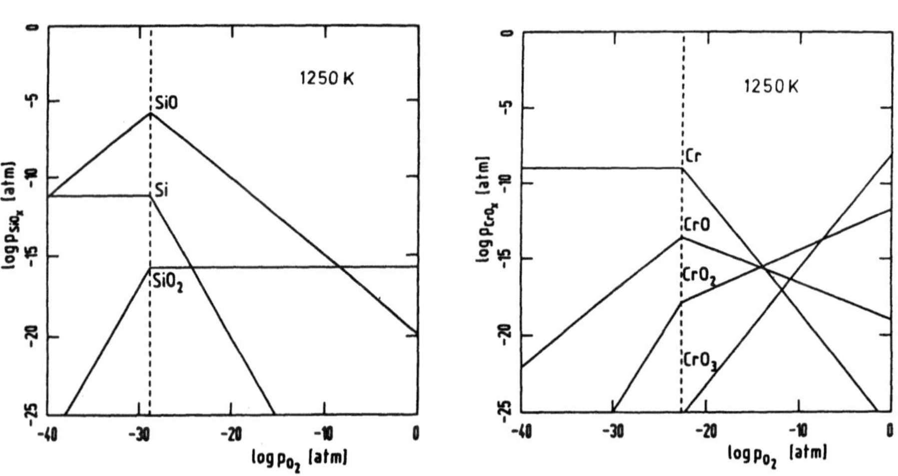

Fig. 2 Vapor pressures of elements and compounds in dependence on the oxygen pressure
a) in the system Si–O b) in the system Cr–O

equilibrium lines for the most stable oxides are in a range of very low oxygen pressures. Chromium oxide, Cr_2O_3, has a very low formation pressure, SiO_2 an even lower and formation pressure is very low for Al_2O_3 , alumina, which is together with ZrO_2, zirconia, one of the most stable oxides.

This means that these oxides can practically not be reduced by gases and they are also very stable concerning conversion to other compounds. However, it will be shown that such reduction is still possible for Cr_2O_3 and that conversion to other compounds also is possible even for ZrO_2 and Al_2O_3. In table 1 data are shown of the equilibrium values for formation pressures of the oxides Cr_2O_3, Al_2O_3 and ZrO_2, given in bar O_2 or in terms of the H_2O/H_2 ratio below which reduction of the oxide is possible in hydrogen. From the pressures it can be seen that it is impossible to decompose the oxides by heating even in ultra high vacuum. With great difficulties Cr_2O_3 can be reduced by H_2,

$$Cr_2O_3 + 3H_2 = 2Cr + 3\,H_2O \tag{2}$$

Table 1 Stability of ceramic oxides, formation resp. dissociation pressure p_{O_2} (in bar) and H_2O/H_2 ratio below which reduction is possible

	Cr_2O_3		Al_2O_3	
°C	pO_2	pH_2O/pH_2	pO_2	pH_2O/pH_2
800°	$1.2\cdot10^{-30}$	$1.5\cdot10^{-6}$	$4.2\cdot10^{-44}$	$3.2\cdot10^{-13}$
1000°	$1.0\cdot10^{-24}$	$1.8\cdot10^{-5}$	$1.8\cdot10^{-35}$	$8.1\cdot10^{-11}$
1200°	$2.1\cdot10^{-20}$	$1.1\cdot10^{-4}$	$3.4\cdot10^{-29}$	$4.5\cdot10^{-9}$

	ZrO_2		$ZrO_2 \longrightarrow ZrPt_3$	
°C	pO_2	pH_2O/pH_2	pO_2	pH_2O/pH_2
800°	$1.1\cdot10^{-43}$	$5.2\cdot10^{-13}$	$1.6\cdot10^{-25}$	$6.0\cdot10^{-4}$
1000°	$2.1\cdot10^{-35}$	$8.9\cdot10^{-11}$	$1.7\cdot10^{-20}$	$2.4\cdot10^{-3}$
1200°	$2.1\cdot10^{-29}$	$3.5\cdot10^{-9}$	$8.5\cdot10^{-17}$	$7.1\cdot10^{-3}$

for this reduction the ratio H_2O/H_2 must be kept below the given ratios in a fast flow of dry hydrogen (see chapter 4). The last rubric gives data on the reaction

$$ZrO_2 + 3\ Pt + 2\ H_2 = ZrPt_3 + 2\ H_2O \qquad (3)$$

in which the intermetallic compound $ZrPt_3$ is formed. This 'Engel-Brewer compound' is very stable /1/ and its formation shifts the reduction equilibrium to the right so that ZrO_2 reduction becomes possible in H_2 at high temperatures. (This reaction sometimes causes problems in the application of solid electrolyte cells with doped ZrO_2 as an oxygen conducting electrolyte and Pt as electrodes and ducts under reducing conditions). Not only with ZrO_2 but also with Al_2O_3, TiO_2 and HfO_2 corresponding reactions are possible under participation of Pt, Pd and Ni /1,2/.

As a further phenomenon mainly controlled by thermodynamics the evaporation of some oxides is to be mentioned. This is demonstrated in Fig. 2 where vapor pressures of compounds are plotted versus oxygen pressure for the system silicon/oxygen and for the system chromium/oxygen /3/4/. In the system silicon/oxygen, the silicon monoxide SiO can be formed which has a rather high vapor pressure, the maximum of this vapor pressure is at the line of coexistence silicon/silicon-dioxide. At higher oxygen pressures, the vapor pressure of SiO decreases. This results from the formation reaction

$$SiO_2 \longrightarrow SiO + 1/2\ O_2 \qquad (4)$$

so the vapor pressure of SiO is proportional to $p_{O_2}^{-1/2}$. The formation and evaporation of SiO is a dangerous process mainly at rather low oxygen pressures, e.g. in dry hydrogen at 1000°C one can evaporate quite a lot of silica by passing a dry hydrogen flow through a quartz tube. The evaporation of chromium oxide Cr_2O_3 is possible by the formation of higher oxides, the most important reaction is the formation of CrO_3 which has a relatively high vapor pressure. For the formation of CrO_3 from Cr_2O_3 the reaction equation is valid:

$$Cr_2O_3 + 3/2\ O_2 = 2\ CrO_3 \qquad (5)$$

So the slope of the line for the vapor pressure in Fig. 2b) has the slope +3/4 and it can easily be seen that the evaporation of CrO_3 will be most dangerous at rather high oxygen pressures. It may be added that chromium oxide can evaporate also under formation of other compounds, e.g. chromium oxihydroxides and chromium chloride or chromium oxichlorides. So this may be a great drawback for the application of chromia as a high temperature ceramic material. Any vapor pressures in the system Al-O at 1250 K are $<10^{-13}$ bar, i.e. much lower than in the systems presented in Fig. 1, the evaporation of alumina is low under any conditions.

2. Solubility of nonmetal atoms in oxides

Solubility of nonmetal atoms such as H, C, N, S, and Cl in oxides is a second point of interest. There have been many studies on hydrogen permeation, diffusion and solubility in oxides in relation to the barrier function of oxide layers on high temperature alloys to the permeation of tritium in the high temperature reactor.
At first I will discuss a study performed in 1968 on the solubility of water vapor in $ZrO_2 \cdot Y_2O_3$, this was one of the very last experimental studies initiated by Carl Wagner /5,6/. It was conceived assuming that the reaction occurs

$$H_2O(g) + V_O^{2+} = 2H_i^+ + O_O \qquad (1)$$

In a ceramic $ZrO_2 \cdot Y_2O_3$ there is a certain amount of oxygen vacancies which is fixed by the concentration of the dopant, in this case yttria. So if such reaction as given occurs, the amount of water vapor dissolved in zirconia should be measurable and be approximately described by the mass action law given in the equation

$$[H_i]^2/p_{H_2O} \cdot [V_O^{2+}] = K \qquad (2)$$

A gas volumetric study of the equilibrium was performed at temperatures between 900 to 1000°C /6/. The material was equilibrated in water vapor at 4.5 to 20 Torr, $ZrO_2 \cdot Y_2O_3$ was used with 8 and 17 wt% Y_2O_3, the equilibration times were in the range 10 to 60 minutes. Indeed, a solubility was observed and the results could be described by the equation

$$[H_i]/(p_{H_2O}(\text{Torr}))^{1/2} \sim 5 \cdot 10^{-6} \qquad (3)$$

The concentration of hydrogen is given in mole fraction, for one bar of hydrogen in the atmosphere a solubility of $1 \cdot 10^{-4}$ moles of water vapor per mole of oxide would result. The concentration of dissolved hydrogen is not temperature dependent and not dependent on the concentration of oxygen vacancies which is constant and is given by the concentration of the yttria. This study was quoted here since it is one of the few in which the solubility of hydrogen in oxide has clearly been demonstrated. There are indeed few studies in which solubility data are given. Many studies have been performed in which the effect of hydrogen has been studied on the conductivity or on the charge of other dopants, e.g. Fe^{3+} in an oxygen lattice can be reduced to Fe^{2+}, or effects on spin resonance or on catalytic activities have been studied. As already mentioned, there are much less reliable studies on permeation, diffusion and solubility. A certain solubility must even be assumed in alumina. This was deduced from conductivity and EMF measurements at α-Al_2O_3 by Kröger and coworkers /7/. These studies have been performed by saturating Al_2O_3 in different H_2O-H_2 mixtures at temperatures between 1100 and 1500°C. The disorder of oxides can be varied with the oxygen pressure and by the use of different water vapor/hydrogen ratios different oxygen pressures are established. Part of the outcome of this study is that the amount of dissolved hydrogen increases with decreasing saturation temperature and increases with increasing hydrogen pressure. This seems very reasonable but the amount of dissolved hydrogen also increases with increasing p_{O_2} and this could not be rationalized fully by Kröger and coworkers. The main point deduced here from this study is the observation by Kröger that oxidation and reduction equilibria must be established even for this very stable oxide Al_2O_3 in a reasonable time at higher temperatures. At 1500°C the equilibration time was about 1 h, upon equilibration the disorder of Al_2O_3 must change somewhat in a very small range of homogeneity which will exist even for Al_2O_3. So oxidation and reduction of Al_2O_3 is possible. It needs very high temperature and it will lead to only very small changes in the stoichiometry which quantitatively may never be detectable.

It looks reasonable that such a small atom such as hydrogen or an even smaller particle like a proton can be dissolved in oxides. However, some authors /8,9/ have claimed a solubility of carbon in oxides, such as MgO, FeO, MnO, Cr_2O_3, Al_2O_3 and they have tried to explain some peculiar kinetic phenomena with effects of dissolved carbon /10,11/. In recent years however, very careful investigations have been performed with radiotracer methods showing that there is virtually no solubility (< 0.01 ppm) of carbon in these oxides /12/. These studies have been conducted by equilibrating the oxides in CO_2-CO mixtures which were tagged with radioactive ^{14}C. After equilibration at high temperatures the radioactivity in the oxide materials was determined by measurements of the radioactivity and by autoradiography and it was confirmed that carbon does not enter the lattice of these oxides but it only enters into pores which were still present in these dense polycrystalline or single crystalline materials. The same result applies for sulfur and for chlorine /13/, also these elements do not enter the lattice of the mentioned oxides. This is an important result, especially for the field of high temperature oxidation and it shows that the dense oxide layer should be protective against active attack of carbon and sulfur from the gas atmosphere. The oxide layer should be protective as long as no molecules can diffuse through pores or cracks of the oxide layer to the metal phase alloy. It may be mentioned that it has recently been shown /14/ that even hydrogen cannot permeate chromia layers, thus Cr_2O_3 seems to have no true solubility for hydrogen. Permeation studies have been performed with chromia covered high temperature alloys in relation to applications in the high temperature reactor and for long time such permeation has been observed but in very recent studies it has been shown that a really dense chromia does not dissolve hydrogen and would not allow permeation of any hydrogen. So many of the previous results on hydrogen permeation diffusion and solubility in oxides should be checked concerning the material which was studied if there are any pores or other microscopic defects in these materials, it may well be that the hydrogen enters or passes through these pores and defects and that there is really no true solubility.

3. Oxidation/reduction of oxide materials

Non-stoichiometric oxides with a certain homogeneity range can be oxidized and can be reduced within their homogeneity range, upon oxidation the cations obtain higher positive valencies and upon reduction they are reduced to the cation in a lower valency state. Certainly the oxides can also be oxidized to other oxide phases in which the cations have higher valency states or they can be reduced to other oxide phases in which the cations have a lower positive valence. At first the oxidation/reduction within the homogeneity range of a rather unstable oxide wustite FeO will be discussed. The Fig. 3

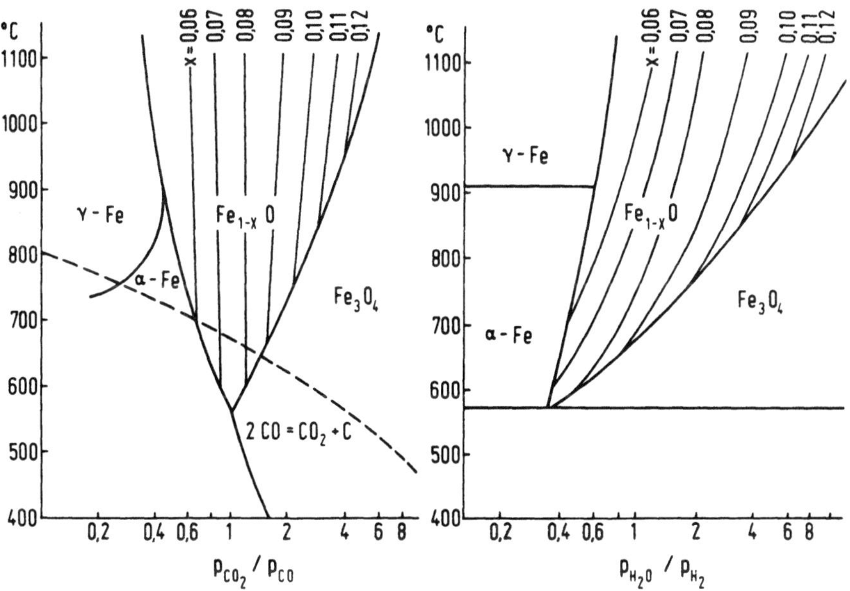

Fig. 3 Stability ranges of metal and oxide phases in the system Fe–O
a) in dependence on the CO_2/CO ratio and temperature
b) in dependence on the H_2O/H_2 ratio and temperature

shows the stability range of FeO in two diagrams, temperature vs. oxygen pressure. In Fig. 3a, the oxygen pressure is given by the CO_2/CO ratio in the gas atmosphere, in Fig. 3b the oxygen pressure is established by the H_2O/H_2 ratio. Within the homogeneity range of FeO rather great changes in oxygen content are possible up to about 15 at%, the conductivity changes and the relations of conductivity and oxygen excess can be described approximately by the mass action law. The oxidation/reduction reactions in CO_2–CO and in H_2O–H_2 have been studied in detail for this oxide which has a rather high reactivity

/15,16,17/. The experimental methods were gravimetry or resistance-relaxation measurements, i.e. the measurement of the conductivity upon an abrupt change in the gas atmosphere, but mainly measurements of isotope exchange reactions in the CO_2-CO and H_2O-H_2 mixtures which will be described in more detail.
At first, a general surview on the oxygen transfer reactions in CO_2-CO and H_2O-H_2 mixtures is presented. Oxygen transfer from CO_2 is given by equation

$$CO_2 = CO + O(ad) \tag{1}$$

For an oxide with high reactivity it can be assumed that this adsorbed oxygen is in equilibrium with lattice oxygen so that there is in practice no difference between the oxygen transfer to the adsorbed state or to the state of lattice oxygen. In a similar way, the oxygen transfer from water vapor hydrogen mixtures can be described

$$H_2O = H_2 + O(ad) \tag{2}$$

But this reaction occurs in two steps, two dissociation steps are necessary to obtain the adsorbed oxygen atom from the water vapor.

$$H_2O = OH(ad) + H(ad) \tag{2.1}$$

$$OH(ad) = O(ad) + H(ad) \tag{2.2}$$

So these reactions have been studied with the aid of tracer exchange reactions on oxides, metals and carbon and the study can be done in gas-solid equilibrium. The great advantage is that the properties of the solid oxides are well defined by the equilibrium with the oxygen pressure of the gas atmosphere. These studies have been proposed by Carl Wagner in 1962, and such studies have been performed by scholars of Carl Wagner in the following years. The state of the oxide is defined by the oxygen activity of the gas atmosphere. In case of reaction (1) the oxygen activity is given by the ratio

$$(p_{CO_2}/p_{CO})_{eq} = K_1 \cdot p_{O_2}{}^{1/2} \tag{3}$$

and this can be defined as the oxygen activity of the CO_2-CO atmosphere. In case of reaction (2) the oxygen activity is given by

$$(p_{H_2O}/p_{H_2})_{eq} = K_2 \cdot p_{O_2}{}^{1/2} = a'_O \tag{4}$$

which can be defined as the oxygen activity in the water vapor hydrogen mixture. In studying the isotope exchange reaction in CO_2-CO, a mixture of $^{12}CO_2$ with $^{14}CO_2$ and ^{12}CO is flown over the solid. The isotope exchange occurs in the reaction steps

$$^{14}CO_2 = {}^{14}CO + O(ad) \tag{5}$$
$$O(ad) + {}^{12}CO = {}^{12}CO_2.$$

These reactions can be summed up to the overall isotope exchange reaction

$$^{14}CO_2 + {}^{12}CO = {}^{14}CO + {}^{12}CO_2. \qquad (6)$$

So by measuring the rate of ^{14}CO formation the rate of the $^{14}CO_2$ decomposition is determined and the rate coefficients are obtained for the oxygen transfer from CO_2. This measurement can be done on any surface and so these rate coefficients can be obtained for any surface, neglecting a small isotope effect. For reaction (5), a rate equation can be formulated

$$dn_{^{14}CO}/Adt = k_1(a_O)\cdot p_{^{14}CO_2} - k_1(a_O)\cdot p_{^{14}CO}\cdot a_O \qquad (7)$$

in which the rate coefficient is written as a function of the oxygen activity. The forward reaction, the decomposition of $^{14}CO_2$ is proportional to its partial pressure, the back reaction is proportional to the partial pressure of $p_{^{14}CO}$ which had been formed and proportional to the activity of oxygen. This rate equation fulfills the equilibrium condition $p_{^{14}CO_2}/p_{^{14}CO} = a_O$ since in equilibrium the ratio of $^{14}CO_2/^{14}CO$ must be equal to the overall ratio CO_2/CO. Such measurements have been done in a flow reactor, passing the radioactive CO_2-CO mixture over the oxide as a catalyst and after that determining the pressure of the formed ^{14}CO. From that measurement the rate coefficient $k_1(a_O)$ can be calculated. In a similar way, the isotope exchange in water vapor/hydrogen mixtures can be measured. A mixture of H_2O with HDO and H_2 is flown over the solid, the isotope exchange occurs in the reaction steps

$$\begin{aligned} HDO &= OH(ad) + D(ad) && (2.1)\\ HDO &= OD(ad) + H(ad) && \\ OD(ad) &= D(ad) + O(ad) && (2.2)\\ OH(ad) &= H(ad) + O(ad) && \\ 2H(ad) + 2D(ad) &= 2HD && (2.3) \end{aligned}$$

So this measurement is not unambiguous. In case that the first dissociation step, the formation of OH resp. OD, is slow and rate determining and the decomposition of OD and OH is fast, the rate of the isotope exchange reaction agrees with the rate of oxygen transfer from water vapor. But in the case that the second dissociation step is slow, most of the isotope exchange would occur by the way of

$$\begin{aligned} H_2 &= 2H(ad)\\ HDO &= OH(ad) + D(ad)\\ H(ad) + D(ad) &= HD\\ H(ad) + OH(ad) &= H_2O \end{aligned}$$

and could be much faster than the rate of oxygen transfer from H_2O. So the measurement of the isotope exchange in water vapor hydrogen mixtures is not unequivocal and it must be checked by other methods, for example resistance-relaxation or gravimetric studies, that indeed the rate of oxygen transfer is measured by this method.

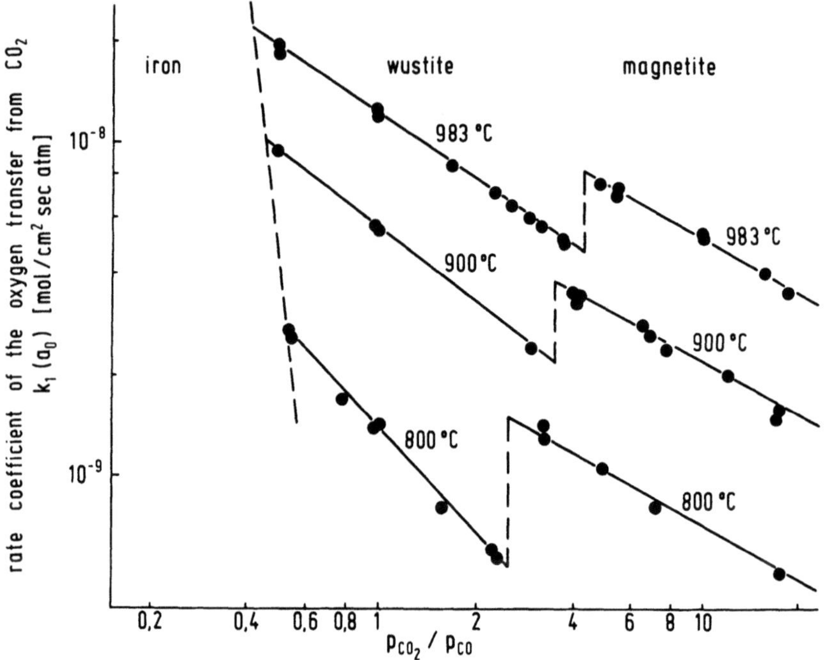

Fig. 4 Rate coefficient of oxygen transfer in CO_2-CO mixtures plotted vs pCO_2/pCO, a) measured using the isotope exchange method on wustite and magnetite, b) comparison of data from the isotope exchange method and resistance-relaxation method measured on wustite at 900°C - showing agreement.

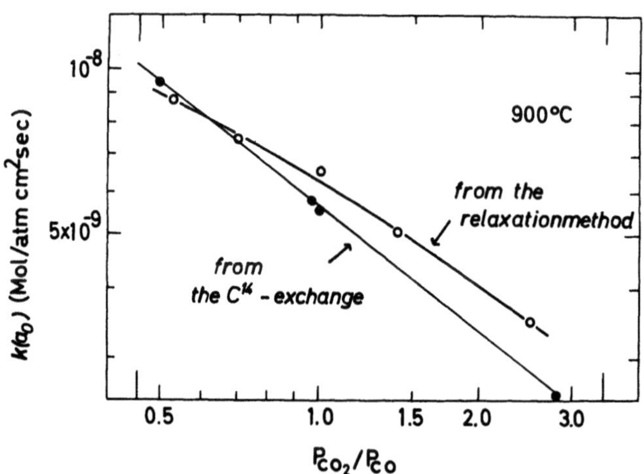

The isotope exchange measurements in CO_2-CO /15/ lead to the result for the rate coefficient

$$k_1(a_O) = k_1 \cdot a_O^{-m}$$

This result has been observed for any solids: metals, oxides and carbons. The exponent m can have values between zero to one. Examples of such results are shown in Fig. 4 for the measurements on wustite and on magnetite. You see that the oxygen transfer decreases with increasing oxygen activity on the solid surface for both phases. Using this result for the rate coefficient the rate equation can be formulated

$$v = dn/Adt = k_1 \cdot a_O^{-m} \cdot p_{CO_2} - k_1 \cdot a_O^{1-m} \cdot p_{CO} \tag{8}$$

The first term describes the oxygen transfer from CO_2 and the second term describes the back reaction of adsorbed oxygen with carbon monoxide. This rate equation fulfills the equilibrium condition $a_O = p_{CO_2}/p_{CO}$. In equilibrium, the oxygen activity on the surface must correspond to the oxygen activity in the gas phase.
The rate law can be rewritten

$$v_1 = k_1 \cdot a_O^{-m} \cdot p_{CO_2} \left(1 - \frac{a_O}{p_{CO_2}/p_{CO}}\right) \tag{9}$$

In this equation one has a kinetic term before the parenthesis

$$k_1 \cdot a_O^{-m} \cdot p_{CO_2} = v_1^0.$$

One might consider this expression as an exchange velocity (as in electrochemical kinetics), the kinetic term contains the dependencies on the partial pressure, the dependence on the state of the solid surface and the rate constant which is also a function of the temperature. In parenthesis, one has a thermodynamic term which describes the distance from equilibrium. At equilibrium, the expression in parenthesis is zero and no reaction occurs. If the oxygen activity on the solid surface is $< p_{CO_2}/p_{CO}$, an oxygen transfer occurs to the solid which corresponds to an oxidation of the solid. If the oxygen activity of the solid surface is higher than p_{CO_2}/p_{CO}, an oxygen transfer will occur from the solid to the gas phase which corresponds to a reduction of the solid. So the equation is a general rate equation for oxygen transfer in CO_2-CO mixtures to and from solid surfaces. Similar rate equations can be written for the oxygen transfer in H_2O-H_2 mixtures. This reaction occurs in two partial reaction steps for which the rate equations can be written

$$v_{2.1} = k_{2.1}(a'_O) \cdot p_{H_2O} - k_{2.1}(a'_O) \cdot p_{H_2} \cdot a'_O \tag{10}$$

$$v_{2.2} = k_{2.2}(a'_O) \cdot p_{H_2O}/p_{H_2}^{1/2} - k_{2.1}(a'_O) \cdot p_{H_2}^{1/2} \cdot a'_O \tag{11}$$

The first rate equation describes the decomposition and formation of the H_2O molecule and the second rate equation describes the decomposition and formation of the OH radical on the solid surfaces. As in the case of the CO_2–CO mixtures, these equations can be rewritten by taking the kinetic terms before the parenthesis

$$v_{2.1} = k_{2.1} \cdot a'_O{}^{-n} \cdot p_{H_2O} \left(1 - \frac{a_O}{p_{H_2O}/p_{H_2}}\right) \quad (12)$$

$$v_{2.2} = k_{2.2} \cdot a'_O{}^{-n} \cdot p_{H_2O}/p_{H_2}^{1/2} \left(1 - \frac{a_O}{p_{H_2O}/p_{H_2}}\right) \quad (13)$$

In equations (12) and (13) the result is introduced from the isotope exchange, from resistance-relaxation measurements and from gravimetric measurements, see Fig. 5, that also these reactions are dependent on the oxygen activity on the solid surface with a negative exponent $0 < n < 1$. The rate equation (12) for step 2.1 shows that in this case the rate is proportional to p_{H_2O} and in relation to the overall pressure it would be proportional to the overall pressure P. In case of the rate equation for step 2.2, the reaction rate is proportional to the quotient $p_{H_2O}/p_{H_2}^{1/2}$, so in case of a rate determining reaction 2.2. the overall pressure dependence of the oxygen in H_2O–H_2 would be proportional to $P^{1/2}$. Certainly, these expressions for $v_{2.1}$ and $v_{2.2}$ are valid again for oxidation and for reduction but it may be mentioned here that they can also be rewritten for reduction in the following forms

$$v_{2.1} = k_{2.1} \cdot a'_O{}^{1-n} \cdot p_{H_2} \left(1 - \frac{p_{H_2O}/p_{H_2}}{a_O}\right) \quad (14)$$

$$v_{2.2} = k_{2.2} \cdot a'_O{}^{1-n} \cdot p_{H_2}^{1/2} \left(1 - \frac{p_{H_2O}/p_{H_2}}{a_O}\right) \quad (15)$$

With these expressions one obtains a positive value for the reduction rate when the oxygen activity on the surface is higher than p_{H_2O}/p_{H_2}.

The dependence of the rate of oxygen transfer on the oxygen activity at oxide surfaces, as given in the expression for the forward reaction in oxygen transfer from CO_2

$$v_1 = k_1 \cdot a_O^{-m} \cdot p_{CO_2} \quad (16)$$

can be interpreted 1. by participation of electrons from the solid in the surface reaction and 2. by a retardation of this forward reaction by adsorbed oxygen. At first, the possible mechanisms upon participation of electrons from the solid are explained. The following mechanisms may be assumed for the decomposition of adsorbed CO_2 on the solid surface

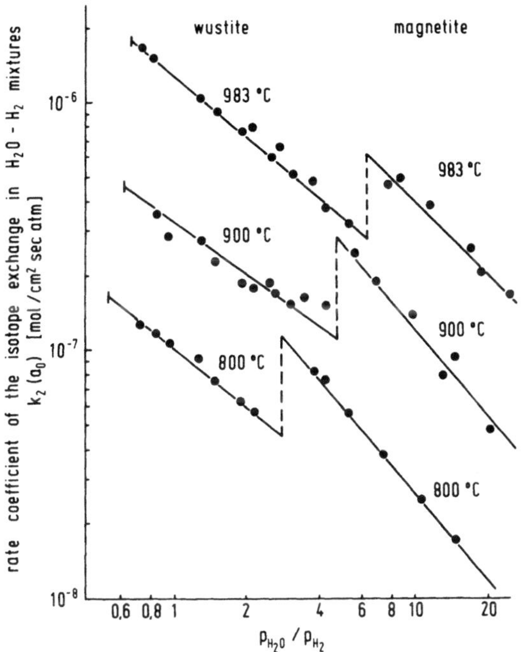

Fig. 5 Rate coefficient of oxygen transfer in H_2O-H_2 mixtures plotted vs pH_2O/pH_2 a) measured using the isotope exchange method and belonging to reaction step (2.1) b) measured using the resistance-relaxation method and belonging to the rate controlling step

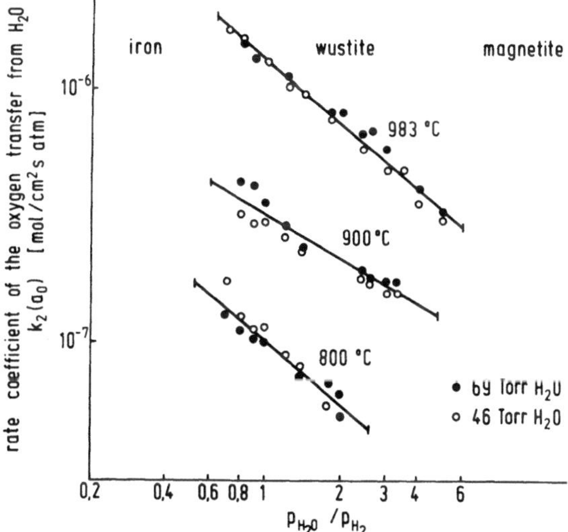

$$CO_2 \longrightarrow CO_2^{\delta-}(ad) \longrightarrow {}^*CO_2^{\delta-}(ad) \longrightarrow CO + O^{\delta-}(ad) \qquad (a) \quad (17)$$
$$CO_2 \longrightarrow CO_2(ad) \longrightarrow {}^*CO_2^{\delta-}(ad) \longrightarrow CO + O^{\delta-}(ad) \qquad (b)$$
$$CO_2 \longrightarrow CO_2(ad) \longrightarrow {}^*CO_2(ad) \longrightarrow CO + O^{\delta-}(ad) \qquad (c)$$

In these equations the activated state of the CO_2 molecule is denoted with an asterisk and the amount of charge transferred, one or two electrons, is denoted with δ. In cases (a) and (b), the electron transfer to $CO_2(ad)$ or to the activated complex $^*CO_2(ad)$, the rate will be dependent on the concentrations of electrons or defect electrons in the solid. In case (c) the electron is transferred to the adsorbed oxygen, in this case the forward reaction rate would be independent on the electron concentration. For p-conductors the surface concentration of charged $CO_2^{\delta-}(ad)$ is given by the equilibrium

$$CO_2 = {}^*CO_2^{\delta-}(ad) + \delta h^+ \qquad (18)$$

here the surface concentration of CO_2 activated or non-activated is proportional to $[h^+]^{-\delta}$. On n-conductors the equilibrium of CO_2 adsorption is described by the following equations

$$CO_2 + \delta e^- = {}^*CO_2^{\delta-}(ad) \qquad (19)$$

and the surface concentration of adsorbed charged CO_2 is proportional to $[e^-]^{\delta}$.

For CO_2 adsorption on a p-conductor, the example wustite is described by the following equations. For the defect equilibrium of wustite the equations are valid

$$O(gas) \longrightarrow O_O^x + V_{Fe}^{2-} + 2h^+ \qquad (20)$$

$$\text{and } [h^+] \ \alpha \ a_O^{1/3}$$

In this case the surface concentration of CO_2 should be proportional to p_{CO_2} and to $a_O^{-\delta/3}$.

As an example for CO_2 adsorption on an n-conductor, zinc oxide is described by the following equations

$$O_O^x = 1/2\ O_2 + Zn_i^+ + e^- \qquad (21)$$

$$\text{and } [e^-] \ \alpha \ a_O^{-1/2}$$

here the surface concentration of the charged CO_2 is proportional to p_{CO_2} again and to $a_O^{-\delta/2}$. Depending on the number δ of electrons participating in the reaction the values are expected

for wustite $m = -1/3 \ldots -2/3$

and for zinc oxide $m = -1/2 \ldots -1$

Such dependencies indeed have been observed for the oxygen transfer from CO_2 on wustite and on zinc oxide. On wustite the exponent m is between -2/3 to -1, see Fig. 4. This indicates a participation of two electrons in the rate determining step.

But additionally, at lower temperatures where the exponent -1 was

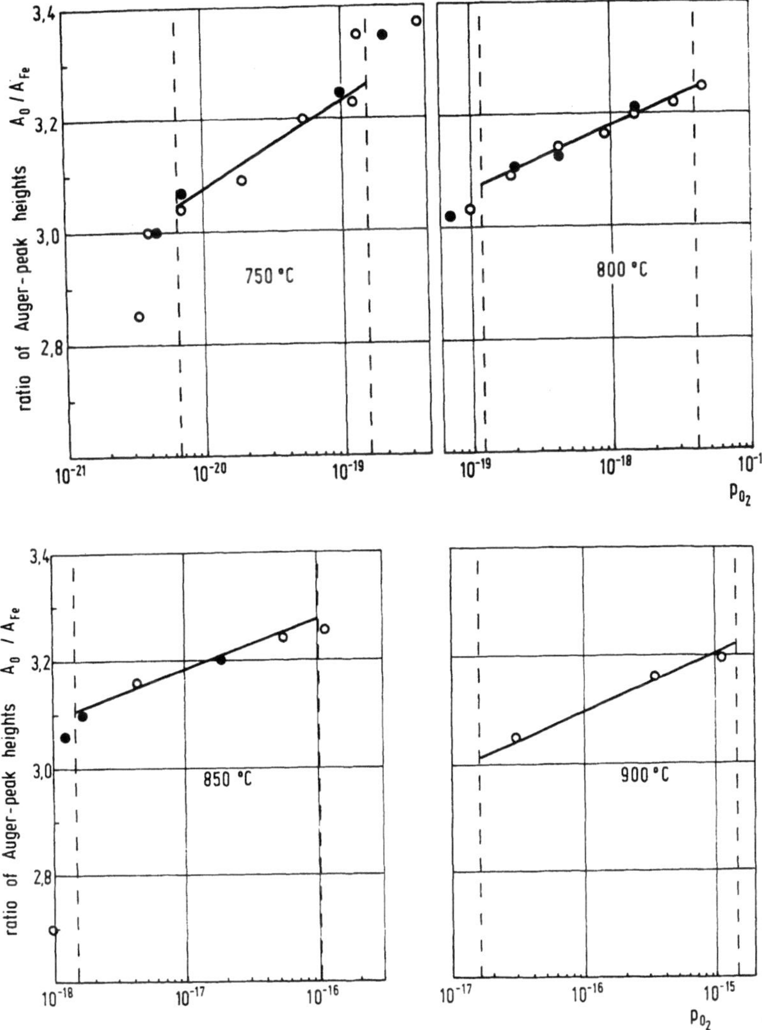

Fig. 6 Results of in situ-AES measurements on wustite surfaces, O/Fe-peak height ratio plotted vs oxygen pressure as established by the EMF in a solid electrolyte cell wustite/ZrO_2•CaO/Cu,Cu_2O within the UHV chamber at different temperatures

observed, a retardation of the forward reaction by adsorbed oxygen plays a role. The presence of adsorbed oxygen in the range of lower temperatures has been confirmed by Auger analysis of wustite surfaces at different temperatures between 700 to 1000°C, see Figs. 6. These studies have been done using an electrochemical cell $FeO/ZrO_2 \cdot CaO/Cu,Cu_2O$ for establishing well defined low oxygen pressures in a wustite film /18/. This wustite film was prepared by oxidation of an iron film deposited on the surface of the zirconia tube used as solid electrolyte in these experiments. At the higher temperatures >800°C the surface composition of the wustite film corresponded exactly to the composition expected according to the well-known non-stoichiometry of wustite /19/. At the lower temperatures, 720°C and 750°C, additional oxygen was observed on the wustite surface, its concentration increasing with increasing oxygen pressure as established by the electrochemical cell. An analysis of the measured data is shown in Fig. 7. Here the total measured excess oxygen is given and the data for the excess lattice oxygen as known and the calculated data for the adsorbed oxygen. So it's reasonable that at the lower temperatures below 700°C adsorbed oxygen also plays a role retarding the forward reaction in the oxygen transfer from CO_2, but on the other hand, the participation of electrons in this oxygen transfer has been confirmed by the isotope exchange measurements.

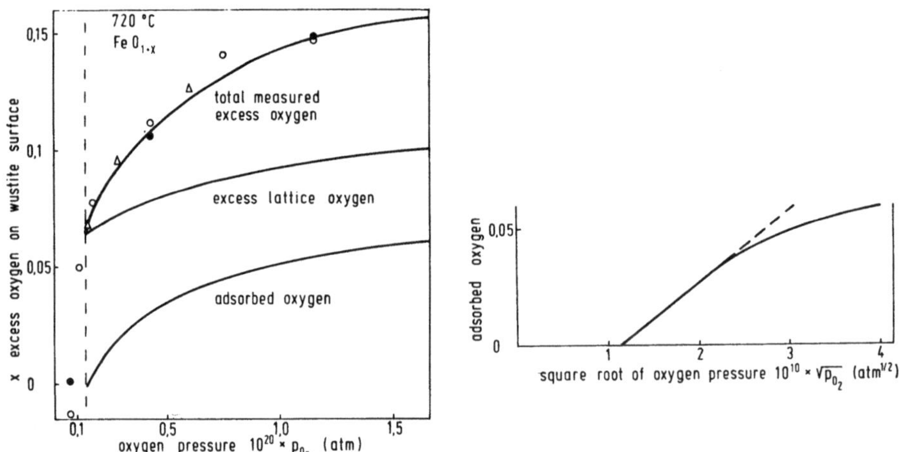

Fig. 7 Analysis of the AES measurement at 720°C,
a) total measured excess oxygen, as composed of the lattice excess oxygen and adsorbed oxygen
b) the surface concentration of adsorbed oxygen plotted vs $\sqrt{pO_2}$

4. Reduction of oxides to metals

In the reduction of oxides in H_2-H_2O mixtures or CO-CO_2 mixtures the reaction steps after the nucleation of the metal phase are

1. removal of O^{2-} by the surface reaction
2. diffusion of cations and electrons through the metals to the metal nuclei
3. uptake of cations and electrons into the metal phase

At first a study of the reduction of FeO (wustite) in H_2-H_2O mixtures

$$H_2 + FeO = Fe + H_2O \tag{1}$$

will be reported in which it becomes clearly evident that the reaction at the three phase interface Fe/FeO/gas plays an important role. The study /20/ has been done at 800°C in a range of $p_{H_2O}/p_{H_2} = 0 \ ... \ 0.3$ bar. The formation and growth of the iron nuclei and crystallites have been observed in the hot stage microscope. A growth of circular disc shaped iron precipitates was observed, see Fig. 8.

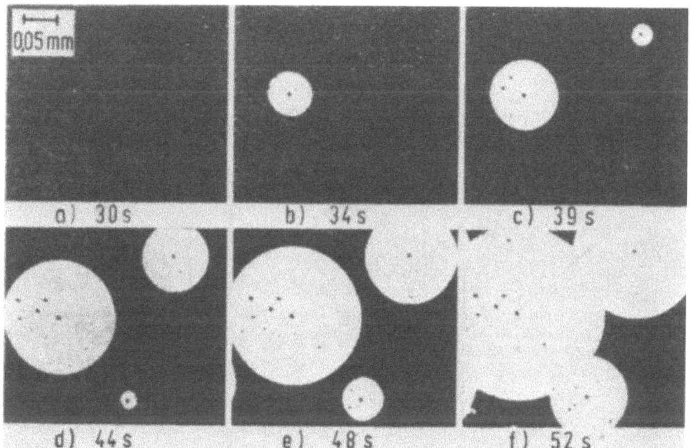

Fig. 8 Observation of wustite reduction in the hot stage microscope. Circular iron platelets growing on the wustite surface in H_2-H_2O at 800°C, photos taken after different reduction times

The increase of the radius of these precipitates is constant with time, i.e. dr/dt = constant. This can be explained by assuming that the removal of oxygen from the system mainly occurs by reaction at the circumference of the iron discs, in an area $2\pi rb = A$, where b is the width of a narrow fringe in which the reaction occurs. Rewriting the expression for the surface reaction by introducing this expression for A and by introducing $dn = 2\pi rh/V_{Fe} \cdot dr$, one obtains

$$\frac{dn}{Adt} = \frac{1}{2\pi rb}\frac{dn}{dt} = \frac{1}{2\pi rb}\frac{2\pi rh}{V_{Fe}}\frac{dr}{dt} = v \tag{2}$$

and it can be easily seen that for such reaction mechanism

$$dr/dt = b \cdot V_{Fe}/h \; v \tag{3}$$

So if the rate of the surface reaction v is constant and only dependent on the partial pressures, indeed the increase of the radius of the precipitates should be linear with time. dr/dt has been measured under different reaction conditions and the result is given in the following equation

$$\frac{dr}{dt} = \frac{bV_{Fe}}{h} \frac{k \cdot p_{H_2}}{1+K_O(p_{H_2O}/p_{H_2})}\left(1 - \frac{p_{H_2O}/p_{H_2}}{a_O}\right) \tag{4}$$

where it can be seen that the rate of reaction is proportional to p_{H_2}. Accordingly, the formation of H_2O from OH(ad) + H(ad) is the rate determining step at the fringe of the Fe disc. Additionally, it can be seen that the iron phase is covered with strongly adsorbed oxygen which retards the surface reaction. The surface reaction can take place only on the part of the surface which is free of oxygen

$$v \propto 1-\theta_O = 1/(1 + K_O \cdot p_{H_2O}/p_{H_2}) \tag{5}$$

Here the oxygen adsorption is described by a Langmuir term for the adsorption equilibrium of O(ad) with the H_2-H_2O mixture. This example clearly shows the important role of the three phase interface: Fe/FeO/gas where the removal of the O^{2-} from the solid phases takes place. It is easy to reduce wustite FeO in gas atmospheres H_2-H_2O or CO-CO_2, it is much more difficult to reduce the more stable oxides like chromium oxide or aluminum oxide. The reduction of Cr_2O_3 is still possible but only at high temperatures and with very dry H_2. This can be seen from the equilibrium data for p_{H_2O}/p_{H_2} at 1073K in table 1. This means that upon reduction in H_2 the water vapor pressure which can be reached is only very low, when this water vapor pressure is established near the Cr_2O_3 surfaces the reaction stops. Only if the water vapor is transported away by diffusion and fast flow of the gas phase the reaction can proceed.

In the reduction by dry H_2 a porous Cr layer is formed on Cr_2O_3 /21/. The reaction steps are

1. phase boundary reaction at the interface Cr/Cr_2O_3
2. transport of the formed H_2O through the porous Cr-layer
3. transport of the H_2O through the convective diffusion boundary layer into the gas atmosphere.

The results at 1173K and $p_{H_2O}/p_{H_2} = 1 \cdot 10^{-5}$ show a linear time dependence of the rate. At low flow rates u the rate is proportional to $u^{1/2}$ and proportional to p_{H_2}. From these results it can be concluded that at low flow rates the diffusion through the boundary layer is rate controlling. This diffusion is given by 1. Fick's law, the diffusion flow is

$$v = j_{H_2O} = 1/3\ \sqrt{Du/L}\ \sqrt[6]{D/\delta}\ \ (c^*_{H_2O} - c_{H_2O}) \tag{6}$$

Here the first factors result from the dependence of the thickness of the diffusion layer δ on the gas flow rate u and the term in parenthesis is the difference between the concentration of H_2O at the place of reaction minus the concentration in the gas flow. The square root dependence on u clearly shows that the gas flow rate under these conditions is rate controlling.
At high flow rates conditions can be obtained under which the phase boundary reaction is rate determining. Under these conditions the rate is independent of u and an expression was observed for the phase boundary reaction

$$v = k \cdot p_{H_2}^{1/2} \cdot f(p_{H_2O}/p_{H_2}) \tag{7}$$

The dependence on the ratio H_2O/H_2 was not fully elucidated but most probably the result is corresponding to

$$v = k \cdot f(a_O) \cdot p_{H_2} \cdot \left(1 - \frac{p_{H_2O}/p_{H_2}}{a_O}\right) \tag{8}$$

This equation indicates that the rate determining step is now the formation of OH by reaction of oxygen in the surface with adsorbed hydrogen (see chapter 3). The following reaction scheme was given by Rahmel /21/.

$$H_2(\text{gas}) = 2H(\text{ad})$$

$$Cr_2O_3 + 2V_{Cr}^{3-} + 3h^+ = 2Cr_{Cr} + 3O^-(\text{ad})$$

$$Cr_{Cr} = Cr$$

$$O^-(\text{ad}) + H(\text{ad}) + h^+ = OH(\text{ad}) \quad \underline{\text{slow}}$$

$$OH(\text{ad}) + H(\text{ad}) = H_2O$$

5. Conversion of Oxides to Other Compounds (Carbides, Intermetallics...)

Some oxides can be converted to carbides under conditions of high carbon activity and low oxygen pressure, especially endangered is chromium oxide. This process plays a role in the conversion of protective Cr_2O_3 layers on Fe-Ni-Cr alloys which are used as cracking tubes in petrochemistry. In the cracking tubes a coke deposit is formed on the inner surface of the cracking tubes which is covered by a chromia layer acting as protection against oxidation and carburization of the material. Below the coke deposit the carbon activity is $a_C = 1$. In the following it is described under which conditions the protective Cr_2O_3 layer can be converted to not protecting chromium carbides /22,23/. The thermodynamics is described in terms of CO_2-CO mixtures for which the oxygen activity is given by equation (1)

$$CO_2 = CO + 1/2\ O_2 \qquad p_{O_2}^{1/2} = K_1 \cdot p_{CO_2}/p_{CO} \tag{1}$$

The carbon activity in such gas mixture is given by equation

$$2CO = C + CO_2 \qquad a_C \cdot p_{CO_2}/p_{CO}^2 = K_2 \tag{2}$$

At $a_C = 1$, the relation is valid

$$p_{CO_2}/p_{CO} = K_2 \cdot p_{CO}.$$

The conversion of chromium oxide to the carbide Cr_3C_2 can take place by the following reaction

$$3\ Cr_2O_3 + 17CO = 2Cr_3C_2 + 13CO_2 \tag{3}$$

$$p_{CO_2}^{13}/p_{CO}^{17} = \exp(-\Delta G_3/RT)$$

Accordingly, the conversion takes place at a value of the CO_2-CO ratio given by

$$\ln(p_{CO_2}/p_{CO}) = \Delta G_3/13RT + 4/13 \ln p_{CO}$$

Similarily, a conversion of chromium oxide to the carbide Cr_7C_3 can occur according to the reaction equation

$$7Cr_2O_3 + 33CO = 2Cr_7C_3 + 27CO_2 \tag{4}$$

$$p_{CO_2}^{27}/p_{CO}^{33} = \exp(-\Delta G_4/RT)$$

For this reaction to occur the ratio p_{CO_2}/p_{CO} must be given by

$$\ln\ (p_{CO_2}/p_{CO}) = -\ \Delta G_4/27 \cdot T + 6/27 \cdot \ln\ p_{CO}$$

Upon heating a material in a CO_2-CO mixture which is in equilibrium with graphite ($a_C = 1$), the ratio p_{CO_2}/p_{CO} is given by equation (2) and it can be seen from Fig. 9 that this ratio decreases with increasing temperature. The plot of p_{CO_2}/p_{CO} vs T, the curve for equation (2), crosses the lines for the reactions (3) and (4) in the points of coexistence of $Cr_2O_3/Cr_3C_2/C$ and $Cr_2O_3/Cr_7C_3/C$, at lower values of p_{CO_2}/p_{CO} or at higher temperatures, the oxide is converted to the respective carbides. So it can be seen from Fig. 9 that with increasing temperature at the carbon activity $a_C = 1$, the chromium oxide should be converted to chromium carbide Cr_3C_2 or Cr_7C_3.

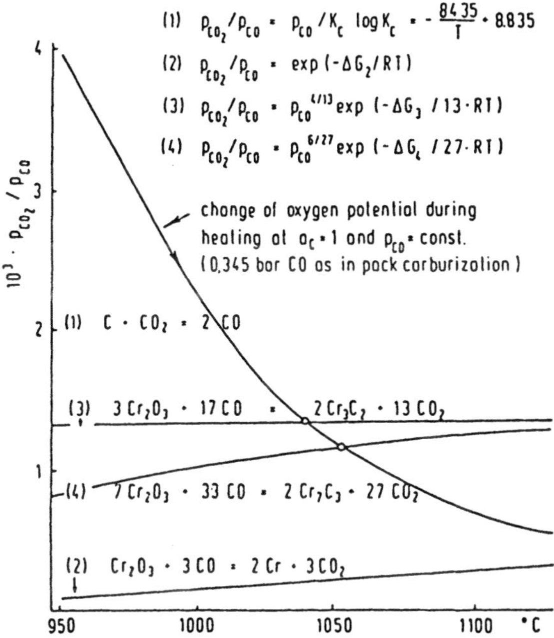

Fig. 9 Thermodynamics of the conversion Cr_2C_3-->chromium carbides at a_C = 1 and constant pCO = 0.345bar. With increasing temperature the ratio pCO_2/pCO decreases and the equilibrium values for the conversion reactions are reached.

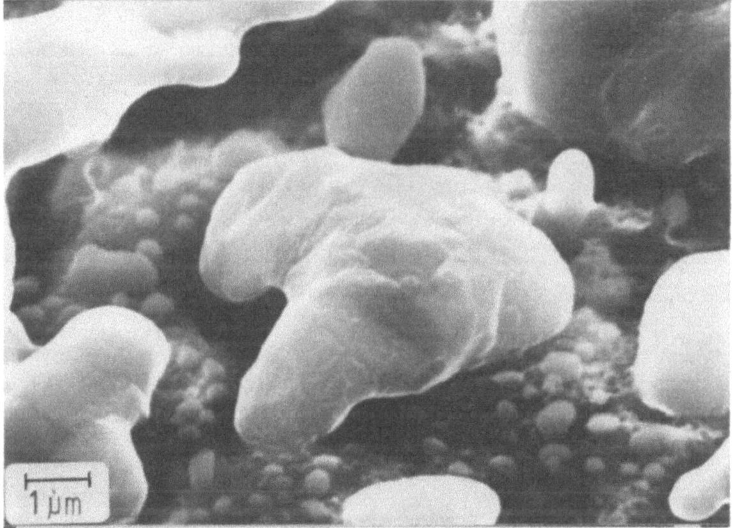

Fig. 10 Scanning electron micrograph of the Cr_2O_3 particles on an Fe-12.5%Cr alloy, partially converted to chromium carbides (which form inside the alloy)

These reactions indeed take place and it's a well-known experience in the petrochemical industry that in the range of these temperatures the chromium layer on the inside of the cracking tubes loses its protective function. So either this range of high temperatures >1100°C must be avoided or other mechanisms for protection must be developed. This can be done by addition of some silicon to the material. Fe-Ni-Cr alloys with 1 to 2% Si develop a SiO_2 layer below the Cr_2O_3 layer which gives protection even at temperatures where the chromia layer is converted to carbide.

The kinetics of the conversion reaction have been studied by Chu and Rahmel /24/ at 1200°C. It has been shown that the conversion of Cr_2O_3 to Cr_7C_3 is rather slow and linearly time dependent. Thus it can be concluded that a phase boundary reaction is rate limiting, the authors have assumed that the reduction of Cr_2O_3 is the rate controlling step. In a later study /25/ the reaction of the Cr_2O_3 layer on a Fe-12.5%Cr alloy was investigated in a flowing CH_4-H_2-H_2O-CO mixture, the Cr_2O_3 layer is metastable in CH_4-H_2O-H_2 which shows that there is only negligible interaction of the oxide with CH_4. However, in this mixture a formation of CO occurs by the reaction of methane CH_4 with H_2O to CO and if enough CO is formed by catalytic action of the oxide surface, a reduction and conversion of the chromia layer to carbides starts. There's evidence that most of the reaction takes place via CO and C dissolved in the metal

$$7\ Cr_2O_3 + 27\ C = 2\ Cr_7C_3 + 21\ CO \tag{5}$$

It can be seen from SEM photos of the reacting surface, see Fig. 10 that the oxide is consumed from its base, i.e. at the three phase boundary metal/oxide gas phase. Most probably, carbon is dissolved from the gas phase into the metal and reacts with the oxide at its contact to the metal phase. So very peculiar shapes of the oxides are generated, mushroom-like or bridge-like shapes are observed, see Fig. 10.

At least, a solid/solid reaction is described which is related to the above oxide/carbide conversion reaction. The reaction of chromium carbide and chromium oxide is of practical interest and has been studied by Maru et al /26/.

$$1/6\ Cr_{23}C_6 + 1/3\ Cr_2O_3 = CO + 9/2\ Cr \tag{6}$$

The reaction has been studied for particles of different size and density compressed in a pellet by thermogravimetry at 1050 to 1100°C in vacuum of 10^{-3} Torr. The first question in this study was if this reaction occurs by direct interaction solid/solid. This possibility was excluded since the overall reaction rate is decreased by a higher density of the particles. In contrast, a higher porosity and lower density of the particles lead to increased reaction rates which indicates that the access of the gases to the reacting solid is important in this reaction. The following reaction mechanism was assumed

$$1/3\ Cr_2O_3 + CO = 2/3\ Cr + CO_2 \qquad a) \qquad (7)$$

$$1/6\ Cr_{23}C_6 + CO_2 = 23/6\ Cr + 2CO \qquad b)$$

This is in agreement with the expression for the rate of reaction which has been observed

$$\text{rate} \propto \frac{k}{r}\left(\frac{p_{CO_2}}{p_{CO}}\right)_{Cr/Cr_2O_3} \qquad (8)$$

where r is the radius of the $M_{23}C_6$ particles and the ratio CO_2/CO is the value for the equilibrium of $CO_2/CO/Cr/Cr_2O_3$. The mechanism of the rate determining step (7b) is given by

$CO_2 = CO + O(ad)$ oxygen adsorption on $Cr_{23}C_6$

$O(ad) + C(carbide) = CO$ slow

So for the overall reaction the reaction step of CO formation from adsorbed oxygen and carbon from the carbide on the carbide surface is rate controlling. The equilibrium of oxygen adsorption on the $Cr_{23}C_6$ is established by oxygen transfer from the CO_2-CO mixture according to step (7a) and the reduction of the Cr_2O_3 by CO leading to CO_2 formation is assumed to be fast. The reaction steps occurring on the carbide surface are similar to that mechanism which takes place upon oxidation of graphite and other carbons in CO_2-CO mixtures:

$$\begin{array}{ll} CO_2 = CO + O(ad) & (a) \\ O(ad) + C = CO & (b) \\ \hline CO_2 + C = 2\,CO & (9) \end{array}$$

Also in this case it can be assumed that the equilibrium of oxygen transfer from the CO_2-CO mixture is established and that the slow step is the reaction of the adsorbed oxygen with the carbon from the graphite or from other forms of carbon /27,28/. Here again, the formalism can be used developed before for the oxygen transfer from CO_2-CO mixtures, see chapter 3, equation (9)

$$\frac{dn_O}{dt} = k_1 \cdot a_O^{-m} \cdot p_{CO_2} - k_1 \cdot a_O^{1-m} \cdot p_{CO} \qquad (10)$$

and the reaction of adsorbed oxygen with carbon is proportional to the oxygen activity on the carbon

$$-\frac{dn_O}{dt} = -\frac{dn_C}{dt} = k \cdot a_O \qquad (11)$$

Both equations can be coupled by the steady state condition that the amount of adsorbed oxygen on the surface does not change in a steady state. This leads to

$$k_1 \cdot a_O^{-m} \cdot p_{CO_2} - k_1 \cdot a_O^{1-m} \cdot p_{CO} = k \cdot a_O \tag{12}$$

and from this steady state condition two limiting cases result:

I. $k_1 \cdot a_O^{-m} \cdot p_{CO_2} \gg k \cdot a_O$

In that case the oxygen activity on the carbon is given by

$$a_O \approx \frac{p_{CO_2}}{p_{CO}} \tag{13}$$

and the rate of carbon consumption is proportional to the p_{CO_2}/p_{CO} ratio. In the second case

II. $k_1 \cdot a_O^{-m} \cdot p_{CO_2} \approx k \cdot a_O$

A more complicated relation results

$$-\frac{dn_C}{dt} \approx k\left(\frac{k_1}{k} p_{CO_2}\right)^{1/1+m} \tag{14}$$

where the carbon gasification is mainly determined by the CO_2-pressure.

At least we come back to reactions mentioned already in the beginning, the reduction of the ceramic oxides ZrO_2 and Al_2O_3 under formation of intermetallic 'Engel-Brewer' compounds, for which the thermodynamics was discussed in the first chapter. Also some studies were performed on the kinetics /29,30/. The conversion of Al_2O_3 to the intermetallic compound Pt_3Al can take place upon contact of Pt with Al_2O_3 in the presence of a very reducing atmosphere. It can be assumed that the hydrogen is dissolved in platinum and reacts at the three phase interface metal/oxide/gas under formation at first aluminum dissolved in platinum and later on formation of the intermetallic compound.
So these are examples that the even more stable oxides, such as Cr_2O_3 and Al_2O_3 can be converted to other compounds by the combined action of a gas and another phase, in the case of chromium the other phase was carbon and in the case of the reduction of Al_2O_3 or ZrO_2 the other phase was platinum.

Conclusions

The solubility of non-metal atoms seems to be well documented for hydrogen in ZrO_2, Al_2O_3, TiO_2 ... however, virtually no carbon, sulfur and chlorine are dissolved in MgO, FeO, Fe_3O_4, Cr_2O_3, Al_2O_3 as was confirmed by radiotracer studies.

The ceramic oxides according to their high thermodynamic stability show very little reactivity in high temperature reactions with gases, this is caused

1. by the strong binding of metal and oxygen atoms M-O,
2. by the low equilibrium pressure of the gaseous reaction products, such as H_2O in the reduction of ceramic oxides by H_2.

Due to the low equilibrium pressures of the reaction gases, in many cases their removal by gaseous diffusion and transport is rate controlling for the overall reaction.

For the phase boundary reactions in oxidation and reduction by CO_2-CO or H_2O-H_2 mixtures general rate equations can be given which were found for more reactive oxides FeO, CoO, NiO, MnO and Fe_3O_4. These rate equations are composed of a term for the forward reaction and a term for the backward reaction. These terms contain dependencies on the partial pressure of the gases and dependencies on the oxygen activity of the solid oxide surface. The dependence on the oxygen activity at the solid surface is caused by participation of electrons from the solid in the oxygen transfer and additionally, by the action of adsorbed oxygen which may block reaction sites on the solid surface. It has been shown that such rate equations apply for the reduction of Cr_2O_3 but it must be still confirmed that similar rate equations are valid also for reactions on TiO_2, Al_2O_3, ZrO_2.

The interface oxide/metal/gas plays an important role in reduction and other reactions on oxides, the metal phase obviously plays a catalytic role or may dissolve components from the gas phase, such as H or C which react with the solids.

Reactions solid/gas can be promoted by the presence of a second solid, e.g of carbon in the conversion of oxides to carbides or for example by Pt in the conversion of oxides to intermetallic compounds. However, generally most reactions do not occur by solid/solid interaction but by intermediate action of gases.

REFERENCES

1. P.J. Meschter, W.L. Worrell: Metallurg. Trans. **7A** (1976) 299
2. P.J. Meschter, W.L. Worrell: Metallurg. Trans. **8A** (1977) 503
3. E.A. Gulbransen, S.A. Jansson: Oxidation of Metals **4** (1976) 181
4. N. Birks, G.H. Meier: 'Introduction to High Temperature Oxidation of Metals' Edward Arnold Publ. London 1983
5. S. Stotz, C. Wagner: Ber. Bunsenges. **70** (1966) 781
6. C. Wagner: Ber. Bunsenges. **72** (1968) 778
7. M.M. El-Aiat, F.A. Kröger: J. Appl. Phys. **53** (1982) 365
8. F. Freund, G. Debras, G. Demortier: J. Amer. Ceramic Soc. **61** (1973) 429
9. F. Freund: Physikalische Blätter **37** (1981) 283
10. P. Kofstad: J. Phys. Chem. of Solids **44** (1983) 129
11. P. Kofstad: Oxidation of Metals **19** (1983) 129
12. I. Wolf, H.J. Grabke: Solid State Commun. **54** (1985) 5
13. D. Bramhoff, H.J. Grabke, R. Lobnig: unpublished results
14. D. Stöver, H.P. Buchkremer, R. Hecker: IRE, KFA Jülich, unpublished results
15. H.J. Grabke: Ber. Bunsenges. phys. Chem. **69** (1965) 48
16. A. Gala, H.J. Grabke: Archiv Eisenhüttenwes. **43** (1972) 463
17. E. Riecke, K. Bohnenkamp: Archiv Eisenhüttenwes. **40** (1969) 717
18. H.J. Grabke, H. Viefhaus: Ber. Bunsenges. phys. Chem. **84** (1980) 152
19. L.S. Darken, R.W. Gurry: J. Amer. Chem. Soc. **67** (1945) 2 68 (1946) 798
20. W. Pluschkell, H. Yoshikoshi: Archiv Eisenhüttenwes. **41** (1970) 715
21. W.F. Chu, A. Rahmel: Metallurg. Trans. **10B** (1979) 401
22. H.J. Grabke, U. Gravenhorst, W. Steinkusch: Werkstoffe u. Korr. **27** (1976) 291
23. J.A. Colwell, R.A. Rapp: Metallurg. Trans. **17A** (1986) 1065
24. W.F. Chu, A. Rahmel: Oxidation of Metals **15** (1981) 331
25. H.J. Grabke, P.C. Prasannan, E.M. Müller: Metallurg. Trans. **17A** (1986) 915
26. Y. Maru, Y. Kuramasu, Y. Awakura, Y. Kondo: Metallurg. Trans. **4** (1973) 2591
27. H.J. Grabke: Ber. Bunsenges. phys. Chem. **70** (1966) 664
28. H.J. Grabke: Carbon **10** (1972) 587
29. A.S. Darling, G.L. Selman, R. Rushfort: Platinum Metals Rev. **14** (1970) 54, **14** (1970) 95, **15** (1971) 13
30. J.T. Klomp: Proceed. of this NATO ASI

GROWTH OF CERAMIC LAYERS FROM VAPOR PHASE

F.TEYSSANDIER
CNRS-Institut de Science et de Génie des Matériaux et Procédés, UP32 Université-Avenue de Villeneuve, 66025 Perpignan Cedex, France

ABSTRACT. Ceramics can be obtained by many techniques and mainly by powder processes such as sintering. They can also be grown from vapor phase with the Chemical Vapor Deposition (CVD) process. Numerous attempts have been made to obtain a better understanding of the mechanisms which involve complex fluid dynamics and surface phenomena but direct evidence is still lacking. Many reviews have dealt with this subject [1-5] and general trends used in the comprehension of the various steps leading to the formation of a ceramic from a gas phase are first presented. Recent developments of the process leading to more complex coatings such as ceramic composites are then discussed. The interest of the thermodynamic calculation used as a predictive tool is finally shown with a few examples dealing with two-phased materials or non-stoichiometric compounds.

1. Introduction

Chemical vapor deposition is a versatile method which allows the growth of a wide variety of materials such as pure elements,compounds, solid solutions or composites on various substrates. A mixture of gases flows around a heated surface where the chemical reactions occur. In the pure CVD process reactions are only thermally activated though this denomination is also used for mixed thermal and physical activation. A distinction is usually made with respect to the total pressure and LPCVD (LP : low pressure) is the common denomination of deposits obtained under reduced pressure. The lowest pressures used in such cases are 10-100 Pa and the mechanisms involved do not fundamentally differ from atmospheric pressure, as the Knudsen flow has not yet been attained in the reactor.

One of the main problems encountered in CVD is the high substrate temperature (> 1000°C) usually required to obtain appreciable reaction rates.This is due to the great stability of the metal bearing molecules which are generally halides (mainly chlorides) and are very stable. In order to cope with this difficulty one can

L.-C. Dufour et al. (eds.), Surfaces and Interfaces of Ceramic Materials, 625–638.

can either use less stable molecules or physically break them in the gas phase to improve the reactivity of the gas mixture.

In the first case, organometallic molecules (OM CVD) lead to a significant decrease in the deposition temperature but these deposits are often poorly crystallized or amorphous and by-products of the reactions may be trapped in the coating, modifying its physical properties.

In the second case the initial molecules are broken in order to obtain more reactive atoms or fractions of starting molecules. This activation is achieved either by plasma PACVD (plasma activated) or by the energetic beam of a laser LECVD (laser enhanced). Assuming that the main feature of a CVD process is the chemical reaction one can also mention the reactive sputtering. This process is at the boundary of CVD and PVD. The reactant gases are activated in a plasma zone before reacting with the species ejected from the target surface.

2. Basic Mechanisms of the Chemical Vapor Deposition Process.

Difficulties in obtaining a comprehensive view of the growth process arise from the number of interdependent parameters involved such as temperature, pressure, composition of the initial gas phase, flowrate, type of reacting chamber ... and the complexity of the various steps. These phenomena are not specific to the CVD process but may also be encountered when ceramics are exposed to reactive gases at high temperature.

The steps of the individual mechanisms involved can be divided into two parts :

- Phenomena occuring in the gas phase, such as bulk transport of the species in a hydrodynamic flow with diffusion mechanisms due to concentration gradients and eventually homogeneous reactions or gravity effects. In a cold wall reactor one must also take into account the thermal diffusion.
- The surface phenomena including adsorption, diffusion, reaction and desorption. In this case a distinction has to be made between the first moment of the deposit when interactions with the substrate take place including epitaxy or nucleation and growth mechanisms and the continuous growth of the coating from the gas phase in steady-state conditions.

This wide variety of relevant scientific domains is one of the difficulties of the CVD process and many assumptions are currently made in order to give a comprehensive description of experimental results.

2.1 GAS PHASE TRANSFERS.

Transfer phenomena in the gas phase were described for a long time by boundary layer models and more recently by a full scale evaluation of the entire set of equations governing the process. These transport phenomena are very important because of their influence on the composition of the gas mixture at the solid-gas interface but multicomponent, diffusive, reactive, three-dimensional, mixed convection systems are quite difficult to represent with fluid dynamics models. Nevertheless real progress has been achieved in this field during the last three or four years. As their description is beyond the scope of this paper some recent reviews may be read on this subject (6-7).

The basic steps and the influence of the main factors will only be presented in the case of the boundary layer hypothesis in order to obtain a better understanding of the coupling between gas phase phenomena and the morphology or the composition of the deposit.

The boundary layer theory (8) is an approximation of the Navier Stokes equation which supposes that the flow around a solid body can be divided into a thin layer in the neighborhood of the body (boundary layer), where frictions play an essential part, and the remaining region, where frictions may be neglected. The validity of the boundary layer hypothesis is questionable in many CVD reactors (7) and the absence of such a layer has been demonstrated in some cases by 3D numerical works.The same assumption is currently made for heat transfer with a thermal boundary layer and for mass transfer with the diffusion boundary layer. The thickness of these three boundary layers are expected to be comparable. If we assume that all chemical reactions take place on the substrate surface and not in the gas-phase, the concentration gradient occurs only across the boundary layer perpendicular to the substrate. The basic mechanistic steps of the CVD process are then generally described in the following way.

1) Forced flow of reactant gases into the reactor.
2) Diffusion of reactant gases through the gaseous concentration boundary layer to the substrate.
3) Surface process including adsorption of the gases onto the substrate, chemical reaction, diffusion.
4) Desorption of species.
5) Diffusion of product gases through the boundary layer.
6) Forced exit of gases.

Steps 3 and 4 deal with surface kinetics and steps 1-2-3-6 control the mass transfer of the gaseous species. Depending on the rate limiting steps, growth rate can be categorized as mass transfer rate-limited or kinetically rate-limited. When plotting the logarithm of the growth rate against T^{-1} (Figure 1), two parts are generally observed which obey the general phenomenological equation $V = K \exp(-Ea/RT)$, where Ea is an apparent activation energy. In the low temperature domain a steep slope is encountered corresponding to a high Ea value typically ranging from 100 to 300 kJ/mole. In this case the growth rate is governed by the surface kinetics including adsorption surface diffusion and reaction. At higher temperatures surface processes increase rapidly and the growth rate becomes weakly dependent on temperature. This part is considered as a region where gaseous diffusion is rate controlling. As long as no modification in the mechanism of the reaction occurs the influence of total pressure is schematically represented in figure 1. Lowering the total pressure reduces the diffusional resistance and the surface kinetic limitation is reached at a higher temperature.

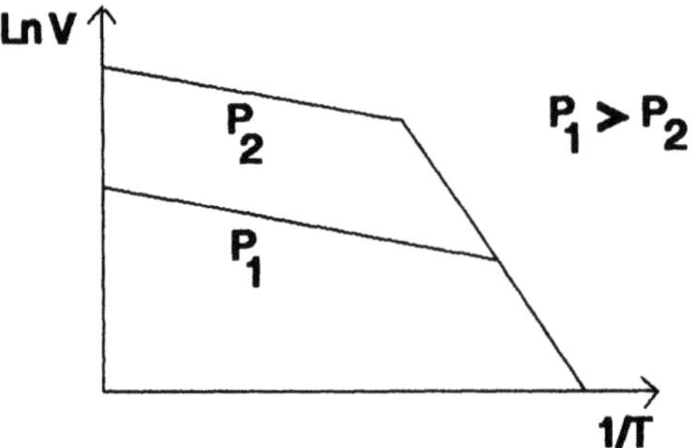

Figure 1: Schematic representation of the experimental growth rate(V) as a function of Temperature (T). P is the total pressure.

The temperature and pressure at which transition from kinetic to diffusion limitation occur have some importance for the structure and the morphology of the deposits, as demonstrated experimentally by coating a smooth surface presenting a groove(9) (Figure 2). Unstable growth is found when it is diffusionally-controlled whereas uniform step coverage is observed under kinetic control. From a more general point of view, gas-phase diffusion can be considered as a destabilizing process leading to surface roughening though surface diffusion tends to smooth out the surface. Competition between mass transport to the surface through the gas phase and surface diffusion sets the scale for surface roughening.

A 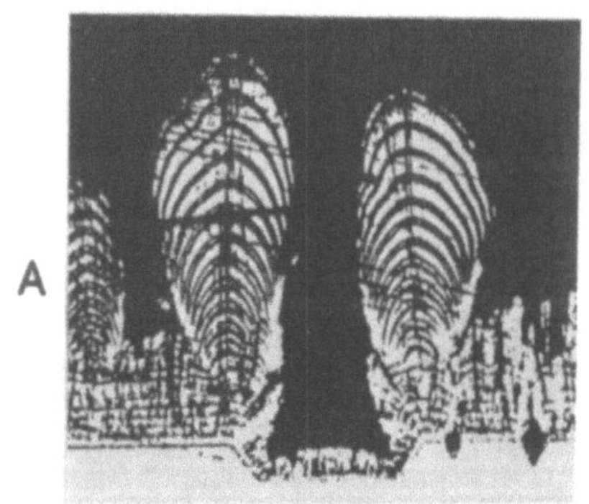

B 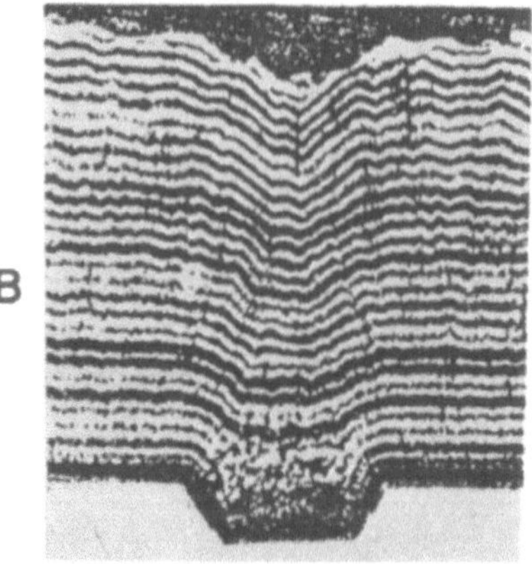

Figure 2: Experimental growth morphologies (9).
A under diffusion limited conditions; B under surface controlled process conditions

2.2 SURFACE PHENOMENA.

A lot of work has been carried out in the case of homoepitaxy (condensed phase on a surface of the same nature) or heteroepitaxy (condensed phase on a different substance but with the same crystallographic system) (10-11).

When the degree of misfit (100 x (b - a)/a, a and b are respectively the lattice parameters of the substrate and the deposit) exceeds 15% to 38% depending on the materials, the growing mode is no longer epitaxy but heterogeneous nucleation. Most ceramics for thermochemical application are grown in such a way. The heterogeneous nucleation models were first developed to represent the condensation process. The model of the cap shaped embryo is generally used. Growth may occur either by direct addition from the vapor or by surface diffusion but this latter process is predominant.

In CVD, supersaturation is so high that the critical cluster radius may be composed of only one or two atoms. Classical thermodynamic variables are no longer valid and an atomistic theory has been developpped (12) in terms of potential energies and partition functions. In the generalization of this theory (13) the idea of critical nucleus is abandoned and the condensation process becomes a kinetic polymerisation-like process.

Due to the number of vapor species involved, application of such theories to pratical cases of CVD growth of a deposit on a polycristalline different substrate is not reasonable. Many phenomena, such as the influence of impurities or preferential nucleation at the grain boundaries, would also have to be considered.

For a given material a wide range of crystal morphologies are observed depending on two main factors : temperature and supersaturation. Reference is usually made to the diagram proposed for physical vapor deposition (figure 3).

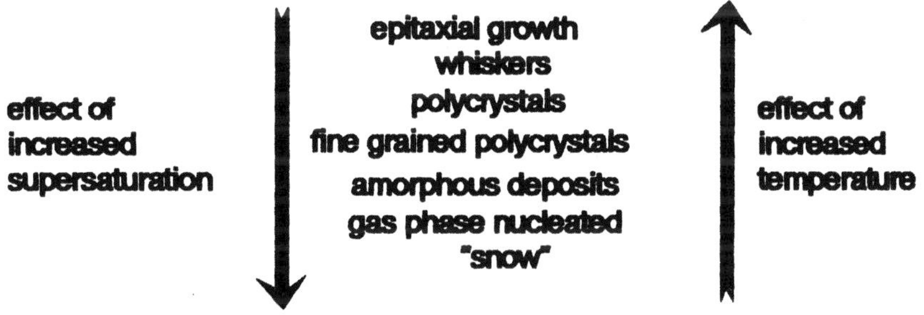

Figure3: The effect of supersaturation and temperature on the structure of materials obtained by PVD process (1).

In the PVD process, which is basically a condensation process, supersaturation is defined as the ratio between the effective pressure of the atoms and their pressure in equilibrium with the solid. In the CVD process one also has to take into account the standard free energy of reaction.

Considering the following dissociation reaction of a metal halide :

$$MX \rightleftharpoons M + X$$

supersaturation is now defined as the concentration of available reaction product in the gas phase relative to the concentration in equilibrium with the solid. If the reaction is exothermic, decomposition of MX is favored at low temperature and the influence of supersaturation is expected to be the opposite to the PVD behavior. As most CVD reactions are endothermic then, according to the modified definition of supersaturation, the observed behavior of CVD systems agrees well with figure 3 if we exclude homogeneous reactions which take place at both high temperature and high supersaturation. This last quantity can be considered as an estimation of the departure from thermodynamic equilibrium of the chemical system

Ceramic layers of mono-phased materials can be obtained with various structure or morphology according to the coating parameters. Appropriate gas mixtures also allow the formation of two or three-phased materials or the control of the composition in non-stoichiometric compounds.

3. Advanced Ceramics Obtained by CVD.

Ceramics for thermomechanical applications are widely used for machining or cutting tools and also for protection against friction and wear (bearings). They are under study for new applications such as blades for high temperature gas turbines, several parts of heat engines (cylinder linings, piston caps, valve faces and seats). The properties required vary with respect to the application but the main features common to ceramics are their great resistance to creep, corrosion, oxidation, erosion and wear. Almost all the phases giving thermomechanical ceramics are known. A lot of work still has to be carried out from a technological point of view in order to increase the reliability of industrial manufacture of such materials. Nevertheless high quality monolithic ceramics can now be prepared and revolutionary improvements are not to be expected in this field. The main problem for the thermomechanical application of ceramics is their low reliability due to their brittleness. Toughness can be substantially enhanced by the use of either fiber composites or dispersed phase composite materials (14). These materials are more promising and the composite concept corresponds to a wide variety of structures and phases which can be classified from a geometric point of view or with respect to their use.

Three classes of fiber composites are distinguished by matrix type : organic matrix, metal matrix, and ceramic (polycrystalline or glass) matrix.

When the matrix is ductile (organic or metallic) it simply holds the reinforcing fibers and transfers the structural load. As long as the fiber-matrix adhesion is good the axial tensile strength of the composite is, to a first approximation, equal to the tensile strengh of the fiber. On the other hand, the tensile strength or the modulus of a ceramic matrix is more or less the same or may be higher than that of the fibers, leading to an improvement in toughness rather than strength. The interfacial strength between matrix and fiber then becomes very important (See article of Professor P.Pirouz) and the rule of mixture no longer accounts for the behaviour of such materials.

Very high toughness can also be obtained from whisker-reinforced ceramics (8,7 MPa $m^{1/2}$ for 20% vol whiskers SiC into alumina (15)) and better properties can be expected from improved dispersion uniformity.

The second family of structural ceramics is composed of dispersed phase composite and essentially the ZrO_2 toughening ceramics (16). The martensitic transformation to monoclinic structure of metastable tetragonal ZrO_2 induced by the stress field of the propagating crack substantially enhances the toughness of two phased ceramics such as partially stabilized zirconia (PSZ) or ZrO_2 into either alumina (ZTA) or Si_3N_4.

The high versatility of the CVD process allows the growth of two-phased materials of various natures.

The composition and grain size can to a large extent be controlled by the deposition parameters. The presence of a dispersed phase can reduce the brittleness of the coating, enhance resistance to oxidation or improve tribological properties, as illustrated for sintered materials. Two-phased systems are difficult to obtain without a third phase by sintering and pollution by adsorbed oxygen at the surface of the initial powder is difficult to avoid.The high uniformity of the dispersion, the possibility of having all the neighbouring phases in equilibrium with each other are some of the interests of ceramics obtained from a gas phase.

The intergrowth of two or three phase equilibria has been studied since the beginning of the seventies. The Ti-Si-C system was first studied from a $TiCl_4$-$SiCl_4$-CCl_4-H_2 mixture (17). On the deposition diagram presented in Figure 4 all binary and ternary phases can be obtained in accordance with the Ti-Si-C phase diagram.

Two or three-phased domains are also deposited and the following trends are observed:

- between the binary phases TiC and β-SiC, non-oriented polycrystalline mixtures with no coherent intergrowth are deposited.
- when ternary phases Ti_3SiC_2 (T_1) or $Ti_5Si_3C_x$ (T_2) are involved, lamellar deposition with a coherent or semi coherent phase boundary are obtained

growth direction : $[110]_{T1}$ || $[110]_{TiC}$

interface : $[001]_{T1}$ || $[111]_{TiC}$

The deposition model proposed by these authors assumes the initial formation of a ternary phase $Ti_5Si_3 + yC_x$ favored because of its large composition range, from which equilibrium phases precipitate.

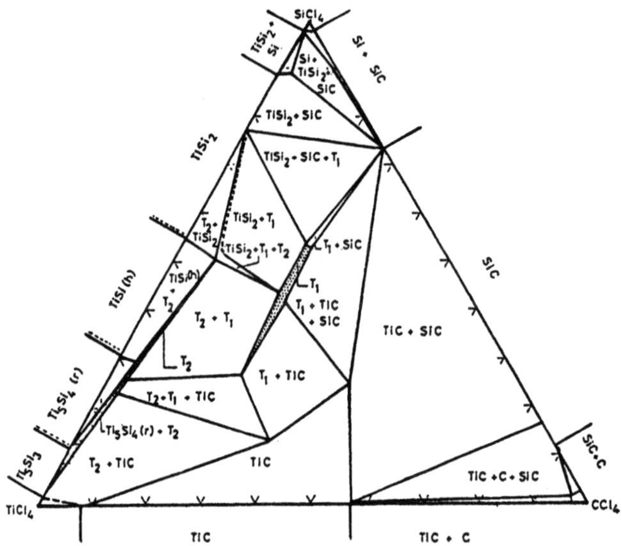

<u>Figure 4:</u> Experimental deposition diagram for the reaction system $TiCl_4$-$SiCl_4$-CCl_4-H_2 at T=1540 k (17)

In a further investigation of the $SiCl_4$-CCl_4-H_2-N_2 system (18), a mixture of crystallized SiC and amorphous Si_3N_4 are supposed to grow from the carburization or the nitridation of the free silicon initially deposited . As a consequence of the stability of N_2 a low amount of CCl_4 kinetically favors the formation of the carbide.

Si_3N_4-TiN composites have been prepared using a mixture of $SiCl_4$-$TiCl_4$-NH_3 and H_2 (19). With respect to the deposition temperature, the structure of the Si_3N_4 matrix ranges from amorphous to the trigonal α-type and the hexagonal β-type, modifying at the same time the shape or the size of the TiN crystallites (20). (figure5)

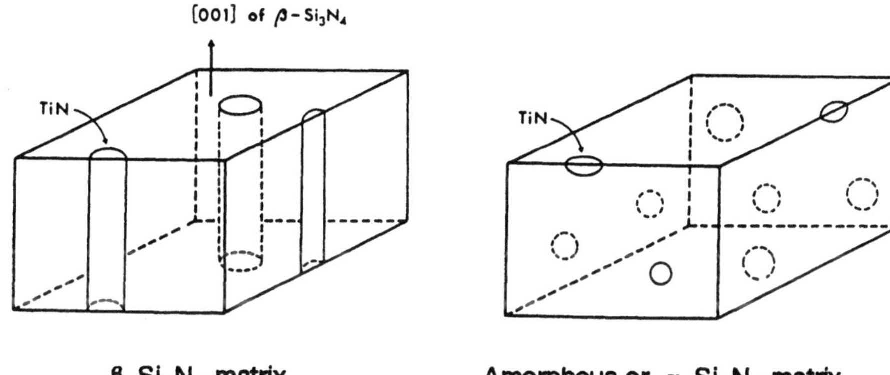

β-Si_3N_4 matrix   Amorphous or α-Si_3N_4 matrix

Figure 5: Schematic representation of the microstructure of the CVD Si_3N_4-TiN composites (20)

The size of TiN crystallites dispersed into amorphous Si_3N_4 was roughly estimated as 3nm from the width of the diffraction peak. This size is larger into the α-type Si_3N_4 with no definite orientation relative to that of Si_3N_4. TiN microstructure has been more detailed (21) in the case of the β-Si_3N_4 matrix. TiN is then present as fibers which can be as long as 1 µm for a diameter of 5nm. From high resolution electron microscopy studies, the fiber axis of TiN can be seen to be always <110> parallel to the [0001] direction of Si_3N_4. The following relations [$\bar{1}$10] || <2$\bar{1}\bar{1}$0> or [$\bar{1}$12] || <1$\bar{1}$00> were found along the short axis of the cross section figure 6 which corresponds to a coincidence between the lattice spacing of TiN and Si_3N_4.

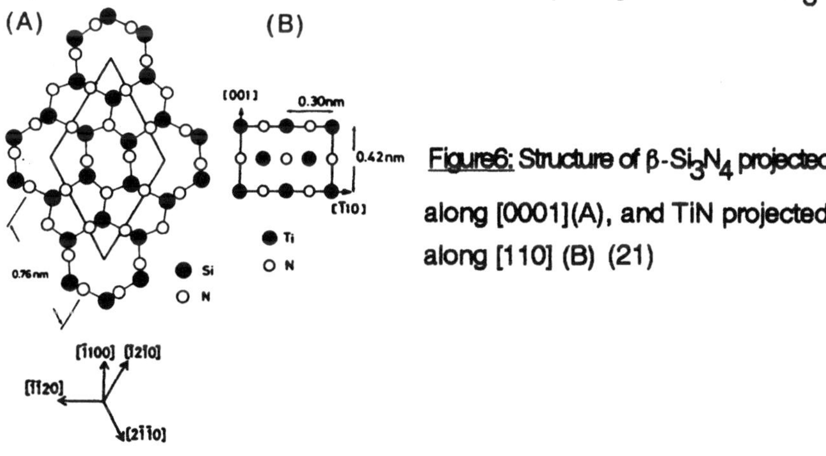

Figure6: Structure of β-Si_3N_4 projected along [0001](A), and TiN projected along [110] (B) (21)

BN-TiN and BN-Si_3N_4 composite have been obtained by CVD (22). In this case 30-50 A° crystalline particles of TiN were observed in the turbostratic BN by high resolution electron microscopy. Their size in the amorphous Si_3N_4 matrix were not determined.

The two-phased AlN-Si_3N_4 mixture can also be mentioned (23). The films are composed of polycrystalline AlN (with grain size ranging from 150 A° to 500 A° as a function of the deposition temperature) and amorphous Si_3N_4.

All these studies show that various two-phased ceramics and crystallites of different size and shape uniformly dispersed into a matrix can be obtained by CVD. Up to now no physical properties and in particular no mechanical behavior, has been determined on such materials.

The increasing number of initial species which are necessary to obtain composite coatings and the quantity of available bearing molecules make the choice of the experimental conditions difficult. A guideline is necessary in order to avoid systematic blind experiments.

4. Thermodynamic Calculation Used as a Predictive Tool.

As already mentioned two domains are observed when plotting the growth rate as a function of temperature. The low temperature domain where surface kinetics are the rate-limiting steps and a high temperature domain where the growth rate is limited by mass transport through the boundary layer above the substrate. In this latter case assuming steady-state conditions, simple equations (24) show that the difference between the surface partial pressure and the equilibrium partial pressure will keep decreasing while increasing temperature. Calculation of the equilibrium behavior of such systems gives valuable predictive information on the composition of the complex gas phase at the solid-gas interface which determine the chemical potential of the various elements in the solid and the possible chemical interactions with the substrate.

Thermodynamic calculation is an attractive tool to predict the influence of temperature, pressure, and initial composition of the gas phase on the nature of the condensed phase at equilibrium (25-26). Some discrepancies between calculated and experimental results may be observed, mainly due to the difference between gas composition at the solid-gas interface and the initial gas mixture as a result of diffusion, homogeneous reactions or condensations. Nevertheless, trends predicted in the behavior of such chemical complex systems are almost always experimentally confirmed.

Several computational programs are available (27-28). The calculation is based on minimizing the component free energies derived from summation over all species which requires all possible species to be included in order to avoid erroneous results. Such calculations have been achieved in the case of ceramics

(VB2 (29), B4C (30), TiN_x (31), TiC_x (32)). More recently they have been applied to ceramic-ceramic composites.

Equilibrium thermodynamic analyses are performed as a function of temperature, total pressure and reactant concentration in order to determine the deposition diagram. These diagrams are usually plotted with respect to the initial concentration of reactant species. This is the case for the calculation of the SiC-Cr_3C_2 domain plotted against CH_3SiCl_3 and $Cr(C_5H_4O_2F_2)_3$ partial pressures in an initial gas mixture also including H_2 (33). Different coordinate systems are also available: in the five component B-Si-Cl-H-N system the Si_3N_4-BN deposition domain is presented as a function of temperature and atomic reactant gas fraction B/(Si+B) (Figure 7)(34).The codeposition of Si_3N_4 and BN is seen to be thermodynamically favored over a wide range of temperature and predominates below 1400 K.

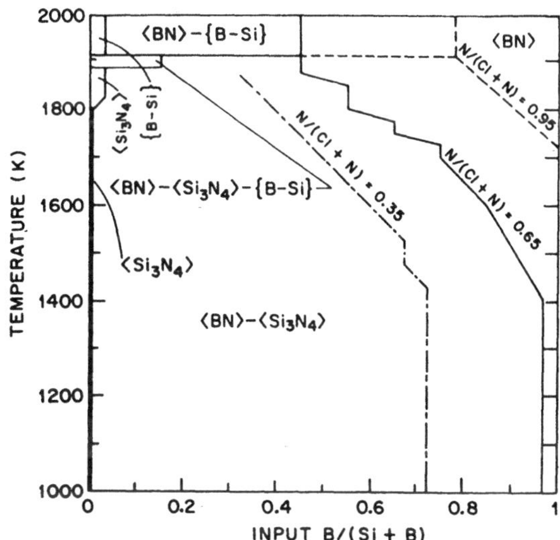

Figure 7: Calculated deposition diagram of Si_3N_4-BN as a function of input B/(Si+B) and Temperature (34).

When dealing only with two-phased domains a more straightforward calculation leads to a rapid investigation of all possible combinations between the initial bearing molecules. In the atomic Ti-Si-C-Cl-H system at constant temperature and pressure, the gas-phase composition in equilibrium with two solid phases is completely detemined by two independent intensive and invariant appropriate variables (atomic abundance ratio) which are used as the coordinate system (35). The representation of the TiC-SiC domain in figure 8 is an intrinsic feature of the chemical system. This diagram allows the determination of the deposition fields against any precursor coordinates without further minimization calculations by resolution of a set of equations.

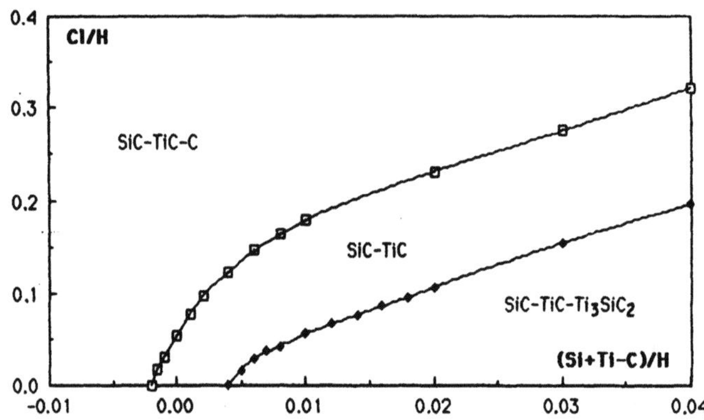

Figure 8: Calculated TiC-SiC deposition domain plotted against variance parameters (35).

5. Conclusion

Many ceramics can be obtained by chemical vapor deposition. Their structure composition and morphologies can be controlled to a large extent according to the coating parameters, mainly temperature, pressure, and initial composition of the gas phase. The growth of composite materials can also be achieved leading to materials of enhanced toughness. The main feature of these ceramics when compared with sintered materials is the uniformity of the crystallite dispersion and the possibility of having all the neighbouring phases in equilibrium with each other. These quasi-equilibrium conditions allow the use of thermodynamic calculation. This latter gives predictive information on the composition of the complex gas phase, the deposition domains with respect to the initial gas mixture and the interactions with the substrate.

Acknowledgments

The author gratefully acknowledge Dr C.Bernard and Dr C.Chatillon from the L.T.P.C.M. of Grenoble for helpfull discussions for the preparation of this lecture.

BIBLIOGRAPHY

1. BLOCHER J.M. Jr., J. Vac. Sci. Technol. **11** 4 (1974) 680.
2. BRYANT W.A. , J. Mater. Sci. **12** (1977) 1285.
3. VANDENBULCKE L. , VUILLARD G. , l'Actualité Chimique, Octobre(1978)
4. AUDISIO S. , l'Actualité Chimique, Septembre (1978) 25.
5. ARCHER N.J. , Phys. Technol., **10** (1979) 152.
6. JENSEN K. F. , in: VAN DEN BREKEL C.H.J., CULLEN G.W., BLOCHER J. , RAI-CHOUDHURY P., ROBINSON Mc. D. (Eds) "Proceedings of the IX International Conference on CVD", Electrochem. Soc., Princeton (Cincinatti USA) (1984)3.
7. ROSENBERGER F. , in: CULLEN G.W. , BLOCHER J.M. (Eds) "Proceeding of the X International Conference on CVD", Electrochem. Soc., Princeton (Honolulu USA) (1987)11.
8. SCHLICHTING H. , "Boundary layer theory", Mac Graw-Hill, (1968).
9. VAN DEN BREKEL C.H.J. in: WAHL G. , BLOCHER J.M. ,VUILLARD G.E. (Eds) "Proceedings of the VIII International Conference on CVD", Electrochem. Soc., Princeton,(Paris FRANCE) (1981)116.
10. KERN R. LE LAY G. METOIS J.J., "Current Topics in Materials Science", Vol.3, KALDIS E. (Ed)(North-Holland, Amsterdam), (1970).
11. GRABOW M.H. GILMER G.H., Surface Science, **194** (1988) 333.
12. WALTON D., Phil. Mag., **7** (1962) 1671.
13. ZINSMEISTER G., Thin Solid Films, **4** (1969) 363.
14. MARSHALL D.B., Ceram. Bull., **66** 2 (1987) 309.
15. BECHER P.B. WEI G.C., J. Am. Ceram. Soc., **67** 12 (1984) C-267.
16. HEUER A.H., J. Am. Ceram. Soc., **70** 10 (1987) 689.
17. NICKL J.J., SCHWEITZER K.K., LUXEMBOURG P. J. Less. Com. Metals, **26** (1971) 335.
18. NICKL J.J., VON BRAUNMÜHL C., J. Less. Com. Metals, **37** (1974) 317.
19. HIRAI T. HAYASHI S., J. Mater. Sci. , **17** (1982) 1320.
20. HAYASHI S., HIRAI T., HIRAGA K., HIRABAYASHI M. J. Mater. Sci., **17** (1982) 3336.
21. HIRAGA K., HIRABAYASHI M., HAYASHI S. , HIRAI T. J. Am. Ceram. Soc, **66** 8 (1983) 539.
22. NAKAE H., MATSUNAMI Y., UNO N., MATSUDA T., HIRAI T., in: CARLSSC J.O. , LINDSTROM J. (Eds) "Proceedings of the V European Conference on CVD", (Upsalla SWEDEN) (1985) 242.
23. LIVINSKY S., IRENE E.A., J. Electrochem. Soc., **125** 2 (1978) 305.
24. BLOEM J., GILLING L.J. "Mechanisms of the chemical vapour deposition of silicon": in KALDIS E.(Eds),Current Topics in Materials Science, vol. 1, North Holland Publishing Company,Amsterdam New-York Oxford (1978) 147.

25. BERNARD C. in: WAHL G. , BLOCHER J.M. ,VUILLARD G.E. (Eds) "Proceedings of the VIII International Conference on CVD", J. Electrochem. Soc. (1981) 3.
26. SPEAR K.E., in: VAN DEN BREKEL C.H.J., CULLEN G.W., BLOCHER J.M. RAI-CHOUDHURY P. , ROBINSON Mc. D. (Eds) "Proceedings of the IX International Conference on CVD", Electrochem. Soc., Princeton (Cincinatti USA) (1984) 81.
27. S.G.T.E. data bank, THERMODATA Domaine Universitaire Grenoble.
28. ERIKSSON G., Chemica Scripta **4** (1973) 193.
29. BESMANN T.M., SPEAR K.E., J. Electrochem. Soc. **112** 11 (1977) 781.
30. DUCARROIR M., BERNARD C. in: BLOCHER J.M. , HINTERMANN H.E. , HALL L.H. (Eds) "Proceedings of the V International Conference on CVD", Electrochem. Soc., Princeton(Fulmer Grange ENGLAND) (1975) 72.
31. TEYSSANDIER F. ,BERNARD C. ,DUCARROIR M., J. Mater. Sci. **23** (1988) 135.
32. TEYSSANDIER F., BERNARD C., DUCARROIR M., J. Electrochem. Soc., **135** 1 (1988) 225.
33. STINTON D.P., LACKEY W.J., LAUF R.J., BESSMANN T.M. Ceram. Eng. Sci. Proc., **5** 7-8 (1984) 668.
34. BESSMANN T.M., J. Am. Ceram. Soc., **69** 1 (1986) 69.
35. TOUANEN M., TEYSSANDIER F., DUCARROIR M., accepted in J. Mater. Sci. let.

REACTIONS OF CERAMIC OXIDES WITH AQUEOUS SOLUTIONS (INCLUDING DISSOLUTION)

Miguel A. Blesa
Departamento Química de Reactores
Comisión Nacional de Energía Atómica
Avenida del Libertador 8250
1429 Buenos Aires, Argentina

ABSTRACT. Traditional ceramic oxides are in general very resistant to attack by water and aqueous solutions; new materials however are not always impervious to this attack and the ensuing degradation places a limit to their durability. In this review both the thermodynamics and the kinetics of the interaction of simple metal oxides with aqueous solutions are described. Massive dissolution, as well as surface reactions of various types are discussed: acid-base, complexation and redox reactions starting at the surface may all give rise to further massive transformations, in some cases with formation of other solid phases, when the conditions are set up outside the stability field of the original phase in the corresponding phase diagram. Potential/pH diagrams are used to discuss the stability of oxides in contact with water, and the kinetic aspects of the transformations are analyzed in terms of the current models of the structure of the interfaces.

1. INTRODUCTION

One of the characteristics of traditional ceramics is the low chemical reactivity towards usual ambient conditions. In particular, porcelain and related materials are extremely resistant to attack by water and dilute aqueous solutions, justifying their use in pottery. Newer ceramic materials, not intended for pottery, are not always impervious to the attack from water solutions, and the ensuing degradation places a limit to their durability (at best), or even to their applications.

The firing procedure involved in ceramics processing in general gives rise to rather passive surfaces whenever such intrinsic property characterizes the chemistry of the constituent compounds. It is well known that the chemical reactivity of solids is highly dependent on the previous history of the sample. One of the reasons, rather trivial, arises from varying surface/volume ratios; heterogeneous reactions proceed at rates that are proportional to the available instantaneous interfacial area S:

L.-C. Dufour et al. (eds.), Surfaces and Interfaces of Ceramic Materials, 639–662.

$$-(dn/dt) = R\,S \qquad (1)$$

In (1), n is the number of moles in the solid phase undergoing the transformation, and R is the specific rate, units mol cm^{-2} s^{-1}. Equation (1) is many times expressed in terms of the fraction of total solid transformed at each time, $f = (w/w_o)$, w standing for mass:

$$(df/dt) = -\frac{1}{w_o}(dw/dt) = -\frac{M}{w_o}(dn/dt) \qquad (2)$$

Thus,

$$(df/dt) = M\,S_s\,R \qquad (3)$$

and the rate increases with the specific surface area S_s of the sample.

A more fundamental question is involved in the changes of reactivity from sample to sample. The reactivity of surface sites towards any type of chemical interaction with the surrounding medium is of course highly dependent on the number and type of bonds linking the site to the solid. Surface reactive sites are many times characterized by reduced coordination numbers ('kinks'). In massive transformations of solids (as opposed, e.g. to catalysis), the same type of defective structures must be available throughout the bulk of the solid, and the concentration of such defects largely determines the reactivity. In any thermodynamically equilibrated sample of metal oxide, there exists a certain concentration of defects of various types; such concentrations are however usually very low, and samples exhibiting an equilibrated defect structure will be in general highly passive. Thermal annealing, or annealing by ionizing radiations, will decrease the concentration of defects towards equilibrium values, and ceramics processing therefore yields in general samples of low reactivity, within the range of reactivity that characterizes each compound. This 'intrinsic' reactivity is determined by the chemical properties of the metallic element, although it should be clear that it is a very elusive magnitude, hardly usable in any other than loose form.

In what follows, we shall try to illustrate the basic ideas involved in the analysis of reactivity of metal oxides. The discussion shall be sorted out in two broad aspects, thermodynamics and kinetics. The presentation shall attempt to give a general outlook of the subject, and we shall not try to make it comprehensive. References to original work in the field shall not be given; rather, the main reviews in the various fields are listed at the end, together with the main numerical data sources.

2. THERMODYNAMICS

2.1. Solubility of Binary Oxides in Aqueous Acids and Bases

The field of thermodynamic stability of a given phase is a limited region of the corresponding phase diagram. The phase diagrams in systems containing metal oxides, water and dissolved chemicals in the fluid phase are of course complex. When the only solutes to be considered are mineral acids or bases, the thermodynamic information is reasonably well represented by the solubility curve. Calculation of solubility curves requires the knowledge of thermodynamic quantities of both the solid oxide and the various hydrolytic species that may be present in solution. For a MO_n oxide, the solubility product quotient Q_s is defined as:

$$MO_n(s) \;+\; 2n\,H^+ \;=\; M^{2n+}{}_{aq} \;+\; n\,H_2O \qquad (4)$$

$$Q_s \;=\; [M^{2n+}] \;/\; [H^+]^{2n} \qquad (5)$$

The square brackets in (5) represent molar concentrations; true constants of course must be written in terms of ionic activities.

The total solubility is given by:

$$S \;=\; \sum_y \sum_x y\,[M_y(OH)_x{}^{(2ny-x)+}] \qquad (6)$$

and may be calculated if the hydrolysis quotients relating the concentrations of all the relevant species to the concentration of M^{2n+} are known. An example of the distribution of species, and the solubility profile is given in Fig. 1.

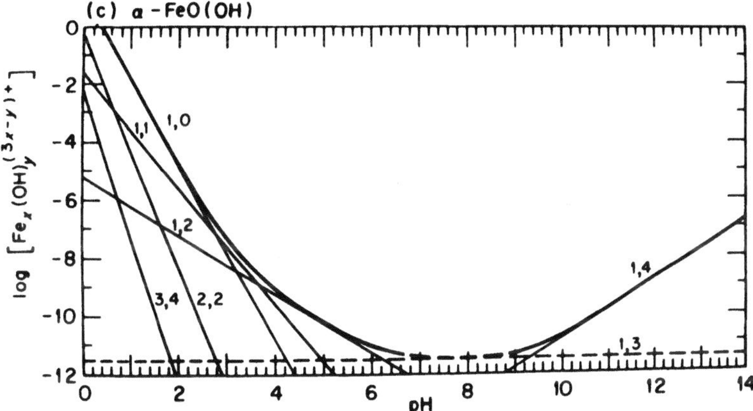

Figure 1: Solubility profile of α-FeOOH at 25 °C as a function of pH. Lines indicate the contribution of various $Fe_y(OH)_x(3y-x)^+$ species and are identified by the pair of values (x,y). From Ref. 27.

In many cases, in acidic media, the main species in solution is the aquoion, and Q_s gives then direct information about the solubility of the oxide. Equation (7) is a very simple relationship between the solubility product constant K_s (i.e., Q_s corrected for activity coefficients) with the charge 2n and the size (represented by the interatomic distance M-O, d in MO_2) of M^{2n+}:

$$\log K_s = -5.6 - 2n\,A - 11.0\,[(2n)^2/d] \qquad (7)$$

A is an empirical constant, that sorts the ions into groups, according essentially on their "hard" or "soft" character; hard, highly non-hydrolyzable ions exhibit high and negative values of A; as the soft character of the ion (linked to the degree of covalency of M-O bonds in the oxide) increases, A values become less negative. The values of A are collected in Table 1.

The hydrolytic behavior of the ions usually imply that at high pH solubility rises through the formation of $M(OH)_{2n+z}^{z-}$ species.In fact, for the more sparingly soluble oxides (i.e., those characterized by high

TABLE I
Values of the parameter A in eqn. (7). From Ref. 26.

$\text{Log } K_{11} = A + 11.0\ (z/d)$

Cation	A
Mg^{2+}, Ca^{2+}, Sr^{2+}, Ba^{2+}, Al^{3+}, Y^{3+}, Ln^{3+}	22.0 ± 0.5
Li^{+}, Na^{+}, K^{+}, Be^{2+}, Mn^{2+}, Fe^{2+}, Co^{2+}, Ni^{2+}, Cu^{2+}, Zn^{2+}, Cd^{2+}, Sc^{3+}, Ti^{3+}, V^{3+}, Cr^{3+}, Fe^{3+}, Rh^{3+}, Ga^{3+}, In^{3+}, Ce^{4+}, Th^{4+}, Pa^{4+}, U^{4+}, Np^{4+}, Pu^{4+}	19.8 ± 1
Ag^{+}, Tl^{+}, Pb^{2+}, Tl^{3+}, Bi^{3+}	15.9 ± 1
Sn^{2+}, Hg^{2+}, Pd^{2+}	∼ 12

charges on the cation and/or less negative A values), the most practical way to enact oxide dissolution is to resort to strongly alkaline solutions. The solubility profile for Zr(IV) is shown in Fig. 2.

When more than one polymorph exists, the relationship between the solubilities of any two phases is given by:

$$\Delta G_{pt} = -2.3\,RT \log (K_{s1}/K_{s2}) \qquad (9)$$

where ΔG_{pt} is the phase transformation Gibbs energy. In the diagrams of Figs. 1 and 2, the log [S] profile for a metastable polymorph would be shifted vertically by ($\Delta G_{pt}/RT$). Thermodynamically, the equilibrium condition between two polymorphs is the same irrespective of the presence or absence of water. Application of the phase rule indicates that the coexistence of two solid phases and the aqueous solution at fixed pressure is possible at one temperature only; at this point, the Gibbs energies of both polymorphs are equal, and the solubilities are identical.

Solubility is sensitive to particle size. Very small crystallites are more soluble, because of the contribution of surface excess Gibbs energy.

2.2. Influence of Redox Potential on Solubility: Pourbaix Diagrams

In the above description, the systems were totally characterized at each temperature by two composition variables, pH and metal ion concentration. Such a description is often incomplete because of the occurrence of redox reactions that can be taken into account through a third variable, the redox potential E. Only in a few cases, like Zr(IV), the redox chemistry is almost irrelevant for our purposes, and within a wide range of E values the information given in Fig. 2 suffices to describe the system.

Complete pictorial description would then require the construction of three dimentional plots E-pH-concentration. The solubility curves are cross sections in the pH-concentration plane, in the most simple cases at constant E values. The Pourbaix diagrams are cross sections in the E-pH plane, at constant concentration values. Some representative Pourbaix diagrams are shown in Figs. 3 and 4.

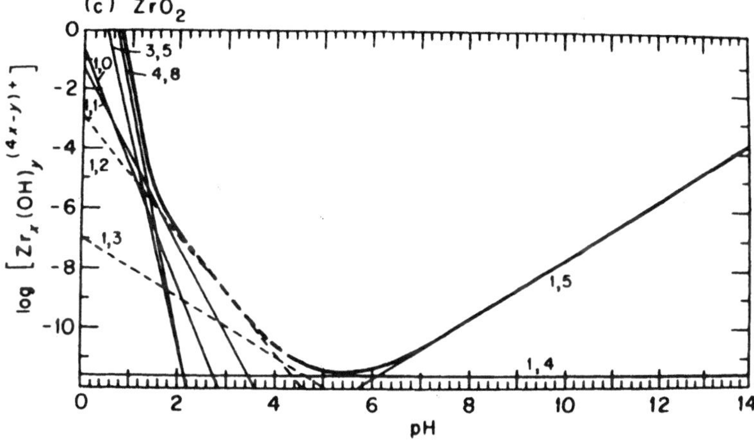

Figure 2: Solubility profile of ZrO_2 at 25 °C as a function of pH. The contributions from various $Zr_y(OH)_x{}^{(4y-x)+}$ are also shown. From Ref. 26.

Whenever the E-pH-concentration conditions are set outside the stability field of a given solid phase, it should transform. The most simple transformation is dissolution, and Fig. 4 illustrate through the arrows some typical changes in solution parameters that may bring about (at least from the point of view of thermodynamics) dissolution of chromium(III) oxide. Acidification (arrow a) pertains to the type of

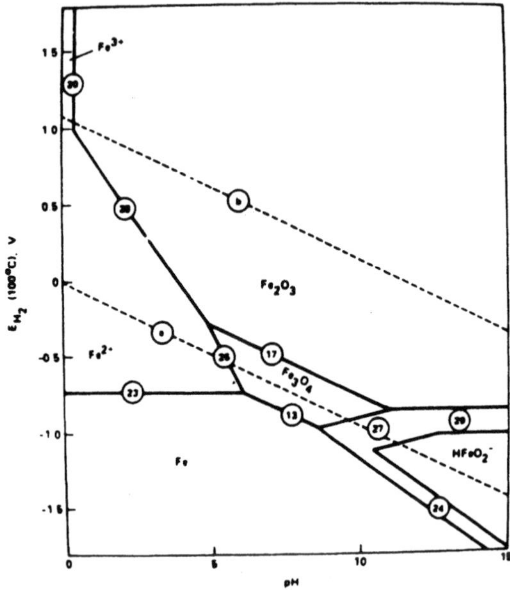

Figure 3: Pourbaix diagram for iron at 100 °C. From Ref. 30.

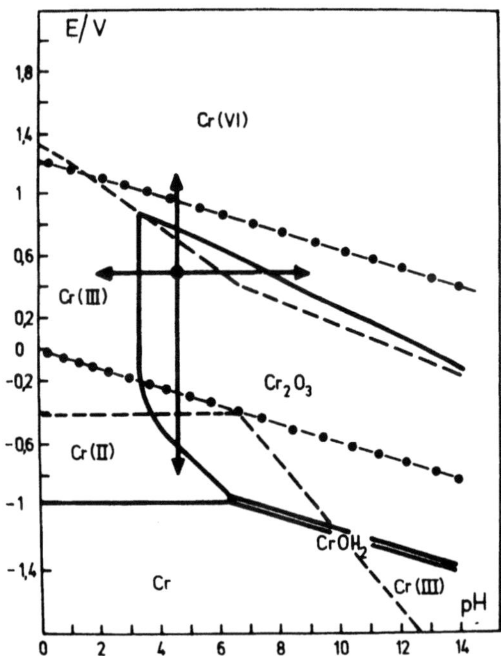

Figure 4: Pourbaix diagram for chromium at 25 °C. From Ref. 29 as quoted in Ref. 1.

dissolution previously discussed; alkalinization in this case however implies an oxidation to CrO_4^{2-}, with net consumption of an oxidant:

$$2\ Cr_2O_3 + 20\ OH^- = 4\ CrO_4^{2-} + 10\ H_2O + 12\ e^- \qquad (10)$$

Vertical arrows represent oxidative and reductive dissolution at constant pH, again with net cosumption of an oxidant or reductant.

A different situation arises when several oxidic phases exist, as in the case of iron (Fig. 3). These oxides are further discussed below.

Redox reactions are very often involved in the corrosive processes of ceramic oxides inmersed in water, because electron transfer reactions are usually fast, and rates of reactions may then be determined by the reactivity of metal ions in their most labile oxidation state (see <u>Kinetics</u>, below).

2.3. Influence of Complexing Agents on Solubility

Whenever complexing molecules or ions, other than H_2O and OH^-, are present in solution, the Pourbaix diagrams in terms of <u>metal ion concentrations</u> may be largely altered. In fact, except for metal oxides immersed in pure water, there are always anions that to a certain extent interact with the metal ions, and thus not all mineral acids are equally corrosive, even from the thermodynamic point of view. Only a few anions are usually 'innocent': perchlorate, perhaps nitrate, and acetate. Thus the information contained in Pourbaix's diagrams is usually incomplete to assess the corrosiveness of a given aqueous solution. Proper description would require the knowledge of the speciation in aqueous solution, including all possible complexes. This can be done if the stability constants are known, but requires the introduction of another variable, complexant concentration. The unified pictorial description would of course be unpractical, and it is more convenient to describe the influence of ligand concentration in more limited cases, and in a

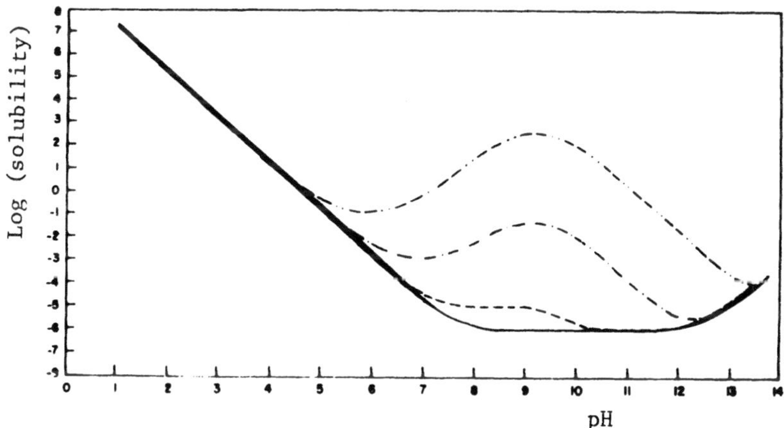

Figure 5: Solubility of copper hydroxide as a function of pH at various ammonia concentrations. Calculated from data in Ref. 31.

parametric form. Fig. 5 shows some calculations of the solubility of copper hydroxide in the presence of various ammonia concentrations. It may be easily seen that the domains of corrosivity may be more complex than expected 'a priori' because of the interplay of all the involved ionic equilibria.

2.4. Solubility of Ternary and More Complex Oxides

These systems in aqueous mineral acids can be adequately described in terms of four components: MO_n, NO_m, H_2O and HCl; at arbitrary fixed pressure temperature and pH the relationship between degrees of freedom F and number of phases P is therefore $F = 3 - P$, and not more than two solid phases may coexist in equilibrium (eg., MNO_{n+m} and MO_n); the concentrations of M^{2n+} and N^{2m+} in the aqueous phase become fixed. If any of these is changed, one of the solid phases should disappear.

Solubility of MNO_{n+m} is given by the equilibrium

$$MNO_{n+m}(s) + 2(n+m)H^+ = M^{2n+} + N^{2m+} + (n+m)H_2O \qquad (11)$$

$$K_s = [M^{2n+}]\,[N^{2m+}]\,/\,[H^+]^{2(n+m)} \qquad (12)$$

Normally, ternary oxides useful in materials science derive from sparingly soluble binary oxides MO_n that combine with other oxide (perhaps highly soluble) in a reaction characterized by a modest $-\Delta G^o$ value. Then, conditions for preferential leaching of NO_y with formation of a second solid phase are often found. We shall not dwell upon this subject any further; Prof. Turner is going to describe in detail one case of much interest, that of titania-based ternary oxides. Suffices it to say that, when more than one solid phases are involved, thermodynamic equilibration is difficult, and the actual behaviour of the material depends strongly on the kinetics of phase transformation.

The description of multi-solid phase systems is also important when the phases are linked to each other through redox reactions:

$$2Fe_3O_4(s) + 2H_2O = 3Fe_2O_3(s) + 2H^+ + 2e^- \qquad (13)$$

Coexistence conditions are set up along the equilibrium line in Pourbaix diagram (Fig. 3), where the ratio $[Fe^{3+}]/[Fe^{2+}]$ is adequate to comply with both solubility equilibria and to establish the correct redox potential.

It should be noted that the chosen example is in fact special, because magnetite itself may be regarded as a mixed Fe(III) - Fe(II) oxide, the solubility of which may be described as in eqn(14), or alternatively by the potential-dependent equilibrium (15):

$$Fe_3O_4(s) + 8H^+ = 2Fe^{3+} + Fe^{2+} + 4H_2O \qquad (14)$$

$$Fe_3O_4(s) + 8H^+ + 2e^- = 3Fe^{2+} + 4H_2O \qquad (15)$$

The link between both descriptions is of course given by the relationship between redox potential and $[Fe^{3+}]/[Fe^{2+}]$ ratios.

Non stoichiometric oxides require a more elaborate thermodynamic formulation, in which the most important descriptor is the change in Gibbs energy with oxygen contents; for MO_{n+x} : $(\partial G^o_{MO_{n+x}} / \partial x)_{P,T}$.

This subject is very important (recall, e.g. high Tc superconducting ceramics), but the description would require more space than available. The excellent review by Heusler (ref. 6) is recommended.

2.5. Surface Reactions: Ionic and Insulating Oxides.

Even within the stability field, metal oxides are not unreactive towards aqueous solutions. The surface of the oxide is in fact rather reactive, except in the case of very perfect surfaces. Metal oxides adsorb water, both as such and dissociatively. The latter process yields surface OH groups that are put in evidence by IR spectroscopy. In a further reaction, the metal oxide particles acquire a net surface charge which is strongly dependent on pH, ionic strength, solution composition, temperature and nature of the oxide. The surface charge generated through ionic adsorption (or surface reactions) gives rise to a potential drop across the metal oxide/solution interface that is important in determining the kinetics of chemical reactions taking place at the interface.

The role of protons as potential determining ions (p.d.i.) was first described by the Nernst equation:

$$\psi_o = 2.3\ k_B T\ e^{-1}\ (pH_o - pH) \qquad (16)$$

where pH_o, the point of zero charge (p.z.c.), is the pH at which both the surface charge and the surface potential are zero; k_B is the Boltzmann constant, T the temperature and e the electronic charge.

In order to derive eqn. 16, it is necessary to assume that the activity of adsorbed protons is independent of surface charge, and this assumption is questionable. The amphoteric properties of metal oxides inmersed in aqueous solutions originate in equilibria involving surface MOH groups:

$$MOH_2^+ = MOH + H^+ \qquad (17)$$

$$MOH = MO^- + H^+ \qquad (18)$$

$$K_{a1}^{int} = (MOH)\ (MOH_2^+)^{-1}\ a_{H^+}\ \exp\ (-e\psi_o/k_BT) \qquad (19)$$

$$K_{a2}^{int} = (MO^-)\ (MOH)^{-1}\ a_{H^+}\ (-e\psi_o/k_BT) \qquad (20)$$

where () denotes surface concentration expressed in number per unit area and a_{H^+} is the activity of protons in bulk solution.
Rearragement of eqns. 19 and 20 leads to 21:

$$\psi_o(mV) = 59.2\ (pH_o - pH) - 29.6\ \log[(MOH_2^+)/(MO^-)] \qquad (21)$$

Eqn. 21 can be regarded as a modified version of the Nernst eqn. (16) that takes into account the dependence of "surface proton activity" with the surface charge. The absolute value of the surface potential should be less than that predicted by eqn. 16, the difference being larger as pH moves apart from pH_o.

The major drawback of modified Nernst equations, such as 21, is that surface concentrations are not experimentally obtainable quantities. Indeed, acid-base potentiometric titrations of metal oxide suspensions only measure the difference of the surface excesses of H^+ and OH^- ions, Γ_{H^+} and Γ_{OH^-}, which is then converted in surface charge density σ_o through

$$\sigma_o = e\,(\Gamma_{H^+} - \Gamma_{OH^-}) \tag{22}$$

Therefore, the use of eqn. 21 requires (except at very low surface charge) an adequate model describing the structure of the electrical double layer to calculate the concentration of the charged surface groups.

Models of the electrical double layer based on the protolytic properties of surface groups, generally called surface ionization or surface dissociation models, have been frecuently used: various colloidal phenomena of metal oxides dispersed other descriptions of the interface include the porous double layer and the gel-like models. The most usual structures of the electrical double layer assumed by surface ionization models are schematically illustrated in Fig. 6.

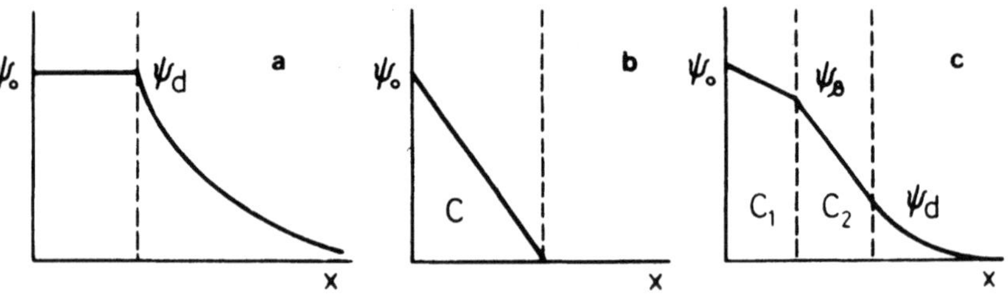

Figure 6: Schematic representation of the metal oxide/solution interface according to various GCSG models: (a) diffuse layer; (b) compact layer-constant capacitance; (c) triple layer

The diffuse layer model is the simplest representation of the metal oxide/solution interface. Its basic assumption is that the interaction between the charged surface and background electrolyte ions is given by the Gouy-Chapman theory. The main drawback of this model is the prediction of too high values of the zeta potential* at variance with the available electrokinetic and colloid stability kinetics experimental evidence.

---

* The zeta potential is the electrostratic potential the characterizes the shear plane dividing the fixed and mobile parts of the interface.

The basic assumption of the compact layer-constant capacitance model is that the potential drops linearly with the distance; adsorbed counter-ions must be located at the plane where the electrostatic potential equals zero. Thus, the model neglects the existence of the electrokinetic phenomena, i.e. $_{d}$ = = 0; a fact that makes the model inappropriate to describe metal oxide/solution systems. This model, however, established the basis to evaluate surface association or surface dissociation equilibrium constants from potentiometric titration data.

The above models are embodied in the triple layer models that represent the metal oxide/solution interface by the classical Gouy-Chapman-Stern-Grahame structure. Among these the model proposed by Yates _et al._ as modified by Davis _et al._, is probably one of the most important contributions to the understanding of metal oxide/solution systems. The model equations are presented below; for further details, the reader is refered to the given references [3,6-9]. The acid-base properties are represented by the surface reactions depicted in equations 17 and 18, and allows the background electrolyte ions to interact with charged surface groups to form surface ion-pairs; ions forming surface ion-pairs are located at the Stern plane. This interaction is represented by mass law equilibria:

$$MOH_2^+ + A^- = MOH_2^+ \ldots A^- \tag{23}$$

$$MO^- + C^+ = MO^- \ldots C^+ \tag{24}$$

$$K_A^{int} = (MOH_2^+ \ldots A^-)\,(MOH_2^+)^{-1}\,(a_A)^{-1}\,\exp(-e\psi_\beta/k_BT) \tag{25}$$

$$K_C^{int} = (MO^- \ldots C^+)\,(MO^-)^{-1}\,(a_C)^{-1}\,\exp(e\psi_\beta/k_BT) \tag{26}$$

Alternatively, eqns. 23 and 24 can be combined with 17 and 18 to yield

$$MOH + A^- + H^+ = MOH_2^+ \ldots A^-;\ (^*K_A^{int})^{-1} = K_A^{int}/K_{a1}^{int} \tag{27}$$

$$MOH + C^+ = MO^- \ldots C^+ + H^+;\ ^*K_C^{int} = K_C^{int}\,K_{a2}^{int} \tag{28}$$

The charge densities in the surface (σ_o), in the Stern plane (σ_β) and in the diffuse region (σ_d) are given by eqns.(29)-(32).

$$\sigma_o = e\,[(MOH_2^+) + (MOH_2^+ \ldots A^-) - (MO^-) - (MO^- \ldots C^+)] \tag{29}$$

$$\sigma_\beta = e\,[(MO^- \ldots C^+) - (MOH_2^+ \ldots A^-)] \tag{30}$$

$$(\sigma_o + \sigma_\beta) = -\sigma_d \tag{31}$$

$$\sigma_d = \varepsilon k_B T \kappa (2\pi e)^{-1} \sinh(-e\psi_d/2k_BT) \tag{32}$$

where ε is the permittivity of the solvent and κ is the inverse of the diffuse layer length:

$$\kappa = (\varepsilon k_B T/8\pi e^2 n_o)^{-1/2} \tag{33}$$

In eqn. 33 n_o is the ionic concentration expressed in number per cm^3.

The charge-potential relationships 34 should also be fulfilled:

$$\psi_o - \psi_\beta = \sigma_o/C_1 \quad ; \quad \psi_\beta - \psi_d = -\sigma_d/C_2 \qquad (34a);(34b)$$

C_1 and C_2 are the integral capacities of the inner and outer Helmholtz planes, respectively. The final equation is the surface mass balance:

$$N_s = (MOH) + (MOH_2^+) + (MO^-) + (MOH_2^+...A^-) + (MO^-...C^+) \qquad (35)$$

The metal oxide/aqueous solution interface is then described by solving the complete set of equations using the appropriate values of the model parameters. These values are either determined experimentally or left as adjustable parameters.

The site binding model accounts for the observed dependence of surface charge with ionic strength and for the usually very low charge of the diffuse layer as compared to σ_o; i.e. $-\sigma_d = (0.1\text{-}0.2)\ \sigma_o$.

The surface acid-base properties of metal oxides inmersed in water are related to the hydrolytic properties of the corresponding metal ion in solution. In principle, the isoelectric point (the pH value at which the oxide particles have null mobility in an electric field) can be determined as the pH value at which the algebraic sum of the charges of dissolved metal ions is zero (i.e. $z_iC_i = 0$, where C_i is the concentration of $M(OH)^{2+}$, $M(OH)_2^+$, $M(OH)_4^-$, etc., and z_i is the corresponding charge number). This relationship between dissolved and surface metal ions determines a parallelism in the pK_a values and also in the corresponding ΔH_a^o and ΔS_a^o. Such comparisons are however not enough to predict pH_o values; a semiempirical procedure based on the properties of the metal ions (charge/radius ratio and electronic configuration) and or the structural properties of the solid (crystal field stabilization energies, crystal structure, etc.) is of limited use; pH_o values and other acid-base properties of metal oxides are better determined experimentally.

Complexing anions bind strongly surface metal ions of metal oxides, in a process that may be described as an ion exchange:

$$MOH + X^- = MX + OH^-; \quad K_X^{int} \qquad (36)$$

Adsorption measurements give values for these surface complexation constants, and also suggest possible stoichiometries and complexation modes from maximum adsorption densities. Thus, one or more OH surface groups, bound to more than one metal ion, may be exchanged for a given polydentate anion. Table II gives some surface equilibrium constants; note that in some cases the choice of the protonation state of the incoming ligand is different, yielding constants that are related to K_X^{int} through the acidity constant of the ligand.

Surface equilibrium constants are related to the constants of the analogous complexation equilibria in solution:

$$MOH^{(n-1)+}\ (aq) + X^-\ (aq) = MX^{(n-1)+}\ (aq) + OH^- \qquad (37)$$

In some simple systems, the relationship is given by eqn.(38):

$$\log K_X^{int} = A \log K_X \quad (38)$$

TABLE II
Surface Equilibrium Constants
(See Ref. 1 for original source)

Surface Equilibrium	$\log K_X^{int}$
$MOH + F^- = MF + OH^-$	- 4.80
$MOH + SO_4^{2-} = MSO_4 + OH^-$	- 5.80
$2\ MOH + SO_4^{2-} = M_2SO_4 + 2\ OH^-$	- 13.5
$MOH + H_4SiO_4 = MSiO_4H_2^- + H_3O^+$	- 3.30
$MOH + H_3PO_4 = MPO_4H_2 + H_2O$	9.50
$MOH + H_3PO_4 = MPO_4H^- + H_3O^+$	5.10
$2\ MOH + H_3PO_4 = M_2PO_4H + 2\ H_2O$	8.50

Of course, no a clearcut distinction between chemisorption and electrostatic adsorption of anions can be made from adsorption measurements. Surface chemical complexation is characterized by high Langmuir constants and large shifts in pH_o and pH_{iep}.

Anions known to bind strongly to the surface of metal oxides include inorganic species like phosphate, and a series of organic anions like oligo- and poly-carboxylates and aminopolycarboxylates. High affinity seems to be related to the availability of more than one coordinating group; whilst the extent of adsorption of simple aminoacids onto crystalline iron oxides has been found variously to be important to negligible, proteins indeed adsorb strongly on metal oxides. Polycarboxylates also exhibit high affinity for metal oxides surfaces, whilst simple carboxylates (formate, acetate) present lower affinities.

In general, two opposite tendencies define the dependence of ΔG^o_{ads} on pH: protonation of the ligand decreases the stability of surface complexes, whilst protonation of the surface promotes adsorption. The resulting shapes of ΔG^o_{ads} profiles include cases of raising $-\Delta G^o_{ads}$ as pH decreases, with some leveling off at low pH, and also cases where a distinct maximum in $-\Delta G^o_{ads}$ is observed.

The affinity of the ligands for the oxide surfaces is low in neutral or alkaline solutions, and acidic media are required to achieve substantial surface complexation. The pH dependence is largely determined by the influence of protons on the chemical reaction itself (see Table 2 for some possible stoichiometries).

2.6. Surface Reactions: Semiconducting Oxides

The classical picture of the semiconductor/electrolyte potential profile is shown in Figure 7. The main portion of potential drop takes place within the oxide, in the space charge region, and thus the potential drop in the solution side is considered to be rather insensitive to bulk semiconductor potential. The extreme view regards the potential drop in the solution side constant as the potential in the semiconductor bulk is varied.

The standard description of the semiconductor/electrolyte interface assumes that the flat-band potential, i.e. the potential at which there is no space charge region nor band bending, is a nernstian function of pH:

$$E_{fb} = E^{o}_{fb} - 0.059 \text{ pH} \qquad (39)$$

E^{o}_{fb} is the flat band potential at pH zero; this parameter is an intrinsic characteristic of each oxide. The potential drop in the solution side of the interface is therefore controlled only by the activity of protons in solution, and not dependent on bulk semiconductor potential. This implies a very high (in principle infinity) capacity for proton adsorption at the surface. Any change in the overall potential drop from bulk solution to bulk semiconductor not related to changes in

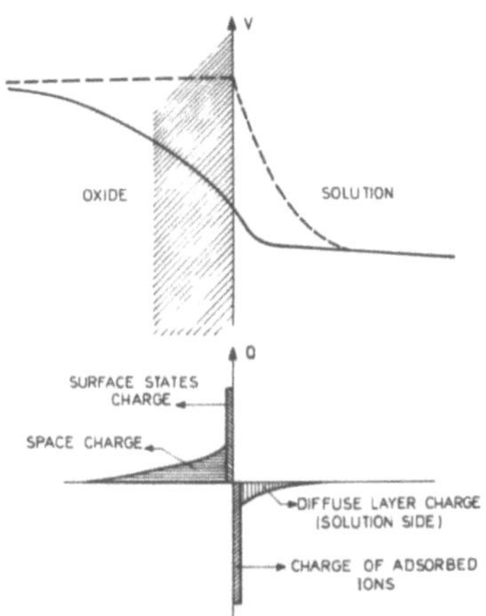

Figure 7: Schematic potential (a) and charge (b) distributions across the semiconducting metal oxide/electrolyte solution interface. In (a), the potential profile across the metal/electrolytic solution interface is shown for comparison (dashed line)

proton activity is wholly located in the space charge region.

The limitations of this description have been questioned and probed by various authors. Thus, Fermi level pinning may take place in various experimental situations; pinning implies that surface potential should accompany bulk solid potential changes. Partial Fermi level pinning should be expected whenever high densities of surface states are present at the interface. Surface trapped states may indeed account for a sizable fraction of the charge in a semiconductor particle, and in amorphous oxides; such surface localized states may be energetically within the band gap, and are very important in defining the chemical reactivity of the oxide in water.

Electron transfer rates between semiconductor and redox couples in solution are determined by the extent of mismatch between the relevant energy level of the solid (conduction or valence band) and of the redox partner in solution, corrected for reorganization energy.

3. KINETICS

3.1. Acid-base dissolution

The surface of a dissolving oxide should develop a charge and potential because of the inherently different rates of transfer of anions and cations: from a hypothetical initial situation $\psi_0 = 0$, faster transfer of, say, anions would leave a positive charge on the surface; the corresponding developed potential difference would assist cation dissolution and hinder anion dissolution, until a steady state characterized by a "free dissolution potential" would be achieved. In this condition, either cation removal or anion removal controls the rate.

The basic electrochemical equations are:

$$jc^+ = jc,o^+ \exp\,(\alpha_+ z_+ F\,\Delta\psi/RT) \qquad (40)$$

$$jc^+ = jc,o^- \exp\,[-(1-\alpha_+)\, z_+ F\,\Delta\psi/RT \qquad (41)$$

where jc^+ and jc^- are the current densities associated to cation dissolution and deposition respectively; jc,o^+ and jc,o^- the current densities corresponding to $\Delta\psi = 0$; α_+ is the electrochemical transfer coeficient. Equivalent expressions hold valid for the current densities ja^+ and ja^- associated to anion transfer.

Transport of cations and anions are statistically independent processes; of course, in "chemical" dissolution experiments, the total current density due to cations $jc = jc^+ + jc^-$ must be equilibrated by the total current density due to anions $ja = ja^+ + ja^-$. Then, for zero net current, $jc = ja$.

The statistical independence of the cation and anion transport processes can be best understood for dissolution experiments carried out in nearly saturated solutions. In true equilibrium , $jc,eq = ja,eq = 0$, or

$jc,eq^+ = -\, jc,eq^-$ and $ja,eq^+ = -\, ja,eq^-$, but $jc,eq^- = ja,eq^-$:
i.e. the exchange currents associated to both type of ions may differ significantly.

The most usual experimental evidence used to probe the validity of dissolution models is the rate dependence on pH. Vermilyea's model predicts:

$$\log R = \log R_o \quad \frac{n\,\alpha_+ z_+}{\alpha_+ z_+ - \alpha_- z_-} \quad pH \tag{42}$$

Where n is the number of protons involved in the phase transfer step that determines the rate. The reaction order depends therefore on the charges of the cations and the actual surface protonation reaction that controls the rate. If the exchange rate of anions is fast enough, in comparison with cation exchange rate, the surface potential may be thermodynamically determined by the proton affinity of the surface. In this case, Diggle modifies eqn.(42) to:

$$\log R = \log R_o - (\alpha_+ z_+)\, pH \tag{43}$$

It is usual to find rate laws of the form:

$$\log R = \log R_o - 0{,}5\ pH \tag{44}$$

Other exponents on $[H^+]$ are also observed, and representation of these values requires accepting ad-hoc values for α_+.

Valverde and Wagner have proposed a general rate expression in terms of interfacial magnitudes. For a MO_n oxide,

$$R = \sum_i \sum_j k_{ij}\, \Gamma_{ij}\, \exp(\alpha_i z_i e\,\psi / k_B T) \tag{45}$$

In eqn. 45, kinks of different reactivity are recognized, and characterized by the number (i) of ligands (complexing agents) bound to them and the number (j) of H^+ associated with neighboring O^{2-} ions. Γ_{ij} represents the surface densities of such kinks, and k_{ij} their specific exchange rate constants. This equation recognizes therefore solid topography, inherent reactivity of kinks, excess surface concentration of H^+ and complexing agents, and potential drop across the interface (ψ) as the factors influencing dissolution rates; all of these interface properties are determined by solid and solution parameters.

Eqn.(45) can be further ellaborated on the basis of the site binding model and with a series of simplifications, assuming only one type of reactive kinks in each experimental condition, the concentration of which is proportional to the surface excess concentration of protons and no specific adsorption of complexing anions. The potential ψ is identified as $\psi_o - \psi_\beta$, given by:

$$\psi_o - \psi_\beta = 2.3\, q(k_B T/e)\,(pH_o - pH) \tag{46}$$

where $q \cong 0.5$. This approximation is valid only for $(pH_o - pH)$ values larger than three units, which is the most usual case. From (45) and (46),

$$R = k'\,(pH_o - pH)\, \exp[2.3\alpha_+ q z_+ (pH_o - pH)] \tag{47}$$

Eqn.(47) is adequate to interpret the empirical rate data.

According to (47), acidity affects the rate through changes in Γ_i and $(\psi_0 - \psi_\beta)$. If wide enough pH ranges are spanned, it is also possible that new sites characterized by larger k' values become more important, and thus it is safer to sort just two factors, $k_i\Gamma_i$ and ψ. The composite parameter $k_i\ \Gamma_i$ is not an inherent magnitude of each oxide, but rather of each batch (also, as mentioned above, $k_i\Gamma_i$ may differ substantially if experiments are performed in solutions of very different composition).

The influence of complexing agents an acid-base dissolution is not straightforward. As discussed before, there is a parallelism between surface complexation and complexation in solution; large K_X^{int} correlates with large K_X, but dissolution rate is not necessarily large: Γ_{Xj} is indeed related to K_X, but k_{Xj} may in fact decrease as K_X increases.

Dissolution in solvents of pH > pH_0 has been scarcely studied. The scanty information seems to indicate that attack by OH^- is more efficient than acid attack for very covalent oxides such as SiO_2, and even also for more amphoretic oxides, such as Cr_2O_3 in the absence of redox reactions.

In the case of semiconducting oxides, it must be recognized that even "simple" acid-base dissolution may imply a redox mechanism, in which minority and majority carriers play a foremost role. In the case of oxides where net oxidation changes may be achieved on the metal ion, redox dissolution is by far more important than the acid-base processes discussed up to now. Thus, the nature and reactivity of kinks in the above models, i.e. $k_i\ \Gamma_i$, may reflect electronic effects that are not obvious in the above treatment. Next, we shall discuss redox dissolution, at the end of which some examples will be presented to illustrate that no a clear cut distinction can be made between acid-base and redox dissolution when semiconductors are considered.

3.2 Redox Dissolution

Expressions such as 45 should suffice for "chemical" (i.e., neither reductive nor oxidative) dissolution. In the more general case however, there are at least two oxidation states of M and each type of ion is characterized by different sets of k_{ij}, Γ_{ij} and z_i values. For the case of M(II) and/or M(III) transfer in the presence of redox reagents in solution, the kinetic scheme is (48) and the corresponding expression for the rate of dissolution is (49):

$$\begin{array}{ccc} M(III) & \xrightarrow{k_{i,j,3}} & M(III)\ (aq) \\ k_{ET}\downarrow\uparrow k_{-ET} & & \\ M(II) & \xrightarrow{k_{i,j,2}} & M(II)\ (aq) \end{array} \qquad (48)$$

$$R = R_2 + R_3 = \sum_i \sum_j k_{i,j,2}\, \Gamma_{i,j,2} \exp(\alpha_{i,2} z_{i,2} e\psi / k_B T) +$$

$$+ \sum_i \sum_j k_{i,j,3}\, \Gamma_{i,j,3} \exp(\alpha_{i,3} z_{i,3} e\psi / k_B T) \quad (49)$$

Steady state dissolution kinetics may reflect rate control by ion phase transfer or by electron transfer as limiting cases. Note that the term $k_{ij}\Gamma_{ij}$ for both types of cations may be largely altered by the transfer of the other cation. In other words, assuming that originally $R_2 >> R_3$, two types of limiting behaviours can be expected: (i) transfer of M(II) shall deplete the surface of $\Gamma_{i,j,2}$ and the ratio $\Gamma_{i,j,2}:\Gamma_{i,j,3}$ shall fall to an adequate low value for which $R_2 = R_3 = 1/2\ R$ (the overall rate is governed by the transfer of the more inert ion), and (ii) transfer of M(II) shall create new and highly reactive M(III) sites through an extensive surface disruption effect; thus $k_{i,j,3}\ \Gamma_{i,j,3}$ increases to high values determined by the transfer rate of M(II) (the overall rate is given by the transfer of the more labile ion).

The rate of transfer of ions across the interface in surface controlled processes is limited by the need to break bonds in the solid. The movement across the interface of a charged species shows a ΔG vs distance profile that is governed by bond-breaking and bond-making reactions along the reaction coordinate. Electron transfer itself may yield species that are vibrationally excited. Thus, it must be recognized that M(II) sites created by electron transfer to M(III) sites may represent highly reactive kinks, for which $k_{i,j,2}$ is substantially larger that the similar values for kinks representing structural M(II) sites; pitting attack is an obvious possibility under these conditions.

Examples of oxides that dissolve faster in reducing media include iron and cobalt(II) oxides. On the other hand, nickel(II), copper(II), uranium(IV) and zinc oxide dissolve faster as E^o of the redox couple increases; Cr_2O_3 is also more readily soluble in oxidizing media.

If a steady state surface composition is reached, the rate corresponding to scheme (48) can be expressed either in terms of R_2 or R_3. It is logical to choose those ions that determine the rate of transfer; i.e. R_2 if the process is reductive, or R_3 if dissolution takes place <u>via</u> M(III). For the first case the steady state condition for Γ_2 is:

$$\Gamma_2 = k_{ET}\, f(a_i)\, \Gamma_3\, [k_{-ET}\, f'(a_i) + k_2 \exp(\alpha_2 z_2 e\psi / k_B T)]^{-1} \quad (50)$$

Thus

$$R = k_2\, k_{ET}\, f(a_i)\, \Gamma_3 \exp(\alpha_2 z_2 e\psi / k_B T)$$
$$[k_{-ET}\, f'(a_i) + k_2 \exp(\alpha_2 z_2 e\psi / k_B T)]^{-1} \quad (51)$$

$f(a_i)$ and $f'(a_i)$ represent adequate functions of dissolved species activities. In the simplest cases, $f(a_i)$ should be function of the activity of the reduced partner only, and $f'(a_i)$ a function of the oxidized species activity only. The kinetic order on these species is not necessarily one, and depends on the characteristics of adsorption pre-equilibria. Eqn. 51 accepts the following limiting simplifications:

a) Electron transfer is fast as compared to dissolution:

$$R = k_2\, K_{ET}\, g(a_i)\, \Gamma_3\, \exp(\alpha_2 z_2 e\, \psi / k_B T) \qquad (52)$$

where $g(a_i)$ represents the quotient $f(a_i)/f'(a_i)$ and $\Gamma_2 + \Gamma_3 = N_s'$, or

$$R = k_2\, K_{ET}\, g(a_i)\, [1 + K_{ET}\, g(a_i)]^{-1}\, N_s'\, \exp(\alpha_2 z_2 e\psi / k_B T) \qquad (53)$$

This expression predicts that the rate should increase with $g(a_i)$ until it reaches a saturation plateau, and this has been indeed found in some cases.

b) Electron transfer is slow and determines dissolution rate.

In this case, Γ_2 falls to such low values that it is not necessary to distinguish between Γ_3 and $\Gamma_2 + \Gamma_3$ as was done in the previous case, and eqn. 51 simplifies to:

$$R = k_{ET}\, f(a_i)\, \Gamma_3 \qquad (54)$$

The rate is now a function of (a_i) rather than of $g(a_i)$. In general $f(a_i)$ is a function only of the activity of the reduced partner of the redox couple, whilst $g(a_i)$ embodies both. The actual form of $k_{ET}f(a_i)$ depends strongly on the mechanism of the electron transfer process.

The electron transfer step is itself complex. It involves prior adsorption, as in the case of iron(III) oxides dissolution by ferrous-carboxylate(L) mixtures:

$$\begin{array}{ccc} Fe^{III}\text{-}L + Fe^{II}\text{-}L\ (aq) & \underset{k_{-diff}}{\overset{k_{diff}}{\rightleftharpoons}} & Fe^{III}\text{-}L \ldots Fe^{II}\text{-}L \\ & & \downarrow k'_{ET}\ (\text{slow}) \\ Fe^{II}\text{-}L\ (aq) + Fe^{III}\text{-}L\ (aq) & \overset{\text{fast}}{\longleftarrow} & Fe^{II}\text{-}L \ldots Fe^{III}\text{-}L \end{array} \qquad (55)$$

The overall rate is controlled by a rather small adsorption equilibrium constant and the kinetic control is defined by a low $k'_{ET}:k_{-diff}$ ratio.

The reducing agents may also be organic molecules or ions; this has been well documented for the dissolution of iron oxides by carboxylates at high temperature, by thioglycollic acid and also by inorganic anions: iodide, thiocyanate, and dithionite.

The equivalent general kinetic scheme for the reductive dissolution by organic reductants is:

$$
\begin{array}{ccc}
M^{III}\text{-OH} + HX & \underset{k_{-HX}}{\overset{k_{HX}}{\rightleftharpoons}} & M^{III}X^{-} + H_2O \\
 & & k_{-ET} \uparrow\downarrow k_{ET} \\
M^{II}OH_2 + X^{\cdot} & \underset{k_{X}}{\overset{k_{-X}}{\rightleftharpoons}} & M^{II}X + H_2O \\
\downarrow k_{PhT} & & \\
M^{II}\ (aq) & &
\end{array}
\qquad (56)
$$

where all the steps may be pH-dependent.

The more important limiting conditions are: (i) adsorption of HX is rate limiting when $k_{HX}[HX] \ll k_{-HX} + k_{ET}$, (ii) electron transfer is rate limiting when $k_{-ET} \cong 0$, $k_{ET} \ll k_{HX}[HX] + k_{-HX}$, and steps following forward electron transfer are fast and irreversible, and (iii) when electron transfer is reversible ($k_{-ET} \neq 0$), where more complex relationships result. When electron transfer is equilibrated, and adsorption of the soluble redox partners is negligible, a simple relationship between the initial rate and E_{HX}, the redox potential of the $X^{\cdot}/HX$ couple is obtained:

$$R = k\ C\ \exp(-eE_{HX}/k_BT)\ [1 + C\ \exp(-eE_{HX}/k_BT)]^{-1} \qquad (57)$$

where k and C are composite constants. Eqn. 57 should be compared with 53; it should be noted that the influence of potential drop across the interface ψ is not considered in 57.

A different analysis of the influence of redox potential can be made on the basis of the mixture potential E^{mix}. This the potential value at which anodic (oxidation of the dissolved reductant) and cathodic (dissolution) currents are equal. Under this hypothesis, the rate of dissolution should be a function of the redox potential for oxide dissolution, the redox potential for the dissolution reagent couple, and the shape of the corresponding polarization curves, as shown in Fig. 8.

The influence of adsorption of surfactant and polymeric anions on the rate of redox dissolution has also been studied. The relationship between adsorption affinity and rate decrease is well substantiated, and may be attributed to either decreased access of oxidant to the surface and/or decreased influence of the oxidant on the surface potential barrier. In the case of dissolution of magnetite by thioglycollic acid, the adsorbed layer prevents the formation of the surface Fe(III) - thioglycollate complexes, rather than blocking phase transfer of dissolving metal ions. On the other hand, there are indications that reductive dissolution proceeding through outer sphere electron transfer is not prevented by adsorbed polyelectrolytes.

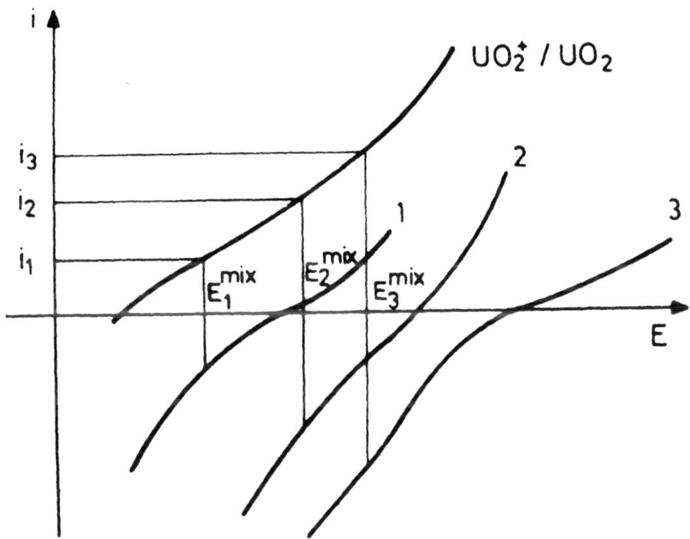

Figure 8: Interpretation of the influence of redox couples on the rate of oxidative dissolution of UO_2, assuming mixed potential conditions. From Ref. 1, taken from Ref. 11.

Reductive dissolution may also take place through the action of microbial metabolites, such as oxalate and pyruvate. Bacterial dissolution of metal oxides is always possibility that should not ve overlooked.

Ionizing electromagnetic radiation is also effective to dissolve iron oxides in adequate media. This is however better considered a reductive process, the reductants being produced by radiolysis of water and/or solute species. On the other hand, prior (γ + n)-irradiation decreases the reactivity through the annealing of reactive sites.

3.3. Semiconductor Oxides: Relationships Between Photodissolution, Acid-Base Dissolution and Redox Dissolution.

Cyclic voltammograms of iron oxides show a cathodic dissolution peak associated to the process:

$$Fe^{III} \xrightarrow{e} Fe^{II}\ (aq) \qquad (58)$$

On the other hand, zinc oxide shows a dissolution peak under anodic polarization (0.65 V vs SCE), corresponding to reaction 58:

$$ZnO \xrightarrow{-2e} Zn^{2+} + 1/2\ O_2 \qquad (59)$$

In agreement, electrons in the conduction band are consumed in the photodissolution of iron oxides, and holes in the valance band are involved in the case of ZnO.

$$ZnO + 2\ h^{+} \longrightarrow Zn^{2+} + 1/2\ O_2 \qquad (60)$$

Reducing agents in the case of iron oxide enhance the rate by either trapping holes or generating localized states that trap the electrons in iron ions, making the oxide prone to dissolution. Addition of iodide to ZnO oxide suspensions, on the other hand, supresses dissolution effectively, as it competes for holes with process 60. Chloride ions, on the contrary, enhance dissolution rates at concentration levels above 0.01 M. This probably reflects again the formation of surface sites that may trap holes effectively in zinc and promote dissolution.

The dissolution of NiO in mineral acids involves holes as in the above ZnO example. The most important factor in determining the reactivity towards nitric acid is the non-equilibrium hole concentration in this p-type semiconductor (i.e. electronic defects). Availability of carriers becomes therefore crucial in semiconductor oxides dissolution. The type and number of surface states also determines to a large extent the rate of dissolution. Photochemically generated (e,h) pairs recombination prevents photodissolution, and this latter process shall become important only when band bending and availability of surface states is adequate. Complexing agents adsorption facilitate photodissolution probably through the creation of adequate reactive sites.

3.4. Phase Transformation Through Dissolution.

After dissolution, precipitation of another less soluble phase may take place. Thus, in many cases, contact with water provides a catalytic pathway for phase transformation, umvailable in dry media. Also possible are direct solid-phase transformations. In either case, the overall process may take the form of an incongruent dissolution, in which the composition of the dissolved fraction is appreciably different of that of the remaining solid. Coherent (e.g. epitactic) layers of the new phase may protect the underlying more reactive material, that can transform further only at rates not higher than the solid phase diffusion of the ions. Chromium dioxide may be used only because such a layer (α-Cr_2O_3) is generated in the manufacturing stage; otherwise it would readily deteriorate by action of moisture. Further details on this wide subject may be found in Ref. 14.

4. FINAL REMARKS

Many of the results described here have been obtained with well characterized simple materials, and therefore the effects of ceramic processing (eg. sintering) are not explored in themselves. Nevertheless, the ideas presented here constitute the basis for any discussion of the corrosion of ceramic oxides by aqueous solutions.

Implicit in the description are ideas like low reactivity of ceramic oxides associated with low exposed surface area and equilibration of defects, but also increased reactivity in grain boundaries and in special phases in multiphase-materials.

ACKNOWLEDGMENTS. A number of co-workers participated actively in the studies of dissolution of metal oxides; their names can be found in the original references. The research projects were partially supported by CONICET and CICPBA. MAB is a member of CONICET.

REFERENCES

General

1. M.A. Blesa, A.E. Regazzoni and A.J.G. Maroto, in "External and Internal Surfaces of Metal Oxides", Eds. L-C. Dufour and J. Nowotny Trans Tech. Publications, Aedermannsdorf, Switzerland (1988)

Dissolution of metal oxides

2. M.A. Blesa and A.J.G. Maroto, J. Chim. Phys. (1986), 83, 757.
3. I.G. Gorichev and N.A. Kipriyanov, Russ. J. Phys. Chem. (1981), 55, 1558.
4. B. Terry, Hydrometallurgy (1983), 11, 315.
5. J.W. Diggle, in "Oxides and Oxides Films", Ed. J.W. Diggle, Vol. 2, Chap. 2, Marcel Dekker, New York, (1973).
6. K.E. Heusler, Electrochim. Acta (1983), 28, 439.
7. M.G. Segal and R.M. Sellers, in Adv. Inorg. Bioinorg. Mech., Vol 3, Sykes, A.G. (Ed.), (Academic Press, London, 1984).
8. I.H. Warren and E.A. Devuyst, Trans. Soc. Mining Eng. AIME (1972), 252, 388.
9. W. Stumm and J.J. Morgan, in "Aquatic Surface Chemistry", Ed. W. Stumm, 2nd. Edition, John Wiley, New York, (1987).
10. A.T. Stone and J.J. Morgan, in "Aquatic Surface Chemistry", Ed. W. Stumm, John Wiley, New York, (1987).
11. M. Spiro, Chem. Soc. Rev. (1986), 15, 141.
12. R.L. Segal, R. St. C. Smart and P.S. Turner, in "Surface and Near-Surface Chemistry of Oxide Materials", Ed. J. Nowotny and L.-C. Dufour, Elsevier, Amsterdam (1987).
13. W.W. Frenier and F.B. Growcock, Corrosion NACE (1984), 40, 663.

Phase transformation of metal oxides in aqueous suspensions

14. M.A. Blesa and E. Matijevic, Adv. Colloid Interface Sci., in press.

Structure of the double layer metal oxide/electrolyte solution

15. G.R. Wiese, R.O. James, D.E. Yates and T.W. Healy, in "International Review of Science, Phys. Chem. Sci. 2", Vol. 6, Ed. J.O.'M. Bockris Butterworths, London, (1976).
16. T.W. Healy and L.R. White, Adv. Colloid Interface Sci. (1978), 9, 30:
17. T.W. Healy, D. Chan and L.R. White, Pure Appl. Chem. (1980), 52, 1207
18. J.A. Davis and J.O. Leckie, in "Chemical Modeling in Aqueous Systems", Chap. 15, Ed. E.A. Jenné, ACS Sym. Ser., 93, ACS, Washington D.C., (1979).
19. F.M.M. Morel, J.R. Yeasted and J.C. Westall, in "Adsorption of Inorganics at Solid-Liquid Interfaces", Chap. 7, Eds. M.A. Anderson and A.J. Rubin, Ann Arbor Sci. Publ., Ann Arbor, (1981).
20. R.O. James, in "Adsorption of Inorganics at Solid-Liquid Interfaces", Chap. 6, Eds. M.A. Anderson and A.J. Rubin, Ann Arbor Sci. Publ., Ann Arbor, (1981).
21. R.O. James and G.A. Parks, in "Surface and Colloid Science", Vol. 12, Chap. 2, Ed. E. Matijevic, Plenum Press, New York, (1982).
22. M.A. Blesa and N. Kallay, (1988), 28, 111.
23. H. Gerischer, in "Physical Chemistry an Advances Treatise", Vol. IXA, Eds. H. Eyring, D. Henderson and W. Jost, Academic Press, New York, (1970).

Adsorption from aqueous solutions

24. P.W. Schindler, in "Adsorption of Inorganics at Solid-Liquid Interfaces", Chap. 1, Eds. M.A. Anderson and A.J. Rubin, Ann Arbor Sci. Publ., Ann Arbor, (1981).
25. C. P. Huang, in "Adsorption of Inorganics at Solid-Liquid Interfaces" Chap. 5, Eds. M.A. Anderson and A.J. Rubin, Ann Arbor Sci. Publ., Ann Arbor, (1981).
26. H. Hol, L. Sigg and W. Stumm, in "Particulates in Water", Chap. 1, Eds. M.C. Kavanaugh and J.O. Leckie, Adv. Chem. Ser. 189, ACS, Washington D.C., (1980).

Thermodynamics data sources

27. R, Mesmer and C.F. Baes, in "The Hydrolysis of Cations", John Wiley, New York, (1970).
28. M. Pourbaix, "Atlas d'Equilibres Electrochimiques a 25°C", Goutier-Villars Editeurs, Paris (1963).
29. M. Pourbaix, A. Pourbaix and X.Z. Yang, Report EPRI-NP-2177, Electrical Power Research Institute, Palo Alto, California, (1981).
30. C.M. Chen, K. Aral and G.J. Theus, Computer-Calculated Potential pH Diagrams to 300°C, Volume 2, Electrical Power Research Institute Report NP-3137, Palo Alto, California (1983).
31. A.E. Martell and R.M. Smith, "Critical Stability Constants", Vols. 1-5, Plenum Press, New York (1974).

DISSOLUTION MECHANISMS OF OXIDES AND TITANATE CERAMICS – ELECTRON MICROSCOPE AND SURFACE ANALYTICAL STUDIES

P.S.TURNER, C.F.JONES, S.MYHRA, F.B.NEALL, D.K.PHAM AND R.ST.C.SMART†
Division of Science and Technology,
Griffith University, Queensland 4111, Australia
† now at School of Chemical Technology,
South Australian Institute of Technology
Ingle Farm, SA 5098, Australia

ABSTRACT. The mechanisms which control the dissolution of simple oxides in aqueous solution are not fully understood, despite substantial research; those for ceramics are even less certain, and may involve "leaching" (preferential release of some species from the solid into solution) rather than "dissolution" (all species released at equal rates). In this paper we consider the reactivity in aqueous solutions of two materials — MgO and $CaTiO_3$ — to illustrate current attempts to establish the detailed mechanisms and rate–determining steps. The importance of using surface analytical and electron microscopy techniques to characterize aspects of the surfaces before and after attack is emphasized. In the case of MgO, there is an initial stage of dissolution related to the restructuring of the initial oxide surface; the dissolution process is controlled by the second protonation reaction. For the perovskite $CaTiO_3$, ion–exchange of the Ca^{2+} from surface sites is accompanied by base–catalyzed hydrolysis of the titanate lattice: there is no significant leaching of the Ca^{2+} from the intact solid unless hydroxylation has occurred. However, crystal defects due to deformation and microstructural characteristics of the material can have major effects on the reactivity of polycrystalline perovskite.

1. Introduction

The reactivity of oxides and ceramics in aqueous environments is important in many situations, from the processing of minerals – where rapid dissolution in dilute solutions at low temperatures is desirable – to the protection of metals via oxide layers, the development of ceramic tiles with extreme corrosion resistance or the weathering of superconducting ceramics.

The broad objectives of the research projects over the past decade, of which the work reported here forms a small part, have been to achieve a more detailed and precise understanding of aqueous attack on solids than can be provided by solution analyses and electrochemical measurements on macroscopic samples, and hence to be able to provide predictive models for situations such as the long–term modelling of the durability of ceramic matrices for immobilization of high level radioactivie waste.

Despite the substantial level of empirical knowledge of the technology of these materials, the precise mechanisms which control the reactivity of even the simplest

L.-C. Dufour et al. (eds.), Surfaces and Interfaces of Ceramic Materials, 663–690.

oxides are not fully understood, as is evident from recent reviews (e.g. Blesa *et al*, 1988;Blesa 1989; Segall, Smart and Turner, 1988). In this paper we present recent results of studies aimed at understanding the fundamental dissolution processes for two contrasting materials – the ionic oxide MgO, and the titanate ceramic $CaTiO_3$, calcium perovskite. These examples may serve to illustrate the difficulties involved in establishing precise knowledge of the controlling reactions at the solid–solution interface, and the important role of techniques such as electron microscopy and surface analysis in such studies.

The reactivity of an oxide or ceramic material in practice will be affected by many factors, in addition to the fundamental reactivity of the perfect crystalline solid under specific solution conditions. There may be a high degree of deformation due to cutting or crushing in preparation, and associated high concentrations of crystal defects (e.g. dislocations) which could greatly affect the number and nature of the reactive surface sites. In polycrystalline materials, especially if significant levels of impurities are present, factors such as grain size, the nature and extent of intergranular films and porosity may be important. However, our experiences in attempting to understand, for example, the hydrothermal durability of the multiphase titanate ceramic Synroc (see Myhra *et al* 1987, 1988a), have convinced us that an understanding of the fundamental processes at the single crystal level is essential before reliable predictive models of the behavior of the "real" materials can be developed.

Against this background we present recent results for the dissolution of MgO in dilute acid solutions, and for perovskite in distilled water and weakly alkaline solutions.

1.1. OXIDE DISSOLUTION KINETICS

There have been several recent reviews concerned with the dissolution of oxides, including the article by Professor Blesa in this volume (Blesa, 1989; see also Gorichev and Kyprianov, 1984; Blesa *et al* 1988; Segall *et al* 1988). Here we are concerned with a restricted class of oxides, the ionic monoxides typified by MgO, and their reactions in dilute solutions of strong monoprotic acids. The oxides are thermodynamically very soluble under these conditions. From detailed studies of the dissolution kinetics, it is possible to determine the factors which control these rates. These factors include:

– adsorption (of protons, hydroxide ions, anions, surfactants, etc);
– proton penetration and ion–exchange;
– dissolution reactions (ion formation, protonation, ion transfer, base–catalysed hydrolysis);
– the surface structures of the solid, including surface modification and the formation of surface phases.

1.1.1. *pH Dependence.* The dissolution of MgO in dilute acid involves the transformation of cations and anions in the solid to cations in solution and water. It may be represented as:

$$Mg^{2+}O^{2-}{}_{(solid)} + 2H^{+} \rightarrow Mg^{2+}{}_{(aq)} + H_2O \qquad (1)$$

or in terms of the two partial reactions:

$$Mg^{2+}O^{2-}{}_{(solid)} + H^+ \rightarrow Mg^{2+}OH^-{}_{(solid)} \quad (2a)$$

$$Mg^{2+}OH^-{}_{(solid)} + H^+ \rightarrow Mg^{2+}{}_{(aq)} + H_2O \quad (2b)$$

If the rate–determining process in these reactions is the protonation of the anion (rather than the transfer of the cation from solid to solution), then the theories of Vermilyea (1966) and Diggle (1973) predict a dependence of rate on hydrogen ion activity. Vermilyea's theory is based on ion transfer control, whereas Diggle incorporated potential control in addition, as described by Blesa (1989) and by Segall *et al* (1988). These predictions are summarised in Table I, and may be tested experimentally through plots of $\log_{10}$(rate) *vs* pH.

TABLE I Predicted slope of log(rate) *vs* pH for ionic monoxides.

Reaction	Vermilyea	Diggle
(1) Double protonation	–1	–1
(2a) 1st protonation	$-^1/_2$	0
(2b) 2nd protonation	$-^2/_3$	$-^1/_2$

In his work on MgO, Vermilyea (1969) claimed to have determined a slope of $-^1/_2$. This was also the conclusion of Segall *et al* (1978), using MgO smoke powders, from measurements of rates using a range of starting pH values and equivalent fractional dissolution of the oxide, in order to compare equally dissolved surface areas. Recently Fruhwirth *et al* (1985) claimed to have established a linear dependence of rate (rather than log(rate)) on pH, in measurements on rotating discs of {100} single crystals of MgO, but we have been unable to reproduce this result using {100} MgO smoke powders.

1.1.2. *Effects of Surface Structure.* The nature of the surface of an oxide may be expected to determine the initial stages, at least, of its reactivity – the kinetic control through hydroxylation, as described above, assumes that the density of reactive sites per unit area is constant, independent of extent of dissolution or of pH. It is well established, both experimentally (Coluccia *et al* 1979, 1981; Jones *et al* 1984b) and theoretically (Colbourne and Mackrodt, 1982, 1983) that the perfect {100} surfaces of MgO are unreactive towards water vapour. The presence of 3– or 4– fold coordinated kink sites and surface steps is necessary for the hydroxylation reaction to proceed. It might therefore be expected that the dissolution rates per unit surface area of MgO prepared in different ways with different surface densities of active sites would differ. This is not, in fact, the case: the rates normalised to

BET surface areas are identical (Jones *et al* 1984a). This effect has been explained in terms of the restructuring of the MgO surfaces on wetting, so that the dissolving surfaces are identical irrespective of the initial surface structures (Jones *et al* 1984a). This restructuring was first observed on MgO smoke samples exposed to dilute acid for times less than that required to dissolve a monolayer (Jones *et al* 1981).

1.2. THE DISSOLUTION OF CALCIUM PEROVSKITE

1.2.1. *Thermodynamic Stability.* The stability of $CaTiO_3$ in an aqueous environment has been considered by Nesbitt *et al* (1981), in terms of the reaction:

$$CaTiO_3 + 2H^+ (aq) \rightarrow Ca^{2+} (aq) + TiO_2 (rutile) + H_2O \qquad (3)$$

$$\text{with } K_{eq} = [Ca^{2+}]/[H^+]^2 = 10^{18.3} \text{ at } 25^0C.$$

Thus $CaTiO_3$ should be thermodynamically stable at 25°C for $[Ca^{2+}] \geq 10^{-4}$ mol dm^{-3} and pH $\geq$ 11 (for example), but would be expected to dissolve with formation of TiO_2 in water at pH 7.

Experimentally, under hydrothermal conditions over the temperature range 100°C – 190°C it has been shown that $CaTiO_3$ does react in water at pH about 7, with the formation of TiO_2 in the form of anatase and brookite (Kastrissios *et al*, 1987). Solution analyses detected only Ca^{2+}, while surface analytical methods indicated the formation of a hydrated TiO_2 layer at the surface of the solid. Neither of these latter analyses can distinguish between two models of the reaction of the perovskite (and other ceramics which have a similar covalently bonded framework lattice enclosing cation species). The first model is that the Ca^{2+} cations are "leached" via ion–exchange and diffusion from the titanate lattice, which does not in itself dissolve but may break down into an amorphous Ti–OH film. The second model involves the dissolution of the whole $CaTiO_3$ lattice, releasing Ca^{2+} into solution and the precipitation of Ti as TiO_2 or as titanium hydroxides.

Electron microscopy provided immediate and unequivocal confirmation that the second model applies for $CaTiO_3$ under hydrothermal conditions. The TiO_2 crystals were observed to nucleate epitaxially at sites spaced across the underlying perovskite. The occasional observation on the lower temperature (~100°C) samples of thin amorphous films consisting of (from EELS analysis) Ti and O supported the idea that Ti and O ions were transported from dissolving lattice sites via surface diffusion of titanium hydroxide ions such as $[Ti(OH)_2]^{2+}$.

The nature of the reaction of $CaTiO_3$ with aqueous solutions at epithermal temperatures (i.e. from ambient up to about 100°C) has now been studied and is reported here (see §4.).

1.2.2. *Ion–Exchange and Base–Catalysed Hydrolysis.* The structure of the titanate ceramics involves a covalently bonded Ti–O framework lattice, with cations located within octahedral interstices. At surface sites, such cations could be readily released to solution, in an ion–exhange reaction with H^+ from solution. If the structure of the solid permits ready diffusion of cations, continued ion–exchange could occur in "leaching" reactions, as in glasses (Paul, 1982; Myhra *at al* 1988a).

It has been established that solution attack on SiO_2 is initiated via a base–catalysed hydrolysis reaction (Paul, 1982). We have proposed (Myhra *et al* 1985) that the titanate lattice is also susceptible to this reaction, which may be written as:

$$\rangle Ti\langle^{O}\rangle Ti\langle + OH^- \rightarrow \rangle Ti\langle^{O} + {}^{OH}\rangle Ti\langle \qquad (4a)$$

$$\rangle Ti/^{O} + H_2O \rightarrow \rangle Ti/^{OH} + OH^- \qquad (4b)$$

Subsequent reaction with OH^- or H_2O leads to further hydroxylation and the formation of hydroxide species such as $Ti(OH)_2^{2+}$, $Ti(OH)^{3+}$, etc, and the formation of TiO_2 by decomposition of these hydroxides.

The overall reaction (equation 3) for dissolution of $CaTiO_3$ may then consist of a series of partial reactions involving ion–exchange, hydrolysis of the surface Ti–O–Ti, formation of Ti hydroxides and of TiO_2. The availability of OH^- in solution would appear to be a controlling factor. It may be noted that the overall reaction consumes H^+ from solution, suggesting that pH would rise during dissolution as Ca^{2+} is released into solution. Equally, the ion–exchange reaction could be controlling if the Ca^{2+} concentration is too high.

2. Materials and Methods

2.1. DISSOLUTION RATE MEASUREMENTS OF MgO

2.1.1. *MgO Samples.* The MgO was prepared in the form of powders in two ways – by burning Mg ribbon in air to form MgO smoke, and by calcining magnesium hydroxide or carbonate in air. The smoke particles are cubic, with remarkably perfect {100} faces, as characterised in many electron micrographs (eg Moodie and Warble 1971; Jones *et al* 1981). The surface area, as determined by numerous B.E.T. N_2 gas adsorption measurements, is 10.0±0.5 $m^2\ g^{-1}$, but the smoke powder shows an unusual size distribution, with a substantial proportion of the mass in a few very large cubes, with dimensions ~1μm compared with median cube edge of ~70 nm (Jones *et al* 1980). The calcined powders have more normal particle size distributions, and consist of rounded cubes, often loosely sintered together (Jones, 1988). The rounded surfaces of such particles have been shown (Moodie and Warble 1971) to consist of {100} facets at unit cell level.

The B.E.T. surface areas and preparation conditions of MgO powders are listed in Table II.

2.1.2. *Dissolution Rates.* The dissolution rate of oxide powders may be determined by measuring the change in solution pH as a function of time after adding the oxide. We have developed a micro computer–controlled system using a rapid response pH electrode with temperature compensation. The output from the pH meter is input to an amplifier with DC offset, so that the required change in pH can be matched to the 12–bit A/D converter. The digitized signals together with calibration data and data concerning the sample entered from the keyboard, are stored on disk for subsequent analysis. A numerical method is used to determine

TABLE II Preparation Conditions and Surface Areas of MgO Powders

Source	Preparation Conditions	B.E.T. area $m^2 g^{-1}$
$Mg(OH)_2$	4h, air, 1000°C	25.7
	4h, air, 1100°C	18.1
$MgCO_3$	4h, air, 1000°C	39.6
	4h, air, 1100°C	33.0
Mg ribbon	Burn in air	10.0

dpH/dt *vs* pH, and hence the dissolution rate per unit surface area of the oxide, corrected for H^+ activity. The details of the system are presented elsewhere (Jones, 1988; Jones *et al*, 1989).

Using this system, dissolution rates can be determined with a precision between repeated runs of a few percent.

2.2. PEROVSKITE

2.2.1. *Samples.* Compacted sintered polycrystalline discs of $CaTiO_3$ were obtained from Cookson Minerals Ltd, U.K. A strontium–doped perovskite, with composition $Ca_{0.8}Sr_{0.2}TiO_3$, was prepared at the ANSTO Lucas Heights Laboratory by hot pressing. Some discs were used as received on one face, with the other face polished (for surface analysis) to a 1μm finish using redistilled ethanol as lubricant; some were cut using a diamond saw, and of these some were finished with 120 grade carborundum paper to give a "rough" surface. Others were polished to a 1μm diamond finish on both sides. All samples were thoroughly cleaned ultrasonically in ethanol.

Powdered samples were prepared by crushing and sieving to produce three size fractions: <10μm, 50–100μm and 200–500μm. These powders were washed ultrasonically in dry ethanol to remove the very fine adhering particles; subsequently it was realised that this washing process did not remove all the fine material.

2.2.2. *Dissolution tests.* The samples were treated in distilled deionised water (DDW) at 20°C in polythene bottles, and at 60°C in teflon containers. The CO_2 in the DDW was removed by boiling and storage under argon. The critical parameter (sample surface area)/(solution volume) (SA/V) was maintained at 0.5 cm^{-1} for most tests, using estimates of the surface area based on sample geometry. In most cases, the solution was changed each hour, the sample washed in ethanol and a fresh aliquot of solution added. The samples were stirred by regular shaking. Concentrations of Ca^{2+} were measured by AAS with detection level 0.001 ppm. Ti^{4+} was tested for using ICPS, although in no case was Ti recorded above the detection limit of 0.03 ppm.

2.3. ELECTRON MICROSCOPY

Transmission electron microscopy at 100 keV and 200 keV was performed using JEOL JEM100C and JEM2000FX instruments, the latter fitted with Tracor Northern TN5400 energy dispensive X–ray spectrometer system. Powders were mounted on holey carbon support films. Images of MgO surface structures were obtained at 100 keV under phase contrast, using the method introduced by Moodie and Warble (1971). Details down to the unit cell (0.4 nm) level could be detected. For the perovskite powder samples, conventional diffraction contrast was used to determine crystal morphology and perfection. High resolution lattice images were obtained at the edges of thin perovskite crystals to reveal the presence of any amorphous layers, and the interface between the perovskite lattice and the surface (or interface with amorphous layers). The Ca/Ti ratio averaged through the thin edges of crystals was determined using EDXS and thin–film quantitative analysis procedures.

2.4 SURFACE ANALYSIS

The $CaTiO_3$ disks were examined in a Perkin Elmer PHI 560 instrument with combined AES/SAM/XPS/SIMS facilities. The XPS spectra were recorded using Al Kα incident X–rays, and a 25 eV analyser window. Peaks were recorded for C 1s, O 1s, Ca 2p, Sr 3d and Ti 2p. To obtain concentration profiles with depth, the samples were eroded using 4 keV Ar^+ ions at 1 μA cm^{-2} over a 9x9 mm^2 scan. XPS and SIMS spectra were recorded at each depth, using 0.1 μA.cm^{-2} 4 keV Ar^+ ions for SIMS. The absolute depths were estimated assuming 0.2 nm per μA.cm^{-2} integrated ion current.

3. Dissolution Kinetics of MgO

3.1 RESULTS

3.1.1. *Initial and Advanced Stages of Dissolution* In previous work on MgO, we reported the observation of an initial rise in $\log_{10}$(rate) with increasing pH, followed by the expected decrease (Jones *et al* 1981). These "initial" and "advanced" stages of dissolution are illustrated in Figure 3.1. The transition between the two stages occurs at different pH values for different initial masses of oxide, for between 5 and 10% dissolution of the original mass. The slope of the advanced stage of these curves, as a measure of the log(rate per unit surface area) *vs* pH relationship, is complicated by the fact that the particle size is decreasing with dissolution, and hence the specific surface area is increasing in an unknown fashion.

In the earlier paper (Jones *et al*, 1981) it was proposed that the initial stage was associated with the stabilization of the solution double layer and the transfer of water molecules through this layer. Thus the rate should be affected by changes to the solution double layer caused by the addition of cations. The more precise experimental results reported here, obtained using the computer–controlled pH analysis system, have led to a re–evaluation of the critical processes.

3.1.2. *Dependence on pH.* In order to resolve the discrepancies between earlier work on the pH dependence of rate, as described in §1.1.1, we have repeated the

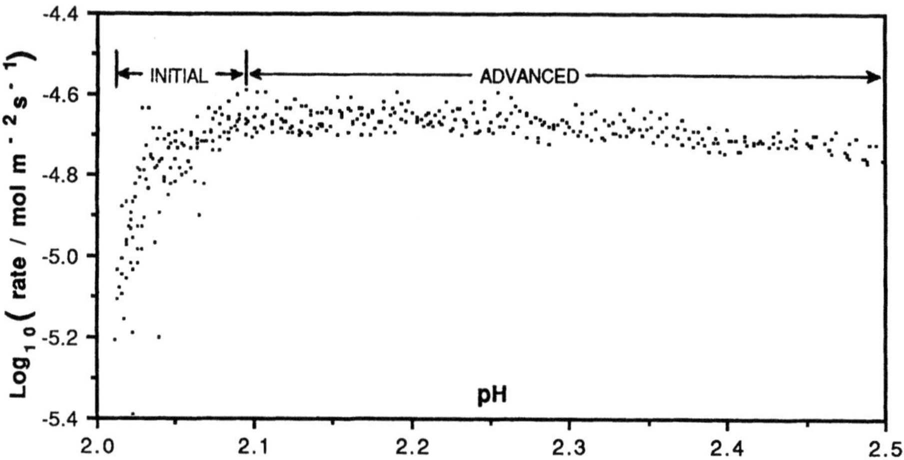

FIGURE 3.1. Dissolution of MgO smoke in pH 2.01 HNO_3 at 30^0C. Points determined from 6 independent repeat experiments as explained in the text. The dissolution data is plotted as Log_{10}(rate/mol m^{-2} s^{-1}) against pH. The "Initial" and "Advanced" stages of dissolution are indicated. In this case, the boundary between these stages occurs for approximately 10% dissolution of the MgO.

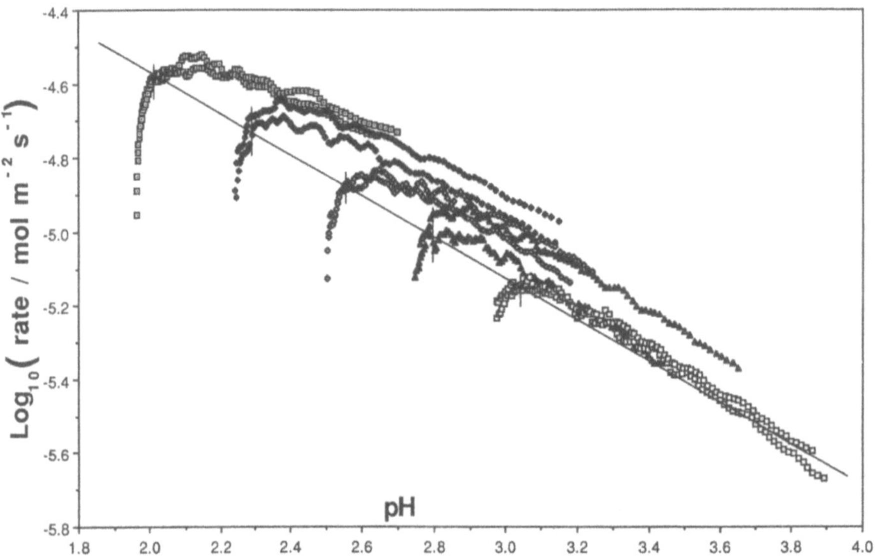

FIGURE 3.2. Duplicated dissolution curves for MgO ex–hydroxide for five values of starting pH. In each case, the mass of powder used was calculated to give a change of 1 unit of pH for 50% dissolution of the sample. The line is drawn through the points corresponding in each case to 5% dissolution (+). The data in Table III was obtained from these curves.

experiments of Segall *et al* (1978) using the computer controlled dissolution system, for MgO prepared by heating the hydroxide in air at 1000°C for 4 h. Five values of initial acid pH were used in the range 2.0 to 3.0, as shown in Figure 3.2. The computer analysis was modified to report the $\log_{10}$(rate) and pH for a given percentage dissolution of the initial mass, to provide data in the form of $\log_{10}$(rate) *vs* pH for equivalent surface areas of partially dissolved oxide. The results are summarised in Table III.

TABLE III Slope of $\log_{10}$(r) pH determined for different % dissolution of MgO

% dissolved	Slope
5	–0.55
10	–0.59
15	–0.60
20	–0.64
25	–0.66
30	–0.68
35	–0.73
40	–0.67

For the range 20% – 40% of total mass dissolved, the slope has mean value 0.68 ± 0.03. This value is consistent with Vermilyea's model if the second protonation is rate determining. As evident in Figure 3.2, the points for 5% dissolved lie on the "initial" part of the log(r) vs pH curve, where the rate is <u>increasing</u> with pH.

3.1.3. *Effects of different source materials on dissolution rates.* The MgO samples produced by the methods described in §2.1.1. not only have significantly different particle morphologies and surfaces structures, but also differ substantially in specific surface areas, as determined by B.E.T. The surface areas for MgO prepared by calcining the carbonate and hydroxide are listed in Table II.

The dissolution rates, measured for these samples and for smoke powder are shown in Figure 3.3. The rates per unit B.E.T. area for smoke and the two ex–hydroxide samples are initially identical, but diverge in the advanced stages, due to the differences in particle size distribution. The ex–hydroxide and ex–carbonate materials show closely similar rates through the advanced stage. The very high initial value for the ex–carbonate sample is believed to be due to segregation of a calcium impurity to the surfaces of the sintered particles (see §3.1.6).

To confirm the previous observation of surface restructuring on wetting, prior to any significant dissolution, a sample of MgO smoke was cooled to <10°C, and transferred rapidly to the electron microscope vacuum, with minimal exposure to the atmosphere. The resulting condensation of water was sufficient to cause restructuring of the initially very smooth {100} surfaces, as shown in Figure 3.4, without dissolution. If dissolution had occurred, a characteristic surface layer of hydroxide would have formed, not evident in the micrograph.

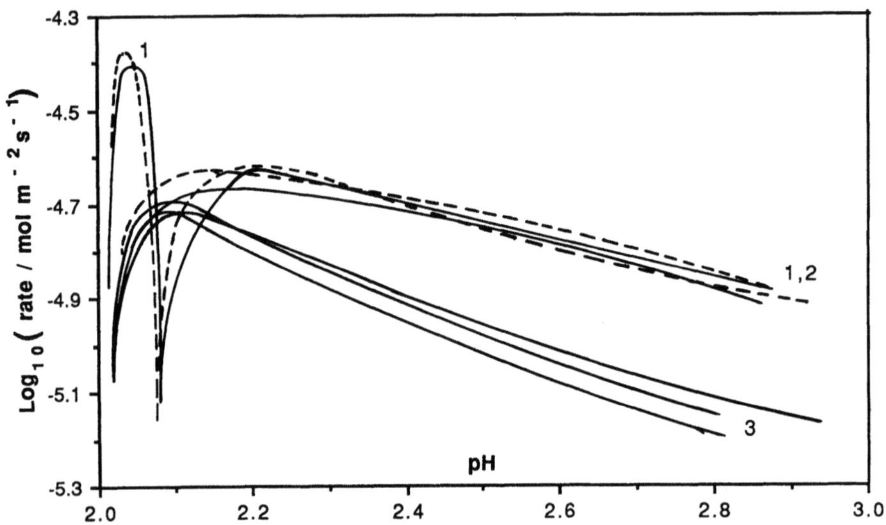

FIGURE 3.3 MgO dissolution curves for calcined and smoke samples. 1: MgO ex—carbonate; 2: MgO ex—hydroxide; (dashed lines: 1100^{0}C; continuous lines: 1000^{0}C); 3: MgO smoke produced by burning in air. See Table II for surface areas of these samples.

3.1.4. *Effects of hydration.* The surfaces of MgO smoke cubes are known to be relatively unreactive to water in the vapour phase. After long periods (>50 h) of exposure to ambient conditions, both IR spectra and electron microscopy observations show that the surface becomes substantially roughened through the hydroxylation of the top layers of the surface (Jones *at al* 1984b).

The dissolution rates for samples treated in HNO_3 within 7 h of preparation and after 65 h exposure to the atmosphere are compared in Figure 3.5. The major difference is the minimal extent of the initial stage of dissolution for the hydrated sample.

3.1.5. *Addition of cations.* As the oxide powder dissolves in the pure acid, the Mg^{2+} concentration rises from zero. In our earlier experiments, we observed some effects on initial dissolution kinetics of Mg^{2+} and other ions added to the original acid. Thus there appeared to be an effect of the dissolving Mg^{2+} on the rates (Jones *et al* 1981). However, repetition of these experiments (Jones, 1988) using the precision rate determination methods has not confirmed these results.

In Figure 3.6 the effects of additions of 10^{-3} mol dm^{-3} NaCl in HNO_3 and HCl dissolution curves for MgO are seen to be negligible. Similarly, the addition of up to 10^{-3} mol dm^{-3} Mg^{2+} as nitrate to HNO_3 had no significant effect. While these added concentrations may not be sufficient to alter the extent of the solution double layer, they do cover the range of [Mg^{2+}] which results from dissolution of the oxide.

3.1.6. *Dissolution of CaO and CaO–doped MgO.* Calcium doped MgO samples, prepared as described in §2.1.2, showed an initial high rate,but advanced rates

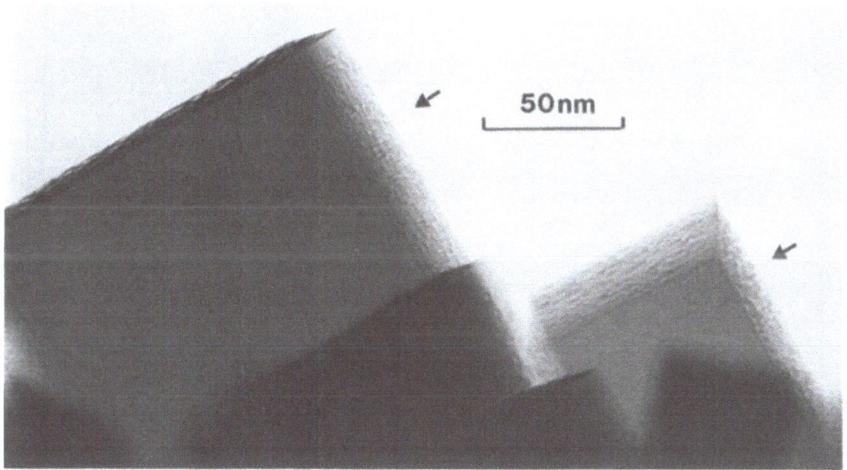

FIGURE 3.4. Transmission electron micrograph showing surface reconstruction of MgO smoke cubes on which water has condensed for a short time before insertion into the microscope vacuum. No significant dissolution has occurred in this time: the drying of $Mg(OH)_2$ would result in a very fine granulated contrast not evident in this image.

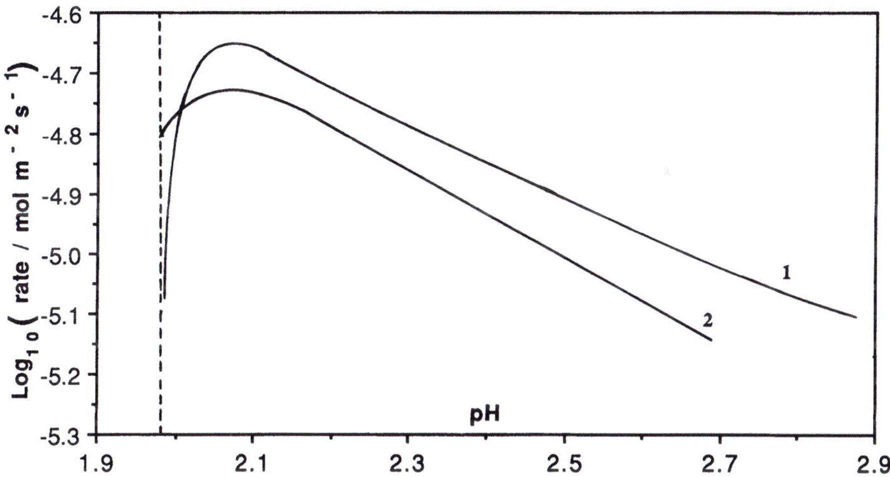

FIGURE 3.5. Effect of hydration prior to dissolution on the initial stage of MgO dissolution.
1: MgO smoke in pH 1.98 HNO_3 at 30.5^0C (average of two runs);
2: Same MgO smoke after 65 hours exposure to moist air, dissolved in pH 1.96 HNO_3 at 30.5^0C (average of two runs). The extent of the initial increasing rate with increasing pH is greatly reduced. Electron microscope examination of sample 2 showed extensive growth of hydroxide across the surface of the smoke particles.

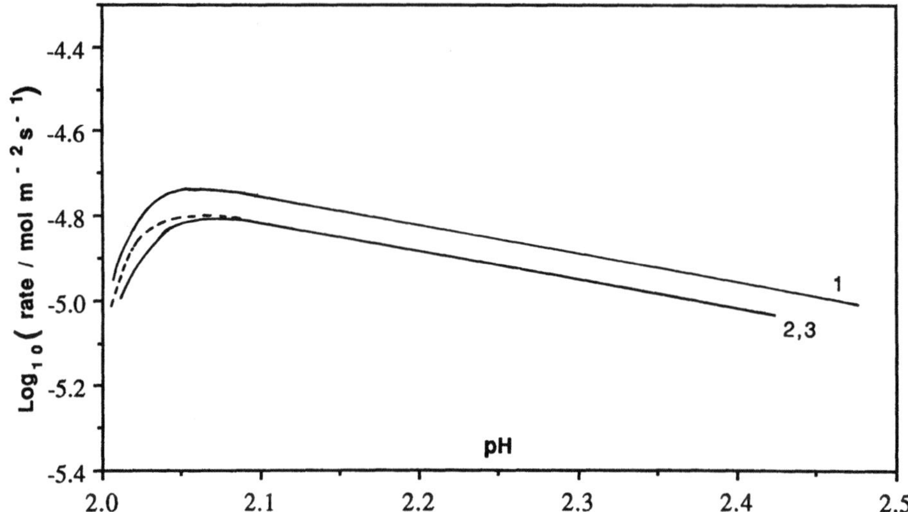

FIGURE 3.6 Effect of added Na^+ on MgO dissolution kinetics. 1: No added cation; 2: 10^{-3} mol dm^{-3} Na^+ as NaCl in HNO_3. 3: 10^{-3} mol dm^{-3} Na^+ as NaCl in HCl (dashed line).

essentially identical to those of MgO (Jones 1988; see Figure 3.7). Since it is believed that Ca^{2+} segregates to the surface of MgO (McCune and Wynblatt, 1983; Colbourn *et al* 1983), samples of CaO were prepared from hydroxide and acetate precursors. The very high reactivity of CaO in air makes such experiments difficult, but it appears to dissolve at least 10 times faster than MgO (Figure 3.8).

3.2. DISCUSSION

3.2.1. *Advanced dissolution kinetics.* The rate at which MgO dissolves in acid solution appears to depend on the hydrogen ion activity to the 2/3 power, in the advanced stages of dissolution. This result suggests that Vermilyea's explanation of the mechanism of ionic oxide dissolution in terms of ion transfer is correct, and that the second protonation reaction is the rate determining process. The results of Fruhwirth *et al* (1985) for {100} MgO, indicating a dependence of rate on pH (rather than log(rate)), might be explicable in terms of a continued restructuring of the {100} surfaces of the single crystal rotating disc used in their experiments, but this could only be tested by TEM examination of the attacked surfaces; our results on {100} MgO smoke cubes do not confirm their claim.

The minor but consistent differences in advanced dissolution rates for MgO smoke compared to MgO prepared by calcination may be understood in terms of the differences in particle size distribution. The smoke has a very asymmetrical size distribution about the median value (Jones *et al* 1980), with significant mass in a few very large crystals. Thus after about 20% dissolution, the mean particle size of the smoke samples will be relatively higher, and the specific surface area relatively lower, compared to the more normal size distributions in the calcined powders. This explains why the rates as plotted in Figure 3.3 are lower for the smoke samples.

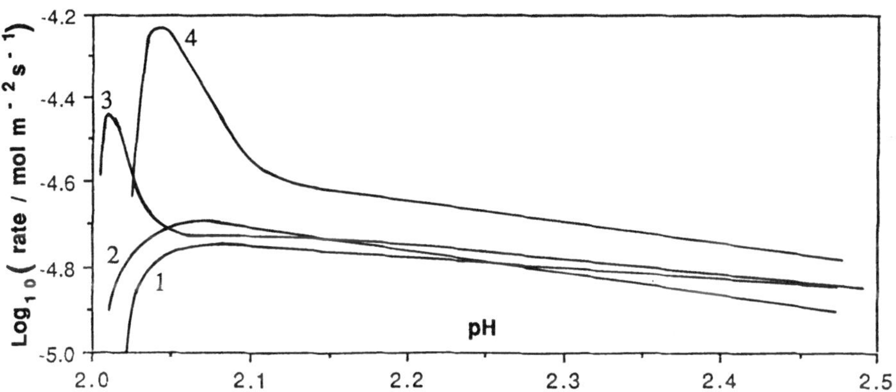

FIGURE 3.7. The effect on dissolution rates of doping MgO with CaO, prepared by co–precipitation; freeze–drying and calcination at 1000^{0}C. 1: no added Ca;
2: 1% CaO; 3: 5% CaO; 4: 10% CaO. There is clear evidence of segregation of CaO to the surface of the MgO.

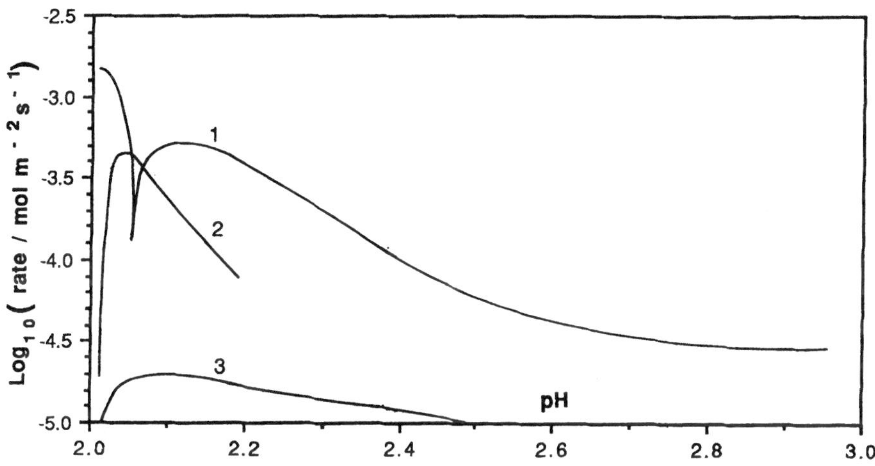

FIGURE 3.8. Dissolution kinetics of CaO in HNO_3 at 30^{0}C. 1: CaO ex–hydroxide calcined at 1100^{0}C for 22 h (surface area 1.4 m^2 g^{-1}); 2: CaO ex–acetate calcined at 1000^{0}C for 24 h (surface area 4.7 m^2 g^{-1}). 3: MgO smoke for comparison. The first peak on sample 1 is believed to be due to a hydroxide layer formed on the powder after preparation. From these preliminary experiments on CaO, it is apparent that CaO dissolves at a rate per unit area at least 10 times faster than MgO under comparable conditions.

3.2.2. *Initial dissolution kinetics.* Our earlier suggestion (Jones *et al* 1981) that the initial stage is controlled by solution double layer parameters is now in doubt, since the addition neither of cations nor of anions (at concentrations equivalent to those of the dissolved species) alters the dissolution curves. The fact that protonation of the MgO surface prior to wetting does reduce the extent of the initial stage (Figure 3.5) suggests that the protonation of the active sites formed by the restructuring on wetting may be involved.

Since smoke samples and MgO ex–hydroxide samples show identical increasing initial dissolution rates per unit B.E.T. area, despite the very different surface concentrations of active (3– and 4– coordinated) sites, we suggest that the restructuring of the surfaces on wetting occurs in a time shorter than that required for the first protonation reaction (equation 2a). The initial rise in rate towards the limit determined by the $[H^+]^{2/3}$ condition (i.e., the second protonation) will then be associated with the rate of increase in the surface density of active protonated O^{2-} sites on the restructured {100} faces (i.e., to the first protonation reaction).

3.2.3. *Surface Reconstruction.* The fact that all MgO samples, irrespective of initial particle size or surface structure, show identical dissolution rates per unit B.E.T. surface area, has been explained in terms of the restructuring of the surface which occurs immediately on immersion in aqueous solution, together with the failure of B.E.T. gas adsorption using N_2 to detect the details of surface structure at the unit cell level (Jones *et al* 1984a). As seen in Figure 3.4, the initially almost perfectly flat {100} faces of the cubic particles are roughened after wetting in a fashion based on {100} facets of order 1–3 nm. This surface provides a very high concentration of active 3– and 4–coordinated sites. As suggested above, it appears that the first protonation reaction may occur subsequent to restructuring. The question as to how and why planar MgO {100} surfaces in water restructure in this way remains to be answered. Surface entropy may be the driving factor.

4. Surface Reactivity of $CaTiO_3$

4.1 RESULTS

Results selected from a large number of dissolution experiments are presented, to illustrate the effects of microstructure and crystal defects on measured Ca^{2+} release rates, and to establish the basic model for reactions at the surface of perovskite crystals. Similar results have been presented recently in the context of perovskite as a component of Synroc (Pham *et al* 1988). Full details of the experimental results are reported by Myhra *et al* (1988b).

4.1.1. *Calcium release from $CaTiO_3$ discs.* The amount of calcium released each hour into fresh aliquots of DDW at 20°C is shown in Figure 4.1, for three discs of $CaTiO_3$ with different surface treatments. All the samples show a steady release of about 0.1 ppm h^{-1} for times beyond about 15 h extending to at least 80 h. For the samples polished on both sides, there is an initial, additional release in the first hour of about 0.1 ppm. The discs with rough surfaces also show an initial additional loss of 0.2–0.4 ppm over the first hour, followed by further release over about 15 h until the steady level of 0.1 ppm h^{-1} is reached.

Based on the geometrical surface area of the discs, and an assumed surface area of 0.25 nm^2 per surface calcium ion in the perovskite, the initial additional release

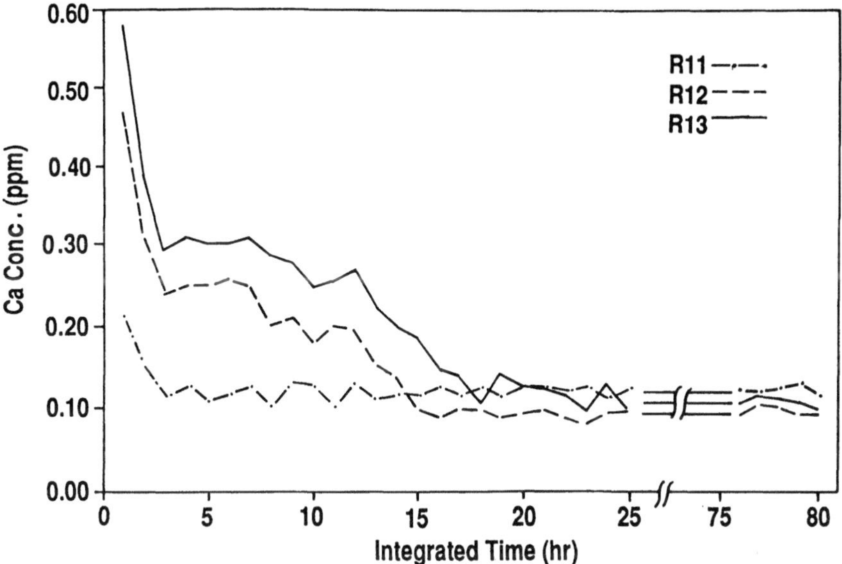

FIGURE 4.1. Release of Ca into solution (ppm h^{-1}) as a function of integrated duration of sequential attack in DDW at 20^0C for three $CaTiO_3$ discs: R11 — polished both sides; R12 — cut one side, polished the other; R13 — cut one side, roughened the other.

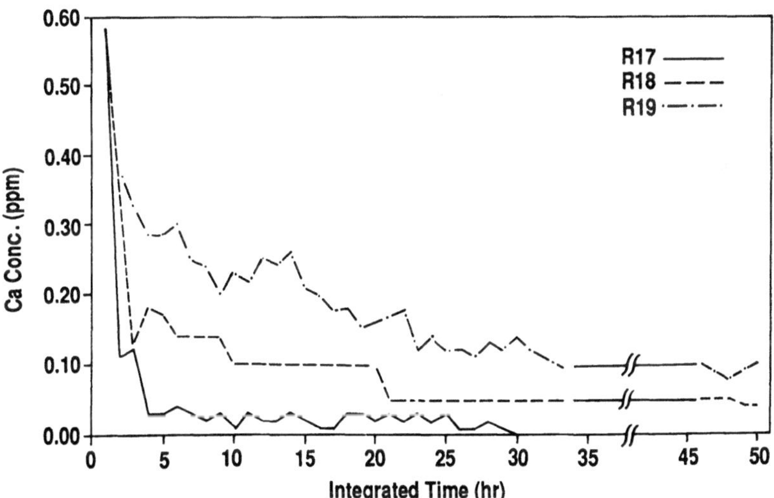

FIGURE 4.2. Release of Ca into solution (ppm/h) as a function of integrated time of sequential attack in DDW at 20^0C on three sized and washed $CaTiO_3$ powders. R17: < 10μm; R18: 50—100μm; R19: 200—500μm.

of 0.1 ppm corresponds to about 8 Ca^{2+} ions per surface site. Given that, even for polished surfaces, the "true" surface area would be substantially higher than the geometrical area, these results indicate an initial loss of Ca^{2+} from the surface sites to a depth of no more than 1 or 2 Ca^{2+} layers.

All these samples showed the same long term release of Ca^{2+}, independent of the initial roughness of the surface.

4.1.2. *Calcium and strontium release from* $Ca_{0.8}Sr_{0.2}TiO_3$. The hot–pressed calcium strontium perovskite showed similar effects, but with a long term rate for Ca^{2+}, normalised against area, almost one half of that from the $CaTiO_3$. Significantly, the Ca^{2+} and Sr^{2+} were always released at rates in the atomic ratio 4:1, equal to the ratio of the Ca/Sr concentration in the solid.

4.1.3. *Calcium release from powdered* $CaTiO_3$ The amount of calcium released into fresh DDW each hour is shown in Figure 4.2 for three powdered samples. The largest size fraction (200–500μm) powder (R10) shows Ca^{2+} release rates very similar to those of the disc samples. The critical sample is the finest powder (< 10 μm – experiment R17), for which a surface area of 1.4±0.3 $m^2\ g^{-1}$ could be determined by the B.E.T. gas adsorption technique. After an initial release of about 0.7 ppm over the first 2 h, a continued loss at $\leq$ 0.03 ppm h^{-1} was observed, reducing to undetectable levels after about 30 h. The initial loss of Ca^{2+} to solution represents 4 ions per surface Ca^{2+} site (relative to B.E.T. surface area).

4.1.4. *Analytical Transmission Electron Microscopy.* Particles from the smallest size fraction powder were examined, and three types were observed. As expected, there were large essentially perfect (1–2μm typically) particles showing little or no defect contrast, with sharp lattice fringes running to the thin edges of crystals (Figure 4.3). The second type consisted of thin flakes showing substantial damage due, presumably, to deformation during crushing (Figure 4.4). Finally there were agglomerates of very fine, thin particles which gave polycrystalline ring diffraction patterns (Figure 4.5). It is apparent that the washing of the powder did not remove these aggregates of very fine (high surface area) particles. In all these samples, EDX spectra from thin particles gave Ca/Ti ratio of 1.0.

After attack in DDW for 50 hours (sample R17, see Figure 4.2), neither the "perfect" particles nor the damaged ones showed any effects. However, the EDX spectra from the clumps of very thin polycrystalline particles showed a loss of calcium, with Ca/Ti $\approx$ 0.8.

After treatment in DDW for some hours, it was observed that the cut and roughened surfaces of the $CaTiO_3$ discs showed a white powder adhering to the surfaces. This could be removed by ultrasonic cleaning in ethanol, for examination by transmission electron microscopy. The effect is illustrated in the scanning EM micrographs in Figure 4.6. The deformation introduced by cutting appears to lead to preferential attack and the fracturing of fine particles from the surface. Some of the resulting powder was found in the solution.

The white powder removed after this treatment of cut perovskite disks showed very similar types of crystal to those found in the crushed powders – some "perfect" fragments, some thin deformed crystals, and polycrystalline aggregates. The latter were depleted in calcium, with Ca/Ti $\approx$ 0.8. Some of the perfect crystal particles showed no evidence for attack, but many had an amorphous film about 10 nm thick, around the edges of the crystals. The crystal lattice fringes extended clearly to the interface with the amorphous film, without distortion (Figure 4.7).

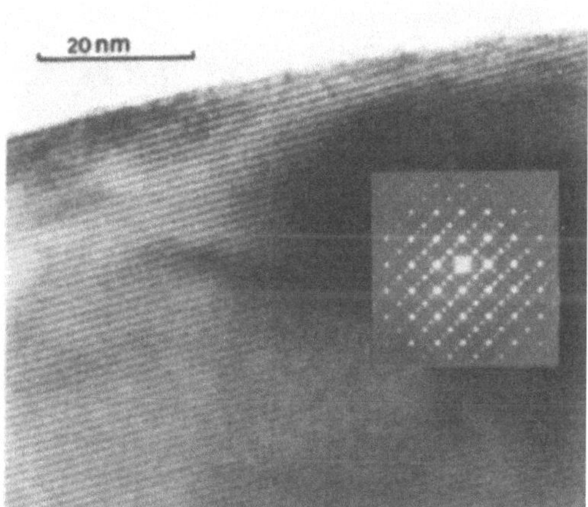

FIGURE 4.3. High resolution TEM image of a thin region of an unattacked $CaTiO_3$ powder particle, showing lattice fringes running to the edge of the crystal.

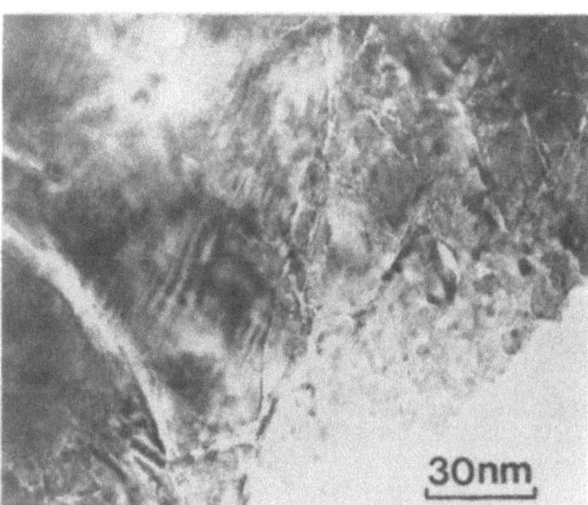

FIGURE 4.4. TEM micrograph of unattacked $CaTiO_3$ powder particle showing damage caused by crushing.

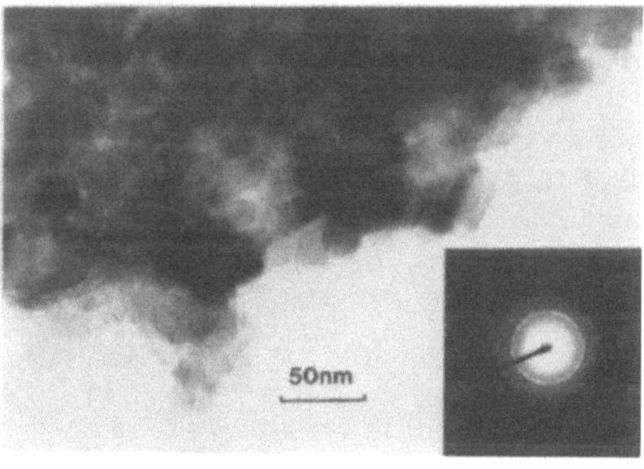

FIGURE 4.5 Agglomeration of very fine $CaTiO_3$ particles which give rise to diffraction rings (inset).

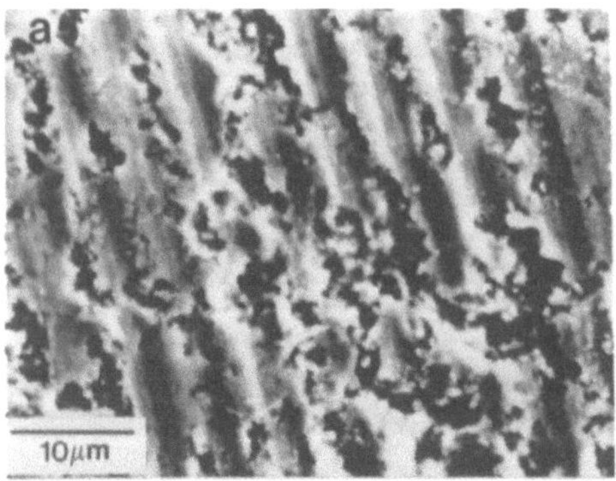

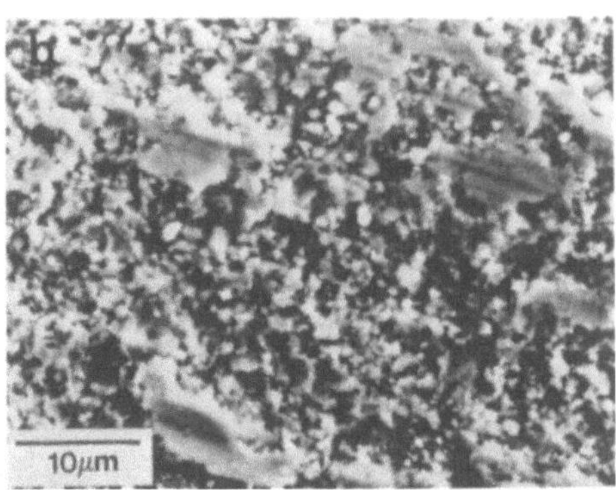

FIGURE 4.6 Scanning electron micrographs of the surface of a $CaTiO_3$ disc; (a) as prepared by cutting with diamond saw and ultrasonic cleaning; (b) after attack in DDW at 20°C for 120h. This attacked surface had a white appearance, which developed after the first few hours. Some particles fractured from the surface during treatment and were found in solution, others could be removed from surfaces such as (b) by ultrasonic cleaning.

Similarly, amorphous films were observed within cracks or boundaries and around the edges of the distorted, defective crystals (Figure 4.8).

4.1.5. *Surface Analytical Measurements.* XPS profiles of $CaTiO_3$ discs before and after sequential treatment in DDW for various times at 20°C are shown in Figure 4.9. In these plots, two features of the depletion of Ca relative to Ti (averaged over the surface) are significant. There is a reduction of Ca/Ti at the surface, to a level of about half the pre–attack ratio, which develops within the first 9h of the attack. The ratio rises sharply to a shoulder at $\Sigma D \simeq 50$ μA min cm^{-2} (10 nm, approximately). The Ca/Ti ratio then rises less abruptly, reaching the bulk value at about 500 μA min cm^{-2} (100 nm).

This profile is established after about 17 cycles of 1–hour attack, and does not change significantly with further attack up to 80 h at this temperature.

In the case of the $Ca_{0.8}Sr_{0.2}TiO_3$ sample, the first effect – depletion at the surface – is observed, but the Ca/Ti and Sr/Ti ratios rise to their bulk values over about 10 nm; there is no significant depletion to greater depths.

SIMS depth profiles for CaOH/Ca and TiOH/Ti reveal relatively high ratios at the surfaces of attacked disks (0.14 and 0.08 respectively), which fall off to the level found in unattacked samples at a depth of about 100 nm.

4.1.6. *Saturation effects.* As noted in §2.2.2, the ratio (SA/V) for the experiments reported above were always about 0.5 cm^{-1} and the calcium ion concentration did not rise above 1 ppm (2.5×10^{-5} mol dm^{-3}). In order to establish the saturation levels of Ca^{2+}, samples of powders of size fraction 50–100 μm were treated in DDW at 25°C for 1 d for various SA/V values, from about 10 cm^{-1} to about 400 cm^{-1}. After 1 d, the solution was replaced with fresh DDW. The resulting Ca^{2+} concentrations are listed in Table IV. The solution pH after 1 day tests ranged between 6.9 and 7.9.

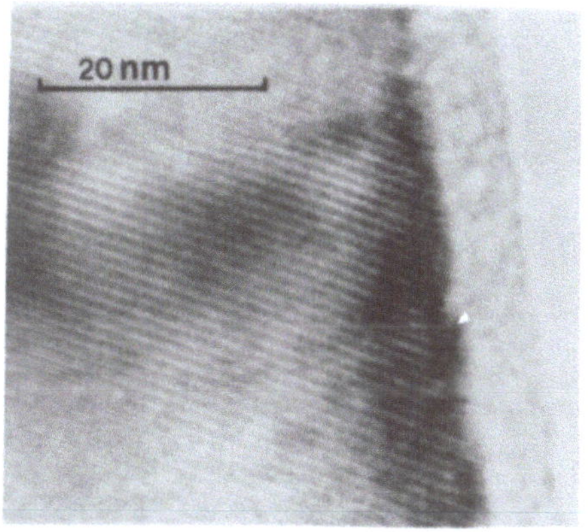

FIGURE 4.7. High resolution TEM micrograph showing amorphous film grown on a "perfect" crystal particle from the white powder washed off an attacked, roughened $CaTiO_3$ disc. Lattice fringes run clearly to the interface between the $CaTiO_3$ and the amorphous film.

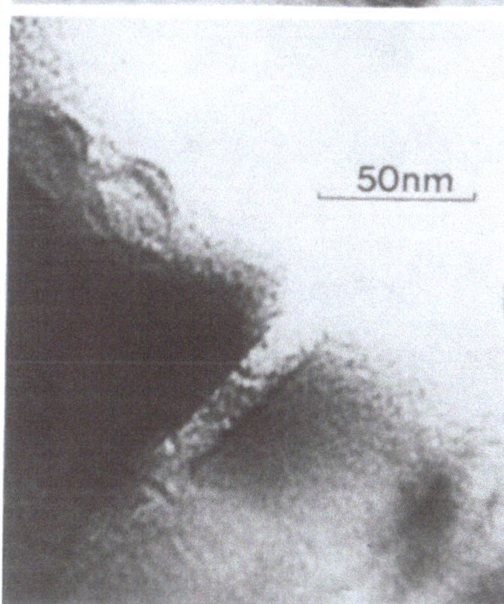

FIGURE 4.8. TEM micrograph showing growth of an amorphous film at the edge and along a boundary in a deformed particle from the white powder sample.

TABLE IV $[Ca^{2+}]$ values under conditions of High SA/V in DDW at 25°C

SA/V	Number of 1 day tests		
	1	2	3
10	2.1	1.0	1.1
100	5.8	4.3	4.2
200	5.7	4.2	4.8
400	5.8	4.6	4.8

Thus at 25°C the saturation effects, within the timescale of this study, have limited the concentration of Ca^{2+} in solution to about 6 ppm (1.5×10^{-4} mol dm^{-3}). We note that the ratio $[Ca^{2+}]/[H^+]^2 < 10^{10}$ in these experiments.

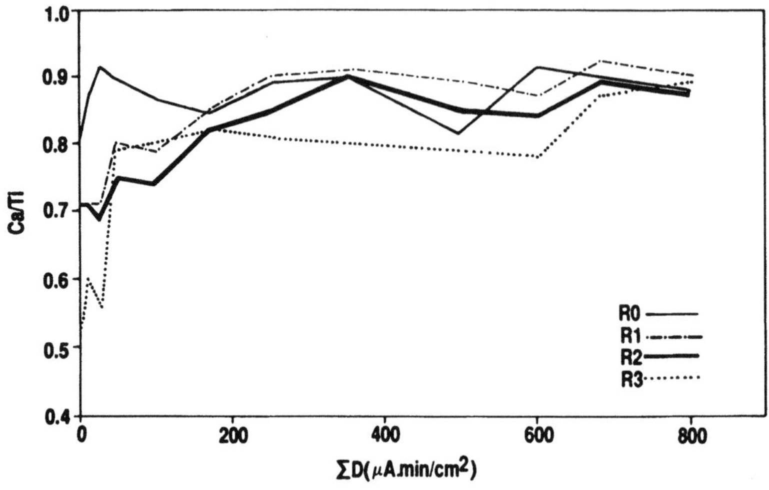

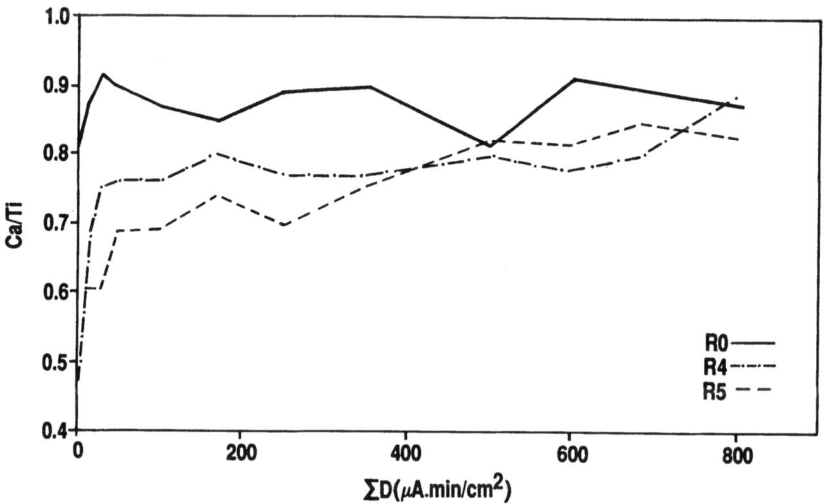

FIGURE 4.9. XPS depth profiles for a polished, unattacked $CaTiO_3$ disc (RO) and for similar surfaces after sequential attack for various periods in DDW at 20^{0}C (R1 – 1 h; R2 – 4 h; R3 – 9 h; R4 – 17 h; R5 – 80 h).

4.1.7. *Effects of Temperature.* The experiments described in §4.1 above were repeated at 60°C. In general, the results were similar to those at 20°C, with higher release rates for Ca^{2+}, and greater depletion of Ca from the solid, as expected. The amorphous layer formed at the surface of $CaTiO_3$ was thicker (about 20 nm). For the purposes of the present paper, no significant new effects were observed at the higher temperature.

4.2. AN HYPOTHESIS AND FURTHER TESTS

In the results from the experiments outlined above, there appeared to be some conflict between surface analytical data – showing substantial depletion of Ca^{2+} within the first 10nm – and electron microscopy, which revealed no surface layers of this dimension on the powder samples. In the case of the smallest fragments, the results were consistent with the removal of the surface layer of Ca^{2+}, together with removal from damaged areas and residual grain boundaries. The observation of amorphous layers on some of the undeformed perovskite fragments removed from the attacked cut surfaces of the discs (the "white powder"), suggested that local conditions of pH at sites with restricted access to the solution (cracks, grain boundaries, pores etc) might be important. Such effects have been invoked to explain attack on glasses. In order to clarify this conflict, two further sets of experiments were designed, based on the following hypothesis: *perovskite dissolves in aqueous environments only under solution conditions for which Ca^{2+} concentration is less than ~10^{-5} mol dm^{-3}; which avoid the thermodynamic stability constraints for $CaTiO_3$ and which provide sufficient OH^- concentration to permit breakdown of the titanate framework via base catalysed hydrolysis.*

4.2.1. *Dissolution of "single crystal" $CaTiO_3$.* In order to avoid the complicating effects of crystal defects or of microstructure, a sample of the <10μm powder fraction was repeatedly ultrasonically washed in a non–polar solvent until no trace of fine suspended material could be detected. TEM observation confirmed that the remaining powder consisted almost entirely of chips of perovskite showing a high degree of crystal perfection. This material was used in six tests under controlled pH (using NH_3/HCl) and [Ca^{2+}] through adjustment of SA/V and/or replacement of solution.

Using the same mass (47 mg) as for experiment R17 (see figure 4.2), there was an initial release of 0.06 ppm Ca^{2+} in the first hour, which is 1/12 of that released in experiment R17. This reduction would be due in part to lower surface area per unit mass for this sample, but is also confirmation that only the Ca^{2+} ions in the solid directly accessible to the surface are released on contact with water – further release of Ca^{2+} from the crystal lattice being prevented until after breakdown of the titanate framework.

The formation of an amorphous TiOH surface layer was taken to be evidence for dissolution of the $CaTiO_3$ lattice.

Of the six samples, only one showed such a film, as summarised in Table V.

These experiments indicate that there is a limited range of [$Ca^{2+}_{(aq)}$] and pH for which dissolution of the $CaTiO_3$ lattice takes place. The earlier TEM observations on the sized powders are confirmed; it is noteworthy that the solution pH was less than 7.5 in all the experiments described in §4.1

Table V The dependence of amorphous film formation on pH and [$Ca^{2+}_{(aq)}$]

pH	[$Ca^{2+}_{(aq)}$]	
	High (~ 5ppm)	Low (<0.5ppm)
7.5	no film	no film
9.5	no film	film
10.5	no film	no film

4.2.2. *The Effect of High [Ca^{2+}].* The second experiment involved the treatment of a $CaTiO_3$ disc under conditions of high [Ca^{2+}]. A polished disc and 2 g of the fine powder were attacked at 25°C for 5 days. The Ca^{2+} concentration was measured at 6.5 ppm, at a pH of 8.5.

The XPS depth profile of this sample revealed calcium depletion to only 2nm. This is close to the detectability limit of the XPS profiling method, and indicates that no significant attack took place during this experiment.

4.3. DISCUSSION

From the results outlined above, a picture of the nature of the attack of aqueous solutions on perovskite emerges, which, taken together with the thermodynamic constraints, implies that the microstructure and the nature and extent of defects in the material can result in the release of Ca^{2+} to solution under conditions of pH in solution which prevent the attack on perfect crystalline perovskite. We consider first several aspects of the experimental observations, and then describe a model for the dissolution of crystalline perovskite. Finally, the effects of crystal defects and of the microstructure of the bulk material are interpreted in terms of the model.

4.3.1. *Estimates of Surface Area.* The surface area of a disc as estimated from its geometry takes no account of contributions to the reactive surface area from surface roughness, pores, etc. Some idea of the ratio of the B.E.T. area of a disc (were this measurable) to geometrical area can be obtained from the 50–100μm size fraction powder. The estimated geometrical area (based on mean of values calculated for cubic and spherical particles) is about 0.02 m^2 g^{-1}; the measured area, by B.E.T. was 0.25 m^2 g^{-1}. Although part of this factor of over 10 is likely to be due to fine particles not removed by ultrasonic cleaning, a ratio of 4 in (B.E.T./geometrical) areas for the discs could be expected.

This would reduce the estimated number of layers of Ca^{2+} removed on contact with DDW to 2.

While such estimates remain very uncertain, it may be argued that the dissolution experiments are consistent with a model in which about one monolayer of Ca^{2+} is removed from the surfaces of the crystals on contact with water, provided that the solution [Ca^{2+}] is not too high.

4.3.2. *Conditions for dissolution of perovskite.* Irrespective of pH in the range 7 – 10.5, the concentration of Ca^{2+} in solution is clearly a controlling factor. No significant dissolution was observed from single crystal particles for $[Ca^{2+}] \geq 10^{-4}$ molar. This concentration was also sufficient to prevent any significant loss from disc samples, as monitored by XPS. At the lower calcium concentration ($\leq 10^{-5}$ molar) dissolution was observed for pH 9.5 but not pH 7.5 or 10.5. Further, the thickness of the amorphous film formed stabilized at about 10nm, irrespective of time of exposure at 20°C.

In Table VI the values of $[Ca^{2+}]/[H^+]^2$ are listed for the solution conditions of Table V. For pH 10.5, the values are approaching those for thermodynamic stability of $CaTiO_3$. It appears that at pH 7.5, the kinetic constraint on dissolution is the low concentration of OH^- ions required to initiate the base catalysed hydrolysis of the titanate lattice. At pH 9.5, a concentration of Ca^{2+} less than $\sim 10^{-5}$ mol dm^{-3} is inadequate to prevent loss of Ca^{2+} to solution, and $[OH^-]$ is high enough to enable breakdown of the Ti–O bonds – dissolution proceeds.

Table VI   Stability parameter $[Ca^{2+}]/[H^+]^2$ (Compare with Table V)

pH	$[Ca^{2+}{}_{(aq)}]$/mol dm^{-3}	
	10^{-4}	10^{-5}
7.5	10^{11}	10^{10}
9.5	10^{15}	10^{14}
10.5	10^{17}	10^{16}

At a given temperature, this dissolution of $CaTiO_3$ continues only until the amorphous film reaches a specific thickness – about 10 nm at 20°C. This effect we interpret in terms of the Ca^{2+} concentration gradient across the film: for the critical thickness, $[Ca^{2+}]$ at the solid–film interface becomes high enough to prevent further loss of Ca^{2+} from the perovskite lattice. The amorphous layer of titanium hydroxides – presumably also hydrated while in solution – acts as a protective layer. The formation of $Ca(OH)^+$ or $Ca(OH)_2$ within the layer (as indicated by SIMS) must serve to restrict the diffusion rates of Ca^{2+} through the film.

4.3.3 *A model for dissolution of $CaTiO_3$ single crystals.* The full set of results outlined above can be explained in terms of a model in which ***either*** the ion–exchange of surface Ca^{2+} ions ***or*** the base–catalysed hydrolysis of the titanate framework can control the overall reaction, and both must occur if dissolution is to take place.

Schematically, we may represent the surface of a $CaTiO_3$ crystal in the following simplified 2–dimensional form:

```
        Ca2+      Ca2+      Ca2+
.....  O         O         O         O......   (surface)
>Ti<Ca  >Ti<Ca  >Ti<Ca  >Ti<Ca
 Ca >O<  Ca >O<  Ca >O<  Ca >O<
>Ti<Ca  >Ti<Ca  >Ti<Ca  >Ti<Ca
```

While this is an artificial picture, it indicates the essential features: the Ti–O covalently bonded lattice, with Ca^{2+} ions within the octahedral interstices, and some Ca^{2+} ions at the free surface. It ignores the state of the surface oxygen atoms, which would probably be hydroxylated after exposure to moist air.

On initial contact with aqueous solution, these surface Ca^{2+} ions will be released into solution by ion exchange with H^+, provided that the solution concentration of Ca^{2+} is not too high. This might be represented as:

```
        H+ H+     H+ H+     H+ H+
.....  O         O         O         O......   (surface)
>Ti<Ca  >Ti<Ca  >Ti<Ca  >Ti<Ca
 Ca >O<  Ca >O<  Ca >O<  Ca >O<
>Ti<Ca  >Ti<Ca  >Ti<Ca  >Ti<Ca
```

If the pH is high enough, there will be sufficient concentration of OH^- to initiate base–catalysed hydrolysis, and the breaking of Ti–O–Ti bonds with the result:

```
                 OH   O-
.....  O                       O         O......   (surface)
>Ti<Ca  >Ti ( Ca  )Ti<Ca  >Ti<Ca
 Ca >O<  Ca  >O<  Ca >O<  Ca >O<
>Ti<Ca  >Ti<Ca  >Ti<Ca  >Ti<Ca
```

(where the surface H^+ have been omitted for clarity). The dangling Ti–O^- bond could be hydroxylated in turn via an H^+ or by reaction with H_2O, releasing an OH^- back to solution. The breaking of the titanate lattice permits the release of further Ca^{2+} via ion exchange:

```
                 OH   OH
.....  O                       O         O......   (surface)
>Ti<Ca  >Ti ( H+ H+ )Ti<Ca  >Ti<Ca
 Ca >O<  Ca  >O<  Ca >O<  Ca >O<
>Ti<Ca  >Ti<Ca  >Ti<Ca  >Ti<Ca
```

If the Ca^{2+} concentration is too high ($\geq 10^{-4}$ mol dm^{-3}), the initial loss of surface Ca^{2+} by ion exchange does not occur, and dissolution is prevented, irrespective of pH. If $[Ca^{2+}]$ is low, the initial loss of surface Ca^{2+} ions will only be followed by dissolution of the solid if the $[OH^-]$ is sufficient to sustain lattice breakdown, with the formation of $Ti(OH)_x$ species. The solubility of titanium hydroxide molecules in aqueous solution is very low. At hydrothermal temperatures, TiO_2 can be nucleated at the surface of perovskite, and the growth of the resulting anatase and brookite crystals provides a sink for $Ti(OH)_x$ species via surface diffusion (Kastrissios *et al*, 1987). However at lower temperatures (< 90°C), TiO_2 does not form, with the result that an amorphous $Ti(OH)_x$ layer forms, incorporating also $Ca(OH)^+$ and/or $Ca(OH)_2$. As this layer grows, diffusion of Ca^{2+} to solution becomes limited, with the result that the effective Ca^{2+} concentration at the dissolving crystal/amorphous–layer interface rises to a level sufficient to prevent further reaction. At 20°C, the resulting protective film thickness is about 10 nm.

4.3.4. *Aqueous attack on polycrystalline perovskite.* The experiments involving aqueous attack at 20°C on polycrystalline $CaTiO_3$ discs show that provided the Ca^{2+} concentration is maintained at a sufficiently low level, dissolution of perovskite at 20°C proceeds to a depth of order 10 nm, as measured by surface analysis. In the case of the hot–pressed calcium strontium perovskite sample, no depletion of Ca^{2+} beyond 10 nm is observed, whereas in the much more porous sintered $CaTiO_3$ discs, the XPS and SIMS depth profiles show further depletion to about 100 nm. These thicknesses are established within about 10 h of attack; further attack to about 100 h does not extend them appreciably.

Since the pH levels in these experiments remain below 8, it is clear that perfect regions of the samples, exposed to the bulk solution pH and [Ca^{2+}] levels, will not be attacked. This is confirmed from the "white powder" observations, in which some crystalline fragments showed no attack, but some showed 10 nm protective films. It must follow that the surface preparation – cutting, grinding and polishing – in these brittle materials results in a high density of surface damage in the form of dislocations, cracks and strained regions at which attack occurs. Locally high pH levels can be associated with such regions of restricted access. The size of the particles (~10 nm) making up the polycrystalline clumps detected in the white powder is an indication of the fine scale of this surface damage. Clear evidence of attack penetrating at cracks or gram boundaries is provided in micrographs such as Figure 4.8.

The solution measurements for the $CaTiO_3$ and $Ca_{0.8}Sr_{0.2}TiO_3$ and for the sized powders, reveal that Ca^{2+} is released at a low but constant rate after the first 10 h of attack independent of surface preparation, in all cases except the very fine powders, which are smaller than the grain size. This suggests that there is continued reaction via grain boundary diffusion, into the bulk of the material. The extent of this, averaged over the exposed surface, is apparently insufficient to be detected in the XPS depth profiles. Further work, including very long–term dissolution experiments, is required to confirm this and to evaluate the extent of grain boundary diffusion over extended times.

5. Conclusions

Several general conclusions may be made on the basis of the two studies presented above.

In both cases, solution measurements of surface reactivity together with results from surface spectroscopies (eg, IR studies of MgO reactivity in water vapour) provided insufficient information to define critical mechanisms of reaction. Transmission electron microscope images of surfaces before and after chemical attack, in conjunction with the surface analytical data, proved essential in both studies in order to distinguish between alternate mechanisms.

The use of macroscopic single crystals for such studies is usually not practical, even when they can be obtained, since they are unsuitable for TEM examination. In order to be able to correlate surface analysis, solution measurements and surface imaging, a range of sample preparations may be necessary, as in the case of perovskite. Considerable care may be required to ensure that conditions of attack are equivalent for different samples.

In both the materials studied, further research is required to confirm the models developed for surface reactions in aqueous electrolytes. Subject to such confirmation, however, the current models provide a valuable basis for extrapolation to similar materials. Thus both MnO and CoO show an initial increase in dissolution rate, suggesting that surface restructuring or wetting may occur for these (semiconducting) oxides (Segall *et al*, 1988). In the case of perovskite, which is the least resistant of the titanate phases of Synroc, (a high level of radioactive waste immobilisation ceramic), the present model may provide an adequate basis for predictions of the long–term stability of the Synroc matrix against possible aqueous attack, and for the design of a suitable packing around the Synroc canisters in a waste repository (Pham *et al*, 1988).

Acknowledgements

The research projects on oxide reactivity and on Synroc have been supported by the Australian Research Grants Committee, the National Energy Research Development and Demonstration Council, the Australian Nuclear Science and Technology Organisation (ANSTO) and the Australian Institute of Nuclear Science and Engineering. The XPS/SIMS analyses were made using the Brisbane Surface Analysis Facility, which is supported by University of Queensland, Griffith University and Queensland Institute of Technology. Electron microscopy was carried out at the Analytical EM Facility supported by Griffith University and Queensland Institute of Technology, and at ANSTO. Discussions with colleagues at ANSTO, particularly Dr. T.J. White (now at University of Melbourne) is acknowledged with thanks. P.S.Turner wishes to thank the S.I.C.Mat–88 organisers for the invitation to present this paper and for financial support.

References

Blesa M.A., Regazzoni A.E. and Maroti A.J.G., 1988, Reactions of Metal Oxides with Aqueous Solutions, in *External and Internal Surfaces in Metal Oxides, Mat Sci Forum*, **29**, (L–C. Dufour, J Nowotny eds).

Blesa M.A., 1989, 'Reactions of Ceramic Oxides with Aqueous Solutions', Proceedings of S.I.C. MAT–88, (this volume).

Colbourn E.A. and Mackrodt W.C., 1982, 'Theoretical aspects of H_2 and CO Chemisorption on MgO Surfaces', *Surf. Sci* **117**, 571–580.

Colbourn E.A. and Mackrodt W.C., 1983, 'Irregularities at the (001) Surface of MgO', *Sol. State Ionics*, **8**, 221–231.

Colbourn E.A., Mackrodt W.C and Tasker P.W., 1983, 'The Segregation of Calcium Ions at the Surface of MgO: Theory and Calculation', *J. Mat Sci.*, **18**, 1917–1924.

Coluccia S., Tench A.J and Segall R.L., 1979, 'Surface Structure and Surface States in MgO Powders', J Chem. Soc. Far. I, **75**, 1769–1779

Coluccia S., Barton A. and Tench A.J., 1981, 'Reactivity of Low Coordination Sites on the Surface of MgO', *J. Chem Soc. Far I.*, **77**, 2203–2207.

Diggle, J.W, 1973, 'Dissolution of Oxide Phases', in *Oxides and Oxide Films* (J.W. Diggle, Ed) Vol.2, Marcel Dekker, New York, 281–386.

Fruhwirth O., Herzog G.W., Hollerer I. and Rachetti A., 1985, 'Dissolution and Hydration Kinetics of MgO', *Surf Tech*, **24**, 301–317.

Gorichev I.G. and Kyprianov N.A, 1984, 'Regular Kinetic Features of the Dissolution of Metal Oxides in Acidic Media', *Russ Chem. Rev.*, **53**, 1039–1061.

Jones C.F., 1988, M Phil. Thesis, Griffith University.

Jones C.F., Segall R.L.,Smart R.St.C. and Turner P.S., 1980, 'Size Distribution of MgO Smoke Particles', *Phil. Mag.* **A42**, 267–270.

Jones C.F., Segall R.L, Smart R.St.C. and Turner P.S., 1981, 'Initial Dissolution Kinetics of Ionic Oxides', *Proc. Roy. Soc. Lond.* **A 374**, 141–153.

Jones C.F., Segall R.L, Smart R.St.C. and Turner P.S. 1984a, 'Surface Structure and the Dissolution Rates of Ionic Oxides', *J. Mat. Sci Letts.* **3**, 810–812.

Jones C.F., Reeve R.A., Rigg R., Segall R.L, Smart R.St.C. and Turner P.S.,1984b, 'Surface Area and the Mechanism of Hydroxylation of Ionic Oxide Surfaces', *J. Chem. Soc. Far. I*, **80**, 2609–2617.

Jones C.F., Smart, R.St.C. and Turner P.S., 1989, in preparation.

Kastrissios, T., Stephenson, M., Turner, P.S. and White, T.J., 1987, 'Hydrothermal Dissolution of Perovskite: Implication for Synroc Formulation', *J. Amer. Cera. Soc.,* **70**[7], C144–146.

McCune R.C. and Wynblatt P., 1983, 'Calcium Segregation to a MgO (100) Surface', *J. Am. Ceram. Soc,* **66**, 111–117.

Moodie A.F. and Warble C.E., 1971, 'Electron Microscope Investigations of MgO Morphology and Surfaces', *J. Cryst. Growth*, **10**, 26–38.

Myhra, S., Segall, R.L., Smart, R.St.C., Stephenson, M., Turner, P.S. and White, T.J., 1985, 'Characterisation of Synroc Microstructure and Durability and its Influence on Final Disposal', pp. 433–444 in *Proceedings of the International Seminar on Radioactive Waste Products.* Ber. Kernforschungsanlage Juelich. Juel–Conf.–54.

Myhra,S., Smart,R.St.C., Turner, P.S.and Neall,F.B.,1987, 'Scientific and Technical Evaluation of Synroc Fabrication Program'. Final Report NERDDC Project. No. 673, Aust. Govt. Publishing Office.

Myhra, S., Smart, R.St.C. and Turner, P.S., 1988a, 'The Surfaces of Titanate Minerals, Ceramics and Silicate Glasses: Surface Analytical and Electron Microscope Studies', *Scanning Microscopy,* **2**, [2], 715–734.

Myhra S., Pham D.K., Smart R.St.C. and Turner P.S., 1988b, 'Scientific and Technical Evaluation of the Synroc Fabrication Program'. Final Report on NERDDP Project No 1026.

Nesbitt, H.W., Bancroft, G.M., Fyfe, W.S., Karkhanis, S.N., Nishijima, A., Shin, S., 1981, 'Thermodynamic Stability and Kinetics of Perovskite Dissolution', *Nature*, **289**, 358–362.

Paul A., 1982, *Chemistry of Glasses*, Chapman and Hall Ltd, London.

Pham D.K., Neall F.B., Myhra S, Smart R.St.C. and Turner P.S. 'Dissolution Mechanisms of $CaTiO_3$ – Solution Analysis, Surface Analysis and Electron Microscope Studies – Implications of Synroc', *Proc. Intl. Conf. on Radioactive Waste Management*, Berlin, October 1988.

Segall R.L., Smart R.St.C. and Turner P.S 1978, 'Ionic Oxides – Distinction between Mechanism and Surface Roughening Effects in the Dissolution of MgO', *J. Chem. Soc. Far I,* **74**, 2907–2912.

Segall, R.L., Smart, R.St.C.and Turner, P.S., 1988, 'Oxide Surfaces in Solution', pp. 527–576 in *Surface and Near–Surface Chemistry of Oxide Materials.* Edited by J. Nowotny and L.C. Dufour. Elsevier Science Publishers B.V., Amsterdam.

Vermillyea D.A., 1966, 'The Dissolution of Ionic Components in Aqueous Media', *J. Electrochem. Soc.*, **113**, 1067–70.

Vermilyea D.A., 1969, 'Dssolution of MgO and $Mg(OH)_2$ in Aqueous Solution', *J. Electrochem. Soc.*, **116**, 1179–83.

THE ROLE OF GRAIN BOUNDARY MODIFICATIONS IN THE THERMAL DECOMPOSITION OF Mn-FERRITES

Jeffrey H. Boy and Gerald P. Wirtz
Department of Materials Science and Engineering
University of Illinois at Urbana-Champaign
105 S. Goodwin - 204 Ceramics Building
Urbana, Illinois 61801 USA

Abstract. Manganese ferrite samples with and without minor (<1 Wt%) additions of SiO_2 and $CaCO_3$ quenched in air from the sintering temperature yielded a single spinel phase. Thermogravimetric analysis in a magnetic field was used to determine Curie temperatures of magnetic phases present. The Curie temperature provides an indication of differences in bulk composition and/or cation distribution between samples. Samples with and without minor additions were subjected to an anneal at temperatures up to 600°C in argon and air. There was little or no change in the Curie temperature for samples annealed in argon. However, samples without additions annealed in air showed a significant change in Curie temperature indicating decomposition of the spinel. This decomposition was suppressed by <1 Wt% SiO_2 and $CaCO_3$ additions indicating that the thermal stability of manganese ferrite is enhanced by the suppression of grain boundary diffusion by minor dopants concentrated at the grain boundaries.

Introduction

The role of grain boundaries in the control of electrical and magnetic properties of ferrites,[1] and in other ceramic material properties, is well documented.[2,3] The modification of the properties of grain boundaries in ferrites by the addition of additives is recognized in the prevention of eddy current losses which are the principal detriment to the high frequency magnetic properties.[4] The simultaneous addition of SiO_2 and $CaCO_3$ added as sintering aides to enhance densification results in their concentration at the grain boundaries. The resulting resistive grain boundaries will form an insulating layer separating the more conductive grains and inhibit macroscopic eddy currents.

The oxidation state and distribution of the cations on the spinel lattice determine the electrical conductivity of the bulk grains in Mn-ferrite microstructures.[5] These conditions are known to be a function of the temperature[6] and oxygen partial pressure.[7] However, the role of the grain boundaries in the kinetic processes occurring in the bulk is little understood. The aim of the present paper is to determine the

L.-C. Dufour et al. (eds.), Surfaces and Interfaces of Ceramic Materials, 691–699.

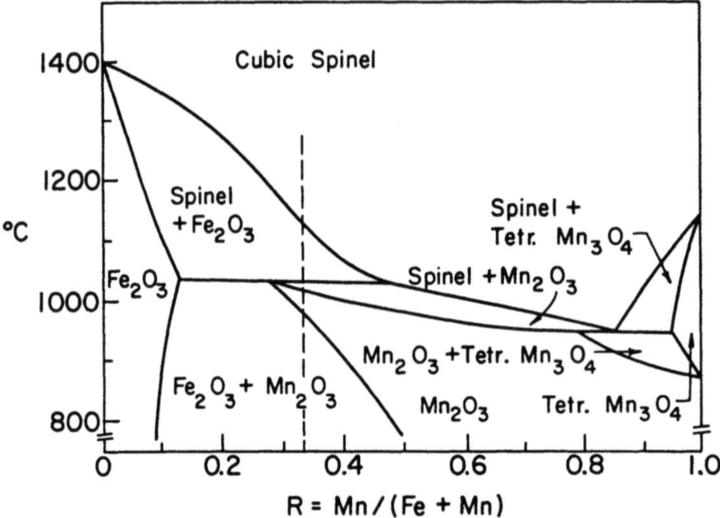

Figure 1. The air isobar for the Mn-Fe-O system.

effect of differences in grain boundaries on the thermal processes occurring in the bulk. Through better understanding of these interactions, it is ultimately hoped to gain better control of the properties of ferrites through control of the grain boundary properties.

The air isobar in the Mn-Fe-O system is shown in Figure 1. The phase diagram of Muan[8] is modified according to Wickham[9] who reported a narrower composition range for the two-phase region containing Mn_2O_3 bixbyite and tetragonal Mn_3O_4 as well as for the cubic spinel and tetragonal Mn_3O_4 two-phase region. The cation fraction corresponding to the $MnFe_2O_4$ composition is shown by the dotted line. At the sintering temperature, typically between 1200°C and 1300°C, the cubic spinel phase is stable. However, equilibrium cooling in air will result in the separation of the spinel into the Fe_2O_3 hematite phase and the Mn_2O_3 bixbyite phase. In order to retain the high temperature spinel phase, a different isobar must be utilized which shifts the stability field of the spinel to lower temperatures. The most common procedure is to cool in an atmosphere containing a lower oxygen partial pressure than air.[10]

Experimental Procedure

Manganese ferrite samples were prepared using standard ceramic processing techniques. Oxides and carbonates of the starting materials were mixed in a ball mill, dried, calcined at 1000°C, remilled, during which equal weight percentages of SiO_2 and $CaCO_3$ were added, uniaxially pressed, and sintered at 1250°C for 4 h. The samples were either sintered and cooled in flowing nitrogen or sintered in air and air-quenched from the sintering temperature. X-ray diffraction analysis, using Cu Kα radiation and a single crystal monochromator, indicated the presence of only the manganese spinel phase in both cases.

The magnetic Curie temperature varies markedly with composition and cation distribution between octahedral and tetrahedral sites for the Mn_3O_4-Fe_3O_4 system. The Curie temperature, therefore, provides a highly sensitive measure of changes in the crystal chemistry of the magnetic phase present. The use of a thermogravimetric analyzer has been shown to provide a ready technique to measure the Curie temperature.[11]

Sintered specimens were placed in the weighing pan of a thermogravimetric analyzer.* A permanent magnet was placed on top of the horizontal furnace assembly of the analyzer. The apparent weight of the sample in the magnetic field of the permanent magnet was recorded as a function of temperature for a heating rate of 10°C/min. When the sample was heated through the Curie temperature, its apparent weight increased due to the loss of spontaneous magnetization. The magnetic Curie temperature was arbitrarily defined as the maximum of the temperature derivative of the apparent weight change.

Results and Discussion

Figure 2 shows a TGA plot for a nitrogen sintered $MnFe_2O_4$ sample without additions. The magnetic transition is sharply defined and the Curie temperature was 289°C. The cation distribution of $MnFe_2O_4$ is known to be dependent on the thermal history of the sample.[12] Differences in the initial Curie temperatures for

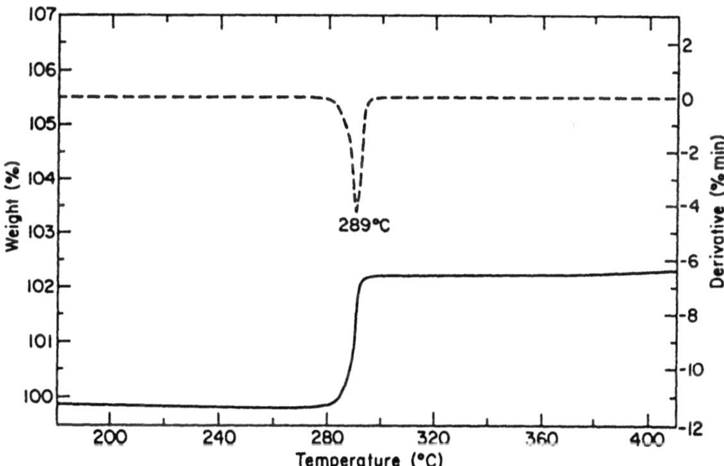

Figure 2. TGA in a magnetic field of $MnFe_2O_4$ showing transition at the magnetic Curie temperature.

*DuPont Model 1090 Thermal Analysis System

TABLE I
MAGNETIC CURIE TEMPERATURE
ARGON ANNEAL

% SiO_2 & $CaCO_3$ Additions	Thermal Anneal	Initial T_c °C	Post-Anneal T_c °C
No addition	500°C/1 h	300	320
0.2	500°C/30 min	281	270
0.3	500°C/1 h	284	284
0.5	500°C/30 min	295	285

samples with different amounts of dopants, Table I, were attributed to differences in the thermal histories due to the difficulty in duplicating the cooling cycle for different firings. The effect of the thermal annealing would be to alter the cation distribution from that initially fixed by the kinetics of the cooling to the equilibrium distribution at the annealing temperature. Following the thermal anneal, the samples were cooled in the furnace of the thermal analysis unit during which additional cation redistribution would be expected to occur. Differences in sample size and geometry would affect the thermal diffusion path and the resultant thermal history of the TGA specimen, and may account for the observed differences in the post anneal Curie temperatures. The magnetic transitions for all samples subjected to a thermal anneal in argon occurred over a very narrow temperature range indicating no decomposition of the magnetic phase.

Figure 3 shows the TGA plot of nitrogen sintered $MnFe_2O_4$ sample without additions which was annealed in air at 470°C. When the sample was initially heated

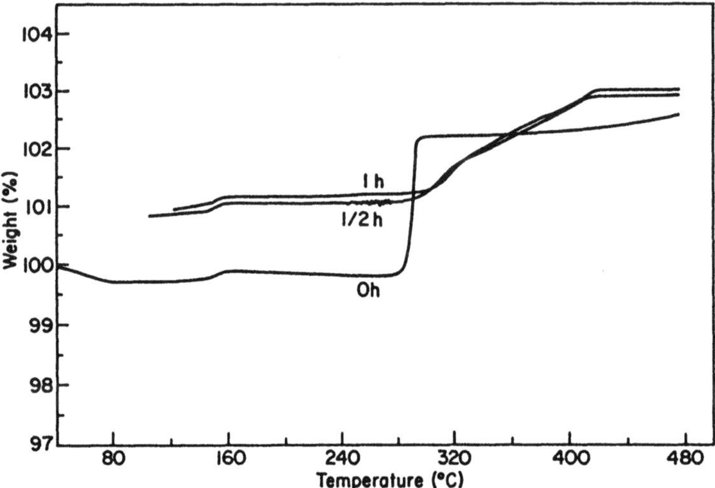

Figure 3. TGA in a magnetic field of a $MnFe_2O_4$ sample with no additions annealed in air at 470°C for a total of 1 h.

in air, the Curie temperature showed a sharp transition at 285°C. During the initial heating at temperatures above the Curie temperature, there was a slight upward bend to the TGA plot indicating that the sample was beginning to gain weight. When the sample was left in the TGA furnace for 30 minutes, the total weight gain was approximately 1%. This weight gain is consistent with the oxidation of the spinel.[13,14] The magnetic transition was measured on cooling and occurred over a wide temperature range corresponding to a distribution of compositions in the magnetic phase. The shift of the magnetic transition to higher temperatures would correspond to an increase in the Fe content of the magnetic phase. This could occur by the diffusion of manganese ions out of the grains, which would leave behind a distribution of compositions with various concentrations of Fe, and a range of Curie temperatures.

Following the thermal analysis, the samples were analyzed by X-ray diffraction. For the $MnFe_2O_4$ sample with no additions annealed in air at 470°C for 1 h, the primary phase was the manganese spinel with bixbyite as a minor phase. For the sample annealed in air at 470°C for 4 h, in addition to the spinel, both bixbyite and hematite were present as minor phases.

The TGA plot of a nitrogen sintered $MnFe_2O_4$ sample without additions which was annealed in air at 600°C for 1 to 4 h is shown in Figure 4. The Curie temperature was 290°C on the initial heating and well defined. During the 1 h anneal at 600°C, there was a total weight gain of approximately 1.5%, due to oxidation of the sample. The Curie temperature measured on cooling showed a broadening of the magnetic transition. This same sample was immediately subjected

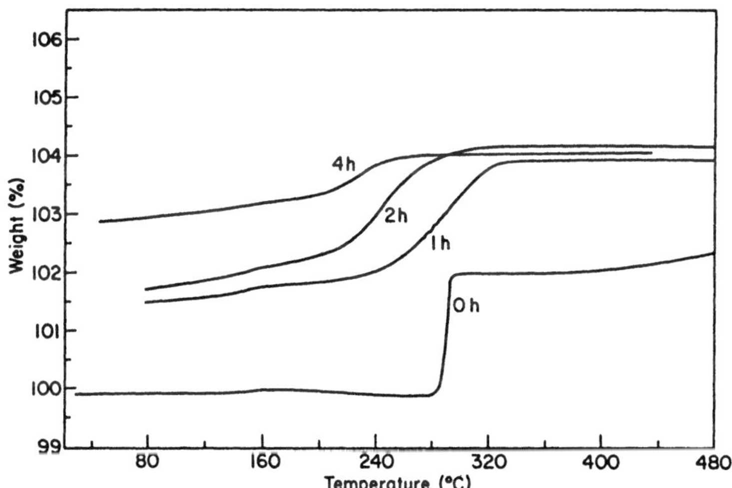

Figure 4. TGA in a magnetic field of a $MnFe_2O_4$ sample with no additions annealed in air at 600°C for a total of 4 h.

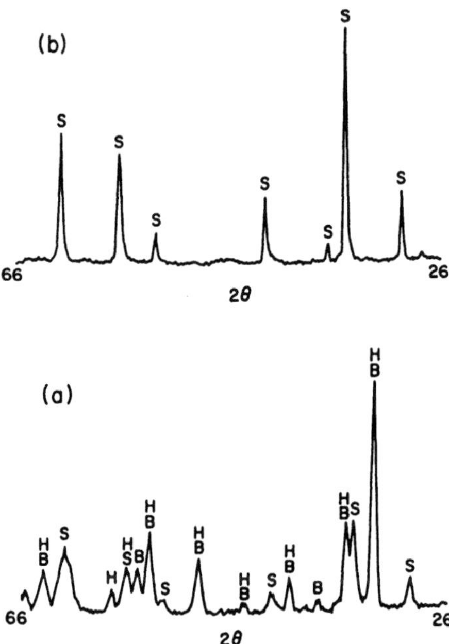

Figure 5. X-ray Diffraction pattern for $MnFe_2O_4$ annealed in air at 600°C for 4 h; (a) no addition, (b) 0.3% SiO_2 and CaO_3 additions (S=Spinel, H=hematite, B=Bixbyite).

to subsequent anneals at 600°C to total 2 and 4 h. The magnetic transition shifted to lower temperatures on cooling. This indicated that the remaining magnetic phase was manganese rich relative to the stoichiometric composition. During the initial heating, there was more than a 2% change in the weight at the Curie temperature. After the 4 h anneal, the weight change associated with the magnetic transition was less than 1%. The decrease in the weight change indicated a decrease in the quantity of the magnetic phase present and was attributed to the decomposition of the spinel into non-magnetic phases.

X-ray diffraction analysis for the sample annealed in air at 600°C for 2 h showed the spinel to be the primary phase with bixbyite a major second phase and hematite a minor phase. For the sample annealed in air at 600°C for 4 h, hematite was the major phase, with spinel as a major second phase and bixbyite as a minor phase, Figure 5a.

During the oxidation of the sample, both manganese and iron are diffusing out of the spinel phase. The shift of the magnetic transition to either higher or lower temperatures is a direct indication of the chemical composition of the remaining magnetic phase. The fact that the remaining magnetic phase is either iron rich or manganese rich, depending on the annealing temperature, could be due to the relative temperature dependence of the diffusion coefficients of the two species.

TABLE II
MAGNETIC CURIE TEMPERATURE
AIR ANNEAL

SiO_2 & $CaCO_3$ Additions Wt%	Thermal Anneal	Initial T_c °C	Post-Anneal T_c °C
0.2	500°C/2 h	284	276
0.3	600°C/2 h	287	287
0.5	600°C/2 h	297	280

The $MnFe_2O_4$ sample containing up to 0.5 Wt% SiO_2 and $CaCO_3$ were also subjected to thermal anneals in flowing air at temperatures up to 600°C for 2 h, Table II. The magnetic transition was sharply defined indicating no decomposition of the magnetic phase during the thermal anneal in air and is shown in Figure 6 for a sample with 0.3% SiO_2 and $CaCO_3$ additions. The differences in the Curie temperatures for samples with different dopant levels were attributed to slight differences in the thermal history of the samples. The suppression of the decomposition of the Mn-ferrite sample with 0.3% SiO_2 and $CaCO_3$ additions, subjected to a thermal anneal in air at 600°C for 4 h, was confirmed by XRD analysis which detected only the spinel phase, Figure 5b. This is in dramatic contrast to samples without additions which decomposed under similar thermal

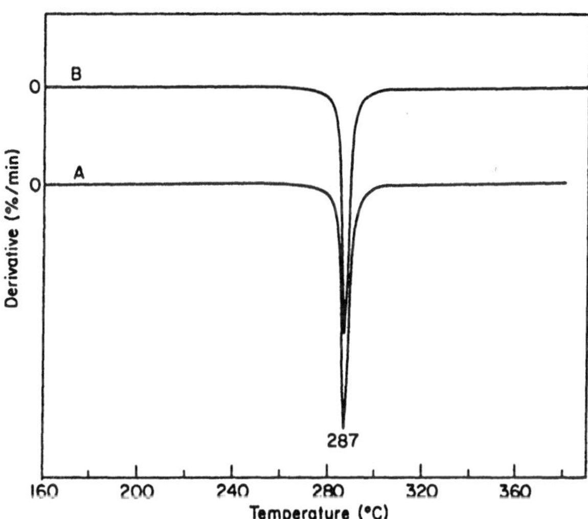

Figure 6. TGA derivative of a $MnFe_2O_4$ sample with 0.3% SiO_2 and $CaCO_3$ additions annealed in air at 600°C; A) initial heating, B) 2 h anneal.

treatments. Therefore, the modification of the grain boundaries in Mn-ferrite, by the addition of Si and Ca, decreases the grain boundary diffusion of oxygen, and inhibits the oxidation and associated decomposition of the ferrite grains when subjected to a post-sintering thermal anneal in air.

Conclusions

Manganese ferrite decomposed when subjected to post-sintering anneals at temperatures below 600°C in air. The addition of less than 1 Wt% of SiO_2 and $CaCO_3$ inhibited this phase decomposition. The role of Si and Ca additives in the control of the thermal stability of ferrites may be as important to the overall electrical properties as the previously recognized role in the control of the electrical conductivity of the grain boundaries and the prevention of eddy current losses.

References

1. C. G. Koops, 'On the Dispersion of Resistivity and Dielectric Constant of Some Semiconductors at Audio Frequencies,' *Physical Review,* **83**, No. 1, 121-124 (1951).

2. 'Grain Boundary Phenomena in Electronic Ceramics,' Advances in Ceramics **Vol 1**, Ed. by L. M. Levinson, The American Ceramic Society, Inc., Columbus, OH (1981).

3. 'Additives and Interfaces in Electronic Ceramics,' Advances in Ceramics **Vol 7**, Ed. M. F. Yan and A. H. Hauer, The American Ceramic Society, Inc, Columbus, OH (1983).

4. K. Ishino and Y. Harumiya, 'Development of Magnetic Ferrites: Control and Application of Losses,' *Ceramic Bulletin*, **66**, No. 10, 1469-1474 (1987).

5. F. K. Lotgering, 'Semiconduction and Cation Valencies in Manganese Ferrite,' *J. Phys. Chem. Solids*, **Vol 25**, 95-103 (1964).

6. S. E. Dorris, 'The Electrical Properties and Cation Distribution of the Fe_3O_4-Mn_3O_4-Co_3O_4 System,' Ph.D. Dissertation, Northwestern University (1986).

7. H. L. Tuller, H. I. Yoo, W. Kehr and R. W. Scheidecker, 'Electrical Property-Phase Equilibria Correlations in Manganese-Zinc Ferrites,' *Proceeding Fourth International Conference of Ferrites, Part 1,*Advances in Ceramics, **Vol 15**, Ed. by F. Y. Young, American Ceramic Society, Columbus, OH, 315-324 (1985).

8. A. Muan and S. Somiya, 'The System Iron Oxide-Manganese Oxide in Air,' *American Journal of Science,* **Vol 260**, 230-240 (March 1962).

9. D. G. Wickham, 'The Chemical Composition of Spinels in the System Fe_3O_4-Mn_3O_4,' *J. Inorg. Nucl. Chem.*, **Vol 31**, 313-320 (1969).

10. J. M. Blank, 'Equilibrium Atmosphere Schedules for the Cooling of Ferrites,' *J. Appl. Phys.*, **Supplement to Vol 32**, No 3, 378s-379s (1961).

11. D. S. Walters and G. P. Wirtz, 'Kinetics of Cation Ordering in Magnesium Ferrite,' *J. Am. Cer. Soc.*, **Vol 54**, No. 11, 563-566 (1971).

12. Z. Jirak, 'Temperature Dependence of Distribution of Cations in $MnFe_2O_4$,' *Czech. J. Phys. B,* **24** 642-647 (1974).

13. T. Yamaguchi and T. Kimura, 'Kinetic Studies on the Precipitation of Hematite from Iron-Rich Spinel Solid Solutions,' *J. Am. Cer. Soc.,* **Vol 59** No. 7-8, 333-335 (1976).

14. M. Dixon, T. S. Stakelon, and R. C. Sundahl, 'Complex Permeability Variations of Iron Excess CoNiZn Ferrites with Annealing Temperature,' *IEEE Trans. on Magnetics,* **Vol Mag 13**, No. 5, 1351-1353 (1977).

TEMPTATIVE MODELING OF SURFACE REACTIVITY WITH OXIDIZING-REDUCING MIXTURES ON RUTILE $TiO_{2-\delta}$

F. MORIN[*] and L.C. DUFOUR[**]
[*]Institut de Recherche d'Hydro-Québec
B.P. 1000, Varennes, QUEBEC, Canada J0L 2P0
[**]Réactivité des Solides, UA23 CNRS,
Université de Bourgogne, B.P. 138, 21004 Dijon, France

Rutile bears some unique features in regard to oxygen transfer with the outer atmosphere, at high temperatures. It possesses very high chemical diffusivity. This diffusivity becomes apparent while performing reequilibration experiments in a final atmosphere of pure oxygen. Conversely, the oxidizing-reducing reaction with CO/CO_2 mixtures at the rutile surface is always a slow process. These two entirely distinct phenomena are clearly illustrated in Fig. 1. c_r represents the degree of reequilibration in the course of a single run and $\tilde{D}_{app}$ is the apparent chemical diffusion coefficient ob-

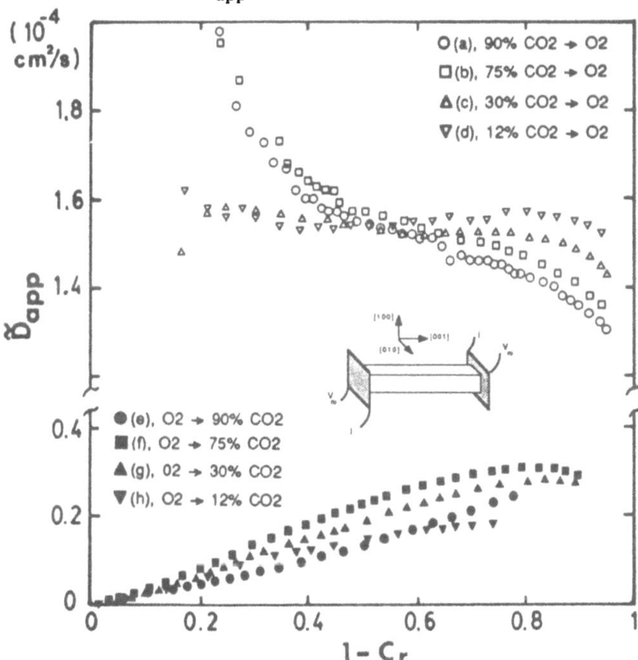

Fig.1 Chemical diffusivity calculated on the basis of a diffusion-controlled process versus the degree of reequilibration. Relaxation experiments in either pure oxygen or CO/CO_2 atmospheres at 1100 °C. 4x4x18 mm^3 TiO_2 single crystal.

L.-C. Dufour et al. (eds.), Surfaces and Interfaces of Ceramic Materials, 701–703.

tained by solving with the usual analytical solution [1,2]. For the latter case (lower curves in Fig.1), the assumption of a constant composition throughout the oxide volume during chemical reequilibration experiments is thus validated. This allows the surface reaction kinetics to be followed readily using this experimental method.

Electrical conductance was used to follow reequilibration kinetics, in CO/CO_2 between 950 and 1150 °C, on two types of $TiO_{2-\delta}$ single crystals initially prepared to preferentially develop either the (001) surface or {100} surfaces. Details of these experiments are extensively reported elsewhere [5]. There is a limited effect of the sample size and orientation on the overall kinetics. Platinum chemically deposited on the initial surfaces exerts a clear promoting effect on the surface reaction, this effect being more pronounced for the reducing reaction. Nevertheless, the catalytic effect of platinum may be seriously inhibited due a likely encapsulation effect. This was corroborated by additional analysis with RHEED, AES and SEM which evidenced facetting of the initial (001) surface.

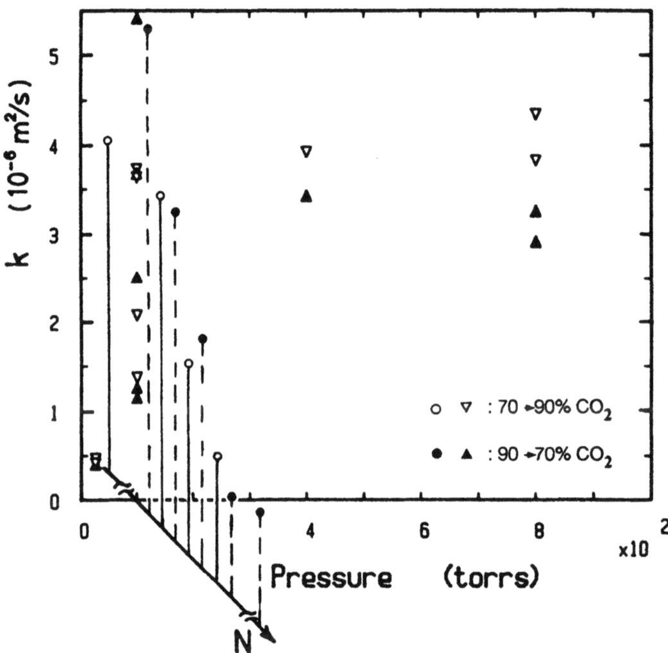

Fig.2 Dependency of the rate constant k ($\propto$ d ln c_r/d t) on the total CO/CO_2 pressure with systematic cycling of a rutile sample. The variability of k for successive runs (N=1, 2, 3,...) at 100 torrs becomes obvious as these points are duplicated on a third, protruding axis (N=0 for an initial oxygen anneal).

A semilogarithmic law expressed as ln $c_r \propto -t$ is rather closely followed in oxidizing-reducing mixtures. Classical models, like those by Wagner or Grabke [3,4], have also been handled in order to verify their sustaining hypotheses. At present, it seems more likely that the slow reaction step would not be between the gas phase and the oxide surface but rather between two different species at the oxide surface.

Finally, a clear and unexpected promoting effect of previous oxygen anneals has been found in the course of reequilibration experiments in CO/CO_2. This is illustrated in Fig. 2. Systematic cycling of a rutile sample was performed between two constant CO/CO_2 ratios for different pressures. Since k was determined at a single degree of reequilibration for all runs, the reported variations are relatively independent of the rate laws under consideration. k tends to stabilize after a certain number of cycles. Taking this phenomenon into consideration should help in removing an appreciable portion of the uncertainties related with many experiments on surface reactivity. Work is presently underway in this direction.

References

1- A.B. Newman, *Trans. AIChE*, **27** (1931) 310
2- H. Dunwald and C. Wagner, *Z. Phys. Chem.*, **B24** (1934) 53
3- C. Wagner in F.S. Pettit, R. Yinger and J.B. Wagner, *Acta Met.*, **8** (1960) 617
4- H.J. Grabke, *Ann. N.Y. Acad. Sci.*, **213** (1973) 110
3- F. Morin and L.C. Dufour, *Solid St. Ionics*, under press

NANOPHASE CERAMICS, MEMBRANES AND ION IMPLANTED LAYERS

A.J. Burggraaf, K. Keiser and B.A. van Hassel
University of Twente, Department of Chemical Technology
Lab. for Inorganic Chemistry, Materials Science and Catalysis
P.O. Box 217, 7500 AE Enschede, The Netherlands

ABSTRACT. Several types of materials with a nanoscale structure will be discussed. The most important synthesis methods are briefly reviewed with a certain focus on sol-gel methods.
Ceramic membranes with a thickness of 1-10μm and a mean pore size which can be varied between 2,5 and 50 nm shall be treated into some detail. A focus will be given to techniques which modify the internal pore structure to decrease the effective pore size or to change the chemical character. The potential for their use in separation processes and in chemical processes is assessed.
Ion implantation as a technique to modify ceramic surfaces will be briefly reviewed and an overview of the changes in surface properties is given.

1. INTRODUCTION AND GENERAL CONSIDERATIONS

Nanocrystalline materials can be thought of as a polycrystalline state of matter in which the size of the individual crystallites is of the order of 1-50 nanometer. In some cases these crystallites form the elementary building units of the microstructure, in other cases they form a substructure of ceramic grains.
Characteristic is the large volume fraction of the atoms located in grain boundaries of interfacial regions. In grains with a size of 5 nm and a thickness of the grain boundary region of 1 nm already 50% of the atoms is located in that grain boundary region. In this region the composition, structure (coordination) and chemical binding is different from that of the bulk. Consequently, quite different properties must be expected for nanocrystalline materials. This will especially be the case when the micro-structural scale is comparable to the physical dimensions that characterise the phenomena (domain size, Debye length in surface region, dimensions of plastic zone around a crack tip), when the property is dependent on a large volume fraction of the interface or on the very small radius of curvature. Some illustrative results are given below.

L.-C. Dufour et al. (eds.), Surfaces and Interfaces of Ceramic Materials, 705–724.

Recently Gleiter [1] showed that properties like density, specific heat and thermal expansion of metals (Fe, Cu, Ti) are in the nanoscale structure quite different from those in the bulk. Nanoscale composites could be produced in Au-CaF, Pb-Al_2O_3 and Au-polyethylene systems. TiO_2 ceramics with a certain ductility at room temperature (under high pressure) were reported. Some examples of exotic non-additive properties which can be obtained in heterogeneous mixtures are given by Yanagida et al. [2]. A heterogeneous (porous) mixture of (acid) TiO_2 and (basic) $MgCr_2O_4$ is electrically much more sensitive to humidity than each of the components. The same holds for mixtures of CuO and ZnO, the hetero contact points giving the material sensitivity to humidity and selectivity for CO oxidation. Some novel micro and nanostructures which can be obtained by advanced processing are discussed by Bender et al. [3]. Interesting results are mentioned in some toughened materials with improved mechanical properties (Al_2O_3-MgO-TiO_2, Al_2O_3-SiO_2-TiO_2 and Al_2O_3-ZrO_2). Meta-stable structures could be obtained from rapid quenching of droplets of the melt phase (obtained by different methods [3]) and by ion implantation [4].
Ceramic membranes with nanoscale structures are reported for monophasic, diphasic and modified materials by Burggraaf et al. [5,6] and for xerogels by Komarmeni et al. [7]. These materials have interesting mass transport properties and especially the combination of permselectivity and (catalytic) reactivity seems very promising. This will be shown later.
In this paper a brief review will be given of the different methods to obtain nanoscale structures in ceramic materials. Some typical results will be mentioned, a more thorough treatment will be postponed to a later paper. The latter sections of this paper will be focussed on recent results in ceramic membranes.

2. SYNTHESIS METHODS AND PRINCIPLES

The most important synthesis methods are summarized below.
Starting with solids, spinodal decomposition of glasses by heat treatment followed by acid leaching results in porous glasses (Vycorglass) with a pore sizes around 4 nm. This technique is discussed too by Hurt and Viechnicki [8] to initiate the nucleation of a crystalline phase in the synthesis of glass ceramics. For the ZrO_2- Al_2O_3-SiO_2 system a spinodal separation wavelength for the resulting two phase mixture down to 2.5-5 nm could be obtained.
A second method makes use of ordering reactions in solid solutions resulting in micro (nano) domain formation. Results are discussed by Burggraaf et al. [9] and by Randall et al. [10]. For the ferro electric perovskite structure $Pb(Sc_{0.5}M_{0.5})O_3$ with M = Nb or Ta the disordered Sc/M sublattice could be ordered by heat treatment starting from domains with a size of about 20 nm which grow during further annealing.
Another example is the formation of nanodomains with pyrochlore

structure in the oxygen ion conductive fluorite structures $(Tb_xGd_{1-x})_2Zr_2O_7$ [11]. The domains have an average diameter ranging from a value smaller than 10 nm for x = 0.2 to 100 nm for $Gd_2Zr_2O_7$. They grow during annealing, have interesting catalytic properties in oxidation reactions and finally form antiphase boundaries with rapid oxygen diffusion pathways.

A second group of methods starts with the nucleation of the new phase in a gas or melt, followed by condensation and/or rapid cooling (quenching). Only a few examples will be given below.

In this way Gleiter et al. [23] reports the evaporation/quenched condensation of metal particles followed by cold compaction of the powder. In the case of Ti, the metal is oxidized before compaction. The final oxide compact should have a particle size of 1-10 nm and behave more or less plastically.

Interesting results are briefly reported by Bender et al. [12] for powders obtained by splat (or anvil) cooled droplets of ceramic melts, followed by a consolidation treatment.

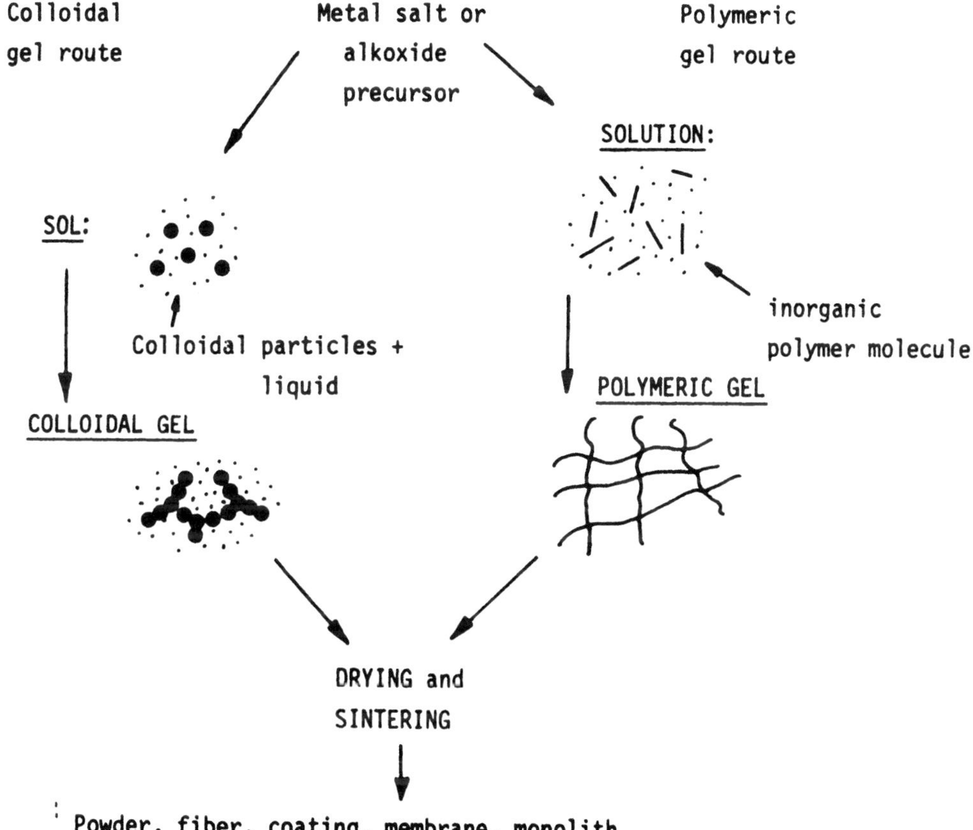

Figure 1. Scheme of sol-gel routes.

This results in metastable phases and a duplex type of micro (nano) structure for Al_2O_3-MgO-TiO_2, Al_2O_3-SiO_2-TiO_2 and Al_2O_3-ZrO_2 materials. Kalonji [13] reports for the eutectic composition in the Al_2O_3-ZrO_2 system a lamellar structure with a spacing of 15-100 nm. The molten droplets are in these cases obtained by plasma or laser-melting of the oxides. In the transformation toughened system ZrO_2-Y_2O_3 a nanoscale substructure is observed within each grain, consisting of the tetragonal prime (T^1) phase and forming antiphase boundaries and very clean grain boundaries [13].
The third group of methods starts with the synthesis of a colloidal suspension (sol) and/or a gel from which are produced powders, coatings, monolithic xerogels and ceramic membranes by subsequent heat treatment and/or crystallisation. There exist many variations and its importance in the production of advanced ceramics is reflected in the rapidly expanding literature and congress activity [14-17].
The main characteristic of this group of methods is its ability to form very small elementary particles (3-10 nm) which in the liquid are loosely bound in larger aggregates (up to 1 μm). During further (appropriate) processing the aggregates can be broken down and a more or less regular and uniform packing results in powder particles or layers with mean pore sizes down to 2.5 nm. The advantages of the method are tabulated below and include the synthesis of:

1. new glasses or ceramic compositions which cannot obtained by conventional processing;
2. novel phase distributions and/or tailored heterogeneity;
3. exceptional (chemically) homogeneous and pure materials;
4. materials with strongly decreased sintering temperature and consequently small grain size;
5. materials with mixed organic and inorganic functionalities.

The sol-gel process can be divided in two main routes which are shown in Figure 1. In both cases a precursor is hydrolysed while simultaneously condensation/polymerisation reactions occur. In the colloidal route the hydrolysis rate is fast by reacting the precursor with excess water and a precipitate of gelatineous hydroxide particles is formed which is peptised in a subsequent step to a stable colloidal suspension. The elementary particle size ranges, depending on the system and the processing, from 3-15 nm and these particles form loosely bound agglomerates with diameters ranging from 5-100 nm. The size of those agglomerates can be decreased e.g. by ultrasonification of the suspension. By increasing the concentration and/or manipulation of the surface (zeta) potential of the sol particles the suspension is transformed now into a gel structure consisting of interlinked 'chains' of particles (see Figure 1). The packing density of such structures can vary from very loose to rather dense depending on the zeta potential. According to Aksay and Schilling [18] colloidal suspensions behave like molecular systems and a phase diagram can be constructed.

In the polymeric route the hydrolysis rate is kept low by adding successively small amounts of water. The precursor is partly hydrolysed, OH-groups attached to metal atoms are formed and these hydroxyl groups give rise to condensation reactions yielding a viscous solution of inorganic-organic polymeric molecules. Further condensation/polymerisation yield a strongly interlinked gel network, containing a certain percentage of organic material. The reactions can be catalysed by acids or bases.

In both routes the gels are dried. This is a critical process step which strongly influences the structure of the obtained xerogel particles and its properties (pore size distribution, porosity, internal surface). In most cases the aim is to obtain soft agglomerates which can be packed more efficiently by subsequent application of pressure or can be dispersed again in solution. Finally a heat treatment (calcining or sintering) converts the hydroxylated oxide to the final chemical and mechanical stable oxidic microstructure.

Binary systems can be obtained e.g. by mixing of sols of different metals [19] or by the cohydrolysation or copolymerisation of mixtures of alkoxides [20,21].

To obtain very intimately dispersed (homogenous) mixtures the polymeric gel route seems to be preferable because here the different metal ions in different precursors are linked before they can separate. The ultimate precursor should be then an alkoxide containing two different metal atoms in one precursor molecule. A process coming close to this ideal situation is reported by Morgan et al. [22]. For the Al_2O_3-ZrO_2 system he obtained an ultimately dispersed system. Powders of solid solutions of ZrO_2 and Y_2O_3 (3 mol%) prepared with a colloidal sol route and with elementary particle diameters of about 7 nm are reproduced by Theunissen et al. [24]. These powders are sintered to dense ceramics with a mean grain diameter of 50 nm. This means that a nanoscale ceramic is produced without pressure sintering [24].

3. CERAMIC MEMBRANES

3.1 Synthesis and Structure.

Ceramic membranes consist essentially of a porous support of a few millimeter thickness, a porous intermediate layer of 10-100 µm thickness with pores of 0.05-0.5 µm and the proper membrane (separation layer) with a thickness of 1-5 µm (Fig. 2). This last layer must have a (very) sharp pore distribution and a mean pore size down to 2.5 nm (Fig. 2). This separation layer can be further modified (Section 4). They can be made in different ways. The most frequently used methods are slibcasting onto the support system with a sol of the membrane material (which is produced by the colloidal route [25-27]) or by film coating with precursor material produced by the polymeric gel route [28,29]. See Figure 1.

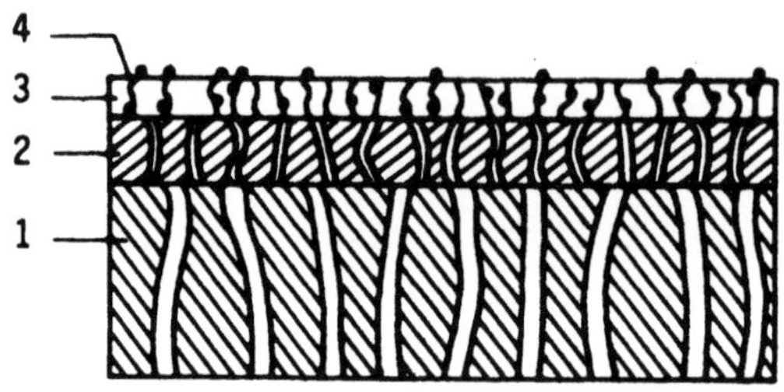

1. porous support (1-15 mu pores)
2. intermediate layer(s) (100-1500 nm)
3. separation (top)layer (3-100 nm)
4. modification of separation layer

1+2	is microfiltration or 'primary' membrane
1+2+3	is ultrafiltration or 'secondary' membrane
1+2+3+4	is hyperfiltration or/and gas separation membrane

Figure 2, Schematic presentation of a multilayer asymmetric membrane system.

During careful drying, capillary forces causes compaction of the gel and a regular packing of particles (size 5-10 nm) is formed. Finally calcining transforms the gel material into oxidic particles and a stable membrane is formed. An essential feature is that the membrane is crack- en defect free. The smaller the elementary particles and the weaker the agglomerates in the sol are, the smaller the pore size is and the sharper its distribution will be. Therefore the sol-gel chemistry is an important step.

In this way supported and non-supported toplayers are produced from γ-Al_2O_3, TiO_2, CeO_2, ZrO_2 and binary combinations of them [25, 26, 30]. Some typical properties are shown in Tables 1 and 2. The non-supported membranes have a thickness which can be varied between 1 and 200µm and have a micro structure which is similar to the supported ones [25, 26, 35]. The elementary particle shape is important. The most frequently produced membranes consist of alumina. It is produced from plate shaped boehmiet particles which are packed into a structure consisting of parallel packed plates (card pack) shown by Burggraaf, Keizer and co-workers [25,26]. In the calcining step the boehmiet is transformed into an isomorphous γ-alumina layer (temp. 450-650°C) or θ-Al_2O_3 layer (higher temp.).

The mean pore diameter amounts about 3 nm at 600°C and this value increases gradually with temperature to 6 nm at 900°C. The time of calcination has some effect and causes a slow increase untill in practice a limiting value is obtained in pore diameters of 3.2 nm after treatment at 600°C for 34 hours.

TABLE 1: Structural properties of non-supported one-component membranes

material	temp./time hrs	°C	pore diam./poro. nm	%	BET-surface m^2/g	cryst.size nm
γ-AlOOH	34	200	2.5*	41	315	≃5x25(-50)#
γ-AlOOH^x	5	300	5.6*	47	131	≃7x300^x
γ-Al_2O_3	34	500	3.2*	50	240	≃5x25(-50)
γ-Al_2O_3^{x}	5	550	6.1*	59	147	≃7x300^x
γ-Al_2O_3	34	800	4.8*	50	154	
θ-Al_2O_3	34	900	5.4*	48	99	
α-Al_2O_3	34	1000	78°	41	15	
TiO_2(a)	3	300	3.8°	30	119	sol:5^b sol:10-40^c
TiO_2(a+r)	3	400	4.6°	30	87	
TiO_2(a^d)	3	450	3.8°	22	80	
TiO_2(r)	3	600	20°	21	10	
CeO_2	3	300	≃2	15	41	sol:5^b
CeO_2	3	400	≃2	5	11	sol:10^c
CeO_2	3	600	nd^f	1	1	

* slit shaped pore model ;° cylinder shaped pore model
\# plate shaped crystals; ^{x} sol prepared at 200°C in autoclave
a anatase phase; r rutile phase; ^{d} titania stabilized with SO_4^{2-}-ions
b primary particle(TEM); ^{f} not detectable
^{c} agglomerated sol particles (10 nm: use of ultrasonic waves,laser scattering)

TABLE 2: Structural properties of non-supported binary composite membrane system

material	temp./time hrs	°C	pore diam./porosity nm	%	BET-surface m^2/g	cryst.size nm
Al_2O_3/CeO_2 (35 wt.%)	3	450	2.4*	39	164	sol:10 and 50^a calc.:5(CeO_2)
Al_2O_3/CeO_2 (35 wt.%)	3	600	2.6	46	133	
Al_2O_3/TiO_2 (30 wt.%)	3	450	2.5	48	260	
Al_2O_3/TiO_2 (65 wt.%)	3	450	2.5	38	220	
Al_2O_3/ZrO_2^{b}	5	450	2.6	43	216	ZrO_2:nd°
Al_2O_3/ZrO_2^{b}	5	750	2.6	44	179	ZrO_2:nd
Al_2O_3/ZrO_2^{b}	5	1000	≥20			ZrO_2:10-15

* slit shaped pore model; ^{a} bimodal distribution of CeO_2 particles
^{b} 17 wt.% ZrO_2; °not detectable,probably ZrO_2 layer on Al_2O_3

This structure seems to be advantageous to obtain rather easy defect free structures. With spherically shaped particles it is considerable more difficult to obtain crack free systems and the processing is much more critical. In the case of γ-Al_2O_3 toplayers the microstructure is build up with penny shaped particles with a size of 25-50 nm (depending on processing) and a thickness of about 3.5-5.5 nm [25,26].
In the case of TiO_2, CeO_2 and ZrO_2 the structure consist of spherically or equi-axed elementary particles with sizes ranging from about 3 nm (CeO_2, particles easily transform to larger particles), 5-7 nm (TiO_2) or 8-9 nm (ZrO_2).
In the case of the binary systems we have a mixture of plate shaped and spherically shaped particles which form crack free, mechanically stable layers with characteristics comparable with γ-alumina membranes as long as the volume contribution of the other component is smaller than about 30-40%. Detailed structural investigations are in progress.
Very important in the synthesis of membranes is the possibility to relax stresses which form during drying and calcination. This is promoted by adding organic (polymer) molecules to the sol and by using very smooth supports. In this way we could produce sandwich structures of several layers.
TiO_2 membranes are reported too by Larbot et al. [28] with a very small thickness (0.7 µm) and pore sizes from 5-20 nm. Electronically conducting RuO_2-TiO_2 binary membranes have been synthesized by Guizard et al. [29]. Pore diameters of 10-20 nm were obtained with membranes

Table 3: Some Modified Nanoscale Ceramic Microstructures

MEMBRANE	DEPOSIT	SHAPE	LOADING	SIZE (nm)
γ-Al_2O_3	Fe or V oxide	monolayer	5-10 %	≈ 0.3 nm
γ-Al_2O_3	MgO/MgOH	particles	5-100%	
γ-Al_2O_3	AL_2O_3/$Al(OH)_3$		5-100%	
γ-Al_2O_3	Ag	particles	5-50 %	≈ 5 nm; > 10 nm
γ-Al_2O_3	CuCl/KCl	multilayer		
γ-Al_2O_3	ZrO_2	layers?		
γ-Al_2O_3	SiO_2(amorphous)	porous props.	5-100%	< 1.0-1.5nm
TiO_2	V oxide			
Al_2O_3-TiO_2	V oxide, Ag			
θ/α-Al_2O_3	ZrO_2-Y_2O_3	multi layers, props	1-100%	- few nm - pore size

from colloidal gels and of less than 5 nm for membranes from polymeric gels. No further microstructural details are given.
These systems show all interesting transport and catalytic properties like surface diffusion and improved selectivity and can be used for influencing chemical reactions and separation processes [35,36]. A review has been recently given by Keizer and Burggraaf [32].

3.2. Modification of membrane structures

The membrane (top layer) structures can be modified further by deposition of material in the internal pore structure from liquids (impregnation, adsorption) or gases. In this case duplex type structures on a scale of 1-2 nm can be obtained. Some results of our current research are summarized in Table 3. Several possible situations are schematically shown in Fig. 3. For silica based xerogels di-phasic nanoscale composites are reported by Komarneni, Roy and co-workers [7]. Typical examples in this case were SiO_2-AgCl, SiO_2-$AlPO_4$, SiO_2-Nd_2O_3 and SiO_2-CdS.
In all these cases the second component is the component which is deposited inside the pore structure of the first one. The substructure inside the pores has a characteristic size equal or smaller than the pore size (which is 3-8 nm).
Monolayer techniques are known from catalyst research and are reviewed briefly by Burggraaf and Gellings [4]. They are applied now by us in membrane systems (Table 3 and Fig.3(a)) using adsorption of metal acetylacetonates from solution, followed by decomposition of the adsorbates. An example is the formation of V oxide monolayers in γ alumina and binary membranes (Table 3). Mg and alumina hydroxides are deposited by a homogeneous precipitation technique inside the pore system and structures and properties are investigated both before and after thermal annealing. In both cases crack-free mechanically stable modified membranes (supported and non-supported) could be obtained with locally narrowed (decreased) pore sizes and changed transport properties [36]. Especially interesting are our preliminary results with silica modification of γ alumina membranes (see Fig. 5).
Modification along this line is reported too by Asaeda et al. [33] starting with γ alumina layers with rather large pores (a few tens of nm) which were subsequently treated by Al isopropoxide and sodium silicate forming an alumina silicate inside the pore system. This system could be used only succesfully after thermal treatment in humid air at low temperature (90°C).
In our MgO modified systems it could be shown by Scanning Auger microscopy that a homogeneous distribution of the MgO within the pores and throughout the toplayer (thickness 3-5μm) could be obtained.
Consequently, the MgO substructure has dimensions smaller than 3-5 nm.
To obtain a very high loading a specially developed technique has to be used [34].
Electrochemical vapour deposition (ECVD) techniques are used too, to vary the shape and the distribution of the nanoscale deposits in the pore system (Table 3, Al_2O_3-(ZrO_2-Y_2O_3) system).

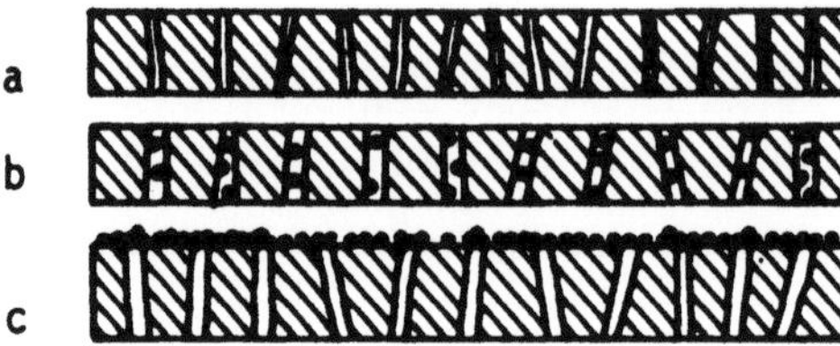

a Homogeneous (multi)layers in the pores
b Plugs in the pores (constrictions)
c Plugs/layers ontop of the pores

Figure 3. Schematic Microstructures of Some Modified Ceramic Membrane Types.

3.3. Transport and separation properties of membranes.

(Porous) ceramic membranes can be used to separate i) solid particles of cells in liquid systems, ii) larger molecules from liquid systems (e.g. proteins from solutions), iii) multicomponent liquids or iv) multicomponent gas or vapour mixtures [6,37,38).
Furthermore they can be used in chemical reactors to influence the conversion or the selectivity [35,43,44]. Most of these applications are just in the R&D stage and are not introduced in the industrial practice. They have, however, a large potential.
Some transport and separation properties of alumina membranes for liquids are described by Leenaars et al. [27]. In this paper we shall focus on gas (vapour) transport properties.
Transport of gases (or vapours) through a porous system is based on four mechanisms:
1. laminar flow
2. Knudsen diffusion
3. surface diffusion
4. capillary condensation /multilayer diffusion.
The first of this mechanisms is non separative of character and has to be eliminated.
A brief review of the transport phenomena involved is given by Keizer and Burggraaf and co-workers [32,34,40]. For interpretation of the experimental results it is very helpful to separate the permeabilities of the support system and the toplayer (the membrane proper). This is described in ref. 32.

For a single gas the total permeability F_o according to the first three mechanisms is described by:

$$F_o = (\varepsilon/\tau)c_1.r.M^{-1/2} + c_2.r^2.P/\eta + c_3D_s \frac{dx_s}{dp}.r^{-1} \qquad (1)$$

with c_1, c_2 and c_3 are constants, ε is the porosity, τ is related with the tortuosity, M = the molecular mass, P is the pressure, η the viscosity of the gas and D_s its surface diffusion coefficient, x_p is the percentage of occupied surface (by adsorption) compared to a monolayer and r is the pore radius. The permeability is defined as the flux per unit of surface and per unit of driving force.

From eq.1 it can be seen that with decreasing pore size first the second term (Poiseuille or laminar flow) vanishes. The Knudsen diffusion flux (first term) is dominant then. With a further decrease in pore size the surface diffusion term becomes dominant, provided some adsorption takes place and a favourable gradient of the adsorbed species across the membrane thickness exist. The transition from dominant Knudsen to dominant surface flow takes, in many cases, place around a pore diameter of 2.5 nm. Furthermore, we see that permselectivity is caused by differences in the molecular mass for the Knudsen flow and with differences in adsorption and surface diffusivity for the surface (diffusion) flow. So with surface flow the interaction between the gas and the solid surface plays a role and surface modification can strongly change the separation behaviour. Finally, it is noted that the Knudsen permeability is independent of the pressure P (note: the flux is not independent of P).

So, we can calculate the selectivity for a bimolecular gas in the Knudsen regime and compare it with the experimental value. A difference between the calculated and experimental value is ascribed to a surface diffusion flux provided no leakage through the membrane occurs. A control on this point is the absence of a pressure dependency of F_o for an indifferent (non adsorbing) gas.

The fourth transport mechanism is that of multilayer adsorption and diffusion followed by capillary condensation with further increasing pressure. Here, the pores are slowly filled (multilayer adsorption) untill a liquid meniscus forms inside the pore (capillary condensation). The liquid "prop" formed condenses at one side and evaporates at the other side. It forms an effective barrier for the condensable gas(es) in the mixture. This type of transport cannot be expressed by one single mathematical equation but needs modelling [41]. The mechanism is very effective with very small pores and vapours and is used e.g. for separating water alcohol mixtures from the vapour phase [33,42].

Examples of the enhancement of the permselectivity of alumina membranes by surface modification are described by Uhlhorn et al. [34]. For a non modified γ alumina membrane the flux ratio's of H_2 and N_2 based on Knudsen diffusion is equal to 3.74. This is measured indeed. After modification of the membrane by depositing very small Ag particles inside the pores (see Table 3) the H_2/N_2 flux ratio increases with decreasing pressure to a value of 8 at a pressure of 80 kPa (T = 298K). This is due to an increasing contribution of the H_2 flux by surface diffusion.

A second example is the permselectivity of CO_2 in CO_2/N_2 mixtures [34,40]. The results for a non modified γ alumina membrane is shown in Fig.4. It can be seen that the experimental flux ratio of CO_2/N_2 is about 30% larger then the calculated one for Knudsen diffusion. This is caused by surface diffusion in a more or less "saturated" adsorption layer (no pressure dependence of the surface flux contrary to the case of H_2 in Ag modified membranes).

The surface contribution to the total flux drastically changes by modification of the alumina membrane with silica. This is shown in Fig.5. Compared with the non modified membrane the flow of indifferent reference gas (N_2 or He) is strongly decreased by a factor 20. This indicates a strong decrease of the product of porosity (ε) and pore size. Adsorption/desorption measurements show a large and pressure independent adsorption starting already at very low relative pressure $P_r=P/P_o$, where P_o is the condensation pressure of N_2 at 77 K(1 atm). This indicates that the pore size is decreased to about one nanometer or smaller!

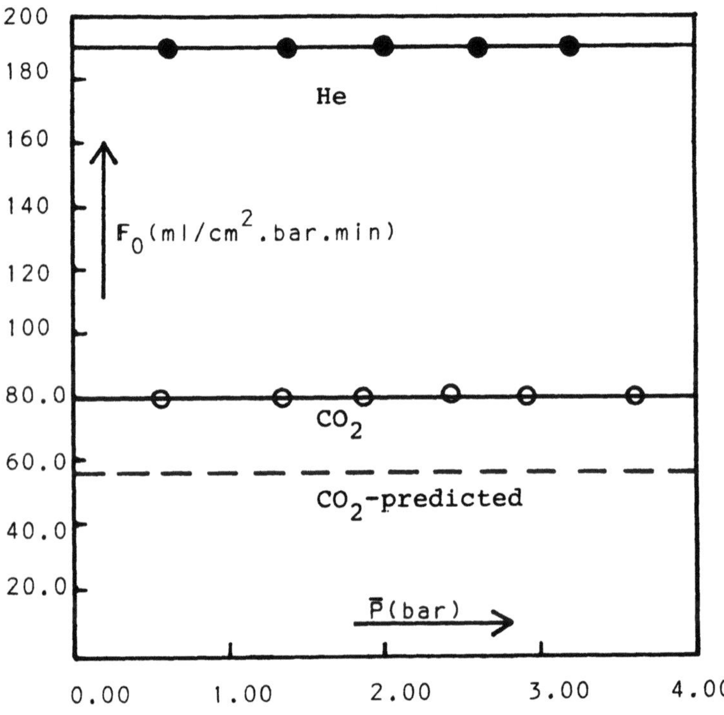

Figure 4. Permeabilities of pure gases through a γ-alumina membrane (top layer only) at 293 K.

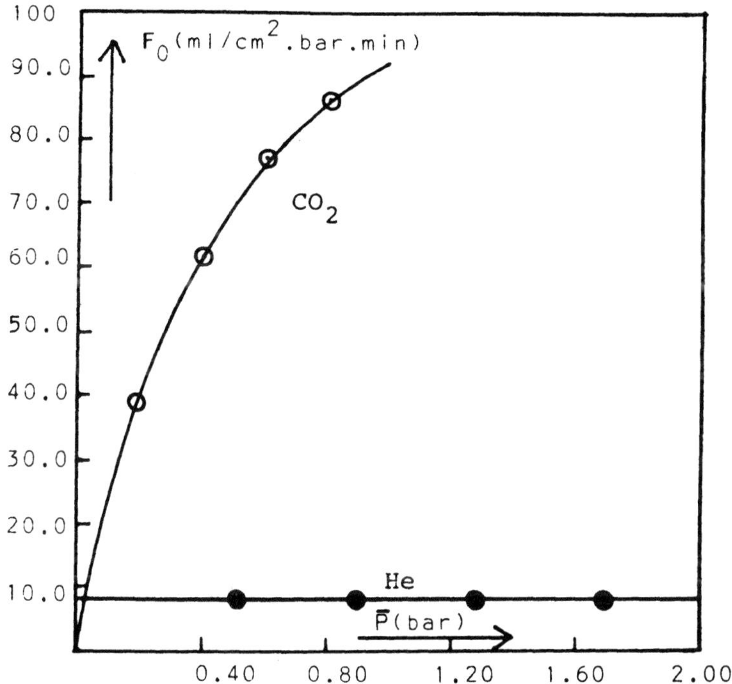

Figure 5. Permeabilities of pure gases through a silica modified γ-alumina membrane (top layer only) at 293 K.

The flux of CO_2 through these very small pores is much less decreased than that of the He (or N_2) and is strongly pressure dependent. At p_{CO_2} = 1 bar and T = 293K the ratio of the respective permeabilities $F_o(CO_2)/F_o(He)$) amounts 9. This implies that almost the complete total flux consist of <u>non</u> Knudsen flow!
Whether this is due to multilayer adsorption and diffusion or to capillary condensation is investigated now (Uhlhorn, Keizer, Burggraaf et al.).

3.4. Catalytically active membranes.

Ceramic membranes can be used to influence chemical reactions in two ways: i) shifting the equilibrium by removing selectively one of the reaction products, e.g. H_2 in the dehydrogenation of butane or propane, ii) by manipulation of the reactant concentration and the location of the catalyst in the membrane pore system. The reactant concentration can be manipulated by selective supply of one of the reactants through the membrane to the reaction place and/or by varying the absolute and partial pressure drops across the membrane.

Application fields are e.g. oxydation or oxidative dehydrogenation processes with improved conversion and selectivity. This is a new field with very interesting possibilities [35,43,44]. Zaspalis et al. [35] showed for example that the dehydrogenation reaction of methanol is strongly different for a normal catalyst bed of γ alumina powder as compared to a catalytic membrane with the same surface area. Furthermore, modification with Ag of this membrane (see Table 3) decreases the optimum reaction temperature from about 420°C to about 280°C. The selectivity to formaldehyde is increased from 18% at 420°C to 25% at 270°C. This is remarkable because at this last temperatures the selectivity to formaldehyde is zero in a non modified membrane!

4. ION IMPLANTATION AS A SURFACE MODIFICATION METHOD

4.1. Ion implantation process

Ion implantation is a technique to modify selectively the surface of metals, semiconductors and insulators [4,45,46]. In the ion implantation process specific atoms are ionised, accelerated and mass selected. Those high energetic ions (5keV-500keV) penetrate in the solid of interest. Due to elastic and inelastic energy losses of the ion in the solid, its penetration is limited. Various computerprograms (e.g. TRIM) [48] have been set up in order to calculate the depth profile of the implanted ion. Surface sensitive techniques like Auger Electron Spectroscopy (AES), X-ray Photoelectron SPectroscipy (XPS), Secondary Ion Mass Spectroscopy (SIMS) and Rutherford Backscattering (RBS) [49,50] are commonly used to determine the depth profiles experimentally.

Due to elastic and inelastic energy losses of the penetrating ion, defects are created in the solid. Depending on the ion dose, ion energy and target temperature, the surface may become amorphous. Due to both structural changes and the chemical nature of the implanted ion the electronic, chemical, optical and mechanical properties of the surface can be modified.

Depending on the process parameters the concentration profile of the dopant ions can be given different shapes, the penetration depth can be varied between some nanometers and about one micrometer and the dopant concentration in the surface can be raised to about 50 at%.

A major advantage of ion implantation as a surface modification technique is the possibility to dope the surface layer of the substrate material with any element, independent of the solubility of this element in the substrate material. This means that one can produce strongly supersaturated metastable compounds or solid solutions and complex micro(nano)structures consisting of very finely dispersed nano precipitates embedded in a matrix material. Some examples are given later.

After the ion implantation process, often a thermal treatment is applied to remove the created defects. Full epitaxial regrowth of the implanted region may occur at this stage [51]. This has been observed for Pb implanted $CaTiO_3$ and $SrTiO_3$, for Yb implanted ZrO_2 [52] and for Al and O implanted Al_2O_3. Depending on the solubility, the thermal treatment may result in the formation of a solid solution [52] or the formation of micro precipitates with nanoscale dimensions [53-56]. Long term anneal treatments will result in a coarsening of the microprecipitates [55], whereas an additional ion implantation may result in a resolution of the precipitates [57,72].

4.2 Changes in the surface properties.

Representative examples will now be discussed of changes in the surface properties of insulators, which are due to the ion implantation process.

Recently, much insight has been obtained in the changes in mechanical properties of ceramics due to ion implantation [58,59]. At low ion doses the hardness of the material tends to increase due to radiation hardening and solid solution hardening. At a critical dose the hardness decreases due to the formation of a subsurface amorphous layer. This amorphous layer extends towards the surface at even higher doses. The final hardness of the surface will be lower than that of the unimplanted layer. Thermal processing of those implanted surface layers may result in the formation of a solid solution or in precipitation hardening.

The optical properties of insulators are modified due to the creation of point defects in the implanted layer. The optical density of certain defect types will increase if those defects are generated due to the elastic and inelastic energy losses of the implanted ion [60]. If the defects are sufficient mobile extended defects may be formed. Thermal treatments may cause precipitation of the dopant with itself or with the matrix. This may result in metallic colloids with a characteristic colloidal absorption band [55,57,60,72]. The change in refractive index which occurs after light ion implantation in $LiNbO_3$ and $LiTaO_3$ has been succesfully applied in producing waveguides on top of those materials [73].

Generally, an increase of the electronic conduction of insulators is observed after the implantation of ions [61,62]. This is due to the formation of charged defects due to both elastic and inelastic energy losses of the implanted ion. Annealing of the material will render the matrix again insulating due to the recombination of those defects. When Fe was implanted in MgO, the high electronic conductivity of the insulating matrix could be retained when the sample was annealed in a vacuum of 10^{-6} Torr at temperatures lower than 225°C [62]. This is explained by a hopping conduction of electrons between Fe ions in the two valent and three valent oxidation state. Annealing at 700°C in a oxygen rich atmosphere, however, resulted in a complete oxidation of the implanted Fe towards the three valent oxidation state, with a corresponding recovery of the insulating properties of the matrix [63].

Apart from electronic conductivity, also ionic conductivity may result from the implantation of ions. Ca implantation in ZrO_2 resulted in an oxygen ion conducting layer [61]. Mg implantation in alumina resulted in an amorphous layer which transformed to a beta alumina like thin film after a solid reaction with sodium aluminates mixtures [64,65]. Recently investigations have been started in order to produce mixed conducting layers (i.e. ionic and electronic conducting layers) on top of an otherwise pure ionic conducting solid electrolyte. For this purpose Fe was implanted in Yttria stabilized zirconia [66,68]. Before annealing a large conductivity was found due to the presence of iron as a mixture of metallic particles and Fe^{2+} and Fe^{3+} ions. Annealing at a temperature of 400°C in a oxygen rich atmosphere resulted in a complete oxidation of the implanted iron towards the three valent oxidation state. Consequently, no large change in the conductivity has been observed with relatively large oxygen partial pressures. An interesting microstructure is however found. If the local Fe concentration exceeds $4{,}5 * 10^{21}$ Fe cm^{-3} than a dispersion of Fe_2O_3 (hematite) particles with a diameter of 3-10 nm in a metastable, strongly supersaturated, solid solution of Fe^{3+} in yttria stabilized zirconia [66].
Ion implantatiaon in superconducting films of $YBa_2Cu_3O_x$ tends to destroy the superconducting properties [69]. This has been used as a method to produce the first superconducting quantum interference device (SQUID) of this material which operated at the temperature of liquid nitrogen.
Some studies of the catalytic activity of ion implanted solids in both gas as liquid phase have been published. Positive changes in catalytical activity, selectivity and stability have been observed. Those catalytic reactions take place at the gas-solid or liquid-solid interface. As the ions are implanted below the surface of the solid, thermal treatments are necessary to increase the surface concentration. Thermal treatments have been useful to increase the Pt surface concentration in Pt implanted Al_2O_3 [70]. This thermal treatment caused the preferential orientation of the Pt (111) face to the surface.
The thermal stability of catalysts may be significantly increased by ion implantation. Abouchacra et al. [55] implanted Ag in MgO. For annealing temperatures up to 800°C the radii of the Ag precipitates varied between 1 and 5 nm. Above 800°C the mean particle radius increased to 25.5 nm (1200°C). Due to the high energy of the implanted Ag (180 keV), the surface concentration is, however, expected to be very low.
To circumvent those thermal treatments often a technique, related to ion implantation and called recoil implantation [71], is used. In the recoil implantation process a thin metal layer is sputtered or evaporated on the surface of the matrix. Subsequently, this metal layer is implanted with high energetic (noble gas) ions, which cause an intermixing of atoms at the interface between the metal and the matrix. As a consequence the mechanical stability of the metal layer is improved whereas a high surface concentration of the catalytic active metal is maintained.

Apart from the thermal stability, also the catalytic activity and selectivity of catalysts can be modified by ion implantation. Pt implanted in RuO_2 resulted in a somewhat higher catalytic activity in the oxidation of methanol than a smooth Pt metal surface itself [71]. A remarkable change in selectivity has been observed in the oxidation of methanol and formic acid. Smooth Pt metal itself did not show any selectivity in the oxidation reaction of those reactants whereas Pt recoil implanted in RuO_2 had a much higher activity in the oxidation of formic acid than in the oxidation of methanol. This difference could be explained in terms of the high dispersity of the Pt particles and a possible metal support interaction.
Ion implantation is a 'line of sight' process. This means that only those areas can be implanted which can be seen by the ion beam. This makes it difficult to prepare catalysts with a high surface area. No experimental studies have been found in literature. Solutions to this problem have been discussed by Wolf [71].

REFERENCES

1. R. Birringer, U. Herr, H. Gleiter, Trans.Jap.Inst.Metall.Suppl., **27** (1986) 43.
2. H. Yanagida, Y. Nakamura, New Materials and New Processes 3 (1985) 13-17.
3. H.A. Bender, R.P. Ingel, W.J. McDonough, J.R. Spann, Advanced Ceramic Materials 1 (1986) 137-144.
4. A.J. Burggraaf, P.J. Gellings, D. Scholten, in High Tech Ceramics, ed. by P. Vincenzini (Elseviers Sc.Publ., Amsterdam 1987), pp. 779-792.
5. A.F. Leenaars, K. Keizer, A.J. Burggraaf in Reverse Osmosis and Ultrafiltration, ed. by S. Sourirayan and T. Matsura, ACS Symposium Series **281** (1985) 57.
6. K. Keizer, A.J. Burggraaf, Science of Ceramics 14 (1987), 83, ed. by D. Taylor, publ. by Institute of Ceramics, Shelton,. Stoke-on-Trent, Staffs, U.K., 1988.
7. S. Komarneni, E. Breval, R. Roy, J.Mat.Science, **21** (1986) 737.
8. J. Hurt, D.J. Viechnicki, Ultra fine grained Ceramics, ed. by J. Burke, N.L. Reed, V. Weiss (Syracuse Univ. Press 1968) pp. 273-293.
9. C.G.F. Stenger, A.J. Burggraaf, Phys.Stat.Sol (a) **61** (1980), 275-285 and 653-664.
10. C.A. Randall, D.J. Barker, R.W. Whatmore, P. Groves, J.Mat.Science **21** (1986) 4456.
11. M.P. van Dijk, F.C. Mylhoff, A.J. Burggraaf, J.Sol.State Chem, **62** (1986) 377.
12. B.A. Bender, R.P. Ingel, W.J. McDonough, J.R. Spann, Advanced Ceramic Materials 1 (1986) 137-144.
13. G. Kalonji, J. McKillick, L.W. Hobbs, in Advances in Ceramics **12** (1984), 816, ed. by N. Claussen, M. Rühle, A.H. Heuer, The American Ceramic Society, Columbus, U.K., 1984.

14.L.L. Hench, D.R. Uhlrich eds., Ultrastructure processing of glasses and composites, Wiley Interscience Publ., N.Y. 1986.
15.L.L. Hench, D.R. Uhlrich eds., Science of Ceramic Chemical Processing,. Wiley Interscience Publ., N.Y. 1986.
16.C.J. Brinker, D.E. Clark, D.R. Uhlrich eds., Better Ceramics through Chemistry, Mat.Res.Soc.Symp.Proc. **32** (1984).
17.R.W. Davidge ed. Novel Ceramic Fabrication Processes, Brit.Cer. Proc. **38** (1986), Institute of Ceramics, Shelton House, Stoke-on-Trent, U.K.
18.J.A. Aksay, C.H. Shilling, see ref. 14 Ch. 34, pp.439-448.
19.R.K. Iler, see ref. 15 pp. 3-20.
20.P.A. Bruggen, A. Mocellin, J.Mat.Sience **21** (1986) 4431.
21.E.A. Pugar, P.D. Morgan, J.Am.Cer.Soc. **69** (1986) C 120.
22.P.E.D. Morgan, H.A. Bump, E.A. Puigar, J.J. Raals, see ref. 15, pp. 327-334.
23.J. Karch, R. Birringer, H. Gleiter, Nature, **330** (1987) 556.
24.G.S.A.M. Theunissen, A.J.A; Winnubst, W.F.M. Groot Zevert, A.J. Burggraaf, Proceedings of the 7th SIMCER, Zirconia '88, 14-17 December 1988, Bologna, Italië, to be published.
25.A.F.M. Leenaars, K. Keizer, A.J. Burggraaf, J.Mat.Science **19** (1984) 1077.
26.A.F.M. Leenaars, A.J. Burggraaf, J.of Colloid and Interface Sc. **105** (1985) 27.
27.A.F.M. Leenaars, A.J. Burggraaf, J.Membrane Science **24** (1985) pp. 245-260 and 260-270.
28.A. Larbot, J. Alary, C. Guizard, J.P. Fabree, N. Idrissi, L. Cot in P. vincenzini (ed.) High Tech Ceramics, Mat.Sc. Monographs 38 (1987) 2259, Elsevier Amsterdam 1987.
29.C. Guizard, N. Idrissi, A. Larbot, L. Cot, see ref. 17, pp.263-274.
30.R.J.R. Uhlhorn, M.H.B.J. Huis in 't Veld, K. Keizer, A.J. Burggraaf, see ref. 6, 'New ceramic membrane materials for use in gas separation applications'.
31.R.J. van Vuren, B.C. Bonekamp, K. Keizer, R.J.R. Uhlhorn, H.J. Veringa, A.J. Burggraaf, see ref. 28, p. 2235.
32.K. Keizer. R.J.R. Uhlhorn, R.J. van Vuren and A.J. Burggraaf, J.Membrane Science **39** (1988), 285.
33.M. Asaeda, Lung Dui Du, M. Fuji, J.Chemical Eng. of Japan **19** (1986) pp. 73-77, 84-85.
34.R.J.R. Uhlhorn, K. keizer, A.J. Burggraaf in Proceedings of the 5th Chemical Symposium of North America, 5-12 June 1988, Toronto, Canada, A.C.S. Symp. Series (1988) in press.
35.V. Zaspalis, K. Keizer, J.G. van Ommen, J.R.H. Ross, A.J. Burggraaf, AIChE Symposium Series, 1989, in press.
36.R.J.R. Uhlhorn, M. Huis in 't Veld, K. Keizer, A.J. Burggraaf, to be published in J. Mat. Science 1989.
37.H.P. Hsieh, AIChE Symposium Series **261** (1988) 84
38.R. Zanetti, H. Short, D. Hunter, S. Ushio, Chemical Engineering, June 1986, 19
39.A.F.M. Leenaars, J.Membrane Science **24** (1985) 261

40.R.J.R. Uhlhorn, K. Keizer, A.J. Burggraaf, J.Membrane Science 1989, in press.
41.K.H. Lee, S.T. Hwang, J.Colloid Interface Science 110 !1986) 544.
42.M. Asaeda, J.Chem.Eng.Japan 14 (1981) 1.
43.R.J. Dellefield, 'High temperature application in Inorganic membranes', to be published in Symposium Series of AIChE, 1989.
44.J.N. Armor, 'Catalysis with perm selective Inorganic Membranes', to be published in J.Applied Catalysis 1989.
45.G. Dearnally, J.H. Freeman, R.S. Nelson, Ion Implantation Defects in Crystalline Solids, 18, North Holland Publishing Company, Amsterdam, 1973.
46.O. Auciello, R. Kelly (eds.) 'Ion bombardment modification of surfaces', fundamentals and applications, beam modification of materials 1, Elsevier Science Publishers BV, Amsterdam, 1984.
47.P. Mazzoldi, G.W. Arnold (eds.), 'Ion beam modification of insulators, beam modification of materials 2, Elsevier Science Publishers BV, Amsterdam, 1987.
48.J.F. Ziegler, J.P. Biersack, U. Littmark, The stopping and range of ions in solids, 1 of the stopping and ranges of ion in matter, J.F. Ziegler (ed.), Pergamon Press, New York, Oxford, Toronto, Sydney, Frankfurt, Tokyo, 1985.
49.M. Grasserbauer, H.J. Dudek, M.F. Ebel, Angewandte Oberflachenanalyse, Springer-Verlag, Berlin, Heidelberg, New York, Tokyo, 1986.
50.Wei-Kan Chu, J.W. Mayer, M.A. Nicolet, Backscattering Spectrometry, Academic Press, New York, 1978.
51.C.W. White, L.A. Boatner, P.S. Sklad, C.J. McHarque, J. Rankin, G.C. Farlow, M.J. Aziz, Nuclear Instruments and Methods in Physics Research B32 (1988) 11-22.
52.C. Cohen, J. Siejka, M. Berti, A.V. Drigo. M. Croset, M.M. Tosic, Radiation Effects 64 (1982), pp.221-231.
53.G. Chassagne, G. Abouchacra, J. Davenas, A. Delmas, C. Dupuy, J. Serughetti, M. Treilleux, Radiation Effects 65 (1982), pp. 119-125 (359-365).
54.J.A. Sawicki, G. Abouchacra, J. Serughetti, A. Perez, Nuclear Instruments and Methods in Physics Research B16 (1986), pp. 355-360.
55.G. Abouchacra, J. Serughetti, Nuclear Instruments in Physics Research B14 (1986), pp. 282-289.
56.A. Perez, Nuclear Instruments and Methods in Physics Research B1 (1984), pp. 621-627.
57.P. Thevenard, M. Treilleux, M.O. Ruault, J. Chaumont, H. Bernas, Nuclear Instruments and Methods in Physics Research B1 (1984), pp. 235-239.
58.P.J. Burnett, T.F. Page, Radiation Effects in Insulators 3, Proceedings of the Third International Conference, Guilford, Surrey, 15-19 July 1985, I.H. Wilson and R.P. Webb (eds.), Gordon and Breach Science Publishers, New York, London, Paris, Montreux, Tokyo, 1986, pp. 123-136.

59.H. Naramoto, C.J. McHarque, C.W. White, J.M. Williams, O.W. Holland, M.M. Abraham, B.R. Appleton,Nuclear Instruments and Methods **209/210** (1983), pp. 1159-1166.
60.M. Treilleux, P. Thevenard, M.O. Ruault, H. Bernas, J. Chaumont, Nuclear Instruments and Methods in Physics Research **B12** (1985) pp. 375-381.
61.J.P. Schnell, M. Croset, D. Dieumegard, G. Vealsco, J. Siejka, Nuclear Instruments and Methods **209/210** (1983) pp. 1187-1191.
62.R. Meaudre, A. Perez, Nuclear Instruments and Methods in Physics Research **B32** (1988) pp.75-78.
63.A. Perez, J. Bert, G. Marest, B. Sawicka, J. Sawicki, Nuclear Instruments and Methods **209/210** (1983), pp. 281-287.
64.Ph. Schnell, G. Velasco, Ph. Colomban, Solid State Ionics **5** (1981), pp. 291-294.
65.J.P. Schnell, G. Velasco, D. Dubreuil, D. Dieumegard, M. Croset, Ph. Colomban, Solid State Ionics **9&10** (1983) pp. 1465-1468.
66.A.J. Burggraaf, D. Scholten, B.A. van Hassel, Nuclear Instruments and Methods in Physics Research **B32** (1988), pp. 32-36.
67.D. Scholten, 'Surface Modification of Yttria Stabilised Zirconia by Ion Implantation', PhD Thesis, University of Twente, Enschede, The Netherlands, 1987.
68.D. Scholten, A.J. Burggraaf, Radiation Effects **97** (1986) 191.
69.G.J. Clark, A.D. Marwick, F. Legoues, R.B. Larbowitz, R. Koch, P. Madakson, Nuclear Instruments and Methods in Physics Research **B32** (1988), pp. 405-411.
70.P. Rabette, A.M. Deane, A.J. Tench, M. Che, Chemical Physics Letters, **2** (1979) pp. 348-352.
71.G.K. Wolf, K. Zucholl, H. Folger, W.E. O'Grady, Nuclear Instruments and Methods 209/210 (1983), pp. 835-840.
72.G. Fuchs, G. Abouchacra, M. Treilleux, P. Thevenard, J. Serughetti, Nuclear Instruments and Methods in Physics Research **B32** (1988), 100-103.
73.E. Glavas, L. Zhang, P.J. Chandler, P.D. Townsend, Nuclear Instruments and Methods in Physics Research **B32** (1988), pp. 45-50.

ACKNOWLEDGEMENTS

R.J.R. Uhlhorn, V. Zaspalis, M.H.B.J. Huis in 't Veld en W. v. Praag are acknowledged for their contributions to parts of the research described in this review. Mrs. F.P. Twilt-van Giessen and Mrs. A. Baveld are thanked for preparing the manuscript.

SURFACE DETERMINED PROPERTIES OF SILICATE GLASSES

Albert A. Kruger
Battelle, Pacific Northwest Laboratories
Chemical Sciences Department
P.O. Box 999
Richland, WA 99352-0999
USA

ABSTRACT. It is normally assumed that glass surfaces are inert; this is why glass is frequently considered as the container material of choice. However, anyone who has ever worked with a vacuum line or looked at the chemical or biological activity of glasses knows otherwise. Furthermore, when treated in specific ways, glass surfaces can be rendered extremely reactive. This presentation will deal with particular treatments that have been shown to greatly modify the chemical activity and mechanical properties of glasses.

Introduction

A problem with the understanding of glasses is that their structures are typically poorly defined. As an amorphous solid, the surface is much less regular and more difficult to describe than those of crystalline solids. Though this text will present results of various analytical techniques, so as to offer a better characterization of the glass surface, the reader is directed to a recent review[1] for a more complete treatment of the prior works.

For many years now it has been thought that the 'best' strength of a glass fiber is dependent upon the largest flaw in the sample taken[2,3]. Griffith's original idea "was that the fracture of a brittle material (glass) would be limited by the largest of a distribution of tiny cracks always present in the material at the time of testing." These distributions are often represented by the Weibull distribution function[4]. However, direct observation of these Griffith-flaws, nor measurement of strengths at these well-known sites has never been reported. This is in spite of the fact that flaw sizes for E–glass fibers testing at 2.8–3.6 GPa should correspond to 120–225 Å dimensions, as calculated from the Griffith relationship. The explanation for the dynamic aging process has been based upon the Inglis stress concentration relationship[5] to describe the flaw growth mechanism, as well as flaw–tip shape, as a function of the temperature, pressure, and corrosive action of water on the components of the glass matrix. Furthermore, the control of the initial steps of the mechanism by alkali ion diffusion was invoked nearly 30 yrs ago. It has subsequently been postulated that the asymmetrical expansion of the glass network, used to model microcrack propagation, results from an interionic diffusion phenomenon. This requires that one to conceptualize the silica backbone as existing in a near–liquid state. This model is consistent neither with dynamic and static fatigue statistics, nor with evidence that interdiffusion is not rate determining. Futhermore, observed activation energies are consistent with a mechanism of interstitial hopping of alkali ions. In spite of these deficiencies, glass

L.-C. Dufour et al. (eds.), Surfaces and Interfaces of Ceramic Materials, 725–735.

technologists continue to cling to the old model.

Recent work in the area of predicting lifetimes of optical glass fibers, as only a single example, has provoked further disagreement between the commonly held reasoning and experimental data. As a result of these discrepancies it was hypothesized that dynamic fatigue is better understood by the reactivity at specific sites on the glass surface rather than by Griffith flaw formalism. In particular, it has been proposed that the rate-determining step of the corrosion mechanism is field–enhanced alkali ion diffusion leading, to the dissolution of the glass network in the presence of water. Therefore, the effects of local variations in charge density on alkali ion migration that occur on the solution side of the glass/water interface play a direct role in the chemical corrosion mechanism.

Materials and Methods

Fibers of monofilament E–glass (SiO_2: 53%, Al_2O_3: 15%, CaO: 23%, B_2O_3: 8%) and with a commonly used insulation glass (SiO_2: 64%, Na_2O: 16%, CaO: 7%, Al_2O_3: 3%) were pulled from Pt crucibles, heated by a resistively heated element, on a standard tower. The E-glass marbles were melted and fined at 1500°C over a 2 h period. The temperature was subsequently lowered (100°C/h) to 1320°C whereupon 10–15 μm diameter fibers were produced at a rate of 16 m/s. Common fiberizing methods were employed for the insulation glass, so as to produce 10–15 μm diameter fibers. Sampling, mounting, fracture testing and accelerated aging were accomplished by the established methods. Diameter measurements were made with the aid of a Ford Video–Micro scaler and scale marker system accurate to ±3%.

Plastic clad silica fibers (Amersil T08 and proprietary UV–curable pre–polymers) were prepared from 13 mm preforms pulled in a CENTORR Series II graphite ring furnace with on–line diameter control (forward laser scattering technique, with an error of $< \pm 1\%$). These 150 μm diameter fibers were pulled at a rate of 1 m/s and were coated in-line so as to yield a 60 μm wall thickness of polymer. As with the E–glass, accelerated aging was accomplished by well-known methods.

Results and Discussion

From a search of the literature it becomes obvious that aqueous attack on glasses can be divided into two very general stages. The first is the selective leaching of mobile ions (*e.g.*, Na^+, Ca^{2+}) from the bulk of the solid through the surface and interface, where the pH remains below that necessary to cause the dissolution of the silica network. The second is a process of hydrolytic depolymerization in solutions containing a high concentration of OH ions (*i.e.*, pH>9.5). However, a silicate glass exposed to an aqueous solution of pH<9 cannot be assumed to be free of a local surface pH that can greatly exceed pH 9, except of course at the isoelectric point where the bulk solution pH and effective surface pH are equal. Thus both of these reactions can occur simultaneously, and can be accompanied by ion exchange, glass hydration, or network hydrolysis. Unfortunately, the extent and relative rates of these reactions are not yet known.

It has been known for at least two decades that when exposed to water, alkali first goes into solution exhibiting square–root time leaching kinetics. Initially it was assumed that this was best explained by a simple diffusion model[6] presuming the applicability of Fick's laws. At longer time periods, the release rate becomes directly proportional to time and this has been attributed to glass dissolution[7]. Activation energies for these two stages of corrosion have been independently determined but seem to yield a common value of *ca.*

80 kJ/mol.

The fact that the diffusion of highly mobile species and the dissolution of a nominally stable silica oxygen network should occur with identical activation energy values in two entirely different reactions is either fortuitous or simply amazing. Also, we must reject the *a priori* assumption that network dissolution occurs at a constant rate, as it does not correlate with experimental observations[8]. Furthermore, the use of the simple thin–film solution of Fick's second law, as given in Equation 1, would not be consistent with the results of leaching experiments such as displayed in Figure 1 where Q(t) is the total alkali ion

$$Q(t) = \int_0^t J(0,\xi)\, d\xi = 2N_0\sqrt{\frac{Dt}{\pi}} \tag{1}$$

concentration in the solution at the surface at time t, N_0 is the uniform free alkali ion concentration in the solid, and D is the diffusion coefficient.

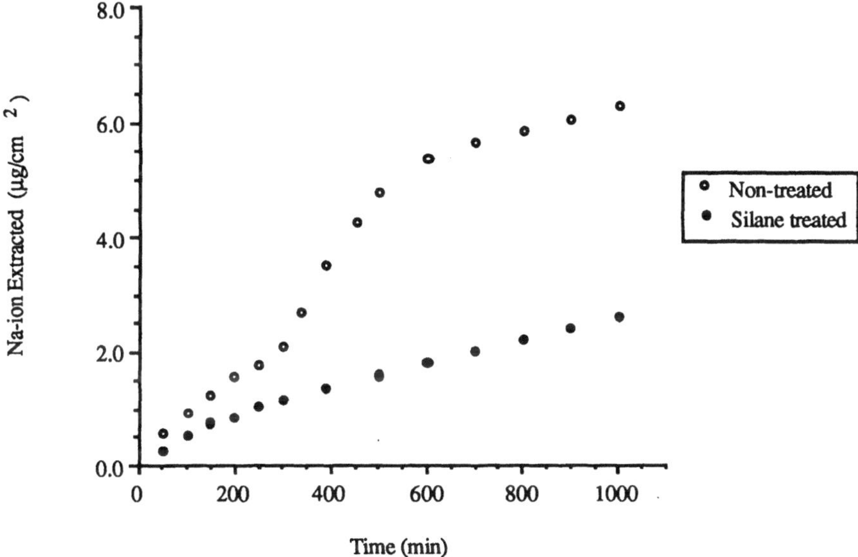

Figure 1. Na+ leaching into unbuffered water at 96°C; (SA/V: 6.7 cm^{-1}).
(Note: For the non-treated glass Na+ extraction proceeds until the surface pH exceeds 9.7, where upon network dissolution occurs. For the treated sample no dissolution effects are observed after this 'critical' value for Na+ extraction has been passed.)

As the diffusion is Fickian in nature the use of concentration–dependent diffusion coefficients and time–invariant concentrations are precluded. This presents us with a number of dilemmas, such as the lack of quantitative agreement with experimental results in describing leaching experiments (using the reduced coordinate; $[x/2(Dt)^{1/2}]$)[9] and diffusion coefficients that are three orders of magnitude higher in the surface region than that in the bulk glass[10].

Examination of the principles of surface potentials would indicate that only under fortuitous circumstances are the surface and the bulk of an ionic and covalently bonded solid at electrostatic equipotential. Furthermore, when glasses are placed in contact with water an electrical charge develops on their surfaces. The presence of this electric double layer (Gouy–Chapman) greatly modifies the various phenomena occurring in the surface region. Therefore, by including a term for the electrostatic potential where the alkali ion velocity $v=(\mu E)$, μ its mobility and E the electrostatic field in the solution of Fick's law, it can be shown that[11]:

$$Q(t) = \frac{2N_0D}{v}\left[\mathrm{erf}\left(\frac{v}{2}\sqrt{\frac{t}{D}}\right)\left(\frac{v^2t}{4D}+\frac{1}{2}\right)+\frac{v}{2}\sqrt{\frac{t}{\pi D}}\,\exp\text{-}\left(\frac{v^2t}{4D}\right)\right] \tag{2}$$

(where the reduced coordinate becomes $[x+vt/2(Dt)^{1/2}]$) such that lim Q(t) for $t\rightarrow 0$ and $t\rightarrow\infty$ are proportional to $\sqrt{t}$ and t, respectively. Thus an expression that gives both qualitative and quantitative agreement with experimental results as does our Equation 2 offers an insight into possible treatments to modify the surface charge so as to render glasses more resistant to corrosion.

Treatment of pure silica fibers[12] (pulled from 13 mm preforms of T08) and the role of these treatments on the leaching experiment[13] have been reported. As can be seen from Figure 2, a properly passivated fiber exhibits what could best be described as a

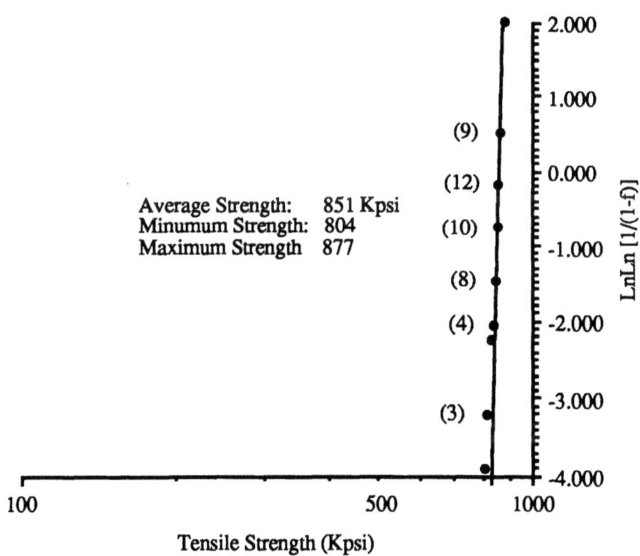

Figure 2. Typical Weibull Failure Probability plots (0.84 m test lengths) for freshly pulled, passivated 150μm diameter T08 silica glass fibers. (*NB*, the numbers in parenthesis represent the number of samples which broke at a given value.)

single–valued strength since the dispersion in the data is equal to the precision of the diameter measurements. Furthermore, when a treatment is applied so as to destabilize the surface, the resulting tensile strength measurements deviate markedly from what is typically observed, as can be seen in Figure 3. The effect of passivation on the static fatigue

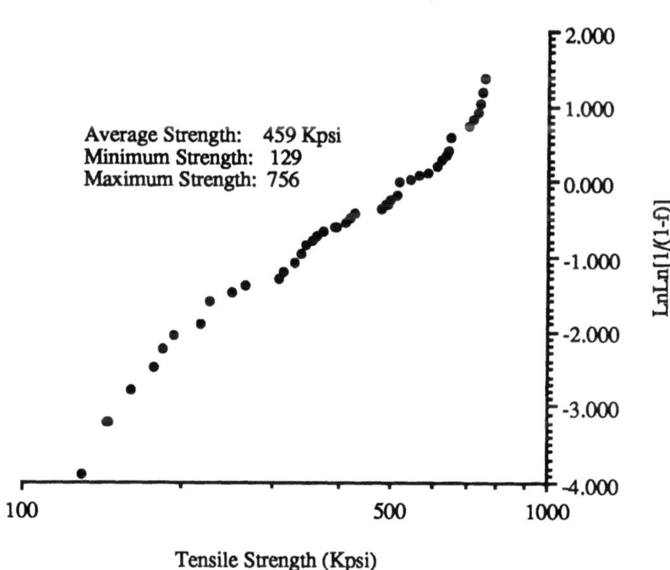

Figure 3. Typical Weibull Failure Probability plot (0.84 m test lengths) for freshly pulled 150μm T08 silica glass fibers with a destabilizing treatment applied.

experiment, aside from the increase of the virgin tensile strength, is that the treated fiber is an order of magnitude more resistant to aging than the non–treated silica[12].

It is interesting to note that despite of the great differences between treated and non-treated samples in both Na^+ extraction experiments and tensile strength tests, no such differences are observed in the surface potential measurements. As seen in Figure 4, there appears to be no difference in the bulk surface Zeta potentials measured by microelectrophoresis between silyted and non-silyted silicas. These results do, however, seem to negate the idea of total uniformity and homogeneity of the surface potential when one considers the vast differences between tensile strengths for treated and non-treated samples. Thus, it is reasonable to believe that the surface potential measurement is so crude that it gives us an average value, but it is the local surface potential (which we are incapable of measuring) that determines the reactivity of a solid surface. This assumption seems justified based upon our results and those obtained at Université Libre de Bruxelles[14,15].

Similar results have been obtained for fibers pulled from E–glass (SiO_2: 55%, CaO: 19%, Al_2O_3: 15%, B_2O_3: 7%), except that no great sensitivity to an increase of virgin tensile strength has been confirmed. As with silica, this low sodium glass was found to respond to passivation and exhibited nearly vertical lines in Weibull plots. The aging experiments showed an order of magnitude increase in resistance to corrosion over non–treated fibers. In addition, aging was not accompanied by any significant change in the Weibull parameter (Figure 5), as was also observed for silica fibers. It must be stated that

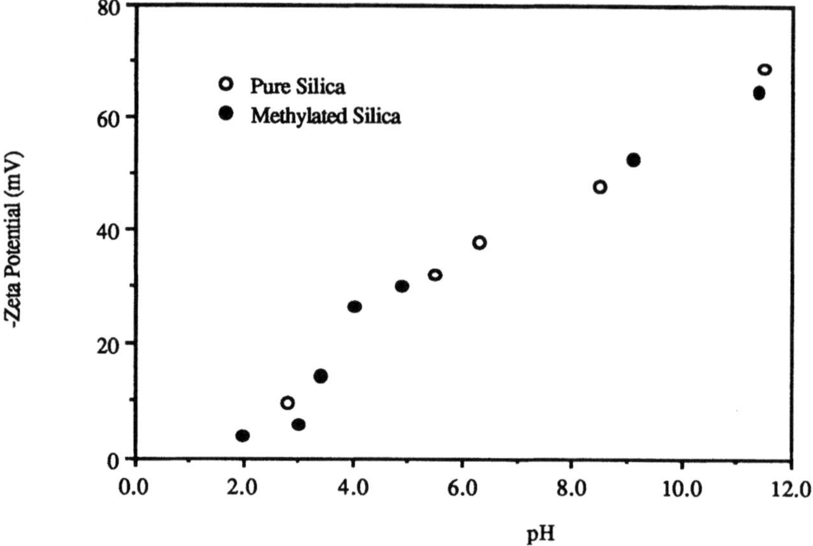

Figure 4. Nominal Zeta potential of pure vitreous silica and methylated silica.

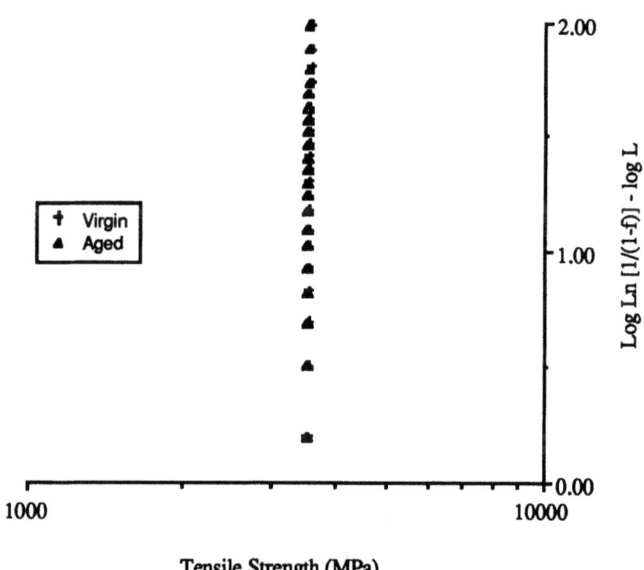

Figure 5. Typical length normalized tensile strength plots for freshly pulled and passivated (+) 10 μm E-glass fibers and identical fibers after accelerated aging (x). (Accelerated aging at 90% RH and 60°C for 20 h.)

in our treatments very few of the surface silanols (< 50%) are converted, given surface silanol concentrations of *ca.* 4.5 OH groups/100^2 Å and maximum packing densities of molecules such as $-Si(CH_3)_3$. However, it is reasonable to assume that the most reactive sites participate preferentially.

It would be expected that in the absence of water or its vapor loss of virgin strength of a non-treated fiber would not be observed. Experiment have been performed both in vacuum (20 mBar), as presented in Figure 6, and in dry nitrogen. As expected the resulting curves are superimposable.

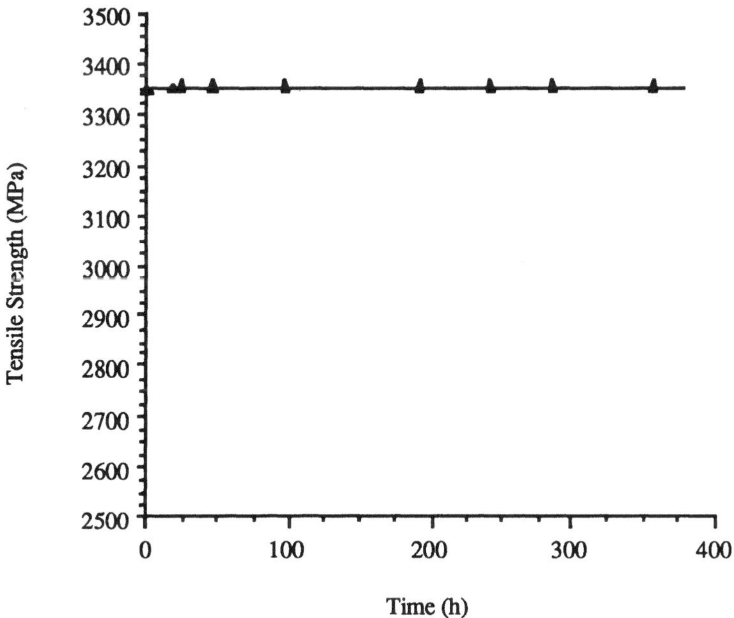

Figure 6. Length normalized tensile strengths as a function of time at reduced pressure.

Accelerated aging of non-treated fibers at 50°C in H_2O and D_2O showed a marked primary isotope effect (virgin: σ_{avg} 3.43 GPa, slope 51.1, D_2O aged for 120 h: σ_{avg} 3.33 GPa, slope 38.9, H_2O aged for 20 h: σ_{avg} 3.17 GPa). Such an effect is consistent with the reaction of water at the glass surface being the rate–determining step. This result cannot, however, be explained by a mechanism that involves the migration of water into, and accompanying swelling of the glass matrix.

In attempting to passivate another commonly used glass composition, that used for insulation fibers (SiO_2: 64%, Na_2O: 16%, CaO: 7%, Al_2O_3: 3%), a rather interesting phenomenon was observed. Firstly, it is obvious that in our chemistry which involves substitution at the H of the SiOH group on the glass surface, that the steric requirements of the organic moieties preclude complete conversion of all the SiOH. So when the insulation fiber was treated to attach a specific R group using a particular organosilane the tensile test gave the Weibull plot shown as Figure 7. It should be noted that this treated glass is nominally stronger (fresh: 2.89 ± 0.3 GPa; aged: 2.19 ± 0.3 GPa) and slightly more resistant to the accelerated aging cycle than the non–treated fiber (fresh: 2.46 ± 0.2 GPa; aged: 2.09 ± 0.3 GPa). Subsequently the R group was attached using a second precursor molecule where the leaving group was much larger such that, the steric requirements for the reaction was also significantly greater. This treatment gave rise to the Weibull plot presented as Figure 8. As is noted only *ca.* 40% of the samples showed a loss of tensile

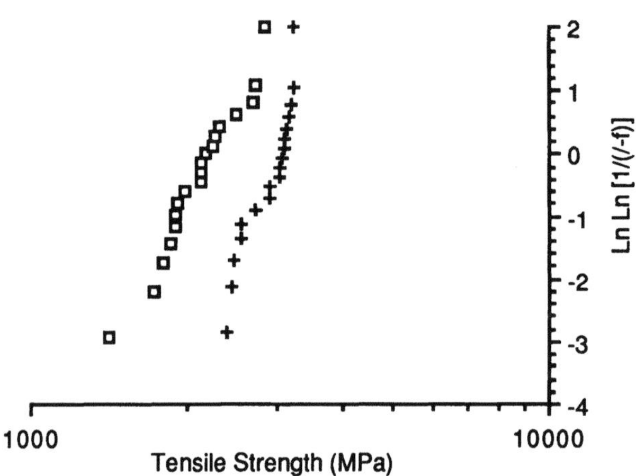

Figure 7. Insulation glass fibers (10μm diameter, 30 mm test length) to bond an organic group 'R' to the surface. Before (+) and after (□) accelerated aging (20 h, 50°C, 90% RH).

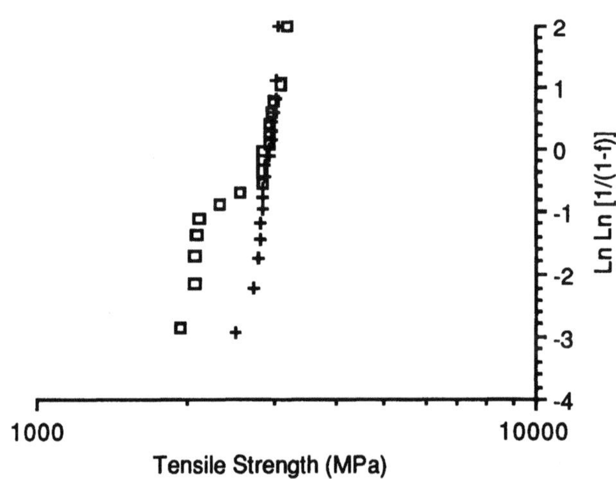

Figure 8. Insulation glass fibers (10μm diameter, 30 mm test length) treated to bond an organic group 'R' to the surface. Before (+) and after (□) accelerated aging (20 h, 50°C, 90% RH).

strength upon aging. Also, the observation of an increase in strength for a certain number of the samples is not consistent with Griffith theory. The virgin fibers were on the average 17% 'stronger' (fresh: 2.89 ± 0.1 GPa) and 26% 'stronger' after aging (aged: 2.64 ± 0.4 GPa). Clearly then this shows that the chemical activity of the surface is not only modified by the attachment of this type of group but also by its concentration. None of these points is particularly surprising in view of chemical knowledge, but such behavior would not be expected from classic corrosion theories.

As had been previously noted "Griffith cracks are not an intrinsic feature of glass, but result from mechanical damage or surface contamination"[16]. The mathematics of Griffith's microflaws predict that a given maximum crack size corresponds to a unique strength. Furthermore, the conventional theory of dynamic and static fatigue based upon fracture mechanics[17] assumes a one–to–one correspondence between maximum flaw size and strength, which is not the case for the samples of these three glasses. Thus use of the Griffith formalism for virgin fibers that display no distribution of strengths would result in flaws that are of a unique size, orientation, and shape. Additionally, fibers for which the expected statistical distribution of flaw sizes (Griffith–flaws) is initially absent and are subsequently aged, give rise to a situation where all flaws must propagate through the solid from the surface at a single velocity. What is being described is the manufacturing of a delta function, which would not be expected to occur in a mechanical process and application of Griffith theory thus becomes a moot point. Thus from our results it is reasonable to suggest that the surface potentials in silica–based glasses overwhelm simple diffusional forces, when considering mechanisms of corrosion processes. It can be then assumed that modifying the surface charge, in particular driving it towards the bulk potential should cause the collapse of the surface alkali accumulation layer thereby reducing the surface bond stresses/rupture probability: this is exactly what has been accomplished for the three silicate glasses herein described. The small distributions of strengths no doubt reflect a local variation in adsorption density (which is related to surface morphology) hence surface potential. Inherent to the field–assisted ion diffusion model as described in Equation 2 is the role of the silanol group upon both the corrosion and possible passivation chemistries. It is believed that the silanol is responsible for the accumulation and distribution of alkali ions in the surface region and that this in turn determines the chemistry of glass over a wide range of temperatures. It is in reality this chemistry that determines the strength and resistance to fatigue of our glass fibers and, as can be imagined, glass in general.

Though it is clear that the bulk of these data were collected on samples exposed to nominally neutral solutions at the outset of the aging experiments. We must not forget that the same local variations in adsorption density, and thus the surface potential which gives rise to the distribution of tensile strengths, can give rise to a local surface pH in excess of 9.5. The corrosion (or aging) process was described as and indeed is a two–fold mechanism. In the case of cementitious matrix materials, the glass, be it fiber or filler, can best be assumed to be in an extremely basic environment. So it is more reasonable to assume that the hydrolytic depolymerization step of the mechanism is the more important. This is not to say that glass cannot be successfully passivated so as to withstand such rigors. Aside from the glasses known to produce what are referred to as alkali resistant fibers, it remains to be seen if a surface treatment of a simpler silicate glass might not provide an equivalent resistance to highly alkaline media.

Conclusions

It has been suggested that Na^+ migration plays an important role in the aging of glass fibers. It was further suggested that if one could affect the migration behavior that the tensile strength of fibers could be altered. The results presented herein are consistent with a mechanism that involves the Na^+ migration in the rate-determining step of dynamic aging of glasses. In addition, we have shown that surface chemistry which modifies the migration behavior of sodium ion has altered the fatigue properties of our samples.

Acknowledgments

We would like to thank our management at Battelle-PNL for the opportunity to present this work. The work on silica fiber optics was realized in the 3M Physics & Materials Group of the Central Research Laboratories and was made easier by the pioneering concepts of W.M. Mularie. A number of the other experiments reported were performed while at Saint–Gobain Recherche and the support of J.P. Causse of Cie Saint-Gobain is gratefully acknowledged. Figures 1 and 4 are reproduced with the permission of Elsevier Science Publishers B.V.

References

1) A.A. Kruger, 'The Role of the Surface on Bulk Physical Properties of Glasses', in *Surface and Near–Surface Chemistry in Oxide Materials*, edited by L–C. Dufour and J. Nowotny, (Elsevier Scientific Publishers BV, Amsterdam, 1988), pp. 413–448.
2) F.T. Pierce, 'Theorems on the Strength of Long and Composite Specimens', *J. Text. Inst.*, **18**, T355 (1926).
3) A.A. Griffith, 'The Phenomena of Rupture and Flow in Solids', *Philo. Trans. Soc. London*, **A221**, 163 (1920).
4) W. Weibull, 'A Statistical Distribution Function of Wide Applicability', *J. Appl. Mech.*, 293 (1951).
5) C.E. Inglis, 'Stress in a plate due to the presence of cracks and sharp corners', *Proc. Inst. Naval Architects*, **55**, 219 (1913).
6) Z. Boksay, G. Bouquet and S. Dobos,'Diffusion Processes in the Surface Layer of Glass' *Phys. Chem. Glasses*, **8**, 140 (1967).
7) R.H. Doremus, 'Chemical durability of glass', in *Treatise on Materials Science and Technology* ,Vol. **17**, edited by M. Tomozawa and R.H. Doremus, (Academic Press, New York, 1979), p. 41–69.
8) R.W. Douglas and T.M. El-Shamy, 'Reaction of glasses with aqueous solutions', *J. Amer. Ceram. Soc.*, **50**, 1 (1967).
9) G.L. McVay and E.H. Farnum, 'Atmosphere Effects on Na Diffusion in Glass', *J. Amer. Ceram. Soc.*, **55**, 275 (1972); 'Anomalous Effects of H_2O and Na Diffusion in Glass' **57**, 43 (1974).
10) F.M. Fowkes and T.E. Burgess, 'Impurity Concentration at "Clean" Oxide Surfaces', in *Clean Surfaces*, edited by G. Goldfinger, (Marcel Dekker, New York, 1970), pp. 351–360.
11) W.M. Mularie, W.F. Furth and A.R.C. Westwood, 'Influence of surface potential on the kinetics of glass reactions with aqueous solutions', *J. Mater. Sci.*, **14**, 2659 (1979).
12) A.A. Kruger and W.M. Mularie, 'Silica glass fibers: modes of degradation and thoughts on protection', in *Optical Fibers in Adverse Environments*, edited by D. Boucher, (SPIE, **404**, Bellingham, WA, 1984), pp. 2–8.

13) A.A. Kruger, B. Ceccaroli and N.M. Harris, 'Chemical Modification of Silica-Based Glass Surfaces', in *Reactivity of Solids*, edited by P. Barret and L-C. Dufour, (Material Science Monographs, **28A**, Elsevier, Amsterdam, 1985), pp. 295–296.

14) M. Prokopowicz-Prigogine and W. Spreutels, 'La Réactivité du Verre I—Le Mécanisme d'Interaction des Molécules d'Acide Fluorhydrique avec la Maille Silicique du Verre', *J. de chimie physique*, **83**,7 (1986).

15) M. Prokopowicz-Prigogine, 'La Réactivité du Verre. II—un Modèle Alternatif du Mécanisme de Désalcalinasation du Verre de Silice', *J. de chimie physique*, **83**, 13 (1986).

16) W.C. LaCourse,'The Strength of Glass', in *Introduction to Glass Science*, edited by L.D. Pye, H.J. Stevens and W.C. LaCourse, (Plenum Press, New York, 1976), pp. 451–512.

17) W.B. Hillig,'Sources of Weakness and the Ultimate Strength of Brittle Amorphous Solids', in *Modern Aspects of the Vitreous State*, edited by J.D. Mackenzie, (Butterworths, London, 1962), pp. 152–194.

IMPORTANCE OF INTERFACIAL STRENGTH ON FRACTURE TOUGHNESS OF BRITTLE MATRIX COMPOSITES

P. PIROUZ, G. MORSCHER, AND J. CHUNG
Department of Materials Science and Engineering,
Case Western Reserve University, Cleveland, OHIO 44106, U.S.A.

ABSTRACT. The role of interfaces in the strength of composites is discussed through a comparison of fracture processes in elastically homogeneoeus materials versus bimaterials. Following this the fracture strength of fibrous composite interfaces are briefly reviewed. Experimental results for the temperature variation of interfacial shear strength in a fibrous composite are presented. A microsructural investigation of the regions where failure occurs is made by scanning and transmission electron microscopy. Based on the microstructural study, a rationale for the failure of the composite under study is offered.

1. Introduction

The high strength of ceramic materials is associated with high brittleness and low values of toughness. On the other hand, metals which exhibit high toughness values suffer from low strength. Thus, it is natural to combine the two materials and make a metal/ceramic composite which, hopefully, exhibits the desirable properties, and does not exhibit any of the undesirable properties, of the two components. This has been successful to a large extent and composites are now available that indeed exhibit relatively high strength and high toughness.

As technology has evolved, however, the need for high temperature applications has meant that the metallic component in the composite may fail either by undue softening with rising temperature, or by oxidation. Therefore, a different class of composites is becoming of increasing importance in which the metallic component is done away with completely and, instead, ceramic materials constitute both components. At first, this may seem to be a strange way of overcoming the problem of low toughness. After all, the ceramic components are, presumably, both brittle, and the combination also should show a brittle behaviour with a low value of toughness. However, the concept of "fracture toughness" as developed for single phase materials no longer holds for such composites.

In the following we shall first briefly review the concept of

L.-C. Dufour et al. (eds.), Surfaces and Interfaces of Ceramic Materials, 737–760.

"fracture toughness" in a monolithic elastically homogeneous material and then discuss the approach usually taken to deal with fracture in multi-phase materials constituting a composite. The discussion is mostly limited to composites made from brittle materials reinforced with high-strength fibers. However, some of the conclusions are more general and apply to other types of composites also. This section of the paper rests heavily on the work of Evans, Marshall, and coworkers and the excellent review papers written by them. Due references to the work of these authors are made in the course of the paper.

The measurement of interfacial shear strength as a function of temperature in a fiber-reinforced composite is then discussed and results presented for a composite made from reaction-bonded Si_3N_4 (RBSN) matrix reinforced by SiC fibers. Finally, microstructural investigation of this composite by means of scanning (SEM) and transmission electron microscopy (TEM) is reported and the results are related to the interfacial shear strength tests.

2. A Brief Review of Fracture Mechanics

2.1. FRACTURE TOUGHNESS OF AN ELASTICALLY HOMOGENEOUS MATERIAL

In an elastically homogeneous material, fracture mechanics relations have been developed for a long time starting with the pioneering work of Griffith [1]. Griffith basically considered the discrepancy between the theoretical cohesive strength of glass and the actual stress required to fracture it. It was known for a long time that the actual stress needed to fracture glass is much less than the theoretical values of bond breaking in this material. Griffith explained this discrepancy by the concentrating effect of a microcrack on the applied stress which magnifies the local stress at the crack tip to a value above that of the theoretical value. This was consistent with the work of Inglis [2] who had earlier shown that the stress concentration near the tip of an elliptical hole having a length of 2c, and a radius of curvature at its tip of ρ, is proportional to $(c/\rho)^{1/2}$ (Fig. 1).

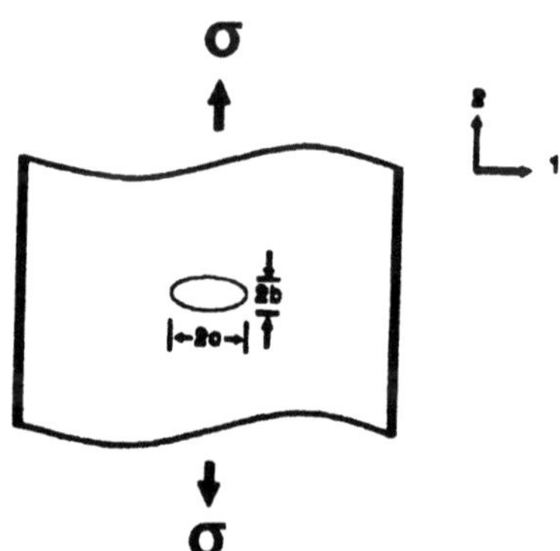

Fig. 1. An elliptical hole in a plate subjected to a uniform tension σ.

More specifically, Inglis showed that the actual stress, σ_{22}, at the crack tip is given by:

$$\sigma_{22} \approx 2(c/\rho)^{1/2}\sigma \qquad ...(1)$$

where σ is the applied stress. Hence the greater the length of the hole, and the sharper its tip, the larger will be its stress magnifying effect.

A sharp crack may be approximated as an elliptical hole with $\rho \to 0$. However, the Inglis relation cannot be used to calculate the stress at the crack tip because as $\rho \to 0$, $\sigma_{22} \to \infty$. To calculate the stress field around a crack tip, the "stress function" must be obtained for the biharmonic equations of linear elasticity theory subject to appropriate boundary conditions [3]. Making simple assumptions, the stress distribution around a crack tip can then be analysed. The analysis shows that the local stress tensor, σ_{ij}, near the crack tip generally depends on the product of the applied stress, σ, and the inverse square root of the distance from the crack tip, r [3]:

$$\sigma_{ij} = \alpha_{ij}\sigma(c/2r)^{1/2} \qquad ...(2)$$

where α_{ij} is a geometrical factor. These equations are valid for regions very close to the crack tip: $\rho < r < c$. Note that, the term $\sigma(c)^{1/2}$ is the same as in Inglis' relation, eqn. (1), and again signifies a magnifying effect dependent on the square root of crack length, c. Irwin [4] called this magnified stress at the crack tip the "Stress Intensity Factor". More precisely, Irwin defined the stress intensity factor, denoted by K, for a sharp elastic crack in an infinitely wide plate as:

$$K = \sigma(\pi c)^{1/2} \qquad ...(3)$$

Eqn. (2) for stress distribution around the crack tip may now be expressed in terms of K:

$$\sigma_{ij} = \alpha_{ij}K(2\pi r)^{-1/2} \qquad ...(4)$$

The dependence of the stress tensor on the distance from the crack tip is in the form of an inverse square root. Note that there is a square root singularity at the crack tip at which point the stress goes to infinity. This singularity is usually avoided by assuming that very close to the crack tip, linear elasticity is not valid and plastic deformations take place.

It should be mentioned that the simplifying assumptions needed to derive eqns. (2) and (4) are: (a) the crack tip in the unstressed state is perfectly sharp, and (b) the crack planes remain free of tractions at all stages of loading. More details of the fracture mechanics of homogeneously elastic materials are given in standard text books, e.g. [3], to which reference should be made.

Griffith's analysis of the crack propagation was based on energetic reasoning. The exposure of crack surfaces implies broken bonds between them and thus an increase in surface energy. The principle of conservation of energy requires that this increase in surface energy be accompanied by a decrease in some other energy. Griffith proposed that it was the elastic strain energy which was the source of the increase in surface energy. The elastic strain energy is released at the crack tip as the crack propagates in order to break the interatomic bonds and create two new surfaces. The Griffith criterion is based on the principles of reversible thermodynamics processes and states that a necessary condition for crack propagation is that the decrease in elastic strain energy be greater than the increase in surface energy. Balancing the two energy terms, he obtained an equation of the following form for the fracture stress, σ_F:

$$\sigma_F=(2E\gamma/\pi c)^{1/2} \qquad ...(5)$$

where E is the Young's modulus and γ is the surface energy of the material.

Equation (5) assumes elastic behavior at all stages of crack propagation. Orowan [5] suggested that very close to the crack tip, elastic behavior no longer holds and instead plasticity occurs in this region and thus the above equation is not valid. He modified the equation by adding to the surface energy, γ, another energy term γ_p this being the plastic work required to extend the crack planes. Orowan's equation for fracture stress, σ_F, is:

$$\sigma_F=[2E(\gamma+\gamma_p)/\pi c]^{1/2} \qquad ...(6)$$

Measurement of γ_p is rather difficult and Irwin [4] later replaced the total energy term with a parameter called the "Crack Extension Force", denoted by G, which is more directly measurable. Irwin defined the crack extension force as:

$$G=\pi c\sigma^2/E \qquad ...(7)$$

G is basically the rate of transfer of energy from the elastic stress field of the cracked structure to the inelastic processes of crack extension. For this reason, G is often called the "Strain Energy Release Rate". Clearly, the higher the applied stress, σ, the greater is the strain energy release rate. In order to get a sufficient transfer of energy to the crack tip to break the bonds and propagate the crack, G must exceed a critical value, G_c, the "Critical Strain Energy Release Rate". This parameter is often used to define the "Fracture Toughness" of the material. Using the critical strain energy release rate, Irwin's modification of Orowan's equation is:

$$\sigma_F=(EG_c/\pi c)^{1/2} \qquad ...(8)$$

A different way of looking at the problem of crack propagation in an elastically homogeneous material is through the concept of the stress intensity factor, K, which is just the magnified stress at the crack tip. Increasing the applied stress increases the stress intensity factor at the crack tip until it exceeds the cohesive strength of the material. The critical value at which this is reached is called the "critical stress intensity factor", K_c, which is another parameter used to define the "fracture toughness" of the material. Thus a necessary condition for fracture of a material is that the applied stress intensity factor, K, be greater than the critical stress intensity factor, K_c, i.e. $K>K_c$.

In dealing with stress intensity factors, the mode by which the external stresses or forces are applied should be taken into account. In general this is classified into three modes, (a) the crack-opening mode, or "mode I", in which case a tensile stress is applied in a direction normal to the faces of the crack, (b) the forward shear mode, or "mode II', in which case a shear stress is applied on the crack faces normal to the crack front, and (c) the parallel shear mode, or "mode III", in which case a shear stress is applied on the crack faces parallel to the crack front. When a load is applied in an arbitrary direction, the stress intensity factor can be resolved into these three components, K_I, K_{II}, and K_{III}. Similarly three critical values of stress intensity factor K_{Ic}, K_{IIc}, and K_{IIIc} can be defined. These are different because of the different way in which atomic bonds are broken during fracture. In mode I, the bonds between atoms on the opposing faces of the crack plane are basically stretched until they break. In mode II and III, however, the bonds are mostly bent until they fail and the faces shear. It should be mentioned that in an elastically homogeneous material, mode I is by far the most important mode of crack propagation. In a brittle solid, a crack always seeks the orientation which minimizes the shear loading, i.e. mode II or III.

We have now defined two parameters for "fracture toughness", K_c and G_c, or two criteria for crack propagation, one in terms of the stress intensity factor, $K>K_c$, and another in terms of the strain energy release rate, $G>G_c$. It can be shown that these two parameters are equivalent and related to each other by the following relationship [3]:

$$G_c=(1/E).[K_{Ic}^2 + K_{IIc}^2 + (1+\nu)K_{IIIc}^2] \qquad ...(9)$$

All the equations are shown for the case of plane stress ($\sigma_{33}=0$). For plane strain ($\varepsilon_{33}=0$, ε being the strain), E should be replaced by $\bar{E}$ for the K_I and K_{II} terms, where:

$$\bar{E}=E/(1-\nu^2) \qquad ...(10)$$

ν being the Poisson ratio. These equations show one advantage of the use of strain energy release rate over the concept of stress intensity factor: the strain energy release rates for different modes are additive.

2.2. FRACTURE TOUGHNESS OF A BIMATERIAL

By definition, a composite consists of two or more components. In this section we shall briefly review the concept of fracture toughness in the simplest composite, i.e. a bimaterial.

It is clear that in a bimaterial, the properties of of both materials should enter into the fracture toughness relationships. More importantly, when two different materials are combined to make a composite, a new entity, an "interface", is created. During the last few years it has become clear that this new entity behaves very differently from its adjoining materials. It has its own structure (two-dimensional), and its own set of mechanical properties. In addition, of course, physical properties such as diffusion rate, corrosion rate, sputtering rate, etc. are very different at an interface than in a bulk material.

Let us now compare the coplanar extension of a single crack from a notch in a monolithic material (Fig. 2a) with that at an interface in a bimaterial (Fig. 2b).

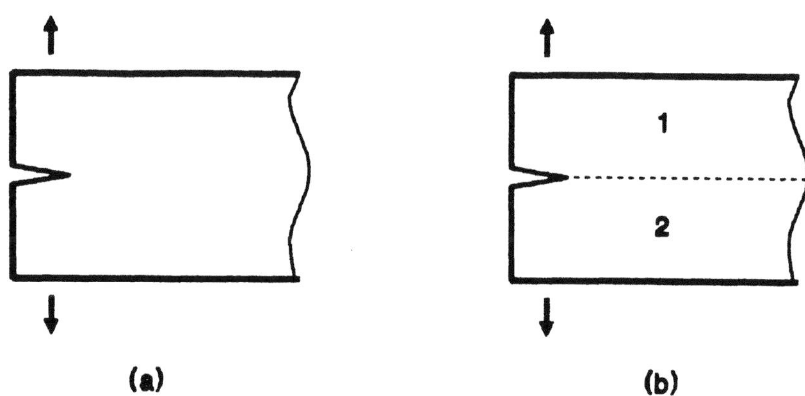

Fig. 2. Application of a tensile stress to a notch (a) in a monolithic material, (b) at an interface in a bimaterial.

In Fig. 2(a), since the load is applied in a direction perpendicular to the crack plane, i.e. in a pure mode I, there are no shearing forces on the crack faces. Hence, one may envisage increasing the applied load, which results in an increasing stress intensity factor K_I at the crack tip, until K_I exceeds K_{Ic} in which case the bonds at the crack tip break and the crack extends. When a load is applied with the same geometry to an interfacial crack in a bimaterial (Fig. 2b), there are not only tensile forces (mode I) present which stretch the bonds at the crack tip, but, in addition, shear forces (mode II and/or mode III) will be present on the crack faces which bend the bonds. In the bimaterial interface, of course, the bonds are interfacial, i.e. they are bonds which connect material 1 to material 2. The reason for the presence of mode II, or mode III, forces when the applied load is normal to the crack faces has to do with the different elastic constants on the two sides of the crack. This is in contrast with the elastically homogeneous material where internal mixed mode loading occurs only under an external mixed mode load.

It was just mentioned that the crack tip fields in a bimaterial interface invariably involve mixed modes. For simplicity we shall consider only mode I and mode II. In order to define a stress intensity factor for an interface crack, it is usual to introduce a complex number $K=K_1+iK_2$. The relative contributions of mode I and mode II can then be shown on an Argand diagram where the ratio K_2/K_1

defines a phase angle ψ where $\tan\psi = K_2/K_1$. Once the phase angle is known, the relative contributions of mode I and mode II loading on the crack tip will be known. Recently, for certain bimaterials, experiments have been devised with known values of phase angle in order to do controlled tests [6].

The discontinuity in the elastic properties across the crack plane has caused a lot of difficulties in the characterization of an interface crack in a bimaterial. The problem was first considered by Williams [7] and followed by England [8], Erdogan [9], Rice and Sih [10], and others. Under the usual boundary condition that the crack surfaces are stress-free, a stress singularity is found, i.e. $\sigma \propto r^{-\lambda}$. In the case of an elastically homogeneous material, λ is real and equal to $-1/2$, i.e. as in eqn. (2) $\sigma \propto r^{-1/2}$. However, it was found that in a bimaterial, λ is a complex number and is given by $\lambda = 1/2 - i\varepsilon$. Thus, the inverse square root singularity is modulated by an oscillating behavior as a function of distance from the crack tip, r. The stress oscillations imply crack surface interpenetration behind the crack tip which is physically inadmissible. However, this behavior is only for regions very close to the crack tip ($r < 10^{-4}c$). It should also be noted that it is possible to obtain stress relationships which do not have the oscillatory behavior but then it is no longer possible to assume that crack surfaces are stress-free.

As a result of mixed mode loading at the crack tip, there is not only a crack opening displacement (due to mode I) but also a rotation of the crack surfaces through an angle ϕ (due to mode II). The relation between the phase angle, ψ, and ϕ is through one of the Dundurs parameters, β [11]:

$$\psi = \phi - [\ln(r)/2\pi].\ln[(1-\beta)/(1+\beta)]$$

Dundurs parameters [12], which have been found to be very useful in describing interfacial stresses, are defined as follows:

$$\alpha = \frac{\mu_1(1-\nu_2)-\mu_2(1-\nu_1)}{\mu_1(1-\nu_2)+\mu_2(1-\nu_1)} \qquad \text{...(11a)}$$

and:

$$\beta = \frac{\mu_1(1-2\nu_2)-\mu_2(1-2\nu_1)}{2[\mu_1(1-\nu_2)+\mu_2(1-\nu_1)]} \qquad \text{...(11b)}$$

where μ is the shear modulus. The Dundurs parameter α is basically the elastic mismatch between material 1 and material 2.

The critical strain energy release rate is not a material constant for a bimaterial independent of the mode of loading, but rather it depends on the testing geometry through the phase angle ψ. Hence to know the fracture toughness of a bimaterial, the variation of G_c with respect to ψ, i.e. $G_c(\psi)$, should be known. In each case, two parameters need to be specified: ψ and G_c. As mentioned before, ψ is determined by the mixity of opening and shearing of the crack surfaces and depends on the modulus mismatch, mode of loading, as well as the crack length. Once the phase angle is known for the particular testing condition, G_c can be specified, e.g. from the $G_c(\psi)$ curve, to determine the debond resistance of the bimaterial interface [13].

3. Fiber Reinforced Composites

3.1. STRONG AND WEAK FIBER/MATRIX INTERFACES

The early works on bimaterials were not directly related to structural composites. They were mostly applications of continuum elasticity, and extension of fracture mechanics from the simpler problems in single phase materials, to the more complex problems of bimaterials. The applications of the fracture mechanics relations to more real problems has accelerated recently with the work of Budiansky, Evans, Hutchinson, Marshall, Rice, etc. (see e.g. [14]). However, one of the earliest problems to be considered was the reinforcement of brittle materials with high strength brittle fibers in which Kelly and coworkers [15,16] did some pioneering work. It was shown that a composite made from a brittle matrix reinforced with high strength fibers increases the toughness. This is to say that the composite undergoes a much larger strain prior to failure as compared to the unreinforced matrix. One of the conditions for this increase in toughness is that the toughness of fibers be greater than that of the matrix. In addition, fibers must be strong enough to remain intact during the propagation of the crack through the matrix.

Let us consider a particular ceramic/ceramic composite which consists of (brittle) fibers set in a (brittle) matrix. The fibers are usually aligned parallel to each other as shown in Fig. 3.

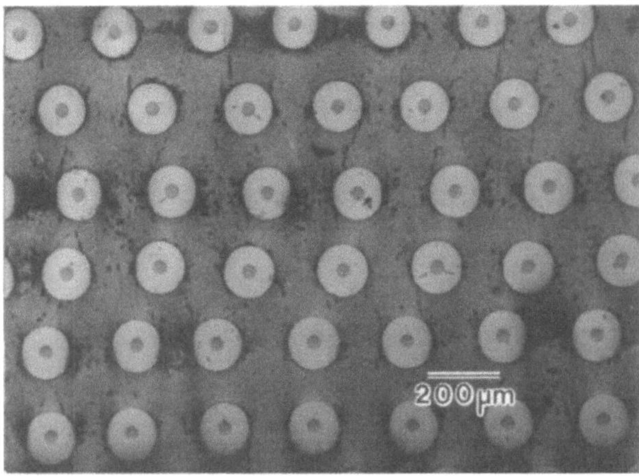

Fig. 3. Cross-sectional view of a typical fiber-reinforced ceramic with unidirectional fibers.

Suppose a crack propagating through the matrix comes across one of the fibers. Two situations may arise which depend on the interfacial strength:

a) Strong Interface: In this case, the crack easily crosses the interface and continues its propagation in the fiber (Fig. 4). Such a case leads to a catastrophic failure of the composite. The toughness of the composite then depends on that of the tougher material.

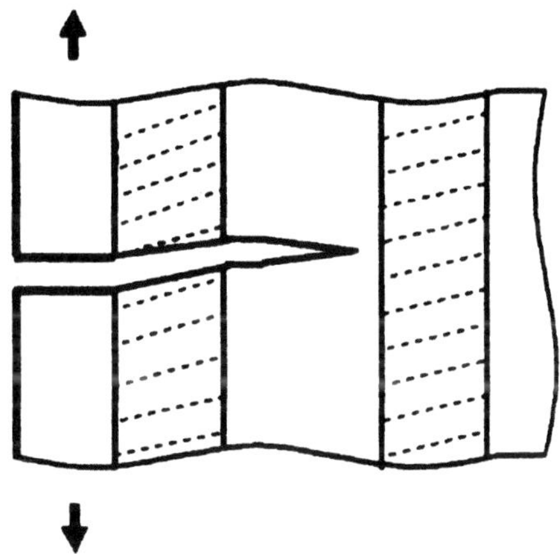

Fig. 4. Propagation of a crack in a composite with a strong fiber/matrix interface.

The stress-strain curve for such a composite looks similar to a homogeneous brittle material (Fig. 5) except, sometimes, for a tail end [13].

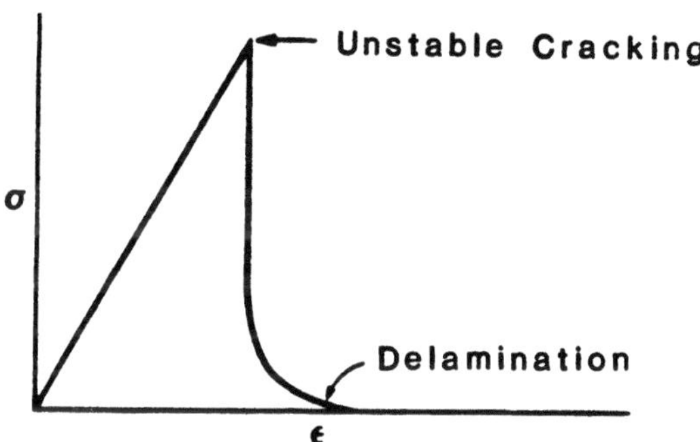

Fig. 5. Stress-strain curve for a fiber-reinforced composite with a strong interface (after [13]).

b) Weak Interface: In this case, the crack does not propagate from the matrix into the fiber but, rather, it chooses the path of least resistance, i.e. fracture occurs at the interface and the fiber debonds from the matrix. After the initiation of the interfacial crack, the debonding may continue in its wake as the crack propagates further in the matrix (Fig. 6) [13].

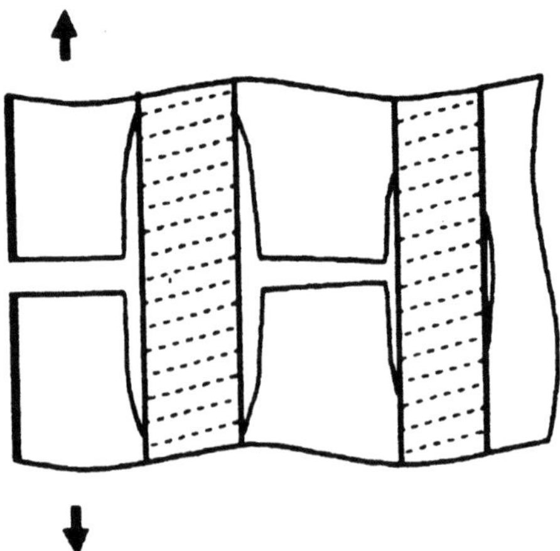

Fig. 6. Propagation of a crack in a composite with a weak fiber/matrix interface (after [13]).

Application of tensile loading to such a composite in a direction parallel to the fibers causes failure in a number of stages (Fig. 7). Initially the composite responds in a linear elastic fashion to the applied load. This linear response holds until a critical stress, σ_m, above which multiple cracking starts occuring in the matrix at regular intervals. From this moment, the stress-strain behavior is non-linear. Despite multiple cracks in the matrix, the composite does not fail because the intact fibers keep the cracked surfaces together.

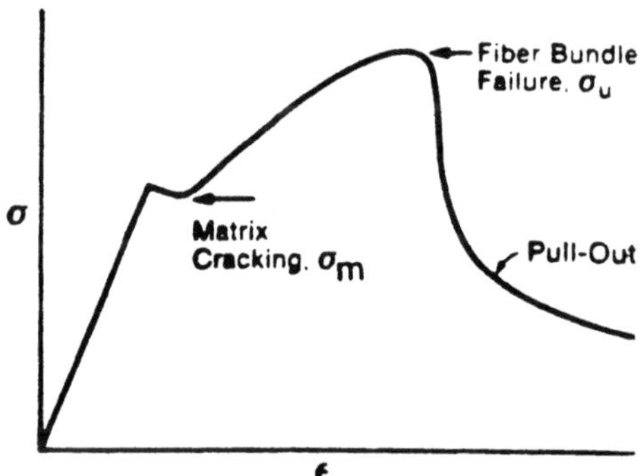

Fig. 7. Stress-strain curve for a fiber-reinforced composite with a weak interface (after [13]).

The intact bridging fibers basically restrain the opening of the crack surfaces. The composite fails at a higher stress, σ_u, when the intact fibers start fracturing, or are pulled out of the matrix. Thus, in this stage, the maximum load-carrying capacity of the composite is determined by the strength of the fibers.

Although composite failure occurs at the ultimate load carrying capacity, σ_u, which may be much greater than the load for the first matrix crack, σ_m, the latter is specially important because this is the first irreversible damage that occurs in the composite. A fracture mechanics analysis for the occurence of the first matrix crack has been given by a number of authors [15-18] and equations have been obtained for σ_m. These will be discussed in section 3.2.

In general, the debonding at the interface is accompanied by frictional sliding of the fiber along the debonded interface. The incidence of interfacial debonding vs. fiber cracking is determined by the relative magnitudes of the strain energy release rates, G, to those of the critical strain energy release rates, G_c. If $G^{int}/G^{fiber} > G_c^{int}/G_c^{fiber}$ then the failure is expected to occur at the interface [19]. The ratio G^{int}/G^{fiber} refers of course to the values of strain energy rate at the relevant phase angle of loading, ψ, for the interface crack.

The interface debonding vs. fiber fracture also depends on the elastic mismatch, α, between the fiber and the matrix [19]. This is the Dundurs parameter α defined in Eqn. 11(a). Rewriting it in terms of the Young's moduli of the fiber and the matrix:

$$\alpha = (\bar{E}_f - \bar{E}_m)/(\bar{E}_f + \bar{E}_m) \qquad ...(12)$$

Once the initial interface debond occurs, further debonding may take place in the crack wake as the crack front propagates further. It is, however, possible that the crack will deviate from the interface into the fiber if the phase angle of loading at the debond changes. So, although the crack starts at an interfacial debond, it may later cross into the fiber. This possibility is due to the changes in crack geometry, residual strain effects, and loading on the fiber during wake debonding which may influence ψ [20]. The change in the phase angle between the crack front and the crack wake has been studied by Evans et al. [21]. For example, in the case of no elastic mismatch (δ=0) at the crack front ψ=45°, whereas in the crack wake ψ increases as the fiber becomes stressed in the crack wake until a maximum steady-state level of ψ=75° is reached. If during this change of ψ, the ratio G^{int}/G^{fiber} ever becomes greater than G_c^{int}/G_c^{fiber}, small kinks near the debond tip will propagate into the fiber and cause its fracture.

3.2. INITIAL MATRIX CRACKING STRESS

In the case, when the strength of the fibers is greater than that of the matrix, Aveston et al. [15,16], Marshall et al. [17], and Budiansky et al. [18] have obtained equations for the first matrix cracking stress, σ_m. The requirement of high fiber strength relative to the matrix strength means that:

$$\sigma_f V_f > \sigma_m \qquad ...(13)$$

In this case, the matrix cracking stress, σ_m, is given by [15-18]:

$$\sigma_m=\{[6(1-\nu^2)K_{int}\tau_{int}V_f^{\,2}E_fE_c]/r(1-V_f)E_m^{\,3}\}^{1/3} \qquad ...(14)$$

where:

σ_m=stress for matrix cracking
τ_{int}=interfacial shear strength
K_{int}=toughness of the interface
r=fiber radius
V_f=volume fraction of fibers.

The subscripts m, f, and c refer to the matrix, fiber and composite respectively.

Note that the matrix cracking stress, σ_m, is independent of the crack size, i.e. it is a damage-tolerant property of the composite [22]. As mentioned before, this equation is based on the assumption that the fiber strength is greater than that of the matrix. Basically, this is because the crack is bridged by the fibers.

Eqn. (13) defines a maximum allowable value of interfacial shear strength, $\tau_{int}=\tau_{crit}$, for a given fiber strength σ_f. For large values of interfacial shear strength, when $\tau_{int}>\tau_{crit}$, fiber failure occurs due to the propagation of the crack across the interface. In the case when the interfacial shear strength is high, e.g. due to strong interfacial bonding, the composite strength is given by [17]:

$$\sigma_c=(E_c/E_m)[1 + (\sigma_fV_f/\sigma_m)^3].[K_{int}/(\pi c)^{1/2}] \qquad ...(15)$$

where c is the crack length. There is an inverse square root dependence on the crack length now and so the composite strength is sensitive to pre-existing cracks. The effect of the fibers is to increase the fracture toughness by the factor $(E_c/E_m)[1 + (\sigma_f f/\sigma_m)^3]$ relative to the monolithic matrix material [17].

According to eqn. (14), for $\tau_{int}<\tau_{crit}$, the matrix cracking stress σ_m increases with increasing τ_{int}. However, there is a maximum value, $\tau_{int}=\tau_{crit}$, above which the crack passes through the fibers and the whole composite fails catastrophically. In this case, the toughening effect of the fibers diminishes as τ_{int} increases further. Hence, some intermediate value of τ_{int} gives the optimum properties to a composite [17].

From the above discussion it is clear that the interfacial shear strength, τ_{int}, is an important parameter and a knowledge of it is essential in determining the strength of a composite. The question is then how is this parameter measured?

3.3. ULTIMATE COMPOSITE CRACKING STRESS

From the above discussion, it can be seen that the maximum load carrying capacity of a composite at which a tensile specimen finally fails is not controlled by the extension of a single crack. Once the matrix fails, the composite is still being held by the intact fibers. This is in line with the discussion in section 2.3 that the usual fracture mechanics criteria, such as fracture toughness, which are based on the extension of a single crack, cannot be used to predict the ultimate load carrying capacity of a composite. The strain to (total) failure of a composite includes the failure of the matrix as well as the fibers bridging the cracked matrix surfaces. The calculation of this ultimate load capacity, σ_u, is a much more complicated parameter which depends on the statistical fracture and pullout of the fibers in the composite [13].

4. Measurement of Fiber/Matrix Interfacial Strength

Direct measurement of the properties of an individual interface between a fiber and the matrix is not simple. However, recently, a simple technique for the measurement of the interfacial shear strength, τ_{int}, has been developed by Marshall [23]. We shall describe the measurement of τ_{int} for a particular fiber-reinforced composite by means of the Marshall technique in this paper. The composite consists of a matrix of reaction-bonded Si_3N_4 (RBSN) and the reinforcements are AVCO fibers which are basically SiC fibers with a carbon core. However, before discussing the results, the Marshall technique will be briefly desribed.

4.1. THE MARSHALL TECHNIQUE

In this technique, a Vickers indenter is used to push on the end of a fiber in a section normal to the fiber axis. The fiber clearly resists this because of possible interfacial bonds that need to be broken, and frictional forces that exist at the interface. Thus, the resisting force of the fiber, as measured by the interfacial shear strength, τ_{int}, may be the stress required to break the interfacial bonds, τ_b, and/or the stress required to overcome the frictional forces at the interface, τ_F. If the debonding strength is greater than that of the frictional resistance, i.e. $\tau_b > \tau_F$, then $\tau_{int} = \tau_b$. In this case, the applied stress on the indenter must be greater than τ_b to debond the fiber from the matrix. Since, $\tau_b > \tau_F$, the applied stress will be more than sufficient to overcome the interfacial friction and push the fiber through the matrix. Conversely, if the frictional resistance is greater than the debonding strength, i.e. $\tau_b < \tau_F$, then $\tau_{int} = \tau_F$ and an applied stress $\tau_b < \tau < \tau_F$ will break the bonds but it will not be sufficient to overcome the interfacial frictional forces and the fiber will not be pushed out.

The magnitude of τ_{int} can be determined from a measurement of the force, F, applied to the fiber and the resulting displacement, u, of the fiber below the composite surface [23] (Fig. 8). The slippage takes place over a certain length 'l' from the surface. There are two cases to consider [23]:

a) The Push-through Technique: In this case, the specimen thickness, t, is less than the slipping distance, l, ($t<l$) and the fiber is actually pushed out of the back surface of the specimen (Fig. 8b). A simple force balance relates the force applied to the indenter, F, to the interfacial shear strength, τ_{int}. Since the latter may vary over the slipping distance, τ_{int} is a function of z:

$$F=2\pi r\int_0^t \tau_{int}(z)\,dz \quad \ldots(16)$$

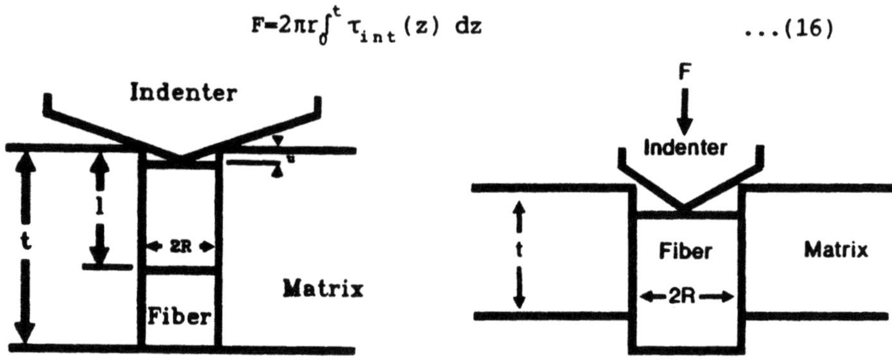

Fig. 8(a) The push-in technique, (b) the push-through technique.

Making the simplifying assumption that $\tau_{int}(z)$ is constant, i.e. $\tau_{int}(z)=\tau_{int}$, we have that [23]:

$$\tau_{int}=F/2\pi rt \quad \ldots(17)$$

In this approximation τ_{int} is the mean interfacial shear strength given by:

$$\tau_{int}=(1/t)\int_0^t \tau_{int}(z)\,dz \quad \ldots(18)$$

Variation of $\tau_{int}(z)$ vs. z has been determined by Grande et al. [24] using finite element techniques. This is shown schematically in Fig. 9 where it can be seen that τ_{int} goes through a maximum at a distance of ~0.8r from the surface.

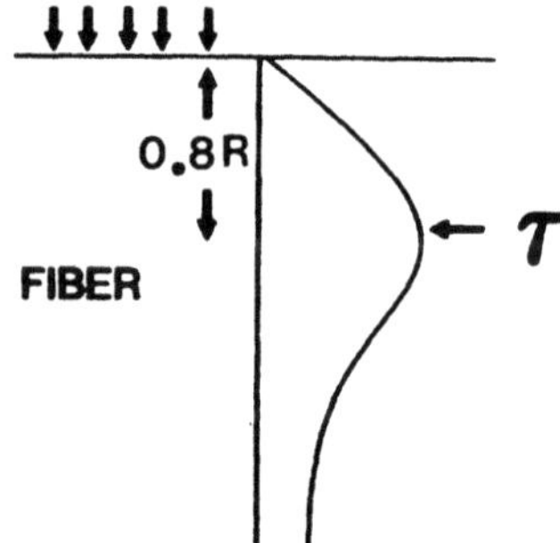

Fig. 9. The schematic variation of $\tau_{int}(z)$ as a function of z, (after [24]).

b) <u>The Push-in Technique:</u> In this case, t>l and the analysis is much more complicated. However, a simplified expression has been given by Marshall [23]:

$$\tau_s=F^2/4\pi^2 ur^3 E_f \quad \ldots(19)$$

The push-in technique requires a measurement of the fiber displacement, u, which is not easy. Using the newly developed nano-indentation machines (e.g. the Nanoindenter [25]), it is possible to plot force-displacement curves for the fiber movement under an applied indenter load. Such tests have been performed by Marshall and Oliver [26] on, for example, the polymer derived SiC Nicalon fibers in a glass matrix. With the ordinary micro-indentation machines, force-displacement curves cannot be obtained. On the other hand, the use of a nano-indentation machine, which allows the acquisition of force-displacement curves in the push-in technique, is limited by the very small loads available. As a result, it cannot be used in many systems for measurement of interfacial shear stress.

One drawback of the Marshall technique should be noted. In any indentation test, the loading is compressive. Hence, in the push-through, or the push-in, test, the fiber expands in the transverse directions owing to the effect of the Poisson ratio. This increases the normal stress and consequently the interfacial friction stress.

It is also clear that the measurement of the interfacial shear stress by this technique depends on the geometry of the fibers. In particular, in a commercial composite, where thousands of fibers are set in the matrix, there are variations in the fiber diameters, in the roughness of the outermost surface of the fibers, in the alignment of the fibers, etc. Hence, the measurement of the interfacial shear strength by this technique is necessarily statistical, and only a mean statistical value is obtained for a particular composite.

4.2. INTERFACIAL SHEAR STRENGTH IN SiC FIBER-REINFORCED/RBSN COMPOSITE

In the present work, the push-through technique has been used to estimate the interfacial shear strength in an AVCO-fiber/RBSN-matrix composite. The geometry of the test is as shown in Fig. 8(b) and the measurements were carried out as a function of temperature in vacuum over the range 25-1300°C. The details of this work are to be published [27] and here only a brief summary of the results will be presented. In the calculation of the interfacial shear strength, the approximate equation (17) has been used.

Fig. 10(a) shows a typical histogram of the results at room temperature. This figure shows the fraction of fibers pushed out at

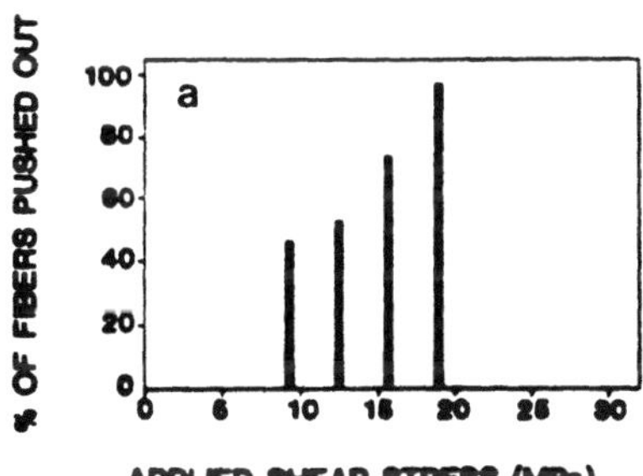

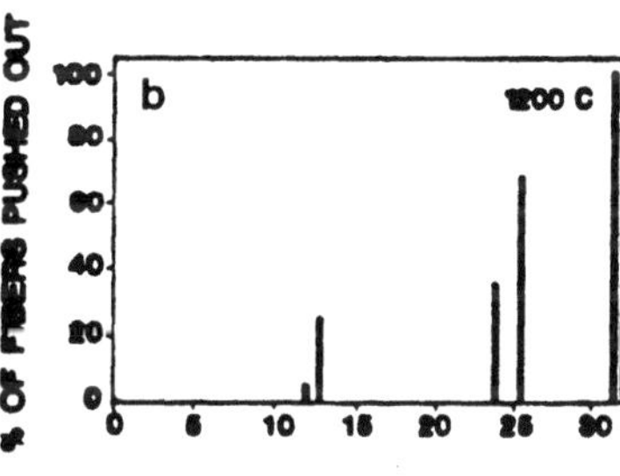

Fig. 10. Histogram of fraction of fibers pushed out vs. applied shear stress in the push-through tests at (a) room temperature, (b) 1200°C.

an applied stress τ. In any such histogram two stresses, τ_l and τ_u, can be defined for the lower bound and the upper bound shear stress respectively. For $\tau<\tau_l$, no fibers can be pushed out whereas for $\tau>\tau_u$ all fibers are pushed out. At any applied stress $\tau_l<\tau<\tau_u$, a fraction of the fibers are pushed out as shown by the histogram. Hence the interfacial shear stress, τ_{int}, defines a range which is a reflection of the statistical nature of the test.

The data in Fig. 10 were obtained from an examination of tested specimens in an optical microscope or a scanning electron microscope (SEM). An SEM micrograph of the back surface of one of the specimens is shown in Fig. 11. Here it can be clearly seen that, for example, a load of 300 g on the indenter has not been sufficient to push out a fiber whereas a load of 500 g has been able to overcome the bonding and/or frictional forces and push the fibers out of the matrix.

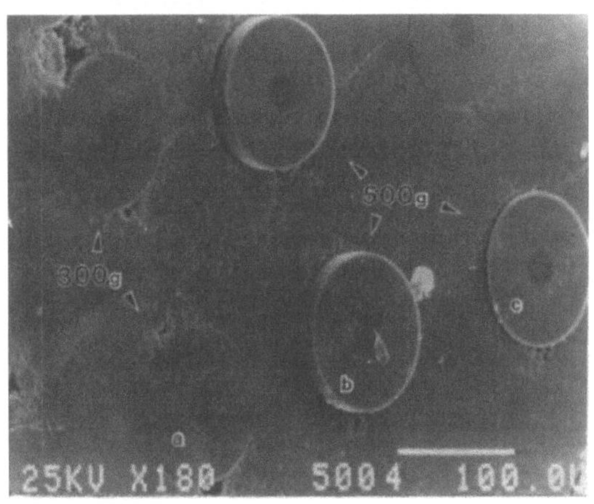

Fig. 11. SEM micrograph of the back surface of a tested specimen. The 500 g load has pushed the fibers out whereas the 300 g load has not.

A histogram of push-through tests carried out at 1200°C is shown in Fig. 10(b). The lower bound, τ_l, and upper bound, τ_u, of the interfacial shear strength increase steadily with temperature indicating that either the bonding strength or the frictional resistance, or both, increase with temperature.

In order to determine whether τ_{int} is controlled by the debonding strength or by the frictional resistance of the fibers another series of tests were carried out. First the fibers were pushed out at a particular temperature requiring a stress of τ_{int}. Subsequently the specimen was turned over and the pushed-out fibers were pushed back in. If $\tau_F>\tau_b$, then the value of τ_{int} measured in the two experiments would be the same. This is only approximately true because the rough surfaces of the fiber and the matrix may smoothen out to a certain extent during the first push-through test so that the frictional resistance would be less in the second push-back test.

On the other hand, if $\tau_F<\tau_b$ the interfacial shear strength measures the stress required to break the bonds and the reverse push test will only measure the frictional resistance, τ_f. Results from measurement of the push-back tests are shown in Fig. 12 at room temperature and at 1200°C.

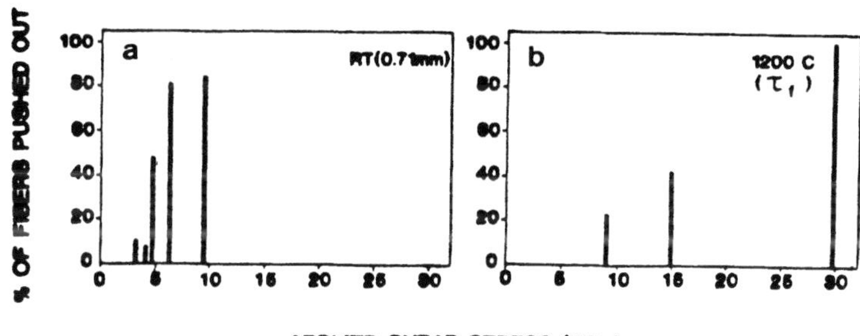

Fig. 12. Histograms of push-back tests at (a) room temperature, (b) 1200°C.

A comparison of Figs. 10 and 12 shows that at room temperature debonding controls the interfacial shear stress and τ_f is only about $\tau_b/2$. At 1200°C, on the other hand, $\tau_{int} \approx \tau_f$ implying a three fold increase in the frictional resistance of the fibers with temperature.

Finally, Fig. 13 shows the variations of interfacial shear stress with temperature. It can be seen that τ_{int} changes from ~15 MPa at room temperature to ~32 MPa at 1300°C for 100% push-out. As mentioned before, it appears that this increase is due to a rise in the frictional resistance of the fibers.

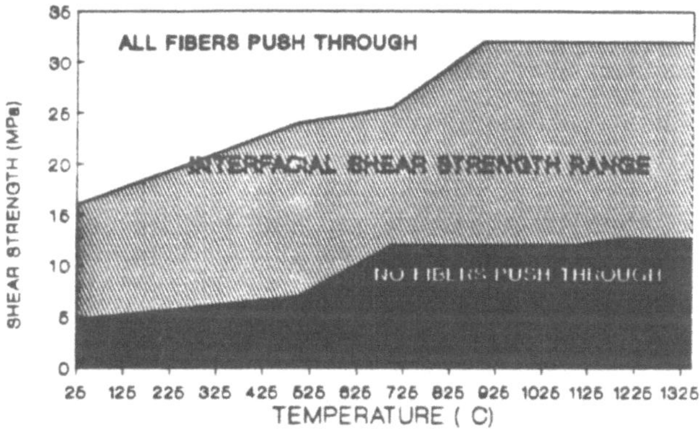

Fig. 13. Interfacial shear strength as a function of temperature.

The increase in the frictional stress of the fibers, σ_f, appears to be due to thermal expansion mismatch effects which increase the interlocking of the rough fiber surfaces into the matrix. This is discussed further in Ref. [27].

5. Microstructural Investigation of the Composite

In order to get a better understanding of the fiber/matrix interfacial failure, a microstructural study of the composite was carried out by SEM and TEM.

SCS AVCO fiber is a complicated composite by itself consisting of many different layers each being a different phase [28,29]. It is produced by chemical vapor deposition (CVD) of a number of different coatings on a carbon fiber which acts as the substrate. A cross-sectional view of the SCS-6 fiber set in the RBSN matrix is shown in Fig. 14.

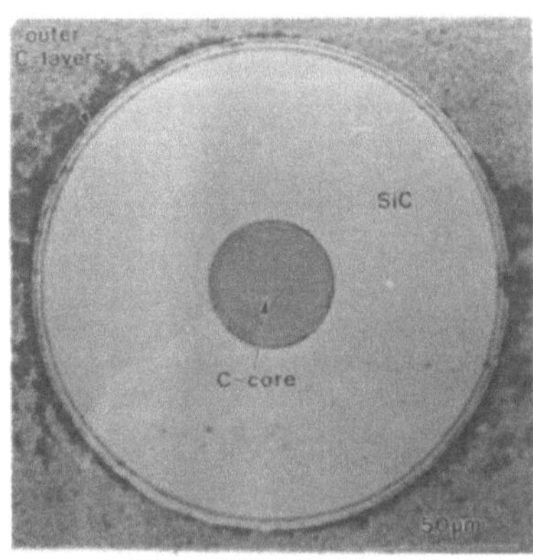

Fig. 14. Cross-sectional optical micrograph of an SCS-6 fiber set in a RBSN matrix.

The fiber consists of a non-graphitic (amorphous) carbon core with a diameter of ~32 μm, a thin coating (~1.5 μm thick) of turbostratic carbon (TC), an inner SiC layer (thickness ~21 μm), an outer SiC coating (thickness ~29 μm), and finally an outermost coating with a thickness of ~4 μm. As shown later, this outermost coating consists primarily of turbostratic carbon with some SiC precipitates. The microstructure of the inner parts of the SCS-6 fiber has been already described in ref. [29]. Here, we shall only focus on the electron microscopy study of the outermost layer, circled in Fig. 14.

An SEM micrograph of a pushed-out fiber is shown in Fig. 15. The fiber push-out can be observed to be due to a slippage at a region about halfway within the outermost ~4 μm coating; this is marked by an arrow. A second type of failure also occurs between the outermost SiC layer and the innermost TC layer of the coating; this is not shown here. It is clear that an understanding of the nature of the outermost coating is of essential importance. The two types of failure do not always occur simultaneously. Roughly 75% of failures occur within the outermost coating, whereas about 25% of failures occur at the interface between SiC and the innermost TC layer of the coating.

Fig. 16 shows a low magnification TEM micrograph of the outermost coating. A narrow white band is observed crossing the mid-section of this layer.

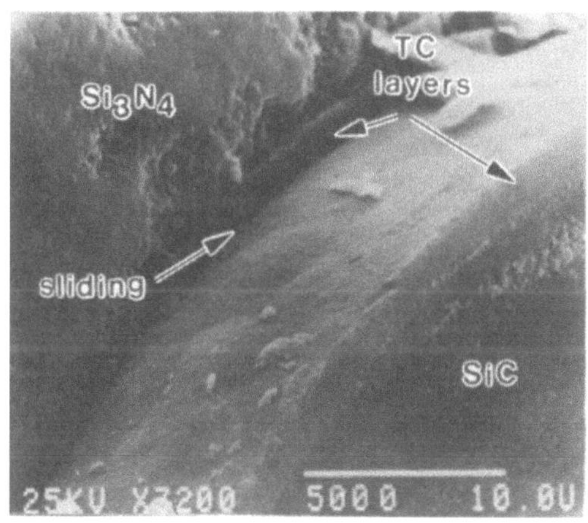

Fig. 15. SEM micrograph of a pushed out fiber.

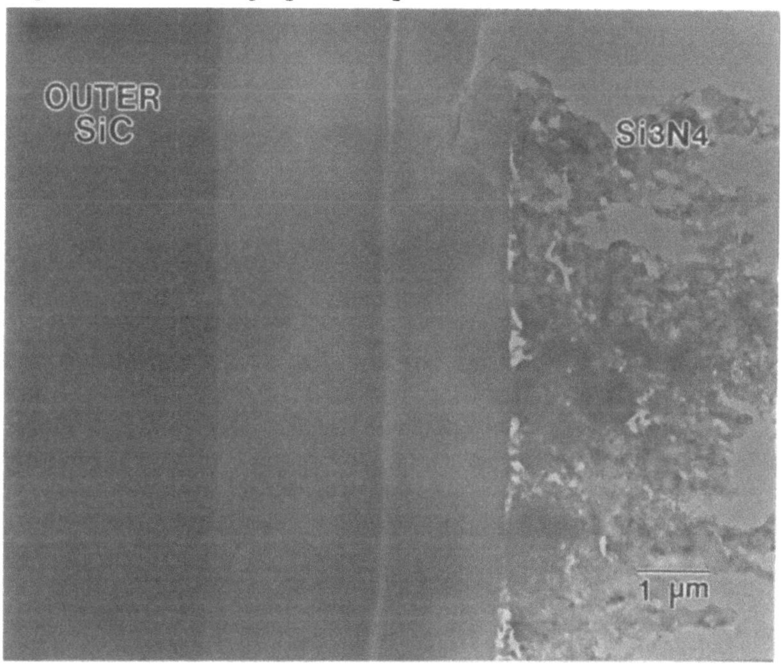

Fig. 16. Low magnification TEM micrograph of the outermost coating of a SCS-6 fiber. Note the narrow bright band at the center of the coating separating two sub-layers.

A higher magnification micrograph of the outermost coating (Fig. 17) shows that it consists of three sub-layers with thicknesses of ~2.4 μm, ~0.1 μm, and ~1.6 μm respectively.

Diffraction patterns from the regions on the two sides of the central narrow band are also shown in Fig. 17 indicating that they are both turbostratic carbon. A high density of precipitates, exhibiting dark contrast in the micrograph (Fig. 17), are embedded in sub-layers

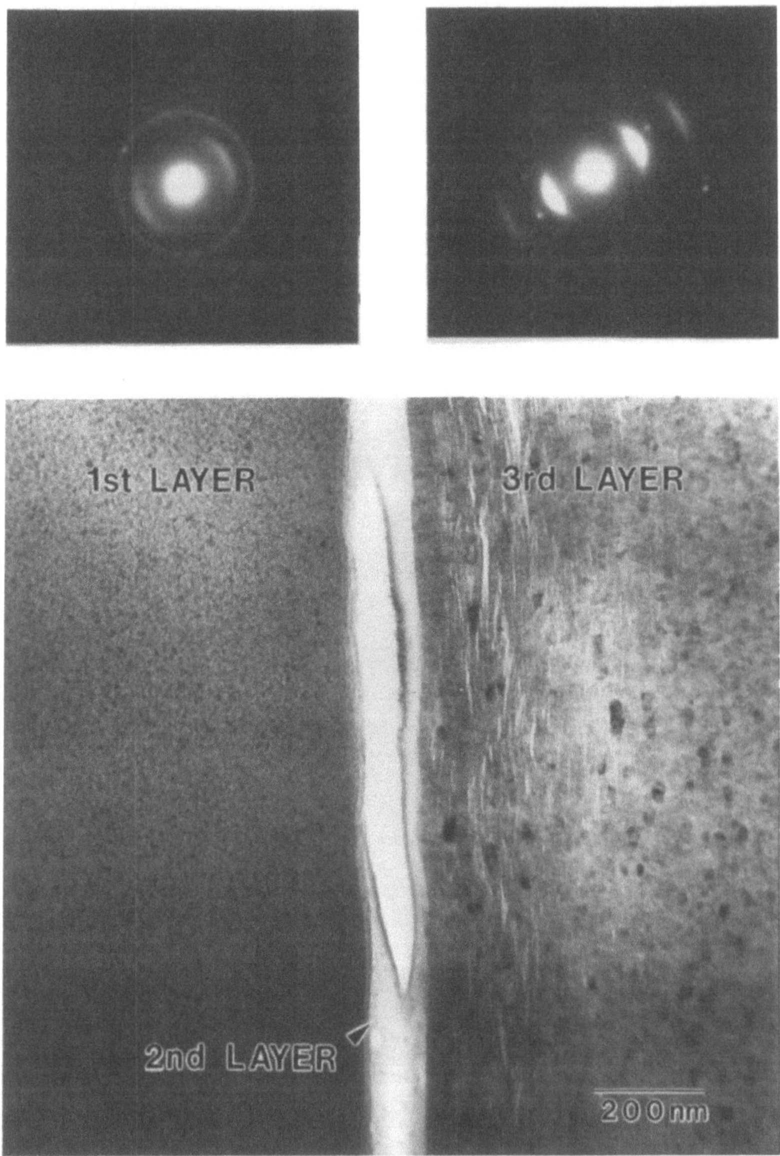

Fig. 17 Higher magnification TEM micrograph of the outermost coating of the SCS-6 fiber showing three distinct sub-layers. The diffraction patterns are from the two sides of the narrow band indicating that the sub-layers are predominantly turbostratic carbon. Note the precipitate-free nature of sub-layer 2.

1 and 3 while the central narrow band (sub-layer 2), separating sub-layers 1 and 3, is free of precipitates showing a uniform bright contrast. Detailed studies show that there are in fact a few precipitates present in sub-layer 2, but the density is very low. Diffraction studies show that the precipitates exhibiting dark contrast in sub-layers 1 and 3 are in fact SiC particles. These are present as weak rings in the diffraction patterns of Fig. 17.

The SiC precipitates start as relatively coarse particulates, with a relatively low density, in sub-layers 1 and 3 closer to the fiber center. They become progressively finer with a higher density in moving out from the fiber center.

Fig. 18 shows a high resolution electron micrograph (HREM) of one of these nanometer-size SiC particles embedded in the turbostratic carbon matrix. It can be seen that the SiC particle is internally twinned.

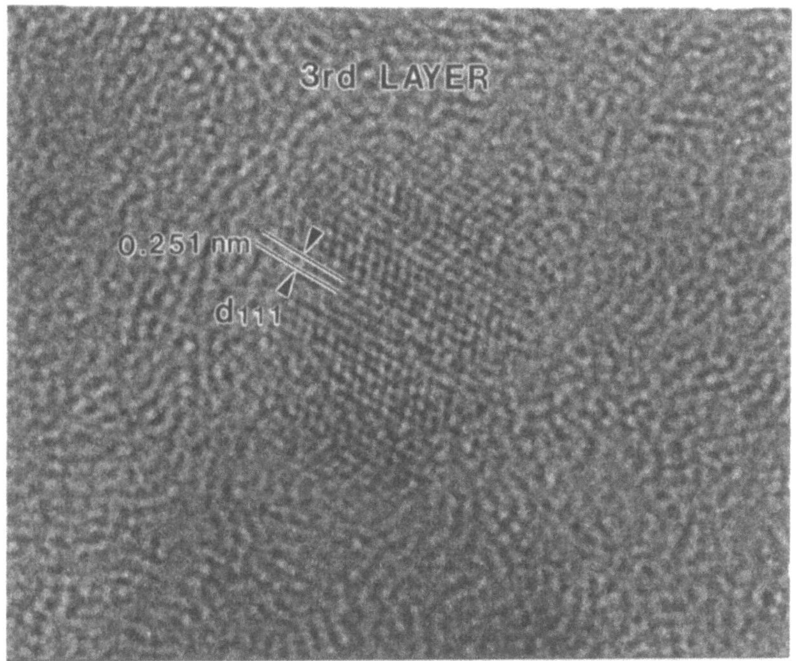

Fig. 18. High resolution TEM micrograph of a SiC particle embedded in turbostratic carbon in the third sub-layer of the coating.

6. Relation of the Fiber Microstructure to the Composite Strength

The picture that emerges from this microstructural study of SCS-6 AVCO fibers is that sub-layers 1 and 3 are micro-composites consisting of a matrix of turbostratic carbon reinforced by SiC particulates. On the other hand, sub-layer 2 is just a layer of turbostratic carbon with hardly any reinforcing particulates. It is therefore the weakest part of the outermost coating of the SCS-6 AVCO fiber. In the case of fiber push-out, where the loading is at the end of a fiber, sub-layer 2, being the weakest region of the fiber, is where the debonding and failure takes place. In addition, however, failure also occurs at the interfacial region between SiC and sublayer 1 indicating that this is also a rather weak region. Closer study of this region [30] shows, in fact, that the graphitic basal planes of the turbostratic carbon in sub-layer 1 are parallel to the {111} planes of SiC. Further away from the interface with the SiC coating, the basal planes of turbostratic carbon in sub-layer 1 are no longer as parallel, but rather become randomly oriented. Also, the density of SiC particulates in sub-layer 1 in this region is much lower than that in regions further away from the fiber center. In fact there is a uniform increase in the density of SiC particles in sub-layer 1 in going away from the outermost SiC coating toward sub-layer 2. As a result of these two features, sub-layer 1 close to the interface with the outermost SiC coating is also a weak region where fiber failure is likely to occur.

The fact that these results are directly relevant to other types of failure in this composite can be seen from Fig. 19 which shows a crack propagating through the composite. Note that the crack advancing from the Si_3N_4 matrix can enter the fiber because of the strong interfacial bonding between Si_3N_4 and sub-layer 3. But, the crack advances only about 2 μm before it reaches sub-layer 2. Here it finds a very weak region as a result of which circumferential debonding takes place within this thin sub-layer. The crack is now unable to advance into the bulk of the fiber. In brief, sub-layer 2 is where fiber pull-out occurs in a SiC/Si_3N_4 composite.

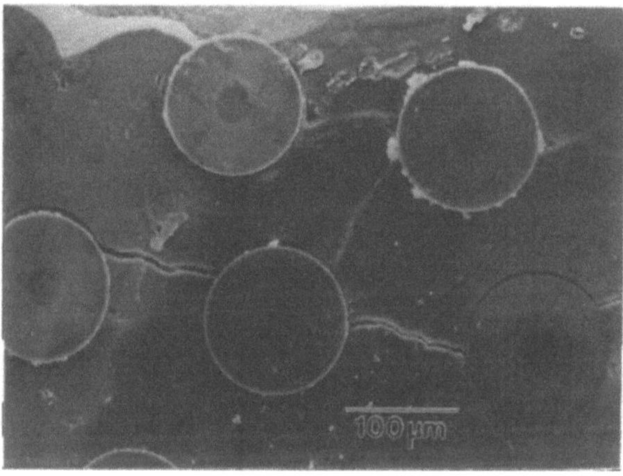

Fig. 19. Crack propagation through the composite during its fracture.

Acknowledgement

The authors are grateful to Dr. D. B. Marshall for permission to use results and figures 5-7 from his papers. Thanks are due to Professors R. Ballarini and A. H. Heuer for useful discussions and their comments. The SiC/RBSN composite was produced by Dr. R. Bhatt of NASA Lewis Research Center. We thank him and Dr. J. DiCarlo for providing us with some of these samples. This work was supported by the University Research Initiative at CWRU grant # N00014-86-K-0773 and NASA grant # NCC 3-73.

References

[1] A. A. Griffith, Philos. Trans. R. Soc. London, **221A**, 163 (1920).
[2] C. E. Inglis, Trans. Inst. Naval Architect. **55**, 219 (1913).
[3] B. R. Lawn and T. R. Wilshaw, Fracture of Brittle Solids, Cambridge University Press, (1975).
[4] G. R. Irwin, in Handbuch der Physik, Vol. VI, (1958). Springer-Verlag, Berlin.
[5] E. Orowan, in "Fatigue and Fracture of Metals", John Wiley & Sons, Inc., New York (1952).
[6] P. G. Charalambides, J. Lund, A. G. Evans, and R. M. McMeeking, J. Appl. Mech. (1989). In press.
[7] M L. Williams, Bull. Seismol. Soc. America, **49**, 199 (1959).
[8] A. H. England, J. Appl. Mech., **32**, 400 (1965).
[9] F. Erdogan, J. Appl. Mech., **32**, 403 (1965).
[10] J. R. Rice and G. C. Sih, J. Appl. Mech., **32**, 418 (1965).
[11] J. R. Rice, J. Appl. Mech., **55**, 98 (1988).
[12] J. Dundurs, J. Appl. Mech., **91**, 650 (1969).
[13] A. G. Evans and D. B. Marshall, in "Fiber-reinforced Ceramics", Ed. K. Mazdiyasni (1989). In press.
[14] Proc. Acta/Scripta Metall. Conf. on "Metal/Ceramic Interfaces", Santa Barbara, California, U.S.A., Jan. 16-18 (1989). Edited by M. Ruhle and A. G. Evans. In press.
[15] J. Aveston, G. A. Cooper, and A. Kelly, in "The Properties of Fiber Composites", Conf. Proc. National Physical Laboratory, IPC Science and Techology Press Ltd., Surrey, England, pp 15-21 (1971).
[16] J. Aveston and A. Kelly, J. Mater. Sci., **8**, 352 (1973).
[17] D. B. Marshall, B. N. Cox, and A. G. Evans, Acta Metall., **33**, 2013 (1985).
[18] B. Budiansky, J. W. Hutchinson, and A. G. Evans, J. Mech. Phys. Solids, **34**, 167 (1986).
[19] A. G. Evans, M. Y. He, and J. W. Hutchinson, "On interface debonding and fiber cracking in brittle matrix composites", University of California, Santa Barbara Internal Report (1988).
[20] M. Y. He and J. W. Hutchinson, Int. J. Fracture (1989). In press.
[21] A. G. Evans and J. W. Hutchinson, Acta Metall. (1989). In press.
[22] D. B. Marshall and A. G. Evans, J. Am. Ceram. Soc., **68**, 225 (1985).
[23] D. B. Marshall, J. Am. Ceram. Soc., **67**, C-259 (1984).

[24] D. H. Grande, J. F. Mandell, and K. C. C. Hong, J. Mater. Sci., **23**, 311 (1988).
[25] J. C. Pethica, R. Hutchings, and W. C. Oliver, Philos. Mag. A, **48**, 593 (1983).
[26] D. B. Marshall and W. C. Oliver, J. Am. Ceram. Soc., **70**, 542 (1987).
[27] G. Morscher, P. Pirouz, and A. H. Heuer, submitted to J. Am. Ceram. Soc., (1989).
[28] S. R. Nutt and F. E. Wawner, J. Mater. Sci., **20**, 1953 (1985).
[29] P. Pirouz, S. C. Farmer, F. Ernst, and J. Chung, in "Interfaces in Polymer, Ceramic, and Metal Matrix Composites", ed. H. Ishida, Elsevier Science Publishing Co., Inc., pp 141-157 (1988).
[30] J. Chung and P. Pirouz. To be published.

PRECIPITATION TOUGHENING AND PRECIPITATION HARDENING IN Y_2O_3-STABILIZED ZrO_2 CRYSTALS

A. Dominguez-Rodriguez
Departamento de Fisica de la Materia Condensada
Facultad de Fisica. Universidad Sevilla
41080 Sevilla, Spain
and
A.H. Heuer
Department of Materials Science and Engineering
Case Western Reserve University
Cleveland, Ohio 44106, USA

ABSTRACT: The literature on precipitation, precipitation toughening, and precipitation hardening in Y_2O_3 partially-stabilized ZrO_2 (Y-PSZ) single crystals is reviewed. Very strong crystals can be obtained at both ambient and elevated temperatures. At room temperature, "classical" transformation toughening appears to be operating, while at elevated temperatures (e.g. 1400°C), potent precipitation hardening results from the difficulty matrix dislocations have in cutting particles; the Burgers vectors of these matrix dislocations are not lattice vectors in the precipitates and the sheared particles are believed to be faulted. Understanding plastic deformation at intermediate temperatures (e.g. 700°C) in crystals deformed in compression remains a task for future work.

1. INTRODUCTION

The polymorphism of ZrO_2 is now well-known|1|. At room pressure, the stable phase has the cubic fluorite structure ($\underline{c}$-ZrO_2) at high temperatures,(melting point (2850°C) to ~2350°C; space group Fm3m), a tetragonally-distorted version of this structure ($\underline{t}$-ZrO_2) is stable at intermediate temperatures (2350°C to ~1170°C; space group $P4_2/nmc$) and an even more distorted form with monoclinic symmetry ($\underline{m}$-ZrO_2) is the equilibrium structure at lower temperatures (space group $P2_1/c$). The three crystal structures are shown in Fig.1.

Much interest has been expended over the years on phase transformations in ZrO_2 and its alloys|2,3|, as polycrystalline and polyphase ceramics find extensive use in advanced structural applications. This obtains because of the good strength and high toughness that is available in suitably-fabricated and heat-treated materials |4|. Some of the

L.-C. Dufour et al. (eds.), Surfaces and Interfaces of Ceramic Materials, 761–776.

same materials also find use as sensors and in other electrolytic applications because of the high ionic conductivity of ZrO_2 alloys |5|.

The polymorphic (c-ZrO_2 → t-ZrO_2) and (t-ZrO_2 → m-ZrO_2) phase transformations have both been studied (see refs. 2 and 3 for example).

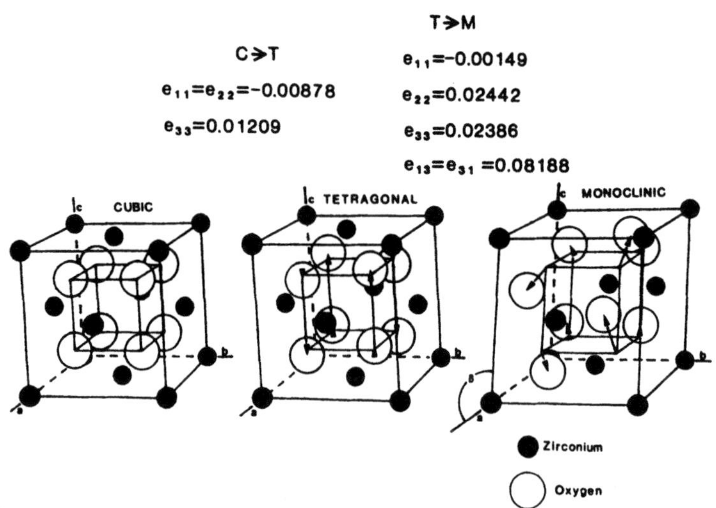

Fig. 1: Crystal structures of c-, t-, and m-ZrO_2. The nonzero components of the transformation strain tensors, ε_{ij}^T, for the (c→t) and (t→m) transformations are shown; the lattice parameter data used to calculate ε_{ij}^T are for pure ZrO_2 from Ref. 3. The arrows in the t and m structures indicate the displacements of the oxygens from their positions in the c phase.

The former reaction is diffusionless and displacive and gives rise to transformation twins and anti-phase domain boundaries (APB'S)|2,6,7|; most of our information derives from studies in Y_2O_3-ZrO_2 alloys |6-9|. The (t→m) reaction is martensitic|3,10|; the transformation start (M_s) temperature varies markedly with alloy composition, and within a given alloy, sensitively depends on the size of the t-ZrO_2 phase. If the M_s temperature is below room temperature, the (t→m) transformation at room temperature can be stress-induced. This phenomenon is at the heart of transformation toughening, which makes certain ZrO_2 alloys so interesting -- if the martensitic phase transformation occurs in the stress field of a propagating crack, the fracture resistance, and hence the strength, the flaw tolerance, and the fracture toughness all increase |11-14|. For historical reasons, single phase cubic alloys containing a sufficient concentration of solutes such as MgO, CaO, or Y_2O_3 to lower the (c→t) and/or (t→m) phase transformations are called fully- stabilized, i.e. Y_2O_3 fully-stabilized ZrO_2 or Y-FSZ. Alloys containing a mixture of phases, e.g. (c + t) or (c + t + m) are called partially-stabilized,e.g. Y_2O_3 partially-stabilized ZrO_2 or Y-PSZ. In many cases of interest, the t-ZrO_2 phase in PSZ's form by precipitation, and we next discuss precipitate morphology in various ZrO_2 systems.

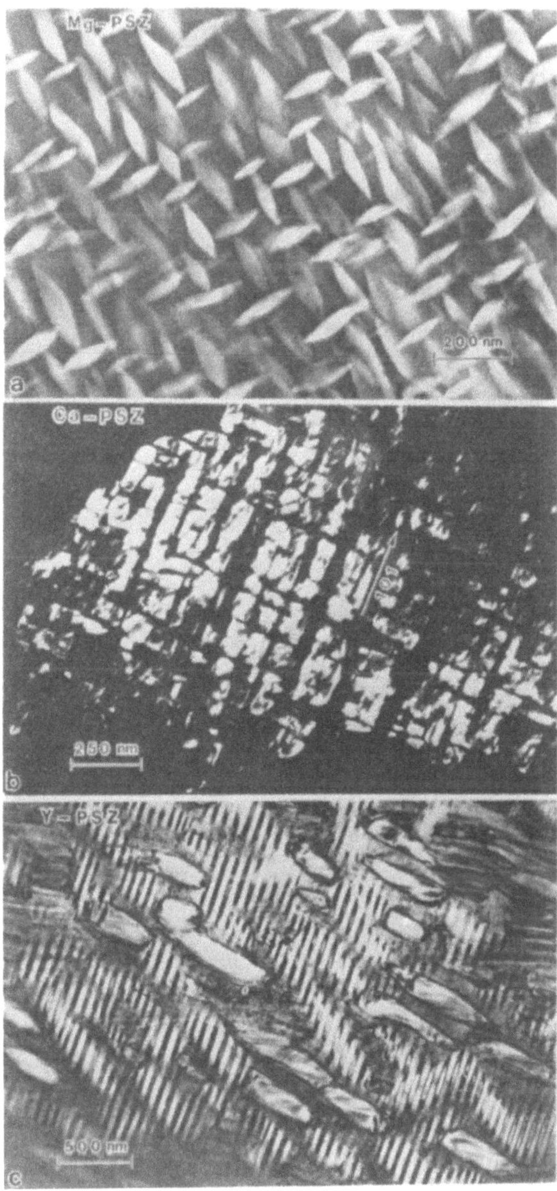

Fig. 2. Solute-lean tetragonal ZrO_2 precipitates in a solute-rich cubic ZrO_2 matrix in a) Mg-PSZ, b) Ca-PSZ, and c) Y-PSZ. Detailed discussion of the differing particle morphologies can be found in refs. 16 and 17.

2: PRECIPITATION IN ZrO_2 ALLOYS

The morphology of t-ZrO_2 precipitates in c-ZrO_2 matrices -- the microstructural feature of most interest in PSZ's intended for structural applications -- is shown in Fig.2 for three common solutes, MgO, CaO. and Y_2O_3. In all cases, these precipitates form by a diffusion-controlled pro-eutectoid precipitation reaction, the kinetics of which must be controlled so that precipitates do not coarsen and thereby increase their M_s temperature above room temperature, These particles are all coherent, as has been demostrated by high resolution electron microscopy (HREM) for Mg-PSZ and Y-PSZ |15|.

The specific particle morpholohy of course depends on minimization of strain and interfacial energies, and has been discussed for these systems by Lanteri, Heuer, et al |16,17|, on the basis of Khachaturyan's theory |18|.

In the following, we restrict our attention to single crystal alloys in the Y_2O_3-ZrO_2 system, which exhibit interesting mechanical properties at both high and low temperatures (other ZrO_2-based alloys, e.g. Mg-PSZ and Ca-PSZ, do not generally exhibit sufficient microstructural stability to permit high temperature use). The phase diagram for this system is shown in Fig. 3 |2,8|.

Before proceeding, however, we point out the interesting implications due to the high solubility of Y_2O_3 in t-ZrO_2. In particular, it is possible to lower M_s so drastically that fully tetragonal polycrystalline ceramics |19,20| (the so-called TPZ's or tetragonal ZrO_2 polycrystals) containing up to ~3m/o Y_2O_3 can be produced. Y-TZP's find widespread use, as they are among the strongest oxide ceramics now known (having room temperature strengths up to ~2GPa) |21|. Such strengths are also possible in single crystals, but only in two-phase precipita-

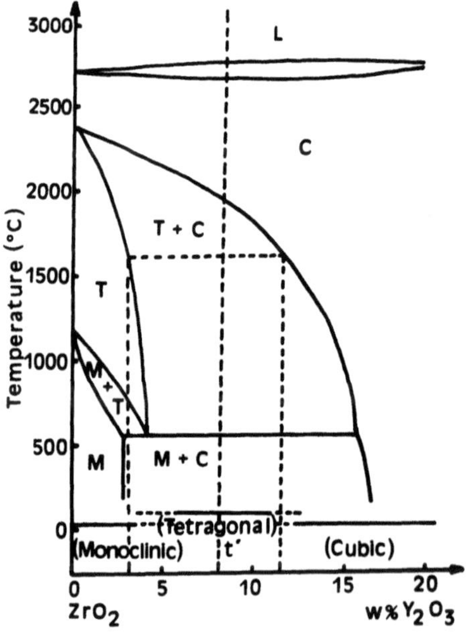

Fig. 3. High ZrO_2 portion of the Y_2O_3-ZrO_2 system (ref.15)

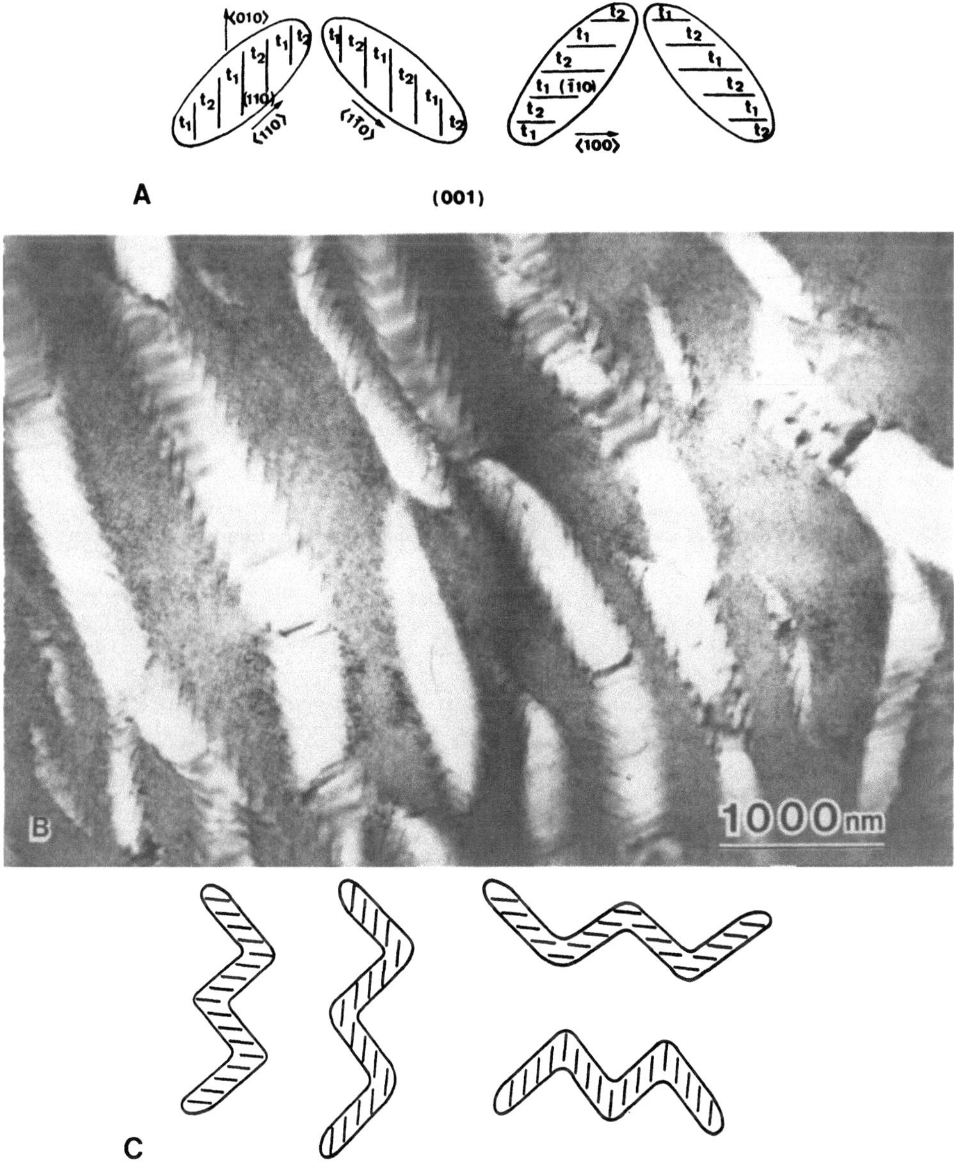

Fig. 4. "Colony" recipitate in Y-PSZ. a) shows a schematic drawing indicating how any one pair of tetragonal ZrO_2 variants (t_1 and t_2 in this case) can give rise to four distinct colony variants. b) shows an electron micrograph of 4.5 m/o alloy heat treated for 150 hours at 1600°C. c) is a schematic drawing revealing how the colonies can grow into elongated accordian-like "fibers".

tion-toughened PSZ's, as will be discussed in the next section. It is thus appropriate to discuss the precipitate microstructure in Y-PSZ in detail.

The $\underline{t}$-ZrO_2 precipitates in Y-PSZ in Fig. 2c are composite and consist of stacks of $\underline{t}$-ZrO_2 plates in twin orientation ($\vec{c}$ axes at nearly 90°), sharing a low energy and coherent {101} plane. Three $\underline{t}$-ZrO_2 variants, t_1, t_2, and t_3, are possible, each having its $\vec{c}$ axis parallel to one of the cube axes. Twelve distinct "colony" precipitates can be formed, with each colony involving t_1/t_2, t_2/t_3, or t_1/t_3 lamellar pairs. For any one pair, say, t_1/t_2, four colony variants are possible, involving two {110} twin planes, say (110) and ($\bar{1}$10), as shown in Fig. 4a. During annealing within the ($\underline{c}$ + $\underline{t}$) two-phase field of Fig.3, both the lamellar spacing and the size of the colonies increases at a rate governed by the diffusion kinetics. The microstructure of a 4.5m/o alloy heated for 150 hrs. at 1600°C is shown in Fig. 4b. The colony precipitates have grown together to form accordian-like "fibers" elongated along <001> by continuous switching of the {101} habit planes, as shown schematically in Fig. 4c.

The colony microstructure is clearly a means of reducing the strain energy of the two-phase system containing coherent precipitates, but certain other crystallographic features are worthy of mention:

i) The $\vec{c}$ axes of the two t variants within any colony are not exactly aligned along the cube axes of the matrix, but are tilted by a small angle α, given by $\tan(\alpha + \Pi/4) = c_t/a_t$ |24| where c_t and a_t are the lattice parameters of $\underline{t}$-ZrO_2.

ii) The two t variants are not present in equal volume fractions within a colony. Rather, the volume fraction of one variant, say t_1, is given by

$$f = -\varepsilon^\circ_{11} \,/\, (\varepsilon^\circ_{33} - \varepsilon^\circ_{11}) \qquad (1)$$

where $\varepsilon^\circ_{11} = (a_t - a_c)/a_c$ and $\varepsilon^\circ_{33} = (c_t - a_c)/a_c$ where a_c is the lattice parameter of $\underline{c}$-ZrO_2. HREM studies gave f= 0.43 (± 0.01) for a 4.5 m/o crystal annealed at 1600°C, whereas eq. 1 predicted f = 0.423 | 9 |.

iii) Lastly, HREM studies show that the colony/matrix interface is facetted, confirming Katchaturyan's prediction|18| that any remaining coherency strain in such a colony microstructure is confined to a small region adjacent the precipitate/matrix interface.

It goes without saying that the microstructure of any composition falling within the two-phase field of Fig. 3 depends on the heat treatment that particular alloy received below the cubic solvus. It is also the case that any alloy can be cooled sufficiently rapidly ("quenched") that the precipitation reaction is suppressed completely. In this case, the cubic matrix can undergo the displacive ($\underline{c} \rightarrow \underline{t}$) phase transformation (discussed earlier) at a critical temperature, T_c, that lies approximately midway between the (c) and ($\underline{c}$ + $\underline{t}$) solvuses of Fig. 3. For a 3 m/o Y_2O_3, T_c of ~1900°C has been determined experimentally |25|. The tetragonal product contains APB's and twins and is referred to as t'ZrO_2 as the solute content is so high that it is virtually impossible to induce the martensitic ($\underline{t}' \rightarrow \underline{m}$) phase transformation (further discussion of $\underline{t}'$-ZrO_2 can be found in ref. 9).

Michel et al |6|, have succeeded in growing tetragonal single crystals by skull melting and rapid cooling of a 3 m/o Y_2O_3 composition.

This important result of the growth of a tetragonal ZrO_2 single crystal (which they refer to as MTZ (metastable tetragonal ZrO_2)) by solid state transformation from a cubic single crystal demonstrates the importance of heat treatment on the microstructure (and resultant mechanical behavior) of ZrO_2 alloys. We begin our discussion of mechanical properties by reviewing the room temperature mechanical properties of these fully tetragonal crystals.

3. MECHANICAL PROPERTIES OF SINGLE CRYSTALS

3.1. Room Temperature Propeties and the Effect of Heat Treatment

Michel et al |6| measured the hardness and indentation fracture toughness, K_c of the 3m/o Y_2O_3 tetragonal ZrO_2 single crystal, and compared it to a 9 m/o Y_2O_3 cubic ZrO_2 single crystal prepared by the same skull melting technique. The hardness of the tetragonal crystal ranged from 11.5 to 13.5 GPa, while that for the cubic crystal was similar, 11.9 to 14.9 GPa. However, the fracture toughness of the tetragonal crystal was much higher than the cubic crystal -- 6 MPa $m^{1/2}$ compared to 1.8 MPa $m^{1/2}$ (± 30% deviation) (the values for the cubic crystals agree with data of Ingel et al |26| (to be discussed next) and with our data. The higher toughness cannot be ascribed to conventional transformation toughening, as the presence of m-ZrO_2 on fracture surfaces could not be detected using x-ray diffraction and Raman spectroscopy. The higher toughness was instead ascribed to the APB domain structure present in the tetragonal single crystals; it was also observed that the fracture surface was very facetted, following only {001} planes of the parent cubic structure. However, a rigorous mechanics-based explanation for the higher toughness of the tetragonal ZrO_2 single crystals remains a task for future work, particularly in the context of the results of Ingel et al|26| and Lankford et al|27| on two-phase crystals of similar composition, to be considered next.

Ingel et al studied crystals containing up to 12 m/o Y_2O_3. As expected from Fig. 3, crystals containing between 8 and 12 m/o Y_2O_3 were cubic, while pure ZrO_2 (0 m/o Y_2O_3) was monoclinic; intermediate compositions were multi-phase, the 2.0 and 2.5 m/o Y_2O_3 compositions containing m, t, and c reflections in x-ray diffraction patterns, while the 3.5 and 4.5 m/o crystals being a mixture of c and t-ZrO_2. Clearly, their skull-melted crystals had been cooled more slowly than those of Michel et al. so that some t-ZrO_2 precipitated from the c-ZrO_2 matrix; the solute content of those precipitates in the 2.0 and 2.5 m/o crystals must have been sufficiently lean that the M_s temperatures was above room temperature. Their 3 m/o crystal contained ~50 v/o t-ZrO_2, which from Fig. 3 suggest an "effective" precipitation temperature of ~1900°C, whereas the 30 v/o t-ZrO_2 they report in the 4.5 m/o crystal suggests an "effective" precipitation temperature of ~1700°C.

The fracture toughness varied markedly with composition. c-ZrO_2 crystals had K_c ≤1.7MPa $m^{1/2}$, which increased to a maximum of 8 MPa $m^{1/2}$ in the 2.5 ,/o Y_2O_3 crystal (Fig. 5). Curiously, the 3 m/o fully tetragonal single crystal of Michel et al. and the 3 m/o two-phase PSZ single crystal of Ingel et al. had nearly the same toughness (Fig. 5). Fracture strengths paralleled the fracture toughness, being a maximum of~1400 MPa

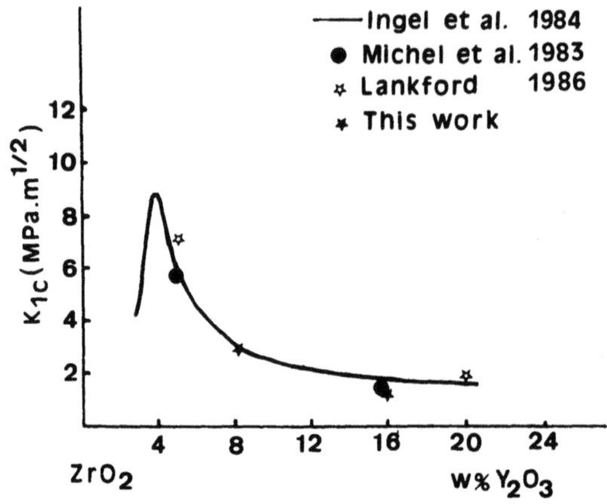

Fig. 5. Room temperature fracture toughness as a function of Y_2O_3 composition in single crystal Y-PSZ.

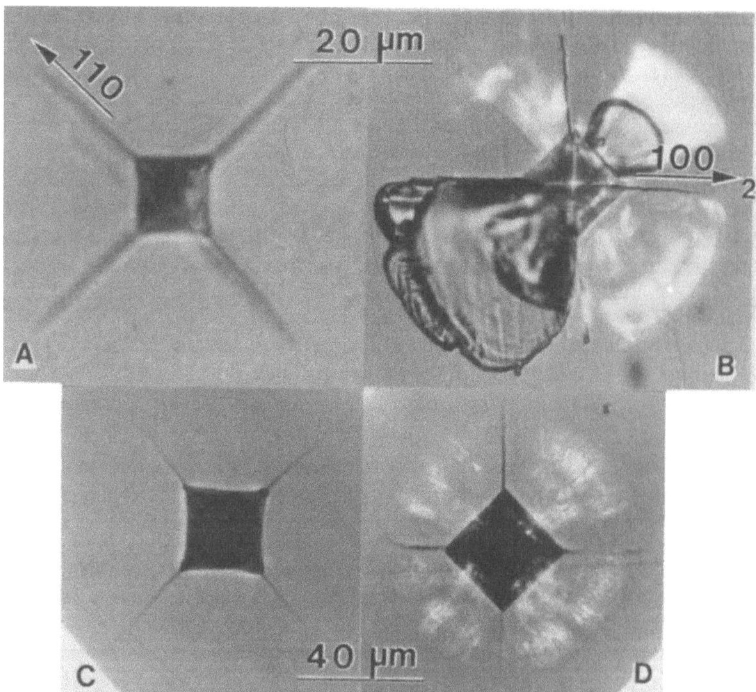

Fig. 6. Microhardness indentation in 9.4 m/o Y-FSZ (2N) ((A) and (B)) and 4.5 m/o Y-PSZ (10N) after heat treatment for 150 hours at 1600°C ((C) and (D)).

at the 2.5 m/o Y_2O_3 composition.

Heat treatment at 1400 - 1500°C for up to 170 hours caused only small changes in fracture toughness, no doubt because of the modest microstructural changes induced by such heat treatment. The fracture strengths of the two phase PSZ crystals were also not sensitive to this heat treatment, although this heat treatment did improve the strength of the 8 and 12 m/o Y-FSZ crystals by improving the surface finish. Heating the 2.5 m/o Y_2O_3 crystal to 1800ºC caused the t-ZrO_2 precipitates to coarsen, such that some had their M_s above room temperature; the heat-treated test specimen, now containing some m-ZrO_2, had a slightly decreased strength of ~1200 MPa, which could be increased to ~2000 MPa by careful surface grinding. It appears that the increased strength of the heat-treated and suface-ground samples was due to the easier transformability of the larger t-ZrO_2 precipitates caused by the 1800ºC heat treatment, and that conventional transformation toughening is operating in these crystals, albeit the magnitude of the transformation zone must be limited. This limited transformability obtains because of the strong stabilizing influence of the Y_2O_3 stabilizer on t-ZrO_2, and the microstructural stability imparted by the colony microstructure. Micrographic evidence of a limited transformation zone in a 4.0 m/o Y_2O_3 crystal heat-treated at 1650°C for 50 hours has been published by Heuer *et al* |9|.Fig. 6 shows microhardness indents and accompanying indent-induced microcracks on a 9.4 m/o Y-FSZ single crystal and a 4.0 m/o Y-FSZ single crystal heated for 150 hours at 1650°C; the decreased cracking due to the increased fracture toughness in the latter crystal is obvious.

3.2. Intermediate Temperature Properties

The temperature dependence of the strength and fracture toughness of the 3.5 and 12 m/o Y_2O_3 compositions has been studied by Ingel *et al*. |26|. Both strength and toughness show a gradual decrease to about 1000°C which was attributed to the decreasing importance of transformation toughening as the equilibrium temperature for the (t→m) phase transformation was approached.However, the strength and toughness of the PSZ single crystal were higher than the FSZ crystal up to 1200ºC, which was attributed to toughening due to some (unspecified) crack/particle interaction arising from structural, elastic, and/or thermal expansion mismatch between precipitate and matrix.

The 2.5 and 12 m/o Y_2O_3 crystals have been studied under compressive deformation at room temperature, 250°C, 700°C, and 1150°C by Lankford *et al*|27|. The behavior was complex, as was the authors' explanation(s). Two different orientations of the PSZ crystals with compression axes along <123> and <100> were studied, as was the <123> orientation for FSZ crystals. The FSZ crystals were brittle at room temperature, with a compressive strength of ~2.2 GPa. With increasing temperature, the strength decreased and yielding occurred, and at 1150°C, 4% plastic strain prior to failure and "jerky" flow was observed.

The behavior of the PSZ crystal was more complicated. The two orientations showed quite different behavior. The <123> orientation yielded at room temperature at a stress of ~2.6 GPa, and then work hardened to a failure stress of ~3.1 GPa at a plastic strain of ~0.005. With increasing temperature, the yield and flow stress decreased, and fracture strain increased (to ~0.06 at 700°C and ~0.07 at 1150°C). The flow was non-hardening and jerky.

The <100> crystal also yielded plastically at room temperature. However, the yield stress was ~2.2 GPa, which as followed by a strain burst of 0.0049, at which elastic deformation resumed until failure at a stress of ~2.5 GPa. At 700°C, yielding occurred in two stages, the first being analogous to that observed at room temperature, and occurring at a stress of ~0.4 GPa; elastic deformation then commenced to a second yield point at ~1 GPa, which was followed by a small yield drop and flow accompanied by smooth monotonic hardening to an ultimate plastic strain >0.15. At 700°C, both yield stresses increased when the strain rate increased from 10^{-5}/sec to 10^{-3}/sec, while the failure strain decreased.

TEM surface replica of <100>crystals deformed at room temperature and 700°C revealed extensive transformation of those precipitates oriented such that their minimum dimension was perpendicular to the compression axis. In both the <100> and <123> crystals tested at intermediate temperatures (250 and 700°C), deformation was by propagation of a Luders band nucleated at one end of the specimen. TEM studies of crystals deformed at 700°C were inconclusive, but dislocations were certainly present.

The explanation for this complex behavior offered by those authors is not easily summarized. At least part of the plastic deformation -- the strain burst in the <100> crystal and the yielding in the <123> crystal -- was attributed to transformation within the t-ZrO_2 precipitates, but the well-know (t → m) phase transformation was specifically discounted, as m-ZrO_2 could not be found in deformed crystals. Rather, this deformation was attributed to a type of ferroelastic domain rotation within particles, a form of transformation plasticity which had been suggested for some TZP's by Virkar and Matsumoto|29| (this explanation replaces an earlier suggestion |27| of deformation via a cubic to rhombohedral phase transformation). In the <100> crystal, the c axes of one set of t variants were all parallel to the loading axis, thus causing the single strain burst, whereas the yielding and work hardening in the <123> crystal were attributed to the compression axis being at an angle to all the $\{100\}_t$ planes, thus causing ferroelasticity over a range of stress. The temperature dependence of this proposed ferroelasticity could arise from thermally-activated migration of "domain" interfaces, or could be due to thermally-activated matrix accommodation of the ferroelastic precipitate shape change, which might involve the punching out of dislocations into the matrix whenever particles transfrom.

As noted above, subsequent to this "pre-yield" deformation, extensive post-yield flow occurred by Luder's band deformation, supplemented by some (t → m) transformation, but the specific dislocation or transformation processes occurring at the Luder front have not been worked out. It is clear that further work is needed to understand the intermediate temperature plastic deformation of these PSZ single crystals.

3.3. High Temperature Plastic Deformation

As with other crystals with the fluorite structure, the operative

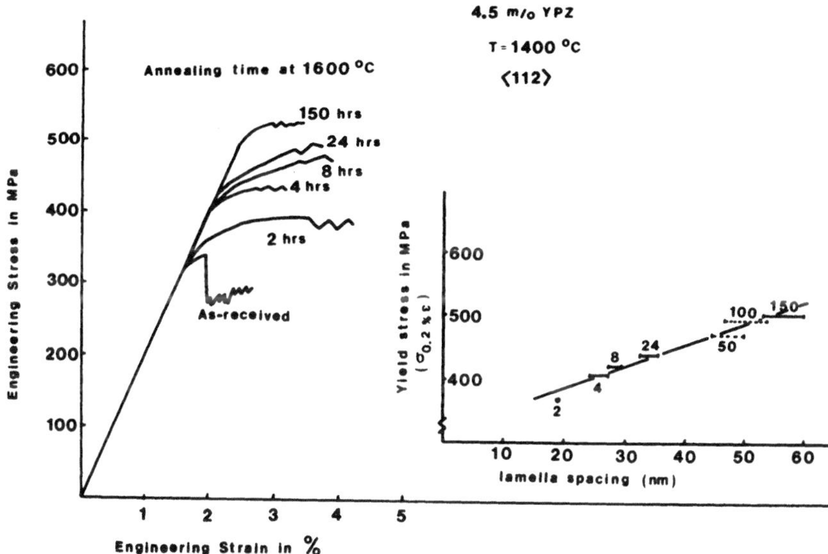

Fig. 7. Stress-strain curves at 1400°C for samples annealed for various times at 1600°C (<112> compression axis). The inset shows the 0.2% strain offset yield stress plotted as a function of lamella spacing; a plot of the latter data as $(l_t - l_o)^3$ vs t, where l_t is the spacing after time t and l_o the spacing in the as-received crystals is linear, indicating that the lamellar coarsening occurs by conventional Ostwald ripening involving bulk diffusion.

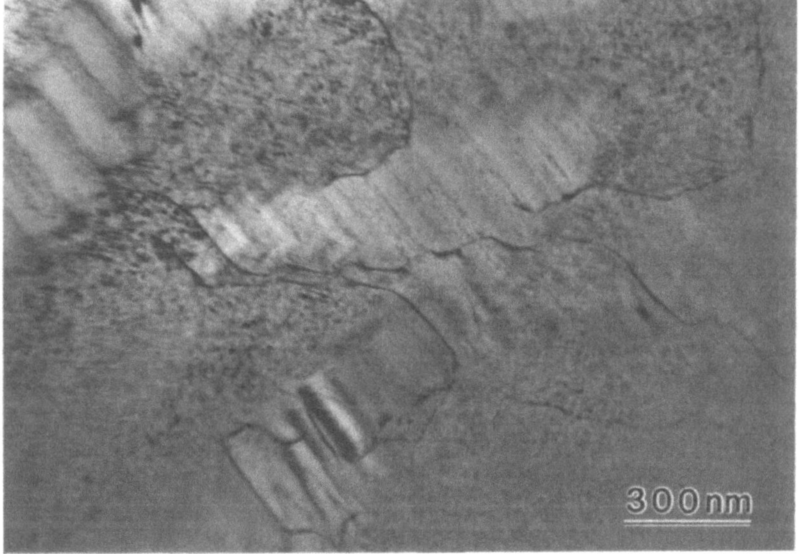

Fig. 8. Transmission electron micrograph of deformed Y-PSZ sample. Note the difficulty that the matrix distinctions have in cutting the particles.

slip systems in c-ZrO_2, when dislocation plasticity is dominant, are $\{001\}\langle 110\rangle$ and $\{\bar{1}11\}\langle 1\bar{1}0\rangle$; the former system generally has the lower critical resolved shear stress (CRSS). We have studied solid solution hardening in compression in Y-FSZ crystals containing 9.4 to 18 m/o Y_2O_3 over a range of temperatures |28| and precipitation hardening at 1400°C using a 4.5 m/o Y-PSZ crystal; we focus on the latter work here |23|.

Crystals were oriented with a $\langle 112\rangle$ compression axis; this orientation gives a Schmid factor of 0.47 for $\{001\}\langle 110\rangle$ slip. As-received crystals, and crystals annealed in air at 1600°C for 2, 4, 8, 24, and 150 hours were tested at 1400°C; the stress-strain curves are shown in Fig. 7, which also includes data showing how the lamellar spacing increases with this annealing. All stress-strain cruves show serrated flow corresponding to the activation of Luders bands, which could be seen by optical microscopy. The flow stress increased with aging time, and increased linearly with the lamellar spacing. The high flow stress of the 150 hour sample, ~550 MPa, is noteworthy, as we know of no other crystals which exhibit such high strengths at this temperature. This potent precipitation hardening derives from the great difficulty matrix dislocations experience in trying to shear particles, as is evident from electron microscopy study of foils of deformed crystals (Fig. 8) This can be undestood by reference to the lattice geometry of t and c-ZrO_2; as will be demonstrated, PSZ's constitute the ceramic equivalent of the γ' in γ microstructure used to strengthen Ni-base superalloys.

The face-centered lattice of c-ZrO_2 is shown in Fig. 9; each lat-ttice point has a ZrO_2 "molecule" associated with it; a Zr at the lattice point (say at 000) and two oxygens (say at ± 1/4 1/4 1/4). The corresponding lattice for t-ZrO_2 is C-centered (this lattice has the same number of ZrO_2 "molecules" as primitive lattice shown by dotted lines); each lattice point has associated with it two "molecules".

Consider a matrix dislocation with 1/2 [110] Burgers vector gliding on (001) of the c-ZrO_2 matrix and encountering a colony precipitate. Aside from any elastic mismatch between t and c-ZrO_2, and ignoring the small difference between a_t and a_c, there will be no particular difficulty in this dislocation entering variant t_3.

The situation is very different with respect to variants t_1 and t_2, however. With the basis appropriate for these variants, the Burgers vector of the same cutting dislocation become 1/2 [011] and 1/2 [101], respectively, neither of which are lattice vectors in t-ZrO_2. How do such partial dislocations shear the colony precipitates? One possibility is for two matrix dislocations- and effective superdislocation with $\vec{b}$ = [011] or [101] to pair, as occurs in ordered alloys such as γ' in γ . We see no evidence for this in TEM, however. Alternatively, a matrix dislocation can dissociate at the particle/matrix interface to become a perfect dislocation within the particle, leaving a second matrix dislocation surrounding the particle at the particle matrix interface. Aside from the difficulty that such a dislocation reaction is energetically very unfavorable, we see no evidence for closed dislocation loops around sheared particles. Finally, consider the consequences of shearing t-ZrO_2 with such a partial dislocation. The sheared particles will be faulted, but the fault energy may be modest. As can be seen from Fig. 1b, the oxygens in t-ZrO_2 are "puckered" up and down from their ideal

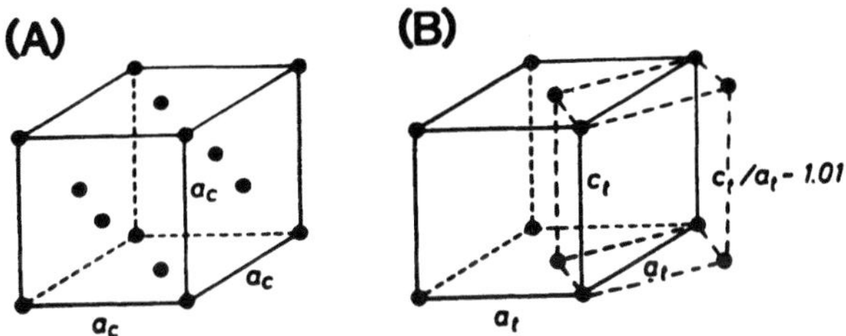

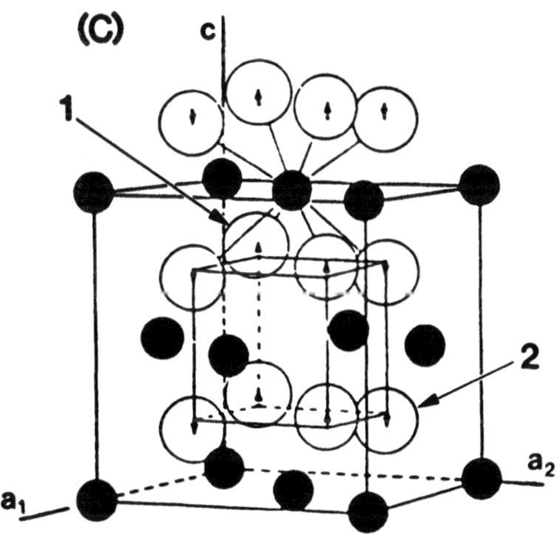

Fig. 9. (a) and (b) show the lattice geometry of c and t-ZrO_2, while (c) shows the actual structure of t-ZrO_2. The oxygen displacement from their ideal 1/4 1/4 1/4 positions are exaggerated in (c). See text for further discussion.

1/4 1/4 1/4 fluorite positions. All along the fault plane, oxygens that should be in an "up" position will be "down" and vice versa. This same type of oxygen misregistry is present across the ubiquitous APB's that form during the ($\underline{c} \rightarrow \underline{t}'$) phase transformation in this same alloy system. Furthermore, it may be possible for the fault to be eliminated during the high temperature deformation by "shuffle" notions, i.e. by self-climb of the fault to a twin or $\underline{c}/\underline{t}$ interface boundary after the dislocation has cut completely through the particle. TEM evidence of such faulted particle is now being sought, but the contrast experiments are complicated by the contrast arising from the particles themselves and the coherency and dislocation strains.

4. SUMMARY

We have discussed precipitation, precipitation-toughening, and precipitation hardening in ZrO_2-rich two-phase single crystal alloys in the Y_2O_3-ZrO_2 system. Coherent precipitates form in this system and assume a "colony" microstructure, which effectively minimizes strain energy during precipitation. Such partially-stabilized ZrO_2's are capable of transformation-toughening and precipitation-hardening, and thus exhibit high strengths at both ambient and elevated temperatures. At room temperature, strengths as high as 2 GPa have been reported, a result of fracture toughnesses up to 8 MPa $m^{1/2}$. This high strength and intermediate toughness implies only a modes transformation zone adjacent to propagating cracks, a result of the strong influence Y_2O_3 can exert on lowering M_s, and on the stability of the colony microstructure.

At 1400°C, these crystals are above their brittle $\rightarrow$ ductile transition. They deform plastically, but with a very high flow stress, ~550 MPa. The potent precipitation hardening is due to the diffuculty matrix dislocations experience in cutting colony particles, as they are partial dislocations within the particles; thus sheared particles must be faulted.

The compresive deformation of these crystals have been studied at intermediate temperatures. Large strains, e.g. 0.06 at 700°C can be obtained. The initial transformation plasticity in these crystals may involve re-orientation of $\underline{t}$-ZrO_2 precipitates, a form of ferroelasticity, but the details of this process and its role in concert with dislocation plasticity in the overall deformation still need to be evaluated.

ACKNOWLEDGEMENT

A. H. Heuer's research on plastic deformation in Y_2O_3-ZrO_2 alloys has been supported by the U.S. Department of Energy under Grant No. DEFG0284ER45110. The collaboration of A.H. Heuer and A. Dominguez-Rodríguez is made possible by the U.S.-Spanish Joint Committee for Scientific and Technological Cooperation under NSF Grant No. CCB8504-029.

REFERENCES

1.- E.C. Subbarao, "Zirconia-an overview" in Advances in Ceramics, Vol.3, Science and Technology of Zirconia (Ed. by A.H. Heuer and L.W. Hobbs) pp 1-24. Am. Ceram. Soc., Columbus, Ohio (1981).

2.- A.H. Heuer and M. Rühle, "Phase transformations in ZrO_2-containing ceramics: I, the instability of c-ZrO_2 and the resulting diffusion-controlled reactions" in Advances in Ceramics, Vol. 12, Science and Technology of Zirconia II (Ed. by N. Claussen, M. Rühle and A. H. Heuer) pp 1-13. Am. Ceram. Soc., Columbus, Ohio (1984).

3.- M. Rühle and A.H. Heuer, "Phase transformations in ZrO_2-containing ceramics: II, the martensitic reaction in t-ZrO_2" ibid. pp14-32.

4.- Ibid. Sections II, III and IV.

5.- Ibid. Section V.

6.- D. Michel, L. Mazerolles and M. Perez y Jorba., J. Mater. Sci. 18, 2618 (1983).

7.- A.H. Heuer, R. Chaim and V. Lanteri., Acta Met. 35, 661 (1987).

8.- H.G. Scott. J. Mater. Sci 10, 1527 (1975).

9.- A.H. Heuer, R. Chaim and V. Lanteri., "Phase transformations and microstructural characterization of alloys in the Y_2O_3-ZrO_2 system" in Science and Technology of Zirconia III. Am. Ceram. Soc., Columbus, Ohio, Oct. (1988)

10.-G.K. Bansal and A.H. Heuer., Acta Met. 20, 1281 (1972); 22, 409 (1974).

11.-R.C. Garvie, R.H. Hanninck and R.T.Pascoe., Nature 258, 235 (1975).

12.-D.L. Porter, A.G. Evans and A.H. Heuer., Acta Met. 27, 1649 (1979).

13.-A.G. Evans and R.M. Cannon., Acta Met. 34, 761 (1986).

14.-A.H. Heuer., J. Am. Ceram. Soc. 70, 689 (1987).

15.-A.H. Heuer, S.P. Kraus-Lanteri, P.A. Labun, V. Lanteri and T.E. Mitchell., Ultramicroscopy 18, 335 (1985).

16.-V. Lanteri, T.E. Mitchell and A.H. Heuer., J. Am. Ceram. Soc. 69, 564 (1986).

17.-K.P.D. Lagerlöf, V. Lanteri and A.H. Heuer., "On precipitate morphology in ZrO_2, α-Al_2O_3 and $FeTiO_3$ matrices" in Tailoring multiphase and Composite Ceramics, Mat. Sc. Research, Vol. 20 (Ed. by R.E. Tressler,

G.L. Messing, C.G. Pantano and R.E. Newnham) pp 239-258. Plenum Press (1986).

18.-A.G. Khachaturyan., "Theory of structural transformations in Solids" John Wiley, 1983.

19.-T.K. Gupta, F.F. Lange and J.H. Bechtold., J. Mater. Sci. 13, 1464 (1978).

20.-M. Rühle, N. Claussen and A.H. Heuer., "Microstructural studies of Y_2O_3-containing tetragonal ZrO_2 polycrystals (Y-TZP)" in Advances in Ceramics, vol. 12, Science and Technology of Zirconia II (Ed. by N. Claussen, M. Rühle and A.H. Heuer) pp 352-370. Am. Ceram. Soc. Columbus. Ohio (1984).

21.-K. Tsukema, K. Veda and M. Shimada., J.Am. Ceram. Soc. 68, C4 (1985).

22.-V. Lanteri, A.H. Heuer and T.E. Mitchell., "Tetragonal phase in the system ZrO_2-Y_2O_3" in Advances in Ceramics, vol 12, Science and Technology of Zirconia II (Ed. by N. Claussen, M. Rühle and A.H. Heuer) pp 118-130. Am. Ceram. Soc. Columbus. Ohio (1984).

23.-A.H. Heuer, V. Lanteri and A. Dominguez-Rodriguez., Acta Met. in press.

24.-V. Lanteri, R. Chain and A.H. Heuer, Ultramicroscopy, 22, 27 (1987).

25.-A. Rouanet., Rev. Int. Htes Temp. Refract. 8, 161 (1971)

26.-R.P. Ingel, D. Lewis, B.A. Bender and R.W. Rice., "Physical, microstructural, and thermomechanical properties of ZrO_2 single crystals". in Advances in Ceramics, Vol 12, Science and Technology of Zirconia II (Ed. by N. Claussen, M. Rühle and A.H. Heuer) pp 408-414. Am. Ceram. Soc. Columbus. Ohio (1984). Idem., J. Am. Ceram. Soc. 65, C-108 (1982). Idem. ibid. C-150.

27.-J. Lankford., J. Mater. Sc. 20, 53(1985). Idem, ibid. 21, 1981 (1986)

28.-A. Dominguez-Rodriguez, K.P.D. Lagerlof and A.H. Heuer., J. Am. Ceram. Soc. 69, 281 (1986). A. Domínguez-Rodriguez and A.H. Heuer., Cryst. Latt. Def. Amorp. Mater. 16, 117 (1987).

29.-A.V. Virkar and R.L.K. Matsumoto., J. Am. Ceram. Soc. 69, C-224 (1986).

RESIDUAL STRESSES IN POROUS PLASMA-SPRAYED ALUMINA COATING ON TITANIUM ALLOY FOR MEDICAL APPLICATIONS

H.CARREROT, J.RIEU
Ecole Nationale Superieure des Mines
42023 Saint-Etienne Cedex 2
FRANCE

A.RAMBERT
S.E.R.F
69152 Décines Cedex
FRANCE

1. Introduction

Hip prosthesis fixation and stabilisation have been improved by biological anchorage. An alumina porous ceramic coating allows bone tissue ingrowth, thus creating strong mechanical bonds.

The alumina porous coating is obtained by plasma-spraying alpha alumina powder on Ti-6Al-4V titanium alloy substrate. The adhesion resistance of the coating is strongly dependent on the ceramic physical and mechanical properties and on residual stresses.

In order to measure the mechanical properties of the coating and the residual stresses, an original method for removing the ceramic layer from the metallic substrate has been developped.

The present study concerns only 0.1 to 0.3 mm coating thicknesses.

2. Coating removing technique

Ceramic coating detachment is obtained by chemical etching of the metallic substrate underlayer by means of a 50% hydrochloric acid solution. The uniform metal dissolution induces less damage in the coating than the thermal shock methods using liquid nitrogen.

3. Physical properties

The coating cristallographic structure has been determined by X-ray diffraction. When 5 to 15 microns diameter stable alpha alumina powder is used, the coating structure is predominantly metastable gamma alumina (1).

The porosity has been determined by mercury porisimetry (2). A maximum value of 40% is dispatched into 20% closed porosity and 20% open porosity. The open porosity consists of two pore size domains: a macroporosity with 30 microns and a microporosity with 0.1 microns radius average.The microporosity is associated with micro-cracks in the surface sprayed droplets.

L.-C. Dufour et al. (eds.), Surfaces and Interfaces of Ceramic Materials, 777–778.

4. Mechanical properties

The removed coating has the shape of a thin plate which is used as a cantilever beam for bending tests. The modulus of elasticity (Young's modulus) is given by:

$$E = \frac{P L^3}{3 f I}$$

where

P = applyed load f = beam deflection
L = beam free length I = moment of inertia

Young's mudulus of 14 and 10 GPa have been found for coatings with respective 30% and 40% porosity.

These values are in good agreement when extrapolating other results (3) where Young's modulus has been reported as a fonction of porosity from 0 to 15%.

5. Residual stresses

When the coating is removed from the substrate, it gets significant curvature demonstrating residual stresses in the coating-substrate system. According to calculated residual stress distribution in plasma-sprayed coatings (4), tensile stresses exist at the surface and compressive ones at the ceramic-metal interface.

The mechanical stress necessary to straight back the removed curved thin ceramic plate leads to a calculated surface residual stress of 4GPa.

This value is low compared to one measured by X-ray diffraction method on the coating-substrate system using (440) diffraction line of gamma alumina. A critical analysis of these different results is proposed.

References:

1 Mc PHERSON R.,Formation of Metastable Phases in Flame-and Plasma- Prepared Alumina,J.Mat.Sc.,8(1973)851-858.
2 AUSKERN A., HORN W.,Capillary Porosity in Hardened Cement Paste,J.of Testing and Eval. JTEVA,1(1973)74-79.
3 FARGEOT D.,Etude des Phases Métastables de l'Alumine Projetée au Chalumeau à Plasma,Thèse 3ème cycle, Limoges(1987)35.
4 MARYNOWSKI C.W.,Variables in Plasma Spraying,Electchem.Techn.,3(1965)109-115.

W-TiN-SiC MATERIALS FOR HIGH TEMPERATURE APPLICATION

L.R. Wolff
Centre for Technical Ceramics
Eindhoven University of Technology
P.O. Box 513, 5600 MB Eindhoven
The Netherlands

ABSTRACT. A composite W-TiN-SiC material, as well as the technology for the manufacture of this material, are being developed at the Centre for Technical Ceramics in co-operation with partners in Italy and the Netherlands. The trilayer material will serve as a multifunctional "hot shell" in a combustion heated Thermionic Energy Converter. This device will operate at 1400°C. The composite material simultaneously serves as a protective confinement of the converter, as an emitter electrode and as an electrical lead. The multitude of requirements induced by these three functions of the hot shell are discussed. A brief outline of the design of the thermionic energy converter is given. The way in which the W-TiN-SiC material is being produced as well as some preliminary results on the evaluation of this material are presented. The properties determined include thermal fatigue and thermoshock resistance, Young's modulus and thermal conductivity. Although the evaluation is still in progress, the results are promising.

1. Introduction

A Thermionic Energy Converter (TEC) is a device which directly (without moving parts) convertes heat into electricity. It is a diode. One electrode of this diode is heated to a sufficiently high temperature (>1200°C) that it will thermally emit electrons. A cooled counterelectrode collects these electrons. Thus a potential difference develops, which can be subjected to an external load. A TEC is a high current density (10 Amps/cm^2), low voltage (0.5 Volt), direct current power source. The principles are explained more extensively elsewhere [1,2].

In the sixties and seventies thermionic energy converters were developed for space applications using a nuclear reactor as a heat source [3-6]. After the energy crisis the western world discontinued this research work alltogether or converted to the development of combustion heated TEC's for terrestrial applications. The USSR, however, continued to develop nuclear thermionic reactors for space applications. Indeed some of these nuclear thermionic systems have

L.-C. Dufour et al. (eds.), Surfaces and Interfaces of Ceramic Materials, 779–787.

been launched into space and are being used in high power radar reconnessance satellites. This became evident during the Falkland crisis when the USSR could produce high resolution radar images of the battle ground. One of these satellites, cosmos 1402, reentered and came down in the Indean Ocean while non nuclear fragments were scattered over Australia. Even today one of these nuclear TEC powered russian satellites, the COSMOS 1900, is on the verge of a reentry and we hope that this will not endanger the human population.

In the USA, Thermo Electron Corporation in Waltham, Ma, was the first to successfully demonstrate a combustion heated TEC for a period of one and a half years [7]. The Thermo Electron Co TEC's had a "hot shell" consisting of a trilayer material (SiC-C-W). The efficiency of their combustion heated TEC was rated at 9% including lead losses.

Since the TEC is essentially a Carnot machine, there should be room for improvement on this efficiency figure. The theoretical limit for a TEC operating between 1400°C and 600°C (Emitter and Collector temperature respectively) would be about 45%.

In the USA the research effort is now once again directed almost exclusively to space nuclear applications of TEC's (Due to the DOD funding of research for President Reagan's Strategic Defence Initiative, S.D.I.). In Europe we directed our research to the terrestrial application of combustion heated TEC's. At Eindhoven University special cermet emitter materials were developed, which promise to eventually raise TEC efficiency [8-10].

In the course of the project on thermionics at Eindhoven it was decided to design and build a combustion heated TEC. This also required the development of a suitable hot shell material, capable of meeting the very severe requirements imposed by the various functions of the hot shell.

2. Design of a Combustion Heated TEC

In order to develop a suitable hot shell material it is necessary to specify all requirements of the hot shell application. To this end a preliminary design of a combustion heated TEC was made (see fig. 1). the main components are the hot shell which serves as a confinement for the TEC, protecting it against the corrosive combustion atmosphere at temperatures up to 1450°C (1700°C combustion temperature). The life time should be five years

The hot shell with, at the hot end, a surface area of 20 cm^2 should accommodate a heat flux of 1000 W (50 W/cm^2). Ten percent of this heat is to be converted into electricity by the TEC. To this end the inner surface of the TEC hot shell serves as an electron emitter. The sleeve of the hot shell serves as an emitter lead for the low voltage (0.5 Volt) direct current of 200 Amps.

Other components of the TEC are the collector electrode which, at an interelectrode distance of less than half a millimeter, will collect the electrons emitted by the emitter, that is the inner surface of the hot shell, and the ceramic seal. This ceramic seal is a metal-ceramic joint, which serves to electrically isolate the emitter from the collector.

Three sapphire ball spacers are meant to keep the emitter and the collector at the desired distance under operating conditions.

The collector is watercooled in order to remain at a collector temperature of 600°C. Watercooling will facilitate efficiency measurements:

$$q = \frac{\text{output power}}{\text{output power + waste heat}}$$

The waste heat can be determined by simply measuring the cooling water flow and the ΔT between ingoing and outgoing water flow.

Finally there is the cesium reservoir which, when kept at a temperature of about 325°C, will maintain an adequate cesium vapour pressure over the liquid cesium in the cesium reservoir. A cesium pressure of the order of one torr is needed inside the TEC in order to reduce the emitter - and collector - workfunction and in order to neutralize the space charge effect.

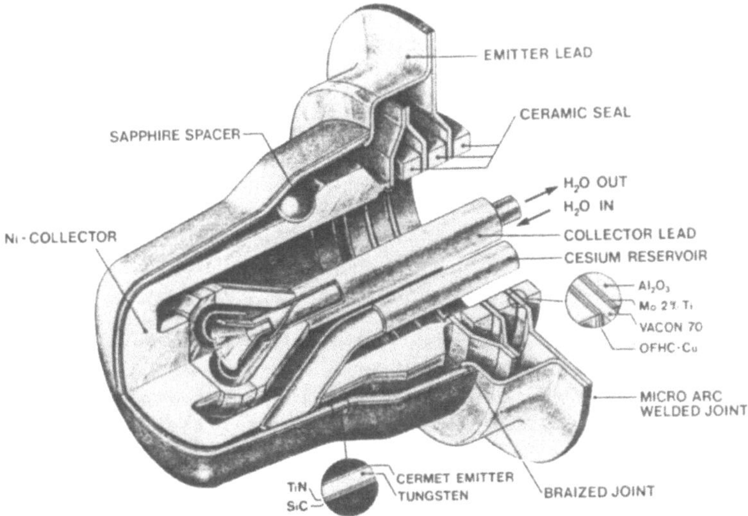

Fig. 1. Preliminary design of a combustion heated TEC by Eindhoven University of Technology.

The concept of this combustion heated TEC required a number of advanced joining techniques to be applied such as: microplasma welding, diffusion bonding, metallizing and vacuum brazing of ceramics to metals......... In this paper, however, we will focus our attention on the composite hot shell material.

Much of the development work on this material was carried out in the context of a European research project in the CEC program on Basic Research in Industrial Technology for Europe (BRITE) in which Eindhoven University of Technology cooperates with industrial partner

Xycarb in Helmond, the Netherlands, with industrial partner ANSALDO in Genova, Italy and with the ENEA institute in Rome, Italy, while Philips in Eindhoven is involved as a subcontractor.

This project, called "Composite Materials for High Temperature Heatexchangers and Thermionic Generators", is in progress and will continue untill May 1989 when several prototypes of this combustion heated TEC will have been subjected to life and performance tests.

3. The Hot Shell Requirements

The various functions of the hot shell in the combustion heated TEC impose a number of requirements on the hot shell composite material:

- Its outside should be corrosion resistant against a combustion atmosphere for a period of five years.
- It should have a sufficiently high E-modulus at 1450°C in order not to deform too much under the load of the pressure difference of one atmosphere over the shell.
- It should have a high emission coefficient in order to absorb the radiative heat from the combustion process at 1700°C.
- It should have a high thermal conductivity at 1450°C.
- It should have a low Coefficient of Thermal Expansion (CTE).
- It should be cesium corrosion resistant on the inside.
- It should be a good thermionic emitter at 1400°C at the inside.
- Its constituents should be chemically stable up to at least 1450°C.
- It should be thermoshock resitant.
- It should be thermal fatigue resistant.
- It should be a good electrical conductor even at high temperatures.
- It should be vacuum-tight.

In our effort to meet all these stringent requirements, we first set out to copy the successful solution developed by Thermo Electron Corporation in Walthem, Massachusetts. Their hot shell consisted of a trilayer material: W-graphite-siliconcarbide [11]. This material was produced by machining a thin walled (1 mm) graphite shell from a bar of this material and subsequently coating it by Chemical Vapour Deposition (CVD) with tungsten and silicon carbide. When we tried to follow this approach, however, we met with difficulties [12]. Due to a mismatch in CTE between the graphite and the CVD silicon carbide, many of the shells cracked upon the SiC CVD even before we could apply the tungsten by CVD. Furthermore we learned that the Thermo Electron approach to the hot shell manufacture was well protected by patents. Finally we were not happy with the fact that carbon and tungsten will react with each other at 1450°C quite readily, forming WC and W_2C reaction layers. This is why we decided to follow an alternative approach.

Siliconcarbide is a suitable candidate for the outside protection of the hot shell because of its corrosion resistance against a combustion atmosphere and its high emission coefficient. Furthermore SiC has a reasonably high thermal conductivity and a rather low CTE.

SiC is a strong material with a high Young's modulus even at high temperature.

Tungsten is a good emitter material. It is cesium resistant. It is a good electrical and thermal conductor. It's CTE matches that of SiC quite well (SiC: $4.50 * 10^{-6}K^{-1}$ versus W: $4.25 * 10^{-6}K^{-1}$). One problem which needed to be solved was that tungsten and silicon carbide will react with each other at elevated temperatures forming W_5Si_3 and W_2C + WC. In order to prevent this reaction, which will cause delamination in a W-SiC composite material [13] we needed to find an adequate diffusion barrier. To this end TiN was selected. TiN is thermodynamically stable versus both tungsten and siliconcarbide up to temperatures of at least 1600°C. Furthermore, when deposited by CVD, an excellent adherance between TiN and both W and SiC can be achieved.

The CTE of TiN is substantially higher than that of W and SiC: $8 * 10^{-6}K^{-1}$. TiN is an electronic conductor. An important property of TiN is that it becomes ductile and metallic in nature above 1000°C. This property reduces the stress induced by the CTE mismatch between TiN and W and SiC.

4. Manufacture of the W-TiN-SiC composite material

A variety of approaches to the manufacture of a W-TiN-SiC composite material was evaluated. Thus it was tried to start from sintered SiC shells of the desired shape and dimensions. A multitude of technical ceramics manufacturers all over the world were approached to produce SiC hot shells according to our specifications. Only a few of them responded by producing prototypes. In one case slipcasting was used as a shaping method. This did not yield the desired shape accuracy. Two other manufacturers used Cold Isostatic Pressing (CIP) as a means of shaping. In all three cases, sintering produced vacuumtight SiC shells. It was not possible to reduce the desired wall thickness to less than 2 mm whereas we would have liked it to be only half a mm. The slibcasted shells were totally unsuitable because they were found to be unround after sintering. The CIPed shells were much better in dimensional accuracy but could only meet the specifications after diamond machining. The cost of the diamond machining alone was prohibitive: $ 1000,-/shell!

Thus we decided to try a different manufacturing approach. We produced a tungsten shell by plasma spraying. To this end a brass mandrel was preheated and subsequently coated with a thin (0.1 mm thick) NaCl coating prior to the tungsten deposition. this facilitated the removal of the W shell from the brass mandrel upon cooling down to room temperature. The salt was removed by soaking in water.

The tungsten shell, thus produced, was still porous and contained substantial amounts of tungsten oxides. Therefore, it was first reduced in hydrogen at a temperature of 1600°C and then sintered to full density (96% theoretical density) by a vacuum heat treatment in a furnace at 2200°C. After this procedure, the tungsten shell with a wall thickness of 0.5 mm was vacuum tight.

The next step was to coat the W shell by CVD with TiN:

$$2\ TiCl_4 + N_2 + 4\ H_2 \underset{850°C}{\rightarrow} 2\ TiN + 8\ HCl$$

A 4 µm thick TiN layer was deposited on the outside of the W shell. The TiN layer is fully dense and follows the rather rough surface of the W shell perfectly. This enhances the adherance of the TiN to the W by mechanicly interlocking.

Finally the shell was CVD coated once again, this time with SiC. To this end we used the pyrolisis reaction:

$$(CH_3)_2\ SiCl_2 \underset{1200°C}{\rightarrow} SiC + CH_3Cl + HCl + H_2$$

Thus a 300 µm thick SiC layer was deposited on the outside of the W-TiN shell. The result is a composite hot shell consisting of 500 µm W - 4 µm TiN - 300 µm SiC. The microstructure of this material is given in fig. 2.

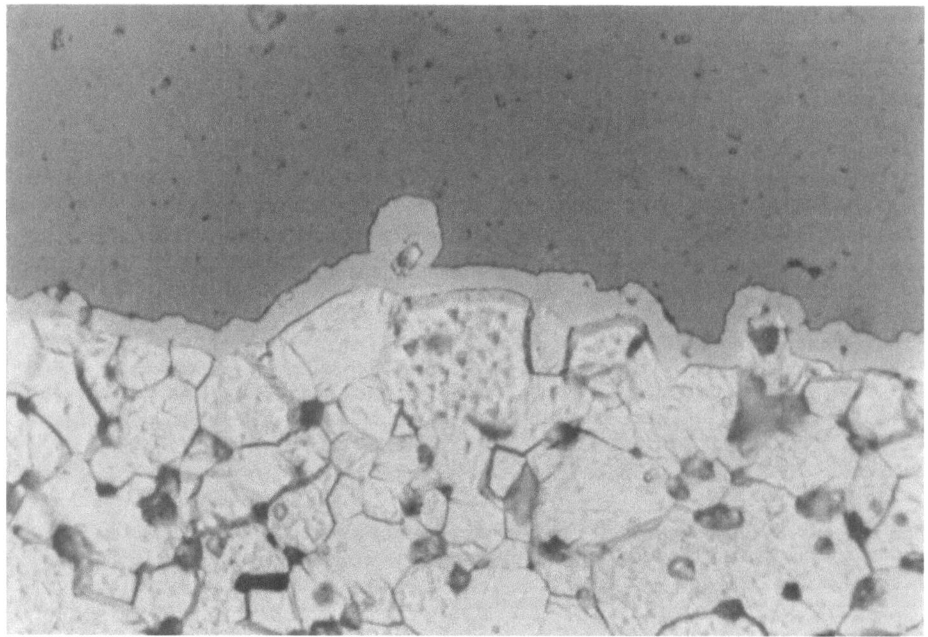

Fig. 2. Microstructure of a plasmasprayed W - CVD TiN - CVD SiC composite material for TEC hot shells developed at Eindhoven University of Technology.

5. Evaluation of the W-TiN-SiC hot shell material

Although evaluation of the W-TiN-SiC hot shell material is still in progress, some preliminary results will be presented here.

When tested at 1400°C in vacuum, the hot shell material remained unchanged after 100 hrs heating. Hundred cycles between room temperature and 1400°C did not affect the microstructure either. The heating and cooling rate during these thermal fatigue tests were approximately 50°C/min.

We subjected the W-TiN-SiC material to water quenching from various temperatures. To this end the material was sealed in an evacuated fused silica capsule. These capsules were heated to the desired temperature and after equilibration were dropped into cold water after which the capsule was broken immediately by a stroke with a hammer. When thermoshock tested in this way the material did not show any sign of delamination even when quenched from temperatures up to 1000°C!

The Young's modulus of the W-TiN-SiC material was measured at the ENSCI in Limoges France by the group of Prof. Gault using a special ultrasonic testing facility [14]. The Young's modulus was found to decrease from 403 GPa at room temperature to 370 GPa at 1200°C. These values are in the same range as those for both tungsten and siliconcarbide. The reduction of the Young's modulus with only 8% over this temperature range indicates that the deformation as a result of the one atmosphere pressure difference between inside (vacuum) and outside of the hot shell will not cause any problem.

The thermal diffusivity of the hot shell material was measured by the laser flash method [15]. The thermal diffusivity was found to be 0.58 cm^2/sec room temperature while it still was 0.155 cm^2/sec at 1400°C. Thermal diffusivity values calculated from separate measurements on the constituent materials resulted in values of 0.60 cm^2/sec at room temp. and 0.160 cm^2/sec at 1400°C. The fact that the measured and calculated values correspond so well shows that the bonding between the constituent materials W, TiN and SiC in the composite material is excellent.

From the thermal diffusivity data the temperature gradient accross the hot shell top can be calculated. The thermal conductivity for the .5 mm W - 4 μm TiN - .3 mm SiC material at 1400°C was calculated to 55 W/mK. At an expected heat flow of 50 W/cm^2 accross the hot shell top this means that the temperature gradient will only be 7°C!

Long term life and corrosion tests of hot shells in a combustion atmosphere as well as life and performance tests of complete combustion heated TEC's still need to be carried out. This is necessary to further validate the W-TiN-SiC hot shell material.

6. Conclusions

- A composite material is being developed based on a combination of three materials: Tungsten, TiN and SiC.

- This material should meet the requirements of a hot shell for a thermionic energy converter operating at 1400°C in a combustion atmosphere.
- Tungsten serves as a thermionic emitter electrode and electrical lead, while SiC serves as a protective corrosion resistant surface layer. TiN serves as a diffusion barrier preventing a reaction between W and SiC.
- A suitable method for the manufacture of the W-TiN-SiC material was developed based on plasmaspraying, reduction and sintering of a freestanding tungsten shell, coated with TiN and SiC by CVD.
- The evaluation of this composite material, while still in progress, shows promise of meeting the many stringent requirements of this advanced high temperature application.

7. Acknowledgements

The work described in this paper is the work of an extensive multinational and multidisciplinary team of scientists and technicians in the Netherlands and Italy. The author thanks them for their enthusiastm and the high quality of their contributions to the project.

Special thanks are due to Prof. Gault of the French ENSCI institute in Limoges for his contribution in the temperature dependent Young's modulus determination on the composite W-TiN-SiC material.

Thanks are also due to Prof. Hasselman of the Virginia Polytechnical Institute and State University and visiting Prof. Suzuki of the Nagoya Institute of Technology in Japan for their contributions in the laser flash thermal diffusivity measurements on various materials used in the development of the W-TiN-SiC material. These measurements were performed at the Virginia Polytech in Prof. Hasselman's laboratory in Blacksburg, Virginia, USA.

8. References

1. G.N. Hatsopoulos and E.P. Gyftopoulos, *Thermionic Energy Conversion*, Vol. I: Processes and Devices, M.I.T. Press, 265 pgs. (1973).

2. G.N. Hatsopoulos and E.P. Gyftopoulos, *Thermionic Energy Conversion*, Vol. II: Theory, Technology, and Application, M.I.T. Press, 661 pgs. (1979).

3. C.D. Sawyer, P.R. Hill, D.R. Wilkins, "Multimegawatt Thermionic Reactor Systems for Space Applications", Proceedings 2nd International Conference on Thermionic Electrical Power Generation, Stresa, Italy p. 171-184 (1968).

4. P. Fiebelmann, H. Neu, C. Rinaldini, "A Heatpipe Thermionic Reactor Concept", Ibid, p. 243-262 (1968).

5. A. Schock, M.J. Abbate and C.L. Eisen, "Out-of-Core Thermionic Power Plant for Manned Space Station", Conference Record of the 1970 Thermionic Conversion Specialist Conference, IEEE, p. 169-178 (1970).

6. V.A. Kuznetsov c.s., "Power Tests of the Topaz-3 Thermionic Reactor", 1975 Thermionic Conversion Specialists Meeting, Eindhoven University of Technology, p. 137-141 (1975).

7. Fred Huffman, "Overview of Terrestrial Thermionics", Proc. 18th IECEC, Orlando, Fa, USA, p. 179-185 (1983).

8. L.R. Wolff, C.J. Heijnen, G.P. v.d. Wouw, "Cr_2O_3, Al_2O_3-Mo as a Thermionic Emitter Material", High Temperatures - High Pressures 13, p. 69-77 (1981).

9. E. Penders-Vankova, G.H.M. Gubbels, L.R. Wolff, "Measuring Thermionic Emission of W-Al_2O_3 and W-$LaCrO_3$ Cermets", High Temperatures - High Pressures 18, p. 689-696 (1986).

10. L.R. Wolff, J.M. Hermans, J. Adriaansen and G.H.M. Gubbels, "Materials for Thermionic Energy Converters", *High Tech Ceramics*, P. Vincenzini ed., Elsevier Science Publ. BV Amsterdam, p. 2397-2405 (1987).

11. G. Miskolczy and D. Lieb, "Cogeneration using a Thermionic Combustor", Proceedings of the 17th IECE, p. 1918-1923 (1982).

12. L.R. Wolff and P.P.J. Ramaekers, "Production of a SiC-C-W hot shell by chemical vapour deposition, High Temperatures - High Pressures 18, p. 135-141 (1986).

13. L.R. Wolff, "Composite Metal-Ceramic Material for High Temperature Energy Conversion Applications", Proc. 14th Science of Ceramics, p. 991-996 (1988).

14. C. Gault, F. Platon and D. Le Bras, "Ultrasonic Measurements of Young's Modulus of Al_2O_3-based Refractories at High Temperatures", Materials Science and Engineering, 74, p. 105-111 (1985).

15. M.A. Bucknam, L.D. Bentsen, J. Makosey, G.R. Angell and D.P.H. Hasselman, "The Measurement of the Thermal Conductivity of Refractories by the Laser-Flash Method", Trans. J. Br. Ceram. Soc., 82, p. 18-23 (1983).

COLLOIDAL FILTRATION OF HYDROPHOBIC ALUMINA. INFLUENCE OF SEDIMENTATION ON FILTRATION KINETICS

Albert P. Philipse and Hubert J. Veringa
National Ceramic Center, NKA
Netherlands Energy Research Foundation, ECN
Westerduinweg 3, P.O. Box 1, 1755 ZG PETTEN,
The Netherlands

ABSTRACT

A method is described for an organic surface modification of α-alumina particles. The modified particles combine to rapidly settling aggregates in water, whereas sedimentation in ethanol is much slower. Suspensions are filtered with a porous mold. A filtration model is introduced for simultaneous filtration and sedimentation, which explains qualitatively the observed filtration kinetics.

1. INTRODUCTION

Colloidal filtration ("slipcasting") is a cheap and relatively reliable ceramic shaping technology to produce mainly thin walled green bodies out of a ceramic suspension. However, attaining a microstructure with a well defined and dense particle packing without defects is considered to be a major problem in slipcasting [1, 2]. One of the reasons is that the starting suspensions, for example α-alumina particles in water, are difficult to handle due to the variety of (often time dependent) parameters such as pH, ionic strength, particle charge, viscosity and average aggregate size. In practice the optimum combination of parameter values to obtain a desired microstructure is often found by a laborious trial-and-error procedure.

At the National Ceramic Center, NKA, filtration on a porous mold is applied to fabricate technical (oxide) ceramics. The filtration process is also studied systematically to deal more adequately with the above mentioned problem. One of the strategies we follow in this study is to investigate the filtration behaviour of colloidal suspensions in non-aqueous solvents.

Aggregation of particles in such suspensions depends among other things on the nature of the solvent. If particles are covered with polymer they will stick together in a solvent in which the polymer is

L.-C. Dufour et al. (eds.), Surfaces and Interfaces of Ceramic Materials, 789–798.

only sparingly soluble. So the suspensions can be used to check for example the influence of aggregates on the fitration kinetics or the microstructure of the filter cake (see also ref. [3]). In this preliminary report we describe a method for preparing a nonaqueous alumina suspension and discuss some first results on the filtration kinetics for this system.

The suspension consists of α-alumina particles dispersed in ethanol. The surface of the particles is modified with an organo-silane compound. The modified particles are easily wetted by ethanol, but not by water. In this latter solvent, large aggregates of particles are formed which settle rapidly, whereas in ethanol sedimentation is slower, indicating smaller aggregates in this solvent. Sedimentation of aggregates significantly influences the filtration kinetics of suspensions. A filtration model is presented which takes this sedimentation effect into account and which explains the filtration results at least qualitatively.

2. PHENOMENOLOGICAL MODEL

This section deals with the filtration kinetics of a suspension in contact with a porous mold. After a brief treatment of a non-sedimenting suspension, the influence of particle sedimentation will be discussed.

No sedimentation

Consider a cylinder of cross sectional area A, placed on a large porous mold and filled with suspension to a height h_0 on time $t = 0$ (fig. 1). Liquid filters into the mold because it experiences a certain pressure difference ΔP due to capillaries in the mold. After a time t the liquid level (which coincides with the suspension level in absence of sedimentation) has travelled a distance h(t) and a particle "cake" of thickness l(t) is formed on the mold surface. This cake is assumed to be homogeneous on a macroscopic scale. Thus with respect to liquid flow, the cake can be characterized with one (specific) liquid permeability k_c [4]. If we further assume that this permeability is much smaller than that of the mold, the cake is the main resistance to flow during the filtration process. Then the pressure drop ΔP occurs across the cake and the pressure gradient driving liquid flow is $\Delta P/l(t)$ where $\Delta P < 0$. Darcy's law [4] reads therefore:

$$v = -\frac{k_c}{\eta}\frac{\Delta P}{l(t)}, \tag{1}$$

where v is the superficial liquid flow velocity in the cake (in m/sec)and η is the liquid viscosity. In what follows we assume that the particle cake is incompressible, i.e. k_c is independent of pressure.

Since the volume flow rate vA equals Adh(t)/dt we have:

$$\frac{d\,h(t)}{d\,t} = -\frac{k_c}{\eta}\frac{\Delta P}{l(t)} \,. \qquad (2)$$

Conservation of the total particle volume gives a relation between the decrease in suspension level, h(t), and the cake thickness, l(t) (see fig. 1):

$$\Phi_s\, A\, h_0 = \Phi_s\, A\, [h_0 - h(t) - l(t)] + \Phi_c\, A\, l(t) \,. \qquad (3)$$

Where:

ϕ_c = particle volume fraction in the cake
ϕ_s = particle volume fraction in the suspension
h_o = initial height of the suspension

Combining Eqs (2) and (3) to eliminate l(t)and solving the resulting equation gives:

$$h^2(t) = -2\, J\, \frac{k_c\, \Delta P}{\eta}\, t \,, \qquad (4)$$

where:

$$J = \frac{(\Phi_c - \Phi_s)}{\Phi_s} \qquad (5)$$

The linear relation between h(t) (or the cake thickness) and $\sqrt{t}$ is a familiar one [4], though certainly not always obeyed in practice. Referring to a more elaborate discussion of this topic [3] we deal here only with one factor giving rise to deviations from eq. (4) and that is sedimentation of particles.

Influence of sedimentation
When a suspension contains large particles or aggregates, leading to a non vanishing sedimentation velocity, suspended particles travel faster downwards than the liquid level, leaving behind a clear layer of liquid (fig. 1). As a result the filtration process can be divided into two parts with different kinetics.

In the first part not all the particles have reached the cake and the filtrating liquid permeates a cake which grows in time. Clearly this cake is thicker than would be the case without sedimentation so the filtration rate will be slower than according to eq. (4).
In the second part of the filtration process all the particles have reached the cake and the remaining liquid permeates a cake of constant thickness. This thickness equals $l = h_0\, \Phi_s/\Phi_c$, which leads on substitution in eq. (2) to:

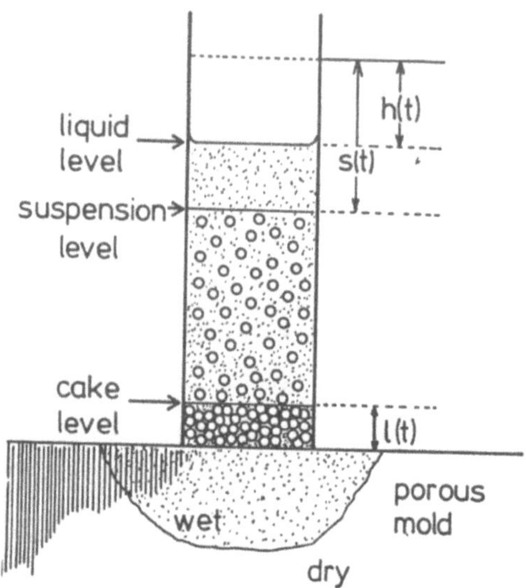

Fig. 1. Colloidal filtration of a suspension with a porous mold. Liquid filters into the mold, leaving behind a filter cake of thickness l(t). The liquid level has travelled a distance h(t) due to filtration. Suspended particles have travelled a distance s(t). When there is no sedimentation of particles, s(t) and h(t) coincide.

$$h(t) = - \frac{\phi_c}{\phi_s h_0} \frac{k_c \, \Delta P}{\eta} t \; . \tag{6}$$

Again the filtration rate is not in accordance with eq. (4).
For the first part of the filtration process, h(t) is calculated as follows. The distances travelled by the liquid level and the suspension level are respectively h(t) and s(t) (see fig. 1). Conservation of the total particle volume gives:

$$h_0 \, \phi_s = \phi_s \, [h_0 - s(t) - l(t)] + \phi_c \, l(t) \; , \tag{7}$$

which combines with eq. (2) to:

$$\frac{d\,h(t)}{d\,t} = - J \, \frac{k_c \, \Delta P}{\eta \, s(t)} \; . \tag{8}$$

The relation between h(t) and s(t) is given by:

$$s(t) = \int_0^t V(t')dt' + h(t) \tag{9}$$

in which V(t) is the settling velocity of the particles (or aggregates) with respect to the liquid. If the particles or aggregates in suspension keep their identity; i.e. do not break up or stick together, the settling velocity is constant, $V(t) = V_0$ and we find from eq. (8) and (9):

$$\frac{d\,h(t)}{dt} = -\,J\,\left(\frac{k_c\,\Delta P}{\eta}\right)\,[V_0 t + h(t)]^{-1}\,. \tag{10}$$

The final solution is:

$$t = \frac{\exp\left(\frac{V_0\,\eta\,h(t)}{-k_c\,J\,\Delta P}\right) - \left(\frac{V_0\,\eta\,h(t)}{-k_c\,J\,\Delta P}\right) - 1}{\frac{V_0^2\,\eta}{-k_c\,J\,\Delta P}} \tag{11}$$

$$= \frac{1}{2}\left(\frac{\eta}{-k_c\,J\,\Delta P}\right) h^2(t) + \frac{V_0}{6}\left(\frac{\eta}{-k_c\,J\,\Delta P}\right)^2 h^3(t) + \ldots. \tag{12}$$

From eq. (12) it is obvious that in absence of sedimentation ($V_0 = 0$) we recover eq. (4). For large h(t), eq. (11) approaches a relationship of the form $h(t) \sim \ln t$, which should be compared with $h(t) \sim \sqrt{t}$ for a non-sedimenting suspension. Note that in the beginning of the filtration process, when the liquid and suspension levels are close together, we should also find that h(t) varies (approximately) with $\sqrt{t}$. Fig. 2 illustrates the total filtration process on the basis of eqs. (6) and (11).

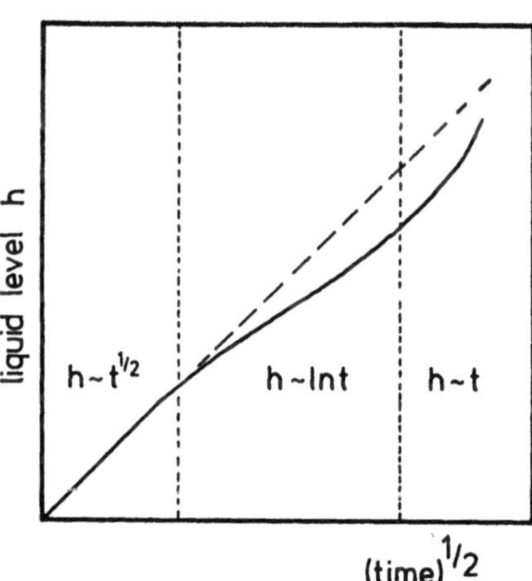

Fig. 2. Sketch of the decrease in liquid level h(t) versus time t for the filtration process of fig. 1, according to eqs. (6) and (11).

3. ORGANIC SURFACE MODIFICATION OF α-ALUMINA

The surface modification of α-alumina particles comprises the treatment of the alumina in an ethanol-ammonia mixture with the silane coupling agent [5] 3-methacryloxypropyltrimethoxysilane, here referred to as TPM:

$$CH_2 = \overset{\displaystyle CH_3}{\overset{|}{C}} - \underset{\displaystyle O}{CO} - CH_2 - CH_2 - CH_2 - Si(OCH_3)_3$$

3-methacryloxypropyltrimethoxysilane (TPM)

Silane coupling agents are well known for their ability to adhere to inorganic substrates [5]. We choose TPM because this silane has proved to be very effective in the coating of colloidal silica spheres with the objective to prepare stable (i.e. non-aggregated) silica dispersions in ethanol [6].

At present we have no detailed information on the deposition of TPM on α-alumina particles. Investigations into the analogous deposition on silica substrates [5] suggest the following scheme. First the methoxygroups of TPM hydrolyse under influence of acid or base. (In this study ammonia is used). Reactive silane triols are formed which readily associate and condensate to oligomers. These oligomers adsorb rapidly on the alumina particles, presumably via hydrogen bonds between hydroxyl groups of TPM and the alumina surface. Part of these bonds may be converted to a chemical bond under loss of a water molecule.

4. EXPERIMENTAL

4.1. Preparation of alumina suspensions

A typical coating experiment was performed as follows. 10 ml 3-methacryloxypropyltrimethoxysilane (Merck, purum grade) was added to a mixture of 10 ml ammonia (Merck, 25%) and 140 ml absolute ethanol. This TPM-solution was mixed with 160 gr alumina powder (CT 2000 SG, average particle diameter in the range 0.5 - 1 µm as checked by SEM). The resulting suspension was milled with alumina balls in a polyethylene vessel for at least two hours. Afterwards the suspension was diluted with absolute ethanol to a concentration of ~ 7 particle volume percent. (This concentration was chosen to compare the filtration results with another filtration study [3] where suspensions of about the same concentration were used). Dilution of the milled suspension with water (also to 7 particle volume percent) induced a rapid settling of particle aggregates on a time scale of minutes. For suspensions in pure ethanol sedimentation was clearly slower.

4.2. Filtration experiments

A filtration experiment was conducted as follows. A glass cylinder (height 74 mm, diameter 12 mm) with a millimeter scale and open at both ends was glued to the surface of a porous ceramic plate (15 x 11 x 3.5 cm). This ceramic plate was developed by ECN originally for filtration purposes [7]. The cylinder is rapidly filled with (previously stirred) suspension using a syringe. Then the height of the suspension level (and also the liquid level in case of a noticeable sedimentation) is measured as a function of time. Prior to a filtration experiment the plate is cleaned with water and ethanol and heat treated at 1000°C for several hours.

5. RESULTS AND DISCUSSION

5.1. Coating procedure

There are two indications for the presence of an organic layer on the alumina particles after treatment with TPM. The first is the sedimentation behaviour of particles in water and in ethanol. The as-received, untreated alumina disperses easily in water. However, after treatment with TPM large particle aggregates are formed in water which settle rapidly; water is a poor solvent for TPM so in this solvent TPM-coated particles tend to stick together. On the other hand, ethanol is a good solvent for TPM so in this solvent a TPM-layer is expected to provide a "steric barrier" against aggregation. Indeed we find that in ethanol the suspension sediments more slowly, indicating smaller aggregates, if any, than in water.
The second indication for the adherence of TPM on alumina is the fact that drying of a suspension yields a powder which is difficult to wet with water, but which is easily wetted by ethanol.

5.2. Filtration experiments

Fig. 3 shows filtration results for a suspension of modified alumina particles in ethanol. The distance h(t) travelled by the liquid level is in the beginning of the filtration process proportional to $\sqrt{t}$ (cf. the leading term in eq. 12). However, at t ~ 2500 sec the filtration curve is seen to deviate from a linear h(t) - $\sqrt{t}$ relationship; the filtration rate slows down. This deviation due to particle sedimentation is more pronounced in fig. 4.
Fig. 4 shows data for modified particles dispersed in water. Settling occurs at a higher rate than in ethanol (compare the s(t) curves in fig. 3 and 4). It is seen in fig. 4 that deviation from linearity starts earlier in the filtration process. This is what we expect, because the larger sedimentation velocity V_0 gives rise to a larger contribution of $h^3(t)$ - and higher order terms - in the filtration equation (12). Further, when in fig. 4 the s(t) curve flattens, i.e. when all the suspended particles have reached the filter cake, the h(t) - $\sqrt{t}$ curve starts to rise. This is in accordance with eq. (6),

which states that for liquid permeation through a cake of constant thickness the liquid level travels at a constant speed so h(t) ~ t.

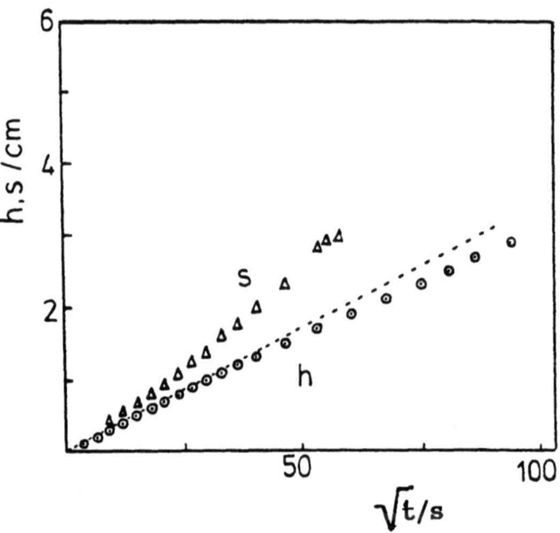

Fig. 3. The decrease in suspension level, s(t), and liquid level, h(t), versus the square root of time t for the filtration of modified alumina in ethanol.

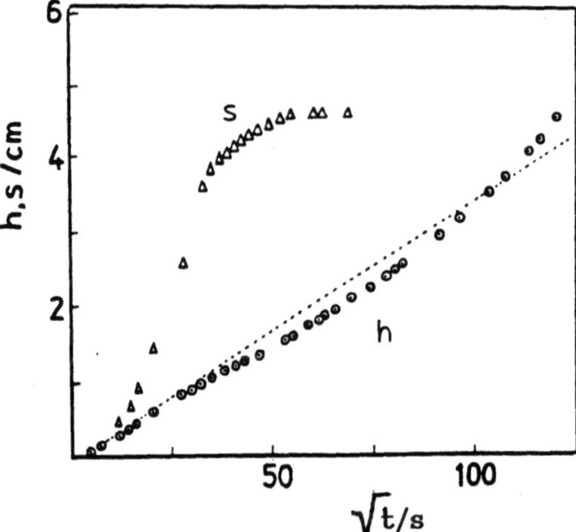

Fig. 4. h(t) and s(t) versus $\sqrt{t}$ for a suspension of modified alumina particles in water. The sedimentation velocity of the particles leads to a deviation from linearity, as explained by the model described in section 2.

To conclude, fig. 4 shows the qualitative features expected from fig. 2.

To illustrate eq. (4), which should apply for a non-sedimenting suspension, we show in fig. 5 filtration data taken from ref. [3] for a suspension of unaggregated silica spheres in ethanol. Referring for further details to ref. [3] we only note here that the supension did not show sedimentation, leading to the expected linear h(t) - $\sqrt{t}$ curve in fig. 5.

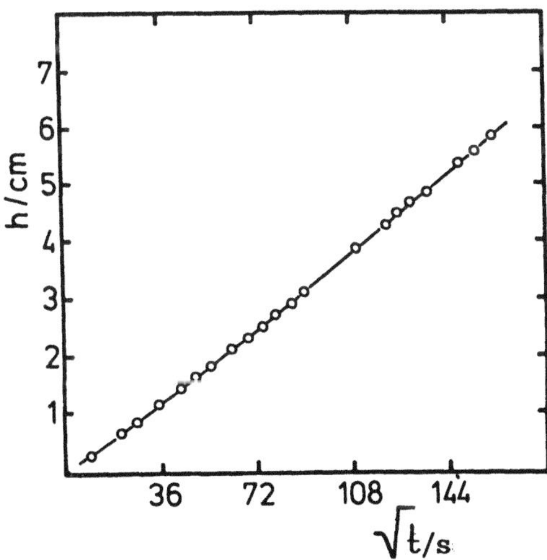

Fig. 5. h(t) versus $\sqrt{t}$ for a non-sedimenting suspension of silica spheres [3], showing the linearity expected from eq. (4).

6. CONCLUSIONS

We conclude that the surface modification of α-alumina with the used organosilane compound leads to particles which combine in water to larger aggregates than in ethanol. The influence of aggregate sedimentation on suspension filtration kinetics can be qualitatively explained with our filtration model.

ACKNOWLEDGEMENTS

André Roskam is acknowledged for help with the experiments and Trudy Zomerdijk for typing the manuscript.
The porous ceramic plate was prepared by Loek Berkeveld and Luci Correia.
This work is part of the research programme on ceramic shaping techniques of the National Ceramic Center, NKA, and its host organization the Netherlands Energy Research Foundation, ECN.

REFERENCES

1. I.A. Aksay, Microstructural Control Through Colloidal Consolidation, Advances in Ceramics-9. Forming of Ceramics, Eds. J.A. Mangles and G.L. Messing, Amer. Ceram. Soc. Columbus, OH, (1984).

2. See also for example Ceramic Transactions, Vol. 1, Ceramic Powder Science II, Ed. G.L. Messing, E.R. Fuller and H. Hausner, Amer. Ceram. Soc. (1988), in particular section IV on Ceramic Powder Forming Processes.

3. A.P. Philipse, B.C. Bonekamp and H.J. Veringa, Colloidal Filtration of Silica and Alumina suspensions, to be published.

4. R.E. Collins, Flow of Fluids through Porous Materials, Reinhold, New York (1961).

5. E. Plueddeman, Silane Coupling Agents, Plenum Press, New York (1982).

6. A.P. Philipse and A. Vrij, Determination of Static and Dynamic Interactions between Monodisperse, Charged Silicaspheres in an Optically Matching, Organic Solvent, J. Chem. Phys. 88, 6459 (1988).

7. L.A. Correia, unpublished results.

SUBJECT INDEX

MIX
Papier aus verantwortungsvollen Quellen
Paper from responsible sources
FSC® C105338

If you have any concerns about our products,
you can contact us on
ProductSafety@springernature.com

In case Publisher is established outside the EU,
the EU authorized representative is:
Springer Nature Customer Service Center GmbH
Europaplatz 3, 69115 Heidelberg, Germany

Printed by Libri Plureos GmbH
in Hamburg, Germany